Classical Mechanics

Classical Mechanics: A Computational Approach with Examples Using Mathematica and Python provides a unique, contemporary introduction to classical mechanics, with a focus on computational methods. In addition to providing clear and thorough coverage of key topics, this textbook includes integrated instructions and treatments of computation.

This newly updated and revised second edition includes two new appendices instructing the reader in both the Python and Mathematica languages. All worked example problems in the second edition contain both Python and Mathematica code. New end-of-chapter problems explore the application of computational methods to classical mechanics problems.

Full of pedagogy, it contains both analytical and computational example problems within the body of each chapter. The example problems teach readers both analytical methods and how to use computer algebra systems and computer programming to solve problems in classical mechanics. End-of-chapter problems allow students to hone their skills in problem solving with and without the use of a computer. The methods presented in this book can then be used by students when solving problems in other fields both within and outside of physics.

It is an ideal textbook for undergraduate students in physics, mathematics, and engineering studying classical mechanics.

Key Features:

- Gives readers the "big picture" of classical mechanics and the importance of computation in the solution of problems in physics
- Numerous example problems using both analytical and computational methods, as well as explanations as to how and why specific techniques were used
- Online resources containing specific example codes to help students learn computational methods and write their own algorithms

A solutions manual is available via the Routledge Instructor Hub and all example codes in the book are available via the Support Material tab, and at the book's GitHub page: https://github.com/vpagonis/Classical_Mechanics_2nd_Edition

Christopher W. Kulp received his PhD in Physics from the College of William and Mary in 2004 and is currently a Professor of Physics at Lycoming College, where he teaches physics at all levels. Dr. Kulp's research interests focus on the fields of nonlinear dynamics and complex systems. He has published more than 20 publications in peer-reviewed journals and conference proceedings and has written two book chapters. More than 10 of his publications have undergraduate co-authors. Much of his work focuses on distinguishing between chaotic and stochastic behaviour in time series data. His current research interests focus on using machine learning to analyse time series and model complex systems.

Vasilis Pagonis is Professor of Physics Emeritus at McDaniel College, Maryland, where he taught undergraduate courses and did research for 36 years. He is currently a Senior Associate Editor of the international journal "Radiation Measurements". His research areas of interest is luminescence dosimetry, and applications of thermally and optically stimulated luminescence (TL and OSL). He has taught courses in classical and quantum mechanics, analog and digital electronics and mathematical physics, as well as numerous general science courses. Dr. Pagonis' resume lists more than 200 peer-reviewed publications in international journals. He is the co-author with Dr Kulp of the textbook "Mathematical methods using Python" (CRC, 2024). He has also co-authored five graduate level books in the field of luminescence dosimetry.

Classical Mechanics

A Computational Approach with Examples Using Mathematica and Python

Second Edition

Christopher W. Kulp and Vasilis Pagonis

CRC Press
Taylor & Francis Group
Boca Raton London New York

CRC Press is an imprint of the
Taylor & Francis Group, an **informa** business

Designed cover image: © Shutterstock

Second edition published 2026
by CRC Press
2385 NW Executive Center Drive, Suite 320, Boca Raton FL 33431

and by CRC Press
4 Park Square, Milton Park, Abingdon, Oxon, OX14 4RN

CRC Press is an imprint of Taylor & Francis Group, LLC

© 2026 Christopher Kulp and Vasilis Pagonis

First edition published by CRC Press 2021
Second edition published by CRC Press 2026

Library of Congress Cataloging-in-Publication Data
Names: Kulp, Christopher W., author. | Pagonis, Vasilis, author.
Title: Classical mechanics : a computational approach with examples using
Python and Mathematica / Christopher Kulp and Vasilis Pagonis.
Description: Second edition. | Boca Raton, FL : CRC Press, 2026. | Includes
bibliographical references and index. |
Identifiers: LCCN 2025005960 | ISBN 9781032594750 (hbk) | ISBN
9781032590516 (pbk) | ISBN 9781003454854 (ebk)
Subjects: LCSH: Mechanics--Mathematics. | Mechanics--Data processing. |
Python (Computer program language) | Mathematica (Computer file)
Classification: LCC QC127 .K85 2025 | DDC 531.0285/53--dc23/eng/20250220
LC record available at https://lccn.loc.gov/2025005960

ISBN: 978-1-032-59475-0 (hbk)
ISBN: 978-1-032-59051-6 (pbk)
ISBN: 978-1-003-45485-4 (ebk)

DOI: 10.1201/9781003454854

Typeset in LM Roman
by KnowledgeWorks Global Ltd.

Publisher's note: This book has been prepared from camera-ready copy provided by the authors.

Dedication

Chris dedicates this book to his wife Gail, mother Linda, and his late father Chester; without their support, this book would not have been possible.

Vasilis dedicates this book to his wife, Mary Jo Boylan, and to his students at McDaniel College.

Contents

PREFACE TO THE SECOND EDITION

We believe that computation should be front and center in a science education. Programming is a core skill for scientists and engineers, that should be taught alongside traditional physics instruction. It is in this spirit that we created this textbook, which merges instruction in Classical Mechanics and programming, into a single presentation. In our approach, computer programming and computer algebra systems are treated as simply one more tool for problem-solving.

WHAT IS NEW IN THIS SECOND EDITION

Since the publication of the first edition of *Classical Mechanics* four years ago, we have received excellent feedback from students and instructors using the book. This second, greatly expanded edition is an attempt to address as many of their suggestions and comments as possible.

Thanks to all the suggestions from our readers, we substantially improved on the first edition by making several important improvements and additions:

- The number of coded examples has been expanded from 98 to 139 solved examples, with all codes now provided in *both* Python and Mathematica in all examples.

- The example problems continue to have a mix of analytical and computational solutions. In many cases, we expanded the discussion in the example problem to include additional instruction on how to use each language.

- Strong emphasis is now placed throughout the book in solving differential equations using the symbolic capabilities of Symbolic Python (SymPy).

- We have added new End-of-Chapter problems and removed some old ones.

- A computer icon now clearly identifies problems whose solution requires a computer.

- We have improved significantly the instructor's solution manual, which now includes codes for *both* Python and Mathematica.

- Two new appendices were added, to give the reader an introduction to Python and Mathematica. This resource was requested by several instructors, so that students can have easy access to this reference material as they work through the book.

- All codes from this textbook are now available for free download. The Python codes are available in the form of Jupyter notebooks, so they can be easily downloaded and run, for example, at the Google Colab. Both Python and Mathematica notebooks can be found at the book's GitHub website:

$$https://github.com/vpagonis/Classical\_Mechanics\_2nd\_Edition$$

- In order to keep the book at a reasonable length, we removed discussions on numerical techniques such as the 4th-order Runge Kutta method for solving ODEs. Readers interested in how these numerical techniques work can consult the many available good books on these topics.

WHY PYTHON AND MATHEMATICA?

Physicists need many different computational tools to solve problems. Python is a powerful, easy-to-use language which has become a mainstay in science.

We also include Mathematica, since it is a world-class computer algebra system. Like Python, it is easy to learn, but Mathematica can often handle complex algebraic manipulations better than Python. Furthermore, Mathematica often requires fewer lines of coding than Python, because of its rich set of commands. By presenting a variety of tools, we are hoping that the student will be able to choose the best tool for solving a particular problem.

The prerequisites for this book are two semesters of introductory physics and two semesters of calculus. The codes in this textbook tend to favor clarity over efficiency. Experienced programmers will certainly find more efficient ways of solving the various example problems in this book. However, we believe that it is much more beneficial for the student to include a few more lines of code for the purpose of clarity, rather than to try to combine multiple lines for a more elegant algorithm. Solving the problems with algorithms that they develop themselves is a beneficial exercise for students, and it is highly recommended.

A NOTE TO THE STUDENTS

One of the goals of this book is to teach you how to use computational tools to solve problems in Classical Mechanics, and physics problems in general. This is an important transferable skill, useful regardless of your career path.

Programming will change the way you think about problems, since writing a program requires you to think procedurally. This will certainly improve your problem-solving skills in your other physics and math classes. Don't hesitate to look online for help. Websites like *stackexchange.com* and generative AI will become your best friend. However, it is important to keep in mind that code posted online or produced by generative AI may contain errors. Debugging those errors is a valuable learning experience. As you solve more programming problems, you will be learning an invaluable skill that will serve you in many ways in your development as a scientist!

A NOTE TO INSTRUCTORS

One of our motivations for writing this book was to better prepare our students for the large variety of careers students pursue after graduating from a physics program. We have found that after graduation, our students are increasingly taking on careers where computation is a critical element. While in college, our students often take at least one computer science course. However, we found they were not always making the connection of how to apply the programming skills they learned in computer science for the purpose of solving problems in physics. Part of the reason is because traditional physics courses are still focused on closed-form solutions and, when computation is used, it is generally focused on using computer algebra systems to perform complicated integrals, with the occasional numerical solution of a differential equation thrown in for good measure.

ACKNOWLEDGMENTS

We thank Danny Kielty at CRC Press for working with us to significantly increase the length of the book, in order to accommodate important additions to this textbook.

Chris Kulp and Vasilis Pagonis
January 2025

1 Foundations of Motion and Computation

So, you have decided to—or are required to—learn the subject of classical mechanics. But, what is *classical mechanics*? Is it fixing old cars? No, but a knowledge of classical mechanics will help you understand how your car works! To understand the term, *classical mechanics* let us first understand the term *classical physics*. Classical physics is the field of physics that doesn't involve either quantum theory or the theory of relativity. *Mechanics* is the branch of physics that deals with the actions of forces on an object that involve motion. So, classical mechanics involves the study of forces on objects that are well-described without using quantum theory and whose motion is nonrelativistic (i.e., cases where relativity is not needed to correctly model the motion). In other words, classical mechanics is the physics of your day-to-day world! An understanding of classical mechanics will help you understand how to build roller coasters, merry-go-rounds, and airplanes. Our knowledge of classical mechanics also allows us to make predictions on the motion of objects like comets and punted footballs. The world you interact with on a daily basis is generally the world of classical mechanics.

At first it might seem that classical mechanics is a dusty old subject that you need to learn in order to get to the "interesting stuff" like quantum mechanics. While the field of classical mechanics is one of the older subjects in physics, it is certainly not dusty! In fact, there is a lot of intriguing current research done in classical systems. For example, in Chapter 13 we will explore the field of nonlinear systems, which occur in many of the natural and social sciences, not only in physics, and we will display interesting types of behaviors including *chaos*.

In this chapter we introduce the basic descriptors of motion in classical mechanics, i.e., position, velocity and acceleration of particles, and show how Mathematica and Python can represent these vector quantities. This is followed by descriptions of the concepts of mass and force and Newton's laws of motion, and by a general description of reference frames. This introductory chapter concludes with an introduction to two important techniques used in computational physics, namely the symbolic and numerical solution of ordinary differential equations (ODEs) in one dimension. A detailed presentation of the solution of such ODEs is found in later chapters.

1.1 BASICS OF CLASSICAL MECHANICS

It is common for classical mechanics to take a central role in many physics problems because of the question it addresses.

The "Fundamental Question" of Classical Mechanics

Given the forces acting on an object and its initial state, what is the resulting motion of that object?

In the rest of this chapter, we will lay down the important foundations for classical mechanics, starting with the next section where we will explore the basic assumptions of classical mechanics.

When thinking about any discipline of science, first ask what are the basic foundations of the field? Sometimes, these foundations are assumptions. For classical mechanics we can start with space and time:

- *Space* serves as a background in which physical processes occur. Unlike in the theory of general relativity, space has no effect on the behavior of physical systems.

- *Time* progresses at the same rate for every observer (unlike in the theories of special and general relativity).

Classical mechanics addresses the problem of predicting an object's motion given the forces acting on the object. For now, we will consider the object to be a point particle. The advantage of working with point particles is that they have no size and no internal dynamics. In addition, point particles do not rotate nor do they deform, further simplifying their dynamics. It turns out that treating objects as point particles can be a very good approximation for describing translational motion. In later chapters, we will study the physics of extended bodies and rotational dynamics, where we will no longer restrict ourselves to working with point particles.

1.2 BASIC DESCRIPTORS OF MOTION

In order to describe the translational motion of a particle, we need three quantities that we will call the basic descriptors of motion: *position, velocity,* and *acceleration.* For rotational motion, we will need three additional quantities: *angular position, angular velocity,* and *angular acceleration,* and we will return to those later in the book.

1.2.1 POSITION AND DISPLACEMENT

The position of a particle is the location of the particle with respect to an origin, and is measured in meters. The meter is defined as the distance light travels in $1/299{,}792{,}458$ seconds. Note that the second was defined in 1967 at the 13th meeting of the International Committee on Weights and Measures. At that meeting the following definition was adopted: "The second is the duration of 9,192,631,770 periods of the radiation corresponding to the transition between the two hyperfine levels of the ground state of the cesium-133 atom."

The position of the particle is typically described by using a vector, whose components consist of the particle's distance from the origin along three perpendicular axes. For example, in Cartesian coordinates the location of a particle can be described using:

$$\mathbf{r} = x\,\hat{\mathbf{i}} + y\,\hat{\mathbf{j}} + z\,\hat{\mathbf{k}} \tag{1.2.1}$$

where $\hat{\mathbf{i}}$, $\hat{\mathbf{j}}$, and $\hat{\mathbf{k}}$ are the Cartesian unit vectors along the x, y, and z axes, respectively. The variables x, y, and z in (1.2.1) give the distance between the particle and the origin along each axis and are called the *components* of the vector. Also note that in (1.2.1) each component is a function of time: $\mathbf{r} = \mathbf{r}(t)$, $x = x(t)$, $y = y(t)$, and $z = z(t)$. In this book, vectors are denoted by boldface font. In Chapter 3, we will discuss vector quantities in more detail.

The vector in (1.2.1) tells us that in order to get to the location of the particle, $\mathbf{r}$, one needs to move a distance x parallel to the x-axis (as denoted by $\hat{\mathbf{i}}$), turn and move a distance y along the y-axis (as denoted by $\hat{\mathbf{j}}$), and then turn and move a distance z parallel to the z-axis (as denoted by $\hat{\mathbf{k}}$). The notation used in this book is that a hat (the ^ symbol) represents a *unit vector*, a vector of length 1. Hence, we can think of $\mathbf{r}$ as the sum of three vectors: $x\,\hat{\mathbf{i}}$, $y\,\hat{\mathbf{j}}$, and $z\,\hat{\mathbf{k}}$, each representing a *displacement* from the origin along one of the axes.

The displacement of the particle, $\Delta \mathbf{r} = \mathbf{r} - \mathbf{r}_0$, is the change of the particle's position from the position, $\mathbf{r}_0$, to the position $\mathbf{r}$. The displacement is found by subtracting the two vectors $\mathbf{r}_0$ and $\mathbf{r}$, component by component,

$$\Delta \mathbf{r} = \mathbf{r} - \mathbf{r}_0 = (x - x_0)\hat{\mathbf{i}} + (y - y_0)\hat{\mathbf{j}} + (z - z_0)\hat{\mathbf{k}} \qquad (1.2.2)$$

A vector quantity contains information about both magnitude (amount) and direction. In this case, the vector $\Delta \mathbf{r}$ tells us how far (magnitude) and in which direction the particle traveled. Example 1.1 shows how to represent graphically position vectors in a Cartesian coordinate system using Python and Mathematica.

Example 1.1: Graphical representation of vectors in Python and Mathematica

Consider the vector $\mathbf{r} = \hat{\mathbf{i}} + \hat{\mathbf{j}} + \hat{\mathbf{k}}$. Plot this vector on a 3D Cartesian coordinate system in Python and Mathematica.

Python Code

In the first line of the code, we import the `matplotlib.pyplot` library and use the alias `plt` to refer to it in the rest of the code. The next two lines use the commands `plt.figure()` and `add_subplot(projection='3d')` to set up a 3D plot.

We define the values of the coordinates of the vector $\mathbf{r}$ to be plotted as `Ax=Ay=Az=1` and plot the arrows for the vectors $\mathbf{r}, \hat{\mathbf{i}}, \hat{\mathbf{j}}, \hat{\mathbf{k}}$ using the `quiver` command. For example, to plot the unit vector $\hat{\mathbf{i}}$ we use the command `ax.quiver(0,0,0,1,0,0)` to plot the arrow from the origin $(0,0,0)$ to the point $(1,0,0)$.

The commands `ax.set_xlabel` and `ax.set_xlimit` set the labels and the limits of the x-axis. The command `ax.text` places appropriate text near the tips of the arrows. The final plot is shown in Figure 1.1a.

```python
import matplotlib.pyplot as plt

fig = plt.figure()
ax = fig.add_subplot(projection='3d')    # set up 3D plot

# plot  vector from (0,0,0) to (1,1,1) using quiver()
ax.quiver(0,0,0,1,1,1,color='k',length=1,arrow_length_ratio=.1)

# plot unit vectors i, j, k
ax.quiver(0,0,0,1,0,0,color='r',length=1,arrow_length_ratio=.2)
ax.quiver(0,0,0,0,1,0,color='b',length=1,arrow_length_ratio=.2)
ax.quiver(0,0,0,0,0,1,color='g',length=1,arrow_length_ratio=.2)

ax.set_xlabel('X')        # label,set limits on x,y,z axes
ax.set_ylabel('Y')
ax.set_zlabel('Z')
ax.set_xlim(0,1.2);
ax.set_ylim(0,1.2);
ax.set_zlim(0,1.2);

ax.text(1,1,1.1,'A',color='black')     # add text A for vector
ax.text(1,0.1,0,'i',color='black')     # add text for vector i
ax.text(0.05,1,0,'j',color='black')    # add text for vector j
ax.text(0,0,1.1,'k',color='black')     # add text for vector k
plt.show()
```

Mathematica Code

We use the commands `Graphics3D` and `Arrow` to create four separate 3D plots for the vectors $\mathbf{r}, \hat{\mathbf{i}}, \hat{\mathbf{j}}, \hat{\mathbf{k}}$. These four plots are combined using the `Show` command. Note the semicolons after each of the four plots, which suppress the graphical output each time. To create for example the vector $\mathbf{r}$ we draw the arrow from the origin $(0,0,0)$ to the point $(1,1,1)$. The text next to each arrow is shown with its own color and with a large font size 22, using for example the options `Red,22`. The final plot is shown in Figure 1.1b.

```
gr1 = Graphics3D[{Arrow[{{0, 0, 0}, {1, 1, 1}}],
Text[Style["A", 22], {.9, .9, 1}]}];
gr2 = Graphics3D[{Arrow[{{0, 0, 0}, {1, 0, 0}}],
Text[Style["i", Red, 22], {.9, 0.1, 0}]}];
gr3 = Graphics3D[{Arrow[{{0, 0, 0}, {0, 1, 0}}],
Text[Style["j", Green, 22], {0.1, .9, 0.05}]}];
gr4 = Graphics3D[{Arrow[{{0, 0, 0}, {0, 0, 1}}],
Text[Style["k", Blue, 22], {0.05, 0.05, .9}]}];
Show[{gr1, gr2, gr3, gr4}]
```

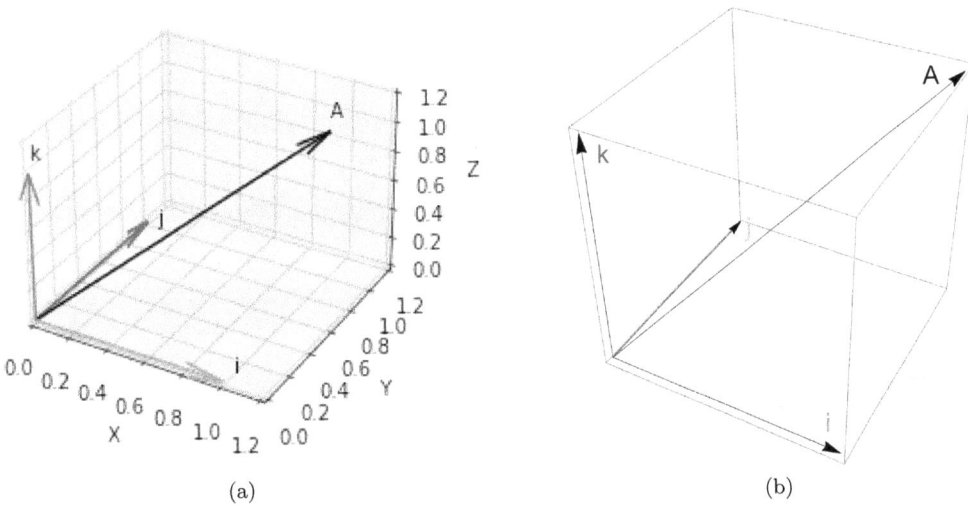

Figure 1.1: Plot of the vector $\mathbf{r} = \hat{\mathbf{i}} + \hat{\mathbf{j}} + \hat{\mathbf{k}}$ and the unit vectors $\hat{\mathbf{i}}$, $\hat{\mathbf{j}}$, $\hat{\mathbf{k}}$ for Example 1.1, using (a) Python and (b) Mathematica.

You may also see the unit vectors in the form $\hat{\mathbf{x}}$, $\hat{\mathbf{y}}$, and $\hat{\mathbf{z}}$, sometimes used to clarify that one is working in the Cartesian coordinate system. In addition, occasionally it will be easier to use a more generic notation for vectors where r_1, r_2, and r_3 are used instead of the components x, y, and z and the unit vectors $\hat{\mathbf{e}}_1$, $\hat{\mathbf{e}}_2$, and $\hat{\mathbf{e}}_3$ are used instead of $\hat{\mathbf{i}}$, $\hat{\mathbf{j}}$, and $\hat{\mathbf{k}}$. This generic notation allows for a more compact method of writing vectors:

$$\mathbf{r} = r_1\hat{\mathbf{e}}_1 + r_2\hat{\mathbf{e}}_2 + r_3\hat{\mathbf{e}}_3 = \sum_{i=1}^{3} r_i\hat{\mathbf{e}}_i \tag{1.2.3}$$

There are other coordinate systems which we will explore later in this book. Each coordinate system will have its own components and unit vectors; however, the basic idea of position and displacement will be the same.

1.2.2 VELOCITY

The velocity **v** of a particle is the particle's displacement (change of position) per unit time and is measured in meters per second (m/s). The instantaneous velocity is found by computing the time derivative of the position vector:

$$\mathbf{v} = \frac{d\mathbf{r}}{dt} = \lim_{\Delta t \to 0} \frac{\Delta \mathbf{r}}{\Delta t} \tag{1.2.4}$$

where $\Delta \mathbf{r} = \mathbf{r}(t + \Delta t) - \mathbf{r}(t)$. All the vectors we typically come across in physics are differentiable, and the limits will exist. Furthermore, the derivative in (1.2.4) will behave like derivatives you have encountered before. Therefore, we can use the standard derivative rules:

$$\frac{d}{dt}(\mathbf{r}_1 + \mathbf{r}_2) = \frac{d\mathbf{r}_1}{dt} + \frac{d\mathbf{r}_2}{dt} \tag{1.2.5}$$

$$\frac{d}{dt}(c\mathbf{r}) = \mathbf{r}\frac{dc}{dt} + c\frac{d\mathbf{r}}{dt} \tag{1.2.6}$$

where c is a scalar function of time. The rules in (1.2.5) and (1.2.6) allow us to distribute the derivative to each vector component. In addition, if the unit vectors are constant, like they are in Cartesian coordinates (but not in other systems!), then we can compute the velocity using (1.2.5):

$$\mathbf{v} = \frac{d\mathbf{r}}{dt} \tag{1.2.7}$$

$$= \frac{d}{dt}\left(x\hat{\mathbf{i}} + y\hat{\mathbf{j}} + z\hat{\mathbf{k}}\right) \tag{1.2.8}$$

$$= \frac{dx}{dt}\hat{\mathbf{i}} + \frac{dy}{dt}\hat{\mathbf{j}} + \frac{dz}{dt}\hat{\mathbf{k}} \tag{1.2.9}$$

or: $\mathbf{v} = v_x\hat{\mathbf{i}} + v_y\hat{\mathbf{j}} + v_z\hat{\mathbf{k}}$ where $v_x = dx/dt$, and so on. Note that there are only three terms (and not six) in (1.2.9) because the Cartesian unit vectors are constant. The *speed* of a particle is the magnitude of its velocity vector. We will discuss how to calculate vector properties in Chapter 3.

Finally, we can further simplify the notation by using dots to denote differentiation with respect to time (i.e., $\dot{x} = dx/dt$). Hence:

$$\mathbf{v} = \dot{\mathbf{r}} = \dot{x}\hat{\mathbf{i}} + \dot{y}\hat{\mathbf{j}} + \dot{z}\hat{\mathbf{k}} \tag{1.2.10}$$

We will find the dot notation to be very useful in the chapters to come.

1.2.3 ACCELERATION

Acceleration, **a**, is the change of velocity per unit time and is measured in meters per second squared (m/s^2). The acceleration of an object is computed similarly to velocity:

$$\mathbf{a} = \frac{d\mathbf{v}}{dt} \tag{1.2.11}$$

$$= \frac{d}{dt}\left(v_x\hat{\mathbf{i}} + v_y\hat{\mathbf{j}} + v_z\hat{\mathbf{k}}\right) \tag{1.2.12}$$

$$= \frac{dv_x}{dt}\hat{\mathbf{i}} + \frac{dv_y}{dt}\hat{\mathbf{j}} + \frac{dv_z}{dt}\hat{\mathbf{k}} \tag{1.2.13}$$

$$= \dot{v}_x\hat{\mathbf{i}} + \dot{v}_y\hat{\mathbf{j}} + \dot{v}_z\hat{\mathbf{k}} \tag{1.2.14}$$

which, like velocity, can be rewritten as $\mathbf{a} = a_x\hat{\mathbf{i}} + a_y\hat{\mathbf{j}} + a_z\hat{\mathbf{k}}$. The acceleration is also the second derivative of the position vector and using the dot notation we can write:

$$\mathbf{a} = \frac{d^2\mathbf{r}}{dt^2} \tag{1.2.15}$$

$$= \ddot{x}\hat{\mathbf{i}} + \ddot{y}\hat{\mathbf{j}} + \ddot{z}\hat{\mathbf{k}} \tag{1.2.16}$$

As we will see in future chapters, the acceleration is very important in classical mechanics. Much of what is discussed in this book is about how to develop an equation for the acceleration, which can then be integrated to get the motion of the system. All measurements of position, velocity, and acceleration, are relative to the frame of reference (choice of origin and axes) from which the measurements are made. We will return to the important idea of reference frames in Section 1.5.

Example 1.2 demonstrates how to obtain the velocity and acceleration vectors $\mathbf{v}$ and $\mathbf{a}$, when the position vector $\mathbf{r}$ is known.

Example 1.2: Evaluating the derivatives of time-dependent vectors

The position of a particle of mass m at time t is given by $\mathbf{r} = a\hat{\mathbf{i}} + b\sin(at)\hat{\mathbf{j}} + ct^2\hat{\mathbf{k}}$ where a, b, c are constants. Evaluate both analytically and using symbolic algebra the velocity $\mathbf{v}$ of the particle, and the acceleration $\mathbf{a}$ at time t.

Solution

The velocity $\mathbf{v}$ of the particle is found from the time derivative of the position:

$$\mathbf{v} = \frac{d\mathbf{r}}{dt} = \frac{dx}{dt}\hat{\mathbf{i}} + \frac{dy}{dt}\hat{\mathbf{j}} + \frac{dz}{dt}\hat{\mathbf{k}}$$

$$\mathbf{v} = ba\cos(at)\hat{\mathbf{j}} + 2ct\hat{\mathbf{k}}$$

The acceleration $\mathbf{a}$ of the particle is found from the time derivative of the velocity:

$$\mathbf{a} = \frac{d\mathbf{v}}{dt} = \frac{dv_x}{dt}\hat{\mathbf{i}} + \frac{dv_y}{dt}\hat{\mathbf{j}} + \frac{dv_z}{dt}\hat{\mathbf{k}}$$

$$\mathbf{a} = -ba^2\sin(at)\hat{\mathbf{j}} + 2c\hat{\mathbf{k}}$$

Python Code

In the beginning of any Python code, we must specify the symbols. In this example we use the `symbols` command to define `a,b,c,t` as real variables.

In the next line we import the library the command `CoordSys3D` from the Symbolic Python (SymPy) library. This command allows us to describe vectors in a Cartesian system.

Here we define a coordinate system `R = CoordSys3D('R')` in which the Cartesian coordinates (x, y, z) are represented internally as the symbols `R.x`, `R.y`, `R.z`. The corresponding unit vectors $\hat{\mathbf{i}}, \hat{\mathbf{j}}, \hat{\mathbf{k}}$, are represented by `R.i`, `R.j`, `R.k`. Note that there is nothing special about choosing the letter `R`, any other letter could have been chosen instead.

The vector $\mathbf{r} = a\hat{\mathbf{i}} + b\sin(at)\hat{\mathbf{j}} + t^2\hat{\mathbf{k}}$ is represented in this coordinate system symbolically as `r = a*R.i + b*sin(a*t)*R.j +c*t**2*R.k`. We evaluate the time derivative of this vector `r` using the command `diff(r,t)`. Similarly the acceleration is evaluated from the time derivative of v using the command `diff(v,t)`.

The output of the Python program is included below the words CODE OUTPUT. You will notice the line `print('-'*28,'CODE OUTPUT','-'*29)` in many of the Python codes in this book, it is used to separate the code from the output generated by the code.

```python
from sympy.vector import CoordSys3D
from sympy import  symbols, diff, sin

print('-'*28,'CODE OUTPUT','-'*29,'\n')

# define symbols
a, b, c, t = symbols('a, b, c, t  ', real=True)

R = CoordSys3D('R')        # define coordinate system R with
                           # unit vectors R.i, R.j, R.k

# define position vector r
r = a*R.i + b*sin(a*t)*R.j +c*t**2*R.k

v = diff(r,t)                       # find v=dr/dt
print('velocity v = ', v)

a = diff(v,t)                       # find a=dv/dt
print('\nacceleration a = ', a)


---------------------------- CODE OUTPUT ----------------------------
velocity v =   (a*b*cos(a*t))*R.j + 2*c*t*R.k

acceleration a =   (-a**2*b*sin(a*t))*R.j + 2*c*R.k
```

Mathematica Code

The first line in the code `Clear["`*"]` clears all previously defined variables in Mathematica, and you will see this command in many of the codes in this book. The position **r** is defined as a Mathematica list and its time derivative is evaluated using the command `D[r,t]`. Similarly the acceleration is evaluated by taking the derivative of the velocity parameter v using `D[v,t]`.

The `Print` command is used to output the results of the code in a manner that is easy to read.

```
Clear["` * "]

r = {a, b * Sin[a * t], c * t^2};

v = D[r, t];

acc = D[v, t];

Print["Position = ", r]

Print["Velocity = ", v]

Print["Acceleration = ", acc]
```
Position = $\{a, b\,\mathrm{Sin}[a\,t], c\,t^2\}$
Velocity = $\{0, a\,b\,\mathrm{Cos}[a\,t], 2ct\}$
Acceleration = $\{0, -a^2\,b\,\mathrm{Sin}[a\,t], 2c\}$

The basic descriptors of motion simply provide a means of describing how an object is moving. If we want to understand why an object moves as it does, then we need to understand the concepts of mass and force.

1.2.4 MASS

The mass of an object is a measure of the object's *inertia*, how strongly the object resists acceleration. For example, a loaded shopping cart is more difficult to accelerate than an empty one. Recall from your introductory physics course that mass is a scalar quantity because in order to describe mass, only a magnitude is needed.

To compare the mass of objects, one could use a beam balance. In order to speak quantitatively about mass, it is helpful to have a standard mass to which all other masses could be compared. Other physical quantities, such as the meter and second, are based on fundamental constants. As mentioned earlier, the meter is based on the speed of light, and the second is based on the ground state hyperfine splitting frequency of cesium-133 atoms, respectively. The definition of the kilogram has recently changed in order to be based on a fundamental constant.

As of May 2019, the kilogram was redefined by setting the Planck constant to be $6.62607015 \times 10^{-34}$ kg m^{2}s^{-1}.

Regardless of how the kilogram is defined, you might be wondering why we can use a beam balance to compare masses. We can use a beam balance because weight is proportional to mass according to the *weak equivalence principle*, which says that *gravitational mass is equal to inertial mass*. Gravitational mass is the mass that determines gravitational forces between objects, whereas the inertial mass determines the acceleration of an object experiencing a given force. At the time of this writing, experiments show that the two masses are equivalent to about one part in 10^{12}, and future experiments are planned for even more precise testing.

1.2.5 FORCE

Like mass, the concept of force is one with which we are intuitively familiar. A force is a push or a pull in a particular direction. You push forward on a shopping cart to move it in the direction you wish to go. However, if someone steps in front of your cart as it is moving, you would pull back on the cart to make it stop. From our everyday experience, we know that forces cause motion and that multiple forces can be acting on an object. For example, if you are in the gym lifting weights, then you are exerting a force to raise a dumbbell. If you let go of the dumbbell, it will fall, demonstrating that the force of gravity is also acting on the dumbbell. Hence, we see that in order to understand the motion of an object, we need to know about all of the forces acting on the object. When accounting for all of the forces, we need to know both the magnitude (or amount) and direction of each force. Similar to displacement, velocity, and acceleration, force is a vector quantity.

The unit of force is the Newton (abbreviated N) and 1 N is the total amount of force needed to provide an acceleration of 1 m/s^2 to a 1 kg mass. We know from everyday experiences that there is a direct linear relationship between force and acceleration. For example, a 4 N net force will cause a 1 kg object to accelerate 4 m/s^2. Of course, the acceleration is caused by the vector sum of the forces. When you hold your cell phone, you are exerting a force upwards that matches the downward force of gravity; hence the two forces are equal in magnitude but opposite in direction, and therefore, their sum is zero. Hence, the cell phone does not accelerate.

We have now laid out all of the tools needed to describe and explain an object's motion. Next, we will discuss the core topic of classical mechanics, Newton's Laws of Motion, which will explain the role of force and mass in determining the motion of an object.

1.3 NEWTON'S LAWS OF MOTION

Isaac Newton (1642–1727) developed both calculus and the foundations for classical mechanics. Newton's book, *Philosophiae Naturalis Principia Mathematica (Mathematical Principles of Natural Philosophy)* was published in 1687 and is considered to be one of the most important works in the history of modern science. In the *Principia*, Newton stated his three laws of motion. Newton's laws of motion are vital tools that allow us not only to explain why objects move the way that they do, but they also provide us with a means of predicting an object's motion. In classical mechanics, the importance of Newton's laws cannot be overstated; they are worth committing to memory. As you solve problems in this book, ask yourself how Newton's laws are involved in the setup and solution to each problem. In this section, we will go through each of the three laws in detail.

1.3.1 NEWTON'S FIRST LAW

Newton's first law of motion is often remembered by students using the phrase, "An object in motion remains in motion, and an object at rest stays at rest unless acted upon by an unbalanced force." The problem with this phrasing is, what is meant by motion? Does motion mean position, velocity, acceleration, or something else? The proper phrasing of physical laws is critical for developing a good understanding of what the law says. The phrasing of the law should use proper physics terminology, not to be confusing, but rather to be clear!

Newton's First Law

A particle's velocity remains constant if the net force acting on the particle is equal to zero.

Let us look carefully at what the first law says. The first law says that a particle's velocity remains constant if the vector sum of the forces (net force) is zero. The term "net force" is important because there can be forces acting on the particle, but if all of those forces sum to zero, then the particle's velocity doesn't change. The first law is sometimes referred to as the law of inertia because it says that an object will continue moving with a constant velocity forever, if there are no net forces acting on it. Inertia is, simply put, an object's resistance to acceleration. Hence, the particle's inertia will ensure that the particle will continue moving in a straight line with a constant velocity or remain at rest (zero velocity), until it experiences a force that will change the particle's speed, direction, or both.

We know that acceleration is the change of velocity and hence, the first law states that a non-zero net force is needed to cause an acceleration.

Finally, the first law provides a definition for the term *equilibrium*. A particle is in equilibrium if the acceleration of the particle is equal to zero. Hence, one condition for equilibrium is that the net force acting on a particle must be zero. We will later see that when we study rotational dynamics, we will also need the net torque to be equal to zero as an additional condition for equilibrium.

1.3.2 NEWTON'S SECOND LAW

Newton's second law of motion is used to find a particle's *equations of motion*, which are equations that give the particle's position, velocity, or acceleration at any point in time. We will be using Newton's second law as the central tool for mathematically describing the motion of a particle throughout this book.

> ### *Newton's Second Law*
>
> A particle's time rate of change of linear momentum, $\mathbf{p}$, is equal to the net force, $\mathbf{F}$, applied to the particle:
>
> $$\mathbf{F} = \dot{\mathbf{p}} \qquad (1.3.1)$$

where the linear momentum of a particle with a mass m and velocity $\mathbf{v}$ is defined to be:

> ### *Linear Momentum of a Single Particle*
>
> $$\mathbf{p} = m\mathbf{v} \qquad (1.3.2)$$

Equation (1.3.1) might not be the way you are used to seeing Newton's second law. If we consider a system with constant mass m, such as a single particle experiencing a non-zero net force, then we have:

$$\mathbf{F} = \frac{d\mathbf{p}}{dt} = \frac{d}{dt}(m\mathbf{v}) \qquad (1.3.3)$$

$$\mathbf{F} = m\mathbf{a} \qquad (1.3.4)$$

and we recover the more familiar form of the second law. However, when studying the motion of an object whose mass m is changing with time, such as in the case of a rocket, the form of the second law presented in (1.3.1) will be the one needed in order to derive the object's equations of motion.

Although we will use the form of Newton's second law $\mathbf{F} = m\mathbf{a}$ most often for calculational purposes in this book, we find it helpful to think of the second law in the form:

$$\mathbf{a} = \frac{\mathbf{F}}{m} \tag{1.3.5}$$

We find this form to be more useful when understanding the concepts behind Newton's second law. Equation (1.3.5) tells us:

- *Acceleration is in the same direction as the net force.* There is no minus sign in front of $\mathbf{F}$ to denote an acceleration in the opposite direction, nor is there any mathematical transformation done to the vector $\mathbf{F}$ to rotate it.

- *The magnitude of the acceleration is directly proportional to the magnitude of the net force.* In other words, net forces with a large magnitude produce larger magnitude accelerations than net forces with small magnitudes. Mathematically, this can be seen because $\mathbf{F}$ appears in the numerator of the fraction in the right-hand side of (1.3.5) and by thinking of the equation as: $|\mathbf{a}| = |\mathbf{F}|/m$.

- *Mass "resists" acceleration.* The mass m appears in the denominator of the fraction in the right-hand side of (1.3.5). Large denominators result in smaller overall fractions when compared to a small denominator with the same numerator (i.e., $1/4 < 1/2$). In other words, with the same given force, objects with a larger mass experience a smaller acceleration, and smaller mass objects experience a larger acceleration.

All three of the above bullet points are contained in the one simple equation (1.3.5)! This is one of the reasons why physicists prefer math as the language for describing the universe. A lot can be said in one simple equation. As a physicist, you should learn how to "read equations" like we did above.

Notice the consistency between the first and second laws. If $\mathbf{F} = 0$, i.e., a zero net force is acting on the particle, then $\mathbf{a} = 0$, and the particle's velocity is not changing, just as stated in the first law. Likewise, we could have restated Newton's first law as: "A particle's momentum remains constant if no external net force acts on the particle."

As we mentioned, the second law will be used to produce differential equations which describe the motion of a particle. As a simple example, we will consider a particle moving in one dimension (along a line) under the influence of a constant force, where $\mathbf{a} = \ddot{x}\hat{\mathbf{i}}$ and $\mathbf{F} = F\hat{\mathbf{i}}$. Recall that the dots above the x denote a second derivative with respect to time. We can use (1.3.5) to write:

$$\ddot{x} = F/m = a \tag{1.3.6}$$

Equation (1.3.6) says that the solution $x(t)$ is a function such that its second derivative is equal to the constant, a. The second-order differential equation (1.3.6) is similar to an algebra problem, except instead of finding a number, you are asked to find a function. Differential equations are the mathematical language used by physicists to describe the motion of a particle. We will solve them both in closed form (analytically) and numerically. Just to whet your appetite, if a is constant, the solution to (1.3.6) is:

$$x(t) = x_0 + v_0 t + \frac{1}{2}at^2 \tag{1.3.7}$$

where $x_0 = x(0)$ and $v_0 = v(0)$ are the initial position and velocity, respectively. You can check this for yourself by taking two derivatives of (1.3.7) and showing that it satisfies (1.3.6) in the case of a constant a. In Chapter 2, we will go through the steps to derive (1.3.7) and other solutions to differential equations.

Finally, we can return to the weak equivalence principle. Suppose a particle is near the surface of the Earth and experiences only the force of gravity (weight W):

$$F = W$$
$$m_i\, a = m_g\, g \tag{1.3.8}$$

where m_i, m_g are the inertial and gravitational masses. Here we have dropped the vector notation, assumed "downward" is the positive direction, and used g for the acceleration due to gravity (9.8 m/s^2). The inertial mass m_i in this equation is the m from Newton's second law, i.e., it is the mass that "resists" acceleration, and was defined as the inertial mass. The mass m_g is the mass that is affected by gravity and was defined as the gravitational mass. As discussed previously in this chapter, the weak principle of equivalence says that $m_i = m_g$, and therefore the masses cancel, hence $a = g$ for a freely falling body near the surface of the Earth. If the weak equivalence principle were not true, then the mass of an object would affect its acceleration due to gravity. So far, physicists have not been able to detect any mass dependence (outside experimental error) on an object's acceleration due to gravity, even in the most sensitive of experiments.

1.3.3 NEWTON'S THIRD LAW

Newton's third law is a statement that discusses the nature of interactions between two particles. In order to state Newton's third law, we need to consider two objects interacting by exerting a force on each other. Let us define $\mathbf{F}_{12}$ to be the force on object 1 exerted by object 2, and $\mathbf{F}_{21}$ is the force on object 2 exerted by object 1. Newton's third law is then stated:

> ### Newton's Third Law
>
> If object 2 exerts a force $\mathbf{F}_{12}$ on object 1, then object 1 exerts a force $\mathbf{F}_{21}$ on object 2 such that:
>
> $$\mathbf{F}_{21} = -\mathbf{F}_{12} \tag{1.3.9}$$

Notice that the minus sign and lack of a scalar multiple in (1.3.9) says that the force exerted by object 2 on object 1 is equal in magnitude (shown by the lack of a scalar multiple which, if present, would change the magnitude), but opposite in direction (denoted by the minus sign) to the force object 1 applies on 2. Many of the interaction forces we will study in this book will be *central forces*. Central forces are forces that act along the line that joins the centers of the two interacting objects, such as gravity and electrostatic forces, and obey Newton's third law. Velocity-dependent forces are not central, and the third law may not apply. An example is the force between two moving electric charges; the Lorentz force is velocity-dependent, and the magnetic force vectors between the two charged particles do not lie along the same line, hence the resulting net forces do not obey Newton's third law.

Let us take (1.3.9) a little further:

$$\mathbf{F}_{21} = -\,\mathbf{F}_{12} \tag{1.3.10}$$
$$m_1 \mathbf{a}_1 = m_2\,(-\mathbf{a}_2) \tag{1.3.11}$$

Not surprisingly, we see that the accelerations of each object are in opposite directions. If we continue manipulating (1.3.11), we find:

$$\frac{m_1}{m_2} = -\frac{a_2}{a_1} \tag{1.3.12}$$

where we have taken the magnitude of the acceleration vectors. Notice that the ratio of the accelerations is inverse to the ratio of the masses. In other words, if $m_1 > m_2$ then $a_1 < a_2$, in order for the right-hand side of (1.3.12) to be a fraction greater than one. As an example, consider the classic scenario of a mosquito hitting the windshield of a moving automobile. Let the mosquito be object 2 and the car be object 1. Clearly, the automobile has more mass than the mosquito, $m_1 > m_2$, and according to (1.3.12), the acceleration of the mosquito is greater than that of the automobile, $a_2 > a_1$. It is a bad day for the mosquito. Another way to interpret (1.3.12) is that an object cannot accelerate without another object accelerating in the opposite direction.

Next, we need to address one other important fundamental concept for classical mechanics: reference frames. Once we have a working understanding of reference frames, we will have finished laying out the foundations of classical mechanics, and we will be ready to do some physics!

1.4   REFERENCE FRAMES

All of the basic descriptors of motion are measured with respect to a reference frame. A reference frame is a choice of origin, spatial and temporal, as well as a set of axes with respect to which all measurements are made. For example, if you are holding your cell phone in a car moving at a constant speed of 60 MPH while your friend is driving, then you observe your phone to be at rest. However, a bystander on the side of the road will observe your phone to be moving at 60 MPH. Who is correct? Both are, because the measurement of velocity depends on the reference frame. In this case, your frame is moving with the car, while the bystander's is at rest on the side of the road (ignoring Earth's motion). By choosing the right reference frame, you may be able to simplify a particular physics problem, in this case a moving phone versus one at rest. For another example, in introductory physics you no doubt worked on problems involving inclined planes. It is well known that choosing one axis to be parallel to the incline greatly simplifies the free-body diagram for the problem. Other changes of reference frame may simply involve changing the time for which $t = 0$.

Another important point to mention about our first example is that the two reference frames are moving relative to one another with a constant velocity. This will make the accelerations measured in each frame the same. Consider the following example, shown in Figure 1.2, with two reference frames S and S', where S' is moving with respect to S with a velocity $u = \dot{d}$, where $d(t)$ is the distance between the two reference frames at a time, t. Note that in this case, primed variables denote the reference frame they belong to, not differentiation. In each frame, an observer is measuring the location of the black dot. The observer in S measures the black dot to be located at x_S. Likewise, the observer in S' measures the black dot to be at $x_{S'}$.

Now suppose the observers in each frame want to communicate to each other the motion of the black dot. To keep things simple, we will consider motion only along the x-direction. The coordinate transformation between the two reference frames:

$$x_S = x_{S'} + d \tag{1.4.1}$$

allows the two frames to consistently describe the motion in each frame. First, we will consider the case where S' is moving at a constant speed relative to S, in other words, $u = \dot{d}$ is constant. Differentiating (1.4.1) with respect to time will give us the transformations of the velocity between coordinate systems:

$$\dot{x}_S = \dot{x}_{S'} + u \tag{1.4.2}$$
$$v_S = v_{S'} + u \tag{1.4.3}$$

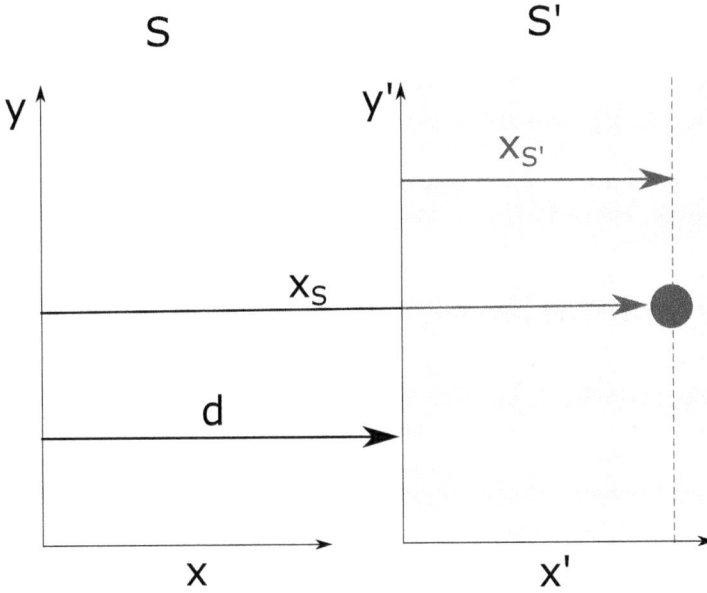

Figure 1.2: The frame S' is moving at a constant velocity with respect to the fixed frame, S. The particle (black dot) is measured to have a position x_S in the reference frame S and $x_{S'}$ in the reference frame S'. The distance between the references frames is d.

where v_S is the velocity measured in S.

Hence, if the observers in each frame wanted to determine if they are consistently measuring the velocity of the particle, then they can insert their measured velocities into (1.4.3) in order to check the other's result. Next, we compute the acceleration transformation:

$$\ddot{x}_S = \ddot{x}_{S'} + \dot{u} \tag{1.4.4}$$

$$a_S = a_{S'} \tag{1.4.5}$$

where we used the fact that u (the relative velocity between frames) is constant. Note that (1.4.5) says that the measured accelerations in the two frames are the same, thus (1.4.5) also tells us that the forces measured on the particle in each reference frame are the same and therefore Newton's laws hold in their typical form in each frame. In other words, if we measured the acceleration of the particle in each reference frame, we would find that the measured acceleration can be accounted for by considering all of the forces acting on the particle. The reference frame S' is called an *inertial reference frame* because it is moving with a constant velocity. Newton's second law holds in inertial reference frames.

Let us now contrast that with the case of S' accelerating with respect to S. In this case, (1.4.3) doesn't change, and (1.4.5) becomes:

$$a_S = a_{S'} + \dot{u} \tag{1.4.6}$$

and $\dot{u} \neq 0$. Now the two measured accelerations are not the same, and therefore the measured forces in each frame are different. In particular, the acceleration of a particle in a noninertial reference frame cannot be accounted for by summing all of the forces acting on the particle. For example, on certain carnival rides that involve rotation, you will experience an apparent outward force which is not caused by any forces acting on you from the ride itself. This is an example of a *noninertial reference frame* where Newton's laws no longer hold in their standard form, because there is an acceleration measured in S that is not apparent in S'.

As we will see in a later chapter, we can modify Newton's laws in order to address the case of noninertial frames. Such modifications involve treating the acceleration $\dot{u}$ as coming from an *inertial force*, a force that is not created by physical interactions, but rather is due to an accelerating frame. While the surface of the Earth typically approximates an inertial frame, the Earth does rotate and revolve around the Sun; hence, the velocity of a reference frame "glued" to the Earth's surface is not constant. There are cases where the noninertial nature of the "glued" frame needs to be accounted for. Such cases include long-distance motion such as missile trajectories and the motion of wind and water currents.

1.5   COMPUTATION IN PHYSICS

An important part of this book is the inclusion of computation in solving problems. When we say "computation" in this book, we mean using a computer to solve physics and mathematics problems, either analytically or numerically. Physicists use computers in a variety of ways to solve problems. In addition to providing instruction in classical mechanics, this book will also provide you instruction on how to use computers to solve physics problems.

Extensive experience in coding is not necessary in order to begin reading this book. In fact, we will mainly rely on commands that are already a part of software packages. That said, you will pick up the coding that you need along the way. In this section, we will provide a motivation for how computing is used in physics and why it is important. Coding has become a fundamental skill for a physicist. We believe that all students should have coding experience before graduating with an undergraduate degree in physics. Hopefully, by the time you have finished this book, you will have a thorough understanding of both classical mechanics and how to use computers to solve physics problems. Think of it as a two-for-one deal!

1.5.1   THE USE OF COMPUTATION IN PHYSICS

To understand how computation is used in physics, we will demonstrate the solution to some problems analytically (also known as "by hand") and using computation. At this time, it is not important to understand all of the physics in the problems, nor is it necessary to understand the computational methods used to solve the problems; we will cover those topics in detail later in this book. The important thing to take away from this section is an appreciation of how computing is used and why it is important.

Let us start with a simple physics problem that can be "solved by hand."

Example 1.3: Solving a simple differential equation

Consider a particle that is moving along a line with a constant acceleration, a. If at time $t = 0$ the particle's velocity is v_0, find the formula for the particle's velocity as a function of time.

Solution:

We know from (1.2.11) that acceleration is the first derivative of velocity with respect to time:

$$\frac{dv}{dt} = a \tag{1.5.1}$$

Next, we separate variables and integrate both sides of the equation. Think of separation of variables as multiplying by dt in order to get all of the $v's$ on one side and the $t's$ on the other. After separation of variables, (1.5.1) becomes:

$$\int_{v_0}^{v(t)} dv' = \int_0^t a \, dt' \tag{1.5.2}$$

where we matched the lower and upper limits on each side. Notice that the lower-limit
on the right-hand side is $t = 0$, and the lower limit on the left-hand side is the value of v
at $t = 0$, similar for the upper limits. In addition, we included a prime on our variables
of integration, to distinguish them from the limits of integration. Finally we integrate,
noting that a is constant:

$$v(t) = v_0 + at \qquad (1.5.3)$$

The result shows that the velocity is a linear function of time, as expected for motion
with a constant acceleration.

Example 1.3 is an example of a problem that is "done by hand", meaning that we were
able to perform the necessary mathematical manipulations to solve the differential equation
(1.5.1) without the aid of a computer. Equation (1.5.3) is called a "closed-form" solution to
the differential equation (1.5.1) because it gives a specific solution consisting of functions
and mathematical operations. What is considered as "closed-form" is somewhat arbitrary
because, for example, a solution in the form of an infinite sum may not be considered in
"closed-form."

As you will see in this book, solving physics problems very often involves finding the
solution to differential equations. Some of these equations have no closed-form solution,
while others are very difficult to find. In those cases, computation can be extremely helpful.

For the purposes of this book, we consider two forms of computation:

- *Symbolic Computation*: involves using a computer to help find closed-form solutions
  to differential equations, integrals, eigenvalues, and more. You enter the equation or
  integral you want solved, and the computer program returns a closed-form solution.
  Recall tables of derivatives and integrals; symbolic computation is a much more sophisticated version of those.

- *Numerical Computation*: provides a list of numerical values representing a solution.
  For example, a numerical solution to (1.5.1) can be thought of as a table of values with
  two columns, t and $v(t)$, where each row contains a specific time t, and the value of v
  at time t. Numerical solutions are often best represented as a graph of ordered pairs
  (and sometimes triplets, depending on the dimension of the problem being solved).

In the next two subsections we demonstrate symbolic and numerical computation, by solving
Example 1.3 using Python and Mathematica.

1.5.2 SYMBOLIC COMPUTATIONS

In this section we present an example of symbolic computation, where we use Python and
Mathematica to solve Example 1.3 symbolically. Since these are the first examples of this
type in this book, we provide detailed explanations of what is going on in the code.

In Python we obtain symbolic general solutions of a differential equation using the
Symbolic Python (SymPy) command `dsolve`. This command has typically three arguments:
the differential equation that we are trying to solve, the function for which the differential
equation is being solved, and the initial conditions of the problem. Let's look at the details
of the Python code.

Python Code for Example 1.3

The Python code for solving the differential equation $dv/dt = a$ from Example 1.3 is shown below.

The first line tells Python to import several commands from the Symbolic Python (SymPy) library. Libraries are written by experienced programmers, and they include functions which can be used by any code that imports the library. In this case, the SymPy library includes functions like `dsolve` , which solve differential equations in closed form.

All of the symbols need to be specified in Python. We use `Symbol('t', real=True, positive=True)` to indicate that t is a real and positive variable. Similarly, `symbols` is used to indicate that a,v0 are real variables. We use the `Function` command to indicate to Python that v[t] is a function.

The general solution of the differential equation is obtained with the SymPy command `dsolve`, whose arguments are the differential equation, the function for which the differential equation is being solved, and the initial conditions of the problem. In this example, the structure of this command is

`dsolve(diff(v(t),t)-a,v(t),ics=initconds)`.

Notice that in Python, the differential equation is written such that the right-hand side is equal to zero, in this example `diff(v(t),t)-a` represents the equation $dv/dt = a$, and only the left-hand side of the differential equation is entered. Here v(t) is the desired solution, and `ics=initconds` specifies the initial conditions of the problem. These conditions are specified in the line `initconds = {v(0):v0}` which represents the given initial condition $v(0) = v_0$.

```
from sympy import Symbol, symbols, Function, dsolve, diff

print('-'*28,'CODE OUTPUT','-'*29,'\n')

# define symbols and the function v
v = Function('v')
t = Symbol('t', real=True, positive=True)
a, v0 = symbols('a, v0', real=True)

# initial conditions for the ODE
initconds = {v(0):v0}

# use dsolve to obtain solution of ODE   dv/dt=a
solution = dsolve(diff(v(t),t)-a,v(t),ics=initconds).rhs

print('The solution is   v(t) = ',solution)

-------------------------- CODE OUTPUT ----------------------------
The solution is   v(t) =   a*t + v0
```

In Mathematica we obtain symbolic general solutions of a differential equation using the command `DSolve`. This command has typically three arguments: the differential equation that we are trying to solve together with the initial conditions of the problem, the function for which the differential equation is being solved, and the independent variable in the solution. Let's look at the details of the Mathematica code.

Mathematica Code for Example 1.3

The Mathematica code for solving the differential equation $dv/dt = a$ from Example 1.3 is shown below. The solution is obtained using the command `DSolve`:

`solution=DSolve[{v'[t] == a, v[0] == v0}, v, t];`.

Notice that the differential equation *and* its initial condition appear as the first argument of the command (in between the curly brackets). Notice also the use of a double equal sign `==`, as opposed to `=` in the equation. Often in programming languages, the single equal sign is used for variable assignment, whereas the double-equal sign is used for logical equivalence.

The code stores the solution generated by `DSolve` in the variable `solution` (notice now the use of a single equal sign). The second argument v in `DSolve` tells Mathematica that v is the function for which we are solving, and the third argument identifies t as the independent variable.

The line of code `speed=v[t]/.solution[[1]]` selects the desired part of the solution of the differential equation, by using the replacement rule `/.solution`.

The `Print` line in the code outputs the solution. As expected, Mathematica's solution is the same as the one we derived in Example 1.1 and in the Python code.

Also note that Mathematica automatically recognizes v0,t as symbols to be manipulated. In addition, the Mathematica syntax `v[t]` allows Mathematica to automatically recognize v as a function of time t. This is a much simpler situation than Python, in which we had to define carefully all the symbols used in the code.

Clear["`*`"]

solution = DSolve[{$v'[t]$==$a, v[0]$==$v0$}, v, t];

speed = $v[t]$/.solution[[1]];

Print["The solution of the ODE is v[t] = ", speed]

The solution of the ODE is v[t] = $at + v0$

In Example 1.3 we see that Mathematica requires a much shorter code to solve this problem than Python. This is because Mathematica is designed, in part, for symbolic manipulations. Computer algebra systems (CAS) like Mathematica are very convenient and easy to use when performing symbolic computations. The disadvantage is that it is proprietary and much more expensive than other options.

Next, we will look at an example of a numeric computation and compare the two languages in that context.

1.5.3   NUMERICAL COMPUTATIONS WITH PYTHON AND MATHEMATICA

In this section we use Python and Mathematica to solve Example 1.3 *numerically*.

In Python we obtain numerical general solutions of a differential equation using the SymPy command `odeint`. This command has typically three arguments: the differential equation that we are trying to solve, the function for which the differential equation is being solved, and the initial conditions of the problem.

In Example 1.4 we solve the simple differential equation $dv/dt = a$, and we get a detailed look at the `odeint` command.

Example 1.4: Numerical solution of a simple differential equation
   Using $a = 9.8$ m/s^2 and an initial velocity of $v(0) = v_0 = 1$ m/s, find and graph the numerical solution to the differential equation from Example 1.3:

$$\frac{dv}{dt} = a \qquad\qquad (1.5.4)$$

Python Code
   We import three libraries, NumPy, SciPy, and the graphics library Matplotlib. From the SciPy library we import only the function `odeint`, which is used to numerically solve the differential equation (1.5.4). After the libraries are imported, we assign the numerical values of the initial condition variable v0 and the acceleration a.
   To solve the differential equation numerically, we create a new function in Python that we called `velderiv(v,t)`, which contains the differential equation we are solving. In Python, function definition is done using the command `def`. The arguments of the function are v and t, and they are included in parentheses following the function name.
   The next few lines contain the actual calculation of the function. Notice we included a local variable `dvdt`, which stays within the function and is equal to the first derivative of v. The last line of the function is `return dvdt`, which returns the value of the variable `dvdt`. Note also that the contents of a Python function are always indented.
   We also need to tell Python for which values of t we will be computing $v(t)$. In this case, `times` is an array which contains the list $[t_0, t_1, \ldots, t_{30}] = [0, 0.1, \ldots, 3.0]$, and hence we will be solving for $v(0)$, $v(0.1)$, $\ldots$, $v(3.0)$. Next, we solve the differential equation using the command `odeint`, which is imported from the `scipy.integrate` library. The `odeint` command in this example is:

```
velocity = odeint(velderiv,v0,times).
```

This command has three arguments: the differential equation (as the user-defined function `velderiv`), the initial value (v0), and the list of times `times`. The result is an array which we called `velocity`, and it contains values $[v_0, v_1, \ldots]$, where $v_i = v(t_i)$. The last group of lines in the program set up the graph of v versus t, and the plot of $v(t)$ is shown in Figure 1.3.

```
import numpy as np
from scipy.integrate import odeint
import matplotlib.pyplot as plt

# initial condition v(0)=1, and acceleration a =9.8 m/s^2
v0 = 1
a = 9.8

# define function velderiv to be called by odeint
def velderiv(v,t):
    dvdt = a
    return dvdt

# times at which to evaluate the numerical solution
times = np.linspace(0,3,30)

# solve the ODE with initial v(0)=v0
velocity = odeint(velderiv,v0,times)

# plot v(t) and label the axes
plt.plot(times,velocity)
plt.ylabel('v(t), m/s')
plt.xlabel('Time, s')
plt.show()
```

Mathematica Code for Example 1.4

After clearing all previously assigned variables with the command `Clear`, we use the `SetOptions` command to set up options for the graphics output. You will see this type of line in most Mathematica codes in this book.

Next we specify the value of the acceleration a, and solve numerically the differential equation in the line:

`NDSolve[{v'[t]==a,v[0]==1},v,{t,0,3}];`

Note that we define the initial value `v[0]` *inside* the Mathematica command `NDSolve`.

Note also that the arguments for `NDSolve` are similar to those of `DSolve`, i.e. the order of the arguments are: equation to be solved and initial values, the function for which we are solving (v), and identification of the independent variable t. This time, however, we needed to specify the range of t values for which v is being solved. The numerical solution is a table of values containing columns t and v(t), so we need to tell the computer when to start finding the solution $(t = 0)$ and when to stop $(t = 3)$, with the last argument `{t,0,3}`.

The final line of the code produces a graph of the solution. As expected, the graph is a line with a y-intercept of 1.0 and slope of 9.8. Everything after `{t,0,3}` in the final line of the code is formatting commands which only affect the visual appearance of the graph.

```
Clear["`*"]

SetOptions[Plot, Frame->True, BaseStyle->{FontSize->16},
PlotRange->All];

a = 9.8;
solution = NDSolve[{v'[t]==a, v[0]==1}, v, {t, 0, 3}];
speed = v[t]/.solution[[1]];

Plot[speed, {t, 0, 3}, FrameLabel->{"time, s", "v(t) m/s"}]
```

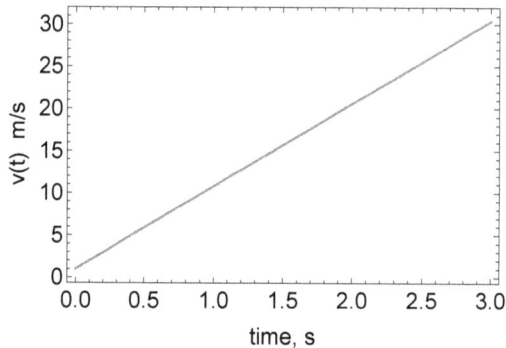

Figure 1.3: Plot of the solution $v(t) = v_0 + a\,t$ of the simple differential equation in Example 1.4.

Again, the Python code is longer than the Mathematica code. Does that make Mathematica better than Python? That answer depends on what you are trying to do. If you are interested in quick development and implementation, then Mathematica might be a very useful tool for you. However, as we mentioned before, Mathematica is proprietary and more expensive than Python. The proprietary nature of Mathematica means that you don't have access to the source code, and this can be a problem if you want to know exactly what the commands are doing. Python is open source software, so if you want to know what is "going on under the hood", you can find out. The open source nature of Python is valuable when doing research, and when you want to identify if unexpected results are due to bugs in the code, or they are due to new science. That said, Mathematica is a language of its own, and you can write your own programs in it. Finally, Python is free, and your budget may dictate your choice. When working on their own research problems, the authors of this book will often use both programs, choosing the most appropriate tool for the particular job at hand.

It should be pointed out that sometimes, a closed-form solution to a differential equation doesn't exist or is difficult to find. In those cases, we rely on numerical solutions to give us the information we need. You may think that numerical solutions are of limited value, but as we will see throughout this book, there is a lot that can be found using numerical solutions (also called numerics). In Chapter 13, we will study nonlinear oscillators, and we will learn how to obtain a lot of information from numerical solutions.

It is easy to walk away from an undergraduate physics education thinking that all physics problems have closed-form solutions. This conclusion arises from the types of problems that undergraduate students solve as part of their education. The truth of the matter is that most problems physicists work on outside of the classroom require numerical solutions. The differential equations governing real-world systems are often complex and not solvable in closed-form. In those cases, numerics may be the only option to gaining any kind of insight into the problem. A physicist with strong computational skills will be well-prepared to tackle a wide variety of problems, not just in classical mechanics, but in any field of physics or engineering.

Furthermore, physicists find themselves working on a variety of problems outside the traditional subfields of physics. Physicists often end up working on problems in finance, economics, biology, climate science, and materials science, just to name a few. Today's physicists are involved in modeling economies, disease propagation, and social networks. These types of problems allow physicists to apply their skills, including strong computational and modeling skills, to interesting problems. It is our hope that this book will not only prepare you well in physics, but also set you on the path of developing strong computational skills, so that you may tackle the exciting problems both inside and outside of physics, wherever your career may take you.

1.5.4 SOME WARNINGS

We will end this section with some warnings. First, the availability of computational tools to solve problems does not mean that you do not need to know how to do math by hand! There are several reasons for this. The first reason is that computer algorithms can and do give wrong answers. Computational algorithms have significantly improved over the decades, and wrong answers from them are not as common as in the early days of computing. However, you need to be able to identify a wrong answer when you see one. A proficiency with mathematics is still needed for identifying wrong answers. Furthermore, it is not unusual for one to perform additional algebraic manipulation to the output of a CAS, in order for the result to be in a more convenient or insightful form. We will see some examples in this book where the output of a CAS will be technically correct, but not as useful for describing the physics of the situation as compared to a result that we would have obtained "by hand."

Second, a computer will provide a solution with no context. The techniques and critical thinking that you learn by doing mathematics will train you in how to understand and interpret computer-generated solutions, so that you can put the solution in the proper context. For example, does your solution have appropriate limits? Does the solution conform to known laws of physics? A person skilled in mathematics and physics is still required to interpret computer-generated solutions (at least for now...). In particular, when one becomes skilled at mathematics, symbolic algorithms should be considered as a tool to help one focus on the physics, without getting encumbered by lots of mathematics. In a sense, you should think of symbolic algorithms as being similar to a calculator. Only after learning arithmetic "by hand," should one begin to use a calculator. In the same sense, mastery of algebra, trigonometry, and calculus, should be a prerequisite for extensive use of symbolic algorithms.

Third, who programs computers to solve math problems? The people who know how to solve the math problems. Hence, we need people who can do math, in order to ensure that we will have future generations of algorithms to assist us.

The final warning comes from the nature of the algorithms presented in this text. The authors do not claim that the codes contained within are the most efficient means of solving the problems in the text. We tended to produce codes that focused on pedagogy over efficiency. We encourage you to rewrite the programs contained within, to see if you can improve them and make them more efficient. By playing around with the codes, you'll get a better understanding of the algorithms and programming in general. Have fun!

1.6 CHAPTER SUMMARY

Motion is described using the *basic descriptors* of motion, displacement, velocity, and acceleration. In Cartesian coordinates the position, velocity, and acceleration of a particle are:

$$\mathbf{r} = x\hat{\mathbf{i}} + y\hat{\mathbf{j}} + z\hat{\mathbf{k}} \tag{1.6.1}$$

$$\mathbf{v} = \dot{x}\hat{\mathbf{i}} + \dot{y}\hat{\mathbf{j}} + \dot{z}\hat{\mathbf{k}} \tag{1.6.2}$$

$$\mathbf{a} = \ddot{x}\hat{\mathbf{i}} + \ddot{y}\hat{\mathbf{j}} + \ddot{z}\hat{\mathbf{k}} \tag{1.6.3}$$

where $\hat{\mathbf{i}}$, $\hat{\mathbf{j}}$, and $\hat{\mathbf{k}}$ are the unit vectors.

When describing the motion of an object, two additional quantities become important, the mass of the object and the net force acting on the object. The inertial mass of an object is a measure of the object's inertia, whereas the gravitational mass of an object is the mass that determines the gravitational force acting on it. However, the weak principle of equivalence says that the inertial and gravitational masses are equal.

A change in an object's motion is caused by a non-zero net external force acting on the object. Newton's laws of motion describe how forces change the motion of an object.

Newton's first law: A particle's velocity remains constant if the net force acting on the object is zero.

Newton's second law: A particle's time rate of change of momentum, dp/dt, is equal to the net force $\mathbf{F}$ applied to the particle :

$$\mathbf{F} = dp/dt = \dot{\mathbf{p}} \tag{1.6.4}$$

Newton's third law: If object 2 exerts a force $\mathbf{F}_{12}$ on object 1, then object 1 exerts a force $\mathbf{F}_{21}$ on object 2 such that:

$$\mathbf{F}_{21} = -\mathbf{F}_{12} \tag{1.6.5}$$

All measurements are made with respect to a reference frame. A reference frame is a choice of origin (spatial x and temporal t), and a set of axes with respect to which all measurements are made.

Two reference frames S, S' are called inertial reference frames when Newton's laws hold, and the forces measured in S' are the same forces as those measured in S.

Two reference frames S, S' are called noninertial reference frames when Newton's laws no longer hold in their standard form, because there is an acceleration measured in S that is not apparent in S'.

Computation is an important tool for physicists. In this book, we will focus on two types of computation, symbolic and numerical. Symbolic computation involves using a computer to help find closed-form solutions to differential equations, integrals, eigenvalues, etc. Numerical computation, or "numerics," provides a list of numerical values representing a solution to differential equations, integrals, eigenvalues, etc.

1.7 END-OF-CHAPTER PROBLEMS

The symbol 🖥 indicates a problem which requires some computer assistance, in the form of graphics, or numerical computation, or symbolic evaluation.

Section 1.2: The Basics of Classical Mechanics

1. 🖥 The position of a particle with mass $m = 0.50$ kg can be described using the vector function:
$$\mathbf{r} = 3a\,t\hat{\mathbf{i}} - 2b\,t^2\hat{\mathbf{j}} + 7c\,t^{-2}\hat{\mathbf{k}}$$
where a, b, c are constants.

   a. What are the SI units of each of the coefficients a, b, c? Assume that time t is measured in seconds and position is measured in meters.

   b. Evaluate the velocity of the particle and the force acting on it at $t = 3$ seconds and for $a = b = c = 1$, both analytically and using the symbolic capabilities of Mathematica and/or Python.

2. 🖥 Consider the position vector $\mathbf{r} = 3\,t\,\hat{\boldsymbol{\rho}}$, where $\hat{\boldsymbol{\rho}} = \cos\phi\,\hat{\mathbf{i}} + \sin\phi\,\hat{\mathbf{j}}$ and $\phi = \phi(t)$ is a scalar function of time t. Compute the velocity and acceleration of the particle, both analytically and using a CAS.

3. 🖥 A particle's position is given by $\mathbf{r}(t) = R\cos(\omega t)\,\hat{\mathbf{i}} + R\sin(\omega t)\,\hat{\mathbf{j}} + (\omega t)\,\hat{\mathbf{k}}$, where R and ω are constants. Describe and plot the path of this particle in a 3D Cartesian coordinate system for the numerical values $R = \omega = 1$ and for times $t = 0$ to $t = 10$ s. What are the particle's velocity and acceleration as functions of time?

4. 🖥 Consider the vectors $\mathbf{A} = \hat{\mathbf{i}} + 2\hat{\mathbf{j}} + 2\hat{\mathbf{k}}$ and $\mathbf{B} = \hat{\mathbf{i}} + \hat{\mathbf{j}} + \hat{\mathbf{k}}$. Plot these two vectors and their sum $\mathbf{A} + \mathbf{B}$ on a 3D Cartesian coordinate system using Python and/or Mathematica.

5. 🖥 Consider the vectors $\mathbf{A} = \hat{\mathbf{i}} + 2\hat{\mathbf{j}}$ and $\mathbf{B} = \hat{\mathbf{i}} + \hat{\mathbf{j}}$. Plot these two vectors and their sum $\mathbf{A} + \mathbf{B}$ on a 2D Cartesian coordinate system using Python and/or Mathematica.

Section 1.3: Newton's Laws of Motion

6. Consider the following applications of Newton's laws of motion:

   a. Using Newton's third law, explain why a gun recoils when it is fired. Explain why the gun's recoil velocity is smaller than that of the bullet.

   b. Consider a universe where the weak equivalence principle was not true. How would Newton's universal law of gravitation be changed?

   c. Using Newton's laws, explain what would happen to the Earth's trajectory if the Sun were to suddenly disappear.

7. Consider the four situations below. Qualitatively describe the velocity and acceleration (as a function of time) of a particle of mass m experiencing each of these forces:

   a. A particle starting at rest at a location $x = A$ from the origin experiences a force $F = -kx$, where x is a position coordinate which represents a displacement from an equilibrium position (where $F_{\text{Net}} = 0$) located at the origin and k is a positive constant.

    b. A particle starting at rest at the origin experiences a force $\mathbf{F} = \frac{c}{t^2+a^2}\hat{\mathbf{i}}$, where t is time and c and a are positive constants.

    c. A particle starting at the origin moving with an initial velocity v_0 in the positive x-direction experiences a force $\mathbf{F} = -bv\hat{\mathbf{i}}$, where v is velocity and b is a positive constant.

    d. An object released from rest at a large height H above the Earth's surface experiences a force $\mathbf{F} = (mg - bv)\hat{\mathbf{j}}$, where g is the acceleration due to gravity, b is a constant, and v is velocity. Assume the positive y-direction is toward the Earth's surface.

Section 1.4 Reference Frames

8. Consider a reference frame S which is at rest, and another reference frame S' whose origin is at a location $\mathbf{r}_0 = ct^2\hat{\mathbf{i}}$ relative to the origin of S, where c is a positive constant. A particle of mass m has a position in S', which described by the vector function $\mathbf{r}' = bt^3\hat{\mathbf{i}}$, where b is a positive constant.

    a. Compute the acceleration of the particle as measured in each frame.

    b. Compute the force on the particle measured in each frame. Why are the two forces different?

9. Consider a reference frame S which is at rest, and another reference frame S', whose origin is at a location $\mathbf{r}_0 = at\hat{\mathbf{i}} + bt\hat{\mathbf{j}}$ relative to the origin in S, where a and b are positive constants and t represents time. Compute the velocity of a particle as measured in S, whose position in S' is described by $\mathbf{r}'(t) = -ct^2\hat{\mathbf{i}}$, where c is a positive constant.

10. Consider a reference frame S which is at rest, and another reference frame S', whose origin is at a location $\mathbf{r}_0 = 3t^2\hat{\mathbf{i}} - 5t^3\hat{\mathbf{j}}$ relative to the origin of frame S, and t represents time. A particle's position in S' is described by the vector function $\mathbf{r}'(t) = -t^2\hat{\mathbf{i}} + 3t^2\hat{\mathbf{j}}$. Find the force measured in each frame if the mass of the particle is m.

11. Consider a reference frame S which is at rest, and another reference frame S', whose origin is at a location $\mathbf{r}_0$ with respect to the origin of S. Is the momentum of the particle $\mathbf{p} = m\mathbf{v}$, as measured in each frame, the same? Show that if S and S' are inertial frames, then the force on the particle, as measured in S' is the same as that measured in S. In other words, $\mathbf{F}' = d\mathbf{p}'/dt = d\mathbf{p}/dt = \mathbf{F}$ where $\mathbf{p}$ $(\mathbf{p}')$ is the momentum of a particle of mass m in S (S'). Is the angular momentum $\boldsymbol{\ell} = \mathbf{r} \times \mathbf{p}$ the same in each frame? How about the torque $\mathbf{N} = \mathbf{r} \times \mathbf{F}$?

12. In the theory of special relativity, we need to alter the transformation equations between two frames. Let S be a rest frame, and S' a frame moving at a speed v relative to S. Suppose that S' moves in the x-direction, in other words, S' moves parallel to the x-axis of the frame S. The theory of special relativity then states that measurements made in S and S' are related by the formulas:

$$x' = \gamma(x - vt)$$
$$y' = y$$
$$z' = z$$
$$t' = \gamma\left(t - \frac{vx}{c^2}\right)$$

where $\gamma = \sqrt{1 - v^2/c^2}$ and c is the speed of light.

a. Consider that two events in S occur at two locations $x = 0$ and $x = a$ at time $t = 0$. Find the times of the two events as measured in S'. Notice that events that are simultaneous in S are not simultaneous in S'. Which event was seen first in S'?

b. ⌨ Show that the space-time interval $(\Delta s)^2 = c^2 t^2 - x^2 - y^2 - z^2$ stays invariant in the system S', i.e., $(\Delta s)^2 = (\Delta s')^2 = c^2 (t')^2 - (x')^2 - (y')^2 - (z')^2$. Do this part of the problem both analytically and using a CAS.

Section 1.5 Computation in Physics

13. ⌨ Solve the following differential equation analytically and numerically using Python and/or Mathematica :

$$\frac{df}{dx} = 5 \sin x - 4e^{-x}$$

with the initial condition $f(0) = 0$. Plot the solution for $x = 0$ to 3 with a step size of 0.1.

14. ⌨ Assign the values 5 and 6 to the variables a and b, respectively. Then compute the following expressions within two decimal points using Python and/or Mathematica:

a. $(a + b) \sin(0.25\pi)$

b. $a\,b\,e^3/(\sinh a)$

c. $\sqrt{7}\,a^b/(a + b\,i)$ where $i = \sqrt{-1}$

15. ⌨ Define an array x (sometimes called a list) which contains the numbers 0 through 2π in steps of 0.01. Define the function $f(x, m, k) = e^{-mx} \cos(2\pi kx)$ of three variables m, k, x. Plot the function $f(x, m, k)$ with $k = 1$ and $m = 0.5$. Label the axes and choose a color and plotting symbol other than the default for your curve. Save the resulting graph to a .jpg or .png file.

16. ⌨ Plot $f(x) = x^2$ and $g(x) = x$ on the same graph, for the range $x = 0$ to $x = 10$. Each function should have its own color, label and plotting symbol in the graph.

17. ⌨ A conditional statement is an important tool for helping computers deal with contingencies. They take the form "if-then-else." Conditional statements allow for computers to do different things, depending on whether or not a condition is true. In other words, if a condition is true, then do one thing, if it is not true (else) then do something different. For example, if a variable x is less than 5 assign the value 1 to the variable a; otherwise, assign the value 0 to a. Write a code in which you assign a value to the variable x and then prints "x is less than 5" if $x < 5$, and prints "x is greater than or equal to 5" if $x \geq 5$.

18. ⌨ Using the conditional *if* statement, described in Problem 17, define the piecewise function

$$f(x) = \begin{cases} 0 & x < 5 \\ x^2 & x \geq 5 \end{cases}$$

evaluate $f(x)$ for $x = 0.5$ and $x = 2$, and plot $f(x)$ for the range $x = 0$ to $x = 3$.

19. 🖥 Computers are very good at repeating tasks over and over again, in a procedure called a *loop*. The basic types of loops are `for` loops, `do` loops, and `while` loops. Using any of these three types of loops, compute 6! (factorial) and display the result.

20. 🖥 The Fibonacci series of numbers x_n is defined by the recursive sequence $x_n = x_{n-1} + x_{n-2}$, starting with the initial values 1,1. Compute the first 17 values of the Fibonacci sequence. You may want to use `for` loops, and you may also want to save your results into an array using the `append()` function in Python, or `Append` in Mathematica, in order to add values to an array.

21. 🖥 Use symbolic computation and the `solve` command in Python and `Roots` command in Mathematica, to solve the following equations for x. Note that you will get complex roots for some of the solutions.

    a. $7x + 5 = 0$

    b. $x^2 - 5x + 2 = 0$

    c. $x^3 + 7x - 5 = 3$

22. 🖥 Use symbolic computation and the `solve` command in Python, and `Solve` in Mathematica, to solve the following system of equations for (x, y):

$$2x - 5y = 7$$
$$x + y = 2$$

23. 🖥 Numerically solve the system of coupled differential equations:

$$\frac{dx}{dt} = 0.6x - 1.2x\,y$$
$$\frac{dy}{dt} = x\,y - y$$

using the initial conditions $x(0) = 40$ and $y(0) = 7$. Plot x as a function of time, y as a function of time, and y as a function of x.

2 Single-Particle Motion in One Dimension

In this chapter, we will examine one-dimensional motion, i.e., motion along a line. A careful study of one-dimensional motion will be a useful foundation for understanding more general motion in higher dimensions. In this chapter, we will give several examples of solving Newton's second law, $F = ma$ in one dimension. We will consider several types of forces: both constant and those which depend on time $F(t)$, velocity $F(v)$, and position $F(x)$. In addition, we will discuss and demonstrate how to use computers to solve ordinary differential equations (ODEs) using computer algebra systems and numerical methods. Finally, we discuss various methods available for numerical evaluation of definite integrals of functions of a single variable.

2.1 EQUATIONS OF MOTION

To begin our study of one-dimensional motion, we first need to make some assumptions about the object whose motion we are examining. One fundamental assumption in this chapter is that the object being studied is a point particle. In order to mathematically describe the motion of a particle under the influence of a force, we need to find the particle's *equations of motion*. The equations of motion of a particle are the equations which describe its position, velocity, and acceleration as functions of time. Equations of motion can be in the form of algebraic equations, or in the form of differential equations.

As we will see, the equations of motion of a particle can be found by solving Newton's second law as a differential equation. In this chapter, we will focus on one-dimensional motion, where the force vector and the particle's displacement are along the same line (but not necessarily in the same direction). Because all vectors in a given problem lay along the same line, we drop the vector notation in all the equations. A negative sign between two quantities will denote vectors that lie in opposite directions along the same line.

Newton's second law in one dimension is:

$$F = ma \tag{2.1.1}$$

Recall that acceleration is the first derivative of velocity v with respect to time and the second derivative of displacement x with respect to time. To solve for the equations of motion, we will think of (2.1.1) as a differential equation, by rewriting (2.1.1) in the following ways:

$$F = m\frac{d^2x}{dt^2} \tag{2.1.2}$$

$$F = m\frac{dv}{dt} \tag{2.1.3}$$

$$F = mv\frac{dv}{dx} \tag{2.1.4}$$

Notice that each of the above equations is a differential equation which can be solved once the net force F acting on the particle is specified. Each of the above equations yields a different equation of motion: (2.1.2) can be solved for $x(t)$, (2.1.3) for $v(t)$, and (2.1.4) for $v(x)$. Equation (2.1.4) comes from the chain rule and is used when the force is dependent upon the particle's position x:

$$m\frac{dv}{dt} = m\frac{dv}{dx}\frac{dx}{dt} = mv\frac{dv}{dx} \tag{2.1.5}$$

Throughout the rest of this chapter, we will solve (2.1.1) for several different cases where forces are constant ($F = F_0$), time-dependent ($F = F(t)$), velocity-dependent ($F = F(v)$), and position-dependent ($F = F(x)$). However, before solving (2.1.1), we will make a few comments about differential equations in general.

2.2   ORDINARY DIFFERENTIAL EQUATIONS

Simply put, an ordinary differential equation (ODE) is an equation that contains the derivative of a function. For example,

$$\frac{dx}{dt} = 7 \tag{2.2.1}$$

is a differential equation that says that the solution to the differential equation $x(t)$ is a function whose first derivative is equal to 7. Of course we know that $x(t) = 7t$ is a solution that works. However, there are an infinite number of other solutions as well, since we can add a constant to $x(t)$ and still have a solution to our ODE. Hence, the so-called general solution is $x(t) = 7t+c$, where c is a constant. This differential equation is simple enough to solve. However, most ODEs are not that simple, and many cannot be solved at all. Before discussing how to solve an ODE, let's first point out a few things about our example.

1. Equation (2.2.1) is called a *first-order ODE*, because the highest derivative in the equation is a first derivative. In general, an nth-order ODE is an ODE whose highest derivative is an nth derivative.

2. The ordinary derivative implies that x is a function of only one variable t, which is the variable of differentiation.

3. The number of arbitrary constants in the general solution of (2.2.1) is equal to the order of the ODE.

To solve (2.2.1), we needed to separate the variables of the equation. Colloquially speaking, this means getting all the terms with x on one side of the equation and all the terms that are either constant or depend on t on the other. This process is called *separation of variables* and is performed by treating the derivative as a fraction and multiplying both sides of (2.2.1) by dt,

$$dx = 7dt \tag{2.2.2}$$

$$\int dx = \int 7dt \tag{2.2.3}$$

$$x = 7t + c \tag{2.2.4}$$

To solve (2.2.1), we carried out the integral after separating variables. Note that both integrals would produce constants of integration, but since both are constants, we can combine them into one arbitrary constant. You can double-check the solution by computing the derivative of $7t + c$, to check that it satisfies (2.2.1).

What happens in the case of second-order ODEs? Second-order ODEs are common in physics. There are many techniques to solve them, but we will demonstrate only one. Consider the ODE,

$$\frac{d^2x}{dt^2} = 7 \tag{2.2.5}$$

Separation of variables does not make sense here, because we typically do not integrate terms like d^2x. However, we can define a new variable v, such that, $v = dx/dt$. Then (2.2.5) becomes,

$$\frac{dv}{dt} = 7 \tag{2.2.6}$$

which we know gives the answer $v(t) = 7t + c_1$, where c_1 is the constant of integration. However, we want $x(t)$, so we use $v = dx/dt$:

$$\frac{dx}{dt} = 7t + c_1 \tag{2.2.7}$$

$$\int dx = \int (7t + c_1)\, dt \tag{2.2.8}$$

$$x(t) = 3.5t^2 + c_1 t + c_2 \tag{2.2.9}$$

where c_2 is the constant of integration obtained by performing the integral in (2.2.8). Hence, we pick up an additional constant of integration in our solution, giving two arbitrary constants for the solution of the second-order ODE (2.2.5). Loosely speaking, we see that the number of arbitrary constants in the solution of an nth-order ODE is equal to n, because we need to do n integrations in order to solve the equation, and each integration produces an arbitrary constant.

Next, we return to (2.2.1). Suppose (2.2.1) was an equation we wanted to use in order to find the position of a particle as a function of time. The infinite number of solutions is not helpful. Which solution describes the actual path taken by the particle? In order to specify the *particular solution* for an ODE, we need to include *initial conditions*, the value of our function at a particular time (normally at $t = 0$). Suppose we know that at $t = 0$, the particle is at a position, $x(0) = 3$. Then we can solve for the arbitrary constant by inserting the initial condition into our general solution,

$$x(0) = (7)(0) + c = 3 \tag{2.2.10}$$

which gives $c = 3$. Our particular solution is then, $x(t) = 7t + 3$. A different initial condition will give a different particular solution.

Now suppose we wanted to find a particular solution to (2.2.5); in that case one initial condition will not be enough, because it will leave out one arbitrary constant. Hence, we will need to specify both $x(t)$ and dx/dt at a particular time (usually $t = 0$). Suppose that $x(0) = 3$ and $v(0) = 1$, where $v = dx/dt$. Then we have:

$$x(0) = 3.5(0)^2 + c_1(0) + c_2 = 3 \tag{2.2.11}$$
$$v(0) = 7(0) + c_1 = 1 \tag{2.2.12}$$

where (2.2.12) is the derivative of the general solution evaluated at $t = 0$. The result is $c_1 = 1$ and $c_2 = 3$, and the particular solution is $x(t) = 3.5t^2 + t + 3$.

In summary, in order to solve for the particular solution of an nth-order ODE, we need n initial conditions. In addition, we can also specify x using knowledge of the value of x at two different times, as opposed to knowing initial values of x and its first derivative. In classical mechanics, it is most common to know the initial conditions of the position and velocity.

However, in other fields, such as electromagnetism and thermodynamics, it is often more common to know the value of a function, say the temperature, at two different locations. In this case, we have what is known as a *boundary value problem,* and the conditions that provide the constants of integration are known as *boundary conditions.*

There are many other properties of ODEs such as linear superposition, that we will explore in this book as we need them. For now, we know enough about ODEs to get started. Let's get back to the physics.

2.3 CONSTANT FORCES

Consider the case where a constant net force, $F = F_0$, acts on a particle with mass m constrained to move along a line. In this case, Newton's second law (2.1.1) gives:

$$\frac{dv}{dt} = \frac{F_0}{m} = a \tag{2.3.1}$$

where a is a constant because F_0 is constant. You may recall (2.3.1) from Example 1.1, Equation (2.3.1) is an example of a first-order differential equation. We can solve this through the process of separation of variables and then integrating the resulting equation. The first step in separating variables is to multiply both sides of the equation by dt:

$$dv = a\,dt \tag{2.3.2}$$

which can be integrated to yield,

$$v(t) = \int a\,dt = at + c_1 \tag{2.3.3}$$

where c_1 is the constant of integration. The constant, c_1, can be found using the initial condition of $v(t_0) = v_0$, where v_0 is the initial velocity at the initial time t_0. When $t_0 = 0$, (2.3.3) gives $v(0) = c_1$ or $c_1 = v_0$. Therefore, the solution to the differential equation (2.3.1) is:

$$v(t) = v_0 + a\,t \tag{2.3.4}$$

Before moving forward, we should mention that there is an alternate method for finding (2.3.3), which can be found by explicitly inserting the initial and final conditions when integrating (2.3.2):

$$\int_{v_0}^{v(t)} dv' = \int_{t_0}^{t} a\,dt' \tag{2.3.5}$$

$$v(t) - v_0 = a(t - t_0) \tag{2.3.6}$$

where we have introduced primes to the variables of integration, in order to distinguish them from the limits of integration. Notice that the lower limit in the left-hand side of (2.3.5) corresponds to the value of $v(t)$ when $t = t_0$, the lower limit of the right-hand side of (2.3.5), and with similar considerations for the upper limits. It is very important that the limits match on both sides of the equation.

Next, we can get an equation for $x(t)$ by writing $v = dx/dt$:

$$\frac{dx}{dt} = at + v_0 \tag{2.3.7}$$

and separating variables, we obtain:

$$\int_{x_0}^{x(t)} dx' = \int_{t_0}^{t} (at' + v_0)\,dt' \tag{2.3.8}$$

Notice how all the time-dependent and constant terms are on the same side of the equation. In this context, separation of variables always involves multiplication and division, never addition and subtraction. Choosing $t_0 = 0$ and performing the integral results in:

$$x(t) = \frac{1}{2}at^2 + v_0 t + x_0 \qquad (2.3.9)$$

Together, equations (2.3.4) and (2.3.9) are the only equations you need to know, in order to solve for the position and velocity of a particle moving in one-dimension while experiencing a constant net force.

Finally, if we want $v(x)$, we can solve (2.1.4):

$$v\frac{dv}{dx} = \frac{F_0}{m} \qquad (2.3.10)$$

$$v\,dv = a\,dx \qquad (2.3.11)$$

$$\int_{v_0}^{v} v'\,dv' = a\int_{x_0}^{x} dx' \qquad (2.3.12)$$

$$v^2 - v_0^2 = 2a(x - x_0) \qquad (2.3.13)$$

where $a = F_0/m$ was used in (2.3.11). We could have also obtained this result by eliminating t between equations (2.3.4) and (2.3.9). The box below summarizes all of the constant force equations, sometimes called the *kinematic equations*. These are equations that you should memorize.

Kinematic Equations for Constant Acceleration a

$$v(t) = v_0 + at \qquad (2.3.14)$$

$$x(t) = x_0 + v_0 t + \frac{1}{2}at^2 \qquad (2.3.15)$$

$$v(x)^2 = v_0^2 + 2a\,(x - x_0) \qquad (2.3.16)$$

In the case of a freely falling particle near the surface of the Earth, we use $a = -g = -9.8\,\mathrm{m/s^2}$ (assuming "down" is in the negative direction), where g is the acceleration due to gravity. The kinematic equations become:

$$v(t) = v_0 - gt \qquad (2.3.17)$$

$$y(t) = y_0 + v_0 t - \frac{1}{2}gt^2 \qquad (2.3.18)$$

$$v^2(y) = v_0^2 - 2g(y - y_0) \qquad (2.3.19)$$

We used y as the position variable, which is common when describing vertical motion. Notice that these are not different equations from the kinematic equations, but simply the kinematic equations with a specific value of a. We now look at some well-known examples of situations in which the acceleration of the system is constant.

Example 2.1: The Atwood machine
   The Atwood machine consists of two masses m_1, m_2 connected by a massless, inextensible string that hangs over a massless, frictionless pulley as shown in Figure 2.1.

Assume that $m_2 > m_1$. Find the acceleration a of the masses and the tension of the string.

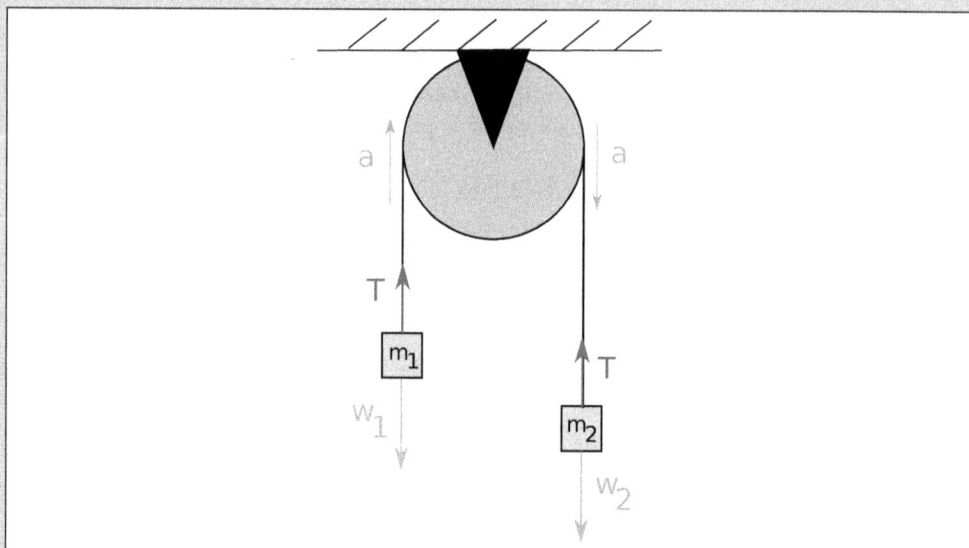

Figure 2.1: Free body diagrams of the two hanging masses of the Atwood machine.

Solution:

We can derive an equation for the acceleration by using force analysis. The only forces we must consider are: tension (T) and weight of the two masses $(W_1 = m_1 g$ and $W_2 = m_2 g$). The tension acting on each mass will be the same, because the tension is constant throughout the string. Because the string is inextensible, the magnitude of the acceleration of each mass will also be the same, although they will be in opposite directions.

To find the acceleration, we need to consider the forces affecting each individual mass. First, we need to define a frame of reference. In this case, we will choose $+y$ is downwards so that the heavier mass, m_2, moves in a positive direction. Newton's second law can be used to derive a system of equations:

$$\left. \begin{array}{l} m_1 g - T = -m_1 a \\ m_2 g - T = m_2 a \end{array} \right\} \qquad (2.3.20)$$

where the right hand side of the top (bottom) equation of (2.3.20) is the net force acting on m_1 (m_2). The upward motion of m_1 has been made explicit by including a minus sign on the right-hand side of the equation describing the forces affecting m_1.

We can solve the system (2.3.20) by subtracting the equations to eliminate T. We can also use a computer algebra system (see below).

Either way, the result is

$$a = g\frac{m_2 - m_1}{m_1 + m_2} \qquad T = \left(\frac{2m_1 m_2}{m_1 + m_2} \right) g \qquad (2.3.21)$$

Once a has been found, we can find the position $x(t)$ and velocity $v(t)$ of the masses using (2.3.15) and (2.3.14). Note that for the position and velocity of m_1, we would need to insert $-a$ in the kinematic equations.

We can check if these results makes physical sense, and also if the units are correct. If the two masses are equal $m_1 = m_2$, these equations give $a = 0$ and $T = m_1 g = m_2 g$. This result makes physical sense, since equal masses mean that the system is stationary ($a = 0$), and the tension T will be equal to the weights of either mass.

In addition, the fraction $(m_2 - m_1)/(m_1 + m_2)$ is dimensionless, so that the units for a in (2.3.21) are the same as the units of g. The units for the fraction $(2m_1 m_2)/(m_1 + m_2)$ are kg, so that the units of T are $kg\,m/s^2$ or Newtons.

Python Code

We solve the system of equations using the `solve` command, which requires that the equations to be solved to have a zero right-hand side. After defining the necessary symbols, we store (2.3.20) into the variables `eq_1` and `eq_2` and use the `solve` command which outputs a Python dictionary. We can identify each solution by using square brackets behind the variable name `soln`, which contains the solutions.

```
from sympy import symbols, solve

print('-'*28,'CODE OUTPUT','-'*29,'\n')

m1, m2, a, g, T = symbols('m1,m2,a,g,T')

eq_1 = m1*g - T + m1*a          # define equations to solve
eq_2 = m2*g - T - m2*a

soln = solve((eq_1,eq_2),(a,T)) #solution to system of equations

print('a = ', soln[a])
print('T = ', soln[T])

--------------------------- CODE OUTPUT ----------------------------
a =   (-g*m1 + g*m2)/(m1 + m2)
T =   2*g*m1*m2/(m1 + m2)
```

Mathematica Code

We define the equations (2.3.20) as the variables `eq1` and `eq2`. Note the use of `==` to demonstrate the equality of the equation. Next, we use the `Solve` command which has two arguments, the equations to be solved and the variables for which the equations are being solved. Notice that the result is a list of replacement rules `/.`, which Mathematica can use for further calculations.

$$\text{eq1} = \text{m1} * g - T == - \text{m1} * a;$$

$$\text{eq2} = \text{m2} * g - T == \text{m2} * a;$$

$$\text{Solve}[\{\text{eq1}, \text{eq2}\}, \{a, T\}]$$

$$\left\{\left\{a \to -\frac{g\,\text{m1} - g\,\text{m2}}{\text{m1}+\text{m2}}, T \to \frac{2g\,\text{m1}\,\text{m2}}{\text{m1}+\text{m2}}\right\}\right\}$$

Example 2.2: Motion on an inclined plane

The shaded box with mass m shown in Figure 2.2 is placed on an inclined plane with angle θ. The coefficient of friction between the mass and the plane is μ. If the mass is released from rest, find the position $x(t)$, velocity $v(t)$, and acceleration a.

Solution:

We will use the standard reference frame for inclined planes that is commonly used for this type of problem; hence we define $+x$ as downward, along the plane. The force of friction is written as $f = \mu N$, where N is the normal force between the mass and the inclined plane. We will obtain the position and speed of the mass as a function of time, both analytically and by using the symbolic capabilities of Python and Mathematica.

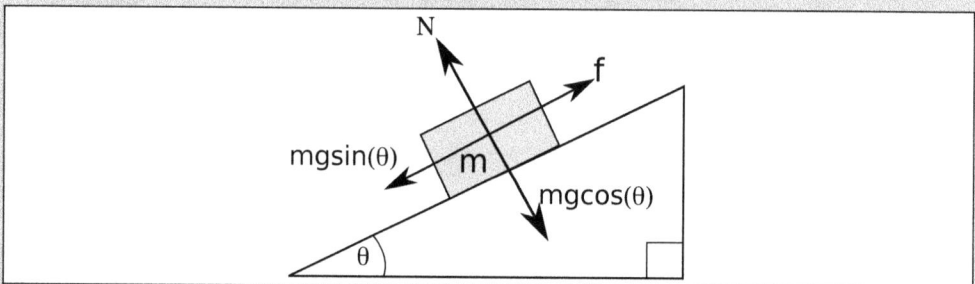

Figure 2.2: Forces on an inclined plane.

The normal force here represents the force applied by the plane against the object (and vice versa). The magnitude of the normal force N can be calculated from the free body diagram, as shown in Figure 2.2, to be $N = mg\cos\theta$, where g is the acceleration due to gravity, and θ is the angle of the inclined surface measured from the horizontal. The component of the weight $W = mg$, which acts along the direction of the plane, is again found from the force diagram to be $F = mg\sin\theta$.

The total force along the x-axis (downward direction along the plane) is:

$$F_{total} = ma = mg\sin\theta - f = mg\sin\theta - \mu mg\cos\theta \tag{2.3.22}$$

and by dividing with the mass m we find the constant acceleration:

$$a = g\sin\theta - \mu g\cos\theta \tag{2.3.23}$$

This acceleration can then be used in (2.3.15) and (2.3.14) to get $x(t)$ and $v(t)$:

$$v(t) = v_0 + (g\sin\theta - \mu g\cos\theta)\,t \tag{2.3.24}$$

$$x(t) = x_0 + v_0 t + \frac{1}{2}\left(g\sin\theta - \mu g\cos\theta\right)t^2 \qquad (2.3.25)$$

Python Code

In the code, we use `dsolve` to solve Newton's law in the form:

$$\frac{d^2x}{dt^2} = a = g\sin\theta - \mu g\cos\theta$$

with the initial conditions `x(0)=x0` and `x'(0)=v0`. We start by defining the parameters g, θ, μ, $x0$, $v0$ as `g,theta,mu,x0,v0,t` using the command `symbols`. The parameter `x` is defined as a function by using the command `Function` in SymPy.

The parameter `initconds` stores the initial conditions. Note that `dsolve` contains now a third parameter `ics=initconds` which tells SymPy to obtain the solution of the differential equation using the given initial conditions. The solution of the ODE is stored in the variable `solnx`, and we select the right-hand side of the solution using the method `.rhs`.

The speed $v(t) = dx/dt$ is evaluated as the time derivative using `diff(solnx,t)`.

```
from sympy import symbols, dsolve, Function, diff, sin, cos
import matplotlib.pyplot as plt
import numpy as np

print('-'*28,'CODE OUTPUT','-'*29,'\n')

g, theta, mu, x0, v0, t = symbols('g, theta, mu, x0, v0, t',real=True)
x = Function('x')                        # define position function x

a = g*sin(theta)-mu*g*cos(theta)         # define acceleration a

# initial conditions x(0)=x0 and v(0)=v0
initconds = {x(0):x0, diff(x(t),t).subs(t, 0): v0}

# use dsolve to solve the ODE x''(t)=a
solnx = dsolve(diff(x(t),t,t)-a,x(t), ics=initconds).rhs

solnv = diff(solnx,t)                    # evaluate v = dx/dt
print(' x(t) = ', solnx)                 # print x(t)

print('\nv(t) = ',solnv)                 # print v(t)

--------------------------- CODE OUTPUT ---------------------------
 x(t) =  g*t**2*(-mu*cos(theta) + sin(theta))/2 + t*v0 + x0

v(t) =  g*t*(-mu*cos(theta) + sin(theta)) + v0
```

Mathematica Code

The Mathematica command `DSolve` is used to solve the differential equation

$$\frac{d^2x}{dt^2} = a = g\sin\theta - \mu g\cos\theta \qquad (2.3.26)$$

In `DSolve` the first argument is a list containing the ODE and the initial conditions, and the result for the position $x(t)$ is stored in the variable **posx**. The speed $v = dx/dt$ is evaluated using the derivative command `D[posx,t]`.

$a = g * \text{Sin}[\theta] - \mu * g * \text{Cos}[\theta];$

$\text{soln} = \text{DSolve}[\{x"[t]{==}a, x[0]{==}x0, x'[0]{==}v0\}, x[t], t];$

$\text{posx} = x[t]/.\text{soln}[[1]];$

$\text{Print}[\text{"The position x(t)=", posx}]$

The position x(t)=$\frac{1}{2}\left(2\,t\,v0 + 2\,x0 - g\,t^2\,\mu\,\text{Cos}[\theta] + g\,t^2\,\text{Sin}[\theta]\right)$

$\text{Print}[\text{"The speed v(t)=", D[posx, t]}]$

The speed v(t)=$\frac{1}{2}(2\,v0 - 2\,g\,t\,\mu\,\text{Cos}[\theta] + 2\,g\,t\,\text{Sin}[\theta])$

2.4 TIME-DEPENDENT FORCES

Now consider a case where a particle constrained to move along a line is experiencing a net force $F(t)$, which is dependent on time. Time-dependent forces arise in a variety of applications, one common occurrence is when a sinusoidal external force acts on the system. In the cases where the net force $F = F(t)$, Newton's second law (2.1.1) gives:

$$\frac{dv}{dt} = \frac{F(t)}{m} \qquad (2.4.1)$$

and by separating variables and integrating from $t' = t_0$ to t we find:

$$v(t) - v(t = 0) = \frac{1}{m}\int_{t'=t_0}^{t} F(t')\,dt' \qquad (2.4.2)$$

By using the initial conditions $v(t = 0) = v_0$:

$$v(t) = v_0 + \frac{1}{m}\int_{t'=0}^{t} F(t')\,dt' \qquad (2.4.3)$$

Once we know $v(t)$, we can find $x(t)$ by integrating $v = dx/dt$ and using the initial conditions $x(t_0) = x_0$ to obtain:

$$x(t) - x_0 = \int_{t'=0}^{t} v(t')\,dt' \qquad (2.4.4)$$

In the next example, we demonstrate how to find the equations of motion for a particle experiencing a time-dependent force. We will obtain the solution both by hand and using Python and Mathematica.

Example 2.3: A decreasing force, F(t)

The force acting on a mass $m =1$ kg is decreasing with time according to $F(t) = c/(t^2 + b^2)$ where t is the elapsed time, and the constants c and b are in units such that F is measured in Newtons. Find the position $x(t)$ and the velocity $v(t)$ with the initial conditions $x(0) = x_0 = 0$ and $v(0) = v_0 = 0$, and with the constants $c =1$ N.s^2, $b = 1$ s.

Solution:

This is the case of a time-dependent force $F = F(t)$, so we can use (2.4.3) with $v_0 = 0$:

$$v(t) = v_0 + \frac{1}{m}\int_{t'=0}^{t} F(t')\,dt' = \frac{1}{m}\int_{t'=0}^{t} \frac{c}{t^2 + b^2}\,dt \tag{2.4.5}$$

The integral is evaluated using the transformation $t = b\tan\theta$ to obtain:

$$v(t) = \frac{c}{m\,b}\tan^{-1}\left(\frac{t}{b}\right) \tag{2.4.6}$$

Does (2.4.6) make sense? To check, we ask what happens when $t \to \infty$? In the limit of $t \to \infty$, $F \to 0$, and we would expect that v becomes constant. In our particular case with $b = 1$ s, as $t \to \infty$, $\tan^{-1}(t) \to \pi/2$, and therefore, our solution $v(t)$ follows our expectation of the velocity approaching a constant. Next, we find $x(t)$ using (2.4.4) with $x_0 = 0$:

$$x(t) - x_0 = \int_{t'=0}^{t} v(t')\,dt' \tag{2.4.7}$$

$$= \frac{c}{m\,b}\int_{t'=0}^{t} \tan^{-1}\left(\frac{t'}{b}\right)\,dt' \tag{2.4.8}$$

$$= \frac{c}{m\,b}\left[t\,\tan^{-1}\left(\frac{t}{b}\right) + \frac{b}{2}\ln\left(\frac{b^2}{t^2 + b^2}\right)\right] \tag{2.4.9}$$

where we used integration by parts.

In this case, as $t \to \infty$, we know that $v \to$ constant, and therefore we would expect $x(t)$ to continue to grow, which, as shown in Figure 2.3, is exactly what our solution does! Figure 2.3 was created using Mathematica.

Notice that the arguments of the logarithm and of the inverse tangent in the solution must be dimensionless. This is is indeed the case, since the constant b has units of time (seconds) as we can see from the given force equation $F(t) = c/(t^2 + b^2)$. It is always a good idea to examine the solution of a problem, and to make sure that the units of the various terms are correct.

Python Code

In the code we use **dsolve** to solve the ODE, specifying its initial conditions using the **ics** option. We use the **textwrap** library so that the output would fit on the page of this book. Notice that the argument for **textwrap.fill** must be a string. Hence the use of **str(solnx)**,70) and the number 70 specifies that the output should be restricted to 70 spaces per line.

In order to plot the symbolic result from SymPy, we use `lambdify` to obtain NumPy arrays for the position and speed variables `position` and `velocity`. The Python method `.subs` is used to substitute the numerical values of the parameters `c,m,x0,b`.

Note also that Python does not combine the logarithmic functions in the output.

```python
from sympy import symbols, dsolve, Function, diff, simplify, lambdify
import textwrap
import numpy as np
import matplotlib.pyplot as plt
print('-'*28,'CODE OUTPUT','-'*29,'\n')

m, c, b, t, x0 = symbols('m, c, b, t, x0',real=True)
x = Function('x')

initconds = {x(0):x0, diff(x(t),t).subs(t, 0): 0}
solnx = dsolve(m*diff(x(t),t,t)-c/(t**2+b**2),x(t), \
    ics=initconds).rhs

solnv = diff(solnx,t)                          # v = dx/dt
print(' x(t) = ', textwrap.fill(str(solnx),70))  # print x(t)
print('\nv(t) = ',simplify(solnv))              # print v(t)

xt = solnx.subs({c:1,m:1,x0:0,b:1})    # substitute values of constants
vt = solnv.subs({c:1,m:1,x0:0,b:1})    # substitute values of constants

position = lambdify(t,xt,'numpy')
velocity = lambdify(t, vt,'numpy')
times = np.linspace(0,10,20)                  # times t=0-10

fig, ax = plt.subplots(nrows=1,ncols=2) # use 2 subplots for x(t),v(t)
ax[0].plot(times, position(times))
ax[0].set_xlabel('time')
ax[0].set_ylabel('position')
ax[1].plot(times, velocity(times))
ax[1].set_xlabel('time')
ax[1].set_ylabel('velocity')
fig.tight_layout()

------------------------------ CODE OUTPUT ----------------------------
 x(t) =  -c*log(b**2 + t**2)/(2*m) + (c*log(b**2) + 2*m*x0)/(2*m) +
c*t*atan(t/b)/(b*m)

v(t) =  c*atan(t/b)/(b*m)
```

Mathematica Code

The Mathematica command `DSolve` is used to solve the differential equation,

$$m\frac{d^2x}{dt^2} = \frac{c}{t^2 + b^2} \tag{2.4.10}$$

This code is very similar to the Mathematica code in Example 2.2. The graphics command GraphicsGrid is used to create two side-by-side plots of $x(t)$ and $v(t)$.

```
SetOptions[Plot, Frame->True, BaseStyle->{FontSize->16}, PlotRange->All];

ODEsolution = DSolve[{m * x"[t]==c/(t^2 + b^2), x[0]==0, x'[0]==0}, x[t], t]
```
$$\left\{\left\{x[t] \to \frac{c\left(2\,t\,\mathrm{ArcTan}\left[\frac{t}{b}\right] + b\,\mathrm{Log}\left[b^2\right] - b\,\mathrm{Log}\left[b^2 + t^2\right]\right)}{2\,b\,m}\right\}\right\}$$

```
position = x[t]/.ODEsolution[[1]]
```
$$\frac{c\left(2\,t\,\mathrm{ArcTan}\left[\frac{t}{b}\right] + b\,\mathrm{Log}\left[b^2\right] - b\,\mathrm{Log}\left[b^2 + t^2\right]\right)}{2\,b\,m}$$

```
velocity = D[position, t]//FullSimplify
```
$$\frac{c\,\mathrm{ArcTan}\left[\frac{t}{b}\right]}{b\,m}$$

```
c = 1; m = 1; b = 1;

GraphicsGrid[

{{Plot[position, {t, 0, 10}, FrameLabel->{"time, s", "x, m"}],

Plot[velocity, {t, 0, 10}, FrameLabel->{"time, s", "v, m/s"}]}}]
```

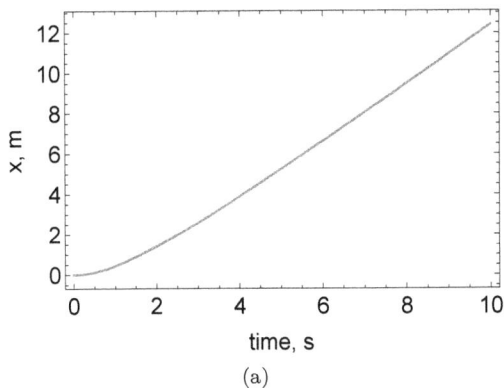

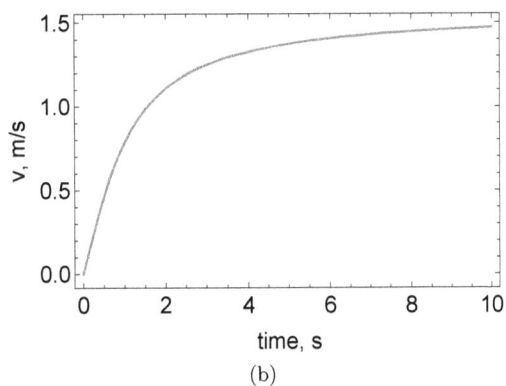

Figure 2.3: Plot of (a) the position and (b) the velocity for Example 2.3. Notice the linear nature of $x(t)$ as t gets large and that the speed $v(t)$ approaches a constant value.

2.5 AIR RESISTANCE AND VELOCITY-DEPENDENT FORCES

Recall that when studying the motion of a particle in free fall, air resistance is often ignored. The force of air resistance depends on the velocity of the object moving through the air. Hence, air resistance is a velocity-dependent force. Here we will learn how to address air resistance when describing the motion of a particle. It should be noted, however, that although we focus on air resistance as an example of a velocity-dependent force, most of our comments will be applicable to any velocity-dependent force.

When speaking about an object moving through the air, two forces are often described, drag and lift. The *drag* force is the component of the air resistance, which is in the opposite direction of the object's velocity (direction of motion). *Lift* forces are the component of the air resistance that is perpendicular to the object's motion. In this section, we will consider cases where the lift is negligible. We will also only consider nonrotating bodies. A rotating body moving through air will experience the *Magnus effect* where the spinning motion of the object drags air faster on one side than the other. The resulting pressure difference on either side of the object causes the object to curve away from its flight path. This is what allows a baseball pitcher to throw a curve ball. There are many interesting videos online which demonstrate the Magnus effect.

In general, the drag force studied in this chapter takes the mathematical form,

$$\mathbf{f} = -f(v)\hat{\mathbf{v}} \tag{2.5.1}$$

where the unit vector $\hat{\mathbf{v}} = \mathbf{v}/v$ is along the direction of the object's velocity (or motion), and the minus sign is included to explicitly show that the direction of the drag force is opposite of the direction of the velocity. The magnitude of the drag force $f(v)$ is a function of velocity which generally takes the form,

$$f(v) = b\,v + c\,v^2 \tag{2.5.2}$$

and therefore has both a linear and a quadratic component in v. For spheres in air, the values of the constants b and c are (Taylor, [1]):

$$\begin{aligned} b &= 1.6 \times 10^{-4} D \quad \text{N·s/m} \\ c &= 0.25 D^2 \quad \text{N·s}^2/\text{m}^2 \end{aligned} \tag{2.5.3}$$

where D is the diameter of the sphere in meters. Fortunately, one usually doesn't need to include both the linear and quadratic terms for air resistance. In order to determine which term, linear or quadratic, is the most important in a given situation, one can compute the ratio:

$$\gamma = \frac{cv^2}{b\,v} = \frac{0.25D^2v^2}{1.6 \times 10^{-4}D\,v} = 1.6 \times 10^3 D\,v \tag{2.5.4}$$

Note that v should be in meters per second, and D in meters. If $\gamma \gg 1$, then the quadratic term dominates, and the linear term can be neglected. If $\gamma \ll 1$, then the linear term dominates, and the quadratic term can be neglected. However, if $\gamma \approx 1$, then both the linear and quadratic terms need to be included. So for a soccer ball with a diameter of 0.22 m, the quadratic term dominates for speeds approximately greater than 0.003 m/s (or 3 mm/s), while for lower speeds the linear term dominates. Both linear and quadratic terms would need to be included in the drag force formula if the soccer ball is traveling near 3 mm/s.

To obtain equations for position and velocity as functions of time, we will consider a generic form for $F(v)$. Newton's second law (2.1.1) gives:

$$\frac{dv}{dt} = \frac{F(v)}{m} \tag{2.5.5}$$

Rearranging the above equation gives:

$$dt = m\frac{dv}{F(v)} \tag{2.5.6}$$

and by integrating from $t' = t_0$ to t , and using the initial conditions $v(t_0) = v_0$ we find :

$$t - t_0 = m \int_{v'=v_0}^{v} \frac{dv'}{F(v')} \tag{2.5.7}$$

By solving (2.5.7), we can find $v(t)$, the velocity as a function of time. Once we know $v(t)$, we can integrate once more using (2.4.4) to find $x(t)$ with the initial condition $x(t_0) = x_0$. We can follow a similar procedure for finding $v(x)$. Newton's second law gives:

$$mv\frac{dv}{dx} = F(v) \tag{2.5.8}$$

After rearranging, we obtain:

$$x - x_0 = m \int_{v'=v_0}^{v} \frac{v'dv'}{F(v')} \tag{2.5.9}$$

Next, we will show some examples involving velocity-dependent forces.

Example 2.4: Air resistance varying linearly with speed
The drag force acting on a mass m depends linearly on the velocity v according to $F(v) = -bv$. Find $x(t)$ and $v(t)$ with the initial conditions $v(0) = v_0$ and $x(0) = x_0$. Plot the solutions using $x_0 = 0$, $v_0 = 1$, mass $m = 1$ and $b = 1$ (all physical quantities in SI units).

Solution:
We use (2.5.7) to find $x(t)$ with $t_0 = 0$:

$$t = m \int_{v'=v_0}^{v} \frac{dv'}{F(v')} = -\frac{m}{b} \int_{v'=v_0}^{v} \frac{dv'}{v'} = -\frac{m}{b} \ln\left(\frac{v}{v_0}\right) \tag{2.5.10}$$

Solving for $v(t)$, we obtain:

$$v(t) = v_0 e^{-\frac{bt}{m}} \tag{2.5.11}$$

This result makes physical sense, as the time t increases, the value of the exponential function $\exp(-bt/m)$ decreases and the speed $v(t)$ gets smaller, until it reaches zero at very large time.
We also find $x(t)$ from (2.4.4):

$$x(t) - x_0 = \int_{t'=t_0}^{t} v(t')\,dt' = \int_{t'=t_0}^{t} v_0 e^{-\frac{bt'}{m}}\,dt' \tag{2.5.12}$$

$$x(t) = x_0 + \frac{m\,v_0}{b}\left(1 - e^{-\frac{bt}{m}}\right) \tag{2.5.13}$$

As the time t increases, the second term $1 - \exp(-bt/m)$ in $x(t)$ increases and approaches 1, so that $x(t)$ at very large times reaches the constant value of $x_0 + m\,v_0/b$ where the motion stops ($v \to 0$).

Python Code

We use the method `xsoln.limit(t,oo)` to find the limit of the variable `solnx` (representing $x(t)$), as $t \to \infty$. The infinity symbol in SymPy is `oo`. As discussed above, these limits are $v(t) \to 0$ and $x(t) \to x_0 + m v_0/b$, as $t \to \infty$.

```python
from sympy import symbols, dsolve, Function, diff,simplify, oo
print('-'*28,'CODE OUTPUT','-'*29,'\n')

t, x0, v0 = symbols(' t, x0, v0')
m, b = symbols('m, b', positive=True)
x = Function('x')

inits = {x(0):x0, diff(x(t),t).subs(t, 0): v0}
solnx = dsolve(m*diff(x(t),t,t)+b*diff(x(t),t),x(t),ics=inits).rhs

print(' x(t) = ', simplify(solnx))        # print x(t)
solnv = diff(solnx,t)                      # v = dx/dt

print('\nv(t) = ',simplify(solnv))         # print v(t)
print('The limit of v(t) at large time t = ',solnv.limit(t,oo))
print('The limit of x(t) at large time t = ',solnx.limit(t,oo))


---------------------------- CODE OUTPUT ----------------------------
 x(t) =   x0 + m*v0/b - m*v0*exp(-b*t/m)/b

v(t) =   v0*exp(-b*t/m)
The limit of v(t) at large time t =  0
The limit of x(t) at large time t =  (b*x0 + m*v0)/b
```

Mathematica Code

The `Collect` command gathers together terms that involve the same powers of the objects in the curly brackets (second argument of `Collect`). It is a useful command for simplifying algebraic terms and for identifying coefficients.

We find the limit of the functions `velocity` and `position` as $t \to \infty$ using the Mathematica `Limit` command with the assumptions $b > 0$ and $m > 0$.

ODEsolution = DSolve[$\{mx''[t] == -bx'[t], x[0] == \text{xo}, x'[0] == \text{vo}\}, x[t], t$];

position = Collect[$x[t]/.$ODEsolution[[1]], $\{\text{v0}, m\}$]

$$\frac{e^{-\frac{bt}{m}} m \left(-\text{vo} + e^{\frac{bt}{m}} \text{vo}\right)}{b} + \text{xo}$$

limitx = Assuming[$b > 0 \&\& m > 0$, Limit[position, $t\text{->}$Infinity]];

Print["The limit of x(t) as t->Infinity = ", limitx]

The limit of x(t) as t->Infinity = $\frac{m\,\text{vo}}{b} + \text{xo}$

velocity = D[position, t]

$$\text{vo} - e^{-\frac{bt}{m}} \left(-\text{vo} + e^{\frac{bt}{m}} \text{vo}\right)$$

limitv = Assuming[$b > 0 \&\& m > 0$, Limit[velocity, $t\text{->}$Infinity]];

Print["The limit of v(t) as t-> ∞ = ", limitv]

The limit of v(t) as t->Infinity = 0

vo = 1; xo = 0; $m = 1$; $b = 1$;

GraphicsGrid[$\{\{$Plot[position, $\{t, 0, 10\}$, FrameLabel $\to \{$"time", x$\}$],

Plot[velocity, $\{t, 0, 10\}$, FrameLabel $\to \{$"time", v$\}]\}\}$]

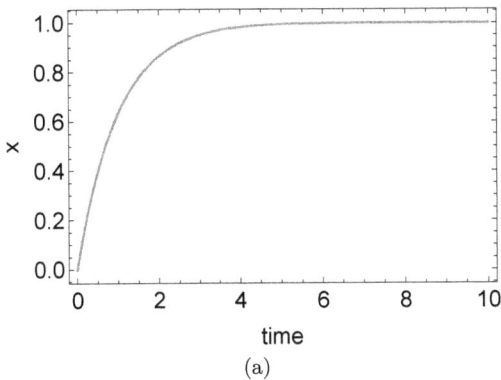

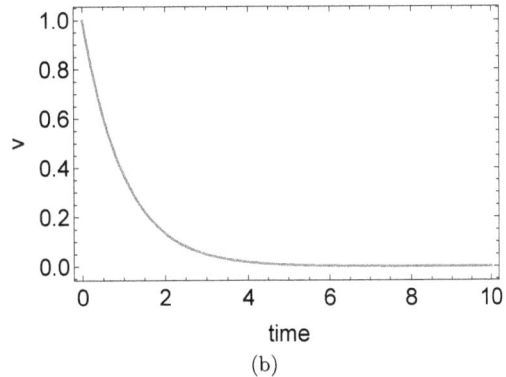

Figure 2.4: (a) Position and (b) velocity as a function of time for Example 2.4. Notice the position asymptotes to $x = 1$ m, and that the velocity decays to zero.

Example 2.5: Air resistance in the presence of gravity
      Consider the case of a falling object near the surface of the Earth. Suppose in this
case, the drag force is linear. Then the net force on the falling body depends on the
velocity v according to

$$F(v) = -b\,v + m\,g \tag{2.5.14}$$

where we have defined $+y$ to be vertically downward so that the body's motion is in
the positive direction. Find the displacement $y(t)$ and the velocity $v(t)$, with the initial
conditions $v(0) = v_0$ and $y(0) = y_0$.

Solution:
From a physical point of view, we expect the mass to initially accelerate and the speed
$v(t)$ to increase due to gravity. As the speed v increases, the force of air resistance $-b\,v$
will also increase, until it becomes equal and opposite to the gravitational force, i.e.,
until $F = -b\,v + m\,g = 0$. When this happens, the mass has reached the terminal
velocity equal to:

$$v_t = \frac{m\,g}{b} \tag{2.5.15}$$

To obtain $v(t)$ we use (2.5.7) with $t_0 = 0$:

$$t = m \int_{v'=v_0}^{v} \frac{dv'}{-b\,v' + m\,g} = -\frac{m}{b} \ln\left(\frac{\frac{m\,g}{b} - v}{\frac{m\,g}{b} - v_0} \right) \tag{2.5.16}$$

Note that we have factored out the constant b from the denominator of the integral
before performing the integration. Solving for $v(t)$, we obtain:

$$v(t) = \frac{m\,g}{b} + \left(v_0 - \frac{m\,g}{b} \right) e^{-\frac{b\,t}{m}} \tag{2.5.17}$$

We also find $y(t)$ from (2.4.4):

$$y(t) - y_0 = \int_{t'=t_0}^{t} v(t')\,dt' = \int_{t'=t_0}^{t} \left[\frac{m\,g}{b} + \left(v_0 - \frac{m\,g}{b} \right) e^{-\frac{b\,t'}{m}} \right] dt' \tag{2.5.18}$$

which gives after integrating:

$$y(t) = y_0 + \frac{m\,g}{b}t + \left(\frac{m^2\,g}{b^2} - \frac{m\,v_0}{b} \right) \left(e^{-\frac{b\,t}{m}} - 1 \right) \tag{2.5.19}$$

When $t \to \infty$ we can substitute $e^{-\frac{b\,t}{m}} \to 0$ in (2.5.17) to obtain the *terminal velocity*:
$v_t = m\,g/b$ as discussed above.
Hence, our solution matches the expected physics. As the object falls, its velocity increases due to the force of gravity. However, at some point, the object is moving fast
enough that the drag force equals that of the object's weight. At that point, the net
force on the object is zero, and the velocity reaches the constant called terminal velocity
v_t for the duration of its fall.

      Figure 2.5 shows the plots of $x(t)$ and $v(t)$ found in Example 2.5. The numerical values
of the parameters are $b = 0.2$, $m = 1$, $x(0) = 0$, $v(0) = 0$, $g = 9.8$ (all in SI units). Notice
that the velocity of the object approaches a constant value as the object falls for a long
period of time.

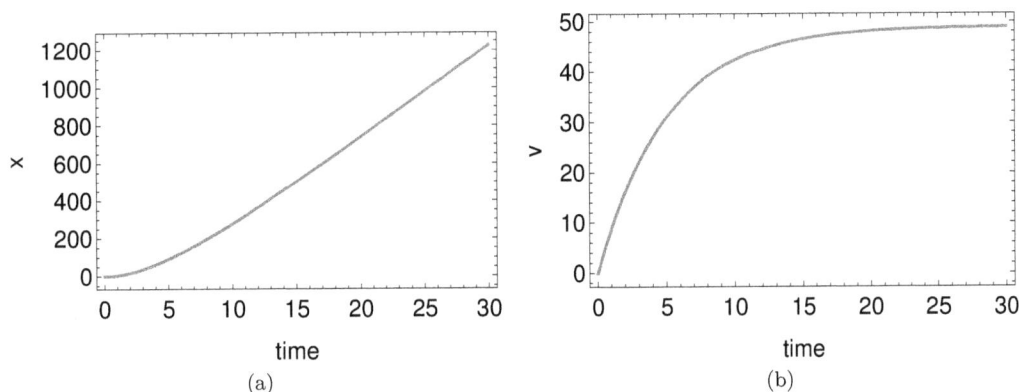

Figure 2.5: Plots of (a) $x(t)$, (b) $v(t)$ from Example 2.5. The plot shows the linear increase of $x(t)$ after terminal velocity is reached (at $v = mg/b = 49$ m/s). See this book's website for the code used to generate this plot.

Example 2.6: Air resistance proportional to v^2

A more realistic drag force for a falling object is a force proportional to the square of the velocity v according to $F(v) = -cv^2 + mg$. Find $x(t)$ and $v(t)$ with the initial conditions $v(0) = 0$ and $x(0) = 0$.

Solution:

The terminal velocity v_t is found in this case by setting $F = -c\,v_t^2 + mg = 0$:

$$v_t = \sqrt{\frac{mg}{c}} \tag{2.5.20}$$

Next, we use (2.5.7) to find $v(t)$ by integrating from the initial velocity v_0 to the final velocity v:

$$t = m \int_0^v \frac{dv'}{-c\,v'^2 + mg} = \frac{m}{c} \int_0^v \frac{dv'}{v_t^2 - v'^2} \tag{2.5.21}$$

We rewrite this equation by factoring the integrand, in order to avoid the presence of imaginary numbers in logarithmic functions in Python. However, this issue is not present in the Mathematica code.

$$t = \frac{m}{2\,v_t\,c} \int_0^v \left(\frac{1}{v_t + v'} + \frac{1}{v_t - v'} \right) dv' \tag{2.5.22}$$

$$t = \frac{m}{2\,v_t\,c} \ln \left(\frac{v_t - v}{v_t + v} \right)$$

After some straightforward algebra, this can be simplified to:

$$t = \frac{m}{c\,v_t} \tanh^{-1} \left(\frac{v}{v_t} \right)$$

We can now easily solve this equation for v to obtain the speed:

$$v(t) = v_t \tanh \left[\frac{c\,v_t}{m} t \right] \tag{2.5.23}$$

We also find $x(t)$ from (2.4.4):

$$x(t) - x_0 = \int_0^t v(t')\, dt' = \int_0^t v_t \tanh\left[\frac{c\,v_t}{m}\, t'\right] dt' \tag{2.5.24}$$

$$x(t) - x_0 = \int_0^t v_t \frac{\sinh\left[\frac{c\,v_t}{m}\, t'\right]}{\cosh\left[\frac{c\,v_t}{m}\, t'\right]}\, dt' = \frac{m}{c} \ln\left(\cosh\left(\frac{c v_t}{m} t\right)\right) \tag{2.5.25}$$

which with $x_0 = 0$ gives:

$$x(t) = \frac{m}{c} \ln\left(\cosh\left(\frac{c v_t}{m} t\right)\right) \tag{2.5.26}$$

Python Code

After defining the parameters in SymPy as positive using the option `real=True,positive=True` in the `symbols` command, we use `integrate` to evaluate the integral in (2.5.22), in order to obtain an expression for the time t. The answer produced for t by SymPy is complex, and contains several logarithmic terms. These terms are combined using the function `logcombine` with the option `force=True`, to ensure that they are combined regardless of the signs of the parameters. We use `solve` to obtain the speed $v(t)$, with the result agreeing with the analytical solution in (2.5.23). Finally `integrate` is used to obtain the position $x(t)$.

```python
from sympy import symbols, integrate,logcombine,solve,cosh, sinh,log
print('-'*28,'CODE OUTPUT','-'*29,'\n')

v, vf, vt, t1, m, c, t= symbols('v, vf, vt, t1, m, c, t',\
 real=True,positive=True)

# evaluate integral after factoring
integral = integrate(1/(v-vt),(v,0,vf))- integrate(1/(v+vt),(v,0,vf))

# combine logarithmic terms from integration
result = logcombine(integral,force=True)*(m/c)/(2*vt)
print('t = ',result )

v = solve(t+result,vf)[0]    # invert equation to find speed v(t)
print('v(t) = ',v)

# integrate to find x(t)
x = integrate(vt*sinh(c*t*vt/m)/cosh(c*t*vt/m),(t,0,t1))
print('x(t) = ',x)

---------------------------- CODE OUTPUT ----------------------------
t =  m*log(-(vf - vt)/(vf + vt))/(2*c*vt)
v(t) =   vt*tanh(c*t*vt/m)
x(t) =   m*log(cosh(c*t1*vt/m))/c
```

Mathematica Code

We use the command **Integrate** to perform the integral. When you perform integrals, you may not always think about the signs of variables. For example you know that the terminal velocity is positive. However, Mathematica makes no such assumptions. In order to perform the integral, and get the above analytical result, we need to tell Mathematica that both the final velocity (**vf** in the code), and terminal velocity, (**vt** in the code) are positive, and that $vf < vt$. Physically, this is making the assumptions: $m, g, k, t, v > 0$ and also $mg > kv^2$ (i.e., the air resistance cannot exceed the weight $W = mg$). The symbol **&&** represents the logical operator AND. These assumptions are provided in the **Assuming** command in the code. So by placing the command **Integrate** inside the **Assuming** command, we are telling Mathematica to integrate with the listed assumptions.

The numerical values of the parameters for the plot in Figure 2.6 are $c = 0.2$, $m = 1$, $x(0) = 0$, $v(0) = 0$, $g = 9.8$ (all in SI units).

```
SetOptions[Plot, Frame->True, BaseStyle->{FontSize->16}, PlotRange->All];

integral = Assuming[vt > 0&&vf > 0&&vf < vt,

Integrate[1/(vt^2 − v^2), {v, 0, vf}]]
```

$$\frac{\text{ArcTanh}\left[\frac{vf}{vt}\right]}{vt}$$

```
Assuming[vt > 0&&t > 0&&m > 0&&c > 0, Solve[integral * m/c==t, vf]];

velocity = First[vf/.%];

Print["The velocity v[t]=", velocity]
```

The velocity v[t]=vt Tanh $\left[\frac{c\,t\,vt}{m}\right]$

```
position = Assuming[m > 0&&c > 0&&vt > 0&&tf > t,

Integrate[velocity, {t, 0, tf}]];

Print["The position x[t]=", position]
```

The position x[t]=$\dfrac{m\,\text{Log}\left[\text{Cosh}\left[\frac{c\,tf\,vt}{m}\right]\right]}{c}$

```
c = 0.2; m = 1; g = 9.8; vt = Sqrt[m * g/c];

Print["Terminal velocity=", vt, " m/s"]

GraphicsGrid[{{Plot[position, {tf, 0, 5}, FrameLabel->{"time, s", "x, m"}],

Plot[velocity, {t, 0, 5}, FrameLabel->{"time, s", "v, m/s"}]}}]
```

Terminal velocity=7.0 m/s

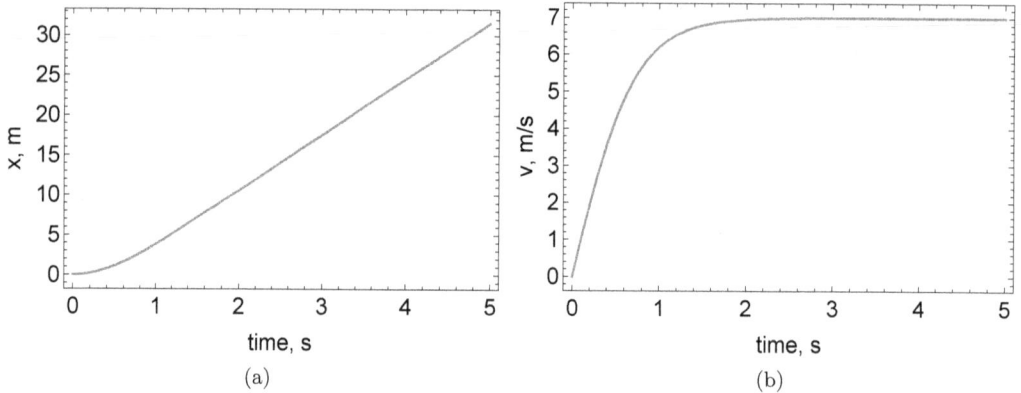

Figure 2.6: (a) Position vs. time and (b) Velocity vs. time graph for Example 2.6. Notice that after $t \simeq 2$ s, the displacement increases linearly with time, and the velocity approaches a constant value of 7 m/s.

A comparison between the velocity of the falling body in the cases of linear air resistance $F(v) = -bv + mg$ and quadratic air resistance $F(v) = -cv^2 + mg$ is shown in Figure 2.7, where $m = 1.0$ kg, $b = 0.2$ Ns/m, and $c = 0.2$ Ns^2/m. The velocity on the y-axis has been normalized by dividing with the corresponding terminal velocity. Notice that the quadratic air resistance (dashed line) leads to the object obtaining terminal velocity in a significantly shorter time than linear air resistance (solid line). This is not surprising; the magnitude of the quadratic drag force will be higher than the linear force for a given velocity, v.

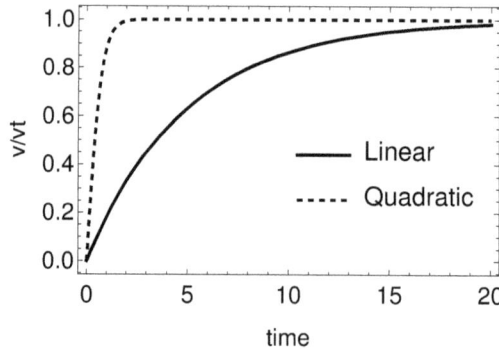

Figure 2.7: Comparison of the velocity of falling body in the presence of linear and quadratic air resistance. The two curves have been scaled by dividing the speed with the terminal velocities, so that the plots produce the same result as $t \to \infty$. The code used to produce this graph can be found on this book's website.

2.6 POSITION-DEPENDENT FORCES

In the cases where the force is a function of the position $F = F(x)$, we can use Newton's second law in the form

$$F(x) = mv\frac{dv}{dx} \qquad (2.6.1)$$

By separating the variables v and x, and using the initial conditions $x = x_0$ and $v = v_0$ when $t = 0$, we obtain:

$$\int_{x_0}^{x} F(x')\, dx' = \int_{v_0}^{v} m\, v'\, dv' = \frac{1}{2} m\, v^2 - \frac{1}{2} m\, v_0^2 \qquad (2.6.2)$$

Solving for the velocity $v(x)$ as a function of the position x:

$$v\,(x) = \frac{2}{m} \sqrt{\int_{x_0}^{x} F(x')\, dx' + \frac{1}{2} m\, v_0^2} \qquad (2.6.3)$$

By substituting $v = dx/dt$, separating the variables, and integrating with the initial condition $x = x_0$ when $t = t_0$, we obtain the relationship between position x and time t:

$$t - t_0 = \frac{m}{2} \int_{x_0}^{x} \left[\frac{dx'}{\sqrt{\int_{x_0}^{x} F(x')\, dx + \frac{1}{2} m\, v_0^2}} \right] \qquad (2.6.4)$$

After performing the integral in equation (2.6.4), one would then try to invert this formula to find $x(t)$, something that is not always easy or possible to do. Often, one relies on computer methods in such cases.

The next example will involve one of the most important problems in all of classical mechanics, simple harmonic motion (SHM). In fact, we will devote a whole chapter to it! As we will see in Chapter 6, SHM is a common model used for small amplitude oscillations. Although SHM is often thought of as a "mass on a spring", it is also a useful model for other small amplitude oscillations including pendulums, small amplitude water waves, and loudspeakers—to name a few. Besides presenting the SHM, the next example also includes a demonstration of using Python to find the solution to a second-order differential equation in closed form.

Example 2.7: Simple harmonic motion
   A mass m is under the influence of a position-dependent force, $F = F(x) = -kx$, which is proportional to the distance x of the mass from the equilibrium position (located at $x = 0$). The mathematical form of F is, of course, the well-known Hooke's law for springs. Find $x(t)$ and $v(t)$ with the initial conditions $v(0) = 0$ and $x(0) = x_0 \neq 0$ at $t = 0$.

Solution:
   In this case (2.6.3) with $v_0 = 0$ becomes:

$$v = \frac{2}{m} \sqrt{-\int_{x_0}^{x} k\, x'\, dx' + \frac{1}{2} m\, v_0^2} = \frac{2}{m} \sqrt{\frac{1}{2} k\, (x_0^2 - x^2)} \qquad (2.6.5)$$

which gives the velocity $v(x)$ as a function of the position x. Next we will find $x(t)$ by using (2.6.4):

$$t = \int_{x_0}^{x} \frac{dx'}{\sqrt{\frac{k}{m}(x_0^2 - x'^2)}} = \sqrt{\frac{m}{k}} \sin^{-1}\left(\frac{x}{x_0}\right)\Bigg|_{x_0}^{x} = \sqrt{\frac{m}{k}} \left[\sin^{-1}\left(\frac{x}{x_0}\right) - \frac{\pi}{2} \right] \qquad (2.6.6)$$

or after some rearranging:

$$x(t) = x_0 \sin\left(\sqrt{\frac{k}{m}}\,t + \frac{\pi}{2}\right) = x_0 \cos\left(\sqrt{\frac{k}{m}}\,t\right) \qquad (2.6.7)$$

The result is that the mass oscillates about the equilibrium position x_0. Hence we find that the natural frequency ω of the oscillator is $\omega = \sqrt{\frac{k}{m}}$.

Python Code

In the code we use the **dsolve** command from the SymPy library.

After importing the necessary commands, we specify that the variables **m, k, t** are real and positive. The command **D** is used in place of **diff** (see the **import** command) and represents the derivative of **x** with respect to time (two **t**'s in the argument means second derivative). The parameter **init** contains the initial conditions for the motion.

If we did not specify that m and k are real and positive, and we had only defined those variables as symbols, Python still solves the differential equation but produces a hyperbolic cosine function (cosh) instead of the cosine function. It is important to remember that it is not unusual to have to do additional algebra on results provided by CAS algorithms, in order to get a result that is useful. This is an additional reason not to forget your math skills and not to rely solely on the computer!

```
from sympy import diff as D
from sympy import symbols, Function, dsolve

print('-'*28,'CODE OUTPUT','-'*29,'\n')

x = Function('x')
t, x0, v0 = symbols('t, x0, v0',real=True)
k, m = symbols('k, m',real=True, positive=True)

# initial conditions
init = {x(0):x0,D(x(t),t).subs(t,0):v0}

#use dsolve to find solution of ODE
soln = dsolve(m*D(x(t),t,t) + k*x(t), simplify=True, ics=init).rhs

# substitute v(0)=0 using .subs
print('The solution is x(t) = ', soln.subs(v0,0))

-------------------------------- CODE OUTPUT ----------------------------------
The solution is x(t) =   x0*cos(sqrt(k)*t/sqrt(m))
```

Mathematica Code

Using **DSolve**, we can find the position as a function of time. Note that in order to isolate the solution, we had to reference the variable **soln**.

$$\text{soln} = \text{DSolve}[\{m * x''[t] == -k * x[t], x[0] == x0, x'[0] == 0\}, x, t];$$

$$\text{position} = x[t]/.\text{soln}[[1]];$$

$$\text{Print}[\text{"x(t)="}, \text{position}]$$

$$\text{x(t)} = \text{x0} \cos\left[\frac{\sqrt{k}\,t}{\sqrt{m}}\right]$$

Not all differential equations can be easily solved in closed form, and some cannot be solved in closed form at all. In cases where closed-form solutions are not possible to obtain, we can solve the differential equation using a numerical method. In the next section, we will discuss a simple method of numerically solving ordinary differential equations.

2.7 EULER'S METHOD OF NUMERICALLY SOLVING ODES

Often, the differential equations derived from Newton's second law cannot be solved in closed form. As we will see in Chapter 13, this is particularly true when the forces are nonlinear in position and/or velocity. However, for now, we will examine numerical methods for solving ODEs which are useful when an ODE cannot be solved in closed form.

For simplicity, let us consider the first-order ODE

$$\frac{dx}{dt} = f(x) \tag{2.7.1}$$

A closed-form method of solving (2.7.1) involves finding an algebraic function $x(t)$ whose derivative is $f(x)$. When finding such a closed form is not possible (or easy), we can rely on numerical methods to get the solution. In this case, we find a list of values $\{x_0, x_1, \ldots, x_{n-1}\}$ corresponding to the set of times $\{t_0, t_1, \ldots, t_{n-1}\}$ such that $x_i = x(t_i)$ where $t_i = idt$ for $i = 0, 1, \ldots, (n-1)$. The variable dt is called the *step size* and will be discussed below. Although we will not obtain a closed-form solution, numerical methods provide useful information about the solution, and their results can be plotted.

In practice, you will often use a numerical ODE solver like `odeint` and `NDSolve` for Python and Mathematica, respectively. These are discussed in the next section and in the two Appendices of this book, and the underlying algorithms can be quite complicated. However, many ODE solvers have similarities with a simple numerical method called Euler's method, which is commonly presented in textbooks. In this method, the derivative is rewritten using Newton's definition of the derivative without the limit

$$\frac{dx}{dt} = \frac{x(t+dt) - x(t)}{dt} \tag{2.7.2}$$

If the chosen step size dt is small enough, (2.7.2) can provide a good approximation of the derivative dx/dt. We can then use (2.7.2) to rewrite (2.7.1) as

$$x(t+dt) = x(t) + f(x(t))dt \tag{2.7.3}$$

Finally, rewriting $x(t+dt)$ as x_i

$$x_i = x_{i-1} + f(x_{i-1})\,dt \tag{2.7.4}$$

If we know $x(t_0) = x_0$, then we can use (2.7.4) to find $x(t)$ for any time $t_i = idt$ which is an integer multiple of the step size dt.

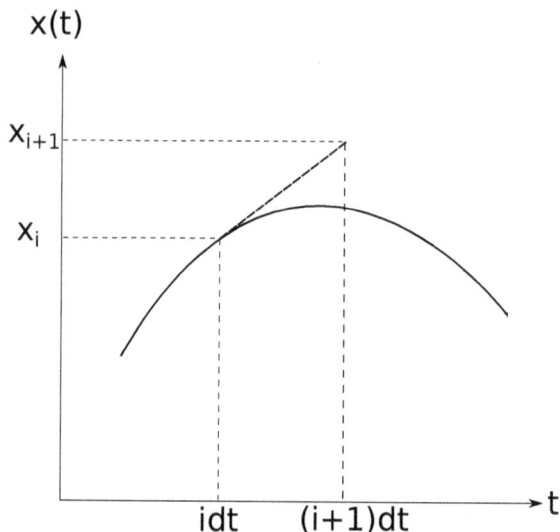

Figure 2.8: The slope $f(x_i)$ (diagonal dashed line) of the true solution (curve) is used to approximate x_{i+1} using the Euler method. Note the mismatch between the value of x_{i+1} and that of the true solution at time $(i+1)dt$.

We can extend Euler's method to second-order ODEs by introducing a new variable $v = dx/dt$. The ODE

$$\frac{d^2x}{dt^2} = f(x, v) \tag{2.7.5}$$

can then be rewritten as the coupled first-order system

$$\left.\begin{aligned} \frac{dx}{dt} &= v \\ \frac{dv}{dt} &= f(x, v) \end{aligned}\right\} \tag{2.7.6}$$

Using Euler's method, we can rewrite (2.7.6) as

$$\left.\begin{aligned} x_i &= x_{i-1} + v_{i-1}dt \\ v_i &= v_{i-1} + f\left(x_{i-1}, v_{i-1}\right)dt \end{aligned}\right\} \tag{2.7.7}$$

The step size dt must be small for Euler's method to provide a good solution. Euler's method works by using the slope $f(x_i)$ of the solution x_i at a time idt for a linear approximation of x_{i+1}, the value of the solution at a time $(i+1)dt$. If the step size is too large, then the slope may not do a good job of approximating x_{i+1} because the true solution $x(t)$ may change significantly over the time interval dt. An example of a poor approximation of x_{i+1} is shown in Figure 2.8. Notice that in Figure 2.8, the slope (diagonal dashed line) gives a reasonable approximation to the solution $x(t)$ (curve) when dt is small. However, for larger values of dt, the difference between the value extrapolated by the slope and $x(t)$ increases. Choosing too small of a step size will compound rounding errors made when computing each Euler step. The best choice of dt is somewhere between these two extremes.

The Euler method is not commonly used in practice. We present it here only to show the reader the basic concept of a numerical ODE solver. Instead, it is common to use more sophisticated algorithms such as fourth-order Runge-Kutta methods with an adaptive step size.

In Example 2.8 we use Python to demonstrate Euler's method. In this example we also demonstrate the use of `for` loops.

Example 2.8: Euler's method

A mass $m = 3.0$ kg is dropped at rest from a height of 20 meters. If the particle encounters a drag force $F_d = -cv^3$ where $c = 2.0$ Ns^3/m^3, use Euler's method to find the position and velocity of the ball as a function of time during the first second of the particle's fall.

Solution:

Using a coordinate system where positive displacements are vertically downward, the ODE corresponding to the particle's equation of motion is

$$m\frac{d^2x}{dt} = mg - cv^3 \tag{2.7.8}$$

Using $v = dx/dt$, we can rewrite the ODE using Euler's method

$$\left.\begin{array}{l} x_i = x_{i-1} + v_{i-1}dt \\ v_i = v_{i-1} + \left(g - \dfrac{c}{m}v_{i-1}^3\right)dt \end{array}\right\} \tag{2.7.9}$$

Python Code

The Python code for Euler's method is below. We begin the code by defining the necessary parameters. Recall that `np.arange` creates a list of values that starts at the first argument, but ends one step (third argument) before the second argument. Hence, we use `t_max + dt` to ensure the list of times ends at $t = 1$. Notice that after defining the necessary parameters, a `for` loop was used to calculate each iteration of Euler's method. Like `np.arange`, `for` ends one step before its second argument. Hence, we need to use `num_steps + 1` as the second argument to ensure x and v are calculated for the full time range. The plots of x and v can be found in Figure 2.9.

```python
import matplotlib.pyplot as plt
import numpy as np

dt, t_max = 0.05, 1        #step size, maximum time
num_steps = int(t_max/dt)  #number of Euler steps
m, g, c = 3.0, 9.8, 2.0    #parameters and coefficients

times = np.arange(0,t_max + dt,dt) # times for x(t), v(t)

def f(v):                  #define the acceleration function
    return (g - c/m*v**3)

x = np.zeros(num_steps+1)  #define x,v arrays
v = np.zeros(num_steps+1)
x[0], v[0] = 0 ,0          # initial conditions

for i in range(1,num_steps+1):
    x[i] = x[i-1] + v[i-1] * dt
    v[i] = v[i-1] + f(v[i-1]) * dt

fig, ax = plt.subplots(nrows=1, ncols=2)
ax[0].plot(times,x)
ax[0].set_xlabel("time, s")
ax[0].set_ylabel("x(t), m")
ax[0].set_title('(a)');
ax[1].plot(times,v)
ax[1].set_xlabel("time, s")
ax[1].set_ylabel("v(t), m/s")
ax[1].set_title('(b)');
fig.tight_layout()
plt.show()
```

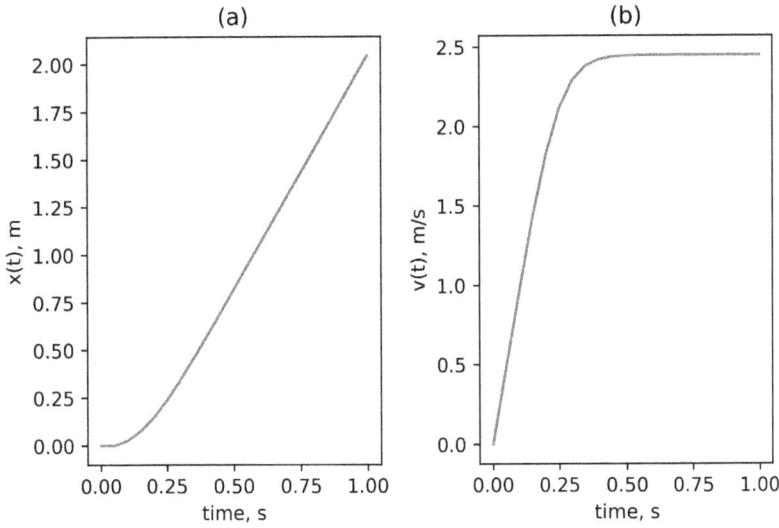

Figure 2.9: Plots of (a) $x(t)$, and (b) $v(t)$ for Examples 2.8 and 2.9. Notice that the position $x(t)$ increases linearly with time after the velocity $v(t)$ becomes constant.

2.8 NUMERICAL ODE SOLVERS IN MATHEMATICA AND PYTHON

In this section, we present two of the main Python and Mathematica built-in algorithms for solving numerically ODEs. Specifically we give examples of using the `odeint` command for Python, and the `NDSolve` command in Mathematica. The algorithms for these commands are more complicated than the Euler method used in the previous section; however most numerical algorithms compute values iteratively using a step size. In other words, they estimate current values from past values, separated in space or time.

In Example 2.9, we repeat the problem of Example 2.8, using the numerical ODE solvers in Mathematica and Python.

Example 2.9: Using Mathematica and Python to numerically solve ODEs
A 3.0 kg particle is dropped at rest from a height of 20 meters. The particle encounters a drag force $F_d = -c\,v^3$ where $c = 2.0$ Ns^3/m^3. Use Mathematica's `NDSolve` command and Python's `odeint` command to find the position and velocity of the ball as a function of time, during the first second of the particle's fall.

Python Code
To use `odeint()` in this example, we need to rewrite $F_d = m\,g - c\,v^3$ as system of two first-order ODEs, similar to Euler's method in Example 2.8:

$$\left.\begin{aligned}\frac{dx}{dt} &= v \\ \frac{dv}{dt} &= g - \frac{c}{m}v^3\end{aligned}\right\} \tag{2.8.1}$$

In the code we define a function `deriv(y,time)` whose arguments are a vector y and the time variable `time`. In our example, the first component y[0] of the vector y represents the position x, and the second component y[1] is the speed $v = dx/dt$. The function deriv evaluates and returns a vector with components $(v, dv/dt)$.

The function odeint(deriv, yinit, t) is called with three arguments. The first argument is the function deriv which contains the information on the first-order ODE to be solved, the second argument is the initial conditions vector yinit, and the third argument is the time variable t.

The line t = np.linspace(0, 1, 20) defines the time interval over which the ODE will be integrated, in this case between $t = 0$ and $t = 1$ s. The variable yinit = (0,0) defines the initial conditions $x(0) = 0$ m, and $v(0) = 0$ m/s. Note that the result of odeint is a matrix, where the first row soln[:,0] is the position $x(t)$, and the second row soln[:,1] is the speed v. The rows are used in the plot commands at the end of the code, with the result shown in Figure 2.9.

```
from scipy.integrate import odeint
import numpy as np
import matplotlib.pyplot as plt

m, g, c = 3.0, 9.8, 2.0    #parameters and coefficients

#define function that returns coupled ODEs
def deriv(y, time):
    return (y[1], g - c/m*y[1]**3)

t = np.linspace(0,1,20)       # time values
yinit = (0,0)                 #initial conditionss

soln = odeint(deriv, yinit, t)  #solve the system of ODEs using odeint

fig, ax = plt.subplots(nrows=1, ncols=2)
ax[0].plot(t,soln[:,0])       #plot position x(t)
ax[0].set_xlabel("time, s")
ax[0].set_ylabel("x(t), m")
ax[0].set_title('(a)');
ax[1].plot(t,soln[:,1])       #plot speed v(t)
ax[1].set_xlabel("time, s")
ax[1].set_ylabel("v(t), m/s")
ax[1].set_title('(b)');
fig.tight_layout()
plt.show()
```

Mathematica Code

The command NDSolve is used to solve the ODE

$$m\frac{d^2x}{dt^2} = mg - cv^3$$

We store the solution found by NDSolve in the variable soln. The first argument of NDSolve is a list which includes the ODE (note the == sign as opposed to =), and also contains the initial conditions. Primes are used to denote derivatives in Mathematica. The second argument of NDSolve is the function being solved for, in this case $x(t)$. The third and final argument is the range of times.

In this example we let Mathematica adjust automatically the integration conditions, to ensure that the solution converges. However, if desirable, the number of integration steps and the integration step size can be specified using the options for NDSolve listed in Mathematica's help files.

When creating the plots, notice that we use the "replace with" command /., in order to specify that x[t] and x'[t] are found using the result of NDSolve stored in soln.

SetOptions[Plot, Frame->True, Axes->False, BaseStyle->{FontSize->16}];

$m = 3.0; g = 9.8; c = 2.0;$

tMax $= 1.0;$

soln = NDSolve[{$m * x"[t]==m * g - c * x'[t]^3, x[0]==0, x'[0]==0$}, x,

{$t, 0,$ tMax}];

GraphicsGrid[{{Plot[$x[t]$/.soln, {$t, 0,$ tMax}, FrameLabel->{"time", "position"}],

Plot[$x'[t]$/.soln, {$t, 0,$ tMax}, FrameLabel->{"time", "position"}]}}]

2.9 EVALUATION OF INTEGRALS WITH NUMERICAL INTEGRATION FUNCTIONS

In this section we demonstrate two commands commonly used for the numerical evaluation of integrals. These are Python's quad and Mathematica's NIntegrate commands.

Example 2.10: Numerical evaluation of integrals
  A particle starting at rest at the origin experiences an acceleration of the form

$$a(t) = a_0 \sin^2(\omega t + \phi) \tag{2.9.1}$$

where $a_0 = 0.3$ m/s^2, $\omega = 1.5$ rad/s, and $\phi = \pi/3$ radians. Use Python's quad and Mathematica's NIntegrate commands to find the particle's velocity at $t = 3.0$ seconds.

Solution:
  To solve this problem, we will need to use (2.4.3):

$$v(t) - v(0) = \int_0^t a(t')dt' \tag{2.9.2}$$

Using $v(0) = 0$ we obtain:

$$v(3) = \int_0^3 a_0 \sin^2(\omega t + \phi) \, dt \tag{2.9.3}$$

Although this integral can be done analytically, we will solve it using **quad** and **NIntegrate**.

Python Code

The command **quad** from SciPy's integrate library is used for numerical calculations of integrals. The function to be integrated **F** is first defined as a lambda function, using the command **lambda**. Lambda functions are useful when a function needs to only be used for a short time in the code. The **quad** command has three arguments, the function to be integrated, the lower limit, and the upper limit of integration. The output is a tuple. The tuple's first argument is the value of the integral, and the second argument is the estimated absolute error on the integration procedure.

```python
from scipy.integrate import quad
import numpy as np

print('-'*28,'CODE OUTPUT','-'*29,'\n')

a0, omega, phi = 0.3, 1.5, np.pi/3

F = lambda t: a0*np.sin(omega * t + phi)**2

integral = quad(F,0,3)

print('The integral = ', integral[0])
print('The absolute error =', integral[1])

---------------------------- CODE OUTPUT ----------------------------
The integral =  0.5430573299677108
The absolute error = 1.974040231812555e-14
```

Mathematica Code

In Mathematica, **NIntegrate** is used to evaluate the definite integral after defining the necessary parameters.

```
a0 = 0.3; ω = 1.5; φ = π/3;

NIntegrate[a0 * Sin[ω * t + φ]^2, {t, 0, 3}]

0.543057
```

2.10   SIMPSON'S RULES FOR NUMERICAL EVALUATION OF INTEGRALS

In this section we show how integrals can be evaluated numerically using a simple method called Simpson's rule. Consider the following integral

$$\int_{x_0}^{x_N} f(x')dx' \tag{2.10.1}$$

where $f(x)$ is a function of a single variable and x_0 and x_N are constant limits of integration. Recall that integrals measure the area between the function $f(x)$ and the x-axis.

In introductory calculus, the *trapezoid rule* approximates the area under $f(x)$ by the area of a trapezoid, whose four corners are the limits of integration on the x-axis and the value of the function at those limits.

$$\int_{x_0}^{x_N} f(x')dx \approx \frac{x_N - x_0}{2}\left[f(x_N) - f(x_0)\right] \tag{2.10.2}$$

Simpson's rule estimates the value of the integral by breaking the region of integration into two steps, by sampling the function $f(x)$ at three points x_0, x_1, and x_N, equally spaced by a distance h. The formula for Simpson's rule is:

$$\int_{x_0}^{x_N} f(x')dx' \approx \frac{h}{3}\left[f(x_0) + 4f(x_1) + f(x_N)\right] \tag{2.10.3}$$

However, both the trapezoid rule and Simpson's rule are often not accurate enough for practical calculation. We can improve upon Simpson's rule by breaking the interval of integration up into even more steps, each separated by a distance h. The result is called the extended Simpson's rule

$$\int_{x_1}^{x_N} f(x')dx' = \frac{h}{3}\left(f_1 + 4f_2 + 2f_3 + 4f_4 + \cdots + 2f_{N-2} + 4f_{N-1} + f_N\right) + O\left(\frac{1}{N^4}\right) \tag{2.10.4}$$

where $f_i = f(x_i)$, $x_i = x_1 + ih$, $h = (x_N - x_0)/N$, and N is the number of steps.

In Example 2.11 we show how to implement the extended Simpson's rule in Python, since this is a good example of using conditional statements.

Example 2.11: The extended Simpson's rule
   Repeat Example 2.10 using the extended Simpson's rule, with the same parameter values.

Python Code
   In the code we calculate the integral using the extended Simpson's rule, and also using SymPy for an exact solution. After defining the force function F, we compute the integral. The coefficients used in the extended Simpson's rule depend on the index of each term in the right-hand side of (2.10.4). The coefficient of f_i is 4 when i is even, and is 2 when i is odd, except for $i = 1$ and $i = N$. In order to handle these dependencies, we used the **if** and **elif** conditional statements. The **if** statement tests whether or not a condition is true and if so, the indented line is executed.
   The first **if** statement tests to see if $i = 1$ or $i = N$. If so, then the term $hF(x)/3$ is added to the sum. Notice that the variable **extended** contains the sum of the terms on

the right-hand side of (2.10.4), and that the command += adds to and overwrites the variable extended. If $i \neq 1$ or N, then the next conditional statement is evaluated.

The next conditional statement is elif, or else-if, which tests to see if i is even by calculating i modulo 2. In Python, the % symbol is used for the modulo calculation. The elif statement provides an additional conditional statement to test after the first if statement, and it is useful when there are multiple possible outcomes for a conditional (in this case i can be even, odd, 1, or N). Finally, if none of the above conditional statements return True, then i must be odd. The else conditional statement tells Python what to execute if no other prior conditionals are true; in this case, the term $\frac{4}{3}hF(x)$ is added to the sum.

```
import numpy as np
from sympy import integrate, pi, sin, symbols
t = symbols('t')

print('-'*28,'CODE OUTPUT','-'*29,'\n')

a0, omega, phi = 0.3, 1.5, np.pi/3

a, b = 0, 3  #lower and upper limit of integration

def F(t):
    return a0*np.sin(omega*t + phi)**2

N = 1000      #number of steps

extended = 0 #parameter stores the result of integration

h = (b-a)/(N-1) #need N steps starting at an index of 0

for i in range(0,N):
    x = a + i*h
    if i==0 or i==(N-1):
        extended += h*F(x)/3.0
    elif i%2 == 0:
        extended += 2.0*h*F(x)/3.0
    else:
        extended += 4.0*h*F(x)/3.0

#exact answer obtained using integrate in SymPy
exact = integrate(a0*sin(omega*t + phi)**2,(t,0,3))

print('The extended Simpson result = ', extended)
print('The exact result = ',exact.subs({t:3}))

------------------------- CODE OUTPUT ----------------------------
The extended Simpson result =  0.5429213205860004
The exact result =  0.543057329967711
```

2.11 CHAPTER SUMMARY

Newton's second law gives the equation of motion for a particle. This equation of motion is a differential equation and can be written in three forms:

$$F = m\frac{d^2x}{dt^2} = m\frac{dv}{dt} = mv\frac{dv}{dx}$$

$F = m\frac{dv}{dt}$ is a *first-order ODE*, because the highest derivative is a first derivative.

$F = m\frac{d^2x}{dt^2}$ is a *second-order ODE*, because the highest derivative is a second derivative.

The number of a*rbitrary constants* in the general solution of an ODE is equal to the order of the ODE.

The method of *separation of variables* is performed by treating the derivative as a fraction, and isolating terms with the dependent variable (e.g., $x(t)$) on one side, and terms involving the independent variable on the other.

When the acceleration $a = $ constant, Newton's second law gives the *kinematic equations*:

$$v(t) = v_0 + at \qquad x(t) = x_0 + v_0 t + \frac{1}{2}at^2 \qquad v(x)^2 = v_0^2 + 2a\left(x - x_0\right)$$

When the force is a function of time only $F(t)$, Newton's second law can be integrated using:

$$v(t) = v_0 + \frac{1}{m}\int_{t'=0}^{t} F(t')\,dt'$$

When the force is a function of speed only $F(v)$, Newton's second law can be integrated using:

$$t - t_0 = m\int_{v'=0}^{v}\frac{dv'}{F(v')}$$

The magnitude of the *air resistance* $f(v)$ has, in general, both a linear and a quadratic component in v:

$$f(v) = bv + cv^2$$

We find the *terminal velocity* for a falling object by setting the air resistance force to be equal to the weight of the object, $f(v) = bv + cv^2 = mg$.

When the force is a function of the position only, $F(x)$, we can integrate Newton's second law in the form:

$$F(x) = ma = mv\frac{dv}{dx}$$

to obtain the function $v(x)$. Then integrate once more the $v(x) = dx/dt$ equation to obtain the position $x(t)$.

An important system in mechanics is the mass-spring system described by *Hooke's Law*:

$$F(x) = m\frac{d^2x}{dt^2} = -kx$$

With the initial conditions $v(0) = 0$ and $x(0) = x_0$ at $t = 0$, this equation has solutions $x(t)$ and $v(t)$ given by:

$$x(t) = x_0 \cos\left(\sqrt{\frac{k}{m}}t\right) \qquad v(t) = dx/dt = -\omega x_0 \sin\left(\sqrt{\frac{k}{m}}t\right)$$

and the mass oscillates about the equilibrium position x_0 with the natural frequency ω of the oscillator:

$$\omega = \sqrt{\frac{k}{m}}$$

Euler's method is a simple numerical technique for integrating differential equations which uses the derivative of the function to approximate future values of the differential equation's solution.

2.12 END-OF-CHAPTER PROBLEMS

The symbol ⌨ indicates a problem which requires some computer assistance, in the form of graphics, or numerical computation, or symbolic evaluation.

Section 2.2: Ordinary Differential Equations

1. ⌨ The number of particles, $N(t)$, in a sample of radioactive materials varies as:

$$\frac{dN}{dt} = -\lambda N \tag{2.12.1}$$

where $\lambda > 0$. Solve this differential equation by hand in the case where $N(0) = N_0$. Plot $N(t)$ as a function of time for various values of λ and various values of N_0. How does the size of λ affect the solution? What happens to the solution if you change N_0?

2. In Chapter 6, we will see the following differential equation:

$$\frac{d^2x}{dt^2} = -\omega^2 x \tag{2.12.2}$$

where $x(t)$ is the position of a particle as a function of time. One method of solving second-order differential equations is by using a trial solution. In this case, the equation says that the second derivative of the function $x(t)$ returns itself with a constant and negative sign. What functions behave this way? Certainly sine and cosine however, an exponential solution $x(t) = Ae^{\lambda t}$ would also work. Insert into the differential equation the exponential function as a trial solution. What values of λ are required in order for $Ae^{\lambda t}$ to be a solution to the differential equation?

3. Consider the linear differential equation $\ddot{x} + \dot{x} + x = f_1(t) + f_2(t)$, where f_1 and f_2 are smooth continuously differentiable functions of time. In this case, the ODE is called a nonhomogeneous differential equation, because there are terms in the ODE which explicitly depend on the independent variable, t. Suppose that you know two functions, $x_1(t)$, which solves the ODE $\ddot{x}_1 + \dot{x}_1 + x_1 = f_1(t)$, and $x_2(t)$, which solves the ODE $\ddot{x}_2 + \dot{x}_2 + x_2 = f_2(t)$. Show that the solution to $\ddot{x} + \dot{x} + x = f_1(t) + f_2(t)$ is $x(t) = x_1(t) + x_2(t)$. Generalize this proof to the solution of the ODE: $\ddot{x} + \dot{x} + x = \sum_{i=1}^{N} f_i(t)$, where you know all $x_i(t)$, each of which solves $\ddot{x}_i + \dot{x}_i + x_i = f_i(t)$.

4. Prove that the result from Problem 3 holds for *any* linear nonhomogeneous ODE.

5. Consider the nonhomogeneous ODE $\ddot{x} + \dot{x} + x = f(t)$, where $f(t)$ is a smooth continuously differentiable function. If $f(t) = 0$, the ODE is called a homogeneous differential equation. Using the result from Problem 3, show that the solution of the nonhomogenous differential equation is $x(t) = x_c(t) + x_p(t)$, where $x_c(t)$ is called the complementary solution, found when $f(t) = 0$, and $x_p(t)$ is called the particular solution and is a solution of the nonhomogeneous ODE, $\ddot{x} + \dot{x} + x = f(t)$. Repeat for *any* linear nonhomogenous ODE.

6. 🖥 Using Mathematica's `DSolve` or Python's `dsolve`, solve the differential equation

$$\frac{df}{dx} = 3 \sin x - 4e^{-2x} \tag{2.12.3}$$

for the initial condition $f(0) = 0$. Plot the solution for $x = 0$ to $x = 3$.

7. Victor slides a puck with mass m up a long ramp with a coefficient of kinetic friction μ. The ramp makes an incline θ with the horizontal. If the puck starts at the bottom of the ramp with an initial velocity of v_0, find how far up the ramp the puck goes before coming to a stop. Find the puck's position and velocity as a function of time.

8. An airplane flying at an altitude H needs to drop supplies at the center of a small village. The plane is flying horizontally with a speed v_0. Neglecting air resistance, how far ahead of the village's center should the pilot drop the supplies in order for the supplies to land in the center of the village?

9. A ball with mass m is thrown up an incline plane with a speed v_0. The plane has an incline of θ with the horizontal, and the ball is thrown at an angle ϕ with respect to the horizontal direction. Neglecting air resistance,

   a. Show that the range of the ball is

$$R = \frac{2v_0^2 \sin{(\phi - \theta)} \cos\phi}{g \cos^2 \theta}$$

   b. Show that the maximum range occurs when $\phi = \frac{\pi}{4} + \frac{\theta}{2}$.

10. 🖥 Consider the double Atwood machine shown in Figure 2.10. Compute the acceleration of the mass m_1 in terms of g, M, and m_2. Ignore friction and assume the pulleys are massless and the strings are massless and inextensible.

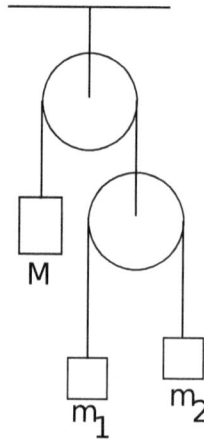

Figure 2.10: A double Atwood machine in Problem 2.10.

Section 2.4: Time-Dependent Forces

11. 🖥 A particle of mass m starts at position x_0 with speed v_0, and experiences the following forces (assume all constants a, b, A and ω are positive). Find $x(t)$ and/or $v(t)$, both analytically and using a CAS.

   a. $F = a + bt$
   b. $F = A\cos(\omega t)$
   c. $F = ate^{-bt}$

12. 🖥 A 1.0 kg particle experiences a force $F = -ae^{-\beta t}$, where $a = 0.5$ N, $\beta = 0.25$ s^{-1}, and the minus sign denotes that the force opposes the particle's velocity. If the particle's initial position and velocity are $x_0 = 0$ m and $v_0 = 10$ m/s, find the position and velocity of the particle as a function of time. Do the problem analytically and using a CAS.

13. Lilly repeats Victor's experiment (from Problem 7) but her ramp starts horizontal and increases its inclination at a rate of ω rad/s. How far along the ramp has Lilly's puck traveled by the time the ramp has an incline of $\pi/3$ radians with respect to the horizontal? The coefficient of friction between the puck and the ramp is μ. Give the result in terms of ω, v_0, μ, and g.

Section 2.5: Air Resistance and Velocity-Dependent Forces

14. 🖥 Find the terminal velocity of a sphere with a 0.25 meter diameter and mass m experiencing a linear and quadratic air resistance, as well as a gravitational force, i.e., $F(v) = bv + cv^2 - mg$. Using numerical methods, find and plot the particle's position and velocity as a function of time if the particle is released from rest at a height $y_0 = H$. To make the plot, use the equation for the coefficients b and c from Section 2.5, and set $m = 1$ kg and $H = 100$ m.

15. 🖥 An object of mass m experiences a force, $F(v) = -cv^2$. What are the SI units of c? If the particle starts at the origin with an initial velocity $v(0) = v_0$, find the particle's position and velocity as a function of time. Use a CAS to verify your results.

16. A ball is thrown upwards with an initial velocity of 10 m/s. The ball has a mass $m = 0.10$ kg and experiences a drag force $F = -bv$ that scales linearly with the ball's velocity with $b = 0.2$:

   a. What are the SI units of b?

   b. How high does the ball travel? Compare your answer to the one you would get in the case of no air resistance.

17. In Example 2.4 we saw that the position and velocity of a falling particle in the presence of linear air resistance are:

$$x(t) = x_0 + \frac{m\,v_0}{b}\left(1 - e^{-\frac{bt}{m}}\right) \tag{2.12.4}$$

$$v(t) = v_0 e^{-\frac{bt}{m}} \tag{2.12.5}$$

By taking the limits of $t \to 0$ and $t \to \infty$, discuss whether these equations make sense for the position and velocity of a particle falling under the forces of gravity and linear drag.

18. An object is thrown vertically upward with an initial velocity v_0.

   a. Find the amount of time it takes for the object to reach its highest point when there is no air resistance.

   b. How much time is required for the object to reach its maximum height if there is linear air resistance $F = -b\,v$?

   c. How about the case of a quadratic air resistance $F = -b\,v^2$? Use a mass $m = 1$ kg, $v_0 = 10$ m/s for the ball's initial velocity, and $b = 0.2$ (SI units) for the coefficients of both types of drag forces.

19. 🖮 Repeat Example 2.5 in this chapter, by using a computer algebra system. The force in this example is given by $F(v) = -bv + mg$. Find $y(t)$ and $v(t)$, and plot the solutions for $m = 1.0$ kg, $b = 0.2$ Ns/m, $x_0 = 0$ m, and $v_0 = 0$ m/s. Using the CAS, take the limit of $v(t)$ as $t \to \infty$. You may need to do a little extra work to manipulate the solution provided by your CAS, so that it looks like the results found in Example 2.5. This is not unusual, and one should expect to need to do some algebraic manipulation of a computer-generated solution, in order to put it into a form that is more easily readable. Another reason why one should not forego their algebra skills and rely solely on computers!

20. Evelyn repeats Victor's experiment from Problem 7. However, this time, the ramp Evelyn is using exerts a force of $F(v) = -cv^2$ on the puck instead of the usual force of friction, μN. If the initial velocity of the puck up the ramp is v_0, how far up the ramp does the puck travel? How much time is required for the puck to reach its maximum height?

21. 🖮 Consider a boat of mass m moving through water experiencing a resisting force, $F(v) = -ae^{bv}$. The boat starts at the origin with a velocity of v_0.

   a. Find the boat's velocity as a function of time, both analytically and using a CAS.

   b. Find the time at which the boat instantaneously comes to a rest.

22. Consider a spherical raindrop with a diameter of 0.5 mm falling through the air at a velocity of 1.0 m/s at time $t = 0$.

    a. Which is the dominant form of air resistance acting on the raindrop, linear or quadratic?

    b. Find an equation that describes the motion of the raindrop and compute the raindrop's terminal velocity.

23. A mass m with an initial velocity of v_0 is sliding along a horizontal surface. The mass experiences a resistance, $F(v) = -cv^{3/2}$. If the mass starts at the origin, what is the maximum distance traveled by the mass?

24. Sarah, who has a mass of 55 kg, parachutes out of an airplane flying at an altitude of 3 km.

    a. How long will it take her to fall if the parachute fails to open and air resistance is ignored?

    b. How long will it take her to fall if the parachute opens? Assume that the parachute opens immediately and that the drag force on Sarah with the parachute open is described by,

$$f = -\frac{1}{2}c_d \rho A v^2 \qquad (2.12.6)$$

    where the drag coefficient for the parachute with an area A=46 m^2 is $c_d = 1.5$. The density of air is $\rho = 1.22$ kg/m^3.

25. ⌨ Consider a falling object whose mass is 1 kg and experiences an air resistance force of $F(v) = -c\,|v|\,v$, where $c = 0.5$ Ns^2/m^2.

    a. Assuming the falling object starts at rest, use a computer to calculate the time it takes the object to fall a distance of 1 km.

    b. Does the velocity of the falling object ever become constant? In other words, is there a terminal velocity for this force?

Section 2.6: Position-Dependent Forces

26. A particle of mass m has a velocity $v(x) = ax^{-n}$. Find the force acting on the particle. If at $t = 0$, the particle is at a position x_0 and moving with a velocity v_0, find the position and velocity of the particle as a function of time.

27. ⌨ Find the velocity $v(x)$ and position $x(t)$ of a particle of mass m starting at rest at the origin experiencing the force $F = a + bx$ where a, b are constants with appropriate SI units. For simplification, use the numerical values $a = 2$, $b = 2$ and mass $m = 2$.

28. ⌨ Compute the value of the acceleration due to gravity as a function of an object's height y above the Earth's surface. Plot your result from $y = 0$ to 25 km. *Hint*: Use Newton's universal law of gravitation:

$$F_g = -\frac{Gm_1 m_2}{r^2} \qquad (2.12.7)$$

where $G = 6.67 \times 10^{-11}$ Nm^2/kg^2 is the universal gravitation constant, m_1 and m_2 are the masses of the interacting bodies, and r is the distance between the centers of m_1 and m_2. Assume that the falling object is a point particle and ignore air resistance.

29. 🖥 As found in Problem 28, the acceleration of a falling object is not constant with height. It turns out that the acceleration varies as:

$$g = \frac{9.8 \text{ m/s}^2}{\left(1 + \frac{y}{R_e}\right)^2} \qquad (2.12.8)$$

where $R_e = 6.37 \times 10^6$ m is the radius of the Earth, y is the altitude of the object above the Earth's surface, and g is in m/s^2. Compute the time it takes for a $m = 1$ kg object with a diameter $D = 1$ m to fall from an altitude of 25 km and land on the Earth's surface, assuming quadratic air resistance. What is the object's final velocity? You will need a computer to solve this problem.

30. A particle of mass m experiences a force $F = cvx$, where $c > 0$ and x, v are the position and speed. If at $t = 0$, the particle is passing through the origin with a velocity $v = v_0$, find the particle's position as a function of time.

Section 2.7: Numerical Methods

31. 🖥 Using Euler's method, solve the ODE from Problem 1 with $N_0 = 10^5$ particles and $\lambda = 0.1$ s^{-1} for $0 \leq t \leq 10$ seconds. Plot the Euler solution and the exact analytical solution $N = N_0 \exp(-\lambda t)$ on the same graph. The function $N(t)$ should decay exponentially. Choose a small enough value of dt such that you get the correct functional form.

32. 🖥 A particle of mass $m = 0.2$ kg is moving horizontally at a velocity of $v_0 = 10$ m/s when it experiences a drag force $F = -0.2v^2$ (where the force is measured in N). Solve for $v(t)$ by hand. Using Euler's method, find $v(t)$ numerically and plot it. Use a small enough value of dt to get the correct result.

33. 🖥 Consider the ODE

$$\frac{d^2x}{dt^2} = -\omega^2 x \qquad (2.12.9)$$

Rewrite the ODE as a coupled system of first-order ODEs using the variable $v = dx/dt$. Use $\omega = 1$ rad/s and Euler's method to solve the coupled ODEs with an initial condition of $x(0) = 1.0$ m and $v(0) = 0$ m/s for the time range $0 \leq t \leq 30$ s. Use $dt = 0.01$ s.

34. 🖥 Consider Problem 33. Repeat the problem, using larger values of dt. What happens to the solution when dt gets too large? Plot your result for different values of dt.

35. 🖥 The Euler-Cromer method produces solutions that are stable for oscillatory systems. Consider an ODE in the form of Newton's second law

$$\frac{d^2x}{dt^2} = a \qquad (2.12.10)$$

where a is an acceleration (not necessarily constant) and we have set the mass $m = 1$ (SI units). Using the variable $v = dx/dt$, (2.12.10) can be rewritten using the Euler-Cromer method as

$$\left.\begin{aligned} v_{n+1} &= v_n + a_n dt \\ x_{n+1} &= x_n + v_{n+1} dt \end{aligned}\right\} \qquad (2.12.11)$$

Note the subscript for v in the second line of (2.12.11). Using Python, write an algorithm which solves (2.12.10) using the Euler-Cromer method with the same initial conditions and parameter values as Problem 33.

36. ⌨ Consider a particle of mass m experiencing a net force $F = -b\sin x$. At $t = 0$, the particle starts at rest. Using either Python or Mathematica, find the particle's position $x(t)$ using a numerical differential equation solver such as odeint or NDSolve. Discuss the particle's motion when $x(0) = \pi/6$, $x(0) = \pi/2$, and $x(0) = 7\pi/8$. Set the values of $m=b=1$ (SI units).

37. ⌨ Consider a particle of mass m experiencing a linear restoring force, $F(x) = -k\,x$, a quadratic drag force $F(v) = -c\,v^2$, plus a sinusoidal external driving force $F(t) = A\cos(\omega t)$, where $k = 3$ N/m, $c = 0.5$ Ns^2/m, $\omega = 1.0$ rad/s, $m = 1$ kg and $A = 2.0$ N. Write the ODE for the resulting equation of motion. If the particle's initial position and velocity are $x(0) = 0$ m and $v(0) = 0$ m/s, find the particle's position and velocity as a function of time using a numerical differential equation solver such as odeint or NDSolve. Describe the motion of the particle.

38. ⌨ The Midpoint Rule is another method of numerically integrating a function. Suppose a function $f(x)$ is defined on the interval $[a, b]$. We subdivide the interval into subintervals of equal length $\Delta x = (b - a)/N$ using the points $x_i = a + i\Delta x$ (where $x_N = b$). We can then approximate the integral as

$$\int_a^b f(x)dx = \sum_{i=1}^N f\left(\frac{x_{i+1} - x_i}{2}\right)\Delta x \qquad (2.12.12)$$

Write a program which uses the Midpoint Rule to solve Example 2.10, in which we evaluated the integral

$$v(3) = \int_0^3 a_0\sin^2(\omega t + \phi)\,dt \qquad (2.12.13)$$

where $a_0 = 0.3$ m/s^2, $\omega = 1.5$ rad/s, and $\phi = \pi/3$ radians.
Compare the result of the Midpoint Rule with the extended Simpson's rule (using the same step size), and by using the **quad** library from SciPy.

39. ⌨ The velocity of a particle (in SI units) can be described using the function

$$v(t) = 2.0e^{-\lambda t}\cos(\omega t)$$

where $\lambda = 0.001\ s^{-1}$ and $\omega = 1.0$ rad/s. Assume the particle starts at the origin.

   a. Evaluate the position of the particle $x(t)$ analytically and using a CAS.

   b. Using a numerical integration technique of your choosing, find the position of the particle at a time $t = 2.0$ seconds.

40. ⌨ The impulse J of a force F is defined as

$$J = \Delta p = \int_{t_0}^t F(t')\,dt \qquad (2.12.14)$$

where p is the momentum of a particle experiencing the force $F(t)$. Suppose a mass

$m = 1.0$ kg experiences a force $F(t) = c/\left(t^3 + b^3\right)$ where $b = 1.0$ s and $c = 2.0$ Ns^3. Using a numerical integration technique of your choice, find the momentum of the particle at a time $t = 2.5$ seconds. Assume the particle is at rest at a time $t_0 = 0$.

3 Motion in Two and Three Dimensions

In the previous chapter, the kinematics and dynamics aspects of particle motion were examined in one dimension by using the scalar position parameter x. In this chapter, we will first describe the motion of particles in two and three dimensions using Cartesian coordinates (x, y, z). Describing motion in two and three dimensions requires the use of vectors. After reviewing the basic properties of vectors and introducing the dot and cross products, we will introduce vector derivatives. Vector derivatives are necessary in order to compute the velocity and acceleration vectors. We then describe polar, cylindrical, and spherical coordinates, and demonstrate how to describe a particle's position, velocity, and acceleration in those coordinate systems. Finally, we conclude the chapter with special vector derivatives commonly used in all fields of physics, such as the gradient, the divergence, and the curl.

3.1 POSITION, VELOCITY, AND ACCELERATION IN CARTESIAN COORDINATE SYSTEMS

In order to describe a particle's motion in two and three dimensions, we will need to use vectors. In this section the main properties of vectors are reviewed, and they are applied to the description of position, velocity, and acceleration of a particle in three dimensions. In Chapter 1, we introduced the concepts of position, velocity, and acceleration using Cartesian coordinates. In this section, we will expand on that discussion. The expanded discussion of the basic descriptors of motion in Cartesian coordinates will prepare us to describe those quantities in other coordinate systems.

A point P_1 in three-dimensional space can be represented by a vector $\mathbf{A}$ from the origin $(0, 0, 0)$ to point P_1 as shown in Figure 3.1. Such a vector $\mathbf{A}$ has three components along the x-, y-, and z-axis, and can be written in the form $\mathbf{A} = [A_x, A_y, A_z]$ or in Cartesian coordinate form as:

$$\mathbf{A} = A_x\hat{\mathbf{i}} + A_y\hat{\mathbf{j}} + A_z\hat{\mathbf{k}} \tag{3.1.1}$$

where $\hat{\mathbf{i}}, \hat{\mathbf{j}}, \hat{\mathbf{k}}$ are the unit vectors along the x-, y-, and z- axis, respectively.

The magnitude A of the vector $\mathbf{A}$ is found from the Pythagorean theorem:

$$A = |\mathbf{A}| = \sqrt{A_x^2 + A_y^2 + A_z^2} \tag{3.1.2}$$

The magnitude of a vector is often thought of as its length and would correspond to a quantity measured from the vector. For example, suppose $\mathbf{A}$ represents the position of a particle located at the point $\mathbf{A} = 3\ m\hat{\mathbf{i}} + 4\ m\hat{\mathbf{j}}$. The magnitude of $\mathbf{A}$ is $|\mathbf{A}| = 5\ m$ and represents the distance between the particle and the origin.

It is often useful to know the direction in which a vector points. This is easily found for simple vectors like $\mathbf{A} = 3\hat{\mathbf{i}}$, which points along the x-axis. However, other vectors like $\mathbf{A} = a\hat{\mathbf{i}} + b\hat{\mathbf{j}}$ do not lie along either the x- or y-axis. In that case, we need a *unit vector* which points in the same direction as the vector $\mathbf{A}$ but has a length of 1. We can obtain a unit vector $\hat{\mathbf{A}}$ in the direction of any vector $\mathbf{A}$, by dividing $\mathbf{A}$ by its magnitude:

$$\hat{\mathbf{A}} = \frac{\mathbf{A}}{|\mathbf{A}|} = \frac{\mathbf{A}}{\sqrt{A_x^2 + A_y^2 + A_z^2}} \tag{3.1.3}$$

Example 3.1: Finding unit vectors.

Find a unit vector along the direction of the vector $\mathbf{a}=[1,\,3,\,-1]$.

Solution:

We write the vector as $\mathbf{a} = a_x\hat{\mathbf{i}} + a_x\hat{\mathbf{j}} + a_x\hat{\mathbf{k}} = 1\hat{\mathbf{i}} + 3\hat{\mathbf{j}} - 1\hat{\mathbf{k}}$. In order to find the unit vector, divide the vector with its magnitude:

$$\hat{\mathbf{a}} = \frac{\mathbf{a}}{\sqrt{a_x^2 + a_y^2 + a_z^2}} = \frac{1\hat{\mathbf{i}} + 3\hat{\mathbf{j}} - 1\hat{\mathbf{k}}}{\sqrt{1^2 + 3^2 + 1^2}} = \frac{1}{\sqrt{11}}\hat{\mathbf{i}} + \frac{3}{\sqrt{11}}\hat{\mathbf{j}} - \frac{1}{\sqrt{11}}\hat{\mathbf{k}}$$

Python Code

As we saw in Chapter 1, vectors in Python can be described using the `CoordSys3D` command imported from the `sympy.vector` library. Within this library we define a coordinate system `C= CoordSys3D('C')` in which the Cartesian coordinates (x, y, z) are represented internally as `C.x`, `C.y`, `C.z`. The corresponding unit vectors $\hat{\mathbf{i}}$, $\hat{\mathbf{j}}$, $\hat{\mathbf{k}}$, are represented by `C.i`, `C.j`, `C.k`. There is nothing special about choosing the letter C; any other letter could have been chosen instead.

The vector $\mathbf{a}=[1,\,3,\,-1]$ is represented as `a = 1*C.i +3* C.j - C.k` and we use the `a.normalize` method to find the unit vector $\hat{\mathbf{a}}$. Note that SymPy does not automatically simplify the term $\sqrt{11}/11$ into $1/\sqrt{11}$.

```
from sympy.vector import CoordSys3D

print('-'*28,'CODE OUTPUT','-'*29,'\n')

C = CoordSys3D('C')          # Cartesian system named C
                             # has unit vectors C.i , C.j , C.k

a = 1*C.i +3* C.j - C.k              # vector a defines desired direction

n = a.normalize()                    # n= normalized vector a

print('The normalized vector a is:\n',n)

--------------------------- CODE OUTPUT ----------------------------
The normalized vector a is:
 (sqrt(11)/11)*C.i + (3*sqrt(11)/11)*C.j + (-sqrt(11)/11)*C.k
```

Mathematica Code

In the Mathematica code, the vector is represented as `A={1,3,-1}` and we use the `Norm(A)` command to find the magnitude of the given vector. We normalize the vector by dividing by its magnitude.

$A = \{1, 3, -1\};$

Print["The normalized vector is", $A/\text{Norm}[A]$]

The normalized vector is $\left\{\frac{1}{\sqrt{11}}, \frac{3}{\sqrt{11}}, -\frac{1}{\sqrt{11}}\right\}$

If we know the coordinates of two points P_1 and P_2 in space, then we can find the vector connecting these two points by subtracting their respective x-, y-, z-components. For example, in Figure 3.1b, let $\mathbf{r}_1$ be the vector from the origin to P_1, and $\mathbf{r}_2$ the vector from the origin to P_2. The vector $\mathbf{r}_{12}$ from P_2 to P_1 can be drawn by the head-to-tail method, and is given by (see Figure 3.1b):

Vector $\mathbf{r}_{12}$ Connecting Two Points in Space

$$\mathbf{r}_{12} = \mathbf{r}_2 - \mathbf{r}_1 \qquad (3.1.4)$$

Now that we have covered some of the basic concepts behind vectors, we can apply them to the basic descriptors of motion. As discussed in Chapter 1, the position vector function $\mathbf{r}(t) = (x(t), y(t), z(t))$ for a particle in three dimensions as described by Cartesian coordinates is written as:

Position Vector

$$\mathbf{r}(t) = x(t)\,\hat{\mathbf{i}} + y(t)\,\hat{\mathbf{j}} + z(t)\,\hat{\mathbf{k}} \qquad (3.1.5)$$

This is a vector-valued function, and $x(t)$, $y(t)$, $z(t)$ are the scalar functions representing the position along the x-, y-, and z-axis at time t, respectively. The vector $\mathbf{r}(t)$ has its tail at the origin, and its head at the position of the particle at time t. The magnitude of the position vector is equal to the distance between the particle and the origin.

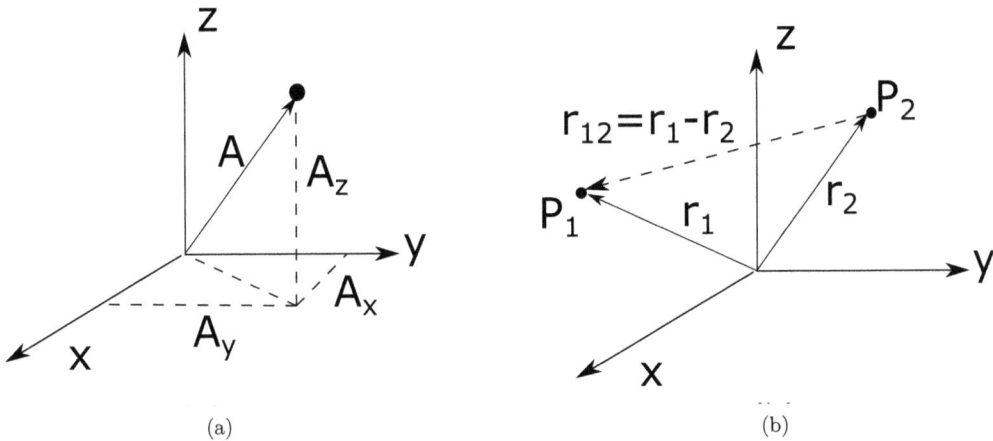

(a)   (b)

Figure 3.1: (a) Cartesian components of vector $\mathbf{A} = A_x\hat{\mathbf{i}} + A_y\hat{\mathbf{j}} + A_z\hat{\mathbf{k}}$; (b) The vector $\mathbf{r}_{12} = \mathbf{r}_2 - \mathbf{r}_1$ connecting two points P_1 and P_2 in space.

The derivative of the position vector is the *velocity* $\mathbf{v}(t)$ of the particle. To find the velocity, we need to know how to differentiate vectors. Figure 3.2 illustrates the derivative of a vector function $\mathbf{r}$. Without loss of generality, $\mathbf{r}$ can be thought of as a position vector; however, what follows is true for any vector. For the derivative of a vector function to exist, the vector function must be a continuous function of the scalar variable t. The vector function $\mathbf{r}(t)$ is shown as the dashed line in Figure 3.2, where the vector $\mathbf{r}$ is shown at times t and $t + \Delta t$. The vector $\Delta \mathbf{r} = \mathbf{r}(t + \Delta t) - \mathbf{r}(t)$ represents the change in the value of the vector function and will decrease in magnitude as $\Delta t \to 0$. Therefore, we can find the derivative using the standard formula:

$$\frac{d\mathbf{r}}{dt} = \lim_{\Delta t \to 0} \frac{\Delta \mathbf{r}}{\Delta t} = \lim_{\Delta t \to 0} \frac{\mathbf{r}(t + \Delta t) - \mathbf{r}(t)}{\Delta t} \tag{3.1.6}$$

Notice that in the limit of $\Delta t \to 0$, the vector $d\mathbf{r}/dt$ becomes tangent to the curve shown in Figure 3.2.

The derivatives of vector functions obey the rules of differentiation that you already know:

$$\frac{d}{dt}(\mathbf{r} + \mathbf{s}) = \frac{d\mathbf{r}}{dt} + \frac{d\mathbf{s}}{dt} \tag{3.1.7}$$

$$\frac{d}{dt}(c\mathbf{r}) = c\frac{d\mathbf{r}}{dt} \tag{3.1.8}$$

where $\mathbf{r}$ and $\mathbf{s}$ are vector functions of the variable t, and c is a scalar constant.

Now that we know how to take the derivative of a vector function, we can compute the velocity vector:

$$\frac{d\mathbf{r}(t)}{dt} = \frac{d}{dt}\left(x(t)\hat{\mathbf{i}} + y(t)\hat{\mathbf{j}} + z(t)\hat{\mathbf{k}}\right) \tag{3.1.9}$$

$$= \frac{d}{dt}\left(x(t)\hat{\mathbf{i}}\right) + \frac{d}{dt}\left(y(t)\hat{\mathbf{j}}\right) + \frac{d}{dt}\left(z(t)\hat{\mathbf{k}}\right) \tag{3.1.10}$$

However, the unit vectors in Cartesian coordinates are constant vectors. Therefore, the velocity vector can be found as:

Velocity Vector

$$\mathbf{v}(t) = \frac{d\mathbf{r}}{dt} = \frac{dx}{dt}\hat{\mathbf{i}} + \frac{dy}{dt}\hat{\mathbf{j}} + \frac{dz}{dt}\hat{\mathbf{k}} \tag{3.1.11}$$

Because the velocity vector is the derivative of the position vector, the velocity vector is going to be always tangent to the path followed by the particle.

The magnitude $v(t)$ of the velocity vector $\mathbf{v}(t)$ is called the *speed*:

$$v(t) = |\mathbf{v}(t)| = \sqrt{\left(\frac{dx}{dt}\right)^2 + \left(\frac{dy}{dt}\right)^2 + \left(\frac{dz}{dt}\right)^2} \tag{3.1.12}$$

Equation (3.1.12) can also be written in terms of the arc length s along the curve C describing the motion, as measured from some initial position along the curve. In order to use the arc length, we need to define an infinitesimal displacement along the curve,

$$d\mathbf{s} = dx\,\hat{\mathbf{i}} + dy\,\hat{\mathbf{j}} + dz\,\hat{\mathbf{k}} \tag{3.1.13}$$

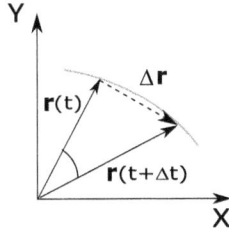

Figure 3.2: The vector function **r** is illustrated at times t and $t + \Delta t$. In the limit $\Delta t \to 0$, the vector $\Delta\mathbf{r}/\Delta t$ becomes the derivative vector function $d\mathbf{r}/dt$, which is tangent to the curve.

The differential distance ds is the magnitude of the infinitesimal displacement and is given by

$$ds = \sqrt{(dx)^2 + (dy)^2 + (dz)^2} \tag{3.1.14}$$

Equation (3.1.12) can now be written in terms of the arc length s, as:

$$v(t) = |\mathbf{v}(t)| = \frac{ds}{dt} \tag{3.1.15}$$

A unit tangent vector $\hat{\mathbf{v}}$ along the curve C describing the motion is obtained by dividing the velocity vector $\mathbf{v}(t)$ by its magnitude v:

$$\hat{\mathbf{v}} = \frac{\mathbf{v}(t)}{v(t)} \tag{3.1.16}$$

Note that the unit tangent vector $\hat{\mathbf{v}}$ points along the direction of the velocity and is always tangent to the curve C. An example of how to calculate the unit tangent vector is given in Example 3.2.

Similarly, the derivative of the velocity vector $d\mathbf{v}/dt$ is the acceleration vector function, which in Cartesian coordinates is:

Acceleration Vector

$$\mathbf{a}(t) = \frac{d\mathbf{v}}{dt} = \frac{d^2\mathbf{r}}{dt^2} = \frac{d^2x}{dt^2}\hat{\mathbf{i}} + \frac{dy^2}{dt^2}\hat{\mathbf{j}} + \frac{d^2z}{dt^2}\hat{\mathbf{k}} \tag{3.1.17}$$

Example 3.2: Finding unit tangent vectors.

The position of a particle along a curve C in 3D space is given by the parametric equations $x = \sin(2t)$, $y = t^2$, $z = 5$, where t represents time, and x, y, z, t are in SI units. (a) Find the magnitude of the velocity and acceleration at time $t = 1$ s. (b) Find a general expression for the unit vector $\hat{\mathbf{v}}$ that is tangent to the curve C at time t.

Solution:

(a) The magnitude of the velocity vector is the speed v:

$$v(t) = \sqrt{\left(\frac{dx}{dt}\right)^2 + \left(\frac{dy}{dt}\right)^2 + \left(\frac{dz}{dt}\right)^2} = \sqrt{[2\cos(2t)]^2 + (2t)^2}$$

The magnitude of the acceleration vector is:

$$a(t) = \sqrt{\left(\frac{d^2x}{dt^2}\right)^2 + \left(\frac{d^2y}{dt^2}\right)^2 + \left(\frac{d^2z}{dt^2}\right)^2} = \sqrt{\left[-4\sin(2t)\right]^2 + (2)^2}$$

At time $t = 1$ s, these magnitudes are:

$$v(1) = \sqrt{\left[2\cos(2)\right]^2 + (2)^2} = 2.2 \text{ m/s}$$

$$a(1) = \sqrt{\left[4\sin(2)\right]^2 + (2)^2} = 4.2 \text{ m/s}^2$$

(b) The unit vector $\hat{\mathbf{v}}$, which is tangent to the curve C at any time t, is found from:

$$\hat{\mathbf{v}} = \frac{\mathbf{v}(t)}{v(t)} = \frac{1}{v(t)}\left(\frac{dx(t)}{dt}\hat{\mathbf{i}} + \frac{dy(t)}{dt}\hat{\mathbf{j}} + \frac{dz(t)}{dt}\hat{\mathbf{k}}\right)$$

By evaluating the derivatives and substituting the speed $v(t)$ from part (a):

$$\hat{\mathbf{v}} = \frac{1}{v(t)}\left(2\cos(2t)\hat{\mathbf{i}} + 2t\hat{\mathbf{j}}\right) = \frac{1}{\sqrt{\left[2\cos(2t)\right]^2 + (2t)^2}}\left(2\cos(2t)\hat{\mathbf{i}} + 2t\hat{\mathbf{j}}\right)$$

The next example shows how to calculate the position vector when one is given an acceleration vector. As you might expect, the problem involves integrating the acceleration.

Example 3.3: Finding position and velocity from acceleration vector.

The acceleration of a particle moving in the Cartesian plane is given by the vector:

$$\mathbf{a}(t) = \sin(2t)\hat{\mathbf{i}} + \cos(2t)\hat{\mathbf{j}}$$

where t is the time, and all physical quantities are measured in SI units. Find the velocity vector $\mathbf{v}(t)$ and the position vector $\mathbf{r}(t)$. Assume that the particle is stationary and is at the origin at time $t = 0$.

Solution:

Since $\mathbf{a}(t) = d\mathbf{v}(t)/dt$, we can find the velocity and position by integration. Just like differentiating a vector, integrating a vector can be done one component at a time:

$$\mathbf{v}(t) = \int \mathbf{a}(t)dt$$

$$= \int \left(\sin(2t)\hat{\mathbf{i}} + \cos(2t)\hat{\mathbf{j}}\right)dt$$

$$= \int \sin(2t)\hat{\mathbf{i}}dt + \int \cos(2t)\hat{\mathbf{j}}dt$$

$$= -\frac{\cos(2t)}{2}\hat{\mathbf{i}} + \frac{\sin(2t)}{2}\hat{\mathbf{j}} + \mathbf{c}$$

where $\mathbf{c}$ is a constant vector of integration, which can be found from the initial conditions at time $t = 0$. Since the particle is stationary at $t = 0$, we must have $\mathbf{v}(0) = 0$, i.e.,

$$\mathbf{v}(0) = -\frac{\cos(0)}{2}\hat{\mathbf{i}} + \frac{\sin(0)}{2}\hat{\mathbf{j}} + \mathbf{c} = 0$$

which gives $\mathbf{c} = \left(\frac{1}{2}\right)\hat{\mathbf{i}}$.

The position is found by integrating the velocity vector:

$$\mathbf{r}(t) = \int \mathbf{v}(t)dt = \int \left(-\frac{\cos(2t)}{2}\hat{\mathbf{i}} + \frac{\sin(2t)}{2}\hat{\mathbf{j}} + \frac{1}{2}\hat{\mathbf{i}}\right)dt$$

$$= -\frac{1}{4}\sin(2t)\hat{\mathbf{i}} - \frac{1}{4}\cos(2t)\hat{\mathbf{j}} + \frac{1}{2}t\hat{\mathbf{i}} + \mathbf{d}$$

where $\mathbf{d}$ is a second constant vector of integration, which can be found from the initial position condition $\mathbf{r}(0) = 0$. Substituting $\mathbf{r}(0) = 0$, we find

$$\mathbf{r}(0) = -\frac{1}{4}\sin(0)\hat{\mathbf{i}} - \frac{1}{4}\cos(0)\hat{\mathbf{j}} + \frac{1}{2}0\hat{\mathbf{i}} + \mathbf{d} = 0$$

which gives $\mathbf{d} = \left(-\frac{1}{4}\right)\hat{\mathbf{j}}$. Again, we have used the fact that the unit vectors $\hat{\mathbf{i}}$ and $\hat{\mathbf{j}}$ in Cartesian coordinates are constant, and therefore can be placed outside of each integral.

The three vector functions $\mathbf{r}(t)$, $\mathbf{v}(t)$, and $\mathbf{a}(t)$ depend on the single parameter t, representing time. If we are given the explicit functions $x(t)$, $y(t)$, $z(t)$ then the path of the moving particle can be plotted in a so-called *parametric plot* in three dimensions using a computer; see Examples 3.4 and 3.5 that follow. A parametric plot uses the functions $x(t)$, $y(t)$, and $z(t)$ to graph a curve that represents the path taken by a particle. Each point in the curve corresponds to the x-, y-, and z-coordinates for various times t.

In the case of two-dimensional motion, if we are given explicit functions $x(t)$ and $y(t)$, one may be able to eliminate the variable t from the simultaneous equations $x = x(t)$, $y = y(t)$. If one of these equations can be solved for t, the expression obtained can be substituted into the other equation to obtain an expression for $y(x)$. In some cases, there is no single equation in closed form that is equivalent to the parametric equations. A similar process can be used in three dimensions in order to get $z(x,y)$. Examples 3.4 and 3.5 show how to create parametric plots in two and three dimensions in Mathematica and Python.

Example 3.4: Parametric equations for motion of a particle in 2D.

Use the appropriate parametric plot commands in Python and in Mathematica to plot the motion of a particle in two dimensions when the (x,y) position is given by the parametric equations:

$$x = \cos(t) \qquad y = \cos(2t)$$

where the parameter t is in the range from 0 to 2π.

Solution:
The time parameter t can be eliminated from the $x(t)$ and $y(t)$ equations using $\cos(2t) = 2\cos^2(t) - 1$ to obtain:

$$y = \cos(2t) = 2x^2 - 1$$

This equation is the equation of parabola on the xy-plane.

Python Code

We import Matplotlib and use `linspace` in NumPy to define the time values **t**. Both the Python and Mathematica codes produce the same plot, and the result of the Mathematica code is shown in Figure 3.3.

```
import numpy as np
import matplotlib.pyplot as plt

t = np.linspace(0,2*np.pi,200)      # define parameter t
plt.plot(np.cos(t),np.cos(2*t))     # parametric plot x=cos(t), cos(t)

plt.xlabel("x")
plt.ylabel("y")
plt.show()
```

Mathematica Code

The equivalent code in Mathematica uses the command `ParametricPlot`.

```
SetOptions[ParametricPlot, BaseStyle->FontSize->16, Frame->True];

ParametricPlot[{Cos[t], Cos[2 * t]}, {t, 0, 2 * Pi}, AxesLabel->{x, y}]
```

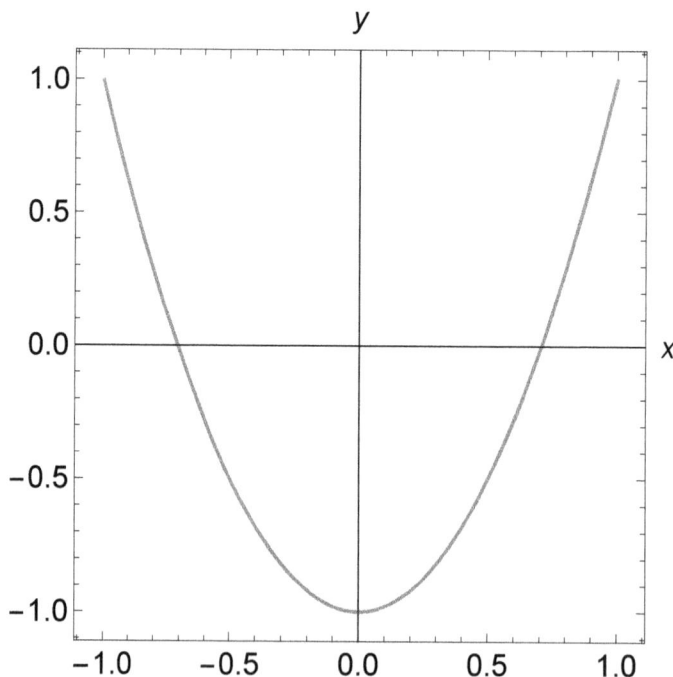

Figure 3.3: Parametric plot of function $(x(t), y(t))$ in two dimensions, produced by the codes in Example 3.4.

Example 3.5: Parametric equations for motion of particle in 3D.

Plot the motion of a particle in three dimensions when the (x, y, z) positions are given by the parametric equations:

$$x = d + a\cos(2t) \qquad y = e + b\sin(2t) \qquad z = ct$$

where the time parameter t is in the range 0 to T. Plot the motion using the numerical values $a = 2$, $b = 4$, $c = 1$, $d = 1$, $e = 2$, $T = 12$ (all quantities in SI units).

Solution:

The time parameter t can be eliminated from the $x(t)$, $y(t)$ equations by solving the equations for $\cos(t)$ and $\sin(t)$ and substituting in the identity $\cos^2(u) + \sin^2(u) = 1$:

$$\left(\frac{x-d}{a}\right)^2 + \left(\frac{y-e}{b}\right)^2 = \cos^2(2t) + \sin^2(2t) = 1$$

This is the equation of an ellipse on the xy-plane with center at $(d,\ e)$. Since $z = ct$, one then expects the motion to be in the form of an elliptical helix moving along the z-axis with a uniform speed c.

Python Code

The code creates a parametric plot of these (x, y, z) equations in three dimensions. Both the Python code and the Mathematica code produce the same plot, shown as Figure 3.4.

```python
import matplotlib.pyplot as plt
import numpy as np

t = np.linspace(0,4,50)    # time values from t=0 to t=12 s

x = 1+2*np.cos(2*t)
y = 2+4*np.sin(2*t)        # define x(t), y(t), z(t)
z = t

fig = plt.figure()         # set up 3D plot and plot
ax = fig.add_subplot(projection='3d')

ax.plot3D(x,y,z)           # plot parametric curve x(t),y(t),z(t)
ax.set_xlabel('x')         # label x,y,z axes
ax.set_ylabel('y')
ax.set_zlabel('z')
plt.show()
```

Mathematica Code

The code in Mathematica uses the **ParametricPlot3D** command, and the graphics options are set up using **SetOptions[ParametricPlot3D]**.

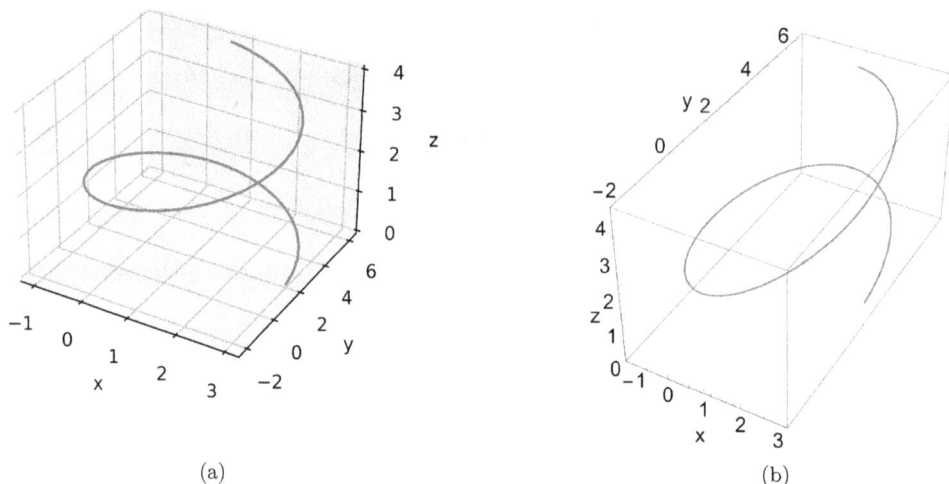

(a)   (b)

Figure 3.4: Parametric plot of function $(x(t), y(t), z(t))$ in three dimensions, produced by (a) Python and (b) Mathematica codes in Example 3.5.

```
SetOptions[ParametricPlot3D, BaseStyle->FontSize->20];

ParametricPlot3D[{1 + 2 * Cos[2 * t], 2 + 4 * Sin[2 * t], t}, {t, 0, 4},

AxesLabel->{x, y, z}]
```

3.2 VECTOR PRODUCTS

There are two methods of multiplying vectors together, the dot product and the cross product. Both methods are very useful in physics.

3.2.1 DOT PRODUCT

Consider two vectors $\mathbf{A} = [A_x,\ A_y,\ A_z]$ and $\mathbf{B} = [B_x,\ B_y,\ B_z]$. The *dot product* or scalar product $\mathbf{A} \cdot \mathbf{B}$ between them is defined as:

$$\mathbf{A} \cdot \mathbf{B} = A_x B_x + A_y B_y + A_z B_z \qquad (3.2.1)$$

and produces a scalar. An alternative definition of the dot product is:

$$\mathbf{A} \cdot \mathbf{B} = AB \cos\theta \qquad (3.2.2)$$

where A and B are the magnitude of the vectors $\mathbf{A}$ and $\mathbf{B}$, respectively, and θ is the angle between the two vectors as shown in Figure 3.5. Notice that the dot product is multiplying the magnitude of $\mathbf{B}$ by the component of $\mathbf{A}$, which is parallel to $\mathbf{B}$. A physical example involving the dot product is the definition of work. The work done by a constant force can be computed as $W = \mathbf{F} \cdot \mathbf{r}$, where a force $\mathbf{F}$ is applied to a particle along a displacement $\mathbf{r}$. The component of the force doing work on the particle is parallel to the particle's displacement.

The dot product is used to compute the work because we need the component of $\mathbf{F}$ parallel to $\mathbf{r}$.

Equation (3.2.2) can be used to obtain the angle between two vectors:

$$\cos\theta = \frac{\mathbf{A}\cdot\mathbf{B}}{AB} = \frac{A_x B_x + A_y B_y + A_z B_z}{\left(\sqrt{A_x^2 + A_y^2 + A_z^2}\right)\left(\sqrt{B_x^2 + B_y^2 + B_z^2}\right)}. \tag{3.2.3}$$

The dot product can also be interpreted geometrically as the *scalar projection* of vector $\mathbf{A}$ on the direction of vector $\mathbf{B}$, as shown in Figure 3.5. This is useful if you want to know the component of the vector $\mathbf{A}$ along the direction of the vector $\mathbf{B}$. First, we create the unit vector $\hat{\mathbf{B}} = \mathbf{B}/|\mathbf{B}|$ pointing in the direction of $\mathbf{B}$. The scalar projection is equal to $A_B = \mathbf{A}\cdot\hat{\mathbf{B}} = |\mathbf{A}|\cos\theta$. The *vector projection* of $\mathbf{A}$ onto $\mathbf{B}$ is, $\text{proj}_{\mathbf{B}}\mathbf{A} = A_B\hat{\mathbf{B}}$.

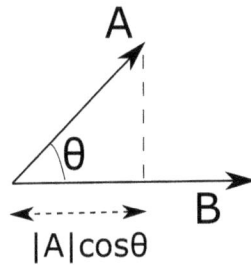

Figure 3.5: The scalar projection of vector $\mathbf{A}$ on the direction of vector $\mathbf{B}$, is $\mathbf{A}\cdot\hat{\mathbf{B}} = |\mathbf{A}|\cos\theta$, where $\hat{\mathbf{B}}$ is the unit vector in the direction of $\mathbf{B}$.

There are several algebraic relationships for the dot products of two vectors $\mathbf{A}$ and $\mathbf{B}$, and most of them are intuitive. The following properties hold if $\mathbf{A}$, $\mathbf{B}$, and $\mathbf{C}$ are real vectors and c is a scalar.

The dot product is commutative:

$$\mathbf{A}\cdot\mathbf{B} = \mathbf{B}\cdot\mathbf{A} \tag{3.2.4}$$

The dot product is distributive over vector addition:

$$\mathbf{A}\cdot(\mathbf{B}+\mathbf{C}) = \mathbf{A}\cdot\mathbf{B} + \mathbf{A}\cdot\mathbf{C} \tag{3.2.5}$$

Likewise, with a scalar c we can show:

$$\mathbf{A}\cdot(c\mathbf{B}) = c\,(\mathbf{A}\cdot\mathbf{B}) \tag{3.2.6}$$

Two non-zero vectors $\mathbf{A}$ and $\mathbf{B}$ are *orthogonal* if and only if

$$\mathbf{A}\cdot\mathbf{B} = 0 \tag{3.2.7}$$

The geometric interpretation of the orthogonality of vectors is that they are perpendicular, i.e., the angle between them $\theta = \pi/2$. Finally, the product rule applies to the dot product:

$$\frac{d}{dt}(\mathbf{A}\cdot\mathbf{B}) = \frac{d\mathbf{A}}{dt}\cdot\mathbf{B} + \mathbf{A}\cdot\frac{d\mathbf{B}}{dt} \tag{3.2.8}$$

Example 3.6: Calculating the dot product.

Find the dot product of the vectors $\mathbf{A} = 3\hat{\mathbf{i}} + 4\hat{\mathbf{j}} - \hat{\mathbf{k}}$ and $\mathbf{B} = \hat{\mathbf{i}} - \hat{\mathbf{j}} - \hat{\mathbf{k}}$ analytically, and also by using appropriate codes in Python and Mathematica.

Solution:

The *scalar* dot product is found by multiplying and adding the coefficients of the unit vectors $\hat{\mathbf{i}}, \hat{\mathbf{j}}, \hat{\mathbf{k}}$:

$$\mathbf{A} \cdot \mathbf{B} = (3\hat{\mathbf{i}} + 4\hat{\mathbf{j}} - \hat{\mathbf{k}}) \cdot (\hat{\mathbf{i}} - \hat{\mathbf{j}} - \hat{\mathbf{k}})$$
$$= (3)(1) + (4)(-1) + (-1)(-1) = 0$$

Python Code

In the Python code, the command `CoordSys3D` is imported from the `sympy.vector` library, and we define a coordinate system C= `CoordSys3D('N')` in which the Cartesian coordinates (x, y, z) are represented internally as `N.x`, `N.y`, `N.z`. The `A.dot(B)` method is used to evaluate the dot product of the two vectors.

```
from sympy.vector import CoordSys3D

print('-'*28,'CODE OUTPUT','-'*29,'\n')

N = CoordSys3D('N')          # Define the reference frame N

A = 3*N.i+4*N.j-1*N.k        # Define vectors A, B in N
B = 1*N.i-1*N.j-1*N.k

dotprod = A.dot(B)

print('The dot product A.B = ',dotprod)

--------------------------- CODE OUTPUT ----------------------------
The dot product A.B =  0
```

Mathematica Code

The Mathematica code uses the command `Dot[A,B]`, and the vectors **A**, **B** are entered in the form of lists of their Cartesian components.

```
A = {3, 4, −1};

B = {1, −1, −1};

Print["The dot product of A.B = ", Dot[A, B]]

The dot product of A.B = 0
```

Example 3.7: Finding unit vectors using dot products.
The position of a particle along a curve C in three-dimensional space is given by the parametric equations $x = t^2$, $y = t$, $z = 5$, where t represents time and all quantities are in SI units. The velocity of the particles is $\mathbf{v}(t)$
(a) Find a unit vector tangent to the curve C at any time t.
(b) Show that the vector $d\hat{\mathbf{v}}/dt$ is perpendicular to the unit vector $\hat{\mathbf{v}}$ at *any* time t.
(c) Find a unit vector $\mathbf{N}$ perpendicular to the unit vector $\hat{\mathbf{v}}$ at time $t = 1$ s.

Solution:
(a) The velocity vector is:

$$\mathbf{v}(t) = \frac{dx(t)}{dt}\hat{\mathbf{i}} + \frac{dy(t)}{dt}\hat{\mathbf{j}} + \frac{dz(t)}{dt}\hat{\mathbf{k}} = (2t)\hat{\mathbf{i}} + (1)\hat{\mathbf{j}} = 2t\hat{\mathbf{i}} + \hat{\mathbf{j}}$$

and the unit vector $\hat{\mathbf{v}}$ tangent to the curve C at any time t is found as in Example 3.2:

$$\hat{\mathbf{v}} = \frac{\mathbf{v}(t)}{v(t)} = \frac{2t\hat{\mathbf{i}} + \hat{\mathbf{j}}}{\sqrt{(2t)^2 + (1)^2}} = \frac{2t\hat{\mathbf{i}} + \hat{\mathbf{j}}}{\sqrt{4t^2 + 1}}$$

At $t = 1$ s we find the unit tangent vector $\hat{\mathbf{v}} = \left(2\hat{\mathbf{i}} + \hat{\mathbf{j}}\right)/\sqrt{5}$.

(b) Since $\hat{\mathbf{v}}$ is a unit vector with a magnitude of 1, the dot product $|\hat{\mathbf{v}}|^2 = \hat{\mathbf{v}} \cdot \hat{\mathbf{v}} = 1$. By differentiating $|\hat{\mathbf{v}}|^2$ with respect to time t, we obtain:

$$\hat{\mathbf{v}} \cdot \frac{d\hat{\mathbf{v}}}{dt} + \frac{d\hat{\mathbf{v}}}{dt} \cdot \hat{\mathbf{v}} = 2\hat{\mathbf{v}} \cdot \frac{d\hat{\mathbf{v}}}{dt} = 0$$

Therefore, the vector $d\hat{\mathbf{v}}/dt$ is perpendicular to the tangent unit vector $\hat{\mathbf{v}}$ *at any time t*, and is found from:

$$\frac{d\hat{\mathbf{v}}}{dt} = \frac{d}{dt}\left[\frac{2t\hat{\mathbf{i}} + \hat{\mathbf{j}}}{\sqrt{4t^2 + 1}}\right] = \frac{(2)\hat{\mathbf{i}} + (-4t)\hat{\mathbf{j}}}{(4t^2 + 1)^{3/2}}$$

(c) From part (b) we obtain at $t = 1$ s: $\frac{d\hat{\mathbf{v}}}{dt} = 5^{-3/2}\left(2\hat{\mathbf{i}} - 4\hat{\mathbf{j}}\right)$, and its magnitude is:

$$\left|\frac{d\hat{\mathbf{v}}}{dt}\right| = \sqrt{\left(\frac{2}{(5)^{3/2}}\right)^2 + \left(\frac{-4}{(5)^{3/2}}\right)^2} = \frac{\sqrt{20}}{(5)^{3/2}}$$

The desired unit vector $\mathbf{N}$ in this perpendicular direction is obtained by dividing $d\hat{\mathbf{v}}/dt$ by its magnitude:

$$\mathbf{N} = \frac{\frac{d\hat{\mathbf{v}}}{dt}}{\left|\frac{d\hat{\mathbf{v}}}{dt}\right|} = \frac{\frac{2\hat{\mathbf{i}} - 4\hat{\mathbf{j}}}{(5)^{3/2}}}{\frac{\sqrt{20}}{(5)^{3/2}}} = \frac{2\hat{\mathbf{i}} - 4\hat{\mathbf{j}}}{\sqrt{20}}$$

3.2.2 CROSS PRODUCT

The *cross product* or *vector product* of two vectors $\mathbf{A}$ and $\mathbf{B}$ is denoted by $\mathbf{A} \times \mathbf{B}$, and it is defined as a vector $\mathbf{C}$ that is perpendicular to *both* vectors $\mathbf{A}$ and $\mathbf{B}$ as shown in Figure 3.6(a). The *direction* of the cross product vector $\mathbf{A} \times \mathbf{B}$ is given by the right-hand rule shown in Figure 3.6(b). The *magnitude* of this vector is equal to the area of the parallelogram formed by vectors $\mathbf{A}$ and $\mathbf{B}$, as shown in Figure 3.6(c).

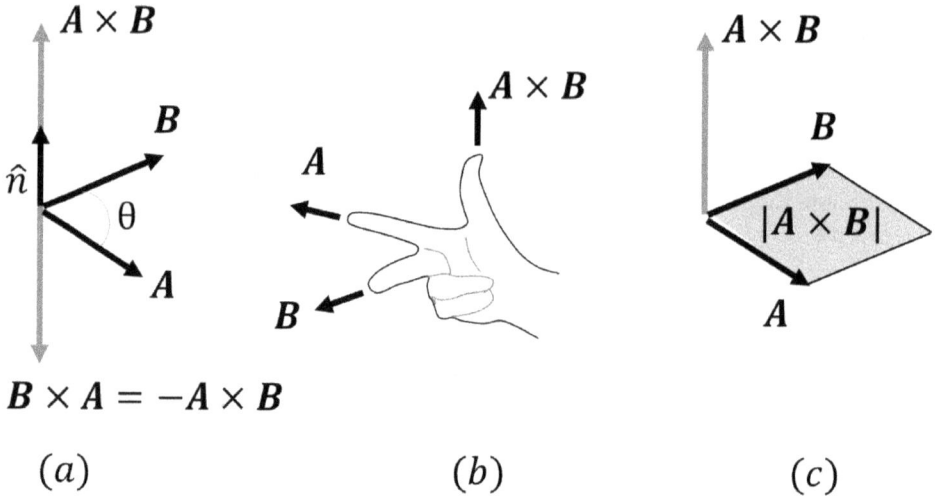

$$A \times B$$

$$B \times A = -A \times B$$

(a)　　　　　　　　　　(b)　　　　　　　　　　(c)

Figure 3.6: (a) The cross product in respect to a right-handed coordinate system; (b) finding the direction of the cross product by the right-hand rule; (c) the area of a parallelogram formed by two vectors **A** and **B**, is equal to the magnitude of the cross product of these two vectors.

The magnitude of the cross product is defined by the formula

$$\mathbf{A} \times \mathbf{B} = a\, b \sin \theta \, \hat{\mathbf{n}} \tag{3.2.9}$$

where θ is the angle between **A** and **B** ($0 \leq \theta \leq \pi$), $a = |\mathbf{A}|$ and $b = |\mathbf{B}|$ are the magnitudes of vectors **A** and **B**, and $\hat{\mathbf{n}}$ is a unit vector perpendicular to the plane containing **A** and **B** in the direction given by the right-hand rule as in Figure 3.6(a). If the vectors **A** and **B** are parallel (i.e., the angle θ between them is either 0 radians or π radians), then the cross product of **A** and **B** is the zero vector **0**.

The magnitude of the cross product is the product of the magnitude of the vector **A**, and the component of **B** perpendicular to **A**. We can compute the torque applied to a door by using the formula $\mathbf{N} = \mathbf{r} \times \mathbf{F}$, where **N** is the torque applied by the force **F**, and **r** is the vector which points from the axis of rotation to the point where the force is applied. For a door, the vector **r** lies along the door, pointing from the hinges to the edge of the door, where the force to open it is applied. We know that the component of the force responsible for the rotation of the door is perpendicular to the door, and that the component of the force parallel to the door is not causing the rotation. Therefore, the cross product is used to compute the torque, because we need the component of the force perpendicular to **r**.

The cross product of two vectors $\mathbf{a} = [a_1,\ a_2,\ a_3]$ and $\mathbf{b} = [b_1,\ b_2,\ b_3]$ can also be represented by the *determinant* of a formal matrix:

$$\mathbf{a} \times \mathbf{b} = \begin{vmatrix} \hat{\mathbf{i}} & \hat{\mathbf{j}} & \hat{\mathbf{k}} \\ a_1 & a_2 & a_3 \\ b_1 & b_2 & b_3 \end{vmatrix} \tag{3.2.10}$$

This determinant can be expanded by using cofactor expansion along the first row to yield:

$$\mathbf{a} \times \mathbf{b} = \begin{vmatrix} a_2 & a_3 \\ b_2 & b_3 \end{vmatrix} \hat{\mathbf{i}} - \begin{vmatrix} a_1 & a_3 \\ b_1 & b_3 \end{vmatrix} \hat{\mathbf{j}} + \begin{vmatrix} a_1 & a_2 \\ b_1 & b_2 \end{vmatrix} \hat{\mathbf{k}} \tag{3.2.11}$$

which gives the components of the resulting cross product vector directly:

$$\mathbf{a} \times \mathbf{b} = (a_2 b_3 - a_3 b_2)\hat{\mathbf{i}} + (a_3 b_1 - a_1 b_3)\hat{\mathbf{j}} + (a_1 b_2 - a_2 b_1)\hat{\mathbf{k}} \qquad (3.2.12)$$

Like the dot product, the cross product has some familiar algebraic identities. The cross product is distributive with addition:

$$\mathbf{a} \times (\mathbf{b} + \mathbf{c}) = \mathbf{a} \times \mathbf{b} + \mathbf{a} \times \mathbf{c} \qquad (3.2.13)$$

Likewise, with a scalar c, we can show:

$$(c\mathbf{a}) \times \mathbf{b} = \mathbf{a} \times (c\mathbf{b}) = c\,(\mathbf{a} \times \mathbf{b}) \qquad (3.2.14)$$

Unlike the dot product, the cross product is anti-commutative, i.e.,

$$\mathbf{b} \times \mathbf{a} = -(\mathbf{a} \times \mathbf{b}) \qquad (3.2.15)$$

This property is easily concluded by considering the right-hand rule; when the two vectors are interchanged, the sign of the cross-product vector is reversed. The product rule for derivatives also applies to the cross product:

$$\frac{d}{dt}(\mathbf{a} \times \mathbf{b}) = \frac{d\mathbf{a}}{dt} \times \mathbf{b} + \mathbf{a} \times \frac{d\mathbf{b}}{dt} \qquad (3.2.16)$$

Example 3.8: Example of cross product with Python and Mathematica.
Find the cross product of the vectors $\mathbf{A} = 3\hat{\mathbf{i}} + 4\hat{\mathbf{j}} - \hat{\mathbf{k}}$ and $\mathbf{B} = \hat{\mathbf{i}} - \hat{\mathbf{j}} - \mathbf{k}$ analytically, and also by using appropriate codes in Python and Mathematica.

Solution:
The cross product is found by using the determinant:

$$\mathbf{A} \times \mathbf{B} = \begin{vmatrix} \hat{\mathbf{i}} & \hat{\mathbf{j}} & \hat{\mathbf{k}} \\ 3 & 4 & -1 \\ 1 & -1 & -1 \end{vmatrix} = \begin{vmatrix} 4 & -1 \\ -1 & -1 \end{vmatrix}\hat{\mathbf{i}} - \begin{vmatrix} 3 & -1 \\ 1 & -1 \end{vmatrix}\hat{\mathbf{j}} + \begin{vmatrix} 3 & 4 \\ 1 & -1 \end{vmatrix}\hat{\mathbf{k}} = -5\hat{\mathbf{i}} + 2\hat{\mathbf{j}} - 7\hat{\mathbf{k}}$$

Python Code
We define the reference frame N using `CoordSys3D('N')`. The A.cross(B) method is used to evaluate the dot product of the two vectors.

```
from sympy.vector import CoordSys3D

print('-'*28,'CODE OUTPUT','-'*29,'\n')

N = CoordSys3D('N')          # Define the coordinate system N

A = 3*N.i+4*N.j-1*N.k        # Define vectors A, B in N
B = 1*N.i-1*N.j-1*N.k

crossprod = A.cross(B)

print('The cross product A.B = ',crossprod)

----------------------- CODE OUTPUT ----------------------------
The cross product A.B =  (-5)*N.i + 2*N.j + (-7)*N.k
```

Mathematica Code

The Mathematica code uses the command `Cross[A,B]`, and the vectors **A**, **B** are entered in the form of lists of their Cartesian components.

$A = \{3, 4, -1\};$

$B = \{1, -1, -1\};$

Print["The cross product of A.B = ", Cross[A, B]]

The cross product of A.B = $\{-5, 2, -7\}$

Example 3.9: Finding unit vectors using cross products.

Find a unit vector perpendicular to both vectors $\mathbf{a} = [1, 3, -1]$ and $\mathbf{b} = [4, -2, 1]$.

Solution:

We first calculate the cross product of the two vectors since we know that it will be perpendicular to both vectors. The cross product is:

$$\mathbf{c} = \mathbf{a} \times \mathbf{b} = \begin{vmatrix} \hat{\mathbf{i}} & \hat{\mathbf{j}} & \hat{\mathbf{k}} \\ 1 & 3 & -1 \\ 4 & -2 & 1 \end{vmatrix} = \begin{vmatrix} 3 & -1 \\ -2 & 1 \end{vmatrix}\hat{\mathbf{i}} - \begin{vmatrix} 1 & -1 \\ 4 & 1 \end{vmatrix}\hat{\mathbf{j}} + \begin{vmatrix} 1 & 3 \\ 4 & -2 \end{vmatrix}\hat{\mathbf{k}} = 1\hat{\mathbf{i}} - 5\hat{\mathbf{j}} - 14\hat{\mathbf{k}}.$$

To find the unit vector, divide this vector by its magnitude. The magnitude of $\mathbf{a} \times \mathbf{b}$ is:

$$|\mathbf{a} \times \mathbf{b}| = \sqrt{1^2 + 5^2 + (-14)^2} = \sqrt{222}$$

The desired unit vector is

$$\hat{\mathbf{c}} = \frac{\mathbf{c}}{\sqrt{c_x^2 + c_y^2 + c_z^2}} = \frac{1\hat{\mathbf{i}} - 5\hat{\mathbf{j}} - 14\hat{\mathbf{k}}}{\sqrt{222}} = \frac{1}{\sqrt{222}}\hat{\mathbf{i}} - \frac{5}{\sqrt{222}}\hat{\mathbf{j}} - \frac{14}{\sqrt{222}}\hat{\mathbf{k}}$$

Python Code

In the Python code, we define a coordinate system `C= CoordSys3D('N')` and the `A.cross(B)` method is used to evaluate the cross product of the two vectors. The `.normalize` method is used to find the unit vector along the direction of the cross product.

```
from sympy.vector import CoordSys3D

print('-'*28,'CODE OUTPUT','-'*29,'\n')

N = CoordSys3D('N')          # Define the reference frame N

A = 1*N.i+3*N.j-1*N.k        # Define vectors A, B in N
B = 4*N.i-2*N.j+1*N.k

crossprod = A.cross(B)

answer = crossprod.normalize()
print('The normalized vector is = \n',answer)

---------------------- CODE OUTPUT ----------------------------
The normalized vector is =
 (sqrt(222)/222)*N.i + (-5*sqrt(222)/222)*N.j + (-7*sqrt(222)/111)*N.k
```

Mathematica Code

In the Mathematica code below, we use the `Cross[A,B]` command to find the cross product, and we use the `Norm(c)` command to find the magnitude of the given vector.

$A = \{1, 3, -1\};$

$B = \{4, -2, 1\};$

$c = \text{Cross}[A, B];$

Print["The cross product of A.B = ", c]

Print["The normalized vector is", $c/\text{Norm}[c]$]

The cross product of A.B = $\{1, -5, -14\}$

The normalized vector is $\left\{ \frac{1}{\sqrt{222}}, -\frac{5}{\sqrt{222}}, -7\sqrt{\frac{2}{111}} \right\}$

A useful mathematical construction involving the cross product is the *triple product*, which is used in two different forms.

The *scalar triple product* of three vectors is defined as

$$\mathbf{a} \cdot (\mathbf{b} \times \mathbf{c}) \tag{3.2.17}$$

Notice that this expression means that we first calculate the cross product of vectors $\mathbf{b}$ and $\mathbf{c}$, followed by the dot product between the vectors $\mathbf{a}$ and $\mathbf{b} \times \mathbf{c}$. The magnitude of the triple scalar product represents the volume V of the parallelopiped with edges $\mathbf{a}$, $\mathbf{b}$, and $\mathbf{c}$. Since these three vectors $\mathbf{a}$, $\mathbf{b}$, $\mathbf{c}$ can be interchanged, the following three scalar triple products are equal to each other and to the volume V:

$$\mathbf{a} \cdot (\mathbf{b} \times \mathbf{c}) = \mathbf{b} \cdot (\mathbf{c} \times \mathbf{a}) = \mathbf{c} \cdot (\mathbf{a} \times \mathbf{b}) = V \tag{3.2.18}$$

The *vector triple product* is the cross product of a vector with the result of another cross product and is related to the dot product by the following identity:

$$\mathbf{a} \times (\mathbf{b} \times \mathbf{c}) = \mathbf{b}(\mathbf{a} \cdot \mathbf{c}) - \mathbf{c}(\mathbf{a} \cdot \mathbf{b}) \tag{3.2.19}$$

The mnemonic "BAC minus CAB" is often used to remember the order of the vectors in the right-hand side of this equation. This formula is used in physics to simplify vector calculations. Another useful identity relates the cross product to the scalar triple product:

$$(\mathbf{a} \times \mathbf{b}) \times (\mathbf{a} \times \mathbf{c}) = (\mathbf{a} \cdot (\mathbf{b} \times \mathbf{c}))\mathbf{a} \tag{3.2.20}$$

Example 3.10: Evaluation of triple products

Use Mathematica and SymPy to demonstrate the triple product identity:

$$\mathbf{a} \cdot (\mathbf{b} \times \mathbf{c}) = \mathbf{b} \cdot (\mathbf{c} \times \mathbf{a})$$

for *any* vectors **a**, **b**, and **c**.

Python Code

In the coordinate system N the vector $a_x\hat{\mathbf{i}} + a_y\hat{\mathbf{j}} + a_z\hat{\mathbf{k}}$ is written as `ax*N.i+ay*N.j+az*N.k`. The `A.cross(B)` and `A.dot(B)` method are used to evaluate the cross and dot product of the two vectors. The identity is tested by using the double equal logical sign `==`, which evaluates here as **True**.

```
from sympy.vector import CoordSys3D
from sympy import symbols
print('-'*28,'CODE OUTPUT','-'*29,'\n')

N = CoordSys3D('N')                    # Define the Cartesian frame N
ax, ay, az = symbols('ax, ay, az')     # Components of vectors a, b, c
bx, by, bz = symbols('bx, by, bz')
cx, cy, cz = symbols('ax, ay, cz')

a = ax*N.i+ay*N.j+az*N.k               # Define vectors a, b, c
b = bx*N.i+by*N.j+bz*N.k
c = cx*N.i+cy*N.j+cz*N.k

product1 = a.dot(b.cross(c))           # Evaluate a.(bxc)
product2 = b.dot(c.cross(a))           # Evaluate b.(cxa)

check=(product1.expand()==product2.expand())  # test with ==
print('\nThe identity    a.(bxc) = b.(cxa)    is: ',check)

------------------------------ CODE OUTPUT ----------------------------

The identity    a.(bxc) = b.(cxa)    is:  True
```

Mathematica Code
By combining the `Dot` and `Cross` commands, the identities are verified in the code below. The two sides of the identities are set equal, by using the Boolean equality sign `==` and Mathematica returns `True` for the logical statements.

```
a = {ax, ay, az};

b = {bx, by, bz};

c = {cx, cy, cz};

Print["The triple product of a.(bxc) = "]

Print[Dot[a, Cross[b, c]]]

The triple product of a.(bxc) =
az(−by cx + bx cy) + ay(bz cx − bx cz) + ax(−bz cy + by cz)

check = (Dot[a, Cross[b, c]]==Dot[b, Cross[c, a]]//Simplify);

Print["The identity a.(bxc)= b.(cxa) is: ", check]

The identity a.(bxc)= b.(cxa) is: True
```

3.3 POSITION, VELOCITY, AND ACCELERATION IN NON-CARTESIAN COORDINATE SYSTEMS

In this section, we discuss how to describe a particle's motion in coordinate systems other than Cartesian coordinates. The new coordinate systems often exploit symmetries in the problem, making the non-Cartesian coordinate systems a more natural means of describing the motion of the particle. For example, the motion of a particle constrained to move along a circle can be more easily described in polar coordinates than it is in Cartesian coordinates.

3.3.1 POLAR COORDINATES

Figure 3.7 shows the position *(x,y)* of a particle located at the point P on the xy-plane. Using polar coordinates, the position can also be described by the angle θ with the x-axis and the radius r representing the distance from the origin.

The relationship between Cartesian *(x,y)* coordinates and the polar coordinates r, θ is:

From Polar to Cartesian Coordinates

$$x = r\,\cos\theta \tag{3.3.1}$$

$$y = r\,\sin\theta \tag{3.3.2}$$

From Cartesian to Polar Coordinates

$$r = \sqrt{x^2 + y^2} \tag{3.3.3}$$

$$\theta = \tan^{-1}(y/x) \tag{3.3.4}$$

We can also write the position vector as a complex number in the following form by using Euler's equations:

$$z = x + iy = r(\cos\theta + i\sin\theta) = re^{i\theta} \qquad (3.3.5)$$

and the complex conjugate of this complex number is:

$$\bar{z} = r(\cos\theta - i\sin\theta) = re^{-i\theta} \qquad (3.3.6)$$

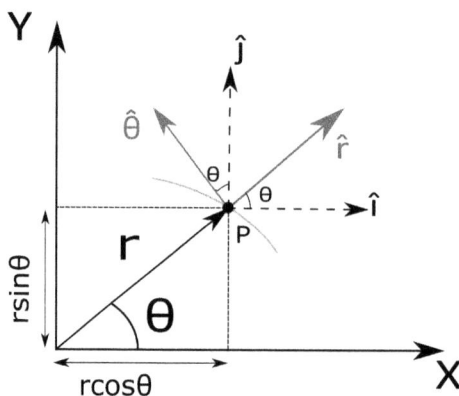

Figure 3.7: The radius r and angle θ locate a particle in polar coordinates, and they also represent a complex number: $z = x + iy = r(\cos\theta + i\sin\theta) = r\,e^{i\theta}$. The two unit vectors $\hat{\mathbf{r}}$ and $\hat{\boldsymbol{\theta}}$ (in blue in the e-book) in polar coordinates change direction as the particle moves in the xy-plane along a curved path (in red in the e-book).

By adding and subtracting the last two equations, we obtain *Euler's formulas* for sine and cosine functions. These equations provide a powerful connection between analysis and trigonometry and should be memorized.

Euler's Identities

$$\cos\theta = \frac{e^{i\theta} + e^{-i\theta}}{2} \qquad (3.3.7)$$

$$\sin\theta = \frac{e^{i\theta} - e^{-i\theta}}{2i} \qquad (3.3.8)$$

$$e^{i\theta} = \cos\theta + i\sin\theta \qquad (3.3.9)$$

$$e^{-i\theta} = \cos\theta - i\sin\theta \qquad (3.3.10)$$

While polar coordinates can be convenient for describing the position of a particle, especially a particle constrained to move along an arc or a circle, the unit vectors require special attention. In Cartesian coordinates, the unit vectors $\hat{\mathbf{i}}$ and $\hat{\mathbf{j}}$ always point along the x- and y-axis respectively, regardless of the particle's location. However, the directions of the unit polar vectors $\hat{\mathbf{r}}$ and $\hat{\boldsymbol{\theta}}$ change with the location of the particle. This can be clearly seen in Figure 3.7.

The unit vector $\hat{\mathbf{r}}$ always points in the radially outward direction. The position vector of the particle moving in the xy-plane can be written in terms of the unit vector $\hat{\mathbf{r}}$ as:

$$\mathbf{r} = r\hat{\mathbf{r}} \qquad (3.3.11)$$

In order to find the velocity and acceleration in polar coordinates, we first need to write the unit vectors $\hat{\mathbf{r}}$ and $\hat{\boldsymbol{\theta}}$ in terms of $\hat{\mathbf{i}}$ and $\hat{\mathbf{j}}$. Using the geometry shown in Figure 3.7, we can write:

$$\hat{\mathbf{r}} = \cos\theta\,\hat{\mathbf{i}} + \sin\theta\,\hat{\mathbf{j}} \tag{3.3.12}$$

$$\hat{\boldsymbol{\theta}} = -\sin\theta\,\hat{\mathbf{i}} + \hat{\boldsymbol{\theta}}\cos\theta\,\hat{\mathbf{j}} \tag{3.3.13}$$

where as expected $|\hat{\boldsymbol{\theta}}| = |\hat{\mathbf{r}}| = 1$, and the two unit vectors are perpendicular to each other, i.e., $\hat{\boldsymbol{\theta}}\cdot\hat{\mathbf{r}} = 0$.

The unit vectors $\hat{\mathbf{r}}$ and $\hat{\boldsymbol{\theta}}$ vary in direction with time, and we can calculate their time derivatives by using the chain rule:

$$\frac{d}{dt}\hat{\mathbf{r}} = \frac{d}{dt}\left[\cos\theta\hat{\mathbf{i}} + \sin\theta\hat{\mathbf{j}}\right] = -\sin\theta\frac{d\theta}{dt}\hat{\mathbf{i}} + \cos\theta\frac{d\theta}{dt}\hat{\mathbf{j}} = \frac{d\theta}{dt}\left[-\sin\theta\hat{\mathbf{i}} + \cos\theta\hat{\mathbf{j}}\right] \tag{3.3.14}$$

By combining the previous two equations, we obtain:

$$\frac{d\hat{\mathbf{r}}}{dt} = \frac{d\theta}{dt}\hat{\boldsymbol{\theta}} \tag{3.3.15}$$

By working in a similar manner, we obtain the time derivative of the unit vector $\hat{\boldsymbol{\theta}}$:

$$\frac{d\hat{\boldsymbol{\theta}}}{dt} = -\frac{d\theta}{dt}\hat{\mathbf{r}} \tag{3.3.16}$$

We can now calculate the velocity vector in polar coordinates by taking the first time derivative of the position vector $\mathbf{r}$:

$$\mathbf{v} = \frac{d\mathbf{r}}{dt} = \frac{d}{dt}\left(r\hat{\mathbf{r}}\right) = \frac{dr}{dt}\hat{\mathbf{r}} + r\frac{d\hat{\mathbf{r}}}{dt} \tag{3.3.17}$$

By substituting (3.3.15) into (3.3.17), we obtain the velocity vector in polar coordinates:

$$\mathbf{v} = \frac{d\mathbf{r}}{dt} = \frac{dr}{dt}\hat{\mathbf{r}} + r\frac{d\theta}{dt}\hat{\boldsymbol{\theta}} \tag{3.3.18}$$

The two components of the velocity in this equation have distinct physical meanings. The first term $\mathbf{v}_r = \frac{dr}{dt}\hat{\mathbf{r}}$ represents the radial component of the velocity, which points along the radius r. The second term $\mathbf{v}_\theta = r\frac{d\theta}{dt}\hat{\boldsymbol{\theta}} = r\omega\hat{\boldsymbol{\theta}}$ is proportional to the angular velocity $\omega = d\theta/dt$, and represents the velocity of the particle along the $\hat{\boldsymbol{\theta}}$ direction. This second component also represents the velocity of uniform circular motion.

From (3.3.18), we obtain the infinitesimal displacement $d\mathbf{s}$ between two points in space in polar coordinates by canceling the differential time dt from both sides of this equation:

$$d\mathbf{s} = dr\hat{\mathbf{r}} + rd\theta\hat{\boldsymbol{\theta}} \tag{3.3.19}$$

Based on this expression, the infinitesimal area element dA in polar coordinates can then be written as:

$$dA = dxdy = rd\theta dr \tag{3.3.20}$$

and we can calculate double integrals of any function $f(x,y)$ in polar coordinates by using the transformation:

$$\iint f(x,y)\,dxdy = \iint f(r\cos\theta, r\sin\theta)\,r\,dr\,d\theta \tag{3.3.21}$$

In summary, the unit vectors in polar coordinates and their derivatives are given by:

Unit Vectors in Polar Coordinates

$$\hat{\boldsymbol{\theta}} = -\sin\theta\hat{\mathbf{i}} + \cos\theta\hat{\mathbf{j}} \qquad (3.3.22)$$

$$\hat{\mathbf{r}} = \cos\theta\hat{\mathbf{i}} + \sin\theta\hat{\mathbf{j}} \qquad (3.3.23)$$

$$\frac{d}{dt}\hat{\mathbf{r}} = \frac{d\theta}{dt}\hat{\boldsymbol{\theta}} \qquad \frac{d}{dt}\hat{\boldsymbol{\theta}} = -\frac{d\theta}{dt}\hat{\mathbf{r}} \qquad (3.3.24)$$

The following example shows how to create polar plots in Python and Mathematica.

Example 3.11: Polar plots using Python and Mathematica.
Create a polar plot of the equation: $r(\theta) = \theta/2\pi$.

Solution:
The following are the Python and Mathematica codes for creating the polar plot. As the angle θ increases from 0 to 4π, the radius increases from 0 to 2.

Python Code
The Python code defines the range of the radius using `linspace` and the polar plot is set up using `subplot(111,projection='polar')` from the `matplotlib` graphics library.

```python
import numpy as np
import matplotlib.pyplot as plt


r = np.linspace(0,2,200)    # define range of values for radius r

# evaluate values of angle theta as function of the radius r
theta = 2*np.pi *r

ax =plt.subplot(111,projection='polar')

ax.plot(theta,r)
ax.set_rticks([0.5, 1, 1.5, 2])  # set values of radial ticks

plt.show()
```

Mathematica Code
The polar plot is created using the `PolarPlot` command, and the graphics options are set up using `SetOptions[PolarPlot]`.

```
SetOptions[PolarPlot, BaseStyle->FontSize->16, Frame->True];

PolarPlot[θ/(2 * Pi), {θ, 0, 4 * Pi}, AxesLabel->{x, y}]
```

Polar coordinates are very useful in describing the motion of celestial bodies around each other, as we will explore in detail in a later chapter. The following example shows how polar coordinates can simplify the mathematical description of the elliptical motion of planets around the Sun.

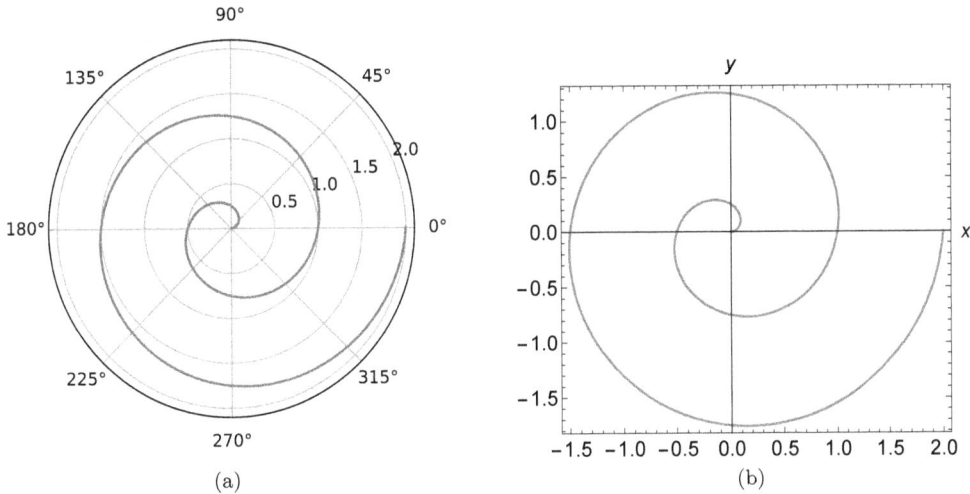

Figure 3.8: Polar plot of function $r = \theta/2\pi$ produced by (a) Python, and (b) Mathematica, in Example 3.11.

Example 3.12: Ellipses in polar coordinates.
   The equation of an ellipse with semi-major axes a, b in Cartesian coordinates is:

$$(x/a)^2 + (y/b)^2 = 1$$

or in parametric form
$$x = a \cos\theta \qquad y = b \sin\theta.$$

(a) Use the symbolic capabilities of Mathematica to show that the general equation for this ellipse in polar coordinates is:

$$r = \frac{ab}{\sqrt{a^2 - c^2 \cos^2\theta}} \tag{3.3.25}$$

where $c = \sqrt{a^2 - b^2}$ (with $a > b$).
   (b) Show that by shifting the x-axis by the quantity $c = \sqrt{a^2 - b^2}$, the equation of the ellipse in polar coordinates simplifies to:

$$r = \frac{b^2/a}{1 + (c/a)\,\cos\theta} = \frac{a\,(1 - e^2)}{1 + e\,\cos\theta} \tag{3.3.26}$$

where $e = c/a = \sqrt{1 - b^2/a^2}$ (with $a > b$) is the *eccentricity* of the ellipse.
   As we will see in Chapter 9, this expression is fundamental in the mathematical description of planetary motion, with the Sun located at one of the foci of this ellipse.
   (c) Plot the above expressions of the two ellipses and in Mathematica using the `PolarPlot` command.

Mathematica Code

(a) The code performs the necessary algebraic steps for deriving (3.3.25), and then it plots the ellipse with parameters $a = 2.2$, $b = 1.1$. After defining the parametric forms $x = a\cos\theta$ and $y = b\sin\theta = b\sqrt{1-\cos^2\theta}$, the code uses the Solve command to find the distance r as a function of $\cos\theta$. The command "replace with" /. is used to define a parameter rellipse for the desired equation $r = r(\theta)$. Since there are two solutions obtained from the Solve command, we chose the positive solution by selecting the second (positive) square root.

```
SetOptions[PolarPlot, BaseStyle->FontSize->16];
x = r * Cos[θ];
y = r * Sqrt[1 − Cos[θ]^2];
sol = Solve[(x/a)^2 + (y/b)^2==1, r]
```

$$\left\{\left\{r \to -\frac{1}{\sqrt{\frac{1}{b^2}+\frac{\text{Cos}[\theta]^2}{a^2}-\frac{\text{Cos}[\theta]^2}{b^2}}}\right\},\left\{r \to \frac{1}{\sqrt{\frac{1}{b^2}+\frac{\text{Cos}[\theta]^2}{a^2}-\frac{\text{Cos}[\theta]^2}{b^2}}}\right\}\right\}$$

```
rellipse = r/.sol[[2]]
```

$$\frac{1}{\sqrt{\frac{1}{b^2}+\frac{\text{Cos}[\theta]^2}{a^2}-\frac{\text{Cos}[\theta]^2}{b^2}}}$$

```
PolarPlot[rellipse/.{a->2.2, b->1.1}, {θ, 0, 2 * Pi}, AxesLabel->{x, y}]
```

(b) The following code carries out the algebra for deriving (3.3.26), and it plots the ellipse. As the polar plot shows, the resulting ellipse from (3.3.26) is shifted along the x-axis by $c = \sqrt{a^2-b^2} = \sqrt{2.2^2-1.1^2} = \sqrt{1.9}$. After some simple manipulation of the Mathematica output, we obtain (3.3.26), which is the commonly used equation for an ellipse in polar coordinates.

```
SetOptions[PolarPlot, BaseStyle->FontSize->16];
c = Sqrt[a^2 − b^2];
x = r * Cos[θ]; y = r * Sqrt[(1 − Cos[θ]^2)];
sol2 = Solve[((x + c)/a)^2 + (y/b)^2==1, r]//FullSimplify;
rellipse2 = r/.sol2[[1]]//FullSimplify;
Print["The equation of the new ellipse is:"]
Print[rellipse2]
```

The equation of the new ellipse is:
$$\frac{b^2}{a+\sqrt{(a-b)(a+b)}\text{Cos}[\theta]}$$

```
PolarPlot[rellipse2/.{a->2.2, b->1.1}, {θ, 0, 2 * Pi}, AxesLabel->{x, y}]
```

Using the derivatives of the unit vectors $\frac{d}{dt}\hat{\boldsymbol{\theta}}$ and $\frac{d}{dt}\hat{\mathbf{r}}$, we can now also calculate the acceleration vector as the second derivative of the position in polar coordinates (see the

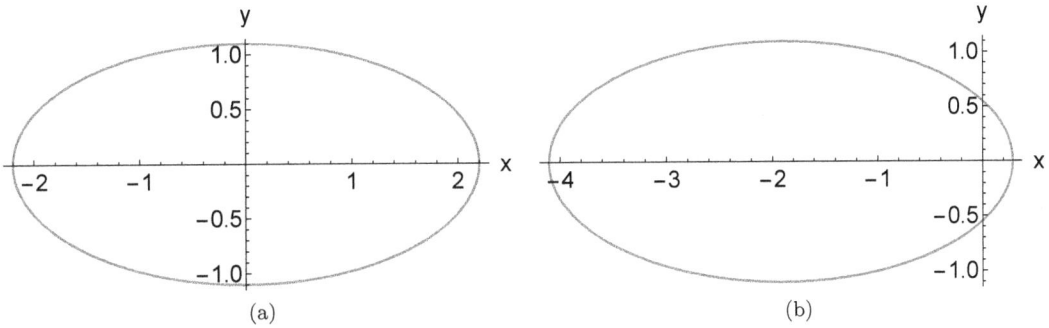

Figure 3.9: Plots of the ellipses produced by Mathematica, for (a) the original ellipse and (b) for the shifted ellipse, in Example 3.12.

End-of-Chapter problems for this section):

$$\mathbf{a} = \frac{d\mathbf{v}}{dt} = \left[\frac{d^2r}{dt^2} - r\left(\frac{d\theta}{dt}\right)^2\right]\hat{\mathbf{r}} + \left[r\frac{d^2\theta}{dt^2} + 2\frac{dr}{dt}\frac{d\theta}{dt}\right]\hat{\boldsymbol{\theta}} \qquad (3.3.27)$$

The two terms in (3.3.27) again have distinct physical meanings. The first term $\mathbf{a}_r = \left[\frac{d^2r}{dt^2} - r\left(\frac{d\theta}{dt}\right)^2\right]\hat{\mathbf{r}}$ represents the radial component of the acceleration, which points along the radius r. The first part of this radial term is $\left[\frac{d^2r}{dt^2}\right]\hat{\mathbf{r}}$ and represents the rate of change of the radial velocity dr/dt. The second part of this radial term is $-r\left(\frac{d\theta}{dt}\right)^2 = -r\omega^2$, and points in the opposite direction from the $\frac{d^2r}{dt^2}$ term, and represents the centripetal acceleration $a_c = v^2/r$.

The second term in (3.3.27)

$$\mathbf{a}_\theta = \left(r\frac{d^2\theta}{dt^2} + 2\frac{dr}{dt}\frac{d\theta}{dt}\right)\hat{\boldsymbol{\theta}} \qquad (3.3.28)$$

also contains two terms; the first term $r\frac{d^2\theta}{dt^2} = r\frac{d\omega}{dt} = r\alpha$ is the angular acceleration along the $\hat{\boldsymbol{\theta}}$ direction. The last term represents the *Coriolis force* and will be discussed in Chapter 10.

In summary, the position, velocity, and acceleration vectors in polar coordinates are given by:

Position, Velocity, and Acceleration in Polar Coordinates

$$\mathbf{r} = r\hat{\mathbf{r}} \qquad (3.3.29)$$

$$\mathbf{v} = \frac{d\mathbf{r}}{dt} = \frac{dr}{dt}\hat{\mathbf{r}} + r\frac{d\theta}{dt}\hat{\boldsymbol{\theta}} \qquad (3.3.30)$$

$$\mathbf{a} = \frac{d\mathbf{v}}{dt} = \left[\frac{d^2r}{dt^2} - r\left(\frac{d\theta}{dt}\right)^2\right]\hat{\mathbf{r}} + \left[r\frac{d^2\theta}{dt^2} + 2\frac{dr}{dt}\frac{d\theta}{dt}\right]\hat{\boldsymbol{\theta}} \qquad (3.3.31)$$

3.3.2 POSITION, VELOCITY, AND ACCELERATION IN CYLINDRICAL COORDINATES

A point (x, y, z) in space can be expressed in *cylindrical coordinates* (ρ, θ, z) as shown in Figure 3.10. Cylindrical coordinates are most useful when the problem to be solved has a cylindrical symmetry. It is useful to remember that cylindrical coordinates (ρ, θ, z) are a direct extension of polar coordinates (r, θ) from two to three dimensions, by adding the z-coordinate.

The transformation from Cartesian to polar coordinates is made by the following relations:

From Cylindrical to Cartesian Coordinates

$$x = \rho \cos \theta \tag{3.3.32}$$

$$y = \rho \sin \theta \tag{3.3.33}$$

$$z = z \tag{3.3.34}$$

From Cartesian to Cylindrical Coordinates

$$\rho = \sqrt{x^2 + y^2} \tag{3.3.35}$$

$$\tan \theta = y/x \tag{3.3.36}$$

$$z = z \tag{3.3.37}$$

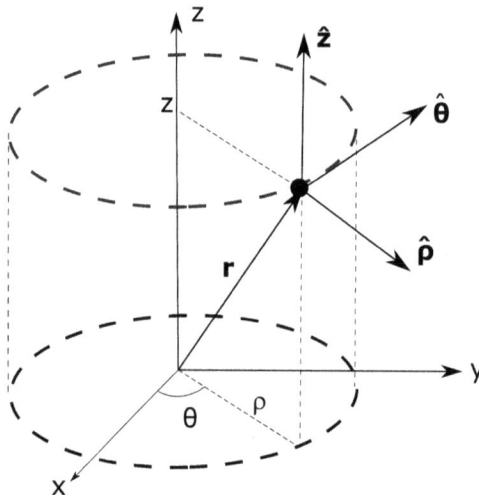

Figure 3.10: Cylindrical coordinates (ρ, θ, z) and the corresponding unit vectors $\hat{\rho}, \hat{\theta}, \hat{z}$.

We can define three unit vectors $\hat{\rho}, \hat{\theta}, \hat{z}$ in cylindrical coordinates, which are shown in Figure 3.10. Since cylindrical coordinates are essentially an extension of polar coordinates in three dimensions, we can write the unit vectors immediately by analogy with the equations in the previous section:

Unit Vectors in Cylindrical Coordinates

$$\hat{\rho} = \cos\theta\,\hat{\mathbf{i}} + \sin\theta\,\hat{\mathbf{j}} \tag{3.3.38}$$

$$\hat{\theta} = -\sin\theta\,\hat{\mathbf{i}} + \cos\theta\,\hat{\mathbf{j}} \tag{3.3.39}$$

$$\hat{\mathbf{z}} = \hat{\mathbf{k}} \tag{3.3.40}$$

Similar to the unit vectors in polar coordinates, the unit vectors in cylindrical coordinates change direction with the location of the particle, with the exception of $\hat{\mathbf{z}}$, which is identical to $\hat{\mathbf{k}}$ and does not change direction. Note also that the cylindrical unit vectors are related to the Cartesian unit vectors by a rotation matrix of the form:

$$\begin{bmatrix} \hat{\rho} \\ \hat{\theta} \\ \hat{\mathbf{z}} \end{bmatrix} = \begin{bmatrix} \cos\theta & \sin\theta & 0 \\ -\sin\theta & \cos\theta & 0 \\ 0 & 0 & 1 \end{bmatrix} \begin{bmatrix} \hat{\mathbf{i}} \\ \hat{\mathbf{j}} \\ \hat{\mathbf{k}} \end{bmatrix} \tag{3.3.41}$$

We can compute the position and velocity vectors in cylindrical coordinates in a manner similar to that used for polar coordinates. The difference is that for cylindrical coordinates there is an additional direction, $\hat{\mathbf{z}}$. However, $\hat{\mathbf{z}}$ does not change direction with position and is therefore constant. Hence, the position, velocity, and acceleration vectors in cylindrical coordinates are given by:

Position, Velocity, and Acceleration Vectors in Cylindrical Coordinates

$$\mathbf{r} = \rho\cos\theta\,\hat{\mathbf{i}} + \rho\sin\theta\,\hat{\mathbf{j}} + z\,\hat{\mathbf{k}} \tag{3.3.42}$$

$$\mathbf{v} = \frac{d\mathbf{r}}{dt} = \frac{d\rho}{dt}\hat{\rho} + \rho\frac{d\theta}{dt}\hat{\theta} + \frac{dz}{dt}\hat{\mathbf{z}} \tag{3.3.43}$$

$$\mathbf{a} = \frac{d\mathbf{v}}{dt} = \left[\frac{d^2\rho}{dt^2} - \rho\left(\frac{d\theta}{dt}\right)^2\right]\hat{\rho} + \left(\rho\frac{d^2\theta}{dt^2} + 2\frac{d\rho}{dt}\frac{d\theta}{dt}\right)\hat{\theta} + \frac{d^2z}{dt^2}\hat{\mathbf{z}} \tag{3.3.44}$$

This section concludes with some additional useful expressions. By analogy to (3.3.19), which was derived for polar coordinates, the infinitesimal displacement $d\mathbf{s}$ between two points in space in cylindrical coordinates can be written as:

$$d\mathbf{s} = d\rho\,\hat{\rho} + \rho\,d\theta\,\hat{\theta} + dz\,\hat{\mathbf{z}} \tag{3.3.45}$$

When computing infinitesimal areas, it is important to keep in mind the specific area on the cylinder that is relevant to your problem. For example, it is common to work with the infinitesimal area element that is on the surface of the body of the cylinder. The infinitesimal area element dA on the surface of a cylinder of radius ρ, can be written as:

$$dA = \rho\,d\theta\,dz \tag{3.3.46}$$

Keep in mind, however, that if one is interested in the top or bottom of the cylinder, then the polar infinitesimal area element is written as $dA = \rho\,d\rho\,d\theta$. An infinitesimal volume element in cylindrical coordinates is given by:

$$dV = dx\,dy\,dz = \rho\,d\rho\,d\theta\,dz \tag{3.3.47}$$

and we can calculate triple integrals of any function $f(x, y, z)$ by using the transformation into cylindrical coordinates:

$$\iiint f(x, y, z)\, dx\, dy\, dz = \iiint f(\rho\cos\theta, \rho\sin\theta, z)\, \rho\, d\rho\, d\theta\, dz \qquad (3.3.48)$$

These triple integrals will become useful in calculating, for example, the center of mass and moment of inertia of an object using cylindrical coordinates.

3.3.3 POSITION, VELOCITY, AND ACCELERATION IN SPHERICAL COORDINATES

In spherical coordinates, we specify the coordinates of every point in space by two angles (θ, ϕ) and by the radial distance r from the origin, as shown in Figure 3.11. It is important to remember that the *polar angle* θ can range between 0 and π, while the *azimuthal angle* ϕ can vary from 0 to 2π. Spherical coordinates are useful for problems involving spherical symmetry, a common occurrence in classical mechanics.

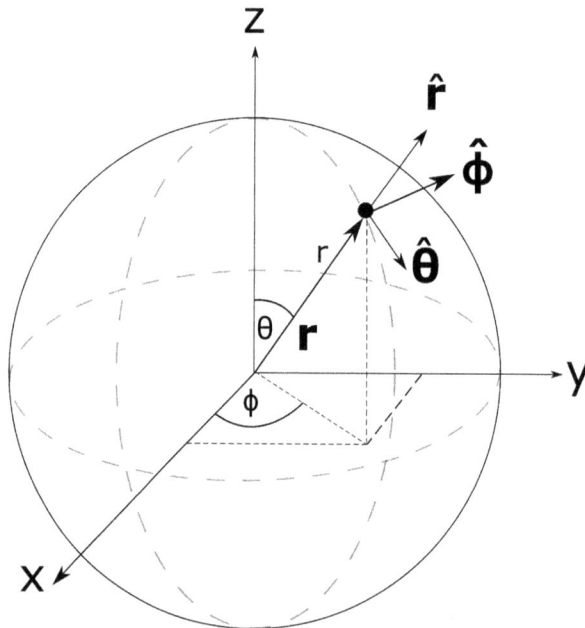

Figure 3.11: Spherical coordinates (r, θ, ϕ) and the corresponding unit vectors. The angle ϕ is defined on the xy-plane and can vary between 0 to 2π, while the azimuthal angle θ can vary between 0 to π *only*.

The Cartesian coordinates x, y, z are related to the spherical coordinates (r, θ, ϕ) by:

From Spherical to Cartesian Coordinates

$$x = r\sin\theta\cos\phi \qquad (3.3.49)$$

$$y = r\sin\theta\sin\phi \qquad (3.3.50)$$

$$z = r\cos\theta \qquad (3.3.51)$$

From Cartesian to Spherical Coordinates

$$r = \sqrt{x^2 + y^2 + z^2} \tag{3.3.52}$$

$$\tan \phi = y/x \tag{3.3.53}$$

$$\cos \theta = \frac{z}{r} = \frac{z}{\sqrt{x^2 + y^2 + z^2}} \tag{3.3.54}$$

Similar to the unit vectors in cylindrical coordinates, the orientation of the spherical coordinate unit vectors $\hat{\mathbf{r}}$, $\hat{\boldsymbol{\theta}}$, $\hat{\boldsymbol{\phi}}$ depends on the position of the particle. We define these three unit vectors as shown in Figure 3.11.

These unit vectors $\hat{\mathbf{r}}$, $\hat{\boldsymbol{\theta}}$, $\hat{\boldsymbol{\phi}}$ for spherical coordinates are related to the Cartesian unit vectors by the following equations:

Unit Vectors in Spherical Coordinates

$$\hat{\mathbf{r}} = \sin \theta \cos \phi \, \hat{\mathbf{i}} + \sin \theta \sin \phi \, \hat{\mathbf{j}} + \cos \theta \, \hat{\mathbf{k}} \tag{3.3.55}$$

$$\hat{\boldsymbol{\theta}} = \cos \theta \cos \phi \, \hat{\mathbf{i}} + \cos \theta \sin \phi \, \hat{\mathbf{j}} - \sin \theta \, \hat{\mathbf{k}} \tag{3.3.56}$$

$$\hat{\boldsymbol{\phi}} = - \sin \phi \, \hat{\mathbf{i}} + \cos \phi \, \hat{\mathbf{j}} \tag{3.3.57}$$

Like the cylindrical coordinate unit vectors, the spherical unit vectors are related to the Cartesian unit vectors by a rotation matrix:

$$\begin{bmatrix} \hat{\mathbf{r}} \\ \hat{\boldsymbol{\theta}} \\ \hat{\boldsymbol{\phi}} \end{bmatrix} = \begin{bmatrix} \sin \theta \cos \phi & \sin \theta \sin \phi & \cos \theta \\ \cos \theta \cos \phi & \cos \theta \sin \phi & - \sin \theta \\ - \sin \phi & \cos \phi & 0 \end{bmatrix} \begin{bmatrix} \hat{\mathbf{i}} \\ \hat{\mathbf{j}} \\ \hat{\mathbf{k}} \end{bmatrix} \tag{3.3.58}$$

In order to calculate velocity and acceleration in spherical coordinates, we need the time derivatives of the unit vectors. They are given by (see End-of-Chapter problems for this section):

$$\dot{\hat{\mathbf{r}}} = \dot{\theta} \, \hat{\boldsymbol{\theta}} + \dot{\phi} \sin \theta \, \hat{\boldsymbol{\phi}} \tag{3.3.59}$$

$$\dot{\hat{\boldsymbol{\theta}}} = -\dot{\theta} \, \hat{\mathbf{r}} + \dot{\phi} \cos \theta \, \hat{\boldsymbol{\phi}} \tag{3.3.60}$$

$$\dot{\hat{\boldsymbol{\phi}}} = -\dot{\phi} \sin \theta \, \hat{\mathbf{r}} - \dot{\phi} \cos \theta \, \hat{\boldsymbol{\theta}} \tag{3.3.61}$$

By using these equations, it can be shown that the velocity vector in spherical coordinates is given by:

$$\mathbf{v} = \dot{r} \, \hat{\mathbf{r}} + r \dot{\theta} \, \hat{\boldsymbol{\theta}} + r \dot{\phi} \sin \theta \, \hat{\boldsymbol{\phi}} \tag{3.3.62}$$

Multiplying both sides of this equation by dt, we obtain the infinitesimal displacement $d\mathbf{s}$ between two points in spherical coordinates:

$$d\mathbf{s} = dr \, \hat{\mathbf{r}} + r \, d\theta \, \hat{\boldsymbol{\theta}} + r \, d\phi \, \sin \theta \hat{\boldsymbol{\phi}} \tag{3.3.63}$$

By multiplying the coefficients of the unit vectors in this equation, the infinitesimal volume element dV in spherical coordinates is found to be:

$$dV = dx \, dy \, dz = r^2 \sin \theta \, dr \, d\theta \, d\phi \tag{3.3.64}$$

We can also calculate triple integrals in spherical coordinates by using the transformation equation:

$$\iiint f(x,y,z)\, dx\, dy\, dz = \iiint f(r\sin\theta\cos\phi, r\sin\theta\sin\phi, r\cos\theta) r^2 \sin\theta\, dr\, d\theta\, d\phi \qquad (3.3.65)$$

Finally, the acceleration vector in spherical coordinates is given by (see End-of-Chapter problems for this section):

$$\begin{aligned}
\boldsymbol{a} = {} & \left(\ddot{r} - r\dot{\phi}^2\sin^2\theta - r\dot{\theta}^2\right)\hat{\mathbf{r}} + \left(r\ddot{\theta} + 2\dot{r}\dot{\theta} - r\dot{\phi}^2\sin\theta\cos\theta\right)\hat{\boldsymbol{\theta}} \\
& + \left(r\ddot{\phi}\sin\theta + 2\dot{r}\dot{\phi}\sin\theta + 2r\dot{\theta}\dot{\phi}\cos\theta\right)\hat{\boldsymbol{\phi}}
\end{aligned} \qquad (3.3.66)$$

Position, Velocity, and Acceleration Vectors in Spherical Coordinates

$$\mathbf{r} = r\hat{\mathbf{r}} \qquad (3.3.67)$$

$$\mathbf{v} = \dot{r}\hat{\mathbf{r}} + r\dot{\theta}\hat{\boldsymbol{\theta}} + r\dot{\phi}\sin\theta\hat{\boldsymbol{\phi}} \qquad (3.3.68)$$

$$\begin{aligned}
\boldsymbol{a} = {} & \left(\ddot{r} - r\dot{\phi}^2\sin^2\theta - r\dot{\theta}^2\right)\hat{\mathbf{r}} + \left(r\ddot{\theta} + 2\dot{r}\dot{\theta} - r\dot{\phi}^2\sin\theta\cos\theta\right)\hat{\boldsymbol{\theta}} \\
& + \left(r\ddot{\phi}\sin\theta + 2\dot{r}\dot{\phi}\sin\theta + 2r\dot{\theta}\dot{\phi}\cos\theta\right)\hat{\boldsymbol{\phi}}
\end{aligned} \qquad (3.3.69)$$

3.4   GRADIENT, DIVERGENCE, AND CURL

There are several important spatial derivatives (i.e., derivatives that are taken with respect to a spatial coordinate) that occur in physics. These are called the gradient, divergence, and curl. In this section we will present these derivatives, along with physical interpretations.

3.4.1   GRADIENT

Consider a scalar function $f(x)$. Suppose we want to know how much the function $f(x)$ changes when we change its argument x by a small amount, say from x to $x + dx$. A Taylor series expansion of $f(x)$ gives:

$$f(x + dx) = f(x) + \frac{df}{dx}dx + \cdots \qquad (3.4.1)$$

We can ignore terms of dx^2 and higher, because dx is very small. Next, we define the differential $df = f(x + dx) - f(x)$ and interpret it as the change of the value of f when we change x by a small amount dx. Then from (3.4.1) we find:

$$df = \frac{df}{dx}dx \qquad (3.4.2)$$

Equation (3.4.2) can be interpreted as follows. The amount df by which f changes when x is changed (by a small amount), is equal to the product of the rate of change of f with respect to x, and the amount by which x is changed (dx). This is similar to the familiar equation $d = vt$, which says the distance traveled by a particle is the product of how fast it is traveling (i.e., the rate v at which the distance changes with time), and of the length of time t traveled.

Next, we can extend (3.4.2) to scalar functions in three dimensions. Let us consider a scalar function $f(x, y, z)$ in Cartesian coordinates. We would like to know how $f(x, y, z)$

changes when we change *each* of its variables by a small amount. A multi-variable Taylor series expansion gives:

$$f(x + dx, y + dy, z + dz) = f(x, y, z) + \frac{\partial f}{\partial x}dx + \frac{\partial f}{\partial y}dy + \frac{\partial f}{\partial z}dz + \cdots \qquad (3.4.3)$$

In this expression, $\frac{\partial f}{\partial x}$, $\frac{\partial f}{\partial y}$, $\frac{\partial f}{\partial z}$ are the partial derivatives of f with respect to (x, y, z) respectively. Similar to the one-dimensional case, we define the differential as $df = f(x + dx, y + dy, z + dz) - f(x, y, z)$ and interpret it as the change of f as we change all of its variables. Using (3.4.3), we can write the differential in dot product form:

$$df = \frac{\partial f}{\partial x}dx + \frac{\partial f}{\partial y}dy + \frac{\partial f}{\partial z}dz \qquad (3.4.4)$$

$$= \left(\frac{\partial f}{\partial x}\hat{\mathbf{i}} + \frac{\partial f}{\partial y}\hat{\mathbf{j}} + \frac{\partial f}{\partial z}\hat{\mathbf{k}} \right) \cdot \left(dx\hat{\mathbf{i}} + dy\hat{\mathbf{j}} + dz\hat{\mathbf{k}} \right) \qquad (3.4.5)$$

Again, we get an equation similar to $d = vt$, where the first term is similar to a rate of change, and the second term is a change in the coordinates. The first term in (3.4.5) can be rewritten as:

$$\left(\hat{\mathbf{i}}\frac{\partial}{\partial x} + \hat{\mathbf{j}}\frac{\partial}{\partial y} + \hat{\mathbf{k}}\frac{\partial}{\partial z} \right) f \qquad (3.4.6)$$

and it describes the rate of change of f in each direction. The operator appearing in (3.4.6) is very important in physics and is called the *del or gradient operator* denoted by ∇:

The Gradient Operator

$$\nabla = \left(\hat{\mathbf{i}}\frac{\partial}{\partial x} + \hat{\mathbf{j}}\frac{\partial}{\partial y} + \hat{\mathbf{k}}\frac{\partial}{\partial z} \right) \qquad (3.4.7)$$

The symbol ∇ is also called *nabla*. The del operator is a vector operator and in (3.4.7), it is written in Cartesian coordinates. As you might expect, the del operator takes different forms in cylindrical and spherical coordinates. We will present the del operator in other coordinates later in this section, but for now we will consider only Cartesian coordinates.

When the del operator is applied to a scalar function f, then ∇f is called the *gradient* of f and the result is a vector function.

The Gradient of a Scalar Function $f(x, y, z)$

$$\nabla f = \left(\frac{\partial f}{\partial x}\hat{\mathbf{i}} + \frac{\partial f}{\partial y}\hat{\mathbf{j}} + \frac{\partial f}{\partial z}\hat{\mathbf{k}} \right) \qquad (3.4.8)$$

We can then write the differential df as a dot product of ∇f and $d\mathbf{r}$:

$$df = \nabla f \cdot d\mathbf{r} \qquad (3.4.9)$$

To better understand the gradient of a scalar field, we will introduce the concept of *level sets of the function f*. The level set of a scalar function f is a curve in two dimensions, or a surface in three dimensions, on which the value of f is constant. Hence, along the curve of a level set we would have $df = 0$. If $d\mathbf{r}$ is a displacement along a level set of f, then:

$$df = \nabla f \cdot d\mathbf{r} = 0 \qquad (3.4.10)$$

and therefore the two vectors ∇f and $d\mathbf{r}$ are perpendicular to each other, i.e., $\nabla f \perp d\mathbf{r}$. We conclude that the gradient of f is perpendicular to the level sets of f.

Example 3.13: The gradient of a scalar field.
Consider the scalar field:

$$f(x,y) = e^{-(x^2+y^2)}$$

(a) Compute and plot the gradient vector of f.
(b) Plot the level sets of this function by using a contour plot to show curves of constant f.

Solution:
The gradient of f can be found by computing:

$$\nabla f = \left(\hat{\mathbf{i}}\frac{\partial}{\partial x} + \hat{\mathbf{j}}\frac{\partial}{\partial y} \right) e^{-(x^2+y^2)}$$

$$= -2xe^{-(x^2+y^2)}\hat{\mathbf{i}} - 2ye^{-(x^2+y^2)}\hat{\mathbf{j}}$$

Python Code
We define the coordinate system R with the unit vectors in this reference frame denoted by `R.i,R.j,R.k` and the Cartesian coordinates x, y, and z denoted by `R.x,R.y,R.z` respectively. The gradient is calculated with the command `gradient(f)` where $f = \exp(-x^2 - y^2)$ is the given scalar function.

In part (b) we use the command `contour(x,y,T,5)` to plot five contours for the function, and `quiver(x,y,x_grad,y_grad)` to draw the arrows for the gradient at points (x,y) on the Cartesian plane. The result is shown in Figure 3.12a.

```
# Code for part (a)
from sympy.vector import CoordSys3D, gradient
from sympy import exp

print('-'*28,'CODE OUTPUT','-'*29,'\n')

R = CoordSys3D('R')        # Define the Cartesian frame R

f = exp(-R.x**2-R.y**2)    # scalar function f(x,y) with x=R.x, y=R.y

gradf = gradient(f)        # evaluate gradient using gradient function

print('The gradient of f(x,y) is:\n', gradf)

------------------------- CODE OUTPUT ----------------------------
The gradient of f(x,y) is:
 (-2*R.x*exp(-R.x**2 - R.y**2))*R.i + (-2*R.y*exp(-R.x**2 - R.y**2))*R.j
```

```
# Code for part (b)
import numpy as np
import matplotlib.pyplot as plt

xyrange = np.linspace(-1,1,20)          # range of x,y values

x, y = np.meshgrid(xyrange,xyrange)

T = np.exp(- (x**2 + y**2))             # scalar function f(x,y)

x_grad = -2*x*np.exp(-(x**2 + y**2))    # x,y components of gradient(f)
y_grad = -2*y*np.exp(- (x**2 + y**2))

plt.contour(x,y,T,5)                    # use contour to plot 5 contours

plt.quiver(x,y,x_grad,y_grad)           # use quiver to plot the arrows
plt.xlabel('x coordinate')
plt.ylabel('y coordinate')
plt.show()
```

Mathematica Code

Notice that Mathematica outputs the gradient as a list, where the first element of the list is the x-component, the second element is the y-component, and the third element is the z-component. Additionally, the command `Grad(f,x,y)` is used to compute the gradient, and Mathematica needs to know what variables are needed for differentiation. Note also that by default, Mathematica computes the gradient in Cartesian coordinates. As we will see below, the gradient is different in cylindrical and spherical coordinates. Mathematica can compute symbolically the gradient in those coordinates as well, when the user specifies which coordinate system to use.

The commands `ContourPlot` and `VectorPlot` are used to create the two sets of plots.

The contour curves shown in Figure 3.12b are the level sets of f, with lighter colors corresponding to greater values of f. The vectors represent the gradient vector field. The size of the arrows corresponds to the magnitude of the gradient at that point. Notice that the vectors are perpendicular to the level sets.

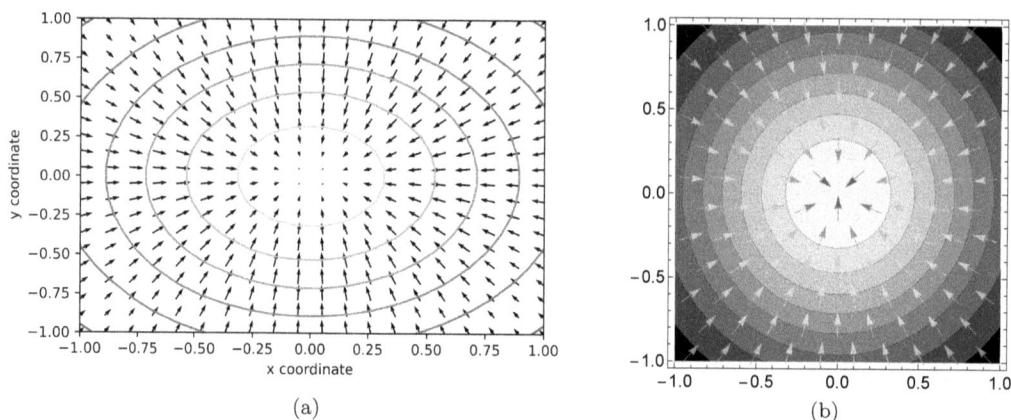

(a)   (b)

Figure 3.12: Plots of the gradient and the contours of a scalar function produced by (a) Python and (b) Mathematica, in Example 3.13.

```
SetOptions[ContourPlot, BaseStyle->FontSize->16];

f = Exp[−(x^2 + y^2)];
gradf = Grad[f, {x, y}]
```

$$\left\{ -2e^{-x^2-y^2}x, \, -2e^{-x^2-y^2}y \right\}$$

```
gr1 = ContourPlot[f, {x, −1, 1}, {y, −1, 1}, ColorFunction->GrayLevel];
gr2 = VectorPlot[gradf, {x, −1, 1}, {y, −1, 1}, VectorStyle->Red, VectorPoints->10];
Show[gr1, gr2]
```

The gradient plays a very important role in classical mechanics. For example, as we will show in Chapter 5, for a particle with potential energy $V(x, y, z)$, the force vector $\boldsymbol{F}(x, y, z)$ acting on the particle is given by the negative gradient of the potential energy:

$$\boldsymbol{F}(x, y, z) = -\nabla V(x, y, z) = -\frac{\partial V}{\partial x}\hat{\mathbf{i}} - \frac{\partial V}{\partial y}\hat{\mathbf{j}} - \frac{\partial V}{\partial z}\hat{\mathbf{k}} \tag{3.4.11}$$

Although we will discuss the relationship between force, energy, and equilibrium states of a mechanical system in more detail in Chapter 5, it is worth noting that the above equation says that forces point in the direction of most rapidly *decreasing* potential energy (because of the minus sign).

In order to get a complete picture of the gradient, we next need to discuss the concept of the *directional derivative*.

The directional derivative df/dn of a scalar field f gives the rate of change of $f(x, y, z)$ in the direction of the unit vector $\hat{\mathbf{n}} = n_x\hat{\mathbf{i}} + n_y\hat{\mathbf{j}} + n_z\hat{\mathbf{k}}$, and is defined as the dot product:

$$\frac{df}{dn} = \hat{\mathbf{n}} \cdot \nabla f = n_x\frac{\partial f}{\partial x} + n_y\frac{\partial f}{\partial y} + n_z\frac{\partial f}{\partial z} \tag{3.4.12}$$

Now we ask the question, in what direction does the scalar field f change most rapidly? We can compute the dot product as:

$$\frac{df}{dn} = \hat{\mathbf{n}} \cdot \nabla f = |\hat{\mathbf{n}}| \, |\nabla f| \cos \theta \qquad (3.4.13)$$

where θ is the angle between $\hat{\mathbf{n}}$ and ∇f. The direction in which the directional derivative is greatest would occur when $\hat{\mathbf{n}} \cdot \nabla f$ is maximal, i.e., when the angle θ equals zero. Therefore, ∇f points in the direction of the most rapid increase of f. Likewise, the direction of greatest *decrease* in the scalar field is $-\nabla f$. As an example, consider a surface whose height above sea level at a point (x, y) is $H(x, y)$. The gradient of H at a point (x, y) is a vector pointing in the direction of the steepest slope, or steepest grade at that point. The steepness of the slope at that point is given by the magnitude of the gradient vector.

Example 3.14: The directional derivative.
　Find the directional derivative of a scalar field $f = x^2 y + xz$ at the point $(1, 2, -1)$ and in the direction of the vector $\mathbf{a} = 2\hat{\mathbf{i}} - 2\hat{\mathbf{j}} + \hat{\mathbf{k}}$.

Solution:
We first obtain the unit vector $\hat{\mathbf{a}}$ in the given direction by dividing $\mathbf{a}$ by its magnitude:

$$\hat{\mathbf{a}} = \frac{\mathbf{a}}{\sqrt{a_x^2 + a_y^2 + a_z^2}} = \frac{2\hat{\mathbf{i}} - 2\hat{\mathbf{j}} + \hat{\mathbf{k}}}{\sqrt{2^2 + 2^2 + (1)^2}} = \frac{2}{3}\hat{\mathbf{i}} - \frac{2}{3}\hat{\mathbf{j}} + \frac{1}{3}\hat{\mathbf{k}}$$

Next find the gradient of f:

$$\nabla f = \hat{\mathbf{i}}\frac{\partial f}{\partial x} + \hat{\mathbf{j}}\frac{\partial f}{\partial y} + \hat{\mathbf{k}}\frac{\partial f}{\partial z} = (2xy + z)\hat{\mathbf{i}} + x^2\hat{\mathbf{j}} + x\hat{\mathbf{k}}$$

Finally we find the directional derivative from its definition as the dot product of the unit vector $\hat{\mathbf{a}}$ and the gradient ∇:

$$\frac{df}{da} = \hat{\mathbf{a}} \cdot \nabla f = (2xy + z)\frac{2}{3} + x^2(-\frac{2}{3}) + x\frac{1}{3}$$

In the final step we substitute the given point $(1, 2, -1)$:

$$\frac{df}{da} = \hat{\mathbf{a}} \cdot \nabla f = \frac{5}{3}$$

Python Code
In this example we show two different libraries which can evaluate the directional derivative.
In the codes we import the `CoordSys3D` command from the `sympy.vector` library. Within this library we define a coordinate system `C= CoordSys3D('C')`.
In the first method we first evaluate the normalized vector $\hat{\mathbf{n}}$, by using the method `a.normalize()` in SymPy. We evaluate the gradient using `gradient(f)`, and then find the dot product of $\hat{\mathbf{n}}$ and ∇f using the method `n.dot(gradient(f))`.
The method `.subs` is used to substitute the given point $(1, 2, -1)$,i.e., $x = 1$, $y = 2$, $z = -1$.

In the second method, we use the dedicated function `directional_derivative(f)`. The result from the two Python methods in this example is the same as the analytical result obtained above.

```python
from sympy import simplify
from sympy.vector import CoordSys3D, gradient, directional_derivative

print('-'*28,'CODE OUTPUT','-'*29,'\n')
C = CoordSys3D('C')          # Cartesian system named C
                             # has unit vectors C.i , C.j , C.k

f = C.x**2 * C.y + C.x * C.z       # scalar field f(x, y, z)=xyz

a = 2*C.i -2* C.j + C.k            # vector a defines desired direction

n = a.normalize()                  # n= normalized vector a

u = n.dot(gradient(f))     # Find the dot product of n and gradient(f)
print('The directional derivative function is:\n',u)

print('\nAt point (1,2,-1) the directional derivative is = ',\
      u.subs({C.x:1,C.y:2,C.z:-1}))

# Shortcut method using the dedicated function directional_derivative

v = simplify(directional_derivative( f, n)) # using dedicated function

print('\nWith method #2, the directional derivative is = ',\
      v.subs({C.x:1,C.y:2,C.z:-1}))

---------------------------- CODE OUTPUT ----------------------------
The directional derivative function is:
 -2*C.x**2/3 + 4*C.x*C.y/3 + C.x/3 + 2*C.z/3

At point (1,2,-1) the directional derivative is =  5/3

With method #2, the directional derivative is =  5/3
```

Mathematica Code

In this code we define the vector **a** as a list, and we normalize it by dividing by its magnitude `Norm(a)`. The command `ResourceFunction["DirectionalD"]` is used to evaluate the directional derivative function. Finally we use the replacement rule `dirderiv/.{x->1,y->2, z->-1}` to substitute the values of the given point $(1, 2, -1)$, i.e., $x = 1$, $y = 2$, $z = -1$.

$a = \{2, -2, 1\};$

$n = a/\text{Norm}[a];$

$\text{Print}[\text{"The normalized vector a is :"}, n]$

The normalized vector a is : $\left\{\frac{2}{3}, -\frac{2}{3}, \frac{1}{3}\right\}$

$\text{dirderiv} = \text{ResourceFunction}[\text{"DirectionalD"}][x\wedge 2 * y + x * z, n, \{x, y, z\}];$

$\text{Print}[\text{"The directional derivative function is :"}]$

$\text{Print}[\text{dirderiv}]$

The directional derivative function is :
$\frac{x}{3} - \frac{2x^2}{3} + \frac{2}{3}(2xy + z)$

$\text{result} = \text{dirderiv}/.\{x\text{->}1, y\text{->}2, z\text{->}-1\};$

$\text{Print}[\text{"The directional derivative at point(1,2,-1) is : "}, \text{result}]$

The directional derivative at point(1,2,-1) is : $\frac{5}{3}$

In order to get the form of the gradient in other coordinate systems, we return to the fact that the total differential df of the scalar function is given by:

$$df = \frac{\partial f}{\partial x}dx + \frac{\partial f}{\partial y}dy + \frac{\partial f}{\partial z}dz = \frac{\partial f}{\partial \rho}d\rho + \frac{\partial f}{\partial \theta}d\theta + \frac{\partial f}{\partial z}dz = \nabla f \cdot d\mathbf{s} \qquad (3.4.14)$$

where in Cartesian coordinates $d\mathbf{s} = dx\,\hat{\mathbf{i}} + dy\,\hat{\mathbf{j}} + dz\,\hat{\mathbf{k}}$, and we used the chain rule for $f(\rho, \theta, z)$. We can use this equation to obtain expressions for the gradient in cylindrical and spherical coordinates, as follows.

The gradient of any scalar function f in cylindrical coordinates (ρ, θ, z) will be of the general form:

$$\nabla f(x, y, z) = (\nabla f)_\rho\, \hat{\boldsymbol{\rho}} + (\nabla f)_\theta\, \hat{\boldsymbol{\theta}} + (\nabla f)_z\, \hat{\mathbf{k}} \qquad (3.4.15)$$

where $(\nabla f)_\rho$, $(\nabla f)_\theta$, $(\nabla f)_z$ are the components of the gradient along the unit vectors $\hat{\boldsymbol{\rho}}$, $\hat{\boldsymbol{\theta}}$, $\hat{z}$ in cylindrical coordinates. From the previous section we recall that $d\mathbf{s}$ in cylindrical coordinates is given by:

$$d\mathbf{s} = d\rho\,\hat{\boldsymbol{\rho}} + \rho\, d\theta\,\hat{\boldsymbol{\theta}} + dz\,\hat{z} \qquad (3.4.16)$$

By evaluating the dot product of (3.4.15) and (3.4.16), and substituting into (3.4.14), we obtain:

$$\frac{\partial f}{\partial \rho}d\rho + \frac{\partial f}{\partial \theta}d\theta + \frac{\partial f}{\partial z}dz = (\nabla f)_\rho\, d\rho + (\nabla f)_\theta\, \rho\, d\theta + (\nabla f)_z\, dz \qquad (3.4.17)$$

By equating the corresponding coefficients of $(d\rho,\ d\theta,\ dz)$ on the two sides of (3.4.17), we find:

$$(\nabla f)_\rho = \frac{\partial f}{\partial \rho} \qquad (\nabla f)_\theta = \frac{1}{\rho}\frac{\partial f}{\partial \theta} \qquad (\nabla f)_z = \frac{\partial f}{\partial z}$$

Therefore by using (3.4.15), the expression of the gradient in cylindrical coordinates is:

$$\nabla f(\rho, \theta, z) = \frac{\partial f}{\partial \rho}\hat{\boldsymbol{\rho}} + \frac{1}{\rho}\frac{\partial f}{\partial \theta}\hat{\boldsymbol{\theta}} + \frac{\partial f}{\partial z}\hat{z} \qquad (3.4.18)$$

The gradient in spherical coordinates can also be obtained with the same method, and is of the form:

$$\nabla f(, \theta, \phi) = \frac{\partial f}{\partial r}\hat{\mathbf{r}} + \frac{1}{r}\frac{\partial f}{\partial \theta}\hat{\boldsymbol{\theta}} + \frac{1}{r\sin\theta}\frac{\partial f}{\partial \phi}\hat{\boldsymbol{\phi}} \qquad (3.4.19)$$

Note that our work above has also given us the del operator in various coordinate systems:

The Gradient in Different Coordinate Systems

$$\nabla = \hat{\mathbf{i}}\frac{\partial}{\partial x} + \hat{\mathbf{j}}\frac{\partial}{\partial y} + \hat{\mathbf{k}}\frac{\partial}{\partial z} \qquad \text{Cartesian Coordinates} \qquad (3.4.20)$$

$$\nabla = \hat{\boldsymbol{\rho}}\frac{\partial}{\partial \rho} + \hat{\boldsymbol{\theta}}\frac{1}{\rho}\frac{\partial}{\partial \theta} + \hat{\mathbf{z}}\frac{\partial}{\partial z} \qquad \text{Cylindrical Coordinates} \qquad (3.4.21)$$

$$\nabla = \hat{\mathbf{r}}\frac{\partial}{\partial r} + \hat{\boldsymbol{\theta}}\frac{1}{r}\frac{\partial}{\partial \theta} + \hat{\boldsymbol{\phi}}\frac{1}{r\sin\theta}\frac{\partial}{\partial \phi} \qquad \text{Spherical Coordinates} \qquad (3.4.22)$$

The next example shows how to evaluate the gradient of a scalar function in cylindrical coordinates using a CAS.

Example 3.15: Evaluating the gradient in cylindrical coordinates.
Evaluate the gradient of the scalar function $f(\rho, \theta, z) = \rho\sin\theta\cos z$ in cylindrical coordinates (ρ, θ, z), both analytically and using Mathematica and the Python functions available in sympy.vector.

Solution:
Using (3.4.21) for cylindrical coordinates:

$$\nabla f = \hat{\boldsymbol{\rho}}\frac{\partial f}{\partial \rho} + \hat{\boldsymbol{\theta}}\frac{1}{\rho}\frac{\partial f}{\partial \theta} + \hat{\mathbf{z}}\frac{\partial f}{\partial z} = \hat{\boldsymbol{\rho}}\sin\theta\cos z + \hat{\boldsymbol{\theta}}\cos\theta\cos z + \hat{\mathbf{z}}\left(-\rho\sin\theta\sin z\right)$$

Python Code
In the Python code we import the CoordSys3D command from the sympy.vector library. Within this library we define a coordinate system c and transform it into cylindrical coordinates, by using transformation='cylindrical'. This transformed system incorporates the cylindrical coordinate parameters (ρ, θ, z) which are represented internally as c.r, c.t, c.z. The corresponding unit vectors $\hat{\boldsymbol{\rho}}$, $\hat{\boldsymbol{\theta}}$, $\hat{\mathbf{z}}$ are represented by c.R, c.T, c.Z respectively.
The gradient within this library is evaluated using gr = gradient(f), and we use gr.coeff() to find the coefficients of the unit vectors in the cylindrical system.

```
# Gradient of a vector in cylindrical coordinates
from sympy import sin, cos
from sympy.vector import CoordSys3D, gradient

import textwrap

print('-'*28,'CODE OUTPUT','-'*29,'\n')

# define Cartesian system named c, with unit vectors c.i, c.j, c.k
# Transform it to system with cylindrical coordinates (rho, theta, z)

# variables are c.r, c.t, c.z for rho, theta, z
# unit vectors c.R, c.T, c.Z
c = CoordSys3D('c', transformation='cylindrical',\
variable_names = list("rtz"),vector_names=list("RTZ") )

scalar = c.r*sin(c.t)*cos(c.z)

print('scalar =',scalar)

gr = gradient(scalar)
print('\nThe gradient of this scalar in cylindrical is:')
print(textwrap.fill(str(gr),63))

print('\nThe rho-component of the gradient is:', gr.coeff(c.R))
print('The theta-component of the gradient is:', gr.coeff(c.T))
print('The z-component of the gradient is:', gr.coeff(c.Z))

--------------------------- CODE OUTPUT ---------------------------
scalar = c.r*sin(c.t)*cos(c.z)

The gradient of this scalar in cylindrical is:
(sin(c.t)*cos(c.z))*c.R + (cos(c.t)*cos(c.z))*c.T +
(-c.r*sin(c.t)*sin(c.z))*c.Z

The rho-component of the gradient is: sin(c.t)*cos(c.z)
The theta-component of the gradient is: cos(c.t)*cos(c.z)
The z-component of the gradient is: -c.r*sin(c.t)*sin(c.z)
```

Mathematica Code

The first line in the Mathematica code defines the function $f(\rho, \theta, z) = \rho \sin\theta \cos z$, and the command `Grad[f,{rho,theta,z},''Cylindrical'']` is used to compute the gradient in cylindrical coordinates.

$f[\text{rho}, \text{theta}, z] = \text{rho} * \text{Sin}[\text{theta}] * \text{Cos}[z];$

$\text{Print}[\text{"The gradient of the function f in cylindrical coordinates is:"}];$

$\text{Grad}[f[\text{rho}, \text{theta}, z], \{\text{rho}, \text{theta}, z\}, \text{"Cylindrical"}]$

The gradient of the f in cylindrical coordinates is:

$\{\text{Cos}[z]\,\text{Sin}[\text{theta}],\ \text{Cos}[\text{theta}]\,\text{Cos}[z],\ -\text{rho}\,\text{Sin}[\text{theta}]\,\text{Sin}[z]\}$

The "derivative" properties of the gradient are shown in the following identity, which is similar to the derivative product rule:

$$\nabla(fg) = f\nabla g + g\nabla f \qquad (3.4.23)$$

where f, g are scalar functions of (x, y, z).

Example 3.16 demonstrates this identity using symbolic algebra methods.

Example 3.16: The gradient of a product of scalar functions.
Prove the following general property of the gradient, for any scalar functions $f(x, y, z)$ and $g(x, y, z)$, by hand and demonstrate using a CAS.

$$\nabla(fg) = f(\nabla g) + g(\nabla f) \qquad (3.4.24)$$

Solution:

$$\nabla(fg) = \frac{\partial(fg)}{\partial x}\hat{i} + \frac{\partial(fg)}{\partial y}\hat{j} + \frac{\partial(fg)}{\partial z}\hat{k}$$

$$= f\frac{\partial g}{\partial x}\hat{i} + g\frac{\partial f}{\partial x}\hat{i} + f\frac{\partial g}{\partial y}\hat{j} + g\frac{\partial f}{\partial y}\hat{j} + f\frac{\partial g}{\partial z}\hat{k} + g\frac{\partial f}{\partial z}\hat{k}$$

$$= f\left(\frac{\partial g}{\partial x}\hat{i} + \frac{\partial g}{\partial y}\hat{j} + \frac{\partial g}{\partial z}\hat{k}\right) + g\left(\frac{\partial f}{\partial x}\hat{i} + \frac{\partial f}{\partial y}\hat{j} + \frac{\partial f}{\partial z}\hat{k}\right)$$

$$= f(\nabla g) + g(\nabla f)$$

Python Code
In the Python code we define the coordinate system C within `sympy.vector`, with unit vectors `C.i`, `C.j`, `C.k`. We define the symbols f, g to be functions of the coordinate parameters `C.x`, `C.y`, `C.z`, and we evaluate the left-hand side (LHS) and right-hand side (RHS) of (3.4.24) using the `gradient()` function.
We verify that the two sides of (3.4.24) are identical by using the logical statement `LHS==RHS`, which returns a `True` value for the equation.

```
from sympy.vector import CoordSys3D, gradient

print('-'*28,'CODE OUTPUT','-'*29,'\n')

C = CoordSys3D('C')      # define Cartesian system, named here C

# The scalar field f(x,y,z) is a function of the Cartesian
# coordinate variables C.x, C.y, C.z

from sympy import symbols, Function
f, g = symbols('f, g', cls=Function)

# Define the scalar fields as fscalar(x,y,z), gscalar(x,y,z),
fscalar = f(C.x, C.y, C.z)
gscalar = g(C.x, C.y, C.z)

# Construct the expression for the LHS: grad(f.g)
lhs = gradient(fscalar*gscalar)

# Construct the expression for the rhs: grad(f).g+grad(g).f
rhs = gradient(fscalar)*gscalar+gradient(gscalar)*fscalar

# Compare the two sides
print('The equation  grad(f.g)=grad(f).g+grad(g).f  is ',lhs==rhs)


-------------------------- CODE OUTPUT ----------------------------
The equation  grad(f.g)=grad(f).g+grad(g).f  is  True
```

Mathematica Code

We use the command `Grad[f[x_,y_,z_]*g[x_,y_,z_]]` to evaluate the $\nabla(fg)$, and `Grad[f[x_,y_,z_]*g[x_,y_,z_]]`, `Grad[g[x_,y_,z_]*f[x_,y_,z_]]` to represent the $\nabla(f)\,g$ and $f\nabla(g)$, respectively. The double equal sign `==` is used to test whether the identity is True or False.

LHS = Grad[f[x_,y_,z_] * g[x_,y_,z_], {x,y,z}];

RHS = Grad[f[x_,y_,z_], {x,y,z}] * g[x_,y_,z_]+

Grad[g[x_,y_,z_], {x,y,z}] * f[x_,y_,z_];

Print["The identity is :", LHS==RHS]

The identity is :True

In vector calculus, the del operator is a vector differential operator. As we have seen, when applied to a scalar field, the del operator produces the gradient of a scalar field ∇f. However, when applied to a *vector field,* the del operator can produce two quantities, the *divergence* of a vector field $\nabla \cdot \mathbf{v}$, and the *curl* of a vector field $\nabla \times \mathbf{v}$. In the next subsections, we will illustrate these new quantities.

3.4.2   DIVERGENCE

The divergence of a vector field $\mathbf{v} = v_x\hat{\mathbf{i}} + v_y\hat{\mathbf{j}} + v_z\hat{\mathbf{k}}$ is a *scalar* function that can be represented as:

$$\nabla \cdot \mathbf{v} = \frac{\partial v_x}{\partial x} + \frac{\partial v_y}{\partial y} + \frac{\partial v_z}{\partial z} \tag{3.4.25}$$

The divergence is obtained by taking a dot product between the del operator and a vector field. The divergence of a vector field is not frequently encountered in classical mechanics, but it does appear prominently in fluid dynamics and in electromagnetism. In this subsection, we provide only a brief overview of the divergence.

The divergence is roughly a measure of a vector field's change in magnitude along the direction in which it points. More accurately, it is a measure of that field's tendency to converge toward or repel from a point, and is sometimes referred to as the *"net outward flow rate"* of a vector field. For example, consider air as it is heated or cooled. The relevant vector field for this example is the velocity of the moving air at a point is pace. If air is heated in a region, it will expand in all directions such that the velocity field points outward from that region and towards regions with lower temperature. Therefore, the divergence of the velocity field in that region would have a positive value, and the region is called a *source*. If the air cools and contracts, the divergence is negative and the region is called a *sink*. In vector calculus, divergence is a vector operator that measures the magnitude of a vector field's source or sink at a given point, in terms of a signed scalar.

In physical terms, the divergence of a three-dimensional vector field is the extent to which the vector field flow behaves like a source or a sink at a given point. It is a local measure of its "outgoingness"–the extent to which there is more "material" exiting an infinitesimal region of space than entering it. If the divergence is non-zero at some point, then there must be a source or sink at that position. Note that when we use the terms source and sink, we are imagining a vector like the velocity vector field of a fluid in motion.

If the divergence of a vector field $\mathbf{A}(x, y, z)$ is zero everywhere in space, this vector field is called *solenoidal*.

Solenoidal Vector Fields A(x,y,z)

$$\nabla \cdot \mathbf{A} = 0 \longleftrightarrow \mathbf{A}(x, y, z) \text{ is a solenoidal vector field} \tag{3.4.26}$$

If $\mathbf{u}$, $\mathbf{v}$ are any vector functions and f is any scalar function, the divergence obeys the following product rule:

$$\nabla \cdot (f\mathbf{v}) = f(\nabla \cdot \mathbf{v}) + \mathbf{v} \cdot \nabla f \tag{3.4.27}$$

However, note that the divergence of a cross-vector product is less intuitive:

$$\nabla \cdot (\mathbf{u} \times \mathbf{v}) = \mathbf{v} \cdot (\nabla \times \mathbf{u}) - \mathbf{u} \cdot (\nabla \times \mathbf{v}) \tag{3.4.28}$$

Example 3.17: The divergence of a vector field.
    Find the divergence of the vector function $\mathbf{v} = xyz\hat{\mathbf{i}} + xyz\hat{\mathbf{j}} + xyz\hat{\mathbf{k}}$ analytically and also by using appropriate codes in Python and Mathematica.

Solution:
    The divergence is a scalar, found by adding the partial derivatives:

$$\nabla \cdot \mathbf{v} = \frac{\partial(x\,y\,z)}{\partial x} + \frac{\partial(x\,y\,z)}{\partial y} + \frac{\partial(x\,y\,z)}{\partial z} = y\,z + x\,z + x\,y$$

The following are the Python and Mathematica codes for the gradient.

Python Code

The command `CoordSys3D('R')` defines the coordinate system R, the unit vectors in this reference frame are denoted by `R.i,R.j,R.k`.

We use the package `physics.vector` from SymPy.

The divergence is calculated with the command `divergence(v)` where $\mathbf{v} = x\,y\,z\,\hat{\mathbf{i}} + x\,y\,z\,\hat{\mathbf{j}} + x\,y\,z\,\hat{\mathbf{k}}$ is the given vector field. We test whether the vector field is solenoidal by using the command `is_solenoidal`. In this example, the function returns the value *False* in Python, so that this field is not solenoidal.

```
from sympy.vector import CoordSys3D, divergence, is_solenoidal

print('-'*28,'CODE OUTPUT','-'*29,'\n')

R = CoordSys3D('R')              # define reference frame R

v = R.x*R.y*R.z*(R.i+R.j+R.k)   # scalar f with x=R.x, y=R.y,z=R.z

divv = divergence(v)            # evaluate divergence using function
print('The divergence of v(x,y,z) is:\n', divv)

test = is_solenoidal(v)
print('\nThe statement   The vector field is solenoidal   is:', test)

---------------------------- CODE OUTPUT ----------------------------
The divergence of v(x,y,z) is:
 R.x*R.y + R.x*R.z + R.y*R.z

The statement   The vector field is solenoidal   is: False
```

Mathematica Code

Notice that Mathematica outputs the divergence as a scalar function, and the vector field is entered as a list. The command `Div[f,x,y]` is used to compute the divergence, and Mathematica needs to know what variables are needed for differentiation. By default, Mathematica computes the divergence in Cartesian coordinates.

$v = \{x * y * z, x * y * z, x * y * z\};$

Print["The divergence of the vector field is: ", Div[$v, \{x, y, z\}$]]

The divergence of the vector field is: $x\,y + x\,z + y\,z$

Note that the divergence takes on a different form in cylindrical and spherical coordinates. To compute the divergence in these coordinate systems, choose the appropriate form of the del operator, and compute the dot product of the appropriate form of the del operator from (3.4.20), (3.4.21), and (3.4.22), and the vector of interest. The results are (see also the problems at the end of this chapter):

The Divergence in Cartesian, Cylindrical, and Spherical Coordinates

$$\nabla \cdot \mathbf{v} = \frac{\partial v_x}{\partial x} + \frac{\partial v_y}{\partial y} + \frac{\partial v_z}{\partial z} \tag{3.4.29}$$

$$\nabla \cdot \mathbf{v} = \frac{1}{\rho} \frac{\partial (\rho\, v_\rho)}{\partial \rho} + \frac{1}{\rho} \frac{\partial (v_\theta)}{\partial \theta} + \frac{\partial v_z}{\partial z} \tag{3.4.30}$$

$$\nabla \cdot \mathbf{v} = \frac{1}{r^2} \frac{\partial \left(r^2\, v_r\right)}{\partial r} + \frac{1}{r \sin\theta} \frac{\partial}{\partial \theta} \left(v_\theta \sin\theta\right) + \frac{1}{r \sin\theta} \frac{\partial (v_\phi)}{\partial \phi} \tag{3.4.31}$$

3.4.3 CURL

The divergence uses the del operator to measure the outflow rate of a vector field. We can also use a form of the del operator to measure the rotational properties of a vector field, called the *curl*. The curl of a vector field $\mathbf{v}(x, y, z) = v_x\hat{\mathbf{i}} + v_y\hat{\mathbf{j}} + v_z\hat{\mathbf{k}}$ is a vector function that can be represented as:

$$\nabla \times \mathbf{v} = \left(\frac{\partial v_z}{\partial y} - \frac{\partial v_y}{\partial z}\right)\hat{\mathbf{i}} + \left(\frac{\partial v_x}{\partial z} - \frac{\partial v_z}{\partial x}\right)\hat{\mathbf{j}} + \left(\frac{\partial v_y}{\partial x} - \frac{\partial v_x}{\partial y}\right)\hat{\mathbf{k}} \tag{3.4.32}$$

From a physical point of view, the curl at a point (x, y, z) is proportional to the on-axis torque to which a tiny pinwheel would be subjected, if it were centered at that point.

The cross-product operation can be written as a pseudo-determinant in Cartesian coordinates:

$$\nabla \times \mathbf{v} = \begin{vmatrix} \hat{\mathbf{i}} & \hat{\mathbf{j}} & \hat{\mathbf{k}} \\ \frac{\partial}{\partial x} & \frac{\partial}{\partial y} & \frac{\partial}{\partial z} \\ v_x & v_y & v_z \end{vmatrix} \tag{3.4.33}$$

If $\mathbf{u}$ and $\mathbf{v}$ are any vector functions and f is any scalar function, the curl also follows a product rule:

$$\nabla \times (f\mathbf{v}) = (\nabla f) \times \mathbf{v} + f\nabla \times \mathbf{v} \tag{3.4.34}$$

However, the curl of a cross product is more complex:

$$\nabla \times (\mathbf{u} \times \mathbf{v}) = (\mathbf{u}\, \nabla \cdot \mathbf{v}) - (\mathbf{v}\, \nabla \cdot \mathbf{u}) + (\mathbf{v} \cdot \nabla)\mathbf{u} - (\mathbf{u} \cdot \nabla)\mathbf{v} \tag{3.4.35}$$

The curl takes a different form in cylindrical and spherical coordinates. To compute the curl in spherical and cylindrical coordinates, choose the appropriate form of the del operator from (3.4.20), (3.4.21), and (3.4.22) and evaluate the cross product between del and the vector of interest. The results are:

The Curl in Cartesian, Cylindrical, and Spherical Coordinates

$$\nabla \times \mathbf{v} = \left(\frac{\partial v_z}{\partial y} - \frac{\partial v_y}{\partial z}\right)\hat{\mathbf{i}} + \left(\frac{\partial v_x}{\partial z} - \frac{\partial v_z}{\partial x}\right)\hat{\mathbf{j}} + \left(\frac{\partial v_y}{\partial x} - \frac{\partial v_x}{\partial y}\right)\hat{\mathbf{k}} \tag{3.4.36}$$

$$\nabla \times \mathbf{v} = \left(\frac{1}{\rho} \frac{\partial v_z}{\partial \theta} - \frac{\partial v_\theta}{\partial z}\right)\hat{\boldsymbol{\rho}} + \left(\frac{\partial v_\rho}{\partial z} - \frac{\partial v_z}{\partial \rho}\right)\hat{\boldsymbol{\theta}} + \frac{1}{\rho}\left(\frac{\partial (\rho v_\theta)}{\partial \rho} - \frac{\partial v_\rho}{\partial \theta}\right)\hat{\mathbf{z}} \tag{3.4.37}$$

$$\nabla \times \mathbf{v} = \frac{1}{r \sin\theta}\left(\frac{\partial}{\partial \theta}(v_\phi \sin\theta) - \frac{\partial v_\theta}{\partial \phi}\right)\hat{\mathbf{r}} + \frac{1}{r}\left(\frac{1}{\sin\theta}\frac{\partial v_r}{\partial \phi} - \frac{\partial}{\partial r}(rv_\phi)\right)\hat{\boldsymbol{\theta}} +$$

$$+ \frac{1}{r}\left(\frac{\partial (rv_\theta)}{\partial r} - \frac{\partial v_r}{\partial \theta}\right)\hat{\boldsymbol{\phi}} \tag{3.4.38}$$

The curl has many uses in physics. For example, as we will study in more detail in Chapter 5, the necessary and sufficient condition for a force $F(x, y, z)$ to be conservative is that its curl equals zero:

Conservative force field F (x,y,z)

$$\nabla \times \mathbf{F} = 0 \longleftrightarrow \mathbf{F}(x, y, z) \text{ is a conservative force field} \qquad (3.4.39)$$

Example 3.18: Calculating the curl using Python and Mathematica.

Find the curl of the vector function $\mathbf{v} = (y^2 z)\hat{\mathbf{i}} + (xz)\hat{\mathbf{j}} + (xy)\hat{\mathbf{k}}$ analytically, and also by using appropriate codes in Python and Mathematica.

Solution:

The curl is found by finding the pseudo-determinant:

$$\nabla \times \mathbf{v} = \begin{vmatrix} \hat{\mathbf{i}} & \hat{\mathbf{j}} & \hat{\mathbf{k}} \\ \frac{\partial}{\partial x} & \frac{\partial}{\partial y} & \frac{\partial}{\partial z} \\ y^2 z & xz & xy \end{vmatrix} = \left(\frac{\partial(xy)}{\partial y} - \frac{\partial(xz)}{\partial z} \right)\hat{\mathbf{i}} + \left(\frac{\partial(y^2 z)}{\partial z} - \frac{\partial(xy)}{\partial x} \right)\hat{\mathbf{j}}$$

$$+ \left(\frac{\partial(xz)}{\partial x} - \frac{\partial(y^2 z)}{\partial y} \right)\hat{\mathbf{k}}$$

$$\nabla \times \mathbf{v} = (x - x)\hat{\mathbf{i}} + (y^2 - y)\hat{\mathbf{j}} + (z - 2yz)\hat{\mathbf{k}} = (0)\hat{\mathbf{i}} + (y^2 - y)\hat{\mathbf{j}} + (z - 2yz)\hat{\mathbf{k}}$$

Python Code

We define the coordinate system `CoordSys3D('R')` and compute the curl using the function `curl(v)` and note that the curl is non-zero, so that this field is not conservative.

To check if a vector field is conservative, we can also use the function `is_conservative`. In this example, the function returns the value **False**.

```
from sympy.vector import CoordSys3D, curl, is_conservative

print('-'*28,'CODE OUTPUT','-'*29,'\n')

R = CoordSys3D('R')                # define reference frame R

v = R.y**2*R.z*R.i+R.x*R.z*R.j+R.x*R.y*R.k

curlv = curl(v)              # evaluate curl using curl function
print('The curl of v(x,y,z) is:\n', curlv)

test = is_conservative(v)
print('\nThe statement   The vector field is conservative   is:', test)

---------------------------------- CODE OUTPUT ---------------------------------
The curl of v(x,y,z) is:
 (R.y**2 - R.y)*R.j + (-2*R.y*R.z + R.z)*R.k

The statement   The vector field is conservative   is: False
```

Mathematica Code

The command `Curl[v,x,y,z]` is used to compute the curl, and we must specify the variables for differentiation. The output is in the form of a list, where the three elements of the list correspond to the x-, y- and z-components of the curl, respectively. The curl is computed in Cartesian coordinates and it is non-zero, which means that the vector field is not conservative.

$v = \{y^2 * z, x * z, x * y\}$;

$\mathrm{curlv} = \mathrm{Curl}[v, \{x, y, z\}]$;

Print["The curl of the vector field is: ", curlv]

The curl of the vector field is: $\{0, -y + y^2, z - 2yz\}$

3.4.4 SECOND DERIVATIVES WITH THE DEL OPERATOR

The Laplace operator or *Laplacian* ∇^2 is a *scalar* operator that can be applied to either vector or scalar fields. The Laplacian is defined as:

$$\nabla^2 = \nabla \cdot \nabla = \frac{\partial^2}{\partial x^2} + \frac{\partial^2}{\partial y^2} + \frac{\partial^2}{\partial z^2} \tag{3.4.40}$$

The Laplacian is ubiquitous throughout modern mathematical physics, appearing in Laplace's equation, Poisson's equation, the heat equation, the wave equation, and the Schrödinger equation, to name a few.

When the del operator acts on the gradient, divergence, and curl of a vector, it produces five possible second derivatives; the use of the scalar Laplacian and vector Laplacian gives two more. Here are these second derivatives created by applying the del operator:

$$\nabla \cdot (\nabla f) \quad \nabla \times (\nabla f) \quad \nabla^2 f \quad \nabla(\nabla \cdot \mathbf{v}) \quad \nabla \cdot (\nabla \times \mathbf{v}) \quad \nabla \times (\nabla \times \mathbf{v})$$

These second derivatives are of interest, principally because they are not always unique or independent of each other. As long as the functions $\mathbf{v}$ and f are well-behaved, two of these identities are always zero:

Two Important Vector Identities

$$\nabla \times (\nabla f) = 0 \tag{3.4.41}$$
$$\nabla \cdot (\nabla \times \mathbf{v}) = 0 \tag{3.4.42}$$

The first identity states that the curl of the gradient of any scalar field is always equal to zero, and the second one states that the divergence of the curl of any vector field is always equal to zero.

Two of the above second derivatives are always equal:

$$\nabla \cdot (\nabla f) = \nabla^2 f \tag{3.4.43}$$

The remaining three vector derivatives are related by the equation:

$$\nabla \times (\nabla \times \mathbf{v}) = \nabla(\nabla \cdot \mathbf{v}) - \nabla^2 \mathbf{v} \tag{3.4.44}$$

Example 3.19: Second derivatives

Consider the function $V(x, y, z) = x + yz + z^2$ in three dimensions. Calculate the gradient, the divergence of the gradient, the curl of the gradient, and the Laplacian of this function.

Solution:

The gradient is

$$\nabla V = \frac{\partial V}{\partial x}\hat{\mathbf{i}} + \frac{\partial V}{\partial y}\hat{\mathbf{j}} + \frac{\partial V}{\partial z}\hat{\mathbf{k}} = (1)\hat{\mathbf{i}} + (z)\hat{\mathbf{j}} + (y + 2z)\hat{\mathbf{k}}$$

The divergence of the gradient is the dot product of the gradient and the del operator:

$$\nabla \cdot (\nabla V) = \frac{\partial}{\partial x}\left(\frac{\partial V}{\partial x}\right) + \frac{\partial}{\partial y}\left(\frac{\partial V}{\partial y}\right) + \frac{\partial}{\partial z}\left(\frac{\partial V}{\partial z}\right) = \frac{\partial^2 V}{\partial x^2} + \frac{\partial^2 V}{\partial y^2} + \frac{\partial^2 V}{\partial z^2}$$

Calculating the second derivatives we obtain: $\quad \nabla \cdot (\nabla V) = 2$
This also represents the Laplacian of V, since

$$\frac{\partial^2}{\partial x^2} + \frac{\partial^2}{\partial y^2} + \frac{\partial^2}{\partial z^2} = \nabla \cdot \nabla = \nabla^2$$

To find the curl of the gradient, we calculate the cross product, $(\nabla \times \nabla V)$. This quantity is always zero, as seen in (3.4.41) for the curl of a gradient.

3.5   CHAPTER SUMMARY

Motion in two and three dimensions is described using vectors. In this book, we use the following conventions for a vector, its magnitude, and a unit vector:

$$\mathbf{A} = A_x\hat{\mathbf{i}} + A_y\hat{\mathbf{j}} + A_z\hat{\mathbf{k}} \qquad A = |\mathbf{A}| = \sqrt{A_x^2 + A_y^2 + A_z^2}$$

$$\hat{\mathbf{A}} = \frac{\mathbf{A}}{|\mathbf{A}|} = \frac{\mathbf{A}}{\sqrt{contour\, A_x^2 curves + A_y^2 + A_z^2}}$$

The dot product between two vectors $\mathbf{A}$ and $\mathbf{B}$ and its derivative are found from the following relationships:

$$\mathbf{A} \cdot \mathbf{B} = A_xB_x + A_yB_y + A_zB_z = AB\cos\theta \qquad \frac{d}{dt}(\mathbf{A} \cdot \mathbf{B}) = \frac{d\mathbf{A}}{dt}\mathbf{B} + \mathbf{A}\frac{d\mathbf{B}}{dt}$$

The cross product between two vectors $\mathbf{A}$ and $\mathbf{B}$ and its derivative are found from:

$$\mathbf{A} \times \mathbf{B} = |\mathbf{A}||\mathbf{B}|\sin\theta\,\hat{\mathbf{n}}$$

$$\mathbf{A} \times \mathbf{B} = \begin{vmatrix} \hat{\mathbf{i}} & \hat{\mathbf{j}} & \hat{\mathbf{k}} \\ A_1 & A_2 & A_3 \\ B_1 & B_2 & B_3 \end{vmatrix} = \begin{vmatrix} A_2 & A_3 \\ B_2 & B_3 \end{vmatrix}\hat{\mathbf{i}} - \begin{vmatrix} A_1 & A_3 \\ B_1 & B_3 \end{vmatrix}\hat{\mathbf{j}} + \begin{vmatrix} A_1 & A_2 \\ B_1 & B_2 \end{vmatrix}\hat{\mathbf{k}}$$

$$\mathbf{A} \times \mathbf{B} = -\mathbf{B} \times \mathbf{A} \qquad \frac{d}{dt}(\mathbf{A} \times \mathbf{B}) = \frac{d\mathbf{A}}{dt} \times \mathbf{B} + \mathbf{A} \times \frac{d\mathbf{B}}{dt}$$

There are several relationships that can simplify expressions with multiple dot or cross products. They are:

The vector triple product: $\quad \mathbf{A} \times (\mathbf{B} \times \mathbf{C}) = \mathbf{B}(\mathbf{A} \cdot \mathbf{C}) - \mathbf{C}(\mathbf{A} \cdot \mathbf{B})$

The scalar triple product: $\quad \mathbf{A} \cdot (\mathbf{B} \times \mathbf{C}) = \mathbf{B} \cdot (\mathbf{C} \times \mathbf{A}) = \mathbf{C} \cdot (\mathbf{A} \times \mathbf{B}) = V$

In addition, we often need to perform derivatives of vectors with respect to spatial coordinates. Those derivatives are the gradient, divergence, and curl, written here in Cartesian coordinates:

$$\nabla f(x, y, z) = \frac{\partial f}{\partial x}\hat{\mathbf{i}} + \frac{\partial f}{\partial y}\hat{\mathbf{j}} + \frac{\partial f}{\partial z}\hat{\mathbf{k}}$$

$$\nabla \cdot \mathbf{v} = \frac{\partial v_x}{\partial x} + \frac{\partial v_y}{\partial y} + \frac{\partial v_z}{\partial z}$$

$$\nabla \times v = \begin{vmatrix} \hat{\mathbf{i}} & \hat{\mathbf{j}} & \hat{\mathbf{k}} \\ \frac{\partial}{\partial x} & \frac{\partial}{\partial y} & \frac{\partial}{\partial z} \\ v_x & v_y & v_z \end{vmatrix} = \left(\frac{\partial v_z}{\partial y} - \frac{\partial v_y}{\partial z}\right)\hat{\mathbf{i}} + \left(\frac{\partial v_x}{\partial z} - \frac{\partial v_z}{\partial x}\right)\hat{\mathbf{j}} + \left(\frac{\partial v_y}{\partial x} - \frac{\partial v_x}{\partial y}\right)\hat{\mathbf{k}}$$

The gradient gives the direction of most rapid change in a scalar field. The divergence measures the magnitude of a vector field's source or sink at a given point. The curl at a point is proportional to the on-axis torque to which a tiny pinwheel is subjected.

The directional derivative df/dn in the direction $\hat{\mathbf{n}} = n_x\hat{\mathbf{i}} + n_y\hat{\mathbf{j}} + n_z\hat{\mathbf{k}}$ is:

$$\frac{df}{dn} = \hat{\mathbf{n}} \cdot \nabla f = n_x\frac{\partial f}{\partial x} + n_y\frac{\partial f}{\partial y} + n_z\frac{\partial f}{\partial z}$$

A force $\mathbf{F}$ will be *conservative* if and only if its curl is zero: $\quad \nabla \times \mathbf{F} = 0$

Second derivatives of vector and scalar fields are also important in mechanics. The Laplacian ∇^2 is a scalar operator that can be applied to either vector or scalar fields:

$$\nabla^2 = \frac{\partial^2}{\partial x^2} + \frac{\partial^2}{\partial y^2} + \frac{\partial^2}{\partial z^2}$$

$$\nabla \times (\nabla f) = 0$$

$$\nabla \cdot (\nabla \times \mathbf{v}) = 0$$

Motion in two and three dimensions can sometimes be best described by non-Cartesian coordinates. Euler's important identities are:

$$\cos\theta = \frac{e^{i\theta} + e^{-i\theta}}{2} \qquad\qquad \sin\theta = \frac{e^{i\theta} - e^{-i\theta}}{2i}$$

$$e^{i\theta} = \cos\theta + i\sin\theta \qquad\qquad e^{-i\theta} = \cos\theta - i\sin\theta$$

Polar coordinates are related to Cartesian coordinates using:

$$x = r\cos\theta \qquad y = r\sin\theta$$

$$r = \sqrt{x^2 + y^2} \qquad \theta = \tan^{-1}(y/x)$$

The unit vectors in polar coordinates are:

$$\hat{\mathbf{r}} = \cos\theta\,\hat{\mathbf{i}} + \sin\theta\,\hat{\mathbf{j}} \qquad\qquad \hat{\boldsymbol{\theta}} = -\sin\theta\,\hat{\mathbf{i}} + \cos\theta\,\hat{\mathbf{j}} \qquad \mathbf{r} = r\hat{\mathbf{r}}$$

The velocity vector in polar coordinates is:

$$v = \frac{d\mathbf{r}}{dt} = \frac{dr}{dt}\hat{r} + r\frac{d\theta}{dt}\hat{\theta}$$

The cylindrical coordinates ρ, θ, z are related to the Cartesian coordinate by:

$$x = \rho \cos\theta \qquad y = \rho \sin\theta \qquad z = z$$

Unit vectors $\hat{\rho}$, $\hat{\theta}$, $\hat{z}$ in cylindrical coordinates *change direction* with the moving particle, and are related to the Cartesian unit vectors by:

$$\begin{bmatrix} \hat{\rho} \\ \hat{\theta} \\ \hat{z} \end{bmatrix} = \begin{bmatrix} \cos\theta & \sin\theta & 0 \\ -\sin\theta & \cos\theta & 0 \\ 0 & 0 & 1 \end{bmatrix} \begin{bmatrix} \hat{i} \\ \hat{j} \\ \hat{k} \end{bmatrix}$$

Spherical coordinates r, θ, ϕ are related to Cartesian coordinates by:

$$x = r\cos\phi\sin\theta \qquad y = r\sin\phi\sin\theta \qquad z = r\cos\theta$$

with unit vectors:

$$\begin{bmatrix} \hat{r} \\ \hat{\theta} \\ \hat{\phi} \end{bmatrix} = \begin{bmatrix} \sin\theta\cos\phi & \sin\theta\sin\phi & \cos\theta \\ \cos\theta\cos\phi & \cos\theta\sin\phi & -\sin\theta \\ -\sin\phi & \cos\phi & 0 \end{bmatrix} \begin{bmatrix} \hat{i} \\ \hat{j} \\ \hat{k} \end{bmatrix}$$

3.6 END-OF-CHAPTER PROBLEMS

The symbol ⌨ indicates a problem which requires some computer assistance, in the form of graphics, or numerical computation, or symbolic evaluation.

Section 3.1: Position, Velocity, and Acceleration in Cartesian Coordinates

1. ⌨ An object with mass $m = 1$ kg has a position vector on the xy-plane given by $\mathbf{r} = (3 + 5\cos t)\hat{i} + (2 + 4\sin t)\hat{j}$, where t is time and all quantities are measured in SI units.

   a. Show analytically that the motion on the xy-plane is an ellipse, and find the equation of the ellipse in Cartesian coordinates (x, y).

   b. Plot the ellipse using Mathematica and/or Python.

   c. On the same plot, show the unit vector tangent to the elliptical path at time $t = \pi/2$ as an arrow.

   d. Repeat (c) for a unit vector perpendicular to the elliptical path at $t = \pi/2$.

2. A particle is moving in space along the curve $(a\cos(\omega t),\ b\sin(\omega t),\ ct^2)$ where t is the time, and a, b, c are constants. All physical quantities are in SI units.

   a. Find expressions for the magnitude of the velocity and acceleration at any point in space.

   b. Find an expression for the unit tangent vector at any point along the motion.

3. The acceleration of a mass moving in two dimensions is given by $\mathbf{a} = (3\hat{\mathbf{i}} - 4\hat{\mathbf{j}}) \, \text{m/s}^2$. At time $t = 0$ the body is located at the origin and is moving with a speed of 1 m/s.

   a. Find the velocity vector $\mathbf{v}(t)$.

   b. Find the position vector $\mathbf{r}(t)$.

4. Find the tangent vector at any point of the curve $(x, \, y, \, z) = (t^3, \, 3t^2 - 1, \, 2t - 5)$, where t represents time, and physical quantities are in SI units.

5. A particle moves along the curve $(x, \, y, \, z) = (t^2, \, t^3, \, \sin t)$. Find the tangential and normal component of the velocity and acceleration vectors at time $t = 1$ s.

6. ⌨ Consider the 3D curve $(x, \, y, \, z) = (t^3, \, 3t^2 - 1, \, 2t - 5)$ where t is time, and all physical quantities are in SI units.

   a. Plot the 3D curve and the tangent vector at $t = 2$.

   b. Plot the angle between the position and velocity vectors as a function of time.

7. ⌨ Create three-dimensional plots of the following objects, using parametric equations. In each case, label the axes and the plots.

   a. A sphere of radius $R = 1$.

   b. A cone of radius $R = 1$ and height $H = 1$.

   c. A cylinder of radius $R = 1$ and height $H = 1$.

   d. An ellipsoid with semi-major axes $a = 1$, $b = 0.6$ and $c = 0.3$.

Section 3.2: Vector Products

8. If the force on a particle is given by $\mathbf{F} = f(r)\hat{\mathbf{r}}$ where r is magnitude of the position vector, $\hat{\mathbf{r}}$ is the unit vector along the position vector $\mathbf{r}$, and $f(r)$ is a scalar function of r, then $\mathbf{F}$ is known as a *central force*. Show that in such cases the vector $\mathbf{r} \times \mathbf{v}$ does not change with time. What is the physical meaning of this result?

9. Identify the cross product and/or dot product equations for the following quantities. In each case identify all the symbols involved in clear sentences.

   a. Mechanical Work

   b. Magnetic Flux

   c. Electric Flux

   d. Lorentz Force

   e. Torque

   f. Angular Momentum

10. ⌨ Consider the vectors $\mathbf{A} = (1, 1, 2)$ and $\mathbf{B} = (1, 1, 1)$.

    a. Plot the vectors $\mathbf{A}$, $\mathbf{B}$ and their cross product vector $\mathbf{A} \times \mathbf{B}$ on a 3D Cartesian coordinate system.

    b. Plot the plane surface function $z = 1 - x - y$ and a vector perpendicular to this plane on a 3D Cartesian coordinate system.

11. Find the area of the triangle defined by the points A $(1, 1, 1)$, B $(2, 3, 0)$, and C $(1, -1, 0)$.

12. If the magnitude of a vector $\mathbf{A}$ is constant, then show that $\mathbf{A}$ and $d\mathbf{A}/dt$ are perpendicular vectors.

13. ⌨ The force on a particle of mass m is given by $\mathbf{F} = 3\mathbf{r}$, where $\mathbf{r}$ is the position vector (x, y, z). Show analytically and using symbolic algebra that the angular momentum $\mathbf{L} = m\,(\mathbf{r} \times \mathbf{v})$ of the body is conserved, i.e., that the derivative $d\mathbf{L}/dt = 0$. Here $\mathbf{v}$ is the velocity of the particle.

14. ⌨ By using the derivative properties of the dot and cross product, show analytically and using Mathematica and/or Sympy that for any vector $\mathbf{v}$ the following identity is true:

$$\frac{d}{dt}\left(\mathbf{v} \cdot \left[\frac{d\mathbf{v}}{dt} \times \frac{d^2\mathbf{v}}{dt^2}\right]\right) = \mathbf{v} \cdot \left(\frac{d\mathbf{v}}{dt} \times \frac{d^3\mathbf{v}}{dt^3}\right)$$

15. If the position of a particle on the xy-plane is given by $\mathbf{r} = \cos(\omega t)\hat{\mathbf{i}} + \sin(\omega t)\hat{\mathbf{j}}$, show that the vector $\mathbf{r} \times \mathbf{v}$ is a constant vector. What is the physical meaning of this result?

16. ⌨ Use the symbolic capabilities of Mathematica and SymPy to demonstrate the following identities for any vectors $\mathbf{A}$, $\mathbf{B}$, $\mathbf{C}$:

$$\mathbf{A} \times (\mathbf{B} \times \mathbf{C}) + \mathbf{B} \times (\mathbf{C} \times \mathbf{A}) + \mathbf{C} \times (\mathbf{A} \times \mathbf{B}) = 0$$

$$(\mathbf{A} \times \mathbf{B}) \cdot [(\mathbf{B} \times \mathbf{C}) \times (\mathbf{C} \times \mathbf{A})] = (\mathbf{A} \cdot [\mathbf{B} \times \mathbf{C}])^2$$

17. ⌨ An object with mass $m = 2$ has a position vector given by $\mathbf{r} = \left(3t^2\,\mathbf{i} - 5\sin t\,\mathbf{j} + 5\cos t\,\mathbf{k}\right)$ where all quantities are measured in SI units and t represents the time. Use the symbolic capabilities of Mathematica and/or SymPy to evaluate the following:

   a. Find the magnitude of the angular momentum vector $\mathbf{L} = \mathbf{r} \times \mathbf{p}$ at $t = \pi$, where $\mathbf{p}$ is the linear momentum of the body.

   b. Find the angle of the position and velocity vectors at time $t = \pi$.

   c. Find the torque $\tau = \mathbf{r} \times \mathbf{F}$ on the body at $t = \pi$, where $\mathbf{F}$ is the force acting on the body.

18. The equation of motion of a particle P with a mass m is given by

$$m\frac{d^2\mathbf{r}}{dt^2} = f(r)\,\hat{\mathbf{r}}$$

   where $\mathbf{r}$ is the position vector of P measured from an origin O, $\hat{\mathbf{r}}$ is a unit vector in the direction $\mathbf{r}$, and $f(r)$ is a function of the distance of P from the origin O.

   a. Show that

$$\mathbf{r} \times \frac{d\mathbf{r}}{dt} = \mathbf{c}$$

   where $\mathbf{c}$ is a constant vector.

   b. Interpret physically the cases $f(r) < 0$ and $f(r) > 0$.

   c. What is the physical meaning of the result in (a)? Note that the angular momentum vector of the particle is defined as the cross product $\mathbf{L} = m\,(\mathbf{r} \times \mathbf{v})$, where $\mathbf{v}$ is the velocity of the particle.

   d. Describe how the results obtained in (a)-(c) relate to the motion of the planets in our solar system.

19. In this chapter the dot and cross products were evaluated using SymPy in Python. An alternative method of evaluating dot and cross products in NumPy is using functions of the form `dot(A,B)` and `cross(A,B)`. Write a Python code to implement the dot and cross product of two vectors in NumPy using functions of this type.

Section 3.3: Position, Velocity, and Acceleration in Non-Cartesian Coordinates

20. A mass m hangs from a string of length L, so that it moves on a horizontal circle. The string makes an angle Θ with the vertical line. Calculate the period T of the rotational motion.

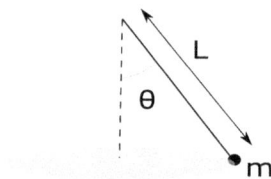

Figure 3.13: A hanging mass moves on a horizontal circle, in Problem 20.

21. Show that the acceleration vector in polar coordinates is given by:

$$\mathbf{a} = \frac{d\boldsymbol{v}}{dt} = \left[\frac{d^2 r}{dt^2} - r\left(\frac{d\theta}{dt}\right)^2\right]\hat{\mathbf{r}} + \left[r\frac{d^2\theta}{dt^2} + 2\frac{dr}{dt}\frac{d\theta}{dt}\right]\hat{\boldsymbol{\theta}}$$

One possible method is by taking the time derivative of the velocity vector

$$\mathbf{v} = \frac{d\mathbf{r}}{dt} = \frac{dr}{dt}\hat{\mathbf{r}} + r\frac{d\theta}{dt}\hat{\boldsymbol{\theta}}$$

and then using the derivatives of the unit vectors $\frac{d}{dt}\hat{\boldsymbol{\theta}}$ and $\frac{d}{dt}\hat{\mathbf{r}}$ derived in this chapter.

22. A point mass m is placed at the top of the frictionless surface of a sphere of radius R, and slides down the surface of the sphere. At what point does the mass leave the spherical surface?

23. 🖥 Consider a mass m moving in the xy-plane under the influence of a force whose polar components are given by the expression $F_\theta = mr\dot{\theta}$ and $F_r = 0$, where (r,θ) are polar coordinates. Show that

$$r = C\dot{\theta}^{-1}$$

$$\dot{r} = \sqrt{A\ln r + B}$$

where A, B, C are constants. You may want to use a CAS for some of the steps in this problem.

24. A wheel of radius R is rolling without slipping on a flat frictionless surface.

a. Show that the path of a point on the circumference of the wheel is a cycloid curve:

$$x = R\theta + R\sin\theta \qquad y = R + R\cos\theta$$

where θ is the angle of rotation of the wheel around its center of mass.

b. Find the velocity and acceleration of this point on the circumference of the rolling wheel.

c. The curvature k at a point of a curve is defined as

$$k = \frac{|\mathbf{a} \times \mathbf{v}|}{|\mathbf{v}|^3}$$

where $\mathbf{a}$ and $\mathbf{v}$ are the acceleration and velocity vectors of the wheel. Find the curvature k at a point located at the top of the cycloid.

25. Find the curvature at the top of the parabolic path of a projectile by assuming no air resistance. The curvature k at a point of a curve is defined as

$$k = \frac{|\mathbf{a} \times \mathbf{v}|}{|\mathbf{v}|^3}$$

where $\mathbf{a}, \mathbf{v}$ are the acceleration and velocity vectors of the object.

26. Find expressions for the velocity and the acceleration vectors in spherical coordinates.

27. ⌨ Use SymPy and/or Mathematica to show that for any vectors $\mathbf{A}, \mathbf{B}, \mathbf{C}, \mathbf{D}$ the following identities are true:

a. $\mathbf{A} \times [\mathbf{B} \times (\mathbf{C} \times \mathbf{D})] = \mathbf{B}\,[\mathbf{A} \cdot (\mathbf{C} \times \mathbf{D})] - (\mathbf{A} \cdot \mathbf{B})\,(\mathbf{C} \times \mathbf{D})$

b. $(\mathbf{A} \times \mathbf{B}) \cdot (\mathbf{C} \times \mathbf{D}) = (\mathbf{A} \cdot \mathbf{C})\,(\mathbf{B} \cdot \mathbf{D}) - (\mathbf{A} \cdot \mathbf{D})\,(\mathbf{B} \cdot \mathbf{C})$

28. ⌨ Show symbolically and analytically that for any two vectors $\mathbf{A}, \mathbf{B}$:

$$\frac{d}{dt}\,(\mathbf{A} \cdot \mathbf{B}) = \left(\frac{d\mathbf{A}}{dt} \cdot \mathbf{B}\right) + \left(\mathbf{A} \cdot \frac{d\mathbf{B}}{dt}\right)$$

where $\mathbf{A}$ and $\mathbf{B}$ are time-dependent vectors.

29. ⌨ Starting from the relationships between the spherical and Cartesian coordinates $x = r\sin\theta\cos\phi$, $y = r\sin\theta\sin\phi$ and $z = r\cos\theta$, show that the arc length in spherical coordinates is given by:

$$ds^2 = dr^2 + r^2\,d\theta^2 + r^2\,\sin^2\theta\,d\phi^2 \qquad (3.6.1)$$

Do this problem both analytically and with the use of a symbolic algebra system.

30. ⌨ Use Mathematica and/or SymPy to solve the following system of equations for the unit vectors $\hat{r}$, $\hat{\theta}$ in polar coordinates:

$$\hat{r} = \cos\theta\,\hat{\mathbf{i}} + \sin\theta\,\hat{\mathbf{j}} \qquad (3.6.2)$$

$$\hat{\theta} = -\sin\theta\,\hat{\mathbf{i}} + \cos\theta\,\hat{\mathbf{j}} \qquad (3.6.3)$$

to obtain the Cartesian unit vectors $\hat{\mathbf{i}}$, $\hat{\mathbf{j}}$ in terms of $\hat{r}$, $\hat{\theta}$.

31. ⌨ Use Mathematica and/or SymPy to differentiate the following expressions for the polar unit vectors $\hat{r}$ and $\hat{\theta}$ with respect to time t,

$$\hat{r} = \cos\theta\,\hat{\mathbf{i}} + \sin\theta\,\hat{\mathbf{j}}$$
$$\hat{\theta} = -\sin\theta\,\hat{\mathbf{i}} + \cos\theta\,\hat{\mathbf{j}}$$

and show that

$$\frac{d\hat{r}}{dt} = \frac{d\theta}{dt}\,\hat{\theta} \qquad\qquad \frac{d\hat{\theta}}{dt} = -\frac{d\theta}{dt}\,\hat{r} \qquad\qquad (3.6.4)$$

Section 3.4: Gradient, Divergence, and Curl

32. ⌨ A particle is located at $\mathbf{r} = -\hat{\mathbf{i}} + 3\hat{\mathbf{j}} - 4\hat{\mathbf{k}}$; find the values of r, θ, ϕ for this vector and the spherical coordinates unit vectors $\hat{r}, \hat{\theta}, \hat{\phi}$ at the position $\mathbf{r}$. In addition, find its components A_r, A_θ and A_ϕ in spherical coordinates, such that $\mathbf{r} = A_r\hat{r} + A_\theta\hat{\theta} + A_\phi\hat{\phi}$.

33. Cylindrical and spherical coordinates are examples of *orthogonal curvilinear systems*. In such coordinate systems, we can obtain expressions for the gradient, divergence and curl by using the following general procedure. Let $\mathbf{r}$ be some point in space and q_i represent the coordinate system, we can define three vectors $\mathbf{e}_i$ which are orthogonal to each other by evaluating the partial derivatives:

$$\mathbf{e}_i = \frac{\partial \mathbf{r}}{\partial q_i} \qquad \mathbf{e}_i \cdot \mathbf{e}_j = 0 \quad (i = 1, 2, 3 \quad i \neq j)$$

These $\mathbf{e}_i$ vectors can be normalized by dividing by their length, and the normalized basis vectors are then:

$$\hat{\mathbf{e}}_i = \frac{\mathbf{e}_i}{h_i} = \frac{\mathbf{e}_i}{|\mathbf{e}_i|}$$

The lengths h_i of the basis vectors $\hat{\mathbf{e}}_i$ are very useful functions known as *scale factors of the coordinates*. In the case of Cartesian coordinates, $q_i = x, y, z$ and

$$\hat{\mathbf{e}}_i = \left(\frac{\partial \mathbf{r}}{\partial x}, \frac{\partial \mathbf{r}}{\partial y}, \frac{\partial \mathbf{r}}{\partial z}\right) = \left(\hat{\mathbf{i}}, \hat{\mathbf{j}}, \hat{\mathbf{k}}\right)$$

and it is easy to verify that $h_i = 1$. The infinitesimal length in this orthogonal coordinate system is found from:

$$ds = \sqrt{d\mathbf{r} \cdot d\mathbf{r}} = \sqrt{h_1^2\, dq_1^2 + h_2^2\, dq_2^2 + h_3^2\, dq_3^2}$$

The corresponding infinitesimal volume elements are calculated from:

$$dV = h_1 h_2 h_3\, dq_1\, dq_2\, dq_3$$

    a. Find expressions for the unit vectors $\hat{\mathbf{e}}_i$ in cylindrical coordinates, and determine the corresponding scale factors.

    b. Find expressions for the infinitesimal arc length ds^2 and infinitesimal volume dV in cylindrical coordinates.

34. Repeat Problem 33 for spherical coordinates, i.e., find the unit vectors $\hat{\mathbf{e}}_i$, the square of the element of arc length ds^2, the corresponding scale factors and the infinitesimal volume dV for spherical coordinates.

35. The gradient of a scalar field f, the divergence and curl of a vector field $\mathbf{F} = (F_1, F_2, F_3)$, and the Laplacian of a scalar field f can be found using the following general expressions:

$$\nabla f = \frac{\hat{e}_1}{h_1}\frac{\partial f}{\partial q_1} + \frac{\hat{e}_2}{h_2}\frac{\partial f}{\partial q_2} + \frac{\hat{e}_3}{h_3}\frac{\partial f}{\partial q_3}$$

$$\nabla \cdot \mathbf{F} = \frac{1}{h_1 h_2 h_3}\left[\frac{\partial}{\partial q_1}(F_1 h_2 h_3) + \frac{\partial}{\partial q_2}(F_2 h_3 h_1) + \frac{\partial}{\partial q_3}(F_3 h_1 h_2)\right]$$

$$\nabla \times \mathbf{F} = \frac{\hat{e}_1}{h_2 h_3}\left[\frac{\partial}{\partial q_2}(h_3 F_3) - \frac{\partial}{\partial q_3}(h_2 F_2)\right] + \frac{\hat{e}_2}{h_3 h_1}\left[\frac{\partial}{\partial q_3}(h_1 F_1) - \frac{\partial}{\partial q_1}(h_3 F_3)\right]$$

$$+ \frac{\hat{e}_3}{h_1 h_2}\left[\frac{\partial}{\partial q_1}(h_2 F_2) - \frac{\partial}{\partial q_2}(h_1 F_1)\right]$$

$$= \frac{1}{h_1 h_2 h_3}\begin{vmatrix} h_1 \hat{e}_1 & h_2 \hat{e}_2 & h_3 \hat{e}_3 \\ \frac{\partial}{\partial q_1} & \frac{\partial}{\partial q_2} & \frac{\partial}{\partial q_3} \\ h_1 F_1 & h_2 F_2 & h_3 F_3 \end{vmatrix}$$

$$\nabla^2 f = \frac{1}{h_1 h_2 h_3}\left[\frac{\partial}{\partial q_1}\left(\frac{h_2 h_3}{h_1}\frac{\partial f}{\partial q_1}\right) + \frac{\partial}{\partial q_2}\left(\frac{h_3 h_1}{h_2}\frac{\partial f}{\partial q_2}\right) + \frac{\partial}{\partial q_3}\left(\frac{h_1 h_2}{h_3}\frac{\partial f}{\partial q_3}\right)\right]$$

where the h_i are the *scale factors* of the basis vectors, $\hat{e}_i$, which were defined in Problem 33.

a. Use the expressions above to obtain the gradient, divergence, and curl in cylindrical coordinates.

b. Obtain the divergence for spherical coordinates.

36. 🖥 Consider the function $V(x, y, z) = x^4 + y^5 + z^6$ in three dimensions. Calculate the following quantities both analytically and using a CAS, and indicate which of these is a vector and which is a scalar.

a. $\nabla\left(\nabla^2 V\right)$

b. $(\nabla V \times \nabla V) \cdot (\nabla V)$

c. $\nabla \times \left(\nabla \times \left(V\hat{\mathbf{i}}\right)\right)$ where $\hat{\mathbf{i}}$ is the unit vector in the x-direction.

d. $\nabla^2(\nabla V)$

e. $\nabla(\nabla V \cdot \nabla V)$

37. 🖥 Given the function $f = y^2 + x^3$ find the following quantities both analytically and using Mathematica and/or SymPy:

a. Curl of the gradient of f

b. Laplacian of f

c. Divergence of the gradient of f

38. 🖥 Consider the function $V(x, y) = x^2 + y^2$ in two dimensions.

a. Graph the equipotential lines $V(x, y) = 1$, $V(x, y) = 4$, and $V(x, y) = 9$ on the $xy-$plane.

    b. Graph the gradient of $V(x, y)$ on the xy-plane.

    c. If $V(x, y)$ is the electric potential on the xy-plane, what does this gradient represent physically?

39. 🖳 Graph the vector fields and contours of constant value for the functions:

    a. $V(x, y) = y\,\hat{\mathbf{i}} + x\,\hat{\mathbf{j}}$

    b. $V(x, y) = -x\,\hat{\mathbf{i}} - y\,\hat{\mathbf{j}}$

    c. $V(x, y) = y\,\hat{\mathbf{i}} + x\,\hat{\mathbf{j}} + z\,\hat{\mathbf{k}}$

40. Prove that $\nabla^2\left(\frac{1}{r}\right) = 0$ where r is the magnitude of the position vector $\mathbf{r}$.

41. 🖳 Prove analytically and using Mathematica and/or Python that:

$$\nabla \cdot (\phi\mathbf{A}) = (\nabla\phi) \cdot \mathbf{A} + \phi\,(\nabla \cdot \mathbf{A})$$

where $\phi(x, y, z)$ is a scalar function, and $\mathbf{A}(x, y, z)$ is a vector function.

42. Prove $\nabla \cdot \left(\mathbf{r}/r^3\right) = 0$, where r is the magnitude of the position vector $\mathbf{r}$.

43. If $\mathbf{v} = \boldsymbol{\omega} \times \mathbf{r}$, prove $\boldsymbol{\omega} = \frac{1}{2}$ curl $\mathbf{v}$ where $\boldsymbol{\omega}$ is a constant vector, and $\mathbf{r}$ is the position vector.

44. 🖳 Find the ∇f if (a) $f = \ln|\mathbf{r}|$, (b) $f = \frac{1}{r}$. Here r is the magnitude of the position vector $\mathbf{r}$. Solve this problem both analytically and with a symbolic algebra system.

45. Show that $\nabla r^n = nr^{n-2}\mathbf{r}$ where r is the magnitude of the position vector $\mathbf{r}$.

46. Maxwell's famous equations concern the magnetic field vector $\mathbf{B}$ and the electric field vector $\mathbf{E}$. If these vectors satisfy the equations:

$$\nabla \cdot \mathbf{E} = 0 \qquad \nabla \cdot \mathbf{B}=0 \qquad \nabla \times \mathbf{E} = -\frac{\partial \mathbf{B}}{dt} \qquad \nabla \times \mathbf{B} = \frac{\partial \mathbf{E}}{dt}$$

then show that $\mathbf{E}$ and $\mathbf{B}$ satisfy the generalized wave equation:

$$\nabla^2\mathbf{E} = \frac{\partial^2\mathbf{E}}{\partial t^2} \qquad\qquad \nabla^2\mathbf{B} = \frac{\partial^2\mathbf{B}}{\partial t^2}$$

47. 🖳 Demonstrate using a symbolic algebra system that for any vector field $\mathbf{A}(x, y, z)$, the following is true:

$$\nabla \times (\nabla \times \mathbf{A}) = -\nabla^2\mathbf{A} + \nabla\,(\nabla \cdot \mathbf{A})$$

4 Momentum, Angular Momentum, and Multiparticle Systems

In this chapter, we focus on two important physical quantities: linear momentum and angular momentum. Under certain conditions, as explained in this chapter, momentum and angular momentum are conserved quantities, meaning that their values are constant during a physical process. The conservation of momentum or angular momentum can then be used to solve problems by comparing the state of a system before a physical process to the state of that same system after this process. In this chapter, we will also study multiparticle systems, continuous mass distributions (i.e., systems that are not point particles), and the center of mass. We develop relationships between the total momentum of a system of particles to the momentum of the system's center of mass. We also develop a relationship between the angular momentum of a system of particles and the angular momentum of the system's center of mass. We show that these relationships also hold for continuous mass distributions. Finally, we discuss numerical integration techniques which will be useful when finding a system's center of mass.

4.1 CONSERVATION OF MOMENTUM AND NEWTON'S THIRD LAW

Recall that Newton's third law states that if Particle 1 exerts a force $\mathbf{F}_{21}$ on Particle 2, then Particle 2 exerts a force $\mathbf{F}_{12}$ on Particle 1 such that:

$$\mathbf{F}_{12} = -\mathbf{F}_{21} \tag{4.1.1}$$

Newton's third law will lead us to an interesting result. Consider two particles interacting via a force, which we will refer to as an "internal force." For example, if our two objects are the Earth and Moon, their gravitational interaction would be the internal force. Furthermore, assume that each particle is experiencing an external force due to other bodies outside of the system. In the case of the Earth and the Moon, that could be the Sun's gravitational force, a body outside the Earth-Moon system. The net force $\mathbf{F}_1$ on Particle 1 is then:

$$\mathbf{F}_1 = \mathbf{F}_{12} + \mathbf{F}_1^{\text{ext}} \tag{4.1.2}$$

where $\mathbf{F}_{12}$ is the force on Particle 1 exerted by Particle 2 (the "internal force"), and $\mathbf{F}_1^{\text{ext}}$ is the net external force acting on Particle 1. Similarly, Particle 2 experiences a net force:

$$\mathbf{F}_2 = \mathbf{F}_{21} + \mathbf{F}_2^{\text{ext}} \tag{4.1.3}$$

Let $\mathbf{p}_1$ and $\mathbf{p}_2$ represent the linear momenta of the two particles, then Newton's second law states that

$$\dot{\mathbf{p}}_1 = \mathbf{F}_1 = \mathbf{F}_{12} + \mathbf{F}_1^{\text{ext}} \tag{4.1.4}$$

and

$$\dot{\mathbf{p}}_2 = \mathbf{F}_2 = \mathbf{F}_{21} + \mathbf{F}_2^{\text{ext}} \tag{4.1.5}$$

Now, we define the total momentum vector as $\mathbf{P} = \mathbf{p}_1 + \mathbf{p}_2$, to be the sum of the momenta of each particle. Then,

$$\dot{\mathbf{P}} = \dot{\mathbf{p}}_1 + \dot{\mathbf{p}}_2 = \mathbf{F}_1^{\text{ext}} + \mathbf{F}_2^{\text{ext}} = \mathbf{F}^{\text{ext}} \tag{4.1.6}$$

Hence, we see that if there are no total net external forces $\mathbf{F}^{\text{ext}} = \mathbf{F}_1^{\text{ext}} + \mathbf{F}_2^{\text{ext}}$ then $\dot{\mathbf{P}} = 0$, or to put it another way:

The Law of Conservation of Linear Momentum

If $\mathbf{F}^{\text{ext}} = 0$, then $\mathbf{P} = \text{constant}$.

The total momentum of a system is conserved if no external forces act on that system. This is an important law in physics, one that we will use to solve problems in cases where the external forces add to zero.

Example 4.1: Inelastic collision between two bodies.

Consider two objects with masses m_1 and m_2 moving with velocities $\mathbf{v}_1$ and $\mathbf{v}_2$, respectively. The two masses collide and stick together, moving away from the collision as one object with a velocity $\mathbf{v}$. Ignoring external forces occurring during the collision, find the velocity of the objects immediately after the collision.

Solution:

This type of collision is called a *perfectly inelastic collision* because the two bodies are stuck together after the collision. Because we can ignore the external forces during the collision, we can use conservation of momentum to solve this problem. The total momentum before the collision is:

$$\mathbf{P}_{\text{initial}} = m_1\mathbf{v}_1 + m_2\mathbf{v}_2$$

and the total momentum after the collision is:

$$\mathbf{P}_{\text{final}} = (m_1 + m_2)\,\mathbf{v}$$

Note that because the two objects stick together after the collision, both masses have the same velocity, hence the right-hand side of $\mathbf{P}_{\text{final}}$ above. Conservation of momentum tells us that $\mathbf{P}_{\text{initial}} = \mathbf{P}_{\text{final}}$ and therefore:

$$\mathbf{v} = \frac{m_1\mathbf{v}_1 + m_2\mathbf{v}_2}{m_1 + m_2}$$

Notice that the final velocity is the weighted average of the initial velocities, where the masses of each object serve as the weights for the average. While this problem may seem to be simple, it is an important application of the conservation of momentum; it is used to solve problems in nuclear physics, astrophysics, and accident reconstruction–to name a few.

Example 4.2: Elastic collision between two bodies.

A particle of mass $m_1 = 0.10$ kg moving with a velocity of $\mathbf{v}_0 = 10.0$ m/s $\hat{\mathbf{i}}$ collides with another particle that is at rest and has a mass $m_2 = 0.20$ kg. After the collision, each particle moves as shown in Figure 4.1, where $\theta = \pi/6$ rad. The collision is *elastic*, meaning that the kinetic energy of each particle is conserved. Find the speed v_1 of mass m_1, and the speed v_2 and angle ϕ for mass m_2 after the collision.

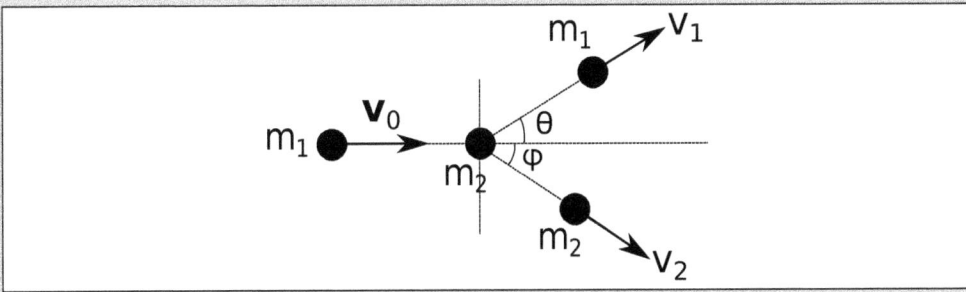

Figure 4.1: Before and after the collision in Example 4.2.

Solution:

Recall that in an elastic collision both the total momentum mv *and* total kinetic energy $1/2mv^2$ are conserved. We will discuss kinetic energy more in depth in the next chapter.

Only mass m_1 is moving before the collision, and the problem states that the velocity is in the x-direction. Therefore,

$$\mathbf{p}_{\text{initial}} = m_1 v_0 \hat{\mathbf{i}}$$

The momentum after the collision involves both particles. By analyzing the velocity vectors $\mathbf{v}_1$ and $\mathbf{v}_2$ into their x and y components, we can write:

$$\mathbf{p}_{\text{final}} = (m_1 v_1 \cos\theta + m_2 v_2 \cos\phi)\,\hat{\mathbf{i}} + (m_1 v_1 \sin\theta - m_2 v_2 \sin\phi)\,\hat{\mathbf{j}}$$

Using conservation of momentum, we equate $\mathbf{p}_{\text{initial}} = \mathbf{p}_{\text{final}}$. In addition, because the collision is elastic, we also equate the initial and final kinetic energies. Together, the two conservation laws lead to three equations:

$$m_1 v_0 = m_1 v_1 \cos\theta + m_2 v_2 \cos\phi$$
$$0 = m_1 v_1 \sin\theta - m_2 v_2 \sin\phi$$
$$\frac{1}{2}m_1 v_0^2 = \frac{1}{2}m_1 v_1^2 + \frac{1}{2}m_2 v_2^2$$

The first two equations represent the total momentum in the x and y directions respectively, and the third equation represents the conservation of kinetic energy. In addition, we know that $v_0 = 10.0$ m/s, $m_1 = 0.10$ kg, and $m_2 = 0.20$ kg. As you might imagine, this problem can involve a lot of algebra. However, a CAS will be able to assist us with that algebra.

Python Code

We use `fsolve` from the SciPy library to solve for v_1, v_2 and ϕ. The `fsolve` command takes a vector function as its input. The components of the vector function `f` are the equations to be solved (with the right-hand side set to zero). We need to give `fsolve` a guess to begin with. We choose 10 and 0 as the initial guesses for v_1 and v_2, because these were the initial velocities of m_1 and m_2. We choose 1 as an initial guess for ϕ because the nearly equal masses would imply the particle would move in similar directions. However, it should be noted that the answer for ϕ given by `fsolve` is very sensitive to the initial guess.

```python
import numpy as np
from scipy.optimize import fsolve

print('-'*28,'CODE OUTPUT','-'*29,'\n')

m1, m2, v0, theta = 0.1, 0.2, 10, np.pi/6

# we use initial guess values of [v1,v2,phi]=[10,0,1]

def f(x):
    #let x[0] = v1, x[1] = v2, x[3] = phi
    return [m1*v0 - m1*x[0]*np.cos(theta) - m2*x[1]*np.cos(x[2]),
    m1*x[0]*np.sin(theta) - m2*x[1]*np.sin(x[2]),
    0.5*m1*v0**2 - 0.5*m1*x[0]**2 - 0.5*m2*x[1]**2]

roots = fsolve(f,[10,0,1])

print('v1 = ', roots[0])
print('v2 = ', roots[1])
print('phi = ', roots[2])

--------------------------- CODE OUTPUT ----------------------------
v1 =   9.34172358964264
v2 =   2.5231131936226365
phi =  1.182656811357155
```

Mathematica Code

We begin the program by defining our known variables, the final direction θ of mass m_1, the initial velocity of $m_1(v0)$, and the values of the masses m_1 and m_2. If you run this code from this book's website, the `Solve` command gives a warning (shown here as a comment line) that some solutions may not be found by Mathematica. This is not uncommon when using a CAS to solve transcendental equations.

However, four solutions are found. The first three solutions include negative values for v_1 and v_2. These are non-physical results because v_1 and v_2 are speeds, and speed is the positive magnitude of the velocity vector. The fourth solution found by Mathematica contains all positive values, which is in agreement with the problem's diagram. Finally, we need to remember that Mathematica, like any calculator, is not respecting significant figures. Therefore, our final answer is: $v_1 = 9.3$ m/s, $v_2 = 2.5$ m/s, and $\phi = 1.2$ rad. As we have said before, the availability of CAS programs is not a substitute for knowing how to do algebra. The reader should verify the solution by hand and by checking the physical meaning of the analytical results.

```
θ = π/6; v0 = 10; m1 = 0.1; m2 = 0.2;

soln = Solve[{m1 * v0 == m1 * v1 * Cos[θ] + m2 * v2 * Cos[φ],

m1 * v1 * Sin[θ] == m2 * v2 * Sin[φ],

0.5 * m1 * v0^2 == 0.5 * m1 * v1^2 + 0.5 * m2 * v2^2}, {v1, v2, φ}];

Print["The four possible solutions are:"]

Do[Print[soln[[i]]], {i, 1, 4}]

(* Solve::ifun : Inverse functions are being used by Solve,

so some solutions may not be found;

use Reduce for complete solution information. *)
```

The four possible solutions are:
$\{v1 \to -3.56822, v2 \to -6.6056, \phi \to 3.00613\}$
$\{v1 \to -3.56822, v2 \to 6.6056, \phi \to -0.135459\}$
$\{v1 \to 9.34172, v2 \to -2.52311, \phi \to -1.95894\}$
$\{v1 \to 9.34172, v2 \to 2.52311, \phi \to 1.18266\}$

Next, we study the case where three or more particles are involved. The question arises: Do we still expect momentum to be conserved when there are many interacting particles? Let's begin by examining the net force acting on a collection of N interacting particles. The net force on the ith particle is:

$$\mathbf{F}_i = \mathbf{F}_i^{\text{ext}} + \sum_{j \neq i} \mathbf{F}_{ij} \tag{4.1.7}$$

where as previously stated $\mathbf{F}_i^{\text{ext}}$ is the net external force acting on the ith particle, and $\mathbf{F}_{ij}$ is the force of interaction between particle j and particle i. The summation term in the equation above denotes the net internal force acting on the particle. The net force acting on the system would then be obtained by summing the above equation over the N particles:

$$\mathbf{F} = \sum_{i=1}^{N} \mathbf{F}_i = \sum_{i=1}^{N} \mathbf{F}_i^{\text{ext}} + \sum_{i=1}^{N} \sum_{j \neq i} \mathbf{F}_{ij} \tag{4.1.8}$$

Equation (4.1.8) can be rewritten as:

$$\mathbf{F} = \sum_{i=1}^{N} \mathbf{F}_i^{\text{ext}} + \sum_{i=1}^{N} \sum_{j > i} (\mathbf{F}_{ij} + \mathbf{F}_{ji}) \tag{4.1.9}$$

If this rewriting is not clear to you, set $N = 3$ and write out the terms in (4.1.8); you will find that you can rewrite the sum as a collection of two terms, as shown previously. Newton's third law tells us that $\mathbf{F}_{ji} = -\mathbf{F}_{ij}$, and therefore the second term in the previous equation is zero leading to:

$$\dot{\mathbf{P}} = \mathbf{F} = \sum_{i=1}^{N} \mathbf{F}_i^{\text{ext}} \tag{4.1.10}$$

The Law of Conservation of Linear Momentum for a System of N Particles

$$\text{If } \quad \sum_{i=1}^{N} \mathbf{F}_i^{\text{ext}} = 0 \quad \text{then} \quad \mathbf{P} = \text{constant} \qquad (4.1.11)$$

In other words, the momentum of the system of particles is conserved if there are no external forces acting on the system, like in the case of two-particle systems.

4.2 ROCKETS

A practical example of conservation of momentum is rocket propulsion. The rocket has a challenge of moving forward without pushing against something. For example, an automobile moves forward by turning its wheels so that it pushes against the ground. However, a rocket in space has nothing to push against. Instead, the rocket solves this problem by ejecting mass, similar to the recoil of a gun. Before a gun is fired, the bullet and the gun are at rest. However, once the bullet is fired, it is propelled forward. In order for the total momentum to be conserved, the gun has to move in the direction opposite to that of the bullet. The velocities of the gun and the bullet are, of course, very different because the gun has more mass than the bullet does.

We begin by considering a rocket flying horizontally and assume that the rocket is not experiencing any external forces. The rocket has momentum $p(t) = mv$ at time t, where m and v are the mass and velocity of the rocket. The momentum of the rocket is measured by an observer in a reference frame that is at rest relative to the rocket. A short time later denoted by $t + dt$, the rocket has expelled a small amount of mass dm', which is moving at a speed u relative to the rocket, as illustrated in Figure 4.2. The speed u is sometimes called the *exhaust speed*. The ejection of mass leads to an increase in the rocket's speed by a small amount dv, as measured in the rest frame. Therefore, the momentum of the system immediately after the mass was ejected is:

$$p(t + dt) = (m - dm')(v + dv) + dm'\,(v - u) \qquad (4.2.1)$$

where the second term above is the momentum of the ejected mass relative to the rest frame.

Figure 4.2: A rocket after ejecting a small amount of mass dm' at a speed u relative to the rocket.

Applying conservation of momentum:

$$
\begin{aligned}
p(t) &= p(t + dt) & (4.2.2)\\
mv &= (m - dm')(v + dv) + dm'\,(v - u) & (4.2.3)\\
m\,dv &= u\,dm' & (4.2.4)\\
m\,dv &= -u\,dm & (4.2.5)
\end{aligned}
$$

where the last line was obtained by noting that the change in mass of the rocket is $dm = -dm'$. Furthermore, we ignored the term $dm'\,dv$ because both dm' and dv are small; therefore their product is negligible. Dividing both sides by dt we obtain:

$$m\,\dot{v} = -u\,\dot{m} \tag{4.2.6}$$

The left-hand side of (4.2.6) is Newton's second law ($F = ma$), and therefore we see that the net force acting on the rocket is equal to $-u\,\dot{m}$, which is sometimes called the *thrust*. Note that $\dot{m} < 0$, so the thrust is positive (and points to the right in Figure 4.2). We can also write (4.2.6) as:

$$dv = -u\frac{dm}{m} \tag{4.2.7}$$

This equation can be integrated by assuming a constant u. If at $t = 0$ the rocket's speed is v_0 and its mass is m_0, we obtain:

$$v - v_0 = u\ln\left(\frac{m_0}{m}\right) \tag{4.2.8}$$

which gives the velocity of the rocket as a function of its mass m. Equation (4.2.8) tells us that in order to make a rocket go as fast as possible, engineers need to create rockets with large exhaust speed u, and with a large initial to final mass ratio m_0/m. The final speed of the rocket will be obtained once all of the fuel is burned. Rockets with multiple stages that are ejected can help maximize m_0/m, so that the rocket can achieve even greater speeds.

Next, we study the case of a rocket moving vertically, experiencing the force of gravity. Our goal is to find an equation which describes the velocity of the rocket. We will use a coordinate system where the positive y-direction is upward, in the opposite direction of the force of gravity. In this case, gravity is an external force acting on the rocket, and therefore:

$$\frac{dp}{dt} = -mg \tag{4.2.9}$$

$$dp = -mgdt \tag{4.2.10}$$

$$p(t+dt) - p(t) = -mgdt \tag{4.2.11}$$

$$mdv + udm = -mgdt \tag{4.2.12}$$

$$m\frac{dv}{dt} + u\frac{dm}{dt} = -mg \tag{4.2.13}$$

In order to obtain an equation for v, we assume a constant *burn rate* α:

$$\alpha = -\frac{dm}{dt} \tag{4.2.14}$$

Using our definition of burn rate, (4.2.13) becomes:

$$\frac{dv}{dt} = -g + \frac{\alpha}{m}u \tag{4.2.15}$$

$$dv = \left(-g + \frac{\alpha u}{m}\right)dt \tag{4.2.16}$$

$$dv = \left(\frac{g}{\alpha} - \frac{u}{m}\right)dm \tag{4.2.17}$$

Integrating the above equation using $v(0) = v_0$ and $m(0) = m_0$ produces:

$$v - v_0 = \frac{g}{\alpha}(m - m_0) + u\ln\left(\frac{m_0}{m}\right) \tag{4.2.18}$$

This equation gives the velocity of the rocket as a function of mass m when the burn rate α is constant. Sometimes (4.2.18) is rewritten by integrating the definition of the burn rate: $m - m_0 = -\alpha t$. In that case, (4.2.18) becomes:

$$v - v_0 = -gt + u \ln \left(\frac{m_0}{m} \right) \tag{4.2.19}$$

The advantage of using (4.2.19) is that one can see two terms which affect the rocket's motion. The first term, $-gt$, is the standard kinematics term for a particle in free fall, and represents the effect of gravity on the velocity. The second term, $u \ln(m_0/m)$, is the same term we obtained for the horizontal motion of the rocket and represents the effect of thrust on the rocket's velocity.

4.3 CENTER OF MASS

Problems involving momentum often involve multiple objects, such as in the case of collisions. In situations where the system of interest involves multiple particles, or the object under study is not a point particle, the concept of *center of mass* becomes useful. Let us begin by considering a system of N point particles. The location $\mathbf{R}$ of the center of mass for a system of particles is defined as:

Center of Mass for a System of Discrete Particles

$$\mathbf{R} = \frac{1}{M} \sum_{i=1}^{N} m_i \mathbf{r}_i \tag{4.3.1}$$

$$X = \frac{1}{M} \sum_{i=1}^{N} m_i x_i \qquad Y = \frac{1}{M} \sum_{i=1}^{N} m_i y_i \qquad Z = \frac{1}{M} \sum_{i=1}^{N} m_i z_i \tag{4.3.2}$$

where $\mathbf{r}_i$ is the position of the ith particle, which has mass m_i. The total mass of the system is $M = \sum m_i$. Equation (4.3.1) is actually three equations, one for each coordinate of $\mathbf{R}$. For example, in Cartesian coordinates the position of the ith particle is $\mathbf{r}_i = x_i \hat{\mathbf{i}} + y_i \hat{\mathbf{j}} + z_i \hat{\mathbf{k}}$ and we can write $\mathbf{R} = X\hat{\mathbf{i}} + Y\hat{\mathbf{j}} + Z\hat{\mathbf{k}}$.

The center of mass is the weighted average location of the mass in the system, with the weights being the masses m_i. The location of the center of mass will be closer to the heavier particles. The next example illustrates this point.

Example 4.3: Center of mass of the Earth-Sun system.
Compute the location of the center of mass of the Earth-Sun system.

Solution:
We choose a coordinate system such that the center of the Sun is at the origin. We will also approximate the Earth's orbit as circular and assume that the plane of the Earth's orbit is constant. Therefore, we can work in polar coordinates and only need to find the distance of the center of mass of the system from the Sun's center. The distance between the Earth and the Sun is approximately 1.5×10^{11} m, and the masses of the Earth and Sun are 5.9×10^{24} kg and 2.0×10^{30} kg, respectively. Therefore, in our

coordinate system, $r_{Sun} = 0$ and $r_{Earth} = 1.5 \times 10^{11}$ m. Inserting these numbers into (4.3.1) gives:

$$\frac{(m_{Earth})\,(r_{Earth})}{m_{Sun} + m_{Earth}} = \frac{(5.9 \times 10^{24}\text{ kg})\,(1.5 \times 10^{11}\text{ m})}{5.9 \times 10^{24}\text{ kg} + 2.0 \times 10^{30}\text{ kg}} = 4.6 \times 10^5\text{ m}.$$

The radius of the Sun is approximately 7.0×10^8 m. Therefore, the center of mass of the Earth-Sun system is well inside the Sun, and it practically coincides with the center of the Sun.

If we want to extend (4.3.1) to continuous distributions of mass, i.e., objects that are not well-described as point particles, then we need to make some changes. One way to approach a continuous mass distribution is by thinking of it as a collection of infinitesimal masses dm. Then the sum over discrete particles in (4.3.1) becomes an integral over infinitesimal mass elements dm. The center of mass equations then become:

Center of Mass for a Continuous Mass Distribution

$$\mathbf{R} = \frac{1}{M} \int \mathbf{r}\,dm \qquad M = \int dm \tag{4.3.3}$$

$$X = \frac{1}{M} \int x\,dm \quad Y = \frac{1}{M} \int y\,dm \quad Z = \frac{1}{M} \int z\,dm \tag{4.3.4}$$

In the equations above, the total mass is found from $M = \int dm$, and like (4.3.1), equation (4.3.3) actually consists of three equations, one for each coordinate. Typically, one does not integrate (4.3.3) using the mass. Most often the integration is done over the length, area, or volume of the object, when the object does not have a uniform density.

The following box summarizes how to calculate the total mass M by using the density of the object in one-, two-, and three-dimensional situations. In the case of one-dimensional objects, we use a linear mass density λ (mass per unit length in units of kg/m). In the case of two-dimensional objects, we use a surface density σ (mass per unit area in units of kg/m^2), and for three-dimensional objects, we use the familiar volume density ρ (mass per unit volume in units of kg/m^3).

Evaluating the Mass dm for 1D, 2D and 3D Objects

$$dm = \lambda\,dx \qquad M = \int \lambda\,dx \tag{4.3.5}$$

$$dm = \sigma\,dA \qquad M = \int \sigma\,dA \tag{4.3.6}$$

$$dm = \rho\,dV \qquad M = \int \rho\,dV \tag{4.3.7}$$

The next two examples demonstrate how to apply (4.3.3) to a continuous mass distribution.

Example 4.4: Center of mass of an isosceles triangle.

A lamina of uniform mass per unit area, σ, is shaped into an isosceles triangle shown below. The triangle has two sides of length a and the base has length $\sqrt{2}a$. Find the center of mass of this triangular lamina.

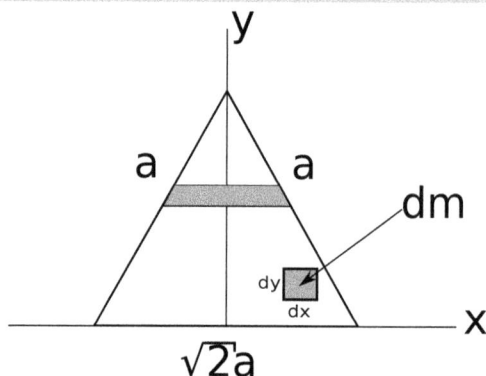

Solution:

First note that due to the symmetry of the triangle, the center of mass lies along the y-axis. Computing the center of mass of a mass distribution involves identifying two components. The first component to identify is the *mass element dm*, shown in the diagram of the triangle. Cartesian coordinates are a natural choice of coordinates for this problem. Therefore, $dm = \sigma dx dy$ will be used for the mass element.

The second component needed to find the center of mass is *the limits of the integral* in (4.3.3). The limits of the integral range over the size of the object. We included a horizontal gray strip near the top of the triangle to illustrate that the width of the triangle varies with the height, with the strip becoming narrower as it moves up the triangle. The right side of the strip is bound by the line $y = a/\sqrt{2} - x$, and the left side is bound by the line $y = a/\sqrt{2} + x$, because the triangle has a height $a/\sqrt{2}$. Therefore, we can consider the left end of the strip to be at $x = y - a/\sqrt{2}$, and the right end to be at $x = a/\sqrt{2} - y$.

The y-coordinate of the center of mass can then be found by computing $Y = (1/M) \int y dm$:

$$Y = \frac{1}{M} \int_{y=0}^{y=a/\sqrt{2}} \int_{x=y-a/\sqrt{2}}^{a/\sqrt{2}-y} y\sigma dx dy$$

$$= \frac{\sigma}{M} \int_0^{a/\sqrt{2}} yx \Big|_{y-a/\sqrt{2}}^{a/\sqrt{2}-y} dy$$

$$= \frac{\sigma}{M} \int_0^{a/\sqrt{2}} 2y \left(\frac{a}{\sqrt{2}} - y \right) dy$$

$$= \frac{\sigma}{M} \left(\frac{ay^2}{\sqrt{2}} - \frac{2y^3}{3} \right) \Big|_0^{a/\sqrt{2}}$$

$$= \frac{a^3}{6\sqrt{2}} \frac{\sigma}{M}$$

To finish the problem, we need to find the total mass M.

$$M = \int dm = \int_{y=0}^{y=a/\sqrt{2}} \int_{x=y-a/\sqrt{2}}^{a/\sqrt{2}-y} \sigma\,dxdy = \sigma \int_{0}^{a/\sqrt{2}} 2\left(\frac{a}{\sqrt{2}} - y\right) dy = \frac{\sigma a^2}{2}$$

By using the value for M found above, we find that:

$$Y = \frac{a}{3\sqrt{2}}$$

We can use a CAS to evaluate the integrals, even if the limits are not constant.

Python Code

We can use the `integrate` command from SymPy to perform both integrals. Notice that in Python unlike Mathematica, the limits of the integral to be done *first* appears as the *second* argument (after the function being integrated).

```python
from sympy import symbols, integrate, sqrt

print('-'*28,'CODE OUTPUT','-'*29,'\n')

x, y, a, sigma = symbols('x, y, a, sigma',real=True)

Y_fcn = y*sigma      # Y component of CM
M_fcn = sigma        # mass

Y_int = integrate(Y_fcn,  (x, y - a/sqrt(2), a/sqrt(2) - y), \
    (y,0,a/sqrt(2)))    # vary y first

M = integrate(M_fcn, (x, y - a/sqrt(2), a/sqrt(2) - y), \
    (y,0,a/sqrt(2)))

print('The mass = ', M)
print('The CM is located at y = ',Y_int/M)

---------------------------- CODE OUTPUT ----------------------------
The mass =  a**2*sigma/2
The CM is located at y =  sqrt(2)*a/6
```

Mathematica Code

Note the order of appearance of the coordinates x, y inside the `Integrate` command.

$Y = \text{Integrate}\left[\sigma * y/m, \left\{y, 0, \frac{a}{\sqrt{2}}\right\}, \left\{x, y - \frac{a}{\sqrt{2}}, \frac{a}{\sqrt{2}} - y\right\}\right];$

Print["The Y-coordinate for the CM is: Y = ", Y]

The Y-coordinate for the CM is: Y = $\frac{a^3\sigma}{6\sqrt{2}m}$

$M = \text{Integrate}\left[\sigma, \left\{y, 0, \frac{a}{\sqrt{2}}\right\}, \left\{x, y - \frac{a}{\sqrt{2}}, \frac{a}{\sqrt{2}} - y\right\}\right];$

Print["The mass M, M = ", M]

The mass M, M = $\frac{a^2\sigma}{2}$

$Y2 = Y/.m \to M;$

Print["Substituting the mass M, Y = ", $Y2$]

Substituting the mass M: Y = $\frac{a}{3\sqrt{2}}$

Example 4.5: The center of mass of a solid cone.

A cone of uniform density ρ is shown in the below figure. The cone has a radius a at the top and a height h. Compute the center of mass of the cone.

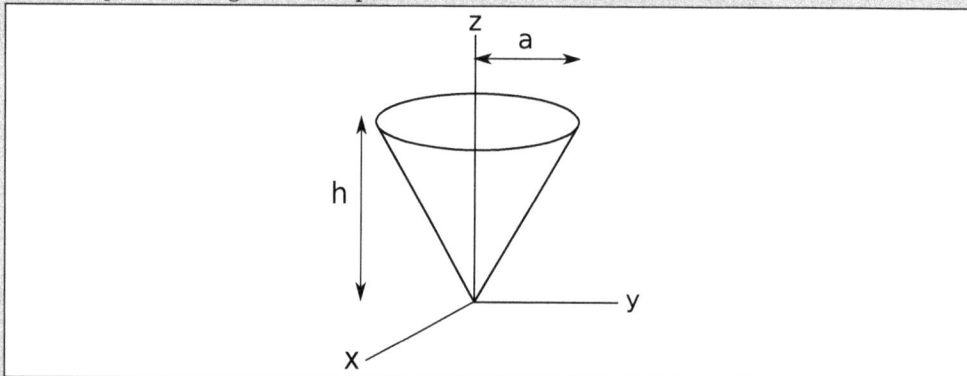

Solution:

To solve this problem, we first need to identify a coordinate system. A cone can be naturally described in cylindrical coordinates. Therefore we use $dm = \rho dV = \rho r dr d\theta dz$. By symmetry, the center of mass is along the z-axis. The θ-coordinate ranges from 0 to 2π, and the z-coordinate ranges from 0 to h.

However, the radius changes with the z-coordinate. The equation of the edge of the cone in the yz-plane in the figure is $r = az/h$, so that the radius r ranges from 0 to az/h. Putting all of this together,

$$Z = \frac{1}{M}\int_{z=0}^{h}\int_{\theta=0}^{2\pi}\int_{r=0}^{az/h} z\rho r\,dr\,d\theta\,dz$$

$$= \frac{2\pi\rho}{M}\int_0^h z\frac{r^2}{2}\bigg|_0^{az/h}\,dz$$

$$= \frac{2\pi\rho a^2}{2h^2 M}\int_0^h z^3\,dz = \frac{\pi\rho a^2 h^2}{4M}$$

Solving for the total mass M:

$$M = \int_0^h \int_0^{2\pi} \int_0^{az/h} \rho r\, dr\, d\theta\, dz$$

$$= 2\pi\rho \int_0^h \frac{r^2}{2}\bigg|_0^{az/h} dz$$

$$= \frac{\pi\rho a^2}{h^2} \int_0^h z^2 dz = \frac{\pi\rho a^2 h}{3}$$

Finally, inserting M into the result for Z, we get:

$$Z = \frac{3}{4}h$$

Again, we can perform the triple integration using CAS.

Python Code

The triple integral is done the same way as the double integral which was evaluated in Example 4.4. Note again the order of integrations, and compare with the Mathematica code. Note again the order of the limits of integration in the **integrate** command.

```
from sympy import symbols, integrate, sqrt, pi

print('-'*28,'CODE OUTPUT','-'*29,'\n')

r, theta, z , rho, h = symbols('r,theta,z,rho,h', real=True)

Z_fcn = z*rho*r     # Y component of CM
M_fcn = rho*r       # mass

Z_int = integrate(Z_fcn, (r, 0, a*z/h), \
    (theta,0,2*pi), (z,0,h))    # vary y first

M = integrate(M_fcn, (r, 0, a*z/h), \
    (theta,0,2*pi), (z,0,h))

print('The mass = ', M)
print('The CM is located at z = ', Z_int/M)

--------------------------- CODE OUTPUT ---------------------------
The mass =  pi*a**2*h*rho/3
The CM is located at z =  3*h/4
```

Mathematica Code

Notice that in order to perform the triple integral in cylindrical coordinates, we needed to explicitly include the equation for $r = az/h$ in the argument of the Integrate command.

$Z = \text{Integrate} \left[\frac{\rho * z * r}{m}, \{\theta, 0, 2\pi\}, \{z, 0, h\}, \{r, 0, az/h\} \right]$;

Print["The Z-coordinate for the CM is: Z = ", Z]

The Z-coordinate for the CM is: $Z = \frac{a^2 h^2 \pi \rho}{4m}$

$M = \text{Integrate}[\rho * r, \{\theta, 0, 2\pi\}, \{z, 0, h\}, \{r, 0, az/h\}]$;

Print["The mass M, M = ", M]

The mass M, $M = \frac{1}{3} a^2 h \pi \rho$

$Z2 = Z/.m \rightarrow M$;

Print["Substituting the mass M, Z = ", $Z2$]

Substituting the mass M: $Z = \frac{3h}{4}$

In the next example, we will use *numerical* methods to find the center of mass of a non-uniform isosceles triangle. Recall that in Chapter 2, we discussed numerical integration techniques.

Example 4.6: The center of mass of a non-uniform isosceles triangle.

Consider the triangle in Example 4.4, but now with a non-uniform density $\sigma = 2y$ (all physical quantities in SI units). Compute the center of mass for the triangle when $a = 1$ m.

Solution:

In this case, the solution is very similar to that of Example 4.4, but now the triangle has a non-uniform density. The limits of integration are the same, with $a = 1$. However, dm is changed due to the non-uniform density. In this case, $dm = \sigma dx dy = 2y dx dy$. Because the density changes only in the y-direction, we still expect that the x-coordinate of the center of mass is zero. The integrals we need to solve are:

$$Y = \frac{1}{M} \int_{y=0}^{y=a/\sqrt{2}} \int_{x=y-a/\sqrt{2}}^{a/\sqrt{2}-y} 2y^2 \, dx dy$$

$$M = \int_{y=0}^{y=a/\sqrt{2}} \int_{x=y-a/\sqrt{2}}^{a/\sqrt{2}-y} 2y \, dx dy$$

We will use Mathematica's and Python's numerical integration commands to solve these integrals.

Python Code

We use `integrate.nquad` from the SciPy library. Notice that we can include the bounds as a function; this allows us to use the non-constant limits of integration necessary to describe the sides of the triangle. In addition to finding Y, we also calculated X to show that $X = 0$, as expected.

Note the square brackets inside the `integrate.nquad` command.

```
from scipy import integrate
import numpy as np

print('-'*28,'CODE OUTPUT','-'*29,'\n')

a = 1.0

def sigma(x,y):
    return 2.0*y

def fx(x,y):
    return x*sigma(x,y)

def fy(x,y):
    return y*sigma(x,y)

def y_bounds():
    return [0,1/np.sqrt(2.0)]

def x_bounds(y):
    return [y-1/np.sqrt(2.0),1/np.sqrt(2.0)-y]

Y = integrate.nquad(fy,[x_bounds,y_bounds])[0]
X = integrate.nquad(fx,[x_bounds,y_bounds])[0]
mass  = integrate.nquad(sigma,[x_bounds,y_bounds])[0]

print('The x-coordinate is: '+str(X/mass))
print('The y-coordinate is: '+str(Y/mass))

------------------------- CODE OUTPUT ----------------------------
The x-coordinate is: 0.0
The y-coordinate is: 0.35355339059327384
```

Mathematica Code

We use `NIntegrate` to numerically evaluate the integrals. We included the result for X to show that it is zero.

```
a = 1.0;

mass = NIntegrate [2 * y, {y, 0, a/√2}, {x, y - a/√2, a/√2 - y}];

Y = NIntegrate [2 * y^2, {y, 0, a/√2}, {x, y - a/√2, a/√2 - y}] / mass;
Print["The Y-coordinate for the CM is: Y = ", Y]
The Y-coordinate for the CM is: Y = 0.353553

X = NIntegrate [2 * y * x, {y, 0, a/√2}, {x, y - a/√2, a/√2 - y}] / mass;
Print["The X-coordinate for the CM is: X = ", X]
The X-coordinate for the CM is: X = 0.
```

4.4 MOMENTUM OF A SYSTEM OF MULTIPLE PARTICLES

Next, consider a system of N point particles with a total mass M. We begin by calculating the net force acting on the whole system. We know from Section 4.1 that the internal forces between particles cancel out, and that the net force acting on the system is the sum of external forces. The net force $\mathbf{F}$ acting on the particles is:

$$\mathbf{F} = \sum_{i=1}^{N} m_i \ddot{\mathbf{r}}_i = \frac{d^2}{dt^2} \sum_{i=1}^{N} m_i \mathbf{r}_i = \frac{d^2}{dt^2} M\mathbf{R} = M\ddot{\mathbf{R}} \qquad (4.4.1)$$

where we used the definition of the center of mass from (4.3.1). Therefore, the system of particles moves like a single particle of mass M, acted upon by the external forces. We can also compute the net momentum of the system of particles:

$$\mathbf{P} = \sum_{i=1}^{N} \mathbf{p}_i = \sum_{i=1}^{N} m_i \dot{\mathbf{r}}_i = \frac{d}{dt} \sum_{i=1}^{N} m_i \mathbf{r}_i = \frac{d}{dt} M\mathbf{R} = M\dot{\mathbf{R}} \qquad (4.4.2)$$

The total momentum of the system is equal to the momentum of the system's center of mass. Again, we can think of the system as a single particle of mass M located at the system's center of mass. The time derivative of the total momentum gives:

$$\dot{\mathbf{P}} = M\ddot{\mathbf{R}} = \mathbf{F} \qquad (4.4.3)$$

This also tells us that the total linear momentum of the system is conserved if there are no net external forces acting on the system.

In summary, we can think of the net external forces as acting on the center of mass of the system. The results of this section also hold for continuous mass distributions, but the summations would need to be replaced by integrals.

When it comes to translational motion, the above result allows us to think of the object or system of particles as a point particle located at the object's center of mass. The mass of the point particle would equal the total mass of the object, or the total mass of the system of particles. Since the external forces act on the center of mass, we can simply follow the motion of the object's center of mass. For example, an American football is not a point particle. However, if the football is punted, its center of mass would follow a parabolic trajectory because the center of mass is being acted upon by the force of gravity, and obeys the equations established for projectile motion in Chapter 3.

Example 4.7: A falling chain.
Consider a chain of length a and uniform density ρ, hanging from one end above a table at time $t = 0$, as shown by the dashed line in the below figure. At time $t = 0$, the end of the chain is released from a height a above the table, and the chain is allowed to fall onto the table. Find the force the table exerts on the chain as a function of the distance x from the originally fixed end.

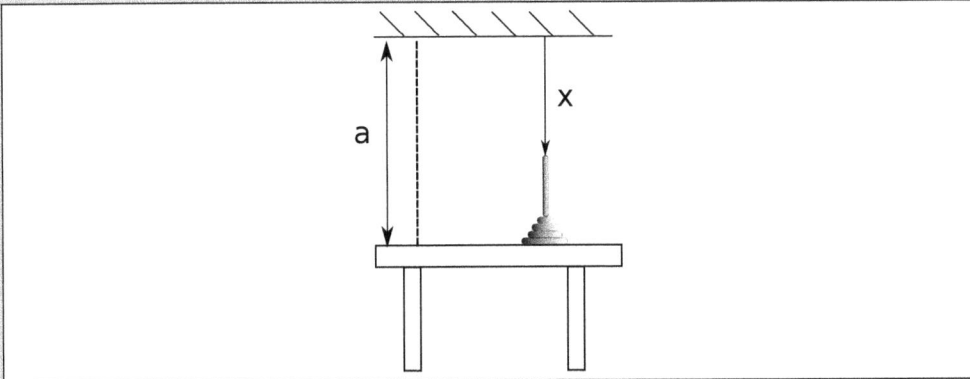

Solution:
The chain can be thought of as a system of particles. The force on the center of mass of the chain is:

$$F_{net} = Mg - F_N$$

where Mg is the weight of the chain and F_N is the force exerted by the table on the chain. We have chosen a coordinate system where $+x$ is in the downward direction, so that the speed of the falling chain is $\dot{x} > 0$. The mass of the falling part of the chain is $\rho(a - x)$, because the length of the chain that has yet to fall is $(a - x)$. The momentum of the falling part of the chain is then:

$$P = mv = \rho(a - x)\dot{x}$$

We know that $\dot{P} = F_{net}$; therefore, the time derivative of the momentum is:

$$\dot{P} = \rho(a - x)\ddot{x} - \rho\dot{x}^2$$

For free fall, we know that $\dot{x} = \sqrt{2gx}$ and $\ddot{x} = g$. By substituting $\dot{x}$ and $\ddot{x}$ in the last equation and collecting terms, we obtain:

$$\dot{P} = \rho g(a - 3x)$$

Finally substituting $\dot{P} = F_{net} = Mg - F_N$ and using the total mass of the chain $M = \rho a$, we find:

$$\rho g(a - 3x) = \rho a g - F_N$$
$$F_N = 3\rho g x$$

As the distance x increases, the force the table exerts on the chain varies from $F_N = 0$ at time $t = 0$, up to $F_N = 3\rho g a$ when the chain has fallen completely on the table.

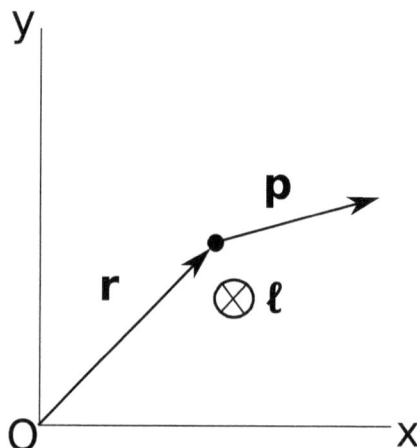

Figure 4.3: A particle of mass m at a location $\mathbf{r}$ relative to the origin O is moving with momentum $\mathbf{p}$, and has an angular momentum $\boldsymbol{\ell} = \mathbf{r} \times \mathbf{p}$ relative to the origin. In this case, the angular momentum of the particle is pointing into the page.

4.5   ANGULAR MOMENTUM OF A SINGLE PARTICLE

Consider a particle of mass m moving with momentum $\mathbf{p} = m\mathbf{v}$ as shown in Figure 4.3. The angular momentum of the particle is defined by:

Angular Momentum

$$\boldsymbol{\ell} = \mathbf{r} \times \mathbf{p} \qquad (4.5.1)$$

Because the position $\mathbf{r}$ of the particle is measured relative to the origin, the angular momentum $\boldsymbol{\ell}$ is also relative to the origin. The linear momentum $\mathbf{p}$ does not depend on the origin, while by contrast the angular momentum vector $\boldsymbol{\ell}$ depends on the choice of origin. The angular momenta of multiple particles can only be compared (and added) if all angular momenta are measured with respect to the same origin. Further, note that the particle does not need to be revolving around the point O, in order for the particle to have an angular momentum relative to O.

The cross product relationship between $\mathbf{r}$ and $\mathbf{p}$ tells us that the direction of angular momentum is found using the right-hand rule. In this case, we can imagine taking the fingers of our right hand and pointing them in the direction of $\mathbf{r}$. Next, we can sweep (or curl) our fingers in the direction of $\mathbf{p}$ resulting in the thumb, which points in the direction of $\boldsymbol{\ell}$, pointing into the page. The $\otimes$ symbol in Figure 4.3 represents the angular momentum vector pointing into the page. If $\boldsymbol{\ell}$ pointed out of the page, we would use a circle with a dot inside to represent the direction of the vector.

If we compute the time derivative of the angular momentum, we obtain:

$$\dot{\boldsymbol{\ell}} = \dot{\mathbf{r}} \times \mathbf{p} + \mathbf{r} \times \dot{\mathbf{p}} = m\left(\dot{\mathbf{r}} \times \dot{\mathbf{r}}\right) + \mathbf{r} \times \dot{\mathbf{p}} = \mathbf{r} \times \dot{\mathbf{p}} \qquad (4.5.2)$$

By using Newton's second law $\dot{\mathbf{p}} = \mathbf{F}$ we find:

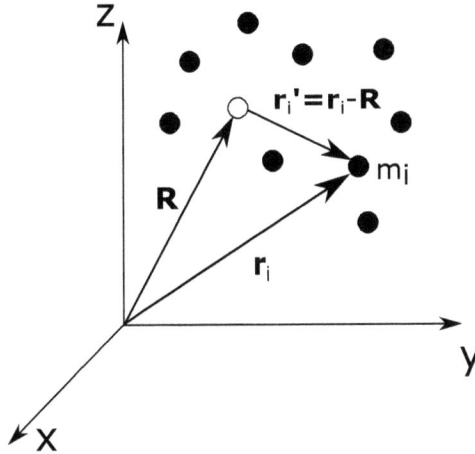

Figure 4.4: A collection of N particles. Each particle m_i is at a location $\mathbf{r}_i$ relative to the origin and a position $\mathbf{r}'_i$ relative to the center of mass. The empty circle is the center of mass of the system.

$$\textit{Newton's Second Law of Rotation}$$

$$\dot{\boldsymbol{\ell}} = \mathbf{r} \times \mathbf{F} = \mathbf{N} \tag{4.5.3}$$

Therefore, the time rate of change of the angular momentum is the torque, $\mathbf{N} = \mathbf{r} \times \mathbf{F}$. Like the angular momentum, the torque is also measured relative to the same origin as the angular momentum. Equation (4.5.3) is sometimes referred to as Newton's second law for rotation, even though rotational motion is not needed in order for the particle to experience a torque, or to have angular momentum.

4.6 ANGULAR MOMENTUM OF MULTIPLE PARTICLES

Next, we consider the total angular momentum (relative to the origin) for a system of discrete particles. Again, the results are the same for a continuous mass distribution. Figure 4.4 illustrates a system of N particles. The mass of the ith particle in Figure 4.4 is m_i, located at a position $\mathbf{r}_i$ relative to the origin, and at a position $\mathbf{r}'_i$ relative to the center of mass (empty circle in Figure 4.4). Notice that $\mathbf{R} + \mathbf{r'_i} = \mathbf{r_i}$, where $\mathbf{R}$ is the location of the center of mass relative to the origin.

We will calculate the total angular momentum (relative to the origin) for the system of particles in Figure 4.4. Each particle has an angular momentum $\boldsymbol{\ell}_i = \mathbf{r}_i \times \mathbf{p}_i$. Therefore, the total angular momentum $\mathbf{L}$ for the system is:

$$\mathbf{L} = \sum_{i=1}^{N} \mathbf{r}_i \times \mathbf{p}_i = \sum_{i=1}^{N} \mathbf{r}_i \times m_i \dot{\mathbf{r}}_i \tag{4.6.1}$$

By substituting $\mathbf{r_i} = \mathbf{R} + \mathbf{r}'_i$, we obtain:

$$\mathbf{L} = \sum_{i=1}^{N} (\mathbf{r}'_i + \mathbf{R}) \times m_i \left(\dot{\mathbf{r}}'_i + \dot{\mathbf{R}} \right) \tag{4.6.2}$$

By expanding the cross product into four terms:

$$\mathbf{L} = \sum_{i=1}^{N} m_i \left[(\mathbf{r}_i' \times \dot{\mathbf{r}}_i') + (\mathbf{r}_i' \times \dot{\mathbf{R}}) + (\mathbf{R} \times \dot{\mathbf{r}}_i') + (\mathbf{R} \times \dot{\mathbf{R}}) \right] \qquad (4.6.3)$$

Of the four terms above, two are of immediate interest:

$$\sum_{i=1}^{N} m_i \left(\mathbf{r}_i' \times \dot{\mathbf{R}} \right) = \left(\sum_{i=1}^{N} m_i \mathbf{r}_i' \right) \times \dot{\mathbf{R}} \qquad (4.6.4)$$

$$\sum_{i=1}^{N} m_i \left(\mathbf{R} \times \dot{\mathbf{r}}_i' \right) = \mathbf{R} \times \frac{d}{dt} \left(\sum_{i=1}^{N} m_i \mathbf{r}_i' \right) \qquad (4.6.5)$$

In each case, the term $\sum m_i \mathbf{r}_i'$ appears. However, $\sum m_i \mathbf{r}_i' = \sum m_i (\mathbf{r}_i - \mathbf{R}) = \sum m_i \mathbf{r}_i - \sum m_i \mathbf{R} = M\mathbf{R} - M\mathbf{R} = 0$ from the definition of the center of mass. Therefore, only two terms survive in (4.6.3), and the total angular momentum is:

$$\begin{aligned}
\mathbf{L} &= \sum_{i=1}^{N} m_i \left[(\mathbf{r}_i' \times \dot{\mathbf{r}}_i') + (\mathbf{R} \times \dot{\mathbf{R}}) \right] \\
&= \sum_{i=1}^{N} (\mathbf{r}_i' \times m_i \dot{\mathbf{r}}_i') + \mathbf{R} \times \sum_{i=1}^{N} m_i \dot{\mathbf{R}}
\end{aligned} \qquad (4.6.6)$$

The final result is the following important equation:

Angular Momentum of a System of Particles

$$\mathbf{L} = \sum_{i=1}^{N} (\mathbf{r}_i' \times \mathbf{p}_i') + \mathbf{R} \times \mathbf{P} \qquad (4.6.7)$$

which says that the total angular momentum of the system is the sum of two terms: the first term is the sum of the angular momenta of the particles about the center of mass ($\sum \mathbf{r}_i' \times \mathbf{p}_i'$); the second term is the angular momentum of the center of mass about the origin ($\mathbf{R} \times \mathbf{P}$).

If we compute the time derivative of $\mathbf{L}$ we find,

$$\dot{\mathbf{L}} = \sum_{i=1}^{N} \mathbf{r}_i \times \dot{\mathbf{p}}_i \qquad (4.6.8)$$

which is the net torque $\mathbf{N}$ acting on the system (recall that $\dot{\mathbf{p}} = \mathbf{F}$). Next, we will examine the net torque acting on the system in detail:

$$\mathbf{N} = \sum_{i=1}^{N} \mathbf{r}_i \times \dot{\mathbf{p}}_i = \sum_{i=1}^{N} \mathbf{r}_i \times \mathbf{F}_i \qquad (4.6.9)$$

where $\dot{\mathbf{p}}_i = \mathbf{F}_i$ is the net force acting on mass m_i. This net force acting on m_i is the sum of external and internal forces; therefore:

$$\mathbf{N} = \sum_{i=1}^{N} \mathbf{r}_i \times \left(\mathbf{F}_i^{\text{ext}} + \sum_{j \neq i} \mathbf{F}_{ij} \right) \tag{4.6.10}$$

$$= \sum_{i=1}^{N} \left(\mathbf{r}_i \times \mathbf{F}_i^{ext} \right) + \sum_{i=1}^{N} \sum_{j \neq i} \left(\mathbf{r}_i \times \mathbf{F}_{ij} \right) \tag{4.6.11}$$

We will now use the same identity as in working with (4.1.8):

$$\mathbf{N} = \sum_{i=1}^{N} \left(\mathbf{r}_i \times \mathbf{F}_i^{\text{ext}} \right) + \sum_{i=1}^{N} \sum_{j>i} \left[(\mathbf{r}_i \times \mathbf{F}_{ij}) + (\mathbf{r}_j \times \mathbf{F}_{ji}) \right] \tag{4.6.12}$$

Using Newton's third law $\mathbf{F}_{ji} = -\mathbf{F}_{ij}$ yields:

$$\mathbf{N} = \sum_{i=1}^{N} \left(\mathbf{r}_i \times \mathbf{F}_i^{\text{ext}} \right) + \sum_{i=1}^{N} \sum_{j>i} (\mathbf{r}_i - \mathbf{r}_j) \times \mathbf{F}_{ij} \tag{4.6.13}$$

Note that if the internal forces $\mathbf{F}_{ij}$ are central forces, then the vector $\mathbf{F}_{ij}$ lies along the line joining m_i and m_j. Therefore, for central forces $\mathbf{F}_{ij}$ lies along the vector $\mathbf{r}_i - \mathbf{r}_j$, and $(\mathbf{r}_i - \mathbf{r}_j) \times \mathbf{F}_{ij} = 0$. In this case, the second term in (4.6.13) is zero, and the net torque is:

$$\mathbf{N} = \sum_{i=1}^{N} \left(\mathbf{r}_i \times \mathbf{F}_i^{\text{ext}} \right) = \sum_{i=1}^{N} \mathbf{N}_i^{\text{ext}} = \mathbf{N}^{\text{ext}} \tag{4.6.14}$$

In other words, the net torque on the system is equal to the sum of the external torques. Similar to the net force, the sum of the internal torques is zero (if the internal forces are central forces).

In summary, similar to the single-particle case, Newton's second law for rotation can be written as,

Newton's Second Law for Rotation

$$\dot{\mathbf{L}} = \mathbf{N}^{\text{ext}} \tag{4.6.15}$$

Hence, the net angular momentum of the system is conserved only if the net external torques are zero. The conservation of angular momentum can then be written as:

The Law of Conservation of Angular Momentum

If $\mathbf{N}^{\text{ext}} = 0$,   then $\mathbf{L} = \text{constant}$

Example 4.8: The total angular momentum of a system of particles.

Consider two particles each with a mass of 1 kg. At time t, Particle 1 is located at $\mathbf{r}_1 = 3\hat{\mathbf{i}} + 2\hat{\mathbf{j}}$ and is moving at a velocity of $\mathbf{v}_1 = 2\,\hat{\mathbf{i}}$ (all physical quantities in SI units). Particle 2 is located at $\mathbf{r}_2 = -3\hat{\mathbf{i}} + 2.0\hat{\mathbf{j}}$ and is moving at a velocity of $\mathbf{v}_2 = -2.0\,\hat{\mathbf{j}}$. Compute the total angular momentum of this system of particles, with respect to the origin.

Solution:

Using (4.5.1) for Particle 1:

$$\boldsymbol{\ell}_1 = m_1\,(\mathbf{r}_1 \times \mathbf{v}_1) = 1.0 \begin{vmatrix} \hat{\mathbf{i}} & \hat{\mathbf{j}} & \hat{\mathbf{k}} \\ 3 & 2 & 0 \\ 2 & 0 & 0 \end{vmatrix} = -4\hat{\mathbf{k}}$$

where we dropped the units in the determinant to simplify the equation. Next, we follow the same steps for Particle 2:

$$\boldsymbol{\ell}_2 = m_2\,(\mathbf{r}_2 \times \mathbf{v}_2) = 1.0 \begin{vmatrix} \hat{\mathbf{i}} & \hat{\mathbf{j}} & \hat{\mathbf{k}} \\ -3 & 2 & 0 \\ 0 & -2 & 0 \end{vmatrix} = 6.0\hat{\mathbf{k}}$$

Finally we add the angular momenta:

$$\mathbf{L} = \boldsymbol{\ell}_1 + \boldsymbol{\ell}_2 = 2.0 \text{ kg m}^2/\text{s}\,\hat{\mathbf{k}}$$

Note that in this case we could add angular momenta, because each of the individual angular momenta was calculated with respect to the same origin.

Python Code

Here we use NumPy to create the vectors and compute the cross products using the `cross(r1,v1)` function.

```
import numpy as np

print('-'*28,'CODE OUTPUT','-'*29,'\n')

m1, m2 = 1, 1

r1 = np.array([3,2,0])
r2 = np.array([-3,2,0])
v1 = np.array([2,0,0])
v2 = np.array([0,-2,0])

l1 = m1*np.cross(r1,v1)
l2 = m2*np.cross(r2,v2)

print('l1 = ', l1)
print('l2 = ', l2)
print('L = ', l1 + l2)

---------------------------- CODE OUTPUT -----------------------------
l1 =  [ 0  0 -4]
l2 =  [0 0 6]
L =  [0 0 2]
```

Mathematica Code

We use the `Cross` command to calculate the angular momenta. Note the use of curly brackets to define vectors as lists.

r1 = {3, 2, 0}; v1 = {2, 0, 0}; m1 = 1.0;

r2 = {−3, 2, 0}; v2 = {0, −2, 0}; m2 = 1.0;

l1 = m1 ∗ Cross[r1, v1];

Print["The angular momentum for particle 1 is: l1 = ", l1]

The angular momentum for particle 1 is: l1 = {0., 0., −4.}

l2 = m2 ∗ Cross[r2, v2];

Print["The angular momentum for particle 2 is: l2 = ", l2]

The angular momentum for particle 2 is: l2 = {0., 0., 6.}

L = l1 + l2;

Print["The total angular momentum is: l1+l2 = ", L]

The total angular momentum is: l1+l2 = {0., 0., 2.}

4.7 CHAPTER SUMMARY

The law of conservation of linear momentum states:

$$\text{If } \mathbf{F}^{\text{ext}} = 0, \quad \text{then } \mathbf{P} = \text{constant}.$$

In an *elastic* collision, both the total momentum *and* total kinetic energy are conserved. In an inelastic collision, only momentum is conserved.

The momentum of a *system of N particles* is conserved if there are no external forces acting on the system:

$$\dot{\mathbf{P}} = \mathbf{F} = \sum_{i=1}^{N} \mathbf{F}_i^{\text{ext}}$$

If $\mathbf{F}^{\text{ext}} = 0$, then $\mathbf{P} = $ constant.

A system of N particles has a center of mass located at $\mathbf{R}$:

$$\mathbf{R} = \frac{1}{M} \sum_{i=1}^{N} m_i \mathbf{r}_i$$

The center of mass for a continuous mass distribution is located at:

$$\mathbf{R} = \frac{1}{M} \int \mathbf{r} \, dm$$

To evaluate the mass dm and M for one-, two-, and three-dimensional objects we use:

$$dm = \lambda dx \qquad M = \int \lambda dx$$

$$dm = \sigma dA \qquad M = \int \sigma dA$$

$$dm = \rho dV \qquad M = \int \rho dV$$

For translational motion, we can think of a system of particles as a point particle located at the object's center of mass, and of the net external forces acting on the center of mass of the system. If M is the total mass and $\mathbf{P}$ is the total momentum:

$$\dot{\mathbf{P}} = M\ddot{\mathbf{R}} = \mathbf{F}$$

The angular momentum $\boldsymbol{\ell}$ of a single particle is the cross product of position vector $\mathbf{r}$ and linear momentum vector $\mathbf{p}$:

$$\boldsymbol{\ell} = \mathbf{r} \times \mathbf{p}$$

and the torque $\mathbf{N}$ is the cross product of position vector $\mathbf{r}$ and force vector $\mathbf{F}$:

$$\dot{\boldsymbol{\ell}} = \mathbf{r} \times \mathbf{F} = \mathbf{N}$$

The angular momentum $\mathbf{L}$ of a system of particles is the sum of two terms, the angular momenta of the particles about the center of mass, and that of the center of mass about the origin $(\mathbf{R} \times \mathbf{P})$:

$$\mathbf{L} = \sum_{i=1}^{N} (\mathbf{r}_i' \times \mathbf{p}_i') + \mathbf{R} \times \mathbf{P}$$

where $\mathbf{R}$ is the location of the center of mass relative to the origin, and $\mathbf{P}$ is the total momentum. If we differentiate $\mathbf{L}$ with respect to time, we find Newton's second law for Rotation:

$$\dot{\mathbf{L}} = \mathbf{N}^{\text{ext}}$$

From Newton's second law for rotation, we can write the law of conservation of angular momentum for a system of particles:

$$\text{If } \mathbf{N}^{\text{ext}} = 0, \text{ then } \mathbf{L} = \text{constant}$$

4.8   END-OF-CHAPTER PROBLEMS

The symbol ⌨ indicates a problem which requires some computer assistance, in the form of graphics, or numerical computation, or symbolic evaluation.

Section 4.1: Conservation of Momentum and Newton's Third Law

1. Prove that after an elastic collision, two particles of equal mass will have velocities that are perpendicular to each other. Assume that one particle is initially moving with velocity $\mathbf{v}_{1i}$, and the other is initially at rest.

2. Consider a perfectly inelastic collision between a particle m_1 initially moving with velocity $\mathbf{v}_0$ and another particle m_2 initially at rest. What fraction of the initial kinetic energy of mass m_1 is lost in the collision?

3. ⌨ A projectile of mass M is launched with a speed v_0 at an angle θ from the horizontal. At the top of its flight, the particle explodes into three pieces. The explosion adds an energy equal to three times the initial kinetic energy of the projectile. Immediately after the explosion, one piece with mass $M/3$ moves vertically upward. A second piece, with mass $M/6$ comes to rest. Find the velocity of each of the three pieces immediately after the explosion. Do this problem both by hand and by using a CAS.

4. A cannon ball with mass m is fired from level ground at a target a distance D from the cannon. The cannon ball is launched with an initial speed v_0 and at an angle of θ with respect to the horizon, such that it has the necessary range to strike the target. At the highest point in its trajectory, the cannon ball breaks up into two pieces of equal mass. One piece lands a distance $D/2$ from the cannon. Where does the other piece fall (as measured from the cannon)?

5. A particle with velocity $\mathbf{v}_0$ and mass m experiences an interaction that causes it to split into two smaller particles with masses $m_1 = m/4$ and $m_2 = 3m/4$. The particle with mass m_1 is observed moving at a speed of $0.01c$ in a direction which is 30 degrees counterclockwise from $\mathbf{v}_0$, and c is a constant. The mass m_2 is moving at a velocity one-third as fast as m_1. Find the direction of m_2 and the magnitude of $\mathbf{v}_0$.

6. Consider a head-on one-dimensional perfectly elastic collision between two particles with mass m_1 and m_2, where initially m_1 is moving at a speed v_0, and m_2 is at rest. Find the ratio of m_2/m_1 for which the kinetic energy of m_2 is maximum.

Section 4.2 Rockets

7. ⌨ Consider a rocket moving horizontally such that the only force acting on the rocket is a linear drag force of the form $F_{\text{drag}} = -bv\hat{\mathbf{v}}$ where b is a positive constant. Using Newton's second law, write the differential equation which describes the velocity of the rocket as a function of time, with the initial condition $v(0) = 0$ at time $t = 0$.

   a. Obtain the analytical solution $v(t)$ by hand.

   b. Using a CAS, solve the differential equation to find the velocity $v(t)$ of the rocket as a function of time. If you use Python, you will notice that the solution in SymPy obtained with `dsolve` may contain complex numbers, while the Mathematica solution contains only real constants. This is an example showing one of the limitations of CAS systems, and it is therefore important that one should compare (when possible) the analytical solutions with the CAS results.

8. A rocket needs to hover just above the surface of Mars which has a gravitational acceleration of $a = 3.8$ m/s^2. The exhaust velocity of the rocket is $u = 2000$ m/s but only 10% of its initial fuel can be burned. How long can the rocket hover?

9. ⌨ A model rocket has a constant burn rate $\alpha = -dm/dt$. Using (4.2.19) derived in this chapter,

$$v - v_0 = -gt + u\ln\left(\frac{m_0}{m}\right)$$

   calculate the height of the rocket as a function of time t. Find the height analytically and by using a CAS.

10. Solve these two rocket problems.

   a. A rocket has an initial mass m and burn rate α. What is the minimum exhaust velocity u needed to allow the rocket to lift-off from the Earth's surface immediately after its engines fire?

   b. The Saturn V rocket had a mass (when fully loaded with fuel) of 2.8 million kg. It consumed fuel at a rate of 20 tons per second. Find the exhaust velocity needed for the Saturn V to lift off from the Earth's surface immediately after its engines fire. Compare your answer to the escape velocity of the Earth.

11. Consider a two-stage rocket flying horizontally far from any other object. During the first phase of flight, the rocket with mass m_0 has exhaust velocity v. At the end of phase 1, the rocket has a mass $m_1 = m_p + m_{f1}$ where m_p is the mass of the payload and m_{f1} is the mass of the fuel container. At burnout, the rocket ejects the mass m_{f1} and begins to burn the fuel stored in the payload. At the end of the second burnout, the rocket's mass is m_2. The exhaust velocity during the second burnout is also u. Find the velocity of the rocket after the second burnout.

Sections 4.3: Center of Mass

12. Consider a two-particle system consisting of masses m_1 and m_2 separated by a distance r. Prove that the center of mass of the two-particle system lies along the line joining the two particles.

13. ⌨ Consider the cone in Example 4.5. Suppose it has a density of $\rho = (3+z)/2$, where the units of the constants are such that ρ is measured in SI units. Using a CAS, find the center of mass of the cone.

14. 🖥 Compute the location of the center of mass of a uniform solid hemisphere of radius R and density ρ. Perform the calculation using a CAS.

15. 🖥 Three particles are located at $\mathbf{r}_1 = 2\mathbf{i} - 2\mathbf{j} + 5\mathbf{k}$, $\mathbf{r}_2 = 5\mathbf{i} - 2\mathbf{k}$, $\mathbf{r}_3 = -\mathbf{i} + \mathbf{j} - \mathbf{k}$, with masses $m_1 = 2$ $m_2 = 5$ m_3. Using either SymPy or Mathematica, find:

 a. The center of mass of the system.

 b. The moment of inertia around the z-axis.

16. 🖥 Using either Mathematica or Python, calculate the center of mass of a lamina with a density of $\sigma = 2(1-x)(1-y^2)$ formed from the area defined by the curves $y = x^2$ and $y = x$, between $x = 0$ and $x = 1$.

17. 🖥 A uniform piece of metal is cut into the shape of the function $y = \sin x$ from $x = 0$ to $x = \pi$. The metal has a density of $\sigma = 2xy$. Compute the location of the center of mass. Solve this problem with a numerical integration algorithm.

18. 🖥 Compute the location of the center of mass of a lamina cut in the shape of $1/3$ of the unit circle with a density of $\sigma = 2xy$. Place one edge of the lamina on the positive x-axis and its "center" at the origin. Solve this problem using a CAS.

19. 🖥 Consider the triangular prism enclosed by the planes $z = 2$, $x = y = z = 0$, and $x + y = 2$. Plot the solid in three-dimensions by hand and using a CAS. Calculate its total mass $M = \iiint \rho \, dV$ when the volume mass density is given by $\rho = x\,y^2$, where all physical quantities are in SI units. Calculate the location of the center of mass of this object along the x-, y- and z-directions, by evaluating the appropriate triple integrals using a CAS.

Section 4.4: Momentum of a System of Multiple Particles

20. Consider a chain in a pile on the floor. The chain has a linear mass density of $\lambda = M/L$, where M is the total mass of the chain and L is the total length of the chain. One end of the chain is pulled horizontally with a constant force F. Assuming no friction is present, compute the position of the end of the chain $x(t)$ as a function of time.

21. A rope of length $2L$ and uniform mass density $\lambda = M/2L$ (where M is the mass of the rope) hangs in equilibrium over a peg in a wall. The rope is given a small perturbation such that it begins to fall freely, but it never leaves the peg during the fall. Find the velocity of the rope as it leaves the peg.

22. 🖥 A chain of length L and uniform mass density $\lambda = M/L$ hangs in a U shape from the ceiling, with its two ends parallel and close together. At time $t = 0$, one end of the chain is released. Find the tension T in the chain at the point of suspension, after the released end has fallen a distance x. You may assume that the end of the chain which is released is experiencing free fall. Plot the ratio of the tension to weight as a function of t/t_f, where $t_f = \sqrt{2L/g}$ is the time it takes for the released end to fall a distance L.

23. A rope of mass M and length L is stretched out along a table such that a length x_0 is hanging over the edge of the table. If the rope is released from rest and neglecting friction, show that the time it takes for the rope to completely slide off of the table is $t = T \cosh^{-1}(L/x_0)$, where $T = \sqrt{L/g}$.

24. ⌨ Consider the rope in Problem 23 in the presence of friction. Find the time it takes for the rope to slide off the table, if μ is the coefficient of friction between the rope and the table. Solve the problem analytically and using a CAS.

Sections 4.5 and 4.6: Angular Momentum of a Single Particle and of Multiple Particles

25. Evidence from ancient corals suggest that 350 million years ago, the Earth took 385 days to revolve around the Sun (Hadhazy [2]). The period of a year hasn't changed, but the Earth was rotating more quickly. What is the magnitude of the effective torque slowing the Earth's rotation over the last 350 million years? Assuming this torque remains constant, how long will it take for the Earth to stop rotating (i.e., to have an angular velocity equal to 0)?

26. Suppose you tie a ball of mass m to the end of a long string of length r_0 and whirl the ball around your head with an angular velocity ω_0, so that the ball travels in a circular path whose plane is parallel to the ground. At time $t = 0$, the string is shortened at a constant rate such that $\dot{r} = -\alpha$. Find the angular velocity ω of the ball as a function of time. Without plotting, qualitatively describe how ω changes with time. Find the time at which the ball's angular velocity is $16\,\omega_0$.

27. Consider a rigid rotating object. By breaking the object up into many small pieces of mass m_i, show that the components of the object's angular momentum parallel to the axis of rotation can be written as $L = I\omega$, such that:

$$I = \sum m_i r_i^2$$

where r_i is the position of the mass element m_i relative to the origin. Recall that I is called the total moment of inertia of the system of particles.

28. ⌨ Consider three particles m_1, m_2, m_3, located at (x_1, y_1), (x_2, y_2), and (x_3, y_3), respectively on the xy-plane. The system of masses rotates around an axis perpendicular to the xy-plane. Find the point P on the xy-plane, about which their total moment of inertia (see Problem 27) is minimized. Show that the result is the system's center of mass. Solve the problem analytically and by using a CAS.

29. Let $\mathbf{L}$ be the angular momentum of a system of discrete particles relative to the origin O. Let $\mathbf{L}'$ be the angular momentum of the same system with respect to another point O'. Find an equation for $\mathbf{L}'$ in terms of $\mathbf{L}$. You will need to consider the vector $\mathbf{r}_{O'}$ which represents the position of O' relative to O.

30. Consider a system of two particles of mass m_1 and m_2 located at positions $\mathbf{r}_1$ and $\mathbf{r}_2$, respectively. Suppose the particles are experiencing the force of interaction:

$$\mathbf{f}_{12} = -k\,(\mathbf{r}_2 - \mathbf{r}_1)$$

where k is a positive constant, and $\mathbf{f}_{12}$ is the force Particle 1 exerts on Particle 2. What is the net internal torque acting on the system? Discuss your result.

31. Consider a system of two particles of mass m_1 and m_2 located at positions $\mathbf{r}_1$ and $\mathbf{r}_2$, respectively. Suppose the particles are experiencing the force of interaction:

$$\mathbf{f}_{12} = -k\,(\dot{\mathbf{r}}_2 - \dot{\mathbf{r}}_1)$$

where k is a positive constant, $\mathbf{f}_{12}$ is the force Particle 1 exerts on Particle 2, and $\dot{\mathbf{r}}_2$, $\dot{\mathbf{r}}_1$ are the velocities of the two masses, respectively. What is the net internal torque acting on the system? Discuss your result.

32. 🖵 Consider two particles each with a mass $m = 1$ (all physical quantities in SI units). Particle 1 is located at $\mathbf{r}_1 = 3\ t\hat{\mathbf{i}} + 2\ t^2\hat{\mathbf{j}} + 6\ \hat{\mathbf{k}}$ and Particle 2 is located at $\mathbf{r}_2 = -3\ t^2\hat{\mathbf{i}} + 2\ t^3\hat{\mathbf{j}} - 4\ t\hat{\mathbf{k}}$, where t represents time. Compute the total angular momentum with respect to the origin of this system of particles. What is the net torque acting on this system? Solve this problem by hand and by using a CAS.

33. 🖵 Consider a system of three particles with masses and positions at time t: mass $m_1 = 1$ is at $\mathbf{r}_1 = 3t^2\mathbf{i} + 2(t-1)\mathbf{j} - 3\cos{(2t)}\,\mathbf{k}$, mass $m_2 = 2$ is at $\mathbf{r}_2 = t\mathbf{i} + 4t^3\mathbf{j} - \exp(-t)\mathbf{k}$, and mass $m_3 = 1.5$ is at $\mathbf{r}_3 = \mathbf{i} + t^2\mathbf{j} - 3\mathbf{k}$. All physical quantities are in SI units. Using a CAS, find:

   a. The center of mass of the system.

   b. The total linear momentum of the system.

   c. The total angular momentum of the system.

   d. The net force acting on the system.

   e. The net torque acting on the system.

5 Energy

In this chapter, we introduce the concepts of energy and conservative forces. We also discuss how energy can be used to describe the motion of a system, and as part of that discussion, we introduce the concept of equilibrium stability. We begin by studying one-dimensional systems because they are easy to visualize, and then extend the concept of work and energy to two and three-dimensional systems, including a discussion of line integrals. We conclude the chapter with a discussion of the energy of multiparticle systems.

5.1 WORK AND ENERGY IN ONE-DIMENSIONAL SYSTEMS

There are multiple ways of solving physics problems. Setting up the equations using Newton's second law requires one to know all of the forces acting on a system. Using Newton's second law leads to a second-order ODE that describes the particle's motion. The challenge with using forces is that forces are vectors, and, in two or three dimensions, we need to solve two or three second-order ODEs, respectively. An alternative approach to solving physics problems involves the concept of energy. Energy is a scalar quantity, and therefore, simpler to work with. The concept of energy is particularly useful when so-called *conservative forces* are acting on a system. In the case of conservative forces, one only needs to compare the total energy of the system before and after an event in order to describe the system's behavior. For example, to find the speed of a rock after it has fallen two meters, all one needs to do is compare the total energy of the rock before and after the fall. In this example, the fall is "the event."

In general, the mathematics of energy requires concepts of vectors and multi-variable calculus, and therefore formally involves higher-dimensional systems. In this section, we are going to present a brief survey of energy and energy conservation for one-dimensional systems. The mathematics for one-dimensional systems is much simpler than that for higher dimensional systems, allowing us to focus more on the concept of energy without getting mired in mathematics. The insights developed here will carry over to the higher-dimensional cases, where the mathematics is more complicated.

Consider the case where the net force $F(x)$ is position-dependent and is not explicitly dependent upon time and velocity. Suppose that this force is exerted on a particle moving from position x_0 to position x, along a line. Furthermore, if we define the positive x-direction to be the same as the direction of the force $F(x)$, then the work done by the force is:

$$W = \int_{x_0}^{x} F(x') \, dx' \tag{5.1.1}$$

where primes are used to distinguish between variables of integration and the limits of integration. The unit of work, the *Joule* is 1 J = 1 N·m. Note that work is negative if the object's displacement is in the opposite direction of the force. For simplicity, we will focus only on positive work right now.

Recall from Chapter 2 that if $F = F(x)$, then we can write:

$$F(x) = mv\frac{dv}{dx} \tag{5.1.2}$$

which can be written as:

$$F(x) = \frac{d}{dx}\left(\frac{1}{2}mv^2\right) \tag{5.1.3}$$

Notice that the right-hand side of the above equation is the derivative of the quantity $T = \frac{1}{2}mv^2$, called the *kinetic energy*:

$$F(x) = \frac{dT}{dx} \tag{5.1.4}$$

Separating variables and integrating:

$$\int_{T_0}^{T} dT' = \int_{x_0}^{x} F(x')dx' \tag{5.1.5}$$

and using the definition of kinetic energy, we find:

Work-Kinetic Energy Theorem

$$\frac{1}{2}mv^2 - \frac{1}{2}mv_0^2 = \int\limits_{x_0}^{x} F(x')dx' \tag{5.1.6}$$

Equation (5.1.6) is called the *work-kinetic energy theorem*, which states that work needs to be done by the force $F(x)$ in order to change the kinetic energy of a system. For example, the force of gravity does work on a falling object, and its kinetic energy increases.

Additional insight into the nature of work can be obtained from (5.1.6). If the change of the particle's kinetic energy ΔT is positive, then positive work is being done by F and the system is gaining kinetic energy. If however $\Delta T < 0$, then $F(x)$ is doing negative work and the particle is losing kinetic energy.

Next, we will add another restriction to $F(x)$. If the work done by $F(x)$ is independent of the path taken by the particle from x_0 to x, then we say that force is *conservative*. In a later section, we will study forces in two and three dimensions and develop a deeper mathematical formalism for conservative forces. However, in one dimension all forces of the form $F = F(x)$ are conservative. Let's consider the force of gravity exerted by an object near the Earth's surface. In this case, the force is the weight $\boldsymbol{w} = -m\,g\,\hat{\mathbf{j}}$, where the negative sign is introduced to denote that the weight points downward, in the negative y-direction. The work done by gravity is,

$$W = \int_{y_0}^{y} -mgdy' = -mg(y - y_0) \tag{5.1.7}$$

where we used y for vertical displacements. Notice that the work done by gravity depends only on the end points of the particle's motion.

Consider a falling rock (in a vacuum, of course!). We noted earlier that the rock will increase its kinetic energy as it falls. Where does that kinetic energy come from? According to (5.1.6), the force of gravity does work on the rock, and (5.1.7) tells us how much work is done by gravity. Another way of thinking of the process is to say that during the fall, the rock's energy is changing from one form into another. While the kinetic energy is the energy associated with the motion of the rock, there is an additional energy associated with the configuration of the rock-Earth system. In this case, the important parameter is the rock's height above the ground. In fact, (5.1.7) supports the idea of configuration-dependent energy. The term $y - y_0$ in (5.1.7) says that the work done by gravity depends on how the distance between the Earth's surface and the rock changes during the motion. That change of distance is a change in the configuration of the rock-Earth system. The energy associated with the configuration of the rock-Earth system, or any system, is called the *potential energy*. The potential energy function $V(x)$ in one-dimension is defined by:

$$F(x) = -\frac{dV(x)}{dx} \tag{5.1.8}$$

Notice that (5.1.8) says that forces act so as to decrease the potential energy along x. Hence, in our rock-Earth example, the force of gravity is decreasing the rock's potential energy and increasing the rock's kinetic energy. Furthermore, this example illustrates that forces transfer one form of energy into another, in this case, potential energy transferring into kinetic energy.

To find the potential energy $V(x)$ for a given force $F(x)$, we need to integrate the force:

$$V(x) - V(x_0) = -\int_{x_0}^{x} F(x')dx' \tag{5.1.9}$$

For example, the gravitational potential energy is:

$$V(y) - V(y_0) = -\int_{y_0}^{y} -mgdy' = mg(y - y_0) \tag{5.1.10}$$

If we choose $y_0 = 0$ (and $V(y_0) = 0$), then we recover the familiar equation, $V = mgy$.

Equation (5.1.9) can be rewritten as $W = -\Delta V$ and tells us that potential energy also measures the ability of the system to do work. By inserting (5.1.8) into (5.1.6) we get:

$$T - T_0 = \frac{1}{2}mv^2 - \frac{1}{2}mv_0^2 = \int_{x_0}^{x} \left(-\frac{dV(x')}{dx'}\right) dx' = V(x_0) - V(x) \tag{5.1.11}$$

or, after simplification:

Conservation of Mechanical Energy

$$T + V(x) = T_0 + V(x_0). \tag{5.1.12}$$

This equation is known as the *conservation of mechanical energy*, stating that the sum of the kinetic and potential energies, called the *total mechanical energy*, is always constant if only conservative forces act on the system. Conservative forces conserve energy!

Conservative forces are very useful in physics as they allow us to work with a scalar quantity, the mechanical energy $E = T + V$. Mechanical energy is a useful way of solving physics problems because instead of needing to keep track of the magnitude and direction of vectors, as one needs to do with forces, all one needs to do is keep track of how the value of the different types of energies are changing as the particle moves. Looking at (5.1.12), we see that solving problems using conservation of energy requires only that we compare one state of the system to another. For example, suppose we drop a rock starting at rest from a height of 5 m above the ground and we want to know the kinetic energy of the rock after it fell 2 m. All we need to know is the initial kinetic energy (which is zero) and the initial and final potential energies in order to obtain the final kinetic energy (and thus the speed). We don't need information about the state of the system during the actual fall!

Equation (5.1.12) also tells us that if the potential energy of the system decreases, its kinetic energy must increase and vice versa, in order to maintain the equality of both sides of the equation. The ability to interpret the meanings behind equations such as (5.1.8) and (5.1.12) is an important skill for physicists.

Example 5.1: Simple harmonic motion and conservation of energy

Consider a mass m on a spring with spring constant k performing simple harmonic motion with amplitude A. The mass is initially at the origin at rest. The force is given by Hooke's Law $F = -kx$ where x is the displacement from equilibrium.

(a) Find the total mechanical energy in terms of A.

(b) Find an expression for the turning points in the motion, i.e., the points at which the velocity v of the particle is zero and the mass turns around, as a function of the total mechanical energy E and spring constant k.

Solution:

(a) We solve this problem using energy conservation. In order to use energy conservation, we need to first find the potential energy associated with the spring. The potential energy associated with the force of the spring which is called the *elastic potential energy* and can be found using:

$$V(x) - V(x_0) = -\int_{x_0}^{x} (-kx')dx' = \frac{1}{2}k\left(x^2 - x_0^2\right) \tag{5.1.13}$$

Therefore, we can write the elastic potential energy as $V(x) = 1/2kx^2$. When the mass reaches the amplitude of oscillation, $x_0 = A$, the mass has traveled as far from equilibrium as it can, and it comes to a stop. Therefore, $v_0 = 0$ and (5.1.12) becomes:

$$T_0 + V(x_0) = T + V(x)$$

$$0 + \frac{1}{2}kA^2 = \frac{1}{2}mv^2 + \frac{1}{2}kx^2$$

Therefore, using $E = T + V(x)$:

$$E = \frac{1}{2}mv^2 + \frac{1}{2}kx^2 = \frac{1}{2}kA^2 \tag{5.1.14}$$

we find that the total mechanical energy in terms of A is $E = \frac{1}{2}kA^2$.

(b) At the turning points x_{TURN} the velocity $v_{\text{TURN}} = 0$ so (5.1.14) gives:

$$E = 0 + \frac{1}{2}k\left(x_{\text{TURN}}\right)^2 = \frac{1}{2}kA^2$$

Therefore, the expression for the turning points in the motion of a simple harmonic oscillator is:

$$x_{\text{TURN}} = \pm A = \sqrt{2E/k} \tag{5.1.15}$$

In the next section, we will discuss turning points in more detail.

What happens if $F(x)$ is not conservative? Friction is an example of a nonconservative force; however, we can still use (5.1.1) to find the work done by friction. Recall that the force of friction $F = \mu F_N$, where μ is the coefficient of kinetic friction and F_N is the normal force, which in this case is constant in magnitude. The work done by friction is then:

$$W_{nc} = \int_{x_0}^{x} -\mu F_N dx' = -\mu F_N\left(x - x_0\right) \tag{5.1.16}$$

where we used W_{nc} to denote that the work is done by a nonconservative force. Note that we explicitly included the minus sign in (5.1.16) because the force of friction opposes the direction of motion, and we consider the displacement to be in the positive direction.

Consider the case of a particle, again constrained to move along a line, experiencing multiple forces, both conservative and nonconservative. The *net work* W_{net} is the total work done by all of the forces. The net work can be written as $W_{\text{net}} = W_c + W_{nc}$, where W_c is the work done by the conservative forces, and W_{nc} is the work done by nonconservative forces. In regards to the work in the above example, $W_{\text{net}} = \Delta T$ and $W_c = -\Delta V$. Therefore,

$$\Delta T = W_{nc} - \Delta V \tag{5.1.17}$$

$$W_{nc} = \Delta T + \Delta V \tag{5.1.18}$$

$$W_{nc} = \Delta E \tag{5.1.19}$$

or, in other words, the work done by nonconservative forces is equal to the change of the system's mechanical energy. So in cases where conservative forces act on the system, kinetic energy gets exchanged for potential energy and vice versa. However, when nonconservative forces act on the system, then mechanical energy can be lost or gained. An object falling through air loses potential energy to both kinetic energy *and* work done by friction heating the object and the surrounding air. However, if we account for the heat, we will find that the total energy is conserved.

In the next section, we will study how to use potential energy to describe the motion of an object.

5.2   POTENTIAL ENERGY AND EQUILIBRIUM POINTS

An equilibrium point is a point in the system's motion where both the net force and net torque are equal to zero. Since in this chapter we are not dealing with rotational motion, we will focus only on the condition that $\boldsymbol{F}_{net} = 0$. Knowing the location of equilibrium points in a system can give us insight into the system's behavior. The location of an equilibrium point x_0 can be found using the potential energy. Equation (5.1.8) tells us that a particle is in an equilibrium state if:

$$\left.\frac{dV}{dx}\right|_{x_0} = 0 \tag{5.2.1}$$

Hence, the local maxima or minima of the potential energy gives the location of the system's equilibrium points, x_0. As we will see, the plot of $V(x)$ can be used to describe the motion of a particle.

Physical systems can be in three types of equilibria: stable, unstable, and neutral. An example of a *stable equilibrium* is shown at the position near $x = 0.7$ mm in Figure 5.1. This corresponds to values for which the potential energy $V(x)$ is a *local minimum*. In other words, $d^2V/dx^2 > 0$ at the equilibrium position. In the case of stable equilibrium, the forces acting on the particle will tend to restore the equilibrium. The restoring nature of the force can be seen in the graph of $V(x)$, by looking at the slope of $V(x)$ on either side of the equilibrium point. Note that the sign of the force is opposite of the sign of the potential energy's slope. Hence a displacement to the right of the equilibrium, where $V(x)$ has a positive slope, is met with a negative (left pointing) force. Similarly, to the left of the equilibrium, $V(x)$ has a negative slope, and therefore a particle displaced to the left of equilibrium experiences a positive (right pointing) force.

An unstable equilibrium is shown at $x = 0$ in Figure 5.1. This corresponds to values for which the potential energy $V(x)$ is a *local maximum*, or in other words, $d^2V/dx^2 < 0$ at the equilibrium position. This means that if the particle is displaced an arbitrarily small distance from the equilibrium state, the force causes it to move even farther away. By studying the slope of $V(x)$ on either side of the equilibrium, we can see that the particle experiences a positive (rightward) force when displaced to the right of equilibrium, and a

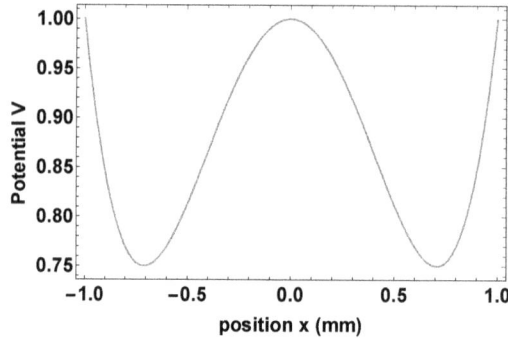

Figure 5.1: Graph of a potential energy function which shows stable equilibrium points near $x = 0.7$ and $x = -0.7$ and an unstable equilibrium at the origin $(x = 0)$.

negative (leftward) force when displaced to the left of the equilibrium position, hence the force acts to move the particle farther from equilibrium.

A final type of equilibrium not shown in Figure 5.1 is called *neutral equilibrium*. A neutral equilibrium corresponds to situations for which the potential energy $V(x)$ is a *constant*, and hence its derivative is zero. In the case of neutral equilibrium, the system will tend to remain in equilibrium if displaced by a small amount.

Example 5.2: Classifying the equilibrium position of a system

Consider the object shown in Figure 5.2. The object consists of a hemisphere with a cylinder on top. The center of the hemisphere is at $y = R$, and when the object is upright, the center of mass of the object is located a height $y = h$ above the point of contact with the ground. The object is tipped an angle θ with respect to the vertical. Find the potential energy of the object when it is tilted. Show that if $h > R$, then the upright equilibrium position is unstable.

Solution:

It is helpful to include a diagram before we perform any calculations.

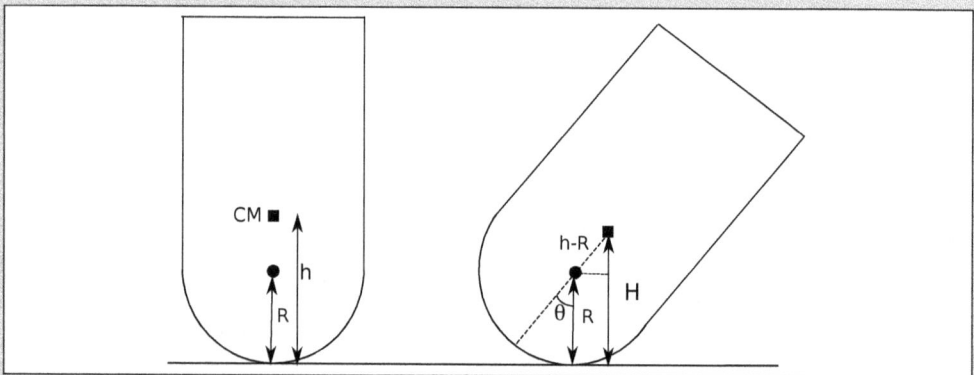

Figure 5.2: The object whose equilibrium is studied in Example 5.2.

The figure on the left shows the object in its equilibrium state (upright). The figure on the right shows the object tilted an angle θ from the vertical. The vertical line passing through the center (circle) and center of mass (square) denotes the equilibrium positions. From the second diagram, we can see that the height of the center of mass is $H = R + (h - R) \cos \theta$. Therefore, the potential energy with respect to the ground is:

$$V = mg\left[R + (h - R) \cos \theta\right] \tag{5.2.2}$$

Next, to get the stability of the equilibrium state, we take two derivatives of V:

$$\frac{dV}{d\theta} = - mg(h - R) \sin \theta \tag{5.2.3}$$

$$\frac{d^2V}{d\theta^2} = - mg(h - R) \cos \theta \tag{5.2.4}$$

Notice that the first derivative $dV/d\theta$ correctly identifies $\theta = 0$ as an equilibrium condition, since $\sin \theta = 0$ at $\theta = 0$. Furthermore, if $\theta = 0$ is inserted into the second derivative, we find that the second derivative is negative when $h > R$. Therefore, the equilibrium at $\theta = 0$ is unstable when $h > R$. Additionally, the upright position is stable if $h < R$, since in that case the second derivative is positive.

Energy can be used to get a qualitative understanding of a particle's motion. Let us consider a particle whose potential energy $V(x)$ is that shown in Figure 5.3. We can imagine the particle's motion being similar to a ball rolling along a "frictionless track" in the shape of $V(x)$. For example, suppose we started the particle at rest at $x = x_1$. This would be similar to holding the ball on the "$V(x)$ track" at x_1. If we released the ball, we would expect it to roll along the track speeding up as it approaches the local minimum, then slowing down and coming to a stop at $x = x_2$, where it would reach its maximum allowed height, $V(x_2)$. The ball would then turn around and head back toward x_1 and repeat the motion. The points x_1 and x_2 are called *turning points*.

While the "frictionless track" described in the above paragraph is useful for visualizing the particle's motion, we will now examine the motion between the turning points x_1 and x_2 in more detail. Again, consider the case where the particle starts at rest at the point, $x = x_1$, in Figure 5.3. At $x = x_1$, the particle has energy E_1 and all of the energy is in the form of potential energy. When released, a force to the right ($V(x)$ has a negative slope at x_1) pushes the ball to the stable equilibrium (local minimum). At the stable equilibrium near $x = 0.5$, the particle would have its highest velocity because it has reached its lowest potential energy. Past the equilibrium point, the force begins to oppose the particle's velocity (note that $V(x)$ has a positive slope in this region). Because the potential energy is increasing, the particle's kinetic energy would decrease, in order to conserve the total mechanical energy. At $x = x_2$, the particle has lost all of its kinetic energy and comes to a stop. However, it is still experiencing a leftward force, and so the particle begins to move back towards x_1, where it will once again come to a stop and move back towards x_2. We see that the resulting motion is an oscillation between the turning points, x_1 and x_2. Both methods of description, the "frictionless track" and the more detailed analysis in this paragraph, tell us that a particle with an initial position of $x_0 = x_1$ and initial velocity $v_0 = 0$, will oscillate about the stable equilibrium reaching an amplitude of x_1 to the left of equilibrium and x_2 to the right of equilibrium.

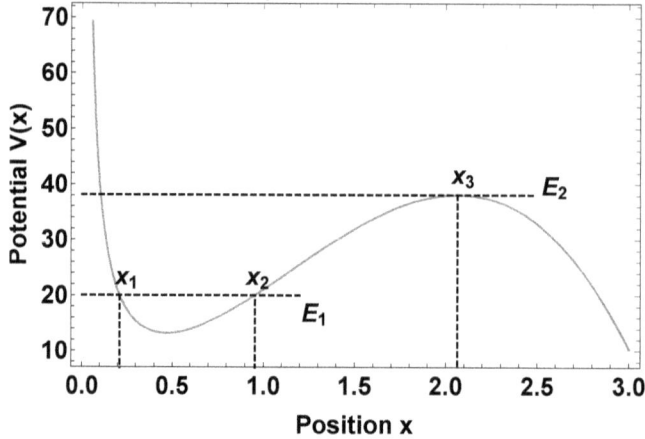

Figure 5.3: Potential energy function demonstrating turning points.

We can, of course, use mathematics to describe this motion. Using energy conservation, $E = T + V = mv^2/2 + V(x)$, we can solve for the velocity as a function of position:

$$v = \pm\sqrt{\frac{2}{m}\left(E - V(x)\right)} \qquad (5.2.5)$$

For a particle with energy $E = E_1$, we can see from Figure 5.3 that $V(x_1) = E_1$ and $V(x_2) = E_1$ and therefore, using (5.2.5), the velocity is zero at the turning points, $x = x_1$ and $x = x_2$. Furthermore, for $x > x_2$ and $x < x_1$, the velocity is a complex number because $V(x) > E$ in those regions. Complex velocities are not physical, and this means that the particle does not have enough energy to access those regions. For a particle with $E = E_1$, the regions $x < x_1$ and $x > x_2$ are called *classically forbidden regions*. According to the physics of classical mechanics, the particle cannot be in classically forbidden regions, because it doesn't have enough energy to access those regions. However, such regions are accessible in quantum mechanical systems.

Using $v = dx/dt$, we can find the position by integrating (5.2.5):

$$t - t_0 = \int_{x_0}^{x} \frac{dx'}{\sqrt{\frac{2}{m}\left(E - V(x')\right)}} \qquad (5.2.6)$$

Equation (5.2.6) can be a difficult way to solve for $x(t)$ in closed form, for all but the most simple of potential energy functions.

We can repeat our qualitative analysis for the motion of a particle starting at rest at the unstable equilibrium x_3 in Figure 5.3. In that case, if the particle is perturbed to the right, it increases its velocity as the potential decreases. The graph ends at $x = 3$, so we cannot tell what will happen to the particle for $x > 3$. If the particle is deflected to the left, then it will increase in speed until it passes through the stable equilibrium, after which it will slow down until it reaches $x \approx 0.1$, before turning around. Mathematically, the particle will then reach the unstable equilibrium coming to a stop.

Notice that both Figures 5.1 and 5.3 have local minima. The motion around local minima corresponds to an oscillatory motion, if the energy of the particle is low enough. Let us understand why that is the case using mathematics. Consider a particle near a stable equilibrium $x = x_0$, hence, the equilibrium position x_0 needs to be a local minimum of $V(x)$. Next, we ask the question, what is the mathematical form of $V(x)$ near the stable equilibrium point? We perform a Taylor expansion in terms of x around an energy minimum ($x = x_0$). A Taylor expansion gives a polynomial approximation of a function near a particular point, and its general form is:

$$V(x) = V(x_0) + (x - x_0)V'(x_0) + \frac{1}{2}(x - x_0)^2 V^{(2)}(x_0) + O(x - x_0)^3 \qquad (5.2.7)$$

where $O(x - x_0)^3$ indicates terms of the order of $(x - x_0)^3$ and higher, and the notations $V'(x_0)$ and $V^{(2)}(x_0)$ indicate the first and second derivatives of V evaluated at $x = x_0$, respectively. If we consider the case where the displacement from equilibrium is small, then $x - x_0$ is small and terms $O(x - x_0)^3$ and higher are negligible. Because $V(x_0)$ is a minimum, $V'(x_0) = 0$, so the linear term drops out:

$$V(x) = V(x_0) + \frac{1}{2}(x - x_0)^2 V^{(2)}(x_0) + O(x - x_0)^3 \qquad (5.2.8)$$

The constant term $V(x_0)$ is arbitrary and thus may also be dropped. This is equivalent to saying that we can set the zero point of the energy to be any value, hence we set $V(x_0) = 0$. Or to put it another way, constant energy terms do not affect the particle's motion, because the force acting on the particle is the derivative of the potential energy, and therefore constant terms do not contribute to the force. With all of this taken into account, we obtain the form of $V(x)$ for the simple harmonic oscillator :

$$V(x) \approx \frac{1}{2}(x - x_0)^2 V^{(2)}(x_0) = \frac{1}{2}k(x - x_0)^2 \qquad (5.2.9)$$

where we set $k = V^{(2)}(x_0)$. Hence, we find that Hooke's Law appears any time we are studying small amplitude oscillations about a stable equilibrium. Thus, given an arbitrary potential energy function $V(x)$ with a local minimum (and a non-vanishing positive second derivative), we can use the solution to the simple harmonic oscillator found in Example 2.7 to provide a solution for small perturbations around the equilibrium point. As the perturbations get larger, however, we need to include higher order terms in the Taylor expansion of $V(x)$, and Hooke's Law no longer applies. These so-called *nonlinear oscillators* can be very difficult to solve. We will study nonlinear oscillators in Chapter 13.

It is worth mentioning that (5.2.9) gives us a method of finding the frequency of small oscillations about the stable equilibrium. Recall that the frequency of oscillations for a particle trapped in a Hooke's Law potential is $\omega = \sqrt{k/m}$. Therefore, the frequency of small oscillations about a stable equilibrium x_0 can be found from:

$$\omega = \sqrt{\frac{k}{m}} = \sqrt{\frac{V^{(2)}(x_0)}{m}} \qquad (5.2.10)$$

Example 5.3: The Lennard-Jones potential.
   The potential energy of the interaction between a pair of neutral atoms or molecules is often described by the Lennard-Jones potential function, also referred to as the 6-12 potential:

$$V(x) = 4\epsilon \left[\left(\frac{\sigma}{x}\right)^{12} - \left(\frac{\sigma}{x}\right)^6 \right] \qquad (5.2.11)$$

where σ, ϵ, x, and V are in appropriate SI units. Setting $\sigma = 1$ and $\epsilon = 1/4$, determine the type of equilibrium possible in this potential, and obtain a parabolic approximation of this potential near the equilibrium point. Solve this problem by hand and by using Python and Mathematica.

Solution:
We find the equilibrium point by setting the derivative of $V(x)$ equal to zero and solving for x:

$$\frac{dV(x)}{dx} = 12x^{-13} - 6x^{-7} = 0$$

$$x = 2^{1/6} = 1.122$$

We can easily verify that this is a stable equilibrium point by examining the second derivative. We can also use Python and Mathematica to expand the function $V(x)$ around the equilibrium point:

$$V(x) = -0.25 + 7.143(x - x_0)^2 \tag{5.2.12}$$

A plot of (5.2.11) and (5.2.12) appear in Figure 5.4. The dashed line is the series approximation near the stable equilibrium, and the solid line is the Lennard-Jones potential. Notice that the approximation fails to provide a reliable estimate of the potential energy the farther the particle moves from the equilibrium position.

Python Code
We begin the code by defining the Lenard-Jones potential, V. We then solve for the equilibrium points. Notice the use of the .subs() method to get a numerical answer from solve. There are many roots, but only one is positive and real. Next, we perform a Taylor series expansion about the value x0 using the SymPy command series. Note that we ask Python to compute up to the third order in the expansion. That ensures we get the quadratic terms.

The method .removeO() removes the order term that Python includes at the end of the series by default. Finally, we use lambdify to convert the SymPy expressions V and expansion into functions that can be plotted, and we create the graph.

```
from sympy import symbols, series, diff,solve, simplify,lambdify,evalf,N
import numpy as np
import matplotlib.pyplot as plt

print('-'*28,'CODE OUTPUT','-'*29,'\n')

x, s, ep = symbols('x, s, ep')

V = 4*ep*( (s/x)**12 - (s/x)**6)

#solve for equilibrium points
x0 = solve(diff(V,x).subs([(s,1),(ep,1/4)]),x)[1]

print('Equilibrium point x0 = ',  N(x0,6))

expansion = simplify(series(V, x, 1.122462, 3).removeO() \
    .subs([(s,1),(ep,0.25)]))

print('\nNear x0: V(x) = ',  expansion.subs([(s,1),(ep,0.25)]).evalf(4))

expansion_plot = lambdify(x,expansion.subs([(s,1),(ep,1/4)]),'numpy')

V_plot = lambdify(x,V.subs([(s,1),(ep,1/4)]),'numpy')

x_values = np.linspace(0.95,2.0,50)

plt.plot(x_values,V_plot(x_values), 'k',label = "Lennard-Jones")
plt.plot(x_values,expansion_plot(x_values),'k--', label='Approximation')
plt.ylim(-0.3,0.5);
plt.xlabel('x')
plt.ylabel('V(x)')
plt.legend()
plt.show()

--------------------------- CODE OUTPUT ----------------------------
Equilibrium point x0 =  1.12246

Near x0: V(x) =  7.143*x**2 - 16.04*x + 8.75
```

Mathematica Code

After defining V, we find the equilibrium position x0 using **Solve** and choose the only positive real root. Next, we compute the Taylor expansion about x_0 using **Series**. We apply the function **Normal** to make the series computable (and to remove the order term). **Chop** replaces real numbers in the expression that are close to zero with the exact value of zero.

```
SetOptions[Plot, Axes->True, BaseStyle->{FontSize->16},

PlotStyle->{{Normal, Black}, {Black, Dashed}}];


σ = 1; ε = 1/4;

V = 4 * ε * ((σ/x)^12 − (σ/x)^6);


x0 = Solve[D[V, x]==0, x][[2]]//N;

Print["Equilibrium point x0 = ", x/.x0]

Equilibrium point x0 = 1.12246

expansion = Series[V, {x, x/.x0, 2}]//Normal//Chop;

Print["Near x0, the potential V(x) = ", expansion//Expand]

Near x0, the potential V(x) = 8.75 − 16.0362x + 7.1433x^2

Plot[{V, expansion}, {x, 0.95, 1.4}, PlotRange->{All, {−0.3, 0.5}},

AxesLabel->{"position,x", "V(x)"}]
```

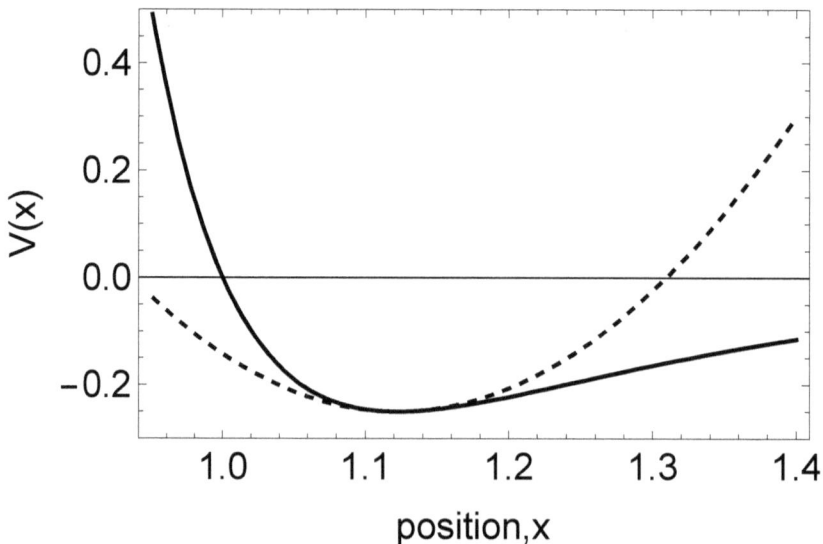

Figure 5.4: Output of the Mathematica code for Example 5.3. The solid line is the Lenard-Jones potential in (5.2.11), and the dashed line is the Taylor series approximation.

Example 5.4: Motion near a stable equilibrium point.

Consider the potential energy illustrated in Figure 5.3, $V(x) = a/x + bx^2 - cx^3$, where $a = 4$ Jm, $b = 25$ J/m^2, and $c = 8$ J/m^3. Numerically solve for the position $x(t)$ as a function of time for a mass of $m = 1$ kg whose initial conditions are $x(0) = 0.21$ m (i.e., very close to x_1 in Figure 5.3), and $\dot{x}(0) = 0$.

Solution:

We begin by differentiating $V(x)$ with respect to x in order to obtain the force $F(x)$:

$$F(x) = -\frac{dV}{dx} = \frac{a}{x^2} - 2bx + 3cx^2$$

The next step is to use Newton's second law in order to create a second-order ODE, which can be solved for $x(t)$:

$$m\ddot{x} = \frac{a}{x^2} - 2bx + 3cx^2 \qquad (5.2.13)$$

We plot the motion using Python and Mathematica. The output of the Mathematica code is shown in Figure 5.5. Notice that the motion is periodic, oscillating between points $x = 0.21$ and $x = 0.95$, as suggested in Figure 5.3.

Python Code

To solve (5.2.13) using `odeint`, we must rewrite it as two first-order equations using the variable, $v = dx/dt = \dot{x}$:

$$\dot{x} = v$$
$$\dot{v} = \frac{1}{m}\left(\frac{a}{x^2} - 2bx + 3cx^2\right)$$

This code parallels many of the other codes in which we used `odeint` to solve an ODE.

```
from scipy.integrate import odeint
import numpy as np
import matplotlib.pyplot as plt

a, b, c, m = 4, 25, 8, 1

inits = [0.21, 0]  # initial conditions

# derivatives function for odeint
def deriv(y,t):
    x, v = y
    dydt = [v, (a/x**2 - 2*b*x + 3*c*x**2)/m ]
    return dydt

times = np.linspace(0, 2.0, 100)

# solve the ode using odeint
soln = odeint(deriv, inits, times)

plt.plot(times,soln[:,0])
plt.xlabel('time (s)')
plt.ylabel('position (m)')
plt.show()
```

Mathematica Code
   We use `NDSolve` to integrate (5.2.13).

SetOptions[Plot, Frame->True, Axes->False, BaseStyle->{FontSize->16}];

$a = 4; b = 25; c = 8; m = 1;$

soln = NDSolve[$\{m * x''[t]$==$a/x[t]^2 - 2 * b * x[t] + 3 * c * x[t]^2, x[0]$==$0.21,$

$x'[0]$==$0\}, x, \{t, 0, 2\}$];

Plot[$x[t]/$.soln, $\{t, 0, 2\}$, FrameLabel->{"time (s)", "position (m)"}]

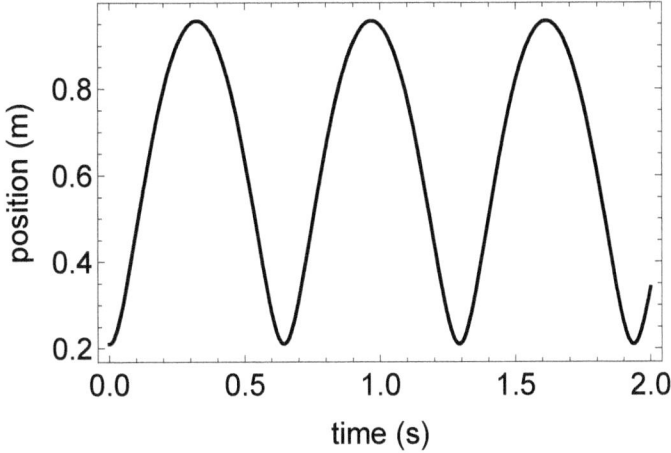

Figure 5.5: Output of the Mathematica code for Example 5.4.

5.3 WORK AND LINE INTEGRALS

Next, we extend the concept of work to higher dimensional systems, i.e., cases where the force $\mathbf{F} = \mathbf{F}(\mathbf{r})$. In this case, the work done by the force $\mathbf{F}$ moving an object between points $\mathbf{r}_1$ and $\mathbf{r}_2$ is generalized from (5.1.1) to be:

Work Done by Force F

$$W(\mathbf{r}_1 \to \mathbf{r}_2) = \int_{\mathbf{r}_1}^{\mathbf{r}_2} \mathbf{F} \cdot d\mathbf{r} \qquad (5.3.1)$$

Notice the dot product between the force $\mathbf{F}$ and the infinitesimal displacement $d\mathbf{r}$ in (5.3.1). The dot product tells us that work calculations involve only the component of the force parallel to the displacement, i.e., the component of the force perpendicular to the displacement does no work. Note that (5.3.1) is also applicable to one-dimensional systems, and we no longer need the restriction that $\mathbf{F}$ is in the same direction as $d\mathbf{r}$.

The right-hand side of (5.3.1) is a *line integral*. In other words, the integral is evaluated along a specific path. The value of W can depend on the path taken by the particle. It is common to rewrite (5.3.1) including the path C followed by the particle:

$$W = \int_C \mathbf{F} \cdot d\mathbf{r} \qquad (5.3.2)$$

In order to calculate integrals of this type, one can break them up into component integrals, by writing out the dot product of the components of the force vector $\mathbf{F} = F_x\hat{\mathbf{i}} + F_y\hat{\mathbf{j}} + F_z\hat{\mathbf{k}}$ and the differential $d\mathbf{r} = dx\hat{\mathbf{i}} + dy\hat{\mathbf{j}} + dz\hat{\mathbf{k}}$, as follows:

$$W = \int_C F_x dx + \int_C F_y dy + \int_C F_z dz \qquad (5.3.3)$$

and then evaluate the individual integrals along the path C. The next several examples demonstrate how to calculate path integrals for the three paths shown in Figure 5.6.

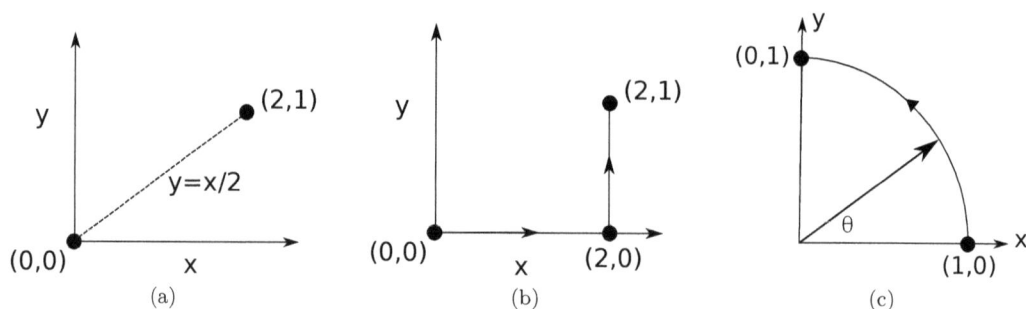

Figure 5.6: Three paths used in Examples 5.5-5.7.

Example 5.5: Calculating the work done by a force along a straight line path.
Find the work done by the force $\mathbf{F} = x\,y\,\hat{\mathbf{i}} - y^2\hat{\mathbf{j}}$ along the straight line path connecting the origin to the point $(2,1)$ as indicated in Figure 5.6a. Do this problem analytically (with and without parameterizing the path) and by using Python and Mathematica.

Solution:
This problem is most easily done in Cartesian coordinates. We write for the work

$$W = \int_C F_x dx + \int_C F_y dy + \int_C F_z dz = \int_C x\,y\,dx - \int_C y^2 dy \qquad (5.3.4)$$

The straight line path shown in the figure can be written as $y = x/2$, and therefore the differentials are related by $dy = dx/2$. The x-coordinate along path AB changes from $x = 0$ to $x = 2$. By substituting these equations for y, dy into the above expression, we end up with simple integrals over the x-axis only:

$$W = \int_0^2 x\,\frac{x}{2}\,dx - \int_0^2 \left(\frac{x}{2}\right)^2 \frac{dx}{2} = \int_0^2 \frac{3}{8}x^2\,dx = 1 \qquad (5.3.5)$$

We can also evaluate the work by parameterizing the path: we introduce the parameter $x = t$ such that t varies from $t = 0$ to $t = 2$. Along the line connecting the origin to the point $(2,1)$ we then have $y = t/2$, and also $dy = dt/2$. The work becomes:

$$W = \int_C F_x dx + \int_C F_y dy = \int_0^2 \left(t^2/2\right) dt - \int_0^2 \left(t/2\right)^2 dt/2 = 1 \qquad (5.3.6)$$

Python Code
In Method #1 we use `integrate` in SymPy to evaluate the two integrals. In the Python code we define the line from the origin to the point $(2,1)$ by using `ParametricRegion((t,t/2),(t,0,2)`, and we store this as the variable `diagonal`. The work using Method #2 is calculated using `ParametricIntegral(F,diagonal)` where F represents the force vector $\mathbf{F} = xy\hat{\mathbf{i}} - y^2\hat{\mathbf{j}}$.

```
from sympy import integrate, symbols
from sympy.vector import  CoordSys3D, ParametricIntegral,ParametricRegion

print('-'*28,'CODE OUTPUT','-'*29,'\n')

x, y, t, x1, y1 = symbols('x, y, t, x1, y1')

C = CoordSys3D('C')      # define Cartesian system, named here C

# Method 1: using two line integrals

intx=integrate(x1*(x1/2), (x1,0,2))

inty=integrate(-y1**2, (y1,0,1))

print('The line integral using two line integrals = ',intx + inty)

# Method 2: using ParametricRegion and ParametricIntegral

x = t
y = t/2

F = x*y*C.i - y**2*C.j   # Force vector with unit vector C.i, C.j

diagonal = ParametricRegion((t, t/2), (t, 0, 2))

W = ParametricIntegral(F, diagonal)
print('The line integral using ParametricIntegral = ',W)
--------------------------- CODE OUTPUT ----------------------------
The line integral using two line integrals =  1
The line integral using ParametricIntegral =  1
```

Mathematica Code

After defining the force **F** as a vector field, we first perform two separate line integrals using the **Integrate** command. Next, we use **LineIntegrate** where the first argument is the vector field. The second argument is the variables written such that they are along the line that joins the origin to the point $(2, 1)$ defined in Mathematica using **Line[0,0,2,1]**.

```
F = {x * y, -y^2};

(* Method 1 : Using two line integrals *)

intx = Integrate[x * x/2, {x, 0, 2}];
inty = Integrate[-y^2, {y, 0, 1}];

total = intx + inty;

Print["The line integral using two line integrals = ", total]
The line integral using two line integrals = 1
(* Method 2 : Using LineIntegrate *)

lineInt = LineIntegrate[F, {x, y} ∈ Line[{{0, 0}, {2, 1}}]];

Print["The line integral using LineIntegrate = ", lineInt]
The line integral using LineIntegral = 1
```

Example 5.6: Calculating the work done by a force along a path consisting of multiple straight-line segments

Find the work done by the force $\mathbf{F} = x\,y\,\hat{\mathbf{i}} - y^2\hat{\mathbf{j}}$ from the previous example along the line path $(0,0) \to (2,0) \to (2,1)$, indicated in Figure 5.6b. Do the problem by hand and by using ParametricIntegral in Python and LineIntegrate in Mathematica. Does the work done by the force depend on the path taken between points $(0,0)$ and $(2,1)$?

Solution:

The work in this case is calculated as the sum of two segments, $(0,0) \to (2,0)$ and $(2,0) \to (2,1)$.

In the first segment we have $y = 0$ and therefore $dy = 0$, while x varies from $x = 0$ to $x = 2$.

$$W = \int_{(0,0)\to(2,0)} F_x dx + \int_{(0,0)\to(2,0)} F_y dy \tag{5.3.7}$$

$$= \int_{(0,0)\to(2,0)} xy dx - \int_{(0,0)\to(2,0)} y^2 dy \tag{5.3.8}$$

$$= 0 + 0 = 0 \tag{5.3.9}$$

The first integral above is zero because $y = 0$ along the path from $(0,0) \to (2,0)$. The second integral is zero because $dy = 0$ along the path.

Along the second segment from $(2,0) \to (2,1)$, we have $x = 2$ and $dx = 0$, while y varies from $y = 0$ to $y = 1$.

$$W = \int_{(2,0)\to(2,1)} F_x dx + \int_{(2,0)\to(2,1)} F_y dy \tag{5.3.10}$$

$$= \int_{y=0}^{y=1} x\,y\,dx - \int_0^1 y^2\,dy \qquad (5.3.11)$$

$$= 0 - \frac{1}{3}y^3\bigg|_0^1 = -\frac{1}{3} \qquad (5.3.12)$$

The first integral above is zero because $dx = 0$ along this segment.

Examples 5.5 and 5.6 show clearly that the work done by this force depends on the path taken between the two points $(0,0)$ and $(2,1)$. Such a force is called a *nonconservative force*; we will see later in this chapter that nonconservative forces have no potential energy associated with them.

Python Code
In the Python code we use the same method as in the previous example, with `ParametricRegion((t,0),(t,0,2)` defining the horizontal line of the integration path with $x = t$ and $y = 0$. Similarly, `ParametricRegion((2,t),(t,0,1)` defines the vertical line of the integration path with $x = 2$ and $y = t$. The work is calculated using `ParametricIntegral(F,line1)` plus `ParametricIntegral(F,line2)`, where F represents again the force vector $\mathbf{F} = x\,y\,\hat{\mathbf{i}} - y^2\hat{\mathbf{j}}$.

```python
from sympy import symbols
from sympy.vector import  CoordSys3D, ParametricIntegral,\
ParametricRegion

print('-'*28,'CODE OUTPUT','-'*29,'\n')

C = CoordSys3D('C')      # define Cartesian system, named here C

t = symbols('t')

# using ParametricRegion and ParametricIntegral

# Force vector
F = C.x*C.y*C.i - C.y**2*C.j

line1 = ParametricRegion((t, 0), (t, 0, 2))

integr1 = ParametricIntegral(F, line1)
print('integral along line 1 = ', integr1)

line2 = ParametricRegion((2, t), (t, 0, 1))
integr2 = ParametricIntegral(F, line2)
print('integral along line 2 = ',integr2)

print('\nWork done is :', integr1+integr2)

--------------------------- CODE OUTPUT ---------------------------------
integral along line 1 =  0
integral along line 2 =  -1/3

Work done is : -1/3
```

Mathematica Code

We break up the path into two steps. The first line integral, stored as `line1` is done along the line joining the origin and the point (2.0). The second line integral, stored as `line2` is done along the line joining (2,0) to (2,1). The final result is the sum of the two integrals.

```
F = {x * y, -y^2};

line1 = LineIntegrate[F, {x, y} ∈ Line[{{0, 0}, {2, 0}}]];

line2 = LineIntegrate[F, {x, y} ∈ Line[{{2, 0}, {2, 1}}]];

Print["The line integral = ", line1 + line2]

The line integral = -1/3
```

Example 5.7: Calculating the work done by a force along a curved path.

Find the work done by the force $\mathbf{F} = x\,y\,\hat{\mathbf{i}} - y^2\hat{\mathbf{j}}$ from Example 5.5 along the unit quarter circle from the point $(1,0)$ to the point $(0,1)$ as indicated in Figure 5.6c. Do the problem by hand and by using `ParametricIntegral` in Python and `LineIntegrate` in Mathematica.

Solution:

This is a problem that is best done in polar coordinates. Let θ be the angle measured counterclockwise with respect to the x-axis, denoted in Figure 5.6c. A particle moving along the quarter circle would have a constant radial coordinate $r = 1$. Therefore, the conversion from Cartesian to polar coordinates would be: $x = \cos\theta$ and $y = \sin\theta$. In order to rewrite the force vector, we need to convert the unit vectors from Cartesian to polar coordinates. To derive the conversion, we start with the unit vectors for polar coordinates from Chapter 3:

$$\left.\begin{array}{l}\hat{\mathbf{r}} = \cos\theta\,\hat{\mathbf{i}} + \sin\theta\,\hat{\mathbf{j}} \\ \hat{\boldsymbol{\theta}} = -\sin\theta\,\hat{\mathbf{i}} + \cos\theta\,\hat{\mathbf{j}}\end{array}\right\} \tag{5.3.13}$$

Next, we multiply $\hat{\mathbf{r}}$ by $\cos\theta$ and $\hat{\boldsymbol{\theta}}$ by $\sin\theta$ and subtract. The result is:

$$\hat{\mathbf{i}} = \cos\theta\,\hat{\mathbf{r}} - \sin\theta\,\hat{\boldsymbol{\theta}} \tag{5.3.14}$$

Similarly, by multiplying the $\hat{\mathbf{r}}$-equation by $\sin\theta$ and the $\hat{\boldsymbol{\theta}}$-equation by $\cos\theta$, and add. The result is:

$$\hat{\mathbf{j}} = \sin\theta\,\hat{\mathbf{r}} + \cos\theta\,\hat{\boldsymbol{\theta}} \tag{5.3.15}$$

By inserting the coordinate transformations for x, y, $\hat{\mathbf{i}}$, and $\hat{\mathbf{j}}$ into the force we obtain:

$$\mathbf{F} = \sin\theta\,\cos(2\theta)\,\hat{\mathbf{r}} - 2\cos\theta\,\sin^2\theta\,\hat{\boldsymbol{\theta}}$$

where we used the trigonometric identity $\cos(2\theta) = \cos^2\theta - \sin^2\theta$. We can use the infinitesimal displacement in polar coordinates, $d\mathbf{r} = \hat{\mathbf{r}}\,dr + \hat{\boldsymbol{\theta}}\,r\,d\theta$, but this simplifies to

$d\mathbf{r} = r\, d\theta\, \hat{\boldsymbol{\theta}}$, because the particle is not moving radially away from the origin (i.e., $r = 1$ and $dr = 0$). After inserting $\mathbf{F}$ and $d\mathbf{r}$ into (5.3.1) and setting $r = 1$, we obtain:

$$W = \int_0^{\pi/2} \left(-2\cos\theta \sin^2\theta\right) d\theta = -2/3 \qquad (5.3.16)$$

Python Code

In the Python code we use the same method as in the previous two examples, with `path=ParametricRegion((cos(t),sin(t),(t,0,pi/2)` defining the quarter circular integration path, and the angle varying from $t = 0$ to $t = \pi/2$. The work is calculated once more using `ParametricIntegral(F,path`, where F represents again the force vector $\mathbf{F} = xy\hat{\mathbf{i}} - y^2\hat{\mathbf{j}}$.

```python
from sympy import  symbols, sin ,cos, pi
from sympy.vector import  CoordSys3D, ParametricIntegral,ParametricRegion

print('-'*28,'CODE OUTPUT','-'*29,'\n')

C = CoordSys3D('C')     # define Cartesian system, named here C

t = symbols('t')

# Force vector
F = C.x*C.y*C.i - C.y**2*C.j

path = ParametricRegion((cos(t), sin(t)), (t, 0, pi/2))

integr = ParametricIntegral(F, path)
print('The integral using ParametricIntegral = ', integr)


----------------------------- CODE OUTPUT -----------------------------
The integral using ParametricIntegral =  -2/3
```

Mathematica Code

Again we use `LineIntegrate`. This time, the path is defined using the command `Circle`, whose arguments are the location of the circle's center $(0,0)$, the circle's radius (1), and the parameter θ values of the path's start and end point (in this case $\theta = 0$ to $\pi/2$).

```
F = {x * y, -y^2};

curve = LineIntegrate[F, {x, y} ∈ Circle[{0, 0}, 1, {0, π/2}]];

Print["The line integral = ", curve]
The line integral = - 2/3
```

5.4 WORK-KINETIC ENERGY THEOREM, REVISITED

Recall that in one dimension, the kinetic energy $T = 1/2mv^2$. In two and three-dimensions, the kinetic energy takes on the form:

$$T = \frac{1}{2}mv^2 = \frac{1}{2}m\mathbf{v} \cdot \mathbf{v} \tag{5.4.1}$$

If we compute the time derivative of T, we get:

$$\frac{dT}{dt} = \frac{1}{2}m(\dot{\mathbf{v}} \cdot \mathbf{v} + \mathbf{v} \cdot \dot{\mathbf{v}}) = m\dot{\mathbf{v}} \cdot \mathbf{v} = \mathbf{F} \cdot \mathbf{v} \tag{5.4.2}$$

$$dT = \mathbf{F} \cdot (\mathbf{v}dt) \tag{5.4.3}$$
$$dT = \mathbf{F} \cdot d\mathbf{r} \tag{5.4.4}$$

After integrating this equation from $\mathbf{r}_0$ to $\mathbf{r}$, we obtain:

The Work-Kinetic Energy Theorem

$$T - T_0 = \int_{\mathbf{r}_0}^{\mathbf{r}} \mathbf{F} \cdot d\mathbf{r} \tag{5.4.5}$$

where T_0 is the value of the kinetic energy at position $\mathbf{r}_0$. Equation (5.4.5) is the work-kinetic energy theorem and is similar to (5.1.6) derived for one-dimensional systems.

5.5 CONSERVATIVE FORCES AND POTENTIAL ENERGY

Recall that for one-dimensional systems, we defined a conservative force $F(x)$, such that the work done by $F(x)$ to move a particle between two points x_0 and x, is independent of the path taken from x_0 to x. This can be easily generalized to two- and three-dimensional systems. A conservative force $\mathbf{F}(\mathbf{r})$ is one that depends only on a particle's position, *and* the work done by $\mathbf{F}(\mathbf{r})$ to move a particle from positions $\mathbf{r}_0$ to $\mathbf{r}$ is independent of the path taken between the points. We can then define the potential energy to be:

$$V(\mathbf{r}) = -W(\mathbf{r}_0 \to \mathbf{r}) = -\int_{\mathbf{r}_0}^{\mathbf{r}} \mathbf{F}(\mathbf{r}') \cdot d\mathbf{r}' \tag{5.5.1}$$

where we have set a reference point $\mathbf{r}_0$ such that $V(\mathbf{r}_0) = 0$. The above definition works for our purposes because we have a conservative force. If the force was not conservative, then the integral in (5.5.1) would also depend on the path taken by the particle and would not be simply a function of $\mathbf{r}$.

Now suppose that a conservative force, $\mathbf{F}(\mathbf{r})$, moves a particle from a point $\mathbf{r}_0$ to the point $\mathbf{r}_2$ along two different paths. One path goes directly from the reference point $\mathbf{r}_0$ to $\mathbf{r}_2$, while the other path goes from $\mathbf{r}_0$ to $\mathbf{r}_1$ and then from $\mathbf{r}_1$ to $\mathbf{r}_2$. Because the work is conservative, the work done along the two paths is the same:

$$W(\mathbf{r}_0 \to \mathbf{r}_2) = W(\mathbf{r}_0 \to \mathbf{r}_1) + W(\mathbf{r}_1 \to \mathbf{r}_2) \tag{5.5.2}$$

Solving for $W(\mathbf{r}_1 \to \mathbf{r}_2)$:

$$W(\mathbf{r}_1 \to \mathbf{r}_2) = W(\mathbf{r}_0 \to \mathbf{r}_2) - W(\mathbf{r}_0 \to \mathbf{r}_1) \tag{5.5.3}$$

$$W(\mathbf{r}_1 \to \mathbf{r}_2) = -V(\mathbf{r}_2) + V(\mathbf{r}_1) \tag{5.5.4}$$

Or, the work done by the conservative force $\mathbf{F}(\mathbf{r})$ is related to the potential energy through the formula:

$$W(\mathbf{r}_1 \to \mathbf{r}_2) = -[V(\mathbf{r}_2) - V(\mathbf{r}_1)] = -\Delta V \tag{5.5.5}$$

Combining equation (5.5.5) with (5.5.1), we see that the work done by a conservative force along a closed-loop path must be equal to zero, since the change of potential energy along that path is zero.

Combining the Work-Kinetic Energy Theorem (5.4.5) with (5.5.1), produces:

$$\Delta T = \int_{\mathbf{r}_1}^{\mathbf{r}_2} \mathbf{F}(\mathbf{r}') \cdot d\mathbf{r}' \tag{5.5.6}$$

$$\Delta T = W(\mathbf{r}_1 \to \mathbf{r}_2) \tag{5.5.7}$$

$$\Delta T = -\Delta V \tag{5.5.8}$$

and therefore:

$$\Delta(T + V) = 0 \tag{5.5.9}$$

which tells us that the total mechanical energy $E = T + V$ remains constant as the particle's position changes from $\mathbf{r}_1$ to $\mathbf{r}_2$.

In the case where nonconservative forces work on the system, (5.1.19) still applies for two- and three-dimensional systems. However, in this case, (5.4.1) is used to compute T, (5.5.1) is used to compute V, and the work done by the nonconservative force $\mathbf{F}_{nc}(\mathbf{r})$ is found using:

$$W_{nc} = \int_{\mathbf{r}_1}^{\mathbf{r}_2} \mathbf{F}_{nc}(\mathbf{r}') \cdot d\mathbf{r}' \tag{5.5.10}$$

We have seen that the potential energy $V(x)$ can be found from the conservative force $\mathbf{F}$, by using (5.5.1). Now suppose we know V and want to find $\mathbf{F}$. We know from our discussion of one-dimensional systems that $F = -dV/dx$; however, how does that generalize to higher dimensions? To derive the relationship between $\mathbf{F}$ and V in three dimensions, we begin by considering an infinitesimal displacement $\mathbf{r} \to \mathbf{r} + d\mathbf{r}$ over which a conservative force $\mathbf{F}(\mathbf{r})$ is acting on a particle. The work done by that force is:

$$dW = \mathbf{F}(\mathbf{r}) \cdot d\mathbf{r} = F_x dx + F_y dy + F_z dz \tag{5.5.11}$$

where we are working in Cartesian coordinates for simplicity, but any other coordinate system can be used. Continuing with the same displacement $\mathbf{r} \to \mathbf{r} + d\mathbf{r}$, the work done by a conservative force can also be computed as:

$$dW = -dV \tag{5.5.12}$$
$$= -[V(\mathbf{r}+d\mathbf{r}) - V(\mathbf{r})] \tag{5.5.13}$$
$$= -\left[V(\mathbf{r}) + \frac{\partial V}{\partial x}dx + \frac{\partial V}{\partial y}dy + \frac{\partial V}{\partial z}dz + \cdots - V(\mathbf{r})\right] \tag{5.5.14}$$
$$= -\left(\frac{\partial V}{\partial x}dx + \frac{\partial V}{\partial y}dy + \frac{\partial V}{\partial z}dz\right) \tag{5.5.15}$$

where $\mathbf{r} = x\hat{\mathbf{i}}+y\hat{\mathbf{j}}+z\hat{\mathbf{k}}$, and in (5.5.15) we kept only linear terms in the multi-variable Taylor series. Comparing the coefficients of dx, dy, and dz in Equations (5.5.11) and (5.5.15), we find:

$$F_x = -\frac{\partial V}{\partial x} \qquad F_y = -\frac{\partial V}{\partial y} \qquad F_z = -\frac{\partial V}{\partial z} \tag{5.5.16}$$

In other words, the force is related to the potential energy through the gradient by:

Potential Energy and Force in Three Dimensions

$$\vec{\mathbf{F}} = -\nabla V(x,y,z) = -\left(\frac{\partial V}{\partial x}\hat{\mathbf{i}} + \frac{\partial V}{\partial y}\hat{\mathbf{j}} + \frac{\partial V}{\partial z}\hat{\mathbf{k}}\right) \tag{5.5.17}$$

$$V = -\int \mathbf{F} \cdot d\mathbf{r} \tag{5.5.18}$$

Notice the minus sign in (5.5.17). The gradient points in the direction of the most rapid increase in a scalar function. In the case of (5.5.17), we see that forces point in the direction of the most rapidly *decreasing* potential energy. Consider the gravitational potential energy as illustrated in Figure 5.7. Points at the same altitude all have the same potential energy. We can think of the dashed lines in Figure 5.7 as surfaces of constant potential energy. As the altitude increases, the potential energy increases. Hence, the direction of the most rapid increase in V is upward ,i.e., perpendicular to the ground (shown as the rectangle). Because $\mathbf{F}_g = -\nabla V$, the force vector points along the direction of the most rapid *decrease* in V, which is downward.

A conservative force $\mathbf{F}$ can be written as the gradient of the potential energy. From Chapter 3, we recall that the curl of a gradient of any scalar function V is always zero:

$$\nabla \times \mathbf{F} = -\nabla \times \nabla V = 0 \tag{5.5.19}$$

Therefore, one method to identify whether or not a force is conservative is to compute its curl. A force $\mathbf{F}$ will be conservative if and only if its curl is zero:

Conservative Force

$$\nabla \times \mathbf{F} = 0 \tag{5.5.20}$$

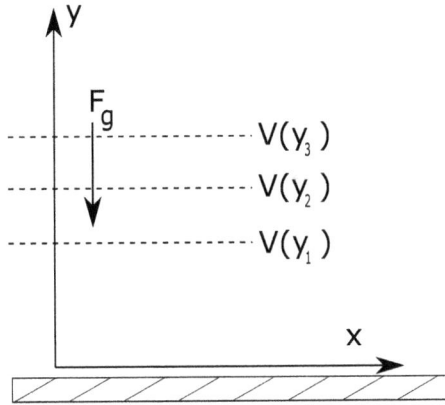

Figure 5.7: Surfaces of constant gravitational potential energy and the direction of gravitational force, $\mathbf{F}_g$.

Example 5.8: Proving a force is nonconservative

Show that the force $\mathbf{F} = xy\hat{\mathbf{i}} - y^2\hat{\mathbf{j}}$ from Examples 5.5-5.7 is not conservative.

Solution:

Calculate the curl of the force:

$$\nabla \times \mathbf{F} = \left(\frac{\partial F_z}{\partial y} - \frac{\partial F_y}{\partial z}\right)\hat{\mathbf{i}} + \left(\frac{\partial F_x}{\partial z} - \frac{\partial F_z}{\partial x}\right)\hat{\mathbf{j}} + \left(\frac{\partial F_y}{\partial x} - \frac{\partial F_x}{\partial y}\right)\hat{\mathbf{k}}$$

with $F_x = xy$, $F_y = -y^2$, $F_z = 0$, this becomes:

$$\nabla \times \mathbf{F} = (0 - 0)\,\hat{\mathbf{i}} + (0 - 0)\,\hat{\mathbf{j}} + (0 - x)\,\hat{\mathbf{k}} = -\,x\hat{\mathbf{k}} \neq 0$$

Hence the force is nonconservative, as we expected based on the results of Examples 5.5-5.7. Below, we include the calculation in Python and Mathematica.

Python Code

We use `CoordSys3D` to define a coordinate system and use the function `curl(F)` from the library `sympy.vector`. We find that the curl is nonzero and the force is nonconservative.

```
from sympy.vector import CoordSys3D, curl

print('-'*28,'CODE OUTPUT','-'*29,'\n')

R = CoordSys3D('R')

F = R.x*R.y*R.i - R.y**2 * R.j

print('The curl of F = ', curl(F))


---------------------------- CODE OUTPUT ----------------------------
The curl of F =   (-R.x)*R.k
```

Mathematica Code

Notice that we defined the force F as a three-dimensional vector to get the correct curl. Otherwise, Mathematica would assume F is a two-dimensional vector field and not give the correct curl.

$F = \{x * y, -y^2, 0\};$

$\text{Print}[\text{"The curl of F} = \text{"}, \text{Curl}[F, \{x, y, z\}]]$

The curl of F $= \{0, 0, -x\}$

Example 5.9: Proving a force is conservative and finding the associated potential energy.

Consider the force $\mathbf{F} = -k\left(y\hat{\mathbf{i}} + x\hat{\mathbf{j}} - z\hat{\mathbf{k}}\right)$.

(a) Show that it is conservative.

(b) Find the potential energy associated with this force.

Solution:

(a) We begin by computing the curl of the force:

$$\nabla \times \mathbf{F} = \begin{vmatrix} \hat{\mathbf{i}} & \hat{\mathbf{j}} & \hat{\mathbf{k}} \\ \frac{\partial}{\partial x} & \frac{\partial}{\partial y} & \frac{\partial}{\partial z} \\ y & x & -z \end{vmatrix}(-k) = -k\left[\hat{\mathbf{i}}(0) + 0\hat{\mathbf{j}} + \hat{\mathbf{k}}(1-1)\right] = 0 \qquad (5.5.21)$$

So **F** is conservative and there exists a potential energy V associated with the force **F**.

(b) Next, we calculate the potential energy V. Recall

$$V = -\int \mathbf{F} \cdot d\mathbf{r} = -k\int y\,dx - k\int x\,dy + k\int z\,dy \qquad (5.5.22)$$

Since the force is conservative, the potential energy V will be independent of the path we choose for evaluating the integral.

We will integrate (5.5.22) from the origin to the point (x, y, z) along the following path:

(1) From the origin to the point $(x, 0, 0)$ along the x-axis.

(2) From the point $(x, 0, 0)$ to the point $(x, y, 0)$ along a line parallel to the y-axis.

(3) From the point $(x, y, 0)$ to the point (x, y, z) along a line parallel to the z-axis.

Along path (1), we have $y = z = 0$, and $d\mathbf{r} = \hat{\mathbf{i}}\,dx$. Therefore the line integral along path (1) is

$$V_1 = -\int_0^x (-ky)\big|_{y=0}\,dx = 0 \qquad (5.5.23)$$

Along path (2), x is constant and we have $z = 0$ and $d\mathbf{r} = \hat{\mathbf{j}}\,dy$. Therefore the line integral along path (2) is

$$V_2 = -\int_0^y (-kx)\big|_{x=\text{constant}}\,dy = kxy \qquad (5.5.24)$$

Along path (3), x and y are constant, and we have $d\mathbf{r} = \hat{\mathbf{k}}\,dz$. Therefore the line integral along path (3) is

$$V_3 = -\int_0^z kz\,dz = -\frac{k}{2}z^2 \tag{5.5.25}$$

The total line integral across the entire path is

$$V = V_1 + V_2 + V_3 = kxy - \frac{k}{2}z^2 \tag{5.5.26}$$

Finally, we check the answer:

$$\mathbf{F} = -\nabla V = \left(-\frac{\partial V}{\partial x}, -\frac{\partial V}{\partial y}, -\frac{\partial V}{\partial z}\right) = -ky\hat{\mathbf{i}} - kx\hat{\mathbf{j}} + kz\hat{\mathbf{k}} \quad \checkmark \tag{5.5.27}$$

We now show an alternative method of calculating the potential energy V in the previous example. In this method we integrate the general expression for the gradient.

Example 5.10: Consider the force $\mathbf{F} = -k\left(y\hat{\mathbf{i}} + x\hat{\mathbf{j}} - z\hat{\mathbf{k}}\right)$ from the previous example. Find the potential energy associated with this force, starting from the expression for the gradient:

$$F = -\nabla V = \left(-\frac{\partial V}{\partial x}, -\frac{\partial V}{\partial y}, -\frac{\partial V}{\partial z}\right)$$

Solution

Since $\mathbf{F}$ is conservative, there exists a potential energy V associated with the force $\mathbf{F}$ such that:

$$V = -\int \mathbf{F} \cdot d\mathbf{r} = -\int F_x dx - \int F_y dy - \int F_y dy$$

$$V = k\int y\,dx + k\int x\,dy - \int z\,dz$$

We must have

$$F = -\nabla V = \left(-\frac{\partial V}{\partial x}, -\frac{\partial V}{\partial y}, -\frac{\partial V}{\partial z}\right)$$

We consider the x-coordinate first:

$$F_x = -\frac{\partial V}{\partial x} = -ky$$

Integrating with respect to x while keeping y, z as constants:

$$V(x,y,z) = -kyx + g(y,z)$$

where $g(y,z)$ is a function to be determined. We must also have:

$$F_y = -\frac{\partial V}{\partial y} = -kx = kx - \frac{\partial g}{\partial y}$$

where we evaluated the $\frac{\partial V}{\partial y}$. Integrating this equation with respect to y while keeping x, z as constants, we obtain:

$$g(x,y) = 2kxy + h(z)$$

where $h(z)$ is a function to be determined. Now we have:

$$V(x,y,z) = -kyx + 2kxy + h(z) = kxy + h(z)$$

We must also have:

$$F_z = -\frac{\partial V}{\partial z} = kz = \frac{\partial h}{\partial z}$$

Integrating with respect to z , we get:

$$h(z) = -\frac{k}{2}z^2 + c$$

where c is an arbitrary constant which we set to $c = 0$. Therefore:

$$V(x,y,z) = kxy - \frac{k}{2}z^2$$

This is of course the same answer we obtained in the previous example, where we used a specific integration path to evaluate V.

5.6 ENERGY AND MULTIPARTICLE SYSTEM

Consider a collection of N discrete particles, each located at $\mathbf{r}_i$ for $i = 1, 2, \ldots, N$, with a center of mass located at $\mathbf{R}$, shown as the empty circle in Figure 5.8. The mass of each particle is m_i and the velocity of each particle is $\mathbf{v}_i$. Note that what follows can be repeated for a continuous mass distribution, but discrete sums will become integrals.

In Figure 5.8 the ith particle is located at a position $\mathbf{r}_i$ relative to the origin, and at a position $\mathbf{r}_i'$ relative to the center of mass. Notice that $\mathbf{R} + \mathbf{r}_i' = \mathbf{r_i}$, where $\mathbf{R}$ is the location of the center of mass relative to the origin.

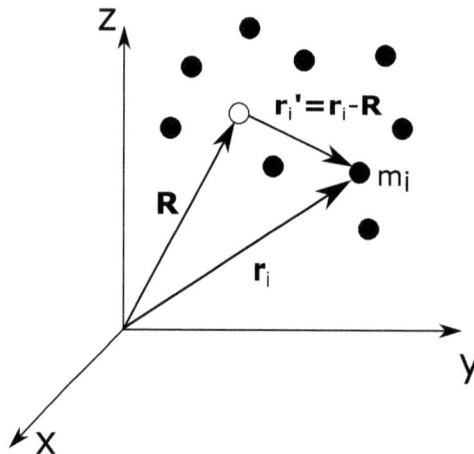

Figure 5.8: A collection of discrete particles. The center of mass is located at $\mathbf{R}$ and is denoted by the empty circle.

First, we calculate the kinetic energy of the system relative to the origin, by using a method similar to our calculation of the total momentum of a system of N particles in

Chapter 4. The total kinetic energy is:

$$T = \sum_{i=1}^{N} \left(\frac{1}{2} m_i \mathbf{v}_i^2 \right) \tag{5.6.1}$$

By using $\mathbf{r_i} = \mathbf{R} + \mathbf{r}_i'$ and $\dot{\mathbf{r}}_i = \dot{\mathbf{R}} + \dot{\mathbf{r}}_i'$ (recall that a dot above a variable means differentiation of that variable with respect to time), we can write $\mathbf{v}_i^2$ as:

$$\mathbf{v}_i^2 = \dot{\mathbf{r}}_i \cdot \dot{\mathbf{r}}_i = \left(\dot{\mathbf{r}}_i' + \dot{\mathbf{R}} \right) \cdot \left(\dot{\mathbf{r}}_i' + \dot{\mathbf{R}} \right) = \left(\mathbf{v}_i' \right)^2 + 2 \left(\dot{\mathbf{r}}_i' \cdot \dot{\mathbf{R}} \right) + \mathbf{V}^2 \tag{5.6.2}$$

where $\mathbf{v}_i' = \dot{\mathbf{r}}_i'$ is the velocity of the ith particle relative to the center of mass, and $\mathbf{V} = \dot{\mathbf{R}}$ is the velocity of the center of mass relative to the origin. Inserting (5.6.2) into (5.6.1) gives:

$$T = \sum_{i=1}^{N} \left[\frac{1}{2} m_i \left(v_i' \right)^2 \right] + \dot{\mathbf{R}} \cdot \frac{d}{dt} \left(\sum_{i=1}^{N} m_i \mathbf{r}_i' \right) + \sum_{i=1}^{N} \left(\frac{1}{2} m_i \mathbf{V}^2 \right) \tag{5.6.3}$$

The middle term in the above equation contains $\sum m_i \mathbf{r}_i'$, which in Chapter 4 was shown to be zero from the definition of the center of mass. Hence, the total kinetic energy becomes:

Kinetic Energy of a System of N Particles

$$T = \frac{1}{2} M \mathbf{V}^2 + \sum_{i=1}^{N} \left[\frac{1}{2} m_i \left(v_i' \right)^2 \right] \tag{5.6.4}$$

The first term in (5.6.4) is the kinetic energy of the center of mass, and the second term corresponds to the kinetic energy of each particle that is moving relative to the center of mass.

For potential energy, we need to consider both external and internal forces and their corresponding potential energies. The potential energy of the particle with mass m_i is:

$$V_i = V_i^{\text{ext}} + \sum_{j \neq i} V_{i,j}^{\text{int}} \tag{5.6.5}$$

where V_i^{ext} is the potential energy due to the external force acting on m_i, and $V_{i,j}^{\text{int}}$ is the potential energy due to the interaction force between m_i and m_j. Examples of such interaction forces could be gravitational, electrostatic, or any other central forces. In general for central forces we have $V_{i,j}^{\text{int}} = V_{i,j}^{\text{int}} (|\mathbf{r}_i - \mathbf{r_j}|)$, meaning that the potential energy depends on the relative distance between the two particles. Therefore the total potential energy of the system is:

$$V = \sum_{i=1}^{N} V_i^{\text{ext}} + \sum_{i=1}^{N} \sum_{j \neq i} V_{i,j}^{\text{int}} (|\mathbf{r}_i - \mathbf{r_j}|) \tag{5.6.6}$$

In the case of rigid bodies, the particles are at fixed distances, and therefore the internal potential energy can be ignored, since it will be constant for each particle. Hence, for a rigid body we will only have to worry about external forces. We will study rigid bodies in more detail in Chapter 11.

5.7   CHAPTER SUMMARY

The work W done by the force $F(x)$ on a particle moving along a line from position x_0 to position x is:

$$W = \int_{x_0}^{x} F(x')\, dx'$$

The *Work-Kinetic Energy Theorem* states that work needs to be done by the force $F(x)$ in order to change the kinetic energy of a system.

$$\frac{1}{2}mv^2 - \frac{1}{2}mv_o^2 = \int_{x_o}^{x} F(x')dx'$$

The *potential energy function* $V(x)$ associated with the force $F(x)$ in one dimension is defined by:

$$V(x) - V(x_0) = -\int_{x_0}^{x} F(x')dx'$$

The force $F(x)$ associated with a potential energy $V(x)$ can be found through differentiation:

$$F(x) = -\frac{dV(x)}{dx}$$

A particle is in an *equilibrium state* at $x = x_0$ if:

$$\left.\frac{dV}{dx}\right|_{x_0} = 0$$

Physical systems can exist in three types of equilibria: stable, unstable, and neutral depending on the sign of d^2V/dx^2 at the equilibrium point $x = x_0$.

The velocity is zero at the *turning points of motion*, so all of the particle's energy is in the form of potential energy.

Hooke's Law appears any time we are studying small amplitude oscillations about a stable equilibrium, by doing a Taylor expansion of the potential $V(x)$:

$$V(x) \approx \frac{1}{2}(x - x_0)^2 V^{(2)}(x_0) = \frac{1}{2}k(x - x_0)^2$$

In two and three dimensions, the work done by the force $\mathbf{F}$ moving an object between points $\mathbf{r}_1$ and $\mathbf{r}_2$ is:

$$W(\mathbf{r}_1 \to \mathbf{r}_2) = \int_{\mathbf{r}_1}^{\mathbf{r}_2} \mathbf{F} \cdot d\mathbf{r}$$

The value of W can depend on the path C taken by the particle:

$$W = \int_C \mathbf{F} \cdot d\mathbf{r}$$

In order to calculate integrals of this type, one can break them up into component integrals:

$$W = \int_C F_x dx + \int_C F_y dy + \int_C F_z dz$$

Potential energy and force in three dimensions are:

$$\mathbf{F} = -\nabla V(x,y,z) = -\left(\frac{\partial V}{\partial x}\mathbf{i} + \frac{\partial V}{\partial y}\mathbf{j} + \frac{\partial V}{\partial z}\mathbf{k}\right) \qquad V = -\int \mathbf{F} \cdot d\mathbf{r}$$

For a *conservative force*: $\nabla \times \mathbf{F} = 0$, and the work $\int_{\mathbf{r}_1}^{\mathbf{r}_2} \mathbf{F} \cdot d\mathbf{r}$ is independent of the path taken between the two points $\mathbf{r}_1$ and $\mathbf{r}_2$.

Conservation of mechanical energy states that the sum of the kinetic and potential energy, called the total mechanical energy E, is always constant only if conservative forces act on the system. In short: Conservative forces conserve energy.

$$E = T + V(x) = T_0 + V(x_0)$$

The kinetic energy of a system of N particles is the sum of the kinetic energy of the center of mass and the kinetic energy of each particle that is moving relative to the center of mass:

$$T = \frac{1}{2}M\mathbf{V}^2 + \sum_{i=1}^{N}\left[\frac{1}{2}m_i\left(v_i'\right)^2\right]$$

In a rigid body the particles are at fixed distances, and the internal potential energy can be ignored, so that we only have to worry about external forces.

5.8   END-OF-CHAPTER PROBLEMS

The symbol ⌨ indicates a problem which requires some computer assistance, in the form of graphics, numerical computation, or symbolic evaluation.

Section 5.1: Work and Energy in One-Dimensional Systems

1. A skier with mass $m = 100$ kg skis down a 20-degree incline. Starting at rest, the skier skis 100 m down the hill and coasts 50 m on level snow before coming to a stop. What is the coefficient of kinetic friction between the skier's skis and the snow?

2. A child slides a toy of mass m along a floor with a frictional coefficient of μ. The initial velocity of the toy is v_0. The toy slides across the floor a distance d before it encounters a spring with spring constant, k. The toy strikes the spring, and the spring compresses a distance Δx before the toy comes to a stop. Find Δx.

3. ⌨ A particle of mass m experiences a force causing it to move in the $+x$-direction with velocity, $v(x) = cxe^{-bx}$, where b and c are positive constants. Do this problem both analytically and using a CAS.

    a. Find the work done by the force if the particle starts at $x = 0$ and travels to $x = a$ $(a > 0)$.

    b. Find the work done when the particle starts at $x = 0$ and travels to $x = \infty$. Note that in this case, infinity is a useful mathematical tool meaning "very far".

    c. Plot the force as a function of the distance x using numerical values $b = c = m = 1$ and explain your answer in part (b).

4. Person A riding in the backseat of a minivan traveling at a velocity u throws a ball of mass m to person B who is sitting in the front seat. The ball is thrown with a velocity v relative to person A.

    a. What is the increase in kinetic energy of the ball as measured by person A?

    b. What is the kinetic energy gain of the ball as measured by a person at rest standing on the side of the road?

    c. How much work did person A do in throwing the ball?

5. 🖥 A particle with mass m experiences a force F, and has velocity $v = a/x$ ($a > 0$) along the $+x$-direction. How much work is done by the force moving the particle from position x_1 to x_2, where $x_2 > x_1$? Do this problem analytically, and by using a CAS.

6. Consider a spherical planet of mass M and radius R. The gravitational potential energy of a particle with mass m a distance r from the center of the planet is:

$$V = -\frac{GMm}{r}$$

where G is the gravitational constant.

    a. What is the escape speed for the planet? In other words, what velocity does the particle need to have at the planet's surface in order to escape the planet's gravitational pull at $r = \infty$?

    b. Calculate numerically the escape velocity for a spacecraft on Mars.

7. 🖥 A mass m is connected to a massless spring and is constrained to move along the x-axis. The coefficient of friction between the floor upon which the mass rests and the mass is μ. The mass is pulled a distance $x = A$ to the right of the equilibrium position and released from rest.

    a. The mass moves to the left, passes the equilibrium position and reaches a turning point. How far to the left of equilibrium is this turning point?

    b. From the turning point in (a), the mass moves to the right, passes equilibrium again and reaches another turning point, which is not at $x = A$. How far to the right of equilibrium is the new turning point?

    c. What value of A results in the mass coming to rest at the equilibrium (without passing through equilibrium) after its initial release?

Section 5.2: Potential Energy and Equilibrium Points in One-Dimensional Systems

8. 🖥 Consider the potential:

$$V = V_0 \left[\left(\frac{x}{a}\right)^2 + \left(\frac{a}{x}\right)^2 \right]$$

where $V_0, a > 0$ and $x > 0$. Find the equilibrium points and determine the stability of each equilibrium point. Plot $V(x)$ and find a Taylor series approximation (to quadratic in x) of V about each equilibrium point. To create the plot, set all constants equal to 1. Do this problem both analytically and using a CAS.

9. 🖥 The equation for the potential energy shown in Figure 5.1 is $V(x) = ax^4 - bx^2 + c$, which is known as a *double-well potential*. In Figure 5.1, $a = b = c = 1$ was used to make the graph:.

a. What are the units for each coefficient a, b, c if V is measured in Joules and x in meters?

b. If a 1.0 kg particle has energy $E = 0.90$ J and starts in the left well, describe the motion of the particle. If there are turning points, find them.

10. ⌨ Consider the double-well potential $V(x) = ax^4 - bx^2 + c$, with $a = b = c = 1$. Find the force acting on the particle. Using a numerical ODE solver, find the position $x(t)$ of the particle with mass $m = 1$ kg as a function of time, when it starts at rest and has a total energy:

   a. $E = 0.85$ J.

   b. $E = 1.0$ J.

   c. $E = 2.0$ J.

11. ⌨ The potential energy function used to make Figure 5.3 is $V(x) = 1/(ax) + bx^2 - cx^3$, where the magnitudes of the coefficients are: $a = 0.25$, $b = 25$, and $c = 8$, and the units are such that V is measured in Joules and x in meters:

   a. What are the units of a, b, and c?

   b. If $E = 20$ J, find the turning points.

   c. Find and classify each of the equilibrium points.

12. ⌨ Consider the potential energy $V(x) = a/x + bx^2 - cx^3$, from Example 5.4 and Figure 5.2.2, where $a = 4$ Jm, $b = 25$ J/m^2, $c = 8$ J/m^3 and mass $m = 1$ kg. Using a numerical differential equation solver, find $x(t)$ if the particle starts at rest at $x = 0.5$ m and has total energy $E = 30$ J.

13. ⌨ Hooke's Law appears as a result of approximating a potential energy function about a local minimum. The next term in the force as a result of the expansion is cubic. Consider a particle experiencing a net force, $F = -kx + \epsilon x^3$.

   a. Find the potential energy $V(x)$ with respect to $x = 0$, find the equilibrium points, and classify their stability.

   b. Plot the potential energy for different values of k and ϵ, to see how the shape of the potential energy graph depends on those parameters. Try both positive and negative values for ϵ. Discuss the motion of the particles in each case.

14. ⌨ The simple plane pendulum consists of a mass m attached at the end of a massless rod of length ℓ. The other end of the massless rod is fixed to a pivot point such that the pendulum is allowed to swing freely in the vertical plane. Find the potential energy of the pendulum as a function of θ, the angle the rod makes with the vertical. Find the equilibrium points and classify their stability. Using $m = 1$ and $\ell = 1$ (in SI units), make a graph of the potential energy and discuss the motion of the pendulum for various values of total energy. Do not assume that the pendulum is limited to small oscillations.

15. Consider a particle of mass m experiencing the force, $\mathbf{F}(x) = -(kx + \epsilon x^2)\hat{\mathbf{i}}$ where $k > 0$ and $\epsilon > 0$. In the cases where there is a stable equilibrium, find the frequency of small oscillations.

16. 🖥 In this chapter we saw Equation (5.2.6) for a particle of mass m and total energy E, moving in a potential $V(x)$:

$$t - t_0 = \int_{x_0}^{x} \frac{dx}{\sqrt{\frac{2}{m}(E - V(x))}}$$

This equation gives $t(x)$, which can be inverted to obtain the position of the particle $x(t)$. Use this equation to find $x(t)$ for a particle whose potential energy is $V(x) = 1/2kx^2$, both analytically and using a CAS.

17. 🖥 Consider the potential energy function:

$$V(x) = \frac{-c^2(x^2 + k^2)}{x^4 + 3c^4}$$

where $c > 0$ and $k > 0$. Find the equilibrium points as a function of x/c. Are the points stable or unstable? Plot $V(x)$ and discuss the possible types of motion of a particle for different values of k/c.

18. 🖥 A particle of mass m is acted upon by a one-dimensional potential $V(x) = 3kx^4/4$. The particle oscillates between two turning points, x_1 and x_2. Find the period of oscillation of a particle with mass $m = 1$ kg, in the case where the total energy is $E = 10$ J and $k = 3.0$ N/m^4. Plot the period of oscillation as a function of total energy and discuss your results.

Section 5.3: Work and Line Integrals

19. 🖥 Analytically, and by using a computer algebra system, compute the work done by the force $\mathbf{F} = xy\hat{\mathbf{i}} - y\hat{\mathbf{j}}$ along the path joining the origin to the point (2,4) along the following paths:

   a. Along the x-axis to the point (2,0), then parallel to the y-axis to the point (2,4).

   b. Along the straight line path connecting the origin to the point (2,4).

   c. Along the path $y = x^2$.

20. 🖥 Analytically, and by using a computer algebra system, compute the work for the force $\mathbf{F} = y\hat{\mathbf{i}} - x\hat{\mathbf{j}}$ along the following closed-loop paths:

   a. The unit square in the first quadrant of the Cartesian plane, with one corner at the origin.

   b. The unit square whose center is at the origin.

   c. The unit circle whose center is at the origin.

   d. The ellipse $\frac{x^2}{4} + \frac{y^2}{25} = 1$.

Section 5.5: Conservative Forces and Potential Energy

21. 🖥 Which of the forces below are conservative? Answer this question analytically and by using a computer algebra system. For each conservative force, find the corresponding scalar potential.

a. $\mathbf{F} = y\hat{\mathbf{i}} - x\hat{\mathbf{j}}$

b. $\mathbf{F} = -k\mathbf{r}$

c. $\mathbf{F} = r \sin\theta\, e^{i\phi}\hat{\mathbf{r}}$ (in spherical coordinates)

d. $\mathbf{F} = kxyz\,\hat{\mathbf{r}}$

e. $\mathbf{F} = \frac{k}{r^3}\hat{\mathbf{r}}$

22. Prove that any central isotropic force $\mathbf{F} = F(r)\hat{\mathbf{r}}$, where $r^2 = x^2 + y^2 + z^2$, is conservative.

23. We showed that a conservative force can be written as $\mathbf{F}(\mathbf{r}) = -\nabla V(\mathbf{r})$, and that when conservative forces act on a system, the total mechanical energy is conserved ($\Delta(T+V) = 0$). Now consider a time-dependent force written as $\mathbf{F}(\mathbf{r},t) = -\nabla V(\mathbf{r},t)$. Is energy conserved for this system? If so, prove it. If not, how is the conservation of energy equation $\Delta(T + V) = 0$, changed?

24. ⌨ Consider the force $\mathbf{F} = -2z^3\hat{\mathbf{i}} + (3 + 2ye^z)\hat{\mathbf{j}} + (y^2 e^z - 6xz^2)\hat{\mathbf{k}}$.

    a. Show that it is conservative, analytically and by using a computer algebra system.

    b. Find the scalar potential associated with this force.

25. Prove that the electrostatic force $\mathbf{F} = kq_1q_2/r^2\hat{\mathbf{r}}$ on a point charge q_1 at position $\mathbf{r}$ due to a fixed charge q_2 at the origin is conservative. The constant $k = 1/4\pi\epsilon_0$ is positive and depends on the permittivity of free space, ϵ_0. Find the potential energy when the charges have the same sign. What happens to the potential energy when the charges have a different sign? Explain your answer.

26. ⌨ For each potential energy function, compute the associated force both analytically and using a computer algebra system:

    a. $V = k\left(x^2 + y^2 + z^2\right)$.

    b. $V = -kr\cos\phi$, where k is a positive constant and r and ϕ are cylindrical coordinates.

    c. $V = -kr\cos\phi + cr^2\sin\theta$, where k and c are positive constants and r, θ, and ϕ are the typical spherical coordinates.

Section 5.6: Energy and Multiparticle Systems

27. ⌨ Consider three particles. Particle 1 has a mass of $m_1 = 3$ and is located at $\mathbf{r}_1 = 3t^2\hat{\mathbf{i}}$ at time t. Particle 2 has a mass of $m_2 = 1$ and is located at $\mathbf{r}_2 = 3t^2\hat{\mathbf{k}}$. Particle 3 has a mass of $m_3 = 2$ and is located at $\mathbf{r}_3 = 4t\,\hat{\mathbf{i}} - 8t^2\,\hat{\mathbf{j}} + 3t^3\,\hat{\mathbf{k}}$. All physical quantities are in SI units. Analytically and by using a computer algebra system, find:

    a. The location of the center of mass for the system.

    b. The kinetic energy of the center of mass.

    c. The kinetic energy of the system of particles.

28. Consider a planet of mass M which is orbited by a Moon of mass $m_1 = m$. The Moon is in a circular orbit with an orbital radius a. A dwarf planet, of mass $m_2 = m$ and with kinetic energy of T_2 approaches the two objects. The dwarf planet gets trapped in the same circular orbit as the Moon, and the Moon is kicked free. Write down the total energy of the three particle system (there will be two kinetic energies and three potential energies). What is the kinetic energy of the Moon after the collision, after it is far away? As usual, take the gravitational potential energy of a particle with mass m at a distance r from a mass M as:

$$V = -\frac{GMm}{r}$$

where G is the gravitational constant.

29. ⌨ Consider a group of three particles: $m_1 = 3$ kg at $\mathbf{r}_1 = 3\hat{\mathbf{i}} - 2\hat{\mathbf{j}} + 3\hat{\mathbf{k}}$, $m_2 = 2$ kg at $\mathbf{r}_2 = -\hat{\mathbf{i}} + 3\hat{\mathbf{k}}$, and $m_3 = 1$ kg at $\mathbf{r}_3 = 7\hat{\mathbf{j}} - 5\hat{\mathbf{k}}$, where each position is measured in meters. If the particles are located far away from any other mass, compute the total gravitational potential energy for this particle configuration. Do this problem analytically and with a computer algebra system.

30. The Virial theorem states that for a gravitationally bound distribution of masses, the average kinetic energy of the system is equal to minus one-half of the total potential energy, $\langle T \rangle = -\frac{1}{2} \langle V \rangle$, where $\langle \, \rangle$ denotes an average over time. Prove that the Virial theorem holds for a single particle of mass m, moving in a circular orbit of radius a around another particle of mass M.

6 Harmonic Oscillations

As discussed in Chapter 5, when a system is displaced only slightly from a stable equilibrium position, it oscillates with harmonic motion about this equilibrium position. In this chapter, we study the problem of harmonic motion. We first introduce the general concepts of *linear ordinary differential equations*, and present the simplest application of these equations for a *linear harmonic oscillator*, with and without damping forces. This is followed by the study of *forced or driven oscillations* under an external force, and the important concepts of amplitude resonance, energy resonance and the associated Q-factor for oscillatory systems. The chapter will conclude by introducing the physical concepts of *phase space* and the *Superposition Principle*, together with the important mathematical technique of *Fourier analysis*.

6.1 DIFFERENTIAL EQUATIONS

In this section, we make some introductory remarks about the mathematical representation of oscillators in general. We will then apply the ideas presented in this section to specific oscillators in later sections. Oscillating systems are of fundamental importance in physics and engineering systems, but interest in their study extends to almost every branch of science. From a mathematical point of view, oscillating systems are often described by *nonhomogeneous linear ordinary differential equations* of the general form:

$$A_n \frac{d^n x}{dt^n} + A_{n-1} \frac{d^{n-1} x}{dt^{n-1}} + .. + A_1 \frac{dx}{dt} + A_0 x = f(t) \tag{6.1.1}$$

where $x = x(t)$ measures the oscillator's displacement from a stable equilibrium, $A_0, \ldots, A_n$ are constants, and $f(t)$ is a function of time t. The *order* of a differential equation is the highest derivative in the equation, and the equation is called *homogeneous* when the function $f(t) = 0$. Equation (6.1.1) is called a *linear differential equation* because it consists of only terms that are linear in $x(t)$ and its derivatives. In general, equations that describe oscillations are *nonlinear* in x and its derivatives. However, as we will see later in this chapter, in the limit of small displacements from equilibrium, the equations of motion usually take a linear form similar to that of (6.1.1).

In this chapter we will first study homogeneous linear second-order differential equations of the form:

Homogeneous Second-Order Linear Differential Equation

$$A_2 \frac{d^2 x}{dt^2} + A_1 \frac{dx}{dt} + A_0 x = 0 \tag{6.1.2}$$

If $x_1(t)$, $x_2(t)$ are two solutions of this equation, then the general solution $x_c(t)$ of (6.1.2) is a linear combination of these two solutions in the form:

Solution of Homogeneous Equation

$$x_c(t) = C_1 x_1(t) + C_2 x_2(t) \tag{6.1.3}$$

---

where C_1, C_2 are two arbitrary constants. These constants C_1, C_2 can be determined by the *initial conditions* of the system at time $t = 0$, which are usually given in the form of the initial position $x(0)$ and the initial speed $v(0) = \dot{x}(0)$. We will find that a homogeneous second-order linear differential equation can be used to describe the motion of a particle experiencing a restoring force (responsible for the oscillation) and a drag force, called a damping force.

When external forces are applied to the oscillator, we will need a *nonhomogeneous* linear second order differential equation to describe the motion. A nonhomogeneous linear second-order differential equation takes the form:

Nonhomogeneous Second-Order Linear Differential Equation

$$A_2 \frac{d^2 x}{dt^2} + A_1 \frac{dx}{dt} + A_0 x = f(t) \tag{6.1.4}$$

In the general theory of differential equations, the solution of this equation is given by the sum of the solution of the *homogeneous equation* $x_c(t)$ from (6.1.3), sometimes called the *complementary solution*, and a *particular solution* $x_p(t)$ of the full nonhomogeneous equation. The particular solution $x_p(t)$ is *any* solution to the equation:

$$A_2 \frac{d^2 x_p}{dt^2} + A_1 \frac{dx_p}{dt} + A_0 x_p = f(t) \tag{6.1.5}$$

The general solution $x(t)$ of a nonhomogeneous second-order differential equation then takes the form:

General Solution of Nonhomogeneous Equation

$$x(t) = x_c(t) + x_p(t) = C_1 x_1(t) + C_2 x_2(t) + x_p(t) \tag{6.1.6}$$

6.2 SIMPLE HARMONIC OSCILLATOR

In this section we study one of the most important problems in classical mechanics, the simple harmonic oscillator (SHO). Simple harmonic motion occurs when an object displaced a small distance from a stable equilibrium experiences a restoring force towards the equilibrium state. An important example of simple harmonic motion is a mass attached to a massless spring with no friction.

6.2.1 EQUATION OF MOTION OF THE SIMPLE HARMONIC OSCILLATOR

Suppose a mass m is attached to a massless spring. When the mass is at the equilibrium position $x = 0$, the spring is neither stretched nor compressed, and the net force acting on the mass is zero. Figure 6.1a shows the system in equilibrium. When the mass is moved from the equilibrium position, the spring exerts a force $F(x)$ which is proportional to the extension or compression of the spring in a direction towards the equilibrium position, as shown in Figures 6.1b and c. The force F is called a *restoring force* because it attempts to return the system back to its equilibrium state.

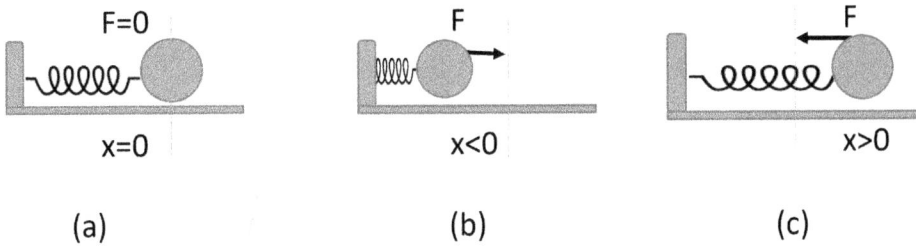

Figure 6.1: Simple harmonic oscillation in a spring-mass system.

For small displacements, we can expand $F(x)$ as a Taylor series about the equilibrium position $x = 0$:

$$F(x) \approx F(0) + \left.\frac{dF}{dx}\right|_{x=0} x + O(x^2) \qquad (6.2.1)$$

We know that $F(0) = 0$, because the restoring force is not exerted when the system is at equilibrium. Likewise, we know that the first derivative of $F(x)$ with respect to x must be negative, in order for $F(x)$ to point towards equilibrium. In other words, the sign of $F(x)$ is opposite that of x. Therefore, we define the *spring constant* as:

$$k \equiv -\left.\frac{dF}{dx}\right|_{x=0} \qquad (6.2.2)$$

We can ignore second-order and higher terms in (6.2.1) because x is small. Therefore, the result is that for small displacements, the restoring force takes on the form of *Hooke's Law*:

Hooke's Law

$$F(x) = -kx \qquad (6.2.3)$$

If $x(t)$ is the location of the mass at a time t and we assume that the restoring force takes the form of Hooke's Law (i.e., small displacements from equilibrium), then we can write Newton's second law, $F = ma = -kx(t)$, as:

Differential Equation for the Simple Harmonic Oscillator

$$\frac{d^2x}{dt^2} + \omega_0^2 x = 0 \qquad (6.2.4)$$

where $\omega_0^2 \equiv k/m$ is a constant. Equation (6.2.4) is the differential equation of motion for the so-called simple harmonic oscillator (SHO), which is an important equation in physics because it is the generic equation of motion for an object experiencing small amplitude oscillations about a stable equilibrium. Simple harmonic motion results when a particle experiences only a restoring force that is proportional to the particle's displacement from equilibrium, i.e., when there are no drag forces or other external forces acting on the particle.

To find an analytical solution to (6.2.4), we need a function $x(t)$ whose second derivative gives itself back after differentiation, with a multiplicative constant and a minus sign. In other words, $x(t)$ must satisfy: $\ddot{x} = -\omega_0^2 x$. Recall that time derivatives can be represented by dots above a function. While one obvious choice is either a sine or a cosine function, we will find that an exponential solution will be more useful (and still equivalent). We look for solutions that have the form $x = Ce^{\lambda t}$, where C is a constant. By substituting this into (6.2.4) we find:

$$\lambda^2 C e^{\lambda t} + \omega_0^2 C e^{\lambda t} = 0 \qquad (6.2.5)$$

or by simplifying:

$$\lambda^2 + \omega_0^2 = 0 \qquad (6.2.6)$$

or

$$\lambda_\pm = \sqrt{-\omega_0^2} = \pm i\, \omega_0 \qquad (6.2.7)$$

and thus λ must be one of the complex numbers $i\,\omega_0$ or $-i\,\omega_0$. The general solution, then, will be a linear combination of the two solutions:

$$x(t) = C_1 e^{i\,\omega_0 t} + C_2 e^{-i\,\omega_0 t} \qquad (6.2.8)$$

Note that in general C_1 and C_2 are arbitrary *complex* constants.

By using Euler's identities $e^{i\,\omega_0 t} = \cos(\omega_0 t) + i \sin(\omega_0 t)$ and $e^{-i\,\omega_0 t} = \cos(\omega_0 t) - i \sin(\omega_0 t)$, we can change the exponentials in this equation into trigonometric functions. We find that the general solution must be of the form:

$$x(t) = C_1 \left[\cos(\omega_0 t) + i \sin(\omega_0 t)\right] + C_2 \left[\cos(\omega_0 t) - i \sin(\omega_0 t)\right] \qquad (6.2.9)$$

This equation can be written in two equivalent mathematical forms:

Two Forms of the Solution $x(t)$ for the SHO

$$x(t) = A \cos(\omega_0 t) + B \sin(\omega_0 t) \qquad (6.2.10)$$

$$x(t) = C \cos(\omega_0 t + \phi) \qquad (6.2.11)$$

$$\text{with} \quad C = \sqrt{A^2 + B^2} \qquad \tan\phi = B/A \qquad (6.2.12)$$

We can re-write (6.2.9) as (6.2.10) by requiring the coefficients A and B to be real numbers. In this case, we require that $C_2^* = C_1$, where C_2^* is the *complex conjugate* of C_2. Complex conjugation changes the sign of the imaginary part of a complex number. Hence, if $c = a + ib$, where a and b are real numbers, then $c^* = a - ib$.

We can also rewrite (6.2.9) as (6.2.11) using the trigonometric identity $\cos(a + b) = \cos a \cos b - \sin a \sin b$.

From the trigonometric forms of the solution of the SHO, we see that the quantity ω_0 represents the angular frequency of the oscillating mass. The angular frequency ω_0 is called the *natural frequency* of the oscillator and is the frequency of small amplitude oscillations.

The Natural Frequency of a Simple Harmonic Oscillator

$$\omega_0 = \sqrt{\frac{k}{m}} \qquad (6.2.13)$$

Returning to Equations (6.2.10) and (6.2.11), A, B, and C are now real constants which can be determined by the initial conditions, which specify the state of the system at a given time (usually at $t = 0$).

Example 6.1 demonstrates the solution of (6.2.4) for the initial conditions $x(0) = x_0$ and $v(0) = x'(0) = v_0$.

Example 6.1: The symbolic solution of the SHO for given initial conditions
Solve (6.2.4) by hand and using Mathematica and SymPy, for the initial conditions $x(0) = x_0$ and $v(0) = x'(0) = v_0$.

Solution:
When solving by hand, we can use (6.2.10) or (6.2.11), where A, B, and C are constants determined by the initial conditions. Inserting the initial condition $x(0) = x_0$ at $t = 0$ into (6.2.10) we find:

$$x(0) = A \cos 0 + B \sin 0 = A = x_0 \qquad (6.2.14)$$

The speed of the mass $v(t)$ is found by taking the derivative of (6.2.10) with respect to time:

$$v(t) = \frac{dx}{dt} = -A\omega_0 \sin(\omega_0 t) + B\omega_0 \cos(\omega_0 t) \qquad (6.2.15)$$

By using the given initial condition $v(0) = v_0$ at $t_0 = 0$:

$$v(0) = \omega_0 \left[-A \sin 0 + B \cos 0 \right] = \omega_0 B = v_0 \qquad (6.2.16)$$

and so $B = v_0/\omega_0$. Therefore, the complete solution for these initial conditions is:

$$x(t) = x_0 \cos(\omega_0 t) + \frac{v_0}{\omega_0} \sin(\omega_0 t) \qquad (6.2.17)$$

Python Code
We can obtain the same result by using `dsolve()` in SymPy. The output of the symbolic evaluation is

```
x(t) = x0*cos(omega*t) + v0*sin(omega*t)/omega
```

In this code we use SymPy to obtain both the general symbolic solution which contains the arbitrary constants C_1 and C_2, as well to obtain the specific solution with the given initial conditions.

```
from sympy import symbols, Function, Derivative as D, dsolve

print('-'*28,'CODE OUTPUT','-'*29,'\n')

# define function x and various symbols
x = Function('x')
omega, t = symbols('omega, t', real=True, positive=True)
x0, v0 = symbols('x0, v0', real=True)

soln = dsolve(D(x(t), t, t) +omega**2*x(t), x(t),\
simplify=True).rhs
print('The general symbolic solution is:\n', 'x(t) =',soln)

initCondits = {x(0): x0, D(x(t),t).subs(t, 0): v0}

soln = dsolve(D(x(t), t, t) +omega**2*x(t), x(t),\
simplify=True,ics=initCondits).rhs

print('\nThe solution with the initial conditions x(0)=x0, v(0)=v0 \
 is:', '\nx(t) =',soln)

--------------------------- CODE OUTPUT ---------------------------
The general symbolic solution is:
 x(t) = C1*sin(omega*t) + C2*cos(omega*t)

The solution with the initial conditions x(0)=x0, v(0)=v0  is:
x(t) = x0*cos(omega*t) + v0*sin(omega*t)/omega
```

Mathematica Code

We use DSolve twice, to obtain both the general symbolic solution which contains the arbitrary constants C_1 and C_2, as well to obtain the specific solution with the given initial conditions.

```
soln1 = DSolve[{x"[t] + ω^2 * x[t]==0}, x[t], t];

Print["The solution of the homogeneous ODE is:"]

Print["x(t) = ", x[t]/.soln1[[1]]]
```

The solution of the homogeneous ODE is:
x(t) = $c_1 \text{Cos}[t\omega] + c_2 \text{Sin}[t\omega]$

```
soln2 = DSolve[{x"[t] + ω^2 * x[t]==0, x[0]==x0, x'[0]==v0},

x[t], t];

Print["The solution of the ODE with initial conditions is:"]

Print["x(t) = ", x[t]/.soln2[[1]]//Simplify]
```

The solution of the ODE with initial conditions is:
x(t) = $\text{x0Cos}[t\omega] + \frac{\text{v0Sin}[t\omega]}{\omega}$

6.2.2 POTENTIAL AND KINETIC ENERGY IN SIMPLE HARMONIC MOTION

Let us write the position and velocity for a mass undergoing simple harmonic motion in the form $x = A\cos(\omega_0 t)$ and $v = -A\omega_0 \sin(\omega_0 t)$. The corresponding potential and kinetic energies are varying with time t, according to: $V(t) = \frac{1}{2}k\,x(t)^2$ and $T(t) = \frac{1}{2}m\,v(t)^2$. The total energy of the system is then:

$$E = T + V = \frac{1}{2}m\omega_0^2 A^2 \sin^2(\omega_0 t) + \frac{1}{2}kA^2 \cos^2(\omega_0 t) \qquad (6.2.18)$$

However, because $\omega_0^2 = k/m$, we get:

$$E = \frac{1}{2}kA^2 \qquad (6.2.19)$$

which is the maximum potential energy of the system. Hooke's Law represents a conservative force; therefore during the oscillation, energy is transferred between kinetic and potential, but their sum always remains the same. At the maximum displacement from the equilibrium position, we have $x = A$, and the potential energy is a maximum and kinetic energy is zero. When the oscillator is at the equilibrium $x = 0$, the kinetic energy is maximum and potential energy is zero. In-between the maximum displacement and equilibrium, the potential and kinetic energies change. For example, as the mass moves towards equilibrium, the potential energy decreases and the kinetic energy increases; however, as mentioned earlier, their sum remains constant.

We can calculate the average kinetic energy $< T >$ in one period of oscillation by taking a time average of T over the period $\tau = 2\pi/\omega_0$ as follows:

$$< T > = \frac{1}{\tau}\int_0^\tau \frac{1}{2}mv^2 dt \;=\; \frac{1}{\tau}\int_0^\tau \frac{1}{2}mA^2\omega_0^2 \cos^2(\omega_0 t)dt \;=\; \frac{1}{4}mA^2\omega_0^2 \qquad (6.2.20)$$

where we used the integral value $\frac{1}{\tau}\int_0^\tau \cos^2(\omega_0 t)dt = \frac{1}{2}$.

Similarly, we can find the average potential energy $< V >$ in one period of oscillation by taking a time average of $V(t)$:

$$< V > = \frac{1}{\tau}\int_0^\tau \frac{1}{2}kx^2 dt \;=\; \frac{1}{\tau}\int_0^\tau \frac{1}{2}kA^2 \sin^2(\omega_0 t)dt \;=\; \frac{1}{4}kA^2 \;=\; \frac{1}{4}mA^2\omega_0^2 \qquad (6.2.21)$$

In conclusion:

$$< V > = < T > = \frac{1}{4}mA^2\omega_0^2 = \frac{1}{4}kA^2 = \frac{1}{2}E \qquad (6.2.22)$$

where we used $k = m\omega^2$ and $E = \frac{1}{2}kA^2$. This equation shows that the time averages of *both* the kinetic and potential energies within one period of oscillation are equal to one-half the total energy E.

Time Averages of Energy

$$< V >_{\text{time}} = < T >_{\text{time}} = \frac{1}{2}E \qquad (6.2.23)$$

This equation is an example of the well-known *Virial theorem* in mechanics, which is a general equation in statistical mechanics relating the average total kinetic energy of a system of particles to its average total potential energy.

6.2.3 THE SIMPLE PLANE PENDULUM AS AN EXAMPLE OF A HARMONIC OS-CILLATOR

The plane pendulum is one of the simplest physical systems which can exhibit harmonic oscillations for small displacements from an equilibrium position. In this section, we calculate the period τ of the simple pendulum for oscillations with a small amplitude and find expressions for the kinetic and potential energies as functions of time.

Figure 6.2 shows a pendulum consisting of a mass m which can swing freely in a vertical plane. A massless rod of length L connects the mass to the pendulum's pivot point. When the mass is displaced sideways from its equilibrium position, it is subject to a restoring force due to gravity. This force will accelerate the mass back toward the equilibrium position. The time for one complete cycle is the period τ.

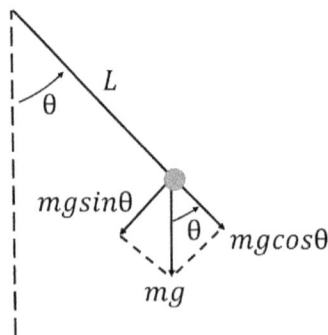

Figure 6.2: Simple plane pendulum.

We can find the equation of motion for the simple plane pendulum by applying Newton's second law to the tangential axis only, $F = -mg\sin\theta = ma$ and $a = -g\sin\theta$. The linear acceleration $a = \frac{d^2s}{dt^2}$, where s is the arc length, which is related to the angle θ by the formula $s = L\theta$. The differential equation for the linear acceleration is $\frac{d^2 s}{dt^2} = L\ddot{\theta}$. Therefore we obtain:

$$L\ddot{\theta} = -g \sin\theta \tag{6.2.24}$$

By substituting $\omega_0^2 = g/L$, we obtain the equation of motion for the simple plane pendulum:

$$\ddot{\theta} + \omega_0^2 \sin\theta = 0 \tag{6.2.25}$$

For small displacements from equilibrium, we can approximate $\sin\theta \approx \theta$, and the equation becomes:

Equation of Motion for Small Pendulum Oscillations

$$\ddot{\theta} + \omega_0^2\theta = 0 \tag{6.2.26}$$

Equation (6.2.26) is exactly similar to the equation for a mass-spring system $\ddot{x} + \omega_0^2 x = 0$, with the angle θ replacing the position x, and with the natural frequency $\omega_0 = \sqrt{k/m}$ being replaced by $\omega_0 = \sqrt{g/L}$. Because the two equations (SHO and pendulum) are mathematically similar, we can repurpose the solution of the SHO in order to solve (6.2.26). The solution to (6.2.26) is therefore given by:

$$\theta(t) = \theta_0 \cos(\omega_0 t + \phi) \tag{6.2.27}$$

where ϕ is a constant phase angle value dependent on the initial conditions. The period of oscillation for small displacements is found using $\tau = 2\pi/\omega_0$ and is:

Period of Oscillation for Small Pendulum Oscillations

$$\tau = 2\pi\sqrt{\frac{L}{g}} \tag{6.2.28}$$

Notice that the period of oscillation is independent of the amplitude of oscillation. This condition holds only for small amplitude oscillations. Once the small angle approximation $\sin\theta \approx \theta$ no longer holds, then the period of oscillation will be related to the amplitude of oscillation. Such a relationship is common in nonlinear oscillators, which we will study in Chapter 13.

The gravitational potential energy of the pendulum is given by the expression:

$$V = mgh = mgL(1 - \cos\theta) \tag{6.2.29}$$

Notice that the above expression is chosen such that $V = 0$ when $\theta = 0$. In other words, the potential energy is zero at the stable equilibrium point. By using the small angle approximation $\cos\theta \simeq 1 - \theta^2/2$, we obtain:

$$V = \frac{1}{2}mgL\theta^2 \tag{6.2.30}$$

The total energy of the pendulum for small angle oscillations is then:

$$E = \frac{1}{2}m\left(\frac{ds}{dt}\right)^2 + \frac{1}{2}mgL\theta^2 = \frac{1}{2}mL^2\left(\frac{d\theta}{dt}\right)^2 + \frac{1}{2}mgL\theta^2 \tag{6.2.31}$$

where $s = L\theta$ was used. Next, we can use (6.2.27) to find the total energy of the pendulum:

$$E = \frac{1}{2}mgL\theta_0^2 \qquad (6.2.32)$$

This equation is of course completely analogous to the equation $E = \frac{1}{2}kA^2$ for a mass-spring oscillator. Notice that mgL appears in (6.2.32) instead of k. The restoring torque acting on the pendulum is $N = -mgL\sin\theta \approx -mgL\theta$ and is linearly proportional to the angular displacement θ of the pendulum (for small displacements), with a coefficient of $k = mgL$. Therefore (6.2.32) is consistent with the results for the spring-mass oscillating system.

6.3 NUMERICAL SOLUTIONS OF THE SIMPLE PENDULUM

The ODE for the simple pendulum cannot be solved in closed form without the use of elliptical functions, and must be solved numerically.

Example 6.2 shows how to numerically evaluate and plot the solution of (6.2.25), using SciPy and Mathematica.

Example 6.2: Numerical solution of the simple pendulum

Integrate numerically and plot the solution of (6.2.25) with $L = 0.5$ m for the initial conditions $\theta(0) = 3$ rad and $\dot{\theta}(0) = 0$ rad/s, using Mathematica and the Scientific Python library SciPy.

Solution:

To solve (6.2.25), we will use the function `odeint()` from the `scipy.integrate` library which numerically solves first-order ODEs of the form:

$$\frac{dy}{dt} = f(y,t) \qquad (6.3.1)$$

where y and t are the dependent and independent variables respectively, and $f(y,t)$ is a known function of the variables y, t.

To use `odeint()` on (6.2.25), we need to rewrite (6.2.25) as a system of two first-order ODEs, a common procedure needed for some ODE solvers. We begin by creating a new variable $\omega = \dot{\theta}$ and (6.2.25) can be rewritten as the system:

$$\left.\begin{aligned} \frac{d\theta}{dt} &= \omega \\ \frac{d\omega}{dt} &= -\frac{g}{L}\sin\theta \end{aligned}\right\} \qquad (6.3.2)$$

Python Code

In the code we define a function `deriv(y,time)` whose arguments are a vector `y` and the time variable `time`. In our example, the first component `y[0]` of the vector `y` represents the angle variable θ, and the second component `y[1]` represents the angular velocity $\omega = d\theta/dt$. The function `deriv` uses (6.3.2) to evaluate and return a vector with components $(\omega, d\omega/dt)$.

The function `odeint(deriv, yinit, t)` is called with three arguments. The first argument is the function `deriv` which contains the information on the first-order ODE to be solved, the second argument is the initial conditions vector `yinit`, and the third argument is the time variable t.

The result of calling `odeint` is an array which is assigned to the variable y. The line `t = np.linspace(0, 10, 100)` defines the time interval over which the ODE will be integrated, in this case between $t = 0$ and $t = 10$ s. The variable `yinit = (3,0)` defines the initial conditions $\theta(0) = 3$ rad, and $\omega(0) = \theta'(0) = 0$ rad/s. The last 6 lines in the code are the graphics commands which plot the two functions $\theta(t)$ and $\omega(t) = \theta'(t)$, as shown in Figure 6.3.

The plot shows clearly that $\theta(t)$ is not a sinusoidal oscillation.

```
from scipy.integrate import odeint

import numpy as np
import matplotlib.pyplot as plt

L = 0.5 # length of pendulum
g = 9.8 # gravitational acceleration

# function to define the ODE for pendulum
def deriv(y, time):
    return (y[1], - (g/L)* np.sin(y[0]))

# define times t
t = np.linspace(0, 10, 100)

# initial conditions theta(0)=3 rad,  theta'(0)=0
yinit = (3, 0)

# solve numerically using scipy odeint() function
y = odeint(deriv, yinit, t)

# plot angle and anglar velocity as functions of time
plt.plot(t, y[:, 0], 'o-',label=r'$\theta$(t)')
plt.plot(t, y[:, 1], '^-',label=r"$\theta$'(t)")

plt.legend(loc='best')
plt.xlabel('Time [s]')
plt.ylabel(r"$\theta$(t), $\theta$'(t)")
plt.title('Numerical solutions for the plane pendulum')
plt.show()

findfont: Font family ['DejaVu Sans Display'] not found. Falling back to
 DejaVu Sans.
```

Mathematica Code

As in previous codes, we use `NDSolve` to obtain the numerical solution of the differential equation with the given initial conditions.

```
SetOptions[Plot, Frame->True, Axes->True,

BaseStyle->FontSize->20, PlotRange->All];


L = 0.5; g = 9.8;

soln = NDSolve[{θ''[t] + g/L * Sin[θ[t]]==0, θ'[0]==0, θ[0]==3},

θ, {t, 0, 10}];

Plot[Evaluate[{θ[t], θ'[t]}/.soln], {t, 0, 10},

FrameLabel->{t, "θ[t], θ'[t]"}, PlotStyle->{Thick, Dashed}]
```

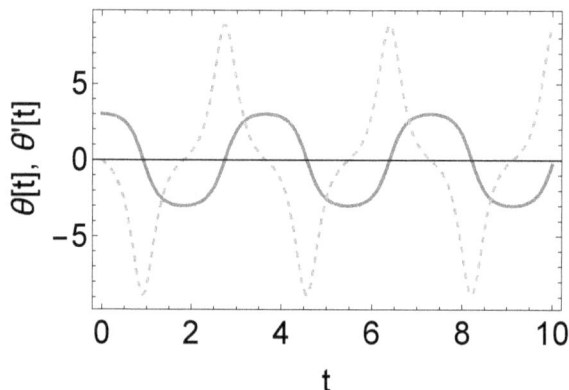

Figure 6.3: Mathematica code output in Example 6.2 for large amplitude oscillations of a simple pendulum, for the angle $\theta(t)$ (solid line) and the angular velocity $\theta'(t)$ (dashed line).

6.4 DAMPED HARMONIC OSCILLATOR

Next, we add a linear resistive force to the simple harmonic oscillator. The resistive force will be modeled similar to the linear drag forces discussed in Chapter 2. The resistive force is sometimes called *damping* and is present, for example, when the oscillator is moving through a fluid such as air. The equation for the damped harmonic oscillator contains an additional friction term which is assumed to be linearly dependent on the speed in the form $F = -bv$, where b depends on the size and shape of the oscillator, as well as other physical parameters such as the type of fluid the oscillator moves through and the coefficient of friction between the oscillator and a supporting surface. The complete equation for the damped harmonic oscillator is obtained now from $F = ma = -kx - bv$. By dividing with the mass m we obtain:

$$\frac{d^2x}{dt^2} + \frac{b}{m}\frac{dx}{dt} + \frac{k}{m}x = 0 \tag{6.4.1}$$

We now define a constant γ called the *damping parameter*, in order to express the amount of damping present in the system:

$$\gamma = b/2m \qquad (6.4.2)$$

and the differential equation becomes:

Equation of Motion for the Damped Harmonic Oscillator

$$\ddot{x} + 2\gamma\dot{x} + \omega_0^2 x = 0 \qquad (6.4.3)$$

where we used again $\omega_0 = \sqrt{k/m}$ for the natural frequency of the undamped oscillator.

Once more we look for solutions that have the form $Ce^{\lambda t}$, where C is a constant. By substituting this into (6.4.3) and dividing all terms by $Ce^{\lambda t}$, we find:

$$\lambda^2 + 2\gamma\lambda + \omega_0^2 = 0 \qquad (6.4.4)$$

This is solved using the quadratic formula:

$$\lambda_\pm = -\gamma \pm \sqrt{\gamma^2 - \omega_0^2} \qquad (6.4.5)$$

This implies a pair of solutions $e^{\lambda t}$, corresponding to the two roots of the quadratic. The general solution is a linear combination of these two solutions:

$$x(t) = C_1 e^{\left(-\gamma + \sqrt{\gamma^2 - \omega_0^2}\right)t} + C_2 e^{\left(-\gamma - \sqrt{\gamma^2 - \omega_0^2}\right)t} = e^{-\gamma t}\left(C_1 e^{\sqrt{\gamma^2 - \omega_0^2}\,t} + C_2 e^{-\sqrt{\gamma^2 - \omega_0^2}\,t}\right)$$
$$(6.4.6)$$

or by introducing the parameter $\omega \equiv \sqrt{\gamma^2 - \omega_0^2}$:

General Solution $x(t)$ for the Damped Harmonic Oscillator $(\gamma \neq \omega_0)$

$$x(t) = e^{-\gamma t}\left(C_1 e^{\omega t} + C_2 e^{-\omega t}\right) \qquad (6.4.7)$$

$$\omega = \sqrt{\gamma^2 - \omega_0^2} = \sqrt{(b/2m)^2 - (k/m)} \qquad (6.4.8)$$

The exact mathematical form of this equation $x(t)$ depends on the numerical values of γ and ω_0, which determine whether the parameter $\omega = \sqrt{\gamma^2 - \omega_0^2}$ is real, imaginary, or zero. Therefore there are three physically different behaviors of the damped harmonic oscillator as follows: overdamped oscillations $(\gamma > \omega_0)$, underdamped oscillations $(\gamma < \omega_0)$, and critically damped oscillations $(\gamma = \omega_0)$.

In the special case of critically damped oscillations, we have $\gamma = \omega_0$, and $\omega = \sqrt{\gamma^2 - \omega_0^2} = 0$. A critically damped system returns to equilibrium as quickly as possible without oscillating. This is often desired for the damping of systems such as doors and car suspensions. In this case, there is only one root of the quadratic equation $\lambda = -\omega_0$. In addition to the solution $x(t) = e^{\lambda t}$, a second solution is given by the function $x(t) = te^{\lambda t}$ (see Problem 26 from the End-of-Chapter problems for this section). The complete solution for a critically damped oscillator is a linear combination of these two solutions:

Solution $x(t)$ for a Critically Damped Oscillator $(\gamma = \omega_0)$

$$x(t) = (C_1 + C_2 t)\, e^{-\omega_0 t} \qquad (6.4.9)$$

where A and B are determined by the initial conditions of the system.

Example 6.3 shows how to obtain the general symbolic solutions (6.4.7) and (6.4.9) using Python and Mathematica.

Example 6.3: The general symbolic solution of the SHO

Solve the ODE for a damped harmonic oscillator using Mathematica and SymPy, for the three cases of overdamped oscillations ($\gamma > \omega_0$), underdamped oscillations ($\gamma < \omega_0$), and critically damped oscillations ($\gamma = \omega_0$). The general solutions contain the arbitrary constants C_1 and C_2.

Python Code

We use `dsolve()` in SymPy. Note that for the critically damped oscillator we use $\gamma = \omega_0$ explicitly in the code.

```
from sympy import symbols, Function, Derivative as D, dsolve

import textwrap

print('-'*28,'CODE OUTPUT','-'*29,'\n')

# define function x and various symbols
x = Function('x')
gamma, t ,omega = symbols('gamma, t ,omega', real=True)

# overdamped and underdamped oscillation
soln = dsolve(D(x(t), t, t) + 2*gamma*D(x(t), t) + omega**2*x(t), x(t),\
    simplify=True).rhs

print('For overdamped and underdamped oscillations the \
solution is : \n\nx(t) =', textwrap.fill(str(soln),63))

# critically damped oscillation:  gamma=omega
soln = dsolve(D(x(t), t, t) + 2*omega*D(x(t), t) +\
            omega**2*x(t), x(t)).rhs

print('\nFor critically damped oscillations, the symbolic \
solution is:\n  x(t) =',soln)

-------------------------- CODE OUTPUT ---------------------------
For overdamped and underdamped oscillations the solution is :

x(t) = C1*exp(t*(-gamma + sqrt(gamma**2 - omega**2))) +
C2*exp(-t*(gamma + sqrt(gamma**2 - omega**2)))

For critically damped oscillations, the symbolic solution is:
  x(t) = (C1 + C2*t)*exp(-omega*t)
```

Mathematica Code

We use DSolve twice to obtain the general solutions for the three cases. The solutions contain the arbitrary constants C_1 and C_2.

Print[

"The general solution for underdamped and overdamped

motion is:"]

Print[DSolve[$\{m * x''[t] + b * x'[t] + k * x[t] == 0\}, x[t], t][[1, 1]]$]

The general solution for underdamped and overdamped motion is:

$$x[t] \rightarrow e^{\frac{1}{2}\left(-\frac{b}{m} - \frac{\sqrt{b^2 - 4km}}{m}\right)t} c_1 + e^{\frac{1}{2}\left(-\frac{b}{m} + \frac{\sqrt{b^2 - 4km}}{m}\right)t} c_2$$

Print["The general solution for critically damped motion is: "]

Print[DSolve[$\{m * x''[t] + b * x'[t] + b\hat{\ }2/(4 * m) * x[t] == 0\}$,

$x[t], t][[1, 1]]$]

The general solution for critically damped motion is:

$$x[t] \rightarrow e^{-\frac{bt}{2m}} c_1 + e^{-\frac{bt}{2m}} t c_2$$

6.4.1 COMPARISON OF OVERDAMPED, UNDERDAMPED AND CRITICALLY DAMPED OSCILLATIONS

In the case of overdamped oscillations we have $\gamma > \omega_0$, which means that the square root appearing in the expression $\omega = \sqrt{\gamma^2 - \omega_0^2}$ is real, and (6.4.7) will be written as the sum of two real exponential functions. The resulting motion is that the system returns to equilibrium by an exponential decay without oscillating. Alternatively, one can write the solution as a linear combination of hyperbolic functions (see Problem 23, in the End-of-Chapter problems for this section):

Solution $x(t)$ for an Overdamped Oscillator

$$x(t) = \left[A_0 e^{\omega t} + A_1 e^{-\omega t}\right] e^{-\gamma t} \qquad (6.4.10)$$

$$x(t) = \left[B_0 \sinh(\omega t) + B_1 \cosh(\omega t)\right] e^{-\gamma t} \qquad (6.4.11)$$

$$\omega = \sqrt{\gamma^2 - \omega_0^2} \qquad \gamma > \omega_0 \qquad (6.4.12)$$

The constants A_0, A_1, B_0, B_1 can be found from the initial conditions of the oscillator.

In the case of underdamped oscillations, we have $\gamma < \omega_0$, and $\omega = \sqrt{\gamma^2 - \omega_0^2}$ is imaginary. The system oscillates at a reduced frequency compared to the undamped case, with the amplitude gradually decreasing to zero. Mathematically, it is preferable to get rid of the imaginary quantities by using the real frequency parameter $\omega_d = \sqrt{\omega_0^2 - \gamma^2}$, expressing the general solution as a linear combination of trigonometric functions (see Problem 24 in the End-of-Chapter problems for this section):

Two Forms of Solutions $x(t)$ for an Underdamped Oscillator

$$x(t) = [A_0 \sin(\omega_d t) + A_1 \cos(\omega_d t)]\, e^{-\gamma t} \qquad (6.4.13)$$

$$x(t) = A e^{-\gamma t} \cos(\omega_d t - \phi) \qquad (6.4.14)$$

$$\omega_d = \sqrt{\omega_0^2 - \gamma^2} \qquad \gamma < \omega_0 \qquad (6.4.15)$$

The solution of the underdamped case oscillates repeatedly through the equilibrium point as the amplitude of oscillation decays to zero. This is unlike the overdamped case, where the oscillator may not pass through equilibrium at all as its amplitude decays to zero. This is an important difference in the behaviors of the two oscillators. Also notice from (6.4.13) and (6.4.14) that the parameter $\omega_d = \sqrt{\omega_0^2 - \gamma^2} = \sqrt{k/m - \gamma^2}$ is the frequency of the underdamped oscillator, which is smaller than the natural frequency $\omega_0 = \sqrt{k/m}$.

The expression $A e^{-\gamma t}$ in (6.4.13) and (6.4.14) represents the decrease of the amplitude as a function of time t for underdamped motion. Notice that the amplitude of oscillation decays exponentially with time. The quantity $\tau_A = 1/\gamma$ has dimensions of time, in order for the argument of the exponential $e^{-\gamma t}$ to be dimensionless, and it characterizes how fast the amplitude of oscillation reaches zero. As a general guideline for exponential decaying functions, the amplitude of oscillation is considered to have reached a value of effectively zero after five *characteristic times*, i.e., after $t \cong 5/\gamma$.

For small damping, i.e., when $\gamma << \omega_0$, we can use the binomial approximation $(1-x)^a \cong 1 - ax$ to obtain:

$$\omega_d = \sqrt{\omega_0^2 - \gamma^2} = \omega_0 \sqrt{1 - (\gamma/\omega_0)^2} \cong \omega_0 \left(1 - \frac{\gamma^2}{2\omega_0^2}\right) \qquad \gamma << \omega_0 \qquad (6.4.16)$$

Example 6.4 illustrates the solution for the equations of motion in a mass-spring system exhibiting various degrees of damping.

Example 6.4: Comparison of overdamped, underdamped and critically damped oscillations

Obtain the analytical solution $x(t)$ of (6.4.3), with parameters $m = 1\,\text{kg}$, $k = 1\,\text{N/m}$ and for three values of the constants $b = 1, 2, 3\,\text{Ns/m}$.

Solution:

For the given values of the parameters, when $b = 1$ the parameter $\omega^2 = \gamma^2 - \omega_0^2 = (b/(2m))^2 - (k/m) = (1/2)^2 - 1 < 0$. Since ω^2 is negative, we expect underdamped motion with frequency $\omega = \sqrt{3}/2$ and $\gamma = b/(2m) = 1/2$. In this case the solution is a linear combination of trigonometric functions $e^{-t/2} \cos(\sqrt{3}/2\,t)$ and $e^{-t/2} \sin(\sqrt{3}/2\,t)$.

For critically damped motion, we use $\gamma = \omega_0 = \sqrt{k/m} = 1$, or $b = 2\sqrt{k\,m} = 2$. In this case the solution is a linear combination of the function $e^{-\omega_0 t}$ and $t\,e^{-\omega_0 t}$, or e^{-t} and $t\,e^{-t}$.

When $b = 3$ the parameter $\omega^2 = \gamma^2 - \omega_0^2 = (b/(2m))^2 - (k/m) = (3/2)^2 - 1 > 0$. Since this quantity is positive, we expect overdamped motion. In this case the solution is a linear combination of exponential functions $e^{\omega t} e^{-\gamma t}$ and $e^{-\omega t} e^{-\gamma t}$. With $\omega = \sqrt{5}/2$ and $\gamma = b/(2m) = 3/2$, the solutions are linear combination of $e^{\sqrt{5}t/2} e^{-3t/2}$ and $e^{-\sqrt{5}t/2} e^{-3t/2}$.

Python Code
We use `dsolve()` in SymPy without specifying the initial conditions. The results from
the code agree with the above discussion for the different b values.

```python
from sympy import symbols, Function, Derivative as D, dsolve, sqrt

print('-'*28,'CODE OUTPUT','-'*29,'\n')

# define function x and various symbols
x = Function('x')
t = symbols('t', real=True)

m, k, b = 1, 1, 3

# overdamped oscillation
soln = dsolve(m*D(x(t), t, t)+b*D(x(t), t) +k*x(t), x(t), \
    simplify=True).rhs

print('\nFor overdamped oscillations the solution :\nx(t) =',soln)

m, k, b = 1, 1, 1
soln = dsolve(m*D(x(t), t, t)+b*D(x(t), t) +k*x(t), x(t), \
    simplify=True).rhs

print('\nFor underdamped oscillations the solution :\nx(t) =',soln)

m, k = 1, 1
b = 2*sqrt(k*m)   # condition for critical SHO
soln = dsolve(m*D(x(t), t, t)+ b*D(x(t), t) + k*x(t), x(t), \
    simplify=True).rhs

print('\nFor critically damped oscillations the solution :\nx(t) =',soln)

--------------------------- CODE OUTPUT ----------------------------

For overdamped oscillations the solution :
x(t) = C1*exp(t*(-3 + sqrt(5))/2) + C2*exp(-t*(sqrt(5) + 3)/2)

For underdamped oscillations the solution :
x(t) = (C1*sin(sqrt(3)*t/2) + C2*cos(sqrt(3)*t/2))*exp(-t/2)

For critically damped oscillations the solution :
x(t) = (C1 + C2*t)*exp(-t)
```

Mathematica Code
Again the analytical solutions from Mathematica are in agreement with the general
form of the analytical solution given in this section.

Print["Solution for underdamped is:"];

$m = 1; b = 1; k = 1;$

DSolve[$\{m * x"[t] + b * x'[t] + k * x[t]==0\}, x[t], t$]

Solution for underdamped is:

$$\left\{\left\{x[t] \to e^{-t/2}c_2\text{Cos}\left[\tfrac{\sqrt{3}t}{2}\right] + e^{-t/2}c_1\text{Sin}\left[\tfrac{\sqrt{3}t}{2}\right]\right\}\right\}$$

Print["Solution for overdamped is: "];

$m = 1; b = 3; k = 1;$

DSolve[$\{m * x"[t] + b * x'[t] + k * x[t]==0\}, x[t], t$]

Solution for overdamped is:

$$\left\{\left\{x[t] \to e^{\left(-\frac{3}{2}-\frac{\sqrt{5}}{2}\right)t}c_1 + e^{\left(-\frac{3}{2}+\frac{\sqrt{5}}{2}\right)t}c_2\right\}\right\}$$

Print["Solution for critically damped is: "];

$m = 1; k = 1; b = 2 * \text{Sqrt}[k * m];$

DSolve[$\{m * x"[t] + b * x'[t] + k * x[t]==0\}, x[t], t$]

Solution for critically damped is:

$$\{\{x[t] \to e^{-t}c_1 + e^{-t}tc_2\}\}$$

A comparison of $x(t)$ for the underdamped, overdamped, and critically damped oscillations is shown in Figure 6.4, where we can easily see that the critically damped oscillator (solid curve) returns to equilibrium in a shorter time than other cases. Note that the same parameter values are used for m, k, $x(0)$, and $v(0)$ to produce all three curves in Figure 6.4, while the value of b is variable.

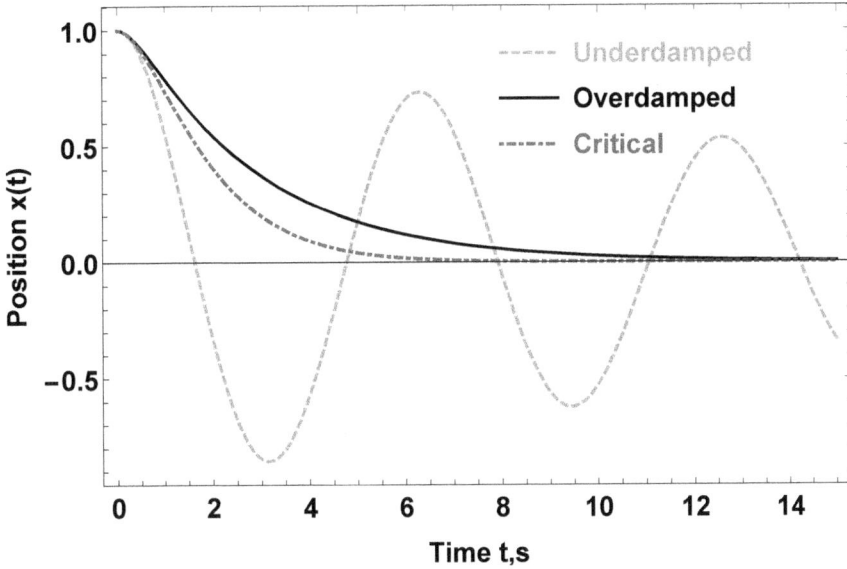

Figure 6.4: Dependence of the harmonic oscillator behavior on the amount of damping, showing underdamped (oscillatory curve), overdamped (solid line) and critically damped (dashed curve) cases.

6.5 ENERGY IN DAMPED HARMONIC MOTION

In this section we derive an analytical expression for the energy E of the damped harmonic oscillator, and we also consider the case of weak damping. The total energy $E = T + V$ of the damped harmonic oscillator is:

$$E = \frac{1}{2}m\left(\frac{dx}{dt}\right)^2 + \frac{1}{2}kx^2 \qquad (6.5.1)$$

For the underdamped case we can write

$$x(t) = A_0 \sin(\omega_d t)\, e^{-\gamma t} \qquad (6.5.2)$$

$$\frac{dx}{dt} = A_0\, e^{-\gamma t}\left[\omega_d \cos(\omega_d t) - \gamma \sin(\omega_d t)\right] \qquad (6.5.3)$$

This expression for the speed $\frac{dx}{dt}$ is rather complicated to work with, and therefore we assume a weak damping situation in which $\gamma << \omega_0$, so that $\omega_d = \sqrt{\omega_0^2 - \gamma^2} \simeq \omega_0$. Using these approximations the term $\gamma \sin(\omega_d t)$ is much smaller that the term $\omega_d \cos(\omega_d t)$ and can be dropped in (6.5.3), to obtain:

$$\frac{dx}{dt} = A_0\, e^{-\gamma t}\left[\omega_d \cos(\omega_d t)\right] \qquad (6.5.4)$$

Substituting $x(t)$ and dx/dt into (6.5.1) we obtain:

$$E = \frac{1}{2}m\left[A_0\, e^{-\gamma t}\omega_d \cos(\omega_d t)\right]^2 + \frac{1}{2}k\left[A_0 \sin(\omega_d t)\, e^{-\gamma t}\right]^2 \qquad (6.5.5)$$

By collecting terms:

$$E = \frac{1}{2}A_0^2\, e^{-2\gamma t}\left[m\omega_d^2 \cos^2(\omega_d t) + k \sin^2(\omega_d t)\right] \qquad (6.5.6)$$

Substituting $\omega_d \simeq \omega_0 = \sqrt{k/m}$ we obtain:

$$E = \frac{1}{2}A_0^2\, e^{-2\gamma t}\left[k\cos^2\left(\omega_d t\right) + k\sin^2\left(\omega_d t\right)\right] = \frac{1}{2}kA_0^2\, e^{-2\gamma t} \tag{6.5.7}$$

The above equation shows that in a situation of weak damping, the total energy decays exponentially with time. The characteristic decay time is $\tau_E = 1/(2\gamma)$. We previously saw in (6.4.13) that the corresponding characteristic decay time for the *amplitude* is $\tau_A = 1/\gamma$.

In summary, the total energy E and the rate of decrease of the total energy dE/dt for the cases of weak damping $\gamma << \omega_d$ (or $\omega_d \simeq \omega_0$) are given by:

Energy and Energy Loss for Underdamped Oscillator, $\gamma << \omega_0$

$$E = \frac{1}{2}kA_0^2\, e^{-2\gamma t} \tag{6.5.8}$$

$$\frac{dE}{dt} = -\gamma k A_0^2 e^{-2\gamma t} = -2\gamma E \tag{6.5.9}$$

The total energy lost within a small time interval Δt can be estimated from:

$$\Delta E \cong \frac{dE}{dt}\Delta t = -2\gamma E \Delta t \tag{6.5.10}$$

where we used (6.5.9). It is customary to define the dimensionless *Quality factor or Q-factor* Q as the ratio $Q = 2\pi|E/\Delta E|$, for a time interval Δt equal to the period $\tau = 2\pi/\omega_d$. By substituting this value of $\Delta t = \tau = 2\pi/\omega_d$ in this definition of Q:

$$Q = 2\pi\left|\frac{E}{\Delta E}\right| = 2\pi\frac{1}{2\gamma\tau} = \frac{\omega_d}{2\gamma} \tag{6.5.11}$$

Q-factor for Underdamped Motion

$$Q = \frac{\omega_d}{2\gamma} \cong \frac{\omega_0}{2\gamma} \tag{6.5.12}$$

In a later section, we will provide a more in-depth discussion of the quality factor, in connection with the concept of energy resonance of a forced harmonic oscillator.

6.6 FORCED HARMONIC OSCILLATOR

Next, we consider the case where an external force $F_0\cos(\omega t)$ is acting on a damped oscillator. Notice that the driving force has an amplitude of F_0 and a frequency of ω, sometimes referred to as the *drive frequency*. This case is referred to as the *forced* or *driven harmonic oscillator*. The net force is then:

$$m\ddot{x} = -kx - b\dot{x} + F_0\cos(\omega t) \tag{6.6.1}$$

The general equation for the driven system is then:

General Differential Equation for a Forced Harmonic Oscillator

$$\ddot{x} + 2\gamma\dot{x} + \omega_0^2 x = D\,\cos(\omega t) \tag{6.6.2}$$

where $\gamma = b/2m$, $\omega_0^2 = k/m$, and $D = F_0/m$. You might be wondering why we limit ourselves to studying driving forces in the form of cosine functions. Clearly not all driving forces in nature are in the form of cosine or sine functions. As we will show later in this chapter, any periodic force (under certain conditions, which are often met in physical problems) can be approximated using a series of sines and cosines. Therefore, the case of cosine or sine driving forces is applicable widely and for many physical systems.

Note that (6.6.2) is a nonhomogeneous ordinary differential equation. As we discussed in the introduction of this chapter on the general theory of differential equations, the solution of nonhomogeneous equations is given by the sum of two parts, the homogeneous solution $x_c(t)$, plus the particular solution $x_p(t)$ of the full nonhomogeneous equation:

$$x(t) = x_c(t) + x_p(t) \tag{6.6.3}$$

We already saw in the previous section that the general solution $x_c(t)$ of the homogeneous equations is given by (6.4.7):

$$x_c(t) = e^{-\gamma t}\left(A_0 e^{\omega t} + A_1 e^{-\omega t}\right) \qquad \omega = \sqrt{(b/2m)^2 - (k/m)} \tag{6.6.4}$$

The solution $x_c(t)$ is commonly referred to as the *transient solution*, because the exponential decay term $e^{-\gamma t}$ causes $x_c(t)$ to decay to zero. After the transient solution has decayed to zero, the only solution left is $x_p(t)$ (assuming that $x_p \neq 0$) and is therefore known as the *steady state solution*. Many physical systems exhibit both a transient and steady state behavior. Physicists and engineers are often (but not always!) more interested in the steady state behavior of the system, because that is the system's long-term behavior.

Before we study the solution $x(t)$ for the full equation (6.6.2), it is instructive to study the case in which damping is absent. When no friction is present, the driven oscillator is described by $b = 0$, so we obtain:

$$\ddot{x} + \omega_0^2 x = D\,\cos(\omega t) \tag{6.6.5}$$

We try a solution of the form $x = A\,\cos\left(\omega t - \phi\right)$, and after we substitute and collect terms we obtain:

$$\begin{aligned} -A\omega^2 \cos\left(\omega t - \phi\right) + \omega_0^2 A\,\cos\left(\omega t - \phi\right) &= D\cos(\omega t) \\ A\left(\omega_0^2 - \omega^2\right)\cos\left(\omega t - \phi\right) &= D\cos(\omega t) \end{aligned} \tag{6.6.6}$$

This equation will have two possible solutions, when $\phi = 0$ and $\cos\left(\omega t - \phi\right) = \cos\left(\omega t\right)$, and when $\phi = \pi$ where $\cos\left(\omega t - \phi\right) = -\cos\left(\omega t\right)$. The amplitude A must be always positive, so we can write the two possible solutions in the form:

$$A = \frac{D}{\omega_0^2 - \omega^2} \qquad \phi = 0 \text{ and } \omega < \omega_0 \tag{6.6.7}$$

$$A = \frac{D}{\omega^2 - \omega_0^2} \qquad \phi = \pi \text{ and } \omega > \omega_0 \tag{6.6.8}$$

and the solution of this driven oscillator with no damping is:

$$x = A\,\cos\left(\omega t\right) = \frac{D}{\omega_0^2 - \omega^2}\cos\left(\omega t\right) \qquad \phi = 0 \text{ and } \omega < \omega_0 \tag{6.6.9}$$

$$x = A\,\cos\left(\omega t\right) = \frac{D}{\omega^2 - \omega_0^2}\cos\left(\omega t\right) \qquad \phi = \pi \text{ and } \omega > \omega_0 \tag{6.6.10}$$

A plot of the amplitude A and of the phase difference ϕ as a function of the external frequency ω is shown in Figure 6.5. We see that the amplitude of the oscillation is maximum when the external frequency $\omega = \omega_0$, and that there is a sharp change in the amplitude and in the phase at this frequency. This is our first demonstration of *amplitude resonance*, where there is a large response from the oscillator when it is driven at a frequency that matches the natural frequency of the oscillator. We will comment on resonance in more detail later in this section.

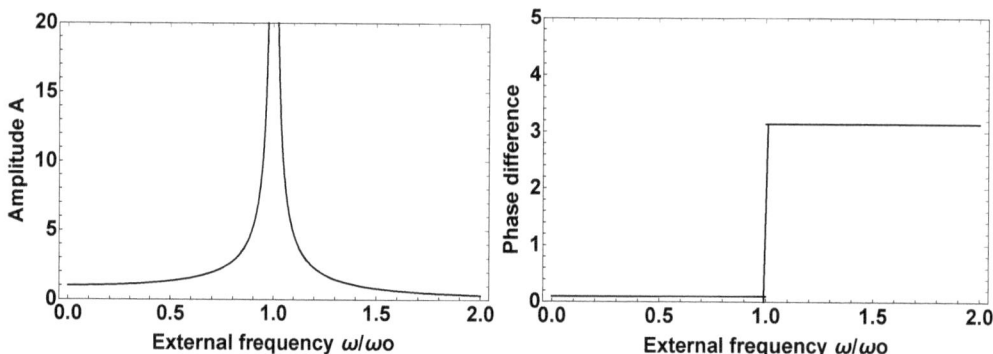

Figure 6.5: Amplitude A and phase difference ϕ as a function of the external frequency ω, in the special case of driven simple harmonic motion with *no damping*.

A more realistic physical situation is of course the case where damping is not zero. For this case, the standard method is to try substituting a solution $x = A\cos(\omega t - \phi)$ in (6.6.2). The choice of a cosine solution is motivated by the fact that the right-hand side of (6.6.2) contains a cosine term with a frequency ω. We need a function $x(t)$ whose derivatives will ultimately give a cosine with the same frequency. We know that the first derivative of the cosine is sine. Therefore we introduce a phase term ϕ, which if chosen properly will allow all of the terms on the left-hand side of (6.6.2) to add up to a cosine, and match the right-hand side. To simplify the mathematics, we write our trial solution as a complex exponential, $x = A\,e^{i(\omega t - \phi)}$, and also replace the term $D\cos(\omega t)$ in this equation with $D\,e^{i\omega t}$. We will carry out the algebra, and then at the end, we will take the real part of the solution, because $\cos(\omega t) = Re\left(e^{i\omega t}\right)$. By substituting and taking the derivatives, we obtain:

$$-A\omega^2 e^{i(\omega t - \phi)} + 2\gamma\omega i A e^{i(\omega t - \phi)} + \omega_0^2 A\,e^{i(\omega t - \phi)} = D\,e^{i\omega t} \tag{6.6.11}$$

By canceling out the $e^{i\omega t}$ from all terms and multiplying by $e^{i\phi} = \cos\phi + i\sin\phi$ on both sides, we obtain:

$$-A\omega^2 + 2\gamma\omega i A + \omega_0^2 A = D\,e^{i\phi} = D\,(\cos\phi + i\sin\phi) \tag{6.6.12}$$

By equating the real parts on the two sides of this equation, and also equating the imaginary parts on the two sides to each other, we obtain:

$$A\left(\omega_0^2 - \omega^2\right) = D\,\cos\phi \tag{6.6.13}$$

$$2\gamma\omega A = D\,\sin\phi \tag{6.6.14}$$

Solving this system of equations for the amplitude A and for the phase angle ϕ, we obtain the particular solution $x_p(t)$ of the nonhomogeneous differential equation (6.6.2) for the driven harmonic oscillator. Finding ϕ from the above equations is simple, all one needs to do is divide (6.6.14) by (6.6.13). Finding A involves squaring each equation and adding them, and recalling that $\sin^2\phi + \cos^2\phi = 1$. The complete solution $x(t) = x_c(t) + x_p(t)$ is then given by:

General Solution $x(t)$ for the Forced Harmonic Oscillator

$$x(t) = x_p(t) + A \cos(\omega t - \phi) \tag{6.6.15}$$

$$v = \frac{dx}{dt} = \frac{dx_p}{dt} - A\omega \sin(\omega t - \phi) \tag{6.6.16}$$

$$A = \frac{F_0/m}{\sqrt{(\omega_0^2 - \omega^2)^2 + 4\gamma^2\omega^2}} \tag{6.6.17}$$

$$\tan\phi = \frac{2\gamma\omega}{\omega_0^2 - \omega^2} \tag{6.6.18}$$

The codes in Example 6.5 numerically solve (6.6.2) using the `odeint` numerical integration method in SciPy, and the `NDSolve` command in Mathematica.

Example 6.5. *Numerical solution for driven harmonic oscillator*
Write Mathematica and Python codes which numerically solve the differential equation for the driven SHO, $\ddot{x} + 2\gamma\dot{x} + \omega_0^2 x = D \cos(\omega t)$. Use the numerical values $m = 1$ mg, $k = 1$ N/m , $b = 15$ Ns/m , $D = 5$ m/s^2 and $\omega = 3$ s^{-1} , from $t = 0$ to $t = 80$ s. Plot both the position $x(t)$ and the velocity $v(t)$. Discuss the behavior of the solution at small times, and at large times.

Solution:
The results are plotted in Figure 6.6, where the left graph is a plot of $x(t)$, and the right graph is a plot of $v(t)$. Notice that in Figure 6.6, the transient behavior is dominant for $t < 40$ s, and that the sinusoidal steady state behavior dominates for $t > 40$ s.

Python Code
In the code, the first component `y[0]` of the vector y represents the position variable x, and the second component `y[1]` represents the derivative dx/dt. The function `odeint` evaluates and returns a vector with components $(x, dx/dt)$. The Python code shows clearly the change of behavior of the solutions $x(t)$ and $v(t)$ as time progresses. The initial part of the functions contains an exponentially decaying transient component. At longer times, the solutions approach the steady state solutions, which oscillate with time with a constant amplitude.

```
from scipy.integrate import odeint
import numpy as np
import matplotlib.pyplot as plt

# SI values of mass, spring constant, force amplitude
b, m, k, Fo, om = 0.2, 1, 1, 5, 3

def deriv(y, time):
    return (y[1], -(b/m)*y[1] - (k/m)*y[0]+Fo/m* np.cos(om*time))

t = np.linspace(0, 80, 400)
yinit = (2, 0)

soln = odeint(deriv, yinit, t)

plt.subplot(1,2,1)
plt.plot(t, soln[:, 0], '-')
plt.ylabel('x(t)')
plt.xlabel('Time [s]')

plt.subplot(1,2,2)
plt.plot(t, soln[:, 1], '-')
plt.ylabel('v(t)')
plt.xlabel('Time [s]')
plt.tight_layout()
plt.show()
```

Mathematica Code

SetOptions[Plot, Frame->True, Axes->True, BaseStyle->FontSize->20];

$m = 1; k = 1; b = .2;$

$d = 5; \omega = 3;$

soln = NDSolve[$\{m * x"[t] + b * x'[t] + k * x[t]==d * Cos[\omega * t], x[0]==2, x'[0]==0\}$,

$x, \{t, 0, 80\}$];

gr1 = Plot[Evaluate[$\{x[t]\}$/.soln], $\{t, 0, 80\}$, FrameLabel->{t, "x[t]"}];

gr2 = Plot[Evaluate[$\{x'[t]\}$/.soln], $\{t, 0, 80\}$, FrameLabel->{t, "x'[t]"}];

GraphicsGrid[{{gr1, gr2}}]

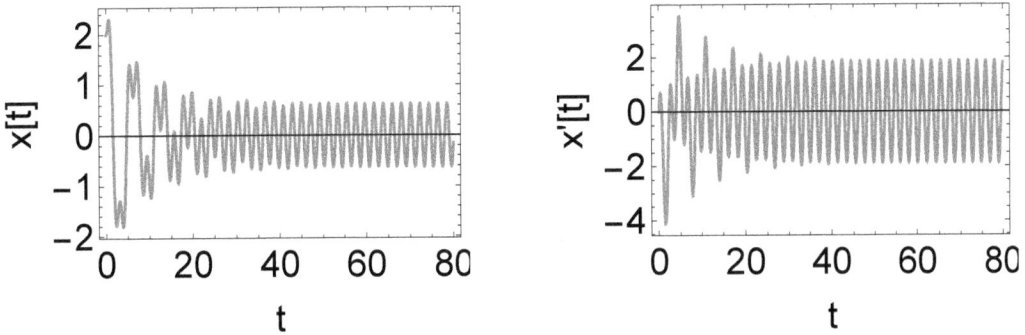

Figure 6.6: The result of running the Mathematica code in Example 6.5 for the externally driven harmonic oscillator. The effect of the *transient* part of the solution is clearly seen for times $t < 40\ s$, while the *steady state* solution is dominant for $t > 40\ s$.

In addition, we can plot the amplitude A and phase angle ϕ as a function of frequency using (6.6.17) and (6.6.18). The plots of the amplitude A and of the phase angle ϕ as a function of the external frequency ω are shown in Figure 6.7. When $\omega << \omega_0$, we find that the phase difference ϕ goes to zero, while for $\omega >> \omega_0$ we find that the phase goes to π (180 degrees), i.e., completely out of phase with the driving force.

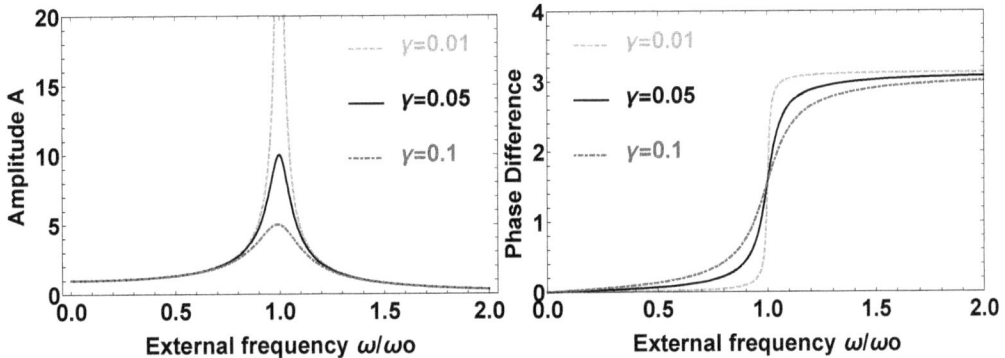

Figure 6.7: Plots of the amplitude $A(\omega)$, and of the phase angle $\phi(\omega)$, in the case of externally driven damped harmonic motion.

One of the most striking results shown in Figure 6.7 is the amplitude $A(\omega)$ as a function of frequency. Notice that similar to the undamped case, amplitude resonance can occur. The damping prevents the amplitude from blowing up to infinity at resonance. Although the small value of $\gamma = 0.01$ results in a large amplitude at resonance, it is not infinite; it simply is beyond the scale of the graph here. Notice that the amplitude resonance occurs when $\omega \approx \omega_0$, because of the small values of γ used to make Figure 6.7. Next, we will compute the exact frequency at which resonance occurs.

The *resonance frequency* ω_r is the value of the drive frequency ω which causes amplitude resonance and is found by computing:

$$\left.\frac{dA}{d\omega}\right|_{\omega=\omega_r} = 0 \tag{6.6.19}$$

From this equation we find that A is a maximum when $\omega = \omega_r = \sqrt{\omega_0^2 - 2\gamma^2}$, where ω_r is the resonant frequency. The corresponding maximum value of A defined as A_{max} is then:

Resonance Frequency and Resonance Amplitude for a Forced Harmonic Oscillator

$$\omega_r = \sqrt{\omega_d^2 - \gamma^2} = \sqrt{\omega_0^2 - 2\gamma^2} \tag{6.6.20}$$

$$A_{max} = \frac{F_0/m}{2\gamma\sqrt{\omega_0^2 - \gamma^2}} \tag{6.6.21}$$

Example 6.6. Amplitude resonance
Use a CAS to obtain the solution $x(t)$ of (6.6.2) and plot the amplitude $A(\omega)$ of the driven oscillation as a function of the external driving frequency ω. Use the numerical values $m = 1\,\text{kg}$, $\omega_0 = 1\,\text{rad/s}$, $F_0 = 1.0\,\text{N}$, $\gamma = 0.1\,\text{rad/s}$.

Solution:
The general solution of the ODE contains the sum of the transient and the steady state solutions. In the Python and Mathematica codes, we select the steady state solution and create a table of the amplitude $A(\omega)$ of the driven oscillation at different frequencies ω.

Python Code
   In the Python code we create a function `findmax(w)` which finds the maximum value of the steady state solution. This function solves the ODE using the `dsolve` function in SymPy, and we select the steady state part of the solution `soln` by using the `.subs` method to set the arbitrary constants to $C_1 = C_2 = 0$.

```python
from sympy import symbols, Function, Derivative as D,\
    dsolve, lambdify, cos, N, Add
import numpy as np
import matplotlib.pyplot as plt

print('-'*28,'CODE OUTPUT','-'*29,'\n')
x = Function('x')
t = symbols('t')
F, wo, gm = 1, 1, .1

#initial conditions x(0)=1  and v(0)=0
initCondits = {x(0): 1, D(x(t),t).subs(t, 0): 0}

def findmax(w):
    # solve the ode to find the full analytical solution soln
    soln = dsolve(D(x(t), t, t)+2*gm*D(x(t), t) +wo**2*x(t)-F*cos(w*t),
      x(t),simplify=True).rhs

# Set the constants C1 and C2 to zero to get the particular solution
    xsteady = soln.subs({'C1': 0, 'C2': 0})
    if (w == 0.1):              # print steady state for w=0.1
        print('\nThe steady state for w=0.1 is:\n  x(t) = ',N(xsteady,3))

    steadyf = lambdify(t, xsteady,'numpy') # lamdify and find max value

    tvals = np.arange(0,70,.02)
    return max(steadyf(tvals))

# omega = external driving frequencies
omega = np.arange(0.1,3,.1)

ampls = [findmax(u) for u in omega]

# plot amplitudes as a function of omega
plt.plot(omega,ampls)
plt.xlabel('Driving frequency, w')
plt.ylabel('Amplitude of steady state, A(w)')
plt.ylim(0,5.4);
plt.show()


-------------------------- CODE OUTPUT ------------------------------

The steady state for w=0.1 is:
  x(t) =  0.0204*sin(0.1*t) + 1.01*cos(0.1*t)
```

Mathematica Code

By using the `DSolve` command as shown in the code, Mathematica produces the analytical solution which clearly contains the sum of $x_c(t)$ and $x_p(t)$. The variable x in the code represents the full solution, while the variable `steady` represents $x_p(t)$. By using the replacement rule `/.Exp[t_]->0`, the two exponential terms in $x_c(t)$ are set to zero, and we are left with the expression `steady` which represents $x_p(t)$. By subtracting the variable `steady` from x, we obtain the variable `trans` which represents the transient solution $x_h(t)$. The parameter u is a list (created by Mathematica's `Table` command) of external frequencies ω and of the corresponding maximum amplitude $A(\omega)$. These maximum values are obtained by using the command `NMaxValue` . Finally the list u is plotted using the `ListPlot` command, and we obtain the desired graph of the maximum amplitude $A(\omega)$ as a function of the external driving frequency ω.

$\text{sol} = \text{DSolve}[\{x''[t] + \text{wo}^2 * x[t] + 2 * \gamma * x'[t] == \text{Fo}/m * \text{Cos}[\omega * t]\}, x[t], t];$

$x = x[t]/.\text{sol}[[1]]//\text{Simplify}$

$e^{-t\left(\gamma+\sqrt{\gamma^2-\text{wo}^2}\right)}c_1 + e^{t\left(-\gamma+\sqrt{\gamma^2-\text{wo}^2}\right)}c_2 + \frac{\text{Fo}\left(\left(-\omega^2+\text{wo}^2\right)\text{Cos}[t\omega]+2\gamma\omega\text{Sin}[t\omega]\right)}{m\left(4\gamma^2\omega^2+(\omega^2-\text{wo}^2)^2\right)}$

$\text{steady} = x/.\text{Exp}[t\_] \to 0$

$\frac{\text{Fo}\left(\left(-\omega^2+\text{wo}^2\right)\text{Cos}[t\omega]+2\gamma\omega\text{Sin}[t\omega]\right)}{m\left(4\gamma^2\omega^2+(\omega^2-\text{wo}^2)^2\right)}$

$\text{trans} = x - \text{steady}$

$e^{-t\left(\gamma+\sqrt{\gamma^2-\text{wo}^2}\right)}c_1 + e^{t\left(-\gamma+\sqrt{\gamma^2-\text{wo}^2}\right)}c_2$

$m = 1; \gamma = 0.1; \text{Fo} = 1; \text{wo} = 1;$

$u = \text{Table}[\{\omega, \text{NMaxValue}[\text{steady}, t]\}, \{\omega, 0, 3, .02\}];$

$\text{ListPlot}[u, \text{FrameLabel} \to \{\text{"Frequency } \omega\text{, Hz"}, \text{"Amplitude A}(\omega)\text{"}\},$

$\text{Frame} \to \text{True}, \text{BaseStyle} \to \{\text{FontSize} \to 16\}, \text{Joined} \to \text{True}]$

$\text{steady} = x/.\text{Exp}[t\_] \to 0/.\omega\text{->}0.1//\text{Simplify}$

$1.00969\text{Cos}[0.1t] + 0.0203978\text{Sin}[0.1t]$

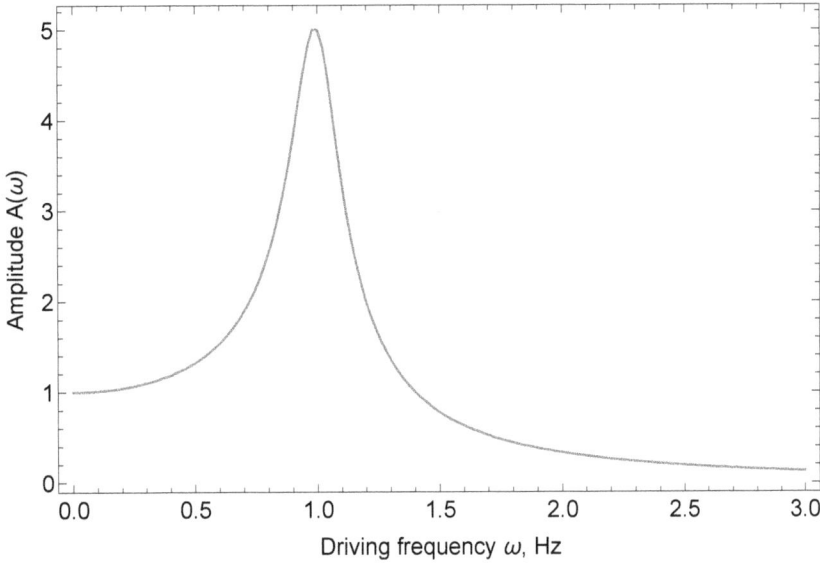

Figure 6.8: The amplitude resonance $A(\omega)$ for a forced harmonic oscillator as a function of the external driving frequency ω, from Example 6.6.

6.7 ENERGY RESONANCE AND THE QUALITY FACTOR FOR DRIVEN OSCILLATIONS

In an earlier section we calculated the quality factor Q for an underdamped oscillator, a concept that is of great importance in physics and engineering. In this section, we will provide a more in-depth derivation of the quality factor, in connection with driven oscillations. This discussion will also lead us to a different kind of resonance, energy resonance.

The quality factor Q is a dimensionless parameter that quantifies the degree of damping and the energy loss present in an oscillator. Higher Q indicates a lower rate of energy loss relative to the stored energy of the resonator, and it means that the oscillations die out more slowly. A pendulum suspended from a high-quality bearing and oscillating in air has a high Q, while a pendulum immersed in oil has a low one. Resonators with high quality factors have low damping, so that they ring or vibrate longer.

To understand Q, we need to find the average mechanical energy of the oscillator. The total mechanical energy of the driven oscillator is found by substituting the general solution $x(t) = A \cos(\omega t - \phi)$, after transients, to obtain:

$$E = T + V = \frac{1}{2}m\dot{x}^2 + \frac{1}{2}kx^2 = \frac{1}{2}mA^2\omega^2 \sin^2(\omega t - \phi) + \frac{1}{2}kA^2 \cos^2(\omega t - \phi) \qquad (6.7.1)$$

By calculating the time average of the $\sin^2(\omega t - \phi)$ term over a period of oscillation, we find that it is equal to $1/2$. Similarly, the time average of the $\cos^2(\omega t - \phi)$ term over a period of oscillation is also equal to $1/2$. Therefore, the time average of the total energy E over a period of oscillation is found from (6.7.1) to be:

$$<E> = \frac{1}{4}mA^2\omega^2 + \frac{1}{4}kA^2 = \frac{1}{4}\left(m\omega^2 + k\right)A^2 \qquad (6.7.2)$$

By substituting the value of the amplitude from (6.6.17) and using $\omega_0^2 = k/m$ we find:

$$<E> = \frac{1}{4}\left(m\omega^2 + m\omega_0^2\right)\frac{(F_0/m)^2}{\left(\omega_0^2 - \omega^2\right)^2 + 4\gamma^2\omega^2} \qquad (6.7.3)$$

and finally:

Average Energy $E(\omega)$ for a Forced Oscillator Over One Period

$$< E > = \frac{F_0^2}{4m} \frac{\omega^2 + \omega_0^2}{\left(\omega_0^2 - \omega^2\right)^2 + 4\gamma^2 \omega^2} \tag{6.7.4}$$

A plot of $< E >$ as a function of the external driving frequency ω is shown in Figure 6.9. Notice that there is a peak similar to the peak we saw for the amplitude graph $A(\omega)$ in Figure 6.8.

Figure 6.9 is illustrating *energy resonance*, and the energy resonance frequency is the drive frequency which maximizes the average energy of the oscillator.

In the context of energy resonance and driven oscillations, the Q factor is defined as the "frequency-to-bandwidth ratio" of the resonator, thus:

$$Q \equiv \frac{\omega_r}{\Delta\omega} \tag{6.7.5}$$

where ω_r is the resonant angular frequency of the peaked graph in Figure 6.9. $\Delta\omega$ in this definition is the full width at half maximum (FWHM) of the energy resonance peak. In order to find $\Delta\omega$, we need to find the two points ω_1 and ω_2 on this graph, which correspond to half of the maximum of the peak.

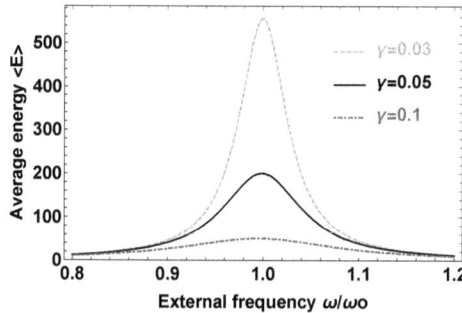

Figure 6.9: Plot of the time averaged energy $< E >$ from (6.7.4), as a function of the ratio ω/ω_0. The value of $\gamma = 0.03$ results in a *lightly damped* harmonic motion, corresponding to a large Q-factor.

To calculate Q, we assume a system with small damping (i.e., a sharp resonance), and we substitute $\omega \simeq \omega_0$ in (6.7.4), except for the term $\left(\omega_0^2 - \omega^2\right)^2$, to obtain:

$$< E > = \frac{F_0^2 \omega_0^2}{2m} \frac{1}{\left(\omega_0^2 - \omega^2\right)^2 + 4\gamma^2 \omega_0^2} \tag{6.7.6}$$

By approximating $\omega_0 \simeq \omega$, the term $\omega_0^2 - \omega^2 = (\omega_0 + \omega)(\omega_0 - \omega) \simeq 2\omega_0 (\omega_0 - \omega)$, and the previous equation becomes:

$$< E > = \frac{F_0^2}{8m} \frac{1}{\left(\omega_0 - \omega\right)^2 + \gamma^2} \tag{6.7.7}$$

Next, taking the derivative of $< E >$ with respect to the driving frequency ω and setting it equal to zero, we find that the maximum of this function occurs at $\omega = \omega_0$. By substituting this value in (6.7.4), the maximum time average of the total energy is given by:

$$< E >_{\text{max}} = \frac{F_0^2}{8m\gamma^2} \tag{6.7.8}$$

Next, we need to find the frequency at which the average energy $< E >$ drops to half its maximum value. We substitute (6.7.8) in (6.7.7) to obtain:

$$\frac{1}{2} < E >_{\text{max}} = \frac{1}{2} \frac{F_0^2}{8m\gamma^2} = \frac{F_0^2}{8m} \frac{1}{(\omega_0 - \omega)^2 + \gamma^2} \tag{6.7.9}$$

which simplifies to:

$$\frac{1}{2\gamma^2} = \frac{1}{(\omega_0 - \omega)^2 + \gamma^2} \tag{6.7.10}$$

The solution to this quadratic equation in ω gives us the two values ω_1 and ω_2 of the driving frequency, at which the time average of the total energy drops to half its maximum value:

$$\omega_1 = \gamma - \omega_0 \qquad \omega_2 = \omega_0 - \gamma \tag{6.7.11}$$

The full width at half maximum is then given by:

$$\Delta\omega = \omega_1 - \omega_2 = 2\gamma \tag{6.7.12}$$

Therefore, from the definition of the Q-factor in (6.7.5), we find the ratio of the resonance frequency ω_0 over the FWHM $\Delta\omega$:

The Quality Factor Q for a Driven Oscillator with Small Damping

$$Q = \frac{\omega_r}{\Delta\omega} = \frac{\omega_0}{2\gamma} \tag{6.7.13}$$

This is the same simple equation we derived for underdamped motion and involves only the coefficients of the second-order differential equation describing most resonant systems, electrical or mechanical. In electrical systems, the stored energy is the sum of energies stored in lossless inductors and capacitors; the lost energy is the sum of the energies dissipated in resistors per cycle. In mechanical systems, the stored energy is the maximum possible stored total energy, i.e., the sum of the potential and kinetic energies at some point in time, while the lost energy is that lost to damping forces such as friction.

Example 6.7. Energy resonance
Use a CAS to obtain the solution $x(t)$ of (6.6.2) and plot the maximum total energy E of the driven oscillation as a function of the external driving frequency ω. Use the numerical values $m = 1\,\text{kg}$, $\omega_0 = 1\,\text{Hz}$, $F_0 = 1.0\,\text{N}$, $\gamma = 0.1\,\text{rad/s}$.

Solution:
We present only the Mathematica code, since the Python code is similar to Example 6.6. The code here evaluates analytically the corresponding speed $v_p(t) = dx_p/dt$, and the total energy $E(t) = \frac{1}{2}mv_p^2 + \frac{1}{2}kx_p^2$. The parameter u is a list of external frequencies

ω and of the corresponding maximum energy $E(\omega)$. We plot the maximum average energy $E(\omega)$ as a function of the external driving frequency ω.

It is notable that the energy resonance $E(\omega)$ shown in this example is much narrower than the corresponding amplitude resonance $A(\omega)$ studied in the previous Example 6.6.

Mathematica Code

```
sol = DSolve[{x"[t] + ωo^2x[t] + 2γx'[t] == Fo/mCos[ωt]}, x[t], t];
x = x[t]/.sol[[1]]//Simplify
```

$$e^{-t\left(\gamma+\sqrt{\gamma^2-\omega o^2}\right)}c_1 + e^{t\left(-\gamma+\sqrt{\gamma^2-\omega o^2}\right)}c_2 + \frac{\text{Fo}\left(\left(-\omega^2+\omega o^2\right)\text{Cos}[t\omega]+2\gamma\omega\text{Sin}[t\omega]\right)}{m\left(4\gamma^2\omega^2+(\omega^2-\omega o^2)^2\right)}$$

```
steady = x/.Exp[t_] → 0
```

$$\frac{\text{Fo}\left(\left(-\omega^2+\omega o^2\right)\text{Cos}[t\omega]+2\gamma\omega\text{Sin}[t\omega]\right)}{m\left(4\gamma^2\omega^2+(\omega^2-\omega o^2)^2\right)}$$

```
en = 1/2msteady^2
```

$$\frac{\text{Fo}^2\left(\left(-\omega^2+\omega o^2\right)\text{Cos}[t\omega]+2\gamma\omega\text{Sin}[t\omega]\right)^2}{2m\left(4\gamma^2\omega^2+(\omega^2-\omega o^2)^2\right)^2}$$

```
m = 1; γ = 0.1; Fo = 1; ωo = 1;
u = Table[{ω, NMaxValue[en, t]}, {ω, 0, 3, 0.02}];
ListPlot[u, PlotRange → All, FrameLabel → {"Driving frequency ω", "Energy E"},
Frame->True, BaseStyle → {FontSize → 16}, Joined → True]
```

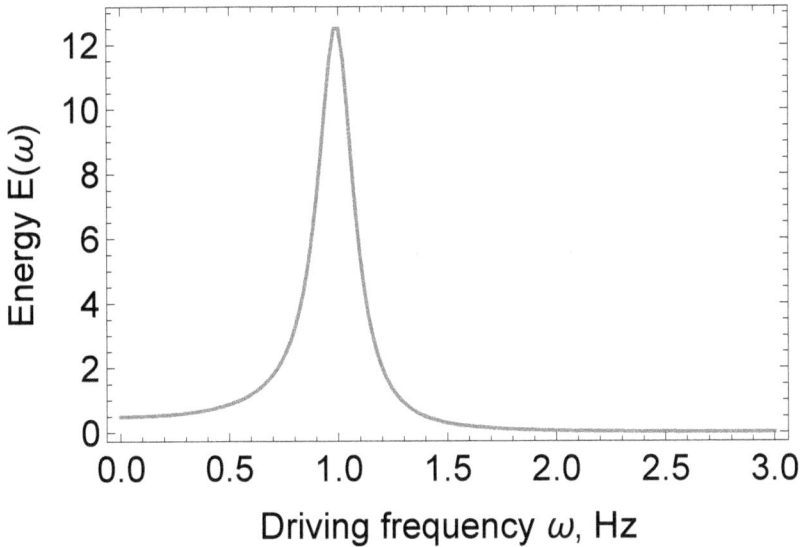

Figure 6.10: The energy resonance $E(\omega)$ for a forced harmonic oscillator as a function of the external driving frequency ω, from Example 6.7.

6.8 ELECTRICAL CIRCUITS

The equation of motion for the forced harmonic oscillator can also be used to describe a driven RLC circuit. In this section, we develop the equations of motion for the RLC circuit and we explore what is sometimes referred to as an *electrical-mechanical analogy*. To begin, consider an RLC circuit as shown in Figure 6.11.

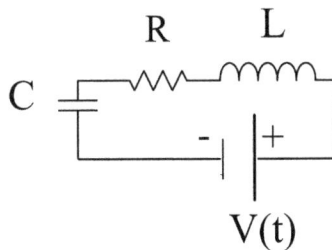

Figure 6.11: RLC series circuit with external voltage $V(t)$, resistance R, inductance L, and capacitance C.

The governing differential equation can be found by substituting into Kirchhoff's voltage law (KVL), the voltage equation for each of the three elements. From the KVL,

$$V(t) + V_R + V_L + V_C = 0 \tag{6.8.1}$$

where $V_R = -IR$ is the voltage drop across the resistor, $V_L = -L\frac{dI}{dt}$ is the voltage drop across the inductor, $V_C = -\frac{Q}{C}$ is the voltage drop across the capacitor, and $V(t)$ is the time varying voltage from the source. Note that we are using the normal conventions that I is the current in the circuit and Q is the charge on the capacitor. Furthermore, L is the inductance of the inductor, R is the resistance of the resistor, C is the capacitance of the capacitor, and the current is $I = \dot{Q}$. For simplicity, we will assume that $V(t) = V_0 \cos(\omega t)$. Substituting these expressions in the KVL equation:

$$L\frac{dI}{dt} + IR + \frac{Q}{C} = V_0 \cos(\omega t) \tag{6.8.2}$$

Dividing by L leads to the second-order differential equation:

General Differential Equation for RLC Series Circuit

$$\frac{d^2 Q(t)}{dt^2} + \frac{R}{L}\frac{dQ(t)}{dt} + \frac{1}{LC}Q(t) = \frac{V_0}{L}\cos(\omega t) \tag{6.8.3}$$

This equation has the exact same mathematical form as equation (6.6.2) for a driven damped harmonic oscillator:

$$\frac{d^2 x}{dt^2} + \frac{b}{m}\frac{dx}{dt} + \frac{k}{m}x = \frac{F_0}{m}\cos(\omega t) \tag{6.8.4}$$

where (6.6.2) is written without the substitution $\gamma = b/2m$.

Because (6.8.3) and (6.8.4) have the same mathematical form, we can reuse the solution of the mechanical system, in order to get the solution of the electrical system. The mathematical similarities imply that the physical behaviors of the two systems are similar. The forced damped harmonic oscillator has two solutions: a transient solution which decays as $t \to \infty$ and a steady state solution consisting of a sinusoidal oscillation with a frequency equal to the drive frequency. Physically, these solutions are describing the displacement of the mass m. Because of the mathematical similarities, we know that the RLC circuit will also have two solutions, a transient which will also decay, and a steady state sinusoidal solution which has the same frequency as the voltage source in the circuit. In the case of the RLC circuit, it is the value of the charge on the capacitor that is oscillating.

Furthermore, we can then draw exact analogies between mechanical systems and electrical systems. These analogies are shown in Table 6.1, and are found by comparing the locations of the variables in (6.8.3) to those in (6.8.4). The simplest analogy is between the displacement x of the mass and the charge Q on the capacitor. A more interesting analogy exists between L and m. The inductance L in (6.8.3) appears in the same location as m in (6.4.3). This implies that L is taking on a role similar to that of mass. This analogy makes physical sense if we recall that the inductance L measures the inductor's ability to resist changes in current ($\dot{Q}$), while the mass m is a measure of inertia, an object's ability to resist changes in velocity ($\dot{x}$).

It is also possible to use the analogies to identify qualitative characteristics of the RLC circuit's behavior. For example, we know that the coefficient of x in the mechanical system is the angular frequency ω_0^2 of oscillation for an undamped oscillator. We can conclude from (6.8.3) that the oscillation frequency of a LC circuit (where $R = 0$) will be given by:

Oscillation Frequency for LC Series Circuit

$$\omega_0 = \frac{1}{\sqrt{LC}} \tag{6.8.5}$$

because $1/LC$ is the coefficient of Q in (6.8.3).

Table 6.1

Analogy between electrical RLC circuit in series and driven harmonic oscillator.

Mechanical System	Series RLC Circuit
Position x	Charge Q
Velocity $\frac{dx}{dt}$	Current $\frac{dQ}{dt}$
Mass m	Inductance L
Spring Constant k	Inverse Capacitance $1/C$
Damping γ	Resistance R
Natural Frequency $\sqrt{\frac{k}{m}}$	$\sqrt{\frac{1}{LC}}$
$m\ddot{x} + \gamma\dot{x} + kx = F$	$L\ddot{Q} + R\dot{Q} + Q/C = V$

Finally, we can derive an expression about the energy of the RLC circuit. We begin by multiplying (6.8.3) by $\dot{Q}$ and L:

$$L\dot{Q}\ddot{Q} + R\dot{Q}^2 + \frac{1}{C}Q\dot{Q} = V_0\cos(\omega t)\dot{Q} \tag{6.8.6}$$

Notice that:

$$\dot{Q}\ddot{Q} = \frac{1}{2}\frac{d}{dt}\left(\dot{Q}^2\right) \quad \text{and} \quad Q\dot{Q} = \frac{1}{2}\frac{d}{dt}\left(Q^2\right) \tag{6.8.7}$$

Substitution of the previous equations into (6.8.6) gives:

$$\frac{d}{dt}\left[\frac{1}{2}L\dot{Q}^2 + \frac{1}{2}\frac{1}{C}Q^2\right] = -R\dot{Q}^2 + V_0\cos(\omega t)\dot{Q} \tag{6.8.8}$$

The $\frac{1}{2}L\dot{Q}^2$ and $\frac{1}{2}\frac{1}{C}Q^2$ terms in (6.8.8) are the energy stored in the inductor and capacitor, respectively. We can see that if $R = 0$ and if there is no voltage source in the circuit ($V_0 = 0$), then the right-hand side of (6.8.8) is zero, so the total energy of the circuit is conserved.

The terms on the right-hand side describe how energy is either lost or gained in the system. Recall that the power dissipated by a resistor is $P = -I^2 R$, which is the first term on the right-hand side of (6.8.8). The second term on the right-hand side of (6.8.8) is describing the rate at which energy is supplied by the variable voltage source. Notice that $V_0\cos(\omega t)\dot{Q}$ is in the form of VI, which is the formula for electric power.

Analogies can be a powerful tool in physics, which can be used to take an understanding of one system, and apply it to another mathematically similar system. As with any analogy (physics-related or not), be careful not to take the analogy too far. Doing so can lead to confusion at best, and to a complete misunderstanding of the system at worst.

6.9 PRINCIPLE OF SUPERPOSITION AND FOURIER SERIES

So far, we have studied the response of an oscillator that is being driven by a continuous periodic force. However, not all driving forces are continuous. Consider a mother pushing her child on a swing. The force applied by the mother is not continuous, but it is periodic with a period τ. How can we mathematically model such forces? We could describe a force that is "off" for a time $\tau/2$ and then constant for a time $\tau/2$ by using a piecewise definition:

$$F(t) = \begin{cases} 0 & 0 \le t < \tau/2 \\ F_0 & \tau/2 \le t \le \tau \end{cases} \tag{6.9.1}$$

however, such a definition could be difficult to work with analytically. We should note that modern CAS can handle such equations in a variety of applications; however, there is much value in learning how to work with such cases analytically. So, the question remains: how do we represent such periodic discontinuous functions in a way that lends itself to analytical solutions? The answer, which we will demonstrate in this section, is that we can represent periodic discontinuous functions as a sum of sines and cosines called a *Fourier series*. This interesting and important result is known as *Fourier's Theorem* and applies to periodic functions that are continuous, or have a finite number of discontinuities over the period τ. Since most functions describing physical processes have only a finite number of discontinuities over an interval $t \in [0, \tau]$, we can consider Fourier's Theorem to be widely applicable.

Fourier's Theorem is only one part of the solution of handling discontinuous periodic drive functions. By representing drive functions as a series of sines and cosines, we need to be able to solve equations of the form:

$$\ddot{x} + 2\gamma\dot{x} + \omega_0^2 x = a_0 + a_1 \cos(\omega t) + a_2 \cos(2\omega t) + \cdots + b_1 \sin(\omega t) + b_2 \sin(2\omega t) + \cdots \tag{6.9.2}$$

In other words, we need to know how to solve linear ordinary differential equations that have multiple terms on the right-hand side. Because the differential equation is linear in x and its derivatives, we can use the powerful concept of the *Principle of Superposition*, which will allow us to break up the differential equation into pieces and then sum the solutions of the individual pieces.

6.9.1 PRINCIPLE OF SUPERPOSITION

Consider a differential equation of the following form:

$$a\frac{d^2x}{dt^2} + b\frac{dx}{dt} + cx = \left(a\frac{d^2}{dt^2} + b\frac{d}{dt} + c \right) x = F(t) \tag{6.9.3}$$

which is very similar to the equations describing oscillations that we have worked with so far. We can rewrite such an equation as $\hat{L}(x) = F(t)$, where the differential operator $\hat{L}$ is defined as:

$$\hat{L} \equiv a\frac{d^2}{dt^2} + b\frac{d}{dt} + c \tag{6.9.4}$$

An operator $\hat{L}$ is linear if it satisfies:

$$\hat{L}(c_1 x_1 + c_2 x_2) = c_1 \hat{L}(x_1) + c_2 \hat{L}(x_2) \tag{6.9.5}$$

where c_1 and c_2 are arbitrary constants, and x_1 and x_2 are two solutions of the differential equation (6.9.3). It is straightforward to show that $\hat{L}$ defined in (6.9.4) is a linear operator.

Next, we consider a differential equation of the form:

$$\hat{L}(x) = c_1 F_1(t) + c_2 F_2(t) \tag{6.9.6}$$

where $\hat{L}$ is defined as in (6.9.4). Suppose we know two functions x_1 and x_2, such that $\hat{L}(x_1) = F_1(t)$ and $\hat{L}(x_2) = F_2(t)$. In other words, x_1 is the solution to (6.9.6) without $F_2(t)$ on the right-hand side, and x_2 is the solution without $F_1(t)$ present. Then:

$$\hat{L}(c_1 x_1 + c_2 x_2) = c_1 \hat{L}(x_1) + c_2 \hat{L}(x_2) = c_1 F_1(t) + c_2 F_2(t) \tag{6.9.7}$$

Therefore, the sum of the individual solutions x_1 and x_2 is the solution to the full differential equation (6.9.6). We can, of course, extend this to an arbitrary number of terms on the right-hand side:

$$\hat{L}(x) = \sum_{i=1}^{N} c_i F_i(t) \tag{6.9.8}$$

which would then have the solution:

$$x(t) = x_c + \sum_{i=1}^{N} c_i x_i \tag{6.9.9}$$

where x_i satisfied $\hat{L}(x_i) = F_i(t)$ and x_c is the complementary solution of the ODE. The ability to add solutions in the above manner is called the *Principle of Superposition* and applies only to linear differential equations. In Chapter 13 we will study nonlinear systems, and the Principle of Superposition will not apply to those systems.

The Principle of Superposition will be critical in dealing with Fourier series representations of discontinuous drive forces. When we represent the discontinuous drive force as a Fourier series, the resulting differential equation will be of the form:

$$\ddot{x} + 2\gamma\dot{x} + \omega_0^2 x = a_0 + a_1 \cos(\omega t) + a_2 \cos(2\omega t) + \cdots + b_1 \sin(\omega t) + b_2 \sin(2\omega t) + \cdots \tag{6.9.10}$$

We now know that in order to solve this type of equation, all we need to do is solve the system of equations:

$$\ddot{x}_i + 2\gamma\dot{x}_i + \omega_0^2 x_i = a_i \cos(i\omega t) \qquad \text{and} \qquad \ddot{x} + 2\gamma\dot{x} + \omega_0^2 x = b_i \sin(i\omega t) \tag{6.9.11}$$

for $i = 0, \ldots, N$. Note that for $i = 0$, $\cos(0) = 1$ and we obtain the solution for the constant term on the right-hand side of the differential equation. Similarly $\sin(0) = 0$ gives the homogeneous solution. While there may be a large number of equations to solve, there are ways that the process can be generalized. For example, notice that in the above case, by changing the value of i, we are simply changing the drive frequency for $i > 0$. Next, we need to learn how to represent a discontinuous periodic function as a Fourier series.

6.9.2   FOURIER SERIES

In this section, we introduce an important result due to French mathematician and physicist Joseph Fourier (1768-1830). Let us consider a function $f(t)$ of the real variable t, which is periodic with a period τ, such that $f(t) = f(t + \tau)$. Fourier's theorem states that, under certain conditions which typically hold in mechanics, this periodic function can be written as the sum of sine and cosine terms in the form of a *Fourier series*:

General Mathematical Form of Fourier Series

$$f(t) = \frac{a_0}{2} + \sum_{n=1}^{N} a_n \cos(n\omega t) + b_n \sin(n\omega t) \tag{6.9.12}$$

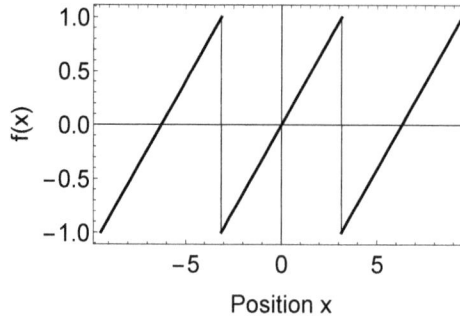

Figure 6.12: The sawtooth function in Example 6.8.

where the lowest angular frequency is $\omega = 2\pi/\tau$, and the higher frequencies are integer multiples $n\,\omega$ of the fundamental angular frequency ω. The coefficients a_n and b_n represent the amplitudes of the various waves, and are known as *Fourier coefficients*. The Fourier coefficients a_n and b_n are computed by using the following general integral relationships:

$$\int_{-\tau/2}^{\tau/2} \cos\left(n\omega t\right)\cos\left(m\omega t\right)dt = \begin{cases} 0 & m \neq n \\ \tau/2 & m = n \neq 0 \end{cases}$$

$$\int_{-\tau/2}^{\tau/2} \sin\left(n\omega t\right)\sin\left(m\omega t\right)dt = \begin{cases} 0 & m \neq n \\ \tau/2 & m = n \neq 0 \end{cases} \qquad (6.9.13)$$

$$\int_{-\tau/2}^{\tau/2} \cos\left(n\omega t\right)\sin\left(m\omega t\right)dt = 0 \qquad \text{for all integers } n \text{ and } m$$

The calculation of a_n and b_n involves multiplying both sides of (6.9.12) with either $\cos(2\pi mt/\tau)$ or $\sin(2\pi mt/\tau)$, and integrating over the whole period. The results of this calculation are known as *Fourier's trick* to many generations of physicists:

Fourier's Trick for Finding the Fourier Amplitude Coefficients

$$a_n = \frac{2}{\tau}\int_{-\tau/2}^{\tau/2} f(t)\cos(n\omega t)dt \qquad (6.9.14)$$

$$b_n = \frac{2}{\tau}\int_{-\tau/2}^{\tau/2} f(t)\sin(n\omega t)dt \qquad (6.9.15)$$

It is not always possible to compute coefficients of all values of n, so a finite Fourier series approximates the original function $f(t)$, and the approximation improves as N increases. The Fourier series is generally presumed to converge everywhere except at discontinuities, where it converges to the mean of the values of $f(t)$ before and after the discontinuity.

Example 6.8 demonstrates how to evaluate and plot the Fourier series of a periodic function $f(x)$, both analytically and using a CAS. Note that in this example the independent variable is x and not the time t.

Example 6.8: Fourier series calculation
Find the Fourier series for the following periodic sawtooth function shown in Figure 6.12, both analytically and using a CAS:

$$f(x) = \frac{x}{\pi}, \quad \text{for } -\pi < x < \pi$$

$$f(x + 2\pi j) = f(x), \quad \text{for } -\infty < x < \infty \text{ and } j \text{ is an integer.}$$

Solution:
    Although in this case the independent variable is x, the function $f(x)$ is periodic in x, so all of the formulas for Fourier series still apply. Next, it is important to identify the period. As can be seen from the graph (and the definition of f), the period is $\tau = 2\pi$. Using the formulas (6.9.14) and (6.9.15), we can compute the Fourier coefficients, with $\omega = 2\pi/\tau = 1$. First we find the a_n coefficients:

$$a_n = \frac{1}{\pi} \int_{-\pi}^{\pi} \frac{x}{\pi} \cos(\tfrac{2\pi n x}{2\pi}) \, dx$$

$$= \left(\frac{1}{\pi}\right)^2 \int_{-\pi}^{\pi} x \cos(nx) \, dx$$

$$= \left(\frac{1}{\pi}\right)^2 \frac{x \sin(nx)}{n} \Big|_{-\pi}^{\pi}$$

therefore, we see that $a_n = 0$ for all values of n, because $\sin(n\pi) = 0$ for all integers n. Next, we compute the b_n coefficients:

$$b_n = \frac{1}{\pi} \int_{-\pi}^{\pi} \frac{x}{\pi} \sin(\tfrac{2\pi n x}{2\pi}) \, dx$$

$$= \left(\frac{1}{\pi}\right)^2 \int_{-\pi}^{\pi} x \sin(nx) \, dx$$

$$= -\frac{2}{\pi n} \cos(n\pi) + \frac{2}{\pi^2 n^2} \sin(n\pi)$$

Again, we use $\sin(n\pi) = 0$ for integer values of n. However, note that $\cos(n\pi) = (-1)^n$. Therefore, we find that:

$$b_n = \frac{2(-1)^{n+1}}{\pi n}$$

The resulting Fourier series is therefore:

$$f(x) = \frac{2}{\pi} \sum_{n=1}^{\infty} \frac{(-1)^{n+1}}{n} \sin(nx)$$

When $x = \pm\pi$, the Fourier series converges to 0, which is the mean of the left and right limits at $x = \pm\pi$.

Python Code
In the code we evaluate the definite integrals for a_n and b_n and specify that n is an integer, by using the option `integer = True` in the `symbols` command. For the plots in Figure 6.13, we define a function `T(numTerms)` which represents the finite summation

of the number of Fourier terms, **numTerms**. A **for** loop is used to evaluate the sum, with the two subplots representing the summation for $N = 10, 30$ terms. As can be expected, as N increases, the sum represents more accurately the sawtooth function.

```
from sympy import symbols, integrate, sin, cos, pi
import matplotlib.pyplot as plt
import numpy as np
print('-'*28,'CODE OUTPUT','-'*29,'\n')


x = symbols('x')          # define symbols
n = symbols('n', integer = True,positive=True)

print("Coefficients an = ",(1/pi**2)*integrate(x*cos(n*x),(x,-pi,pi)))
print("Coefficients bn = ",(1/pi**2)*integrate(x*sin(n*x),(x,-pi,pi)))

# function to evaluate Fourier f(x) by summing the series terms
def T(numTerms):
    suma = [0]*len(x)             # initialize the sum
    for n in range(1,numTerms):
        suma = suma+(2/np.pi)*(-1)**(n+1)/n*np.sin(n*x)
    return suma


# positions along the x-axis
x = np.linspace(0,2*np.pi,200)
x1 = np.linspace(0,np.pi,100)
x2 = np.linspace(np.pi,2*np.pi,100)


# call function to evaluate f(x) for different numTerms
plt.subplot(1,2,1)
numTerms = 10
plt.plot(x, T(numTerms))
plt.plot(x1,x1/np.pi)
plt.plot(x2,-1+(x2-np.pi)/np.pi)
plt.text(5,1,'N = '+str(numTerms))
plt.ylabel('f(x)')


plt.subplot(1,2,2)
numTerms = 30
plt.plot(x, T(numTerms))
plt.plot(x1,x1/np.pi)                    # plot the two parts of f(x)
plt.plot(x2,-1+(x2-np.pi)/np.pi)
plt.text(5,1,'N = '+str(numTerms))
plt.tight_layout()
plt.show()

-------------------------- CODE OUTPUT ----------------------------
Coefficients an =  0
Coefficients bn =  -2*(-1)**n/(pi*n)
```

Mathematica Code

We use the `PieceWise` command to define the function $f(x)$, and we use the `Table` command to create appropriate lists for the plots of the series approximations and of $f(x)$.

$f = \text{Piecewise}[\{\{x/\text{Pi}, x > -\text{Pi}\&\&x < \text{Pi}\}, \{(x - 2 * \text{Pi})/\text{Pi}, x > \text{Pi}\&\&x < 3 * \text{Pi}\}\}];$

$\text{Plot}[f, \{x, -\pi, 3 * \pi\}, \text{PlotRange->All}];$

$T = 2 * \text{Pi};$

$\text{numTerms} = 30;$

$\text{Print}[\text{"ao="}, \text{ao} = (2/T) * \text{Integrate}[f, \{x, 0, T\}]];$

$\text{an:=}(2/T) * \text{Integrate}[f * \text{Cos}[2 * n * \text{Pi} * x/T], \{x, 0, T\}];$

$\text{bn:=}(2/T) * \text{Integrate}[f * \text{Sin}[2 * n * \text{Pi} * x/T], \{x, 0, T\}];$

$\text{Print}[\text{"an-Fourier coefficients="}, \text{an}];$

$\text{Print}[\text{"bn-Fourier coefficients="}, \text{bn}];$

ao=0
an-Fourier coefficients=$\frac{-1 + \text{Cos}[2n\pi] + 2n\pi \text{Sin}[n\pi]}{n^2\pi^2}$
bn-Fourier coefficients=$\frac{-2n\pi \text{Cos}[n\pi] + \text{Sin}[2n\pi]}{n^2\pi^2}$

$\text{anList} = \text{Table}[\text{an}/.n \rightarrow m, \{m, 1, \text{numTerms}\}];$

$\text{bnList} = \text{Table}[\text{bn}/.n \rightarrow m, \{m, 1, \text{numTerms}\}];$

$\text{Plot}\left[\left\{f, \text{ao}/2 + \sum_{n=1}^{\text{Length[anList]}} (\text{anList}[[n]] * \text{Cos}[2 * n * \text{Pi} * x/T]) + \right.\right.$
$\left.\left.\sum_{n=1}^{\text{Length[bnList]}} (\text{bnList}[[n]] * \text{Sin}[2 * n * \text{Pi} * x/T])\right\}, \{x, -T/2, T/2\}\right]$

6.9.3 EXAMPLE OF SUPERPOSITION PRINCIPLE AND FOURIER SERIES

To conclude this section, we present an example of a problem where a damped harmonic oscillator is driven by a discontinuous periodic force. The following problem ties together the concepts of the Principle of Superposition and Fourier Series.

Consider the following discontinuous periodic force $F(t)$:

$$F(t) = mA\frac{t}{\pi} \quad \text{for} - j\pi < x < j\pi \tag{6.9.16}$$

where m is the mass of the oscillator, j is an integer, and A has units of acceleration per unit time. This is essentially the same function from Example 6.8. It may seem strange to have m as a term in the formula of the driving force, but it will simplify the algebra later.

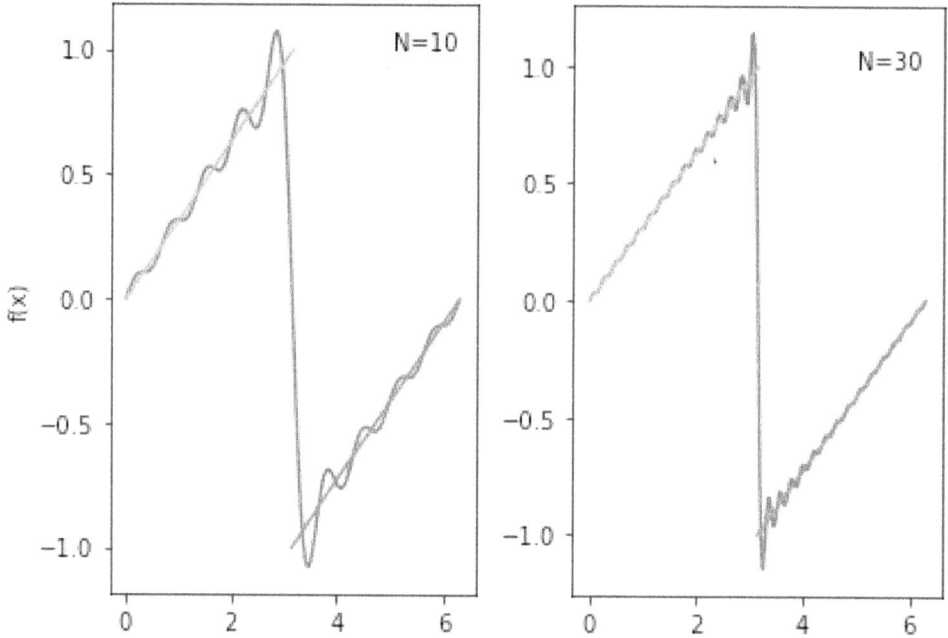

Figure 6.13: Results from the Python code in Example 6.8, showing a comparison of the original function (solid line) and the approximation using the Fourier series for $N = 10, 30$ terms (wavy line).

The driving force in (6.9.16) acts on a damped harmonic oscillator with damping parameter γ and natural frequency ω_0, whose equation of motion is then:

$$\ddot{x} + 2\gamma\dot{x} + \omega_0^2 x = \frac{F(t)}{m} \tag{6.9.17}$$

In order to solve for $x(t)$ using the analytical methods described in this chapter, we need to use the Fourier series for $F(t)$. From Example 6.8, we know that:

$$F(t) = \frac{2mA}{\pi} \sum_{n=1}^{\infty} \frac{(-1)^{n+1}}{n} \sin(n\omega t) \tag{6.9.18}$$

where we changed the independent variable from x to t, included the multiplicative constant A, and explicitly included ω into the Fourier series. Note that $\omega = 1$ because $\tau = 2\pi$. So the resulting differential equation to solve is:

$$\ddot{x} + 2\gamma\dot{x} + \omega_0^2 x = \frac{2A}{\pi} \sum_{n=1}^{\infty} \frac{(-1)^{n+1}}{n} \sin(n\omega t) \tag{6.9.19}$$

However, the work done in this chapter had cosine terms on the right-hand side of the differential equation. In order to use the solutions developed in this chapter, we will need to use the identity: $\sin(a) = \cos(a - \pi/2)$. Doing so results in (6.9.19) becoming:

$$\ddot{x} + 2\gamma\dot{x} + \omega_0^2 x = \frac{2A}{\pi} \sum_{n=1}^{\infty} \frac{(-1)^{n+1}}{n} \cos(n\omega t - \pi/2) \tag{6.9.20}$$

To solve (6.9.20), we consider individual equations of the form:

$$\ddot{x}_n + 2\gamma \dot{x}_n + \omega_0^2 x_n = \frac{2A}{\pi} \frac{(-1)^{n+1}}{n} \cos(n\omega t - \pi/2) \qquad (6.9.21)$$

We use trial solutions of the form $x_n(t) = A_n \cos(n\omega t - \pi/2 - \phi_n)$ and follow the procedure outlined for the solution of the driven damped harmonic oscillator. The result is:

$$A_n = \frac{2A(-1)^{n+1}}{n\pi \sqrt{\left(\omega_0^2 - (n\omega)^2\right)^2 + 4\gamma^2 (n\omega)^2}} \qquad (6.9.22)$$

$$\tan \phi_n = \frac{2\gamma n\omega}{\omega_0^2 - (n\omega)^2} \qquad (6.9.23)$$

Notice that the above solution is the same as (6.6.17) and (6.6.18), however the constant coefficient is now $2A(-1)^{n+1}/n\pi$, and the drive frequency is $n\omega$ instead of ω. Finally, we can use the Principle of Superposition to find the solution to (6.9.20):

$$x(t) = \frac{2A}{n\pi} \sum_{n=1}^{\infty} \frac{(-1)^{n+1}}{\sqrt{\left(\omega_0^2 - (n\omega)^2\right)^2 + 4\gamma^2 (n\omega)^2}} \cos(n\omega t - \pi/2 - \phi_n) \qquad (6.9.24)$$

where the phase angles ϕ_n are defined in (6.9.23).

6.10 PHASE SPACE

Phase space diagrams are a useful concept in physics and other disciplines. A phase space diagram usually plots the position $x(t)$ of a particle on the x-axis and its corresponding momentum $p = m v(t)$ on the y-axis. Alternatively, one can also use the speed $v(t)$ on the y-axis, instead of the momentum $p = m v(t)$. The result is a graph that gives a qualitative description of the particle's motion. Phase space diagrams are important in analyzing nonlinear systems, and we will discuss the phase space diagrams of nonlinear systems in Chapter 13. For now, we will explore the phase space diagrams associated with the harmonic oscillator.

Let us consider an undamped simple harmonic oscillator whose position and speed are described by:

$$x(t) = A \cos(\omega_0 t) \qquad (6.10.1)$$
$$v(t) = dx/dt = -A\omega_0 \sin(\omega_0 t) \qquad (6.10.2)$$

We can eliminate the time t in these equations by using the trig identity $\sin^2(\omega_0 t) + \cos^2(\omega_0 t) = 1$:

$$\frac{x^2}{A^2} + \frac{v^2}{A^2 \omega_0^2} = 1 \qquad (6.10.3)$$

This clearly represents an ellipse on the xv-plane. We say that the phase space representation of a simple harmonic oscillator is an ellipse, with semi-major axis equal to the amplitude of oscillation $a = A$, and the semi-major y-axis equal to $b = A\omega_0$. We know that for the SHO, $E = \frac{1}{2}kA^2$ and $\omega_0^2 = k/m$. Therefore, we can rewrite (6.10.3) as:

$$\frac{x^2}{2E/k} + \frac{v^2}{2E/m} = 1 \qquad (6.10.4)$$

This is an ellipse with semi-major axis equal to the amplitude of oscillation $a = \sqrt{2E/m}$ and the semi-major y-axis equal to $b = \sqrt{2E/k}$. Examples of harmonic oscillators with different energies E are shown in Figure 6.14. In this case, we see that ellipses with larger area correspond to the trajectories with larger energies.

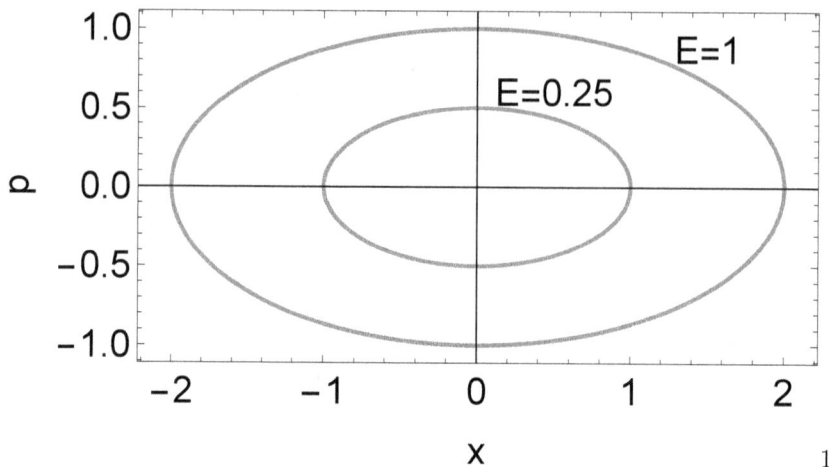

Figure 6.14: Phase space of simple harmonic oscillator

Phase space diagrams can tell us something about the motion of the system. The curve in a phase diagram is sometimes referred to as a *trajectory in phase space*. An image of all possible trajectories in a phase space is usually called a *phase portrait* or *phase diagram*. Each point along a trajectory gives the system's position and momentum. The presence of closed-loop trajectories in a phase space diagram is evidence of oscillatory motion. A closed-loop trajectory tells us that after some time τ, the system returns to its initial position and momentum. The elliptical trajectories in Figure 6.14 tell us that the SHO displays oscillatory motion (which we already knew!).

Next, we will look at the phase space diagram of an underdamped harmonic oscillator. In this case, we write:

$$x(t) = Ae^{-\gamma t}\cos(\omega_d t - \phi) \tag{6.10.5}$$

$$v(t) = dx/dt = -A\omega_d e^{-\gamma t}\sin(\omega_d t - \phi) - A\gamma e^{-\gamma t}\cos(\omega_d t - \phi) \tag{6.10.6}$$

where ω_d is the driving frequency. It can be difficult to visualize the trajectories directly from the above equations. Therefore, we can make the following change of coordinates, in order to simplify the algebra:

$$u = \omega_d x, \qquad w = \gamma x + v \tag{6.10.7}$$

Then the coordinates become:

$$u = \omega_d Ae^{-\gamma t}\cos(\omega_d t - \phi) \tag{6.10.8}$$

$$w = -\omega_d Ae^{-\gamma t}\sin(\omega_d t - \phi) \tag{6.10.9}$$

Next, we convert to polar coordinates: $\rho^2 = u^2 + w^2$ and $\theta = \omega_d t$. Squaring and adding u and w gives:

$$\rho = \omega_d Ae^{-(\gamma/\omega_d)\theta} \tag{6.10.10}$$

We see that the trajectory in polar coordinates corresponds to a curve whose distance ρ from the origin is decreasing as θ increases. In other words, the trajectory is an inward spiral. Notice that the decay rate of ρ is proportional to γ, as we would expect. Furthermore, the inward spiral for the underdamped harmonic oscillator should not be a surprise, because we know that underdamped motion consists of oscillatory motion whose amplitude decays exponentially in time.

Suppose we did not have the insight to use polar coordinates. How would we have created the phase space diagram? We can use a parametric plot of x and v. The codes for creating the underdamped oscillator's phase space diagram are demonstrated in Example 6.9.

Example 6.9: Plotting the phase space of an oscillator
Plot the phase space diagram for the underdamped harmonic oscillator with position $x(t) = Ae^{-\gamma t}\cos(\omega_d t - \phi)$ and numerical values $A = 1$, $\omega_d = 1$, $\gamma = 0.1$, $\phi = 0$ (all in SI units).

Solution:
We use a parametric plot with $x(t) = Ae^{-\gamma t}\cos(\omega_d t - \phi)$ and $y(t) = dx/dt$.

Python Code
In the code we define the time parameter `t` and define $x(t)$ and $v(t)$ as NumPy expressions, which are plotted in MatPlotLib using `plt.plot`. The graph is shown in Figure 6.15.

```python
import matplotlib.pyplot as plt
import numpy as np

A, omega, phi, gamma = 1, 1, 0, 0.1
t = np.linspace(0,100,500)

x = A*np.exp(-gamma*t)*np.cos(omega*t-phi)

# v = dx/dt
v = -A*(gamma*np.cos(omega*t - phi) +\
        omega*np.sin(omega*t - phi))*np.exp(-gamma*t)

plt.plot(x,v)
plt.ylabel('v')
plt.xlabel('x')
plt.show()
```

Mathematica Code
In Mathematica we find the speed $v(t)$ using `v=D[x,t]`, and we can use directly the `ParametricPlot[x,v]` command to create the phase space plot in Figure 6.15.

$A = 1; \omega\mathrm{d} = 1; \phi = 0; \gamma = 0.1;$

$x = A * \mathrm{Exp}[-\gamma * t] * \mathrm{Cos}[\omega\mathrm{d} * t - \phi];$

$v = D[x, t];$

$\mathrm{ParametricPlot}[\{x, v\}, \{t, 0, 100\}, \mathrm{FrameLabel} \to \{\mathrm{x}, \mathrm{v}\}, \mathrm{Frame} \to \mathrm{True},$

$\mathrm{BaseStyle} \to \{\mathrm{FontSize} \to 16\}, \mathrm{PlotRange}\text{->}\mathrm{All}]$

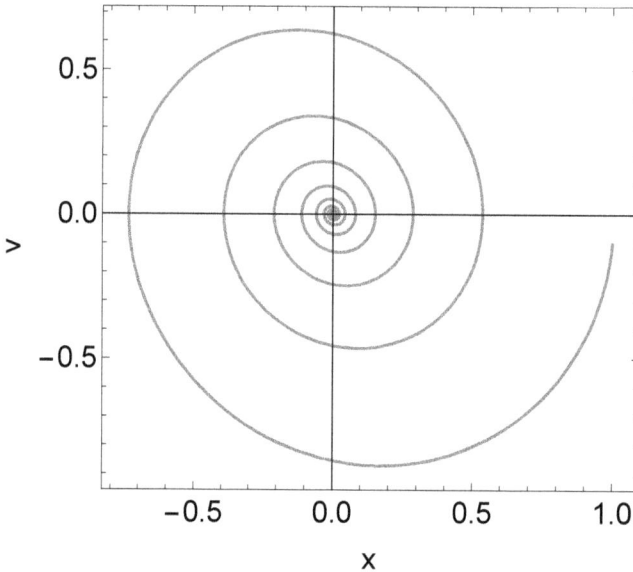

Figure 6.15: Phase space diagram for the underdamped oscillator, in Example 6.9.

The above algorithm can be used to generate the phase diagram for any system whose position $x(t)$ is known. While the phase space diagram does not provide new information about the behavior of the harmonic oscillator, we will find in Chapter 13 that phase space diagrams will become a critical tool for understanding the possible behaviors of a nonlinear system. In Chapter 13, we will develop a method of finding a system's phase space diagram without solving analytically the system's equations of motion. It is often the case in nonlinear systems that the equations of motion cannot be solved analytically. Therefore the phase space diagram may be the only tool available in getting information about the behavior of the system.

6.11 CHAPTER SUMMARY

A spring-mass system is an example of simple harmonic motion and described by:

$$F(x) = -kx \qquad \frac{d^2x}{dt^2} + \omega_0^2 x = 0 \qquad \omega_0 = \sqrt{\frac{k}{m}}$$

and has the equivalent solutions:

$$x(t) = A \cos(\omega_0 t) + B \sin(\omega_0 t) \qquad x(t) = C \cos(\omega_0 t + \phi)$$

and total kinetic energy

$$E = T + V = \frac{1}{2} mv^2 + \frac{1}{2} kx^2$$

A simple plane pendulum is also an example of simple harmonic motion when the angle of oscillation is small. The differential equation is:

$$\ddot{\theta} + \left(\frac{g}{l}\right) \sin\theta = 0$$

In the limit of small oscillations $\theta_0 \ll 1$, the solution of this equation is:

$$\theta(t) = C \cos(\omega_0 t + \phi) \qquad \omega_0^2 = \frac{g}{l} \qquad \tau = 2\pi \sqrt{\frac{l}{g}}$$

The general equations describing the damped harmonic oscillator are:

$$\frac{d^2x}{dt^2} + 2\gamma \frac{dx}{dt} + \frac{k}{m} x = 0 \qquad x(t) = e^{-\gamma t}\left(A_0 e^{\omega t} + A_1 e^{-\omega t}\right) \qquad \omega = \sqrt{\gamma^2 - \omega_0^2}$$

where the constant $\gamma = b/2m$ is the *damping parameter*. There are three possible types of solutions, depending on whether $\omega = \sqrt{\gamma^2 - \omega_0^2}$ is real, imaginary, or zero: (a) overdamped oscillations ($\gamma > \omega_0$), (b) underdamped oscillations ($\gamma < \omega_0$), and (c) critically damped oscillations ($\gamma = \omega_0$ or $\omega = 0$). The corresponding solutions are:

$$x(t) = \left[A_0 e^{\omega t} + A_1 e^{-\omega t}\right] e^{-\gamma t} \qquad \omega = \sqrt{\gamma^2 - \omega_0^2} \qquad \gamma > \omega_0 \text{ Overdamped}$$

$$x(t) = A e^{-\gamma t} \cos(\omega_d t - \phi) \qquad \omega_d = \sqrt{\omega_0^2 - \gamma^2} \qquad \gamma < \omega_0 \text{ Underdamped}$$

$$x(t) = (A + Bt) e^{-\omega_0 t} \qquad\qquad\qquad \gamma = \omega_0 \text{ Critically Damped}$$

The energy and energy loss for weakly damped oscillator are:

$$E = \frac{1}{2} kA_0^2 e^{-2\gamma t} \qquad \gamma \ll \omega_d \qquad \frac{dE}{dt} = -2\gamma E \qquad (6.11.1)$$

The dimensionless Q-factor expresses the degree of damping present in an oscillator, with higher Q indicating a lower rate of energy loss:

$$Q = \frac{\omega_0}{2\gamma}$$

The forced damped oscillator is described by the amplitude $A(\omega)$ and phase angle $\phi(\omega)$, both are functions of the external driving frequency ω:

$$\ddot{x} + 2\gamma \dot{x} + \omega_0^2 x = F_0 \cos(\omega t) \qquad x(t) = x_c(t) + A \cos(\omega t - \phi)$$

$$A = \frac{F_0/m}{\sqrt{(\omega_0^2 - \omega^2)^2 + 4\gamma^2\omega^2}} \qquad \tan\phi = \frac{2\gamma\omega}{\omega_0^2 - \omega^2}$$

where $x_c(t)$ is one of the three possible solutions of the homogeneous equation. This oscillator exhibits the phenomenon of resonance with resonance frequency and resonance amplitude given by:

$$\omega_r = \sqrt{\omega_d^2 - \gamma^2} = \sqrt{\omega_0^2 - 2\gamma^2} \qquad A_{max} = \frac{F_0/m}{2\gamma\sqrt{\omega_0^2 - \gamma^2}}$$

There is an analogy between the charge $Q(t)$ in electrical RLC circuits and the position $x(t)$ in mechanical oscillators:

$$\frac{d^2 Q(t)}{dt^2} + \frac{R}{L}\frac{dQ(t)}{dt} + \frac{1}{LC}Q(t) = 0 \qquad \frac{d^2 x}{dt^2} + \frac{b}{m}\frac{dx}{dt} + \frac{k}{m}x = 0$$

The general mathematical form of a Fourier series for any periodic function with period $\tau = 2\pi/\omega$ is:

$$f(t) = \frac{a_0}{2} + \sum_{n=1}^{N} a_n \cos(n\omega t) + b_n \sin(n\omega t)$$

The Fourier coefficients can be found using Fourier's trick:

$$a_n = \frac{2}{\tau}\int_{-\tau/2}^{\tau/2} f(t)\cos(n\omega t)dt \qquad b_n = \frac{2}{\tau}\int_{-\tau/2}^{\tau/2} f(t)\sin(n\omega t)dt$$

Phase space diagrams are important in analyzing nonlinear systems. They are plots of the position $x(t)$ of a particle on the x-axis and its corresponding speed $v(t)$ (or momentum) on the y-axis.

6.12   END-OF-CHAPTER PROBLEMS

The symbol ⌨ indicates a problem which requires some computer assistance, in the form of graphics, numerical computation, or symbolic evaluation.

Section 6.1-6.3: Differential Equations and the Simple Harmonic Oscillator

1. ⌨ A mass m kg moves with simple harmonic motion along the x-axis with a frequency ω_0. At time $t = 0$ the mass is stationary, and at time $t = 1$ second it is moving with a speed V. Solve this problem both analytically and using a CAS.

   a. Find the position, velocity, and acceleration as a function of time t.

   b. Find the amplitude and period of the motion.

2. Starting from the general expression $x(t) = A\cos(\omega_0 t) + B\sin(\omega_0 t)$ for a harmonic oscillation, show that an equivalent form of this equation is given by the expressions: $x(t) = C\cos(\omega_0 t + \phi)$ where $C = \sqrt{A^2 + B^2}$ $\tan\phi = B/A$.

3. A mass m is attached to a spring of spring constant k and moves with underdamped harmonic motion along the x-axis, with the air resistance given by $F = -bv$.

   a. Show that the difference in times between two successive maxima is given by:

   $$\Delta t = \frac{2\pi}{\sqrt{\omega_o^2 - \gamma^2}}$$

where $\gamma = b/2m$ and $\omega_0 = \sqrt{k/m}$.

    b. Show that the *ratio* of the amplitudes of two successive maxima of the position is given by:

$$R = e^{\gamma \Delta t}$$

where Δt is the expression in part (a).

4. A mass m is attached to a vertical spring of spring constant k, which at equilibrium stretches by an amount a. At time $t = 0$ a force $F(t) = A \sin(\omega t)$ is applied to the mass, where $\omega = \sqrt{k/m}$. Describe qualitatively the position $x(t)$ and the speed $v(t)$ of this mass as a function of time t. You may neglect any damping present in the system.

5. ⌨ A spring has a spring constant of $k = 8\,\text{N/m}$ and a mass $m = 0.5\,\text{kg}$ is attached to it. At time $t = 0$ a force $F = 6\sin(4t)$ (in N) is applied to the mass. Find an analytical expression for the position $z(t)$ at time t, with the initial conditions $z(0) = 0$ and $z'(0) = 0$.

6. ⌨ The differential equation for an oscillating system is:

$$\frac{d^2 z}{dt^2} + 16z - 160\cos(6t) = 0$$

with the initial conditions $z(0) = 0$ and $\frac{dz}{dt}(0) = 0$. Find and plot the solution $z(t)$ for several periods and interpret the results.

7. A cylinder of mass m and radius R floats with its axis vertically and partially submerged in a liquid of density ρ. The cylinder is given a small vertical displacement downwards and is released. Find the period of oscillation.

8. A mass m moves in a region of the xy-plane where the force is given by $\boldsymbol{F} = -k\,x\,\hat{\mathbf{i}} - k\,y\,\hat{\mathbf{j}}$.

    a. Show that depending on the initial conditions of the problem, the motion will be an ellipse, a parabola or a hyperbola.

    b. Find the potential energy of this oscillator.

9. Consider a mass m moving on a vertical wire inside a gravitational field. The shape of the wire is a cycloid, described by the equations: $x = a\theta - a\sin\theta$ and $y = -a + a\cos\theta$.

    a. Find the speed of the mass at the bottom of the cycloid wire, if it is released from rest at the origin $(0,0)$ at time $t = 0$.

    b. Find the period of oscillation.

10. A mass m is on a frictionless table and is connected to two fixed points on opposing walls, by two springs of equal natural length, of negligible mass and spring constants k_1 and k_2, respectively. The mass is displaced horizontally and then released. Prove that the period of oscillation is $\tau = 2\pi\sqrt{m/(k_1 + k_2)}$.

11. A spring having a spring constant k and negligible mass, has one end fixed on an inclined plane of angle θ and a mass m at the other end, as shown in Figure 6.16. If the mass m is pulled down a distance x below the equilibrium position in a direction parallel to the inclined plane and released, find the displacement from the equilibrium position at any time if (a) the incline is frictionless, (b) the incline has coefficient of friction μ.

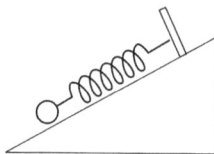

Figure 6.16: A mass attached to a spring on an inclined plane, from Problem 6.11.

12. A particle moves with simple harmonic motion along the x-axis. At times t_0, $2t_0$, $3t_0$ it is located at $x = a, b, c$, respectively. Prove that the period of oscillation is $T = 2\pi t_0 / \left[\cos^{-1} \left(\frac{a+c}{2b} \right) \right]$.

13. Two equal masses m are connected by springs having equal spring constant k, as shown in Figure 6.17, so that the masses are free to slide on a frictionless table. The walls to which the ends of the springs are attached are fixed.

    a. Set up the differential equations of motion of the masses, and find the possible frequencies of oscillation, also called the normal frequencies.

    b. Suppose that the first mass is held at its equilibrium position while the second mass is given a displacement of magnitude $a > 0$ to the right of its equilibrium position. The masses are then released. Find the position of each mass at any later time.

    c. Plot the positions of the two masses as a function of time on the same graph, and discuss the motion. Use the numerical values $k = m = a = 1$ in SI units.

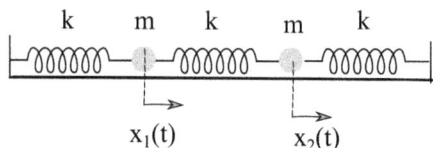

Figure 6.17: Two coupled masses connected to fixed ends, from Problem 6.13.

14. Two equal masses m are on a horizontal frictionless table as shown in the Figure 6.18, are connected by springs of equal spring constants k. The end of one spring is attached to a wall, and the masses are set into motion.

    a. Set up the equations of motion of the system.

b. Find the possible frequencies of oscillation, also called the normal frequencies, and plot the motion of the two masses. Use a numerical value of the frequency $\omega = \sqrt{k/m} = 1$ rad/s.

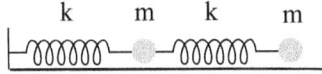

Figure 6.18: Two masses joined by a horizontal spring from Problem 6.14.

15. Two masses m_1 and m_2 are attached to the ends of a spring of constant k, which is on a horizontal frictionless table as shown in Figure 6.19. If the masses are pulled apart and then released, prove that they will vibrate with respect to each other with period $T = 2\pi\sqrt{\mu/k}$ where $\mu = M_1 M_2/(M_1 + M_2)$ is the reduced mass of the system.

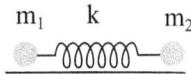

Figure 6.19: Two masses attached to a spring on a frictionless table, from Problem 6.15.

16. ⌨ Find the frequencies of vibration for the system of particles of equal masses m connected by springs with the same spring constant k, as indicated in Figure 6.20.

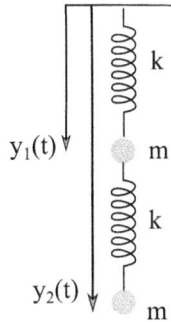

Figure 6.20: Two coupled masses hanging from vertical springs, from Problem 6.16.

17. ⌨ A pendulum has length $L = 1$ m, mass $m = 0.1$ kg and the gravitational acceleration is $g = 9.8$ m/s^2. Let $\theta(t)$ represent the angle of the pendulum from the equilibrium position.

a. Plot the angle $\theta(t)$, the angular velocity $\omega = d\theta/dt$, the kinetic energy T, the potential energy V and the total kinetic energy $E = T + V$ as functions of time t, when the pendulum is released from rest at a small angle of 3 degrees from equilibrium.

b. Plot the motion of the pendulum when it is released from rest at a small angle of 3 degrees, and at a very large angle of 170 degrees from equilibrium. How is the oscillation different in the two cases?

c. Find the time averages $< T >$ and $< V >$ over a complete period of oscillation for the two angles of release in part (b), and compare them with the total mechanical energy E. Discuss your result.

18. 🖵 A pendulum has length $L = 1\,\mathrm{m}$, mass $m = 0.1\,\mathrm{kg}$, and it is released from rest at an angle of 3 degrees from equilibrium. Use a value of $g = 9.8\,\mathrm{m/s^2}$ for the gravitational acceleration.

a. Find the tension T of the string as a function of time t.

b. At what value of time t is the tension a maximum, and at what value is it a minimum? Do these results make sense?

19. In this chapter, we saw that for small amplitude oscillations, the period of a pendulum is $\tau = 2\pi\sqrt{L/g}$.

a. Show that in the more general case of oscillations with any amplitude θ_0, the period is given by the expression:

$$\tau = \sqrt{8L/g} \int_0^{\theta_0} \frac{d\theta}{\sqrt{\cos\theta - \cos\theta_0}}$$

b. By expanding the integrand in a series of the angle θ, show that the integration can be performed term-by-term to obtain the following expression:

$$\tau = 2\pi\sqrt{L/g}\left(1 + \frac{\theta_0^2}{16}\right)$$

Section 6.4: Damped Harmonic Oscillator

20. 🖵 We have studied the ODE for the damped oscillator $\ddot{x} + 2\gamma\dot{x} + \omega_0^2 x = 0$. The corresponding ODE for a damped pendulum oscillating at small angle θ is $\ddot{\theta} + 2b\dot{\theta} + \omega_0^2\,\theta = 0$. A pendulum has length $L = 1\,m$, mass $m = 0.1\,kg$, and is released from rest at a small angle of 3 degrees from equilibrium.

a. If the damping force is such that the motion of the pendulum is critically damped, find the maximum angular speed of the pendulum.

b. Plot the angle $\theta(t)$ and the angular speed $\omega = d\theta/dt$ for the motion.

21. A spring system is underdamped, and the frequency of oscillation when there is no damping is three times the frequency of oscillation with damping. Find the ratio of successive maxima in the oscillations.

22. A spring system is critically damped, and the frequency of oscillation when there is no damping is 100 Hz. If the initial conditions of the spring are $x = x_0$ and $v = 0$ at time $t = 0$, find the displacement at $t = 10$ s.

23. Show that the two expressions $x(t) = [A_0 e^{\omega t} + A_1 e^{-\omega t}] e^{-\gamma t}$ and $x(t) = [B_0 \sinh(\omega t) + B_1 \cosh(\omega t)] e^{-\gamma t}$ are equivalent for the case of overdamped motion, where $\omega = \sqrt{\gamma^2 - \omega_0^2}$ and $\gamma > \omega_0$. What is the mathematical relationship between the constants A_0, A_1 and the constants B_0, B_1?

24. Starting from the general equation $x(t) = (A_0 e^{\omega t} + A_1 e^{-\omega t}) e^{-\gamma t}$ where $\omega = \sqrt{\gamma^2 - \omega_0^2}$, derive

$$x(t) = A e^{-\gamma t} \cos(\omega_d t - \phi)$$

for the case of underdamped motion, where $\omega_d = \sqrt{\omega_0^2 - \gamma^2}$ and $\gamma < \omega_0$. What is the mathematical relationship between the constants A_0, A_1 and the constants A, ϕ?

25. Using the approximation $\gamma << \omega_0$, compute the general position $x(t)$ for the case of underdamped motion. What is the resulting equation in the limit of $\gamma \to 0$?

26. Show that $x(t) = t e^{\lambda t}$ is a solution of the equation $\ddot{x} + 2\gamma \dot{x} + \omega_0^2 x = 0$, in the case of critically damped oscillations.

Section 6.5: Energy in Damped Harmonic Motion

27. 🖥 A pendulum has length $L = 1$ m, mass $m = 0.1$ kg and the gravitational acceleration is $g = 9.8 \, \text{m/s}^2$. The air resistance on the pendulum is given by $F = -bv$ where $v = v(t)$ is the speed of the mass m, and $b = 0.1 \, Ns/m$ is the damping coefficient. The pendulum is released from rest at an angle of 3 degrees from equilibrium.

Find the kinetic energy T and potential energy V as a function of time t, when the pendulum is released from rest at an angle of 3 degrees from equilibrium. Plot T, V and the sum $T + V$ on the same graph .

Sections 6.6: Forced Harmonic Oscillator

28. 🖥 The position $x(t)$ of a forced harmonic oscillator satisfies the equation

$$\frac{d^2 x}{dt^2} + 6\frac{dx}{dt} + 8x = 10 \cos(2t)$$

At time $t = 0$ the particle is at the origin and at rest.

a. Find the transient and steady state solutions for the position and velocity of the oscillator as functions of time t. Do this by hand, and also by using a CAS.

b. Plot the potential energy, the kinetic energy and the total energy on the same graph.

c. Find the rate of change of the kinetic energy dT/dt as a function of time, and plot this quantity over many *periods of oscillation*. Discuss the shape of the resulting graph at different intervals of time t.

29. The position $x(t), y(t)$ of a particle on the xy-plane is given by $x(t) = A \cos(\omega_1 t + \phi_1)$ and $y(t) = B \cos(\omega_2 t + \phi_2)$. Prove that if the ratio ω_1/ω_2 is a rational number, then the particle moves in a closed curve.

30. 🖥 The position $x(t), y(t)$ of a particle on the xy-plane satisfies the equations

$$\frac{d^2x}{dt^2} = -8y$$

$$\frac{d^2y}{dt^2} = -2x$$

At time $t = 0$, the particle is at the position $(1, 0)$ and at rest. Find the position and velocity as a function of time t. Do this problem by hand, and also by using a CAS.

31. 🖥 A particle with mass $m = 1$ kg moves on the xy-plane where the potential energy function is $V(x, y) = x^2 + y^2 + xy + 3$. The particle is initially at the point $(1, 1)$ and moves with a speed of 2 m/s along the positive x-axis.

   a. Find the position as a function of time t.

   b. Find the period of the motion.

Sections 6.7-6.8: Energy Resonance and the Quality Factor for Driven Oscillations and Electrical Circuits

32. 🖥 An electrical circuit consists of a 10 kΩ resistor, a 80 mH inductance, a 10 μF capacitor, and an AC voltage of the form $V = 100\cos(10t)$ (where V is measured in Volts and time in seconds) connected in series.

   a. Find the current $I(t)$ in the circuit at any time t. Plot this function and identify the steady state part and the transient part of the graph.

   b. Plot the electrical energy stored in the capacitor as a function of time, and discuss the shape of this graph.

   c. Find the quality factor Q in this circuit.

   d. Find the average power $P = VI$ dissipated in the resistor R, between $t = 0.2$ s and $t = 0.4$ s.

33. Give examples of mechanical and electrical systems which have a very low Q-factor and a very high Q-factor. What is the physical meaning of these numbers for the oscillators they describe?

Sections 6.9: Superposition Principle, Fourier Series

34. 🖥 Use the symbolic capabilities of Mathematica or SymPy to:

   a. Find analytical expressions for the Fourier coefficients for the periodic square wave function shown below.

   b. Plot the sum of the first 10 terms in the Fourier series and discuss how the plot compares with the periodic square wave.

   c. Repeat part (b) for the first 100 terms of the Fourier series and discuss the results.

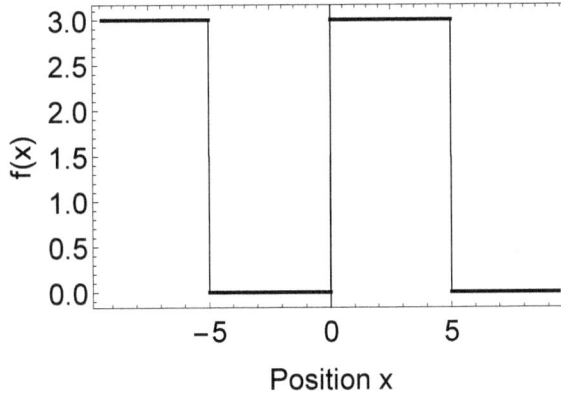

Figure 6.21: Piecewise continuous function $f(x)$, from Problem 6.35.

Section 6.10: Phase Space

Several problems on the concept of phase space can be found in Chapter 13.

7 Calculus of Variations

In the next chapter, we will develop a new way of finding a system's equations of motion. The new method will involve a branch of mathematics called the *calculus of variations*, which we introduce in this chapter. The calculus of variations focuses on finding the extrema of a mathematical object called a *functional*. The foundations of the calculus of variations were laid by Isaac Newton and Gottfried Wilhelm Leibniz (1646–1716) and was further developed by the brothers Jakob and Johann Bernoulli (1655–1705 and 1667–1748, respectively). The first major contributions to the calculus of variations were made by Leonhard Euler (1707–1783), Joseph-Louis Lagrange (1736–1813), and Pierre-Simon Laplace (1749–1827). In this chapter, we will introduce the calculus of variations and use it to solve problems related to physics.

7.1 MOTIVATION FOR LEARNING THE CALCULUS OF VARIATIONS

Up to this point in the book, we have primarily found a particle's equation of motion by solving a second-order differential equation based on Newton's second law. This method can be straightforward when the forces acting on the system are mathematically simple and the problem is best described in Cartesian coordinates. However, many problems in physics require the use of coordinate systems other than Cartesian coordinates. In Chapter 3, we computed velocities and accelerations in polar, cylindrical, and spherical coordinates. We found that those quantities took on forms that are much more complicated than when they are expressed in Cartesian coordinates. Because Newton's second law uses acceleration, problems involving the description of motion in spherical or cylindrical coordinates can be quite messy and difficult to solve.

An additional complication with Newton's second law comes from the fact that it uses force to describe motion. While there are many mathematically simple forces which can act on a particle, such as weight, the normal force, and the linear and quadratic drag forces; forces of constraint can either be very complicated or completely unknown. Examples of a force of constraint would be a particle constrained to move along the surface of a sphere or a bead sliding along a curved wire. Even a rolling object can involve constraint forces. If the forces of constraint are unknown, it may not be possible to use Newton's second law to find the equations of motion.

Granted that one knows all of the forces acting on the particle, Newton's second law can be used to describe the motion of particles even in the cases described above. However, the mathematics can be quite difficult and messy. We need another way to formulate equations of motion which will allow to us to more easily deal with non-Cartesian coordinate systems and with constraint forces, even in cases where the force of constraint is unknown.

One such alternative formulation involves the use of a *variational principle*. A variational principle involves finding a function which either minimizes or maximizes a particular quantity, which is itself a function. The *calculus of variations* provides a means of finding the function which extremizes the quantity of interest. For example, a chain that is suspended at both ends will take on a shape that minimizes the chain's gravitational potential energy. In this case, the shape of the chain is the "function to be found" and finding that function (the shape of the chain) involves minimizing the chain's gravitational potential energy, which is itself a function. The chain's gravitational potential energy is a function of the chain's shape. Hence, the chain's gravitational potential energy is a "function of a function," because the chain's gravitational potential energy function depends on the chain's shape function. In

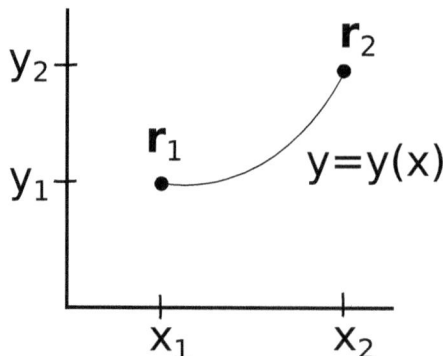

Figure 7.1: A function $y(x)$ which describes a path between two points, $\mathbf{r}_1 = (x_1, y_1)$ and $\mathbf{r}_2 = (x_2, y_2)$.

this case, the calculus of variations can be used to find the function that corresponds to the shape of the chain that minimizes the chain's gravitational potential energy. This will be analogous to finding the value that minimizes a function using the methods of elementary calculus.

In this chapter, we will present the calculus of variations and use it to demonstrate the solution to a few mathematical problems, all of which are relevant to physics. In the next chapter, we will present a method which allows us to formulate a system's equations of motion using variational principles and the calculus of variations.

7.2 SHORTEST DISTANCE BETWEEN TWO POINTS—SETTING UP THE CALCULUS OF VARIATIONS

We will begin our development of the calculus of variations by finding the shortest distance between two points on a plane. The setup of this problem will provide a concrete example of what we are trying to achieve with the calculus of variations. Let's begin by considering two points on the Cartesian plane as shown in Figure 7.1.

The quantity to be minimized is the length L of the curve joining the points $\mathbf{r}_1 = (x_1, y_1)$ and $\mathbf{r}_2 = (x_2, y_2)$. The length can be found by integrating the differential distance $ds = \sqrt{dx^2 + dy^2}$ along a small part of the curve:

$$L = \int_{\mathbf{r}_1}^{\mathbf{r}_2} ds \tag{7.2.1}$$

where $\mathbf{r}_1$ and $\mathbf{r}_2$ denote the location of the first and second point, respectively. It is common to factor out the dx^2 term from ds in order to obtain:

$$L = \int_{x_1}^{x_2} \sqrt{1 + y'(x)^2}\, dx \tag{7.2.2}$$

where we used:

$$dy = \frac{dy}{dx} dx = y'(x) dx \tag{7.2.3}$$

In the introductory paragraphs of this chapter we said, "[the] calculus of variations provides a means of finding the function which extremizes the quantity of interest." In this case, the "quantity of interest" is the length L of the curve $y(x)$ joining the points $\mathbf{r}_1$ and $\mathbf{r}_2$. Notice that the length of the curve depends on the function $y(x)$ (or more specifically, its derivative), hence our "quantity of interest" is a function of the function $y(x)$ and, therefore, we can write the length of the curve as $L = L(y)$. The problem specifically states that we want to find the path which minimizes the distance between the points $\mathbf{r}_1$ and $\mathbf{r}_2$. In other words, we want to find the function $y^*(x)$ such that $L(y^*)$ is the minimum value of L. As we will see, the calculus of variations will provide us with a means of how to find $y^*(x)$.

Before showing how to solve the above problem, it will be useful to establish some notation. Notice that (7.2.2) takes the form:

General Functional for the Calculus of Variations

$$J(y) = \int_{x_1}^{x_2} f\left[y(x), y'(x); x\right] dx \tag{7.2.4}$$

In the case of (7.2.2), the integrand $f[y(x), y'(x); x] = \sqrt{1 + y'(x)^2}$ is a function of only $y'(x)$. The use of square brackets denotes the functional nature of f, i.e., that, loosely speaking, its arguments are functions. The semicolon is used to separate out the independent variable in the argument of f from the functions. In other words, f depends on functions $y(x)$ and $y'(x)$ but can also contain terms of those functions' independent variable, x. It should be noted that there are no hard rules for these two notations, square brackets and semicolons, parentheses and a comma could have been used instead. Many of the problems from the calculus of variations that we will study will involve integrands that are functions of $y(x)$, $y'(x)$, and x. Later on in this chapter, we will explore cases where the integrand depends on multiple functions $y_1(x), y_2(x), \ldots, y_n(x)$, their derivatives, and an independent variable x.

The quantity J is sometimes referred to as a *functional*. A functional is a mathematical object that takes a function as an input and returns a scalar as an output. In the case of (7.2.2), the functional L can be rewritten as $L(y)$, and takes the equation of the curve $y(x)$ as an input and returns as the length of the curve (a scalar) as the output.

Next, we will develop a method for finding the function $y(x)$ that is an extremum of the functional $J(y)$. Or, continuing with our example, we will learn how to find the curve $y^*(x)$ that minimizes the path length between two points on a plane.

7.3 FIRST FORM OF THE EULER EQUATION

Let us return to introductory calculus for a moment. A common problem from introductory calculus was that of finding a number x^*, which is an extremum of $f(x)$. This is done by using the condition:

$$\left.\frac{df}{dx}\right|_{x=x^*} = 0 \tag{7.3.1}$$

In other words, the condition for the value x^* to be an extremum of f is that the first derivative of f must be zero at x^*. Equivalently, the rate of change in the function f is zero at the extremum. It may not surprise you that in order to find the function $y(x)$ that is an extremum of the functional J, we will also perform a derivative and identify a condition for that derivative to be equal to zero. The condition that leads to a zero derivative will also give us the means of finding the extremum, $y(x)$. Note that for clarity of notation, we will drop the asterisk representing the extremum function. Furthermore, the point x^* is sometimes called a *stationary point* because infinitesimal displacements from x^* do not

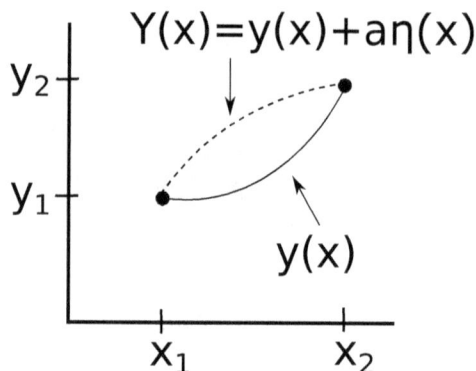

Figure 7.2: The function $\eta(x)$ perturbs $y(x)$, the extremum of (7.3.2), resulting in the functional J not being stationary. The parameter a describes the size of the perturbation from the extremum $y(x)$.

change the value of $f(x)$ since the slope of $f(x)$ is zero at x^*. Likewise, in the case of the functional J, the extremum function $y(x)$ makes J *stationary*. In other words, infinitesimal changes in the function $y(x)$ do not change the value of J.

Continuing with the introductory calculus problem, recall that one can easily distinguish between a local maximum and a local minimum by calculating the second derivative of $f(x)$ and evaluating it at the stationary point. As it turns out, determining if $y(x)$ maximizes or minimizes J can be very difficult. Fortunately for applications in mechanics, it is only necessary to know that $y(x)$ makes a certain functional stationary and whether $y(x)$ maximizes or minimizes the functional will not matter.

Let us now consider the problem of finding the function $y(x)$ that makes the functional

$$J(Y) = \int_{x_1}^{x_2} f\left[Y(x), Y'(x); x\right] dx \qquad (7.3.2)$$

stationary. It will be helpful to think of the "shortest distance between two points on a plane" example from the previous section as we work through the following derivation. However, keep in mind that J is not necessarily a path length. Let us return to Figure 7.1 for a visualization; however, this time we will start with the true stationary function $y(x)$ and vary it by adding a small perturbation $\eta(x)$ such that $\eta(x_1) = \eta(x_2) = 0$. A perturbation is a small additive deviation from a quantity. Hence, the perturbation results in a function $Y(x) = y(x) + a\eta(x)$, shown in Figure 7.2, which no longer makes the functional J stationary because of the perturbation. The parameter a provides a measure of the amount that $y(x)$ is perturbed.

The functional J becomes:

$$J = \int_{x_1}^{x_2} f[Y, Y'; x] dx \qquad (7.3.3)$$

$$= \int_{x_1}^{x_2} f[y + a\eta, y' + a\eta'; x] dx \qquad (7.3.4)$$

In other words, $J = J(a)$ because the integration will "remove" J's dependence on x. In effect, we have reduced our problem to a function of a single variable a, for which we know how to find its extrema by using introductory calculus. Because we know that the extremum exists when $a = 0$, we need to show that $\partial J/\partial a$ equals zero when $a = 0$. As we will see, the result of this work will be a condition for which $y(x)$ makes J stationary.

Next we take the derivative of J with respect to a:

$$\frac{\partial J}{\partial a} = \frac{\partial}{\partial a} \int_{x_1}^{x_2} f[Y, Y'; x] dx \tag{7.3.5}$$

$$= \int_{x_1}^{x_2} \left(\frac{\partial f}{\partial Y} \frac{\partial Y}{\partial a} + \frac{\partial f}{\partial Y'} \frac{\partial Y'}{\partial a} \right) dx \tag{7.3.6}$$

Using $Y = y + a\eta$,

$$\frac{\partial Y}{\partial a} = \eta \quad \text{and} \quad \frac{\partial Y'}{\partial a} = \frac{d\eta}{dx}$$

Therefore,

$$\frac{\partial J}{\partial a} = \int_{x_1}^{x_2} \left(\frac{\partial f}{\partial Y} \eta + \frac{\partial f}{\partial Y'} \frac{d\eta}{dx} \right) dx \tag{7.3.7}$$

Next, we integrate the second term in (7.3.7) by parts using:

$$u = \frac{\partial f}{\partial Y'} \qquad dv = \frac{d\eta}{dx} dx$$

$$du = \frac{d}{dx} \frac{\partial f}{\partial Y'} dx \qquad v = \eta$$

such that,

$$\int_{x_1}^{x_2} \frac{\partial f}{\partial Y'} \frac{d\eta}{dx} dx = \frac{\partial f}{\partial Y'} \eta(x) \Big|_{x_1}^{x_2} - \int_{x_1}^{x_x} \frac{d}{dx} \frac{\partial f}{\partial Y'} \eta(x) dx \tag{7.3.8}$$

The first term on the right-hand side of (7.3.8) is zero because $\eta(x_1) = \eta(x_2) = 0$ as can be seen in Figure 7.2. This leaves the second term remaining in (7.3.8), which will be inserted into (7.3.7) to obtain

$$\frac{\partial J}{\partial a} = \int_{x_1}^{x_2} \left(\frac{\partial f}{\partial Y} - \frac{d}{dx} \frac{\partial f}{\partial Y'} \right) \eta(x) dx \tag{7.3.9}$$

Next, we need to evaluate (7.3.9) at $a = 0$ so that $Y = y + a\eta = y$, resulting in

$$\frac{\partial J}{\partial a} = \int_{x_1}^{x_2} \left(\frac{\partial f}{\partial y} - \frac{d}{dx} \frac{\partial f}{\partial y'} \right) \eta(x) dx = 0 \tag{7.3.10}$$

where we have set the derivative equal to zero to ensure that J is stationary.

Equation (7.3.10) is of the form, $\int h(x)\eta(x)dx = 0$, and must be true for any continuous function $\eta(x)$. It can be shown that if $\int h(x)\eta(x)dx = 0$ for any $\eta(x)$, then $h(x) = 0$ for all x. The end result is the *Euler equation*:

The Euler Equation

$$\frac{\partial f}{\partial y} - \frac{d}{dx} \frac{\partial f}{\partial y'} = 0 \tag{7.3.11}$$

which provides an ODE for the function $y(x)$ that makes J stationary. Equation (7.3.11) is also sometimes referred to as the *first form* of the Euler equation, and it provides the promised condition for finding the extremum y because, as we will see, (7.3.11) will produce a differential equation for $y(x)$.

Next, we use the Euler equation to find the path of the shortest distance between two points on the Cartesian plane.

Example 7.1: The shortest path between two points on the Cartesian plane

Find the path that has the shortest length between two points (x_1, y_1) and (x_2, y_2) on the Cartesian plane, both analytically and using a CAS.

Solution:

We start with (7.2.2),

$$L = \int_{x_1}^{x_2} \sqrt{1 + y'(x)^2} \, dx$$

where we see that $f = \sqrt{1 + y'(x)^2}$. Next, we compute the derivatives needed for the Euler equation:

$$\frac{\partial f}{\partial y} = 0 \quad \text{and} \quad \frac{\partial f}{\partial y'} = \frac{y'}{\sqrt{1 + (y')^2}}.$$

Because the first term of the Euler equation is zero, we have

$$\frac{d}{dx}\left(\frac{y'}{\sqrt{1 + (y')^2}}\right) = 0$$

or, in other words, the term within the parentheses must be constant in x. Setting that constant equal to c produces

$$\frac{y'}{\sqrt{1 + (y')^2}} = c$$

which leads to:

$$y' = \frac{c}{\sqrt{1 - c^2}}$$

The term on the right-hand side of the above equation is also a constant which we define to be m. As promised, the Euler equation leads to a differential equation from which y can be found. It is easy to check that the differential equation $y' = m$ results in:

$$y(x) = mx + b$$

where b is the constant of integration. Hence, the path with the shortest distance between two points in the Cartesian plane is a straight line. No surprise. Computer algebra systems sometimes have algorithms for handling problems involving the calculus of variations. The following codes create the Euler equation for this example.

Python Code

We use the command `euler_equations` from the SymPy library to produce the Euler equation for the problem. The arguments of `euler_equations` consist of the functional f, the dependent function $y(x)$ (the function we ultimately want to find), and the independent variable x. The command `euler_equations` outputs a list of ODEs, which is why we need the index in the variable `euler`. The command `dsolve` can then be used to solve the resulting Euler equation for $y(x)$, with the expected result of $y(x) = c_1 + c_2 x$.

```python
from sympy import diff, symbols, Function, euler_equations, sqrt,\
    dsolve, simplify

print('-'*28,'CODE OUTPUT','-'*29,'\n')

x = symbols('x')
y = Function('y')(x)

f = sqrt(1 + diff(y,x)**2)

euler = simplify(euler_equations(f, y, x)[0])

print('The Euler equation is\n')
print(euler.lhs,'=',euler.rhs)

soln = dsolve(euler,y).rhs

print('\nThe solution y(x) = ', str(soln))

--------------------------- CODE OUTPUT ----------------------------
The Euler equation is
Derivative(y(x), (x, 2))/(Derivative(y(x), x)**2 + 1)**(3/2) = 0

The solution y(x) =   C1 + C2*x
```

Mathematica Code

Similar to Python, Mathematica also has libraries which are called packages. The command `<<` tells Mathematica to import the VariationalMethods package, which contains various functions that can be used to solve problems involving the calculus of variations. The `EulerEquations` command creates the Euler equation for f. Finally, we use `DSolve` to solve the differential equation that results from the Euler equation. The result is the equation of a line.

<< VariationalMethods`

EulerEquations[Sqrt[1 + y'[x]^2], y[x], x]

$-\dfrac{y''[x]}{(1+y'[x]^2)^{3/2}} == 0$

soln = DSolve$\left[-\dfrac{y''[x]}{(1+y'[x]^2)^{3/2}} == 0, y[x], x\right]$;

y[x]/.soln

$\{c_1 + xc_2\}$

7.4 SECOND FORM OF THE EULER EQUATION

There is another form of the Euler equation which is useful when the integrand of J does not explicitly depend on x, or in other words, $\partial f/\partial x = 0$. Note that even with this condition for the partial derivative $\partial f/\partial x = 0$, the derivative df/dx is not necessarily zero. We begin by computing this derivative:

$$\frac{df}{dx} = \frac{\partial f}{\partial y}\frac{dy}{dx} + \frac{\partial f}{\partial y'}\frac{dy'}{dx} \tag{7.4.1}$$

$$\frac{df}{dx} = y'\frac{\partial f}{\partial y} + y''\frac{\partial f}{\partial y'} \tag{7.4.2}$$

$$y''\frac{\partial f}{\partial y'} = \frac{df}{dx} - y'\frac{\partial f}{\partial y} \tag{7.4.3}$$

In addition, we can compute:

$$\frac{d}{dx}\left(y'\frac{\partial f}{\partial y'}\right) = y''\frac{\partial f}{\partial y'} + y'\frac{d}{dx}\frac{\partial f}{\partial y'} \tag{7.4.4}$$

Finally, we insert (7.4.3) into (7.4.4),

$$\frac{d}{dx}\left(y'\frac{\partial f}{\partial y'}\right) = \left(\frac{df}{dx} - y'\frac{\partial f}{\partial y}\right) + y'\frac{d}{dx}\frac{\partial f}{\partial y'} \tag{7.4.5}$$

$$= y'\left[\frac{d}{dx}\frac{\partial f}{\partial y'} - \frac{\partial f}{\partial y}\right] + \frac{df}{dx} \tag{7.4.6}$$

The term in the square brackets in (7.4.6) is Euler's first equation, and if $y(x)$ makes J stationary, then that term must be zero. Therefore we have:

$$\frac{d}{dx}\left(f - y'\frac{\partial f}{\partial y'}\right) = 0 \tag{7.4.7}$$

Using the same kind of argument as we did in Example 7.1, we can set the term inside the parentheses equal to a constant and therefore we obtain the second form of the Euler equation:

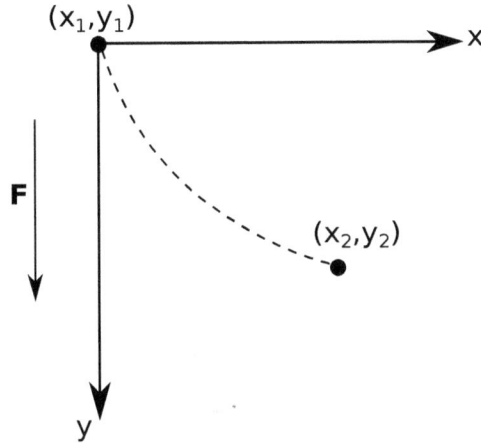

Figure 7.3: The solution of the brachistochrone problem will provide the formula for the dotted path which minimizes the time of travel for a particle moving between the two points (x_1, y_1) and (x_2, y_2) in a constant conservative force field, $\mathbf{F} = mg\hat{\mathbf{j}}$.

The Second Form of the Euler Equation

$$f - y'\frac{\partial f}{\partial y'} = \text{constant} \tag{7.4.8}$$

The second form of the Euler equation is often useful when the dependent variable, x does not explicitly appear in f, i.e., $\partial f/\partial x = 0$. An example of such a case is finding the mathematical form of a geodesic on a sphere, as described later in this chapter.

7.5 SOME EXAMPLES OF PROBLEMS SOLVED USING THE CALCULUS OF VARIATIONS

In this section, we will demonstrate the solution to three problems using the calculus of variations. In the first problem, we will study the so-called brachistochrone problem, which is a well-known problem in the calculus of variations. The second problem will involve computing the shortest path between two points on the surface of a sphere. The third problem will involve finding the minimal surface of revolution between two points.

7.5.1 THE BRACHISTOCHRONE PROBLEM

The first problem we will solve is the classic problem known as the *brachistochrone* problem which was solved by Johann Bernoulli (1667-1748) in 1696. The brachistochrone problem involves finding the path that minimizes the time of travel for a particle moving between two points (x_1, y_1) and (x_2, y_2) in a conservative force field as shown in Figure 7.3. We will consider a particle of mass m starting at rest at the point (x_1, y_1). The particle experiences a constant conservative force, which for our problem will be the force of gravity $\mathbf{F} = mg\hat{\mathbf{j}}$, and we will define the positive y-direction to be downwards.

The time of travel between the two points is:

$$t = \int_{(x_1,y_1)}^{(x_2,y_2)} \frac{ds}{v} \tag{7.5.1}$$

In our previous example, we used $ds = \sqrt{1 + y'(x)^2}\, dx$; however, we are going to find it more convenient to use $ds = \sqrt{1 + x'(y)}\, dy$, where $x' = dx/dy$. Next, we need to find the speed v in order to use the Euler equation. We can use the fact that $\mathbf{F} = mg\hat{\mathbf{j}}$ is a conservative force field to write

$$\frac{1}{2}mv^2 = mgy \tag{7.5.2}$$

at any point along the path between the two points. The above equation, of course, is simply the conservation of energy, recalling that the particle's initial velocity is zero. Therefore, $v = \sqrt{2gy}$, and we have:

$$t = \frac{1}{\sqrt{2g}} \int_{y_1}^{y_2} \sqrt{\frac{1 + x'^2}{y}}\, dy \tag{7.5.3}$$

The conservation of energy introduces y, the independent variable, in our integrand. The use of y as an independent variable instead of x, means that the first form of the Euler equation becomes:

$$\frac{d}{dy}\frac{\partial f}{\partial x'} = \frac{\partial f}{\partial x} \tag{7.5.4}$$

because now we are going to find the function $x(y)$ that makes (7.5.3) stationary. Using

$$f = \sqrt{\frac{1 + x'^2}{y}} \tag{7.5.5}$$

we see that f is independent of x, and and therefore $\partial f/\partial x = 0$. We can then equate the partial derivative $\partial f/\partial x'$ to a constant that we will call $1/\sqrt{2a}$,

$$\frac{\partial f}{\partial x'} = \frac{x'}{\sqrt{y\,(1 + x'^2)}} = \frac{1}{\sqrt{2a}} \tag{7.5.6}$$

where a is a constant. The choice of using $1/\sqrt{2a}$ as our constant will simplify the mathematics later on in the calculation. Next, we will square (7.5.6) and solve for x' to obtain,

$$x' = \frac{dx}{dy} = \sqrt{\frac{y}{2a - y}} \tag{7.5.7}$$

which is a differential equation whose solution will give us the function $x(y)$ that makes (7.5.3) stationary. To solve for $x(y)$, we need to solve the integral:

$$x = \int \sqrt{\frac{y}{2a - y}}\, dy \tag{7.5.8}$$

which can be done using the substitution:

$$y = a\,(1 - \cos\theta) \tag{7.5.9}$$

The result of the integration is that,

$$x = a(\theta - \sin\theta) \tag{7.5.10}$$

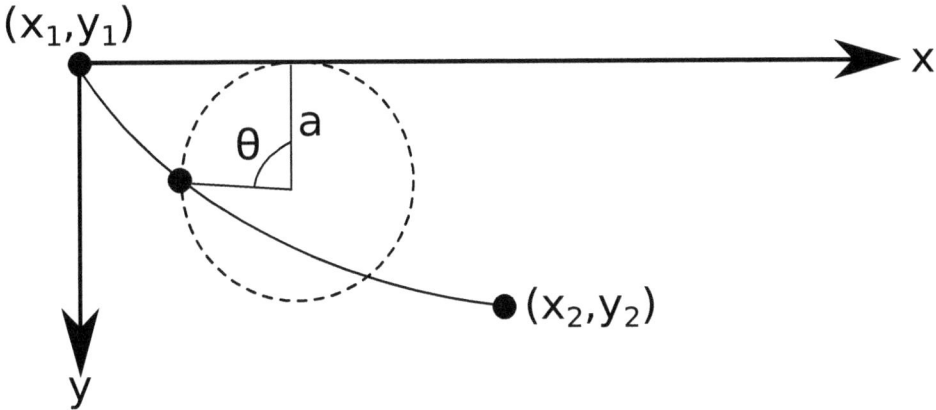

Figure 7.4: The cycloid is the path that makes (7.5.3) stationary. The path is parameterized by the variable θ.

where we set the integration constant to zero by choosing the point (x_1, y_1) to be the origin. In other words, $x = 0$ when $\theta = 0$. The equations (7.5.9) and (7.5.10) use the parameter θ in order to describe the path that makes (7.5.3) stationary. The path described by (7.5.10) and (7.5.9) is called a *cycloid*. A cycloid is the curve traced out by the point on the rim of a circular wheel of radius a, as it rolls along a horizontal path without slipping as shown in Figure 7.4.

The cycloid is the path that extremizes (7.5.3), and although we will not prove it, it is the path that minimizes the time of travel between the two points (x_1, y_1) and (x_2, y_2).

The next example shows how we can arrive at the equations for the cycloid using a CAS and how to plot it. The example also shows some of the limitations of a CAS when dealing with complex integrals, especially when using SymPy.

Example 7.2: The cycloid
Integrate (7.5.8) using a CAS, to obtain the analytical form $x(y)$.
Using $a = 2$, plot one and a half periods of the cycloid produced by equations (7.5.9) and (7.5.10).
Solution:
See the codes below for carrying out the integration, and for creating the plot of the cycloid.

Python Code
We perform the integration using the `integrate` command in SymPy. However, note that the answer produced by SymPy contains complex numbers; this is is because SymPy does not have a fully developed method for including detailed assumptions for the variables in the code.

No special command is needed to create a parametric plot in Python. We simply define the x and y coordinates and the parameter, `theta`.

```
from sympy import symbols,  integrate, sqrt
import numpy as np
import matplotlib.pyplot as plt
import textwrap
wrapper = textwrap.TextWrapper(width=70)


print('-'*28,'CODE OUTPUT','-'*29,'\n')

y1, a, y = symbols('y1, a, y',real=True,positive=True)

x = integrate(sqrt(y1/(2*a-y1)),(y1,0,y))

print('The solution of the ODE is\n\nx = ',wrapper.fill(str( x)))

# Parametric plot
theta = np.linspace(0,3*np.pi,100)

a = 2
x = a*(theta - np.sin(theta))
y = -a*(1 - np.cos(theta))

plt.plot(x,y)
plt.xlabel('x')
plt.ylabel('y')
plt.show()

-------------------------- CODE OUTPUT ------------------------------
The solution of the ODE is

x =   2*I*a*asinh(sqrt(2)*sqrt(-2*a + y)/(2*sqrt(a))) + pi*a +
I*sqrt(y)*sqrt(-2*a + y)
```

Mathematica Code

The Mathematica code performs better than the SymPy code above, since we can specify in detail the conditions that must be satisfied by the variables `a,y` using the `Assumptions` command.

The code obtains the analytical equation for $x(y)$, which does not contain complex numbers. Furthermore, by substituting the parametric form

$$y = a\left(1 - \cos\theta\right)$$

into the analytical equation $x(y)$, we obtain the parametric equation for x as

$$x = a(\theta - \sin\theta)$$

The `ParametricPlot` command in Mathematica is used for the plot.

```
x = Integrate[Sqrt[y1/(2 * a − y1)], {y1, 0, y},

Assumptions->{Element[{a, y, y1}, Reals],

a > 0 && y1 > 0 && y > 0 && 2 * a − y > 0 && 2 * a − y1 > 0}]//

Simplify;

Print["x(y) = ", x]
```

$$x(y) = -\sqrt{(2a - y)y} + 2a \operatorname{ArcSin}\left[\frac{\sqrt{\frac{y}{a}}}{\sqrt{2}}\right]$$

```
(*Substitute parametric y = a * (1 − Cos[θ]) *)


yparametric = a * (1 − Cos[θ]);

xparametric = Assuming[θ > 0 && θ < Pi && a > 0,

x/.{y->yparametric}//Simplify];


Print["Parametric equation for x = ", xparametric]

Print["Parametric equation for y = ", yparametric]
```

Parametric equation for x = $a(θ − \operatorname{Sin}[θ])$
Parametric equation for y = $a(1 − \operatorname{Cos}[θ])$

```
SetOptions[ParametricPlot, Frame->True, Axes->False,

BaseStyle->{FontSize->16}];


a = 2;


ParametricPlot[{xparametric, −yparametric}, {θ, 0, 3π},

Frame->True, BaseStyle->{FontSize->16},

FrameLabel->{x, y}]
```

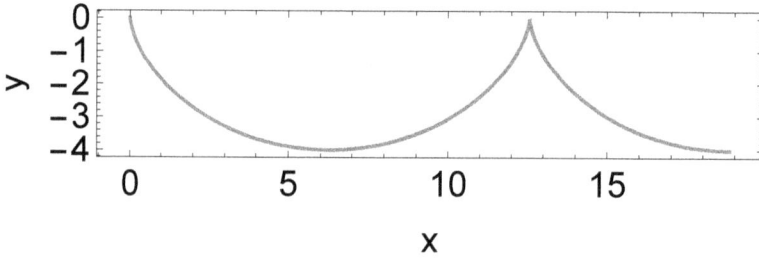

Figure 7.5: Cycloid plot for the Mathematica code in Example 7.2.

7.5.2 GEODESICS

A *geodesic* is the shortest possible line between two points on a curved surface. In this example, we will find the geodesic on a sphere of radius R. We begin by noting that the infinitesimal displacement along the surface of the sphere ($dr = 0$) is:

$$ds = R\sqrt{d\theta^2 + \sin^2\theta\, d\phi^2} = R\sqrt{\theta'^2 + \sin^2\theta}\, d\phi$$

where we used $\theta' = d\theta/d\phi$. In this case, we are going to find the geodesic as a curve $\theta(\phi)$. If we use the Earth as our sphere, then finding $\theta(\phi)$ is similar to describing the latitude of the points along the geodesic as a function of their longitude.

The length integral we are attempting to make stationary is,

$$L = \int ds = \int_{\phi_0}^{\phi} R\sqrt{\theta'^2 + \sin^2\theta}\, d\phi \tag{7.5.11}$$

The variable R will not affect the process of minimization. Therefore,

$$f = \sqrt{\theta'^2 + \sin^2\theta} \tag{7.5.12}$$

Because $\partial f/\partial\phi = 0$, we can use the second form of Euler's equation (7.4.8),

$$f - \theta'\frac{\partial f}{\partial\theta'} = a \tag{7.5.13}$$

which gives

$$\sqrt{\theta'^2 + \sin^2\theta} - \theta'\frac{\partial}{\partial\theta'}\left(\sqrt{\theta'^2 + \sin^2\theta}\right) = a \tag{7.5.14}$$

where a is a constant.

Your next thought might be that we solve (7.5.14) for θ' in order to get a differential equation which would produce $\theta(\phi)$. However, it will be mathematically easier if we solve the inverse function $d\phi/d\theta = \theta'^{-1}$. The result is,

$$\frac{d\phi}{d\theta} = \frac{a\csc^2\theta}{\sqrt{1 - a^2\csc^2\theta}}$$

which when integrated by separation of variables yields

$$\phi = \sin^{-1}\left(\frac{\cot\theta}{\beta}\right) + \alpha \tag{7.5.15}$$

where $\beta^2 = (1 - a^2)/a^2$ and α is a constant of integration. Finally, we can rewrite (7.5.15) as:

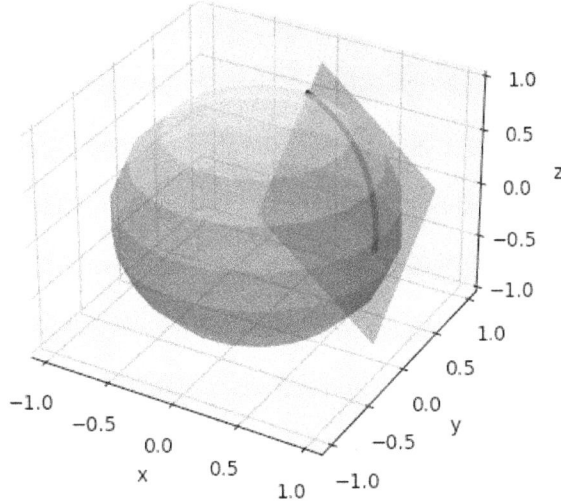

Figure 7.6: The geodesic (black line) on the sphere is the intersection of the plane and the surface of the sphere. The code used to create this figure can be found on this book's website and in the book by Pagonis and Kulp [3].

$$\cot\theta = \beta\sin(\phi - \alpha) \qquad (7.5.16)$$

Strictly speaking, with (7.5.16) we are finished as we have $\theta(\phi)$, but it is difficult to interpret this result. In order to clarify our result, we will multiply (7.5.16) by $R\sin\theta$ on both sides and expanding $\sin(\phi - \alpha) = \cos\alpha\sin\phi - \sin\alpha\cos\phi$,

$$R\cos\theta = \beta\cos\alpha\,(R\sin\theta\sin\phi) - \beta\sin\alpha\,(R\sin\theta\cos\phi) \qquad (7.5.17)$$

Notice that the terms inside the parentheses in (7.5.17) are the coordinate transformations:

$$\left.\begin{aligned} x &= R\sin\theta\cos\phi \\ y &= R\sin\theta\sin\phi \\ z &= R\cos\theta \end{aligned}\right\} \qquad (7.5.18)$$

with constant coefficients, $A \equiv \beta\cos\alpha$ and $B \equiv \beta\sin\alpha$. Therefore, in Cartesian coordinates, (7.5.17) can be expressed as:

$$z = Ay - Bx \qquad (7.5.19)$$

which is the equation of a plane passing through the center of the sphere as shown in Figure 7.6.

Because we are restricted to the surface of a sphere of radius R, the geodesic is the path made by the intersection of the plane with the sphere's surface. Hence, the geodesic is the *great circle* that lies at the intersection of the plane and the surface of the sphere. The great circle between two points is the path often used for air and sea navigation, because it

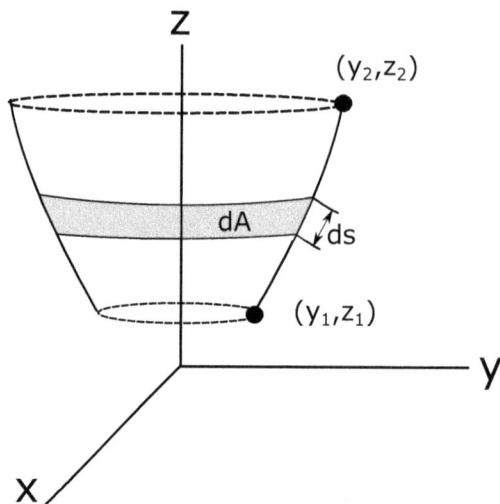

Figure 7.7: The surface of revolution formed by a line joining the points (y_1, z_1) and (y_2, z_2).

provides the path with the shortest distance between those points. The code used to create Figure 7.6 can be found on this book's website.

7.5.3 MINIMUM SURFACE OF REVOLUTION

Consider two parallel circular wire rings that were dipped in soapy water. The soap film between the rings will then form a surface. What is the shape of that surface? The answer can be found by finding the minimum surface of revolution formed by a line joining the two rings. Why does the soap film take the shape of the minimum surface of revolution? The answer has to do with the free energy of the soap film, which is $E = \sigma A$, where σ is the film tension and A is the area of the film. In equilibrium the film's energy is minimized. Hence, minimizing the area is the same as minimizing the energy of the film.

Consider two points (y_1, z_1) and (y_2, z_2) as shown in Figure 7.7. We can generate a surface of revolution by revolving a line that joins (y_1, z_1) and (y_2, z_2) about the z-axis. We want to find the equation of the line joining (y_1, z_1) and (y_2, z_2), which minimizes the area of the surface of revolution.

We begin by writing down the equation for the area of the surface of revolution. The area is found by integrating the shaded area dA over the whole surface. Note that Figure 7.7 shows only a part of the infinitesimal area dA, and that dA actually extends around the surface. We can think of dA as being the surface area of a cylinder, which has area $2\pi rh$ where r is the radius of the cylinder and h is the height of the cylinder. In this case, r would be the distance y from the z-axis to the curve defining the surface of revolution. The height of the cylinder ds is the differential distance along the curve $z(y)$, $ds = \sqrt{dy^2 + dz^2}$. Therefore, our functional for the area is:

$$A = \int_{y_1}^{y_2} 2\pi y \sqrt{1 + z'^2} \, dy \qquad (7.5.20)$$

We want to find $z(y)$ that minimizes the area of the surface of revolution. Therefore, we factored dy out of ds to produce (7.5.20), where $z' = dz/dy$. Next, we insert $f = y\sqrt{1 + z'^2}$ into the first form of the Euler equation (7.3.11) and $\partial f/\partial z = 0$; therefore

$$\frac{d}{dy}\left(\frac{y\,z'}{\sqrt{1+z'^2}}\right) = 0 \qquad (7.5.21)$$

$$\frac{yz'}{\sqrt{1+z'^2}} = a \qquad (7.5.22)$$

where a is a constant. Solving (7.5.22) for z' we obtain,

$$\frac{dz}{dy} = \frac{a}{\sqrt{y^2 - a^2}} \qquad (7.5.23)$$

After separating variables and integrating, we obtain,

$$z = a\,\cosh^{-1}\left(\frac{y}{a}\right) + b \qquad (7.5.24)$$

where b is a constant of integration. Note that both a and b can be determined once the points (y_1, z_1) and (y_2, z_2) are specified. We can solve (7.5.24) for y to obtain,

$$y = a\cosh\left(\frac{z-b}{a}\right) \qquad (7.5.25)$$

The next example shows how to visualize this result.

Example 7.3: The soap film problem

Plot the surface of the soap film that connects two circular rings. Assume that each ring is parallel to the xy–plane. Furthermore, each ring is centered on the z–axis and has a radius of 3 units. One ring is located at $z = 1$, and the other is located at $z = -1$, where the z coordinate is measured in the same units as the radius.

Solution:

The soap film will take on the shape of the minimum surface of revolution between the two rings. We can use (7.5.25),

$$y(z) = a\cosh\left(\frac{z-b}{a}\right)$$

as the curve that describes the shape of the film's surface in the yz–plane. The curve will have end points $(y_1, z_1) = (3, -1)$ and $(y_2, z_2) = (3, 1)$.

The first step in the solution is to use the end points to solve for a and b in (7.5.25). Note that if we insert the end points into (7.5.25), we find that $y(-1) = y(1)$. Furthermore, cosh is an even function. Together, the condition that $y(-1) = y(1)$ and the even nature of the cosh function require that $b = 0$.

Next, we find a by numerically solving $y(1) = 3$,

$$a\cosh\left(\frac{1}{a}\right) = 3 \qquad (7.5.26)$$

which must be solved using numerical approximations. The following code numerically solves (7.5.26) for a and plots the surface of revolution.

Python Code

We use the command **newton** from the SciPy library to perform Newton's method of finding roots. The plot is created by revolving the curve $a\cosh(z/a)$ around the z-axis between the values of $z = -1$ and $z = 1$. Notice the use of trig functions to create the effect of the revolution when defining the variables X and Y.

```python
from scipy.optimize import newton
import numpy as np
import matplotlib.pyplot as plt

print('-'*28,'CODE OUTPUT','-'*29,'\n')

def f(a):
    return a*np.cosh(1/a) - 3

root1 = newton(f, 0.1)
root2 = newton(f,1)

print('The two roots are located at a = ',round(root1,4),\
    ' and ', round(root2,4))

#next, we create the plot using u and v
#to establish the plot range values for x and y

u = np.linspace(-1,1,100)
v = np.linspace(0, 2*np.pi, 100)
U,V = np.meshgrid(u,v)

Y = root2*np.cosh(U/root2)*np.sin(V)
X = root2*np.cosh(U/root2)*np.cos(V)
Z = U

fig = plt.figure()
ax = fig.add_subplot(111, projection='3d')
ax.plot_surface(X, Y, Z)
ax.set_xlabel('x')
ax.set_ylabel('y')
ax.set_zlabel('z')
plt.show()

-------------------------- CODE OUTPUT ----------------------------
The two roots are located at a =  0.3536  and  2.8209
```

Mathematica Code

The option **Reals** in **NSolve** instructs Mathematica to produce only real solutions for a. Notice that there are two solutions that were stored in the variable **params**. There are two curves that extremize the functional (7.5.20) from which (7.5.26) was derived. The two curves are plotted in the yz-plane as shown in Figure 7.8.

We can plot the surface using Mathematica's `RevolutionPlot3D` command, as shown in the code below. Although we will not prove stability, we will use $a = 2.82089$ because that is the value of a associated with the minimal surface of the soap film. When $a = 0.35363$, the soap film is not minimal and is unstable.

The default for `RevolutionPlot3D` is to plot a function $f_z(t)$ which describes the height of the curve above the xy-plane, where t is the radius from the z-axis. The command then rotates the curve of f_z about the z-axis. However, our function describes the height of the curve above the xz-plane. In order to get the shape we expect, we need to tell Mathematica to revolve about what it believes to be the x-axis by using the option, `RevolutionAxis` $\rightarrow \{1,0,0\}$. This is the type of technical issue one occasionally encounters when working with software. If you had plotted (7.5.25) using the defaults, you would have obtained a plot that was clearly incorrect. This is another reminder that one can't simply use computer algorithms as black boxes. You need to make sure that you understand the algorithm and check to see that the output makes sense.

```
SetOptions[RevolutionPlot3D, BaseStyle->{Black, Bold, FontSize->18}];

params = NSolve[a * Cosh[1/a] == 3, a, Reals]

{{a → 0.353638}, {a → 2.82089}}

RevolutionPlot3D[a * Cosh[z/a]/.params[[2]], {z, −1, 1},
RevolutionAxis → {1, 0, 0}, ViewPoint → {1.3, −2.4, 2.},
AxesLabel → {x, y, z}]
```

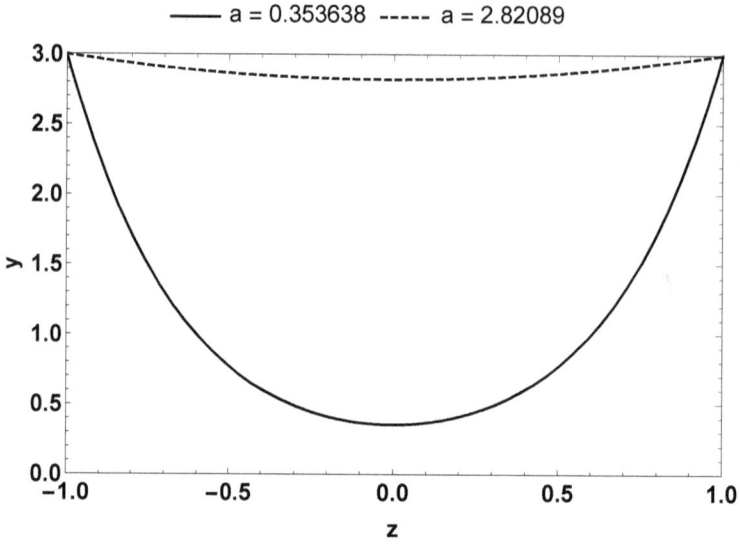

Figure 7.8: The soap film profile for $a = 0.353638$ (solid curve) and $a = 2.82089$ (dashed curve), from Example 7.3.

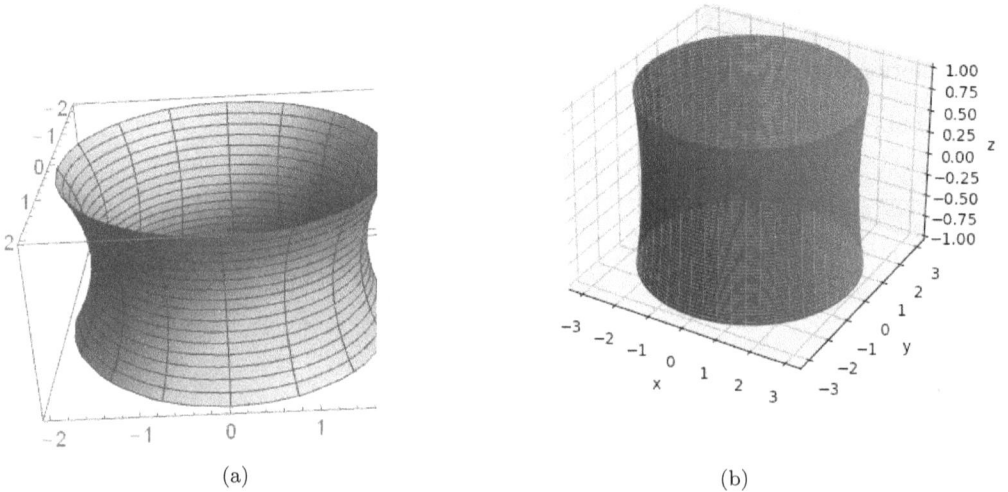

Figure 7.9: The surface of the soap film as produced by (a) Mathematica and (b) Python, from Example 7.3.

7.6 MULTIPLE DEPENDENT VARIABLES

In classical mechanics, it is common for f to depend on multiple dependent variables $y_i(x)$ all with the same independent variable. In this case, f takes the form,

$$f\left[y_1(x),\ y_1'(x),\ y_2(x),\ y_2'(x),\ \ldots,\ y_n(x),\ y_n'(x);\ x\right] \qquad (7.6.1)$$

where there are n dependent variables y_i. In this case, it is often convenient to write $f = f[y_i, y_i'; x]$, where $i = 1, 2, \ldots, n$.

We want to find a set of functions $\{y_i(x)\}$ $(i = 1, 2, \ldots, n)$ that makes the functional

$$J = \int f\left[y_i(x), y_i'(x); x\right] dx \tag{7.6.2}$$

stationary. To find $\{y_i(x)\}$, we follow the work done in Section 7.2. We write $Y_i(\alpha, x) = y_i(x) + \alpha \eta_i(x)$, insert Y_i into (7.6.2), differentiate the resulting J with respect to α, and then set that derivative equal to zero. The result is,

$$\frac{\partial J}{\partial \alpha} = \int_{x_1}^{x_2} \sum_{i=1}^{n} \left(\frac{\partial f}{\partial Y_i} - \frac{d}{dx}\frac{\partial f}{\partial Y_i'} \right) \eta_i(x) dx \tag{7.6.3}$$

Each perturbation $\eta_i(x)$ is independent of one another. Therefore, in order for the derivative in (7.6.3) to be zero at $a = 0$, we must have,

Multi-variable Euler Equation

$$\frac{\partial f}{\partial y_i} - \frac{d}{dx}\frac{\partial f}{\partial y_i'} = 0 \tag{7.6.4}$$

In other words, we will have an Euler equation for each dependent function y_i. As before, each Euler equation will produce an ordinary differential equation. However, these ODEs are often coupled and can be difficult to solve in closed form.

For an example of f depending on multiple dependent variables, we can reconsider the first problem we studied, the shortest path between two points on the Cartesian plane. When we first solved that problem, we assumed $y = y(x)$. However, if we wanted to consider all possible paths, then we would need to consider curves which are parameterized using a variable t,

$$x = x(t) \qquad \text{and} \qquad y = y(t)$$

The above generalization allows us to look at curved paths, such as spirals in the Cartesian plane that cannot be written as $y = y(x)$. For example, one of the simplest parameterizations is that for the unit circle, where $x(t) = \cos t$ and $y(t) = \sin t$.

The infinitesimal displacement along a parameterized path in the Cartesian plane is $ds = \sqrt{dx^2 + dy^2}$, which can be rewritten and inserted into (7.2.1),

$$L = \int_{t_1}^{t_2} \sqrt{x'(t)^2 + y'(t)^2} \, dt \tag{7.6.5}$$

where $x' = dx/dt$ and $y' = dy/dt$. Using $f = \sqrt{x'^2 + y'^2}$, and the Euler equations:

$$\frac{d}{dt}\frac{\partial f}{\partial x'} - \frac{\partial f}{\partial x} = 0 \qquad \text{and} \qquad \frac{d}{dt}\frac{\partial f}{\partial y'} - \frac{\partial f}{\partial y} = 0$$

we find,

$$\frac{x'}{\sqrt{x'^2 + y'^2}} = a \qquad \text{and} \qquad \frac{y'}{\sqrt{x'^2 + y'^2}} = c \tag{7.6.6}$$

where a and c are constants. Note that we used the fact that $\partial f/\partial x = \partial f/\partial y = 0$, in order to set $\partial f/\partial x'$ and $\partial f/\partial y'$ equal to the constants a and c, respectively. By dividing the two equations in (7.6.6), we get,

$$\frac{y'}{x'} = \frac{dy}{dx} = \frac{c}{a} = m \tag{7.6.7}$$

Integrating (7.6.7), we find $y = mx + b$, where b is the constant of integration. In other words, we again get the equation for a line.

7.7 CHAPTER SUMMARY

The calculus of variations provides us a means of finding a function $y(x)$ that extremizes the functional:

$$J = \int_{x_1}^{x_2} f[y(x), y'(x); x]dx$$

Common problems in the calculus of variations involve finding the path of the shortest distance between two points, finding the path that minimizes the time of travel between two points, and finding the minimum surface area for a surface of revolution. The function $y(x)$ can be found using Euler's equation:

$$\frac{d}{dx}\frac{\partial f}{\partial y'} - \frac{\partial f}{\partial y} = 0$$

Euler's equation will produce a differential equation, which can be solved for $y(x)$. An alternative form of the Euler's equation is,

$$f - y'\frac{\partial f}{\partial y'} = \text{constant}$$

which can be used when $\partial f/\partial x = 0$.

Problems involving finding the extrema of a functional which depends on several dependent variables $\{y_i(x)\}$, for $i = 1, 2, \ldots, n$, of the form,

$$J = \int_{x_1}^{x_2} f[y_i(x), y_i'(x); x]dx$$

can be solved using multiple Euler equations,

$$\frac{d}{dx}\frac{\partial f}{\partial y_i'} - \frac{\partial f}{\partial y_i} = 0$$

The result is typically a system of n-coupled ordinary differential equations, which can be solved, either numerically or in closed form, for each $y_i(x)$.

7.8 END-OF-CHAPTER PROBLEMS

The symbol ⌨ indicates a problem which requires some computer assistance, in the form of graphics, or numerical computation, or symbolic evaluation.

Section 7.2: Shortest Distance Between Two Points—Setting Up the Calculus of Variations

1. Fermat's principle states that the path taken between two points by a light ray is the path that can be traversed in the least time. Consider a light ray traveling between Points 1 and 2 in the Cartesian plane. The points are in a medium with a spatially varying index of refraction, $n(x, y)$. Show that the path taken by the light ray must minimize the functional,

$$t = \int_1^2 \frac{n(x, y)}{c} \sqrt{1 + y'(x)^2} dx \tag{7.8.1}$$

where $y' = dy/dx$ and the index of refraction is defined as the ratio:

$$n = \frac{c}{v}$$

where c is the speed of light in vacuum, and v is the speed of light inside the medium.

2. What functional gives the length of the path that joins two points on the surface of a cylinder with radius R? Use cylindrical coordinates (r, θ, z) and assume that the curve can be written as $\theta(z)$.

Sections 7.3 and 7.4: First and Second Form of the Euler Equation

3. ⌨ Find the path that makes the functional,

$$J(y) = \int_0^1 x^2 \sqrt{1 - y'} dx,$$

stationary with $y' = dy/dx$. The path must connect the points $(0, 0)$ and $(1, 1)$. Solve this problem analytically and by using Python or Mathematica.

4. Find the shortest path between two points on the surface of a cylinder with radius R. As in Problem 2, set up the necessary functional by using cylindrical coordinates (ρ, θ, z) and assume that the curve can be written as $\theta(z)$.

5. Consider the light ray (solid line) reflecting off of the surface of a plane mirror at the point (x_r, y_r) as shown in Figure 7.10. The surface of the mirror is in the xz-plane, and the light ray is in the xy-plane. Using Fermat's principle from Problem 1, prove the law of reflection. In other words, show that the angle of incidence θ_1 is equal to the angle of reflection θ_2 in Figure 7.10. Although not necessary, you may find it helpful to set $x_1 = 0$ and $y_r = 0$ as suggested in Figure 7.10. To solve the problem, compute the time it takes to travel along the path in Figure 7.10, then show that the time is a minimum when $\theta_1 = \theta_2$. You may start by assuming that the point (x_r, y_r) is in the same vertical plane as (x_1, y_1) and (x_2, y_2). If you were not to make that assumption, then explain how you would need to change your solution in order to prove the law of reflection.

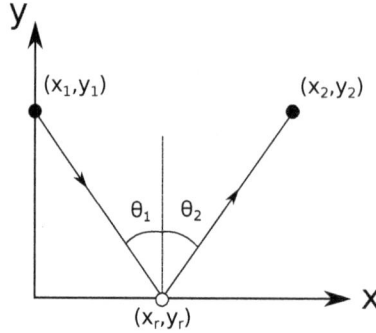

Figure 7.10: The path of a light ray in the xy-plane reflecting off a mirror in the xz-plane, in Problem 7.5.

6. Consider the light ray shown in Figure 7.11. The light ray starts in a medium with a spatially uniform index of refraction n_1 and, at the point (x_r, y_r) enters a new medium with a spatially uniform index of refraction n_2. Using Fermat's principle from Problem 1 to minimize the time of travel between the two points (x_1, y_1) and (x_2, y_2), prove Snell's law, $n_1 \sin \theta_1 = n_2 \sin \theta_2$ and that the point of refraction (x_r, y_r) is coplanar with the points (x_1, y_1) and (x_2, y_2). As before, the index of refraction is defined as the ratio $n = \frac{c}{v}$, where c is the speed of light in vacuum and v is the speed of light inside the medium.

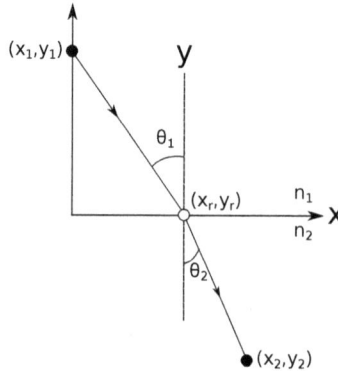

Figure 7.11: The path of a light ray in the xy-plane leaving a medium with an index of refraction n_1, and entering a medium with an index of refraction n_2, in Problem 7.6.

7. 🖥 Find the extremum of the functional,

$$J = \int_0^1 \left(y'^2 - y^2 + x \right) dx$$

subject to the two boundary conditions: $y(0) = 1$, $y'(0) = -1$ and where $y'(x) = dy/dx$. Solve this problem analytically and by using Python or Mathematica.

8. 🖥 Show that the extrema of the functional,

$$\int \sqrt{x^2 + y^2}\sqrt{1 + y'^2}\,dx$$

are

$$x^2 \cos\alpha - 2xy \sin\alpha - y^2 \cos\alpha = \beta$$

where $y' = dy/dx$ and α, β are constants.
Hint: Rewrite the functional in polar coordinates and use a CAS to find Euler's equations and solve the resulting ODE.

9. 🖥 Find the geodesics on the surface of the cone whose equation in cylindrical coordinates is $z = c\rho$.

10. Imagine a ball rolling down the first half of the first loop of a cycloid (from $x = 0$ to $x = \pi a$), defined by the equations:

$$y = a(1 - \cos\theta) \qquad x = a(\theta - \sin\theta)$$

Show that the time of travel from $x = 0$ to $x = a$ is equal to $\pi\sqrt{a/g}$.

11. For a little bit of foreshadowing of upcoming chapters, consider the functional,

$$J(x) = \int_{t_0}^{t} \left(\frac{1}{2}mx'^2 - \frac{1}{2}kx^2\right) dt'$$

where m and k are positive constants, $x' = dx/dt'$ and t' is the independent variable (the prime after the t does not denote a derivative). Find the function $x(t)$ that makes the functional stationary. Does this look familiar?

Section 7.6 Multiple Dependent Variables

12. Show that the shortest distance between two points in three-dimensional space is a straight line. You may want to consider writing x, y, and z in parametric form.

13. Repeat Problem 4; however, this time do not assume that $\theta = \theta(z)$. Instead, you should parameterize the curve such that $z = z(t)$ and $\theta = \theta(t)$, where t is the parameter of the curve.

14. 🖥 Find the shortest path on the conical surface $z = z_0 - c\sqrt{x^2 + y^2}$, where z_0 and c are constants. Use a CAS to find the solution of the differential equation resulting from Euler's equation.

15. 🖥 Find the extrema of the functional,

$$J(y, z) = \int_0^{\pi} \left(y'^2 + z'^2 + yz\right) dx$$

where $y' = dy/dx$ and $z' = dz/dx$, subject to the conditions, $y(0) = 0$, $y(\pi) = 1$, $z(0) = 0$, and $z(\pi) = 1$. Solve this problem using Mathematica or Python.

8 Lagrangian and Hamiltonian Dynamics

In this chapter, we present an alternative way of formulating physics problems. This new method was first published by French-Italian mathematician Joseph-Louis Lagrange (1736–1813) in 1788 and is therefore called *Lagrangian dynamics*. Lagrange's formulation uses a function called the *Lagrangian*, which is found using the particle's energy. As we will see, a particle will move in such a way as to make the integral of the Lagrangian with respect to time stationary. This integral, called an *action integral*, is a functional (a mathematical formula that maps a function to a number). Hence, the particle's path can be described using a function that is an extremum of a functional (in this case, the action integral). In the last chapter, we learned how to find the functions that are the extrema of a functional by using the Euler equation. In this chapter we will learn how the Euler equation can be used to find a system's equations of motion.

One advantage of the Lagrangian formulation is that it is the same, regardless of the coordinate system being used. Lagrangian dynamics uses the Euler equation to derive a particle's equation of motion and the Euler equation takes the same form regardless of the coordinate system being used. As we saw in Chapter 3, Newton's second law takes different forms depending on the coordinate system, because the formula for the acceleration depends on the coordinate system. Another advantage of Lagrangian dynamics is that it eliminates the need of knowing constraint forces. A bead constrained to move along a wire is a common example involving constraint forces. The force keeping the bead on the wire is often unknown, or very difficult to find. Lagrangian mechanics doesn't need to know the formula of the constraint force in order find the particle's motion.

The final "advantage" we mention is that Lagrangian mechanics is simply elegant! As we will see, visual inspection of the Lagrangian can reveal what quantities are conserved in the system. We will also learn how to use the Lagrangian to derive another quantity called the *Hamiltonian*, which can also be used to derive a particle's equations of motion. The Hamiltonian will provide us with a new way of thinking about the phase space, where we can use pictures to describe the possible behaviors of a system. The Hamiltonian's value extends well beyond the field of classical mechanics and is the central quantity used to describe a particle's wave function in quantum mechanics.

8.1 INTRODUCTION TO THE LAGRANGIAN

In this section, we motivate the material in the rest of the chapter by looking at a basic example, the simple harmonic oscillator (SHO). Recall that a model system for the SHO is a mass m connected to one end of a horizontal massless spring with a spring constant k. The other end of the spring is connected to a support. The mass is connected to the spring and slides along a frictionless surface. We will use the variable x to measure the location of the mass relative to equilibrium. In this case, we know that the kinetic and potential energies can be written as,

$$T = \frac{1}{2}m\dot{x}^2 \quad \text{and} \quad V = \frac{1}{2}kx^2 \tag{8.1.1}$$

respectively. We know that for the SHO,

$$F = -\frac{\partial V}{\partial x} = -kx \tag{8.1.2}$$

is the force acting on the mass. Therefore, using Newton's second law, $F = m\ddot{x}$, we can write the equation of motion for the SHO as,

$$m\ddot{x} + kx = 0 \tag{8.1.3}$$

Recall that normally we define $\omega_0^2 = k/m$ for the SHO, but for now we will leave the equation of motion as it appears above.

There is another method of finding (8.1.3) that doesn't directly involve Newton's second law. Notice that,

$$\frac{d}{dt}\left(\frac{\partial T}{\partial \dot{x}}\right) = \frac{d}{dt}(m\dot{x}) = \dot{p} = m\ddot{x} \tag{8.1.4}$$

Therefore, we can obtain (8.1.3) by writing,

$$\frac{d}{dt}\left(\frac{\partial T}{\partial \dot{x}}\right) + \frac{\partial V}{\partial x} = 0 \tag{8.1.5}$$

Equation (8.1.5) is similar to the Euler equation in appearance, except that the functions appearing in each derivative are different. We can make (8.1.5) look like the Euler equation by defining a new function, the *Lagrangian*:

$$\mathcal{L} \equiv T - V \tag{8.1.6}$$

Notice that V does not depend on $\dot{x}$, and T does not depend on x; therefore $\partial T/\partial x = 0$ and $\partial V/\partial \dot{x} = 0$, and we can write:

$$\frac{d}{dt}\left(\frac{\partial T}{\partial \dot{x}}\right) = \frac{d}{dt}\left(\frac{\partial \mathcal{L}}{\partial \dot{x}}\right) \tag{8.1.7}$$

and

$$\frac{\partial V}{\partial x} = -\frac{\partial \mathcal{L}}{\partial x} \tag{8.1.8}$$

And therefore we obtain from (8.1.5) the *Euler-Lagrange equation*:

The Euler-Lagrange Equation for One Degree of Freedom

$$\frac{d}{dt}\left(\frac{\partial \mathcal{L}}{\partial \dot{x}}\right) - \frac{\partial \mathcal{L}}{\partial x} = 0 \tag{8.1.9}$$

For the SHO, the Lagrangian is

$$\mathcal{L} = \frac{1}{2}m\dot{x}^2 - \frac{1}{2}kx^2 \tag{8.1.10}$$

Inserting (8.1.10) into (8.1.9), we obtain (8.1.3). Hence we see that the equation of motion for the SHO can be obtained by using the Euler-Lagrange equation.

We see that $\mathcal{L} = \mathcal{L}(x, \dot{x}; t)$ and (8.1.9) implies that the integral $\int \mathcal{L}(x, \dot{x}; t)dt$ is stationary along the path $x(t)$ of the oscillator. In other words, the path of the oscillator is the one that makes $\int \mathcal{L}(x, \dot{x}; t) dt$ stationary. We will formalize this statement in the next section. However, the important point here is that we did not use Newton's second law to derive the equation of motion for the SHO. Instead of asking what forces are acting on the system (as is done using Newton's second law), we asked what behavior makes the integral of the Lagrangian stationary. The Lagrangian depends only on the kinetic and potential energies of the system. Note, however, that the Lagrangian is **not** the total mechanical energy of the system. The total mechanical energy is the *sum* of the kinetic and potential energies, while the Lagrangian is the *difference* between them.

Although we have demonstrated the Euler-Lagrange equation for the SHO, as we will see in later sections, the Euler-Lagrange equation can be used to find the equation of motion for any system. Before generalizing the work done in this section to other systems, we need to discuss coordinate systems and the concepts of generalized coordinates and degrees of freedom.

8.2 GENERALIZED COORDINATES AND DEGREES OF FREEDOM

In Chapter 3 we saw that it is sometimes useful to describe a system using coordinates other than Cartesian coordinates. One challenge of using non-Cartesian coordinates with Newton's second law is that one has to find the acceleration in the new coordinate system. The new form of the acceleration is often mathematically more complicated than in Cartesian coordinates. The result is that the form of Newton's second law heavily depends on the coordinate system used. Because the Euler-Lagrange equations take the same form for every coordinate system, it is in general easier to work with non-Cartesian coordinate systems when using the Lagrangian.

We will use the term *generalized coordinates* in order to describe any coordinate system that completely specifies the state of the system. Spherical and plane-polar coordinates are examples of generalized coordinate systems; however, generalized coordinates can take many forms. You may find it useful to work with generalized coordinates which have dimensions of energy, angular momentum, or no dimensions at all. It is customary to write generalized coordinates as $(q_1, , q_2, \ldots)$ or just as q_j (with $j = 1, 2, \ldots$ implied but not always stated).

How many generalized coordinates do we need to completely describe the state of a system? This is an important question because it will determine the number of Euler-Lagrange equations we need in order to solve for a system's behavior. Let's consider a system of N particles. In order to describe each particle, we will need N vectors of the form $\mathbf{r} = x\hat{\mathbf{i}} + y\hat{\mathbf{j}} + z\hat{\mathbf{k}}$, with three components each, x, y, and z. Hence, we will need $3N$ quantities to describe the position of each particle and, therefore, the state of the system. Next, suppose that there are constraints on the particle's motion. Each constraint will relate one quantity to another. For example, each particle may be constrained to move on the xy-plane. Regardless of their specific nature, constraints can relate two or more of the quantities to each other, or they can impose a value for a particular coordinate (e.g., $z = 0$). In other words, not all $3N$ quantities are necessarily independent. If there are m equations of constraint, then there are $s = 3N - m$ *degrees of freedom* for the system. The number of degrees of freedom is the number of independent quantities that are needed to specify the state of the system. In other words, we will need s independent generalized coordinates in order to describe the state of the system. As we will see, the Euler-Lagrange equations will produce s second-order differential equations which, when solved, will give $q_j(t)$ for $j = 1, 2, \ldots, s$.

The following example will help clarify constraints and degrees of freedom. Consider the simple plane pendulum shown in Figure 8.1.

In the case of the simple plane pendulum there is one particle, the mass m, and therefore $N = 1$. Generally speaking, the position of the mass would require three coordinates,

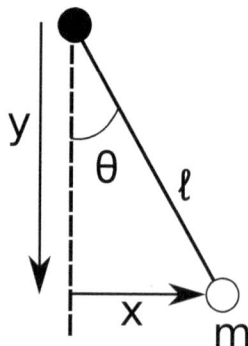

Figure 8.1: A simple plane pendulum composed of a mass m attached to a massless rigid rod of length ℓ which makes an angle θ with the vertical (dashed line).

x, y, and z. However, because we are working with a plane pendulum, we have our first equation of constraint, $z = 0$. Next, we know that the mass is constrained to move along a circle of radius ℓ. Therefore, the next equation of constraint is $x^2 + y^2 = \ell^2$. We see that we have two equations of constraint; therefore, we have $s = 3\,(1) - 2 = 1$ degree of freedom. To describe the system, we would like to work with a generalized coordinate system. The most obvious one to choose is polar coordinates. In this case, we will choose $q = \theta$, where θ is the angle between the pendulum and the dashed line in Figure 8.1. We can write the Cartesian coordinates in terms of the generalized coordinate θ using:

$$x(q) = \ell \sin q \qquad \text{and} \qquad y(q) = -\ell \cos q \tag{8.2.1}$$

Note that the origin of the coordinate system is located at the pendulum's support, denoted by the black-filled circle in Figure 8.1.

To find the Lagrangian, we need to rewrite the kinetic and potential energies using the generalized coordinate q. In Cartesian coordinates, the velocity of the particle is $\mathbf{v} = \dot{x}\hat{\mathbf{i}} + \dot{y}\hat{\mathbf{j}}$, where

$$\dot{x} = \ell \dot{q} \cos q \qquad \text{and} \qquad \dot{y} = \ell \dot{q} \sin q \tag{8.2.2}$$

Using $T = \frac{1}{2}m\mathbf{v} \cdot \mathbf{v}$ for the pendulum's kinetic energy and $V = mgy$ for its gravitational potential energy, we obtain

$$T = \frac{1}{2}m\left(\dot{x}^2 + \dot{y}^2\right) = \frac{1}{2}m\ell^2\dot{q}^2 \tag{8.2.3}$$

$$V = mgy = -mg\ell \cos q \tag{8.2.4}$$

The resulting Lagrangian, $\mathcal{L} = T - V$, is:

$$\mathcal{L}\,(q, \dot{q}) = \frac{1}{2}m\ell^2\dot{q}^2 + mg\ell \cos q \tag{8.2.5}$$

Notice that the Lagrangian is a function of both the generalized coordinate q and the *generalized velocity* $\dot{q}$. The generalized velocity is the term used to describe the first derivative of the generalized coordinate with respect to time.

Although we have not demonstrated that it is true for cases other than the simple harmonic oscillator, we can say that the pendulum's motion is the path $q(t)$ which makes the integral $\int \mathcal{L}\,(q, \dot{q})\,dt$ stationary. Therefore, the Euler-Lagrange equation,

$$\frac{d}{dt}\left(\frac{\partial \mathcal{L}}{\partial \dot{q}}\right) - \frac{\partial \mathcal{L}}{\partial q} = 0 \tag{8.2.6}$$

will produce the equation of motion. Inserting (8.2.5) into (8.2.6) gives the simple plane pendulum equation,

$$\ddot{q} + \frac{g}{\ell}\sin q = 0 \tag{8.2.7}$$

8.3 HAMILTON'S PRINCIPLE

Now that we have developed an understanding of generalized coordinates, it is time that we justify using the Euler-Lagrange equations to derive a system's equations of motion. This justification was done by Irish mathematician, Sir William Rowan Hamilton (1805–1865), in papers published in 1834 and 1835 ([4],[5]). Hamilton's Principle states:

Hamilton's Principle

The path taken by a particle from the point $\mathbf{r}_1$ to the point $\mathbf{r}_2$ during the time interval t_1 to t_2 is the one that makes the action integral stationary:

$$S = \int_{t_1}^{t_2} \mathcal{L}dt \tag{8.3.1}$$

Hamilton's principle is a cornerstone not just of classical physics, but also of quantum physics. From Chapter 7, we know that the condition for S to be stationary is

The Euler-Lagrange Equation

$$\frac{d}{dt}\left(\frac{\partial \mathcal{L}}{\partial \dot{q}_j}\right) - \frac{\partial \mathcal{L}}{\partial q_j} = 0 \tag{8.3.2}$$

where q_j is the jth generalized coordinate, $j = 1, 2, \ldots, s$, and s is the number of degrees of freedom in the system.

How can we be certain that Hamilton's principle produces the actual path that the particle takes? We will demonstrate this using a system with one degree of freedom, but this result is easily generalized to s degrees of freedom.

Let us consider a particle of mass m moving along a line. The particle's location along the line is described by the coordinate q. Therefore, the Lagrangian $\mathcal{L}$ is

$$\mathcal{L} = \frac{1}{2}m\dot{q}^2 - V(q) \tag{8.3.3}$$

To obtain the path $q(t)$ that makes the action integral S stationary, we insert (8.3.3) into (8.3.2):

$$\frac{d}{dt}(m\dot{q}) = -\frac{\partial V}{\partial q} \tag{8.3.4}$$

The result is an ODE for the path $q(t)$. The right-hand side of (8.3.4) is the q component of the force acting on the particle. Likewise, the left-hand side of (8.3.4) is the time derivative of the momentum. Therefore, we see that the path $q(t)$ that makes the action integral S stationary is the one that also solves Newton's second law. Therefore, Hamilton's principle does provide a means of finding a system's equation of motion. Note that the above work is not a proof of Hamilton's principle, but rather a *demonstration* of how it is capable of producing a system's equation of motion.

Before we move on to some examples of using the Lagrangian to find the equation of motion for a system, it is important that we point out some additional terminology that will become useful later. First, notice $V = V(q_j)$ and $T = T(\dot{q}_i)$. Hence, in our example

$$\frac{\partial \mathcal{L}}{\partial \dot{q}_j} = m\dot{q}_j \quad \text{and} \quad \frac{\partial \mathcal{L}}{\partial q_j} = -\frac{\partial V}{\partial q_j} \tag{8.3.5}$$

The term $\partial \mathcal{L}/\partial \dot{q}_j$ is sometimes called the jth component of the *generalized momentum* although it is not always a linear momentum. Likewise, $\partial \mathcal{L}/\partial q_j$ is sometimes called the jth component of the *generalized force*; again it is not always a force, but rather it acts like one. With these terms in mind, (8.3.4) can be reworded as: "the generalized force is equal to the time rate of change of the generalized momentum." It is a "generalized" Newton's second law!

8.4   EXAMPLES OF LAGRANGIAN DYNAMICS

In this section, we present several example problems. We have already seen two examples, the simple harmonic oscillator in Section 8.1 and the simple plane pendulum in Section 8.2. Each problem in this section will present a particular new challenge. First, we will start with an example in three dimensions. The second example will examine a system of two particles with one constraint. In the first two examples, we will analytically find the Lagrangian and the Euler-Lagrange equation. This will allow us to focus on the details of concepts such as degrees of freedom and constraints.

The third and fourth examples will use coordinates other than Cartesian coordinates. The fifth and final example in this section will involve constraints on velocities. The last three examples will contain a computational component, so that we may see how to use Mathematica and SymPy to assist with the solution of problems in Lagrangian mechanics.

Example 8.1: The Euler-Lagrange equations in three dimensions
Consider a particle of mass m moving in three dimensions. In Cartesian coordinates the kinetic energy can be written as,

$$T = \frac{1}{2}m\left(\dot{x}^2 + \dot{y}^2 + \dot{z}^2\right) \tag{8.4.1}$$

The particle experiences only conservatives forces and has a potential energy is $V = V(x, y, z)$. Write down the Euler-Lagrange equations for the particle in Cartesian coordinates.

Solution:
This problem has three degrees of freedom because it consists of one particle ($n = 1$) and no constraints ($m = 0$). The Lagrangian is,

$$\mathcal{L} = \frac{1}{2}m\left(\dot{x}^2 + \dot{y}^2 + \dot{z}^2\right) - V(x, y, z) \tag{8.4.2}$$

Therefore, we are attempting to find functions $x(t)$, $y(t)$, and $z(t)$ that make the action integral

$$S = \int \mathcal{L}(x, y, z, \dot{x}, \dot{y}, \dot{z}) dt$$

stationary.

Because there are three degrees of freedom, i.e., three functions that need to be found in order to make S stationary, there will be three Euler-Lagrange equations. Before finding the Euler-Lagrange equations for this problem, first note that:

$$\left.\begin{aligned}
\frac{d}{dt}\left(\frac{\partial \mathcal{L}}{\partial \dot{x}}\right) &= \frac{d}{dt}(m\dot{x}) = m\ddot{x} \\
\frac{d}{dt}\left(\frac{\partial \mathcal{L}}{\partial \dot{y}}\right) &= \frac{d}{dt}(m\dot{y}) = m\ddot{y} \\
\frac{d}{dt}\left(\frac{\partial \mathcal{L}}{\partial \dot{z}}\right) &= \frac{d}{dt}(m\dot{z}) = m\ddot{z}
\end{aligned}\right\} \qquad (8.4.3)$$

By equating the terms in parentheses in each line above, we see that $\partial \mathcal{L}/\partial \dot{x}$ is the x-component of the particle's momentum (similar for the $\dot{y}$ and $\dot{z}$ derivatives). In this case, the generalized momentum is the same as the linear momentum. Likewise,

$$\left.\begin{aligned}
\frac{\partial \mathcal{L}}{\partial x} &= -\frac{\partial V}{\partial x} = F_x \\
\frac{\partial \mathcal{L}}{\partial y} &= -\frac{\partial V}{\partial y} = F_y \\
\frac{\partial \mathcal{L}}{\partial z} &= -\frac{\partial V}{\partial z} = F_z
\end{aligned}\right\} \qquad (8.4.4)$$

We can see that $\partial \mathcal{L}/\partial x$ is the x-component of the force acting on the particle (similar for the y and z derivatives). In this case, the generalized force is equal to the force acting on the particles. Inserting the derivatives into the Euler-Lagrange equation gives:

$$\frac{d}{dt}\left(\frac{\partial \mathcal{L}}{\partial \dot{x}}\right) - \frac{\partial \mathcal{L}}{\partial x} = 0 \qquad \Rightarrow \qquad m\ddot{x} = F_x$$

$$\frac{d}{dt}\left(\frac{\partial \mathcal{L}}{\partial \dot{y}}\right) - \frac{\partial \mathcal{L}}{\partial y} = 0 \qquad \Rightarrow \qquad m\ddot{y} = F_y$$

$$\frac{d}{dt}\left(\frac{\partial \mathcal{L}}{\partial \dot{z}}\right) - \frac{\partial \mathcal{L}}{\partial z} = 0 \qquad \Rightarrow \qquad m\ddot{z} = F_z$$

The result of the Euler-Lagrange equations is simply Newton's second law.

While Example 8.1 may seem elementary, it points out a few important things. First, the number of Euler-Lagrange equations will equal the number of degrees of freedom in the system. Second, Example 8.1 demonstrates the connections between the generalized momentum and generalized force and the physical quantities momentum and force with which you are already familiar. Of course, to progress further in this problem and find $x(t)$, $y(t)$, and $z(t)$, you would need to know $V(x, y, z)$.

Example 8.2: The Atwood machine

Two masses, m_1 and m_2 are joined by a massless inextensible string of length ℓ on an Atwood machine. The Atwood machine has a pulley with mass M and a radius R and the pulley is frictionless. If the string does not slip with respect to the pulley, find the acceleration of the masses m_1 and m_2.

Solution:

We begin by drawing a diagram so that we may define the coordinates needed to describe the state of the Atwood machine.

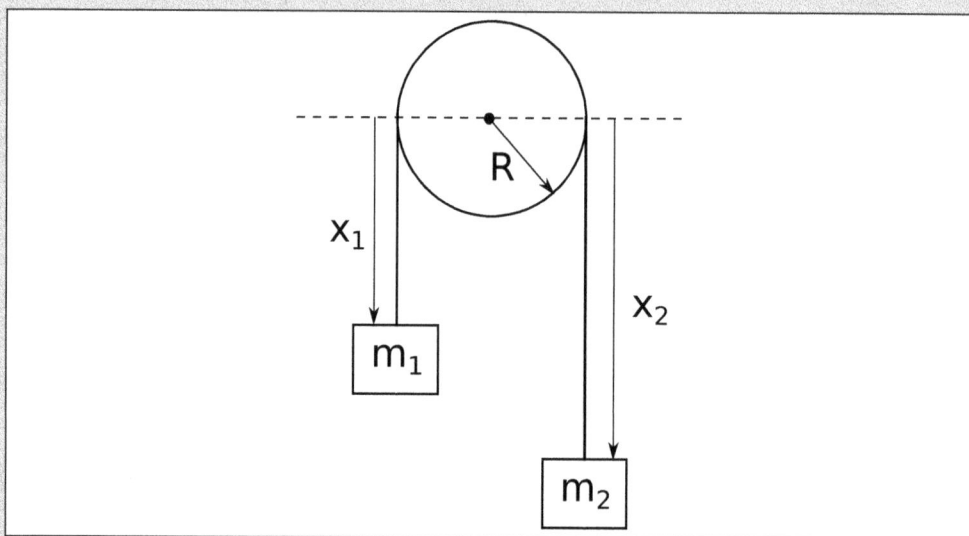

Figure 8.2: The Atwood machine for Example 8.2.

We use the horizontal dashed lined passing through the center of the Atwood machine as the origin for our coordinate system. The coordinates x_1 and x_2 measure the distance from the center of the Atwood machine to masses m_1 and m_2, respectively. Because the string is inextensible, the two coordinates x_1 and x_2 are not independent. For example, as m_1 falls down, m_2 rises and vice versa. Therefore, there is one equation of constraint among the coordinates:

$$\ell = x_1 + x_2 + \pi R \tag{8.4.5}$$

The last term in the constraint is the length of the string that wraps around the pulley. Of course an additional constraint is that the masses are each constrained to move only up or down along a line; therefore, $y_1 = y_2 = z_1 = z_2 = 0$, which are actually four constraints (one on each coordinate) for a total of five constraints on the system ($m = 5$). Therefore, with two particles in the system $N = 2$, we have $s = 3(2) - 5 = 1$ degree of freedom for the problem. We will rewrite the coordinates x_1 and x_2 in terms of the generalized coordinate x using:

$$x_1 = x \tag{8.4.6}$$
$$x_2 = \ell - x - \pi R \tag{8.4.7}$$

Notice that the velocities have the relationship, $\dot{x}_2 = -\dot{x}_1$, and the accelerations have the relationship, $\ddot{x}_2 = -\ddot{x}_1$, which is what we would expect in this situation.

Next, we will find the kinetic and potential energies of the system. Notice that the pulley has mass, but the string does not. Therefore, the total kinetic energy includes the kinetic energy of m_1, m_2, and the rotational kinetic energy of the pulley. Recall that rotational kinetic energy takes the form $T_{rot} = \frac{1}{2}I\omega^2$, where $I = \frac{1}{2}MR^2$ is the rotational inertia for the pulley, and $\omega = \dot{x}/R$ is the angular velocity of the pulley. Therefore, the total kinetic energy for the system is:

$$T = \frac{1}{2}m_1\dot{x}_1^2 + \frac{1}{2}m_2\dot{x}_2^2 + \frac{1}{2}I\omega^2 \tag{8.4.8}$$

$$= \frac{1}{2}m_1\dot{x}^2 + \frac{1}{2}m_2\dot{x}^2 + \frac{1}{2}I\left(\frac{\dot{x}}{R}\right)^2 \tag{8.4.9}$$

The potential energy takes the form of the gravitational potential energy. Measuring upward as the positive direction, we have:

$$V = -m_1 g x_1 - m_2 g x_2 \tag{8.4.10}$$

$$= -m_1 g x - m_2 g \left(\ell - x - \pi R\right) \tag{8.4.11}$$

The resulting Lagrangian is:

$$\mathcal{L} = \left(\frac{1}{2}m_1 + \frac{1}{2}m_2 + \frac{1}{2}\frac{I}{R^2}\right)\dot{x}^2 + m_1 g x + m_2 g \left(\ell - x - \pi R\right) \tag{8.4.12}$$

Because $s = 1$, there is only one Euler-Lagrange equation for this system:

$$\frac{d}{dt}\left(\frac{\partial \mathcal{L}}{\partial \dot{x}}\right) = \frac{\partial \mathcal{L}}{\partial x} \tag{8.4.13}$$

$$\frac{d}{dt}\left[\left(m_1 + m_2 + \frac{I}{R^2}\right)\dot{x}\right] = (m_1 - m_2)g \tag{8.4.14}$$

$$\left(m_1 + m_2 + \frac{I}{R^2}\right)\ddot{x} = (m_1 - m_2)g \tag{8.4.15}$$

The generalized acceleration is:

$$\ddot{x} = \frac{(m_1 - m_2)g}{\left(m_1 + m_2 + \frac{M}{2}\right)} \tag{8.4.16}$$

where we used $I = \frac{1}{2}MR^2$. From the generalized acceleration, we can get the acceleration of each mass because $\ddot{x}_1 = \ddot{x}$ and $\ddot{x}_2 = -\ddot{x}$.

Example 8.2 demonstrates several ideas. First, drawing a picture of the system can help identify constraints on the coordinates. Second, this example reminds us to carefully identify all kinetic and potential energies in the system. It would be easy to overlook the rotational kinetic energy of the pulley if you thought the pulley was massless, a common assumption in introductory physics. Finally, Example 8.2 shows us that although it might be useful to describe the motion in generalized coordinates (in this case x), the solution of the problem needs to address what is being asked by the problem. In this case, we needed to make sure that we showed how to express the acceleration of each mass in terms of the generalized coordinate.

Example 8.3: Pendulum supported by a rotating disk
A pendulum is suspended from a massless disk with a radius R that is rotating with a constant angular velocity ω as shown in Figure 8.3. The pendulum consists of a mass m attached to one end of a rigid massless rod of length ℓ. The other end of the rod is attached to the edge of the rotating disk. Find the equation of motion of the mass m analytically and using Mathematica and Python. Using a numerical ODE solver, find $\theta(t)$ for $t = 0$ to $t = 20$ seconds using the values $R = 0.2$ m, $\omega = 2.5$ rad/s, $\ell = 1.0$ m, $g = 9.8\ m/s^2$, and initial conditions $\theta(0) = 1$ rad and $\dot{\theta}(0) = 0$ rad/s.

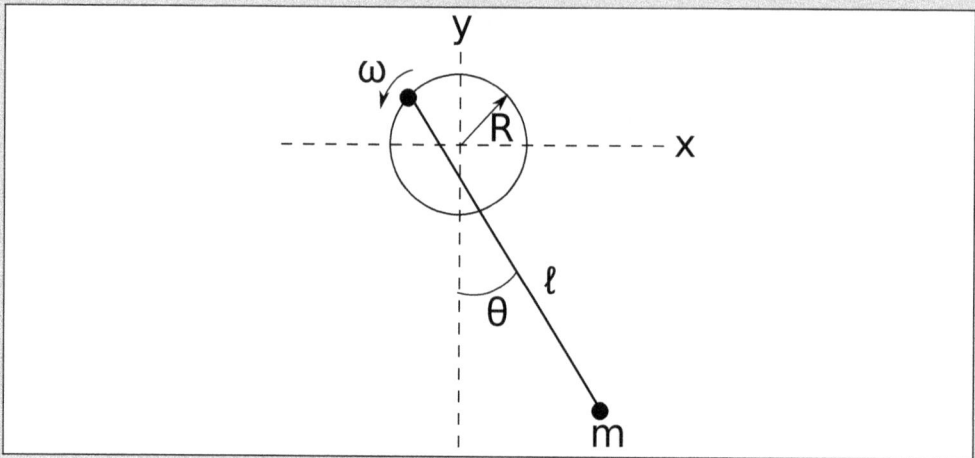

Figure 8.3: The rotating pendulum for Example 8.3.

Solution:
The coordinate system is defined in the figure where the coordinate axes are the dashed lines. The mass is constrained to move in the xy-plane, hence $z = 0$ is our first equation of constraint. There is an additional constraint which may not be immediately obvious. However, if we write the location of the mass in terms of polar coordinates, then it will become clear that this problem has only one degree of freedom. The coordinates of the mass can be written as:

$$x = R\cos(\omega t) + \ell\sin\theta \qquad \text{and} \qquad y = R\sin(\omega t) - \ell\cos\theta \qquad (8.4.17)$$

In other words, we chose the origin to be the center of the rotating disk, and the location of the support (which is attached to the disk) appears as the first term in the equations for x and y. We see that x and y are a function of the generalized coordinate θ. The x and y component of the velocities are:

$$\dot{x} = -R\omega\sin(\omega t) + \ell\dot{\theta}\cos\theta \qquad \text{and} \qquad \dot{y} = R\omega\cos(\omega t) + \ell\dot{\theta}\sin\theta \qquad (8.4.18)$$

Next, we find the kinetic and potential energies. The kinetic energy $T = \frac{1}{2}mv^2$ can be written using the velocity of the mass in the Cartesian plane, $\mathbf{v} = \dot{x}\hat{\mathbf{i}} + \dot{y}\hat{\mathbf{j}}$,

$$T = \frac{1}{2}mv^2 = \frac{1}{2}m\mathbf{v}\cdot\mathbf{v} = \frac{1}{2}m\left(\dot{x}^2 + \dot{y}^2\right)$$

The pendulum is in the Earth's gravitational field; therefore, the potential energy is $V = mgy$. Inserting (8.4.17) and (8.4.18) into our equations for the kinetic and potential energies results in the Lagrangian (after some algebra, see Problem 9 in the End-of-Chapter Problems),

$$\mathcal{L} = \frac{1}{2}m\left(R^2\omega^2 + \ell^2\dot{\theta}^2 + 2R\ell\omega\dot{\theta}\sin(\theta - \omega t)\right) - mg\left(R\sin(\omega t) - \ell\cos\theta\right) \quad (8.4.19)$$

Inserting (8.4.19) into

$$\frac{d}{dt}\left(\frac{\partial\mathcal{L}}{\partial\dot{\theta}}\right) - \frac{\partial\mathcal{L}}{\partial\theta} = 0$$

produces the equation of motion:

$$\ddot{\theta} + \frac{g}{\ell}\sin\theta = \frac{R\omega^2}{\ell}\cos(\theta - \omega t) \quad (8.4.20)$$

Notice that (8.4.20) is the simple pendulum equation with an additional cosine term. The cosine term is the drive term associated with the rotating disk.

The codes below show how to find (8.4.20) and plot its solution $\theta(t)$. A graph of $\theta(t)$ is shown in Figure 8.4.

Python Code

This code follows the same procedure as the examples in Chapter 7. Note that we use the variable q instead of θ. Notice also that we needed to redefine the ODE as the function `deriv` so that it can be solved using `odeint`. The Python result agrees with the analytical solution.

```
from sympy import diff, symbols, Function, euler_equations, nsimplify,\
sin, cos
import numpy as np
from scipy.integrate import odeint
import matplotlib.pyplot as plt
import textwrap

print('-'*28,'CODE OUTPUT','-'*29,'\n')

wrapper = textwrap.TextWrapper(width=60)

D, t, m, g, w, R, l = symbols('D,t,m,g,w,R,l',positive=True)
q = Function('q')(t) # we use q instead of theta

L = 1/2*m*(R**2*w**2 + l**2*diff(q,t)**2 + \
    2*R*l*w*diff(q,t)*sin(q-w*t)) - \
        m*g*(R*sin(w*t) - l*cos(q))

euler = nsimplify(euler_equations(L,q,t)[0].expand().lhs)
euler_string = wrapper.fill(str(euler))

print('The Euler equation is:\n')
print(euler_string, ' = 0')

# Next, we integrate the equation of motion numerically
R, w, l, g = 0.2, 2.5, 1.0, 9.8

def deriv(y, t):
    q, v = y    #let v = dq/dt
    dydt = [v, R*w**2/l*np.cos(q-w*t)-g/l*np.sin(q)]
    return dydt

ics = [1,0]
t = np.linspace(0,20,100)
soln = odeint(deriv, ics, t)

plt.plot(t,soln[:,0])
plt.xlabel('time')
plt.ylabel('theta')
plt.show()

--------------------------- CODE OUTPUT ------------------------------
The Euler equation is:
R*l*m*w**2*cos(t*w - q(t)) - g*l*m*sin(q(t)) -
l**2*m*Derivative(q(t), (t, 2))  = 0
```

Mathematica Code

This code follows the same procedure as in Chapter 7. Notice that by using a re-placement in the variable `ode`, we can directly use the output of `EulerEquations` in `NDSolve`.

```
SetOptions[Plot, Axes->False, Frame->True, BaseStyle->{FontSize->16}];

<< VariationalMethods`
```

$$L = 1/2 * m * (R\text{^}2 * \omega\text{^}2 + l\text{^}2 * \theta'[t]\text{^}2 + 2 * R * l * \omega * \theta'[t] * \text{Sin}[\theta[t] - \omega * t]) -$$
$$m * g * (R * \text{Sin}[\omega * t] - l * \text{Cos}[\theta[t]]);$$

```
euler = EulerEquations[L, θ[t], t]//FullSimplify
```

$$lm\left(-R\omega^2\text{Cos}[t\omega - \theta[t]] + g\text{Sin}[\theta[t]] + l\theta''[t]\right) == 0$$

```
ode = euler/.{R->0.2, ω->2.5, l->1.0, g->9.8};

soln = NDSolve[{ode, θ[0]==1, θ'[0]==0}, θ[t], {t, 0, 20}];

Plot[θ[t]/.soln, {t, 0, 20}, FrameLabel->{"time", "θ"}]
```

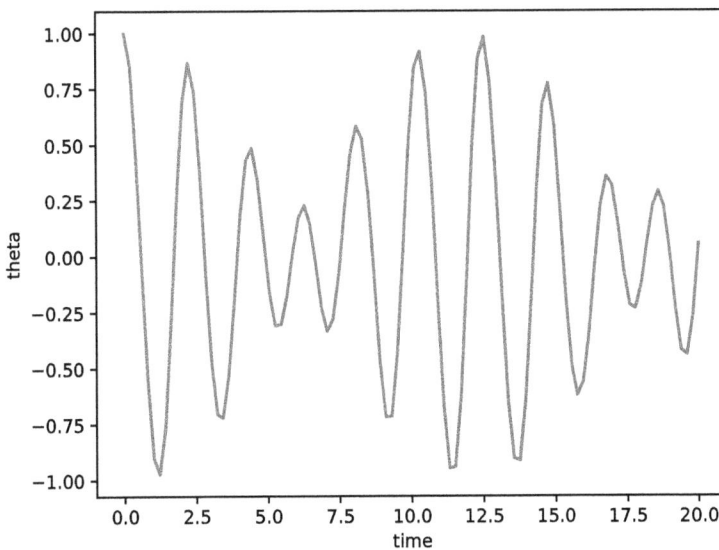

Figure 8.4: The solution $\theta(t)$ from Example 8.3 as produced by the Python code. Notice that $\theta(t)$ behaves similar to a beat wave as the pendulum interacts with the rotating disk.

Example 8.3 is an example of a problem which appears to have two degrees of freedom, but it can actually be described using only one. Another important point demonstrated in Example 8.3 is that it is necessary to be careful with kinetic energies. While often we focus more on potential energies, this is a case where the kinetic energy is a bit more difficult. It was important to include all of the components of the velocity in order to correctly obtain the kinetic energy. Finally, if you solve Problem 9, you will see the value in using trigonometric identities to cast the final solution in terms of something that is easier to interpret.

Example 8.4: Bead on a wire

A bead of mass m slides along a wire near the surface of the Earth. The wire is bent in the shape of the parabola $z = ar^2$, where a is a positive constant and (r, θ, z) are the usual cylindrical coordinates. When the wire rotates with an angular velocity ω about the z-axis, the bead travels in a circle of radius R parallel to the xy-plane, as shown in Figure 8.5. Find the value of a in terms of ω and g, the acceleration due to gravity. Solve this problem analytically and use Mathematica and Python to find the Euler-Lagrange equation.

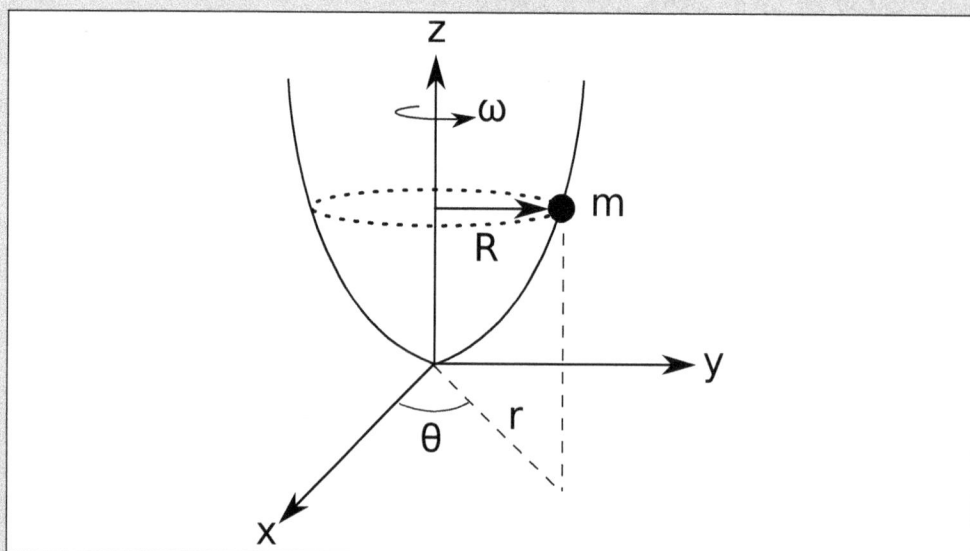

Figure 8.5: A bead of mass m free to move on a rotating parabolic wire from Example 8.3.

Solution:

The system consists of one particle $N = 1$, and in order to find the number of degrees of freedom s, we will need to find the number of constraints. Cylindrical coordinates are a natural choice for this problem. In those coordinates, we have two constraints. The first constraint is that $z = ar^2$ because the bead is constrained to move along the wire. The second constraint is due to the rotation of the wire. We know that the wire is rotating about the z-axis at a constant angular velocity; therefore, $\theta = \omega t$. Hence, we are left with one degree of freedom, r.

Next we find the kinetic and potential energies. Using the velocity in cylindrical coordinates, $\mathbf{v} = \dot{r}\hat{\mathbf{r}} + r\dot{\theta}\hat{\boldsymbol{\theta}} + \dot{z}\hat{\mathbf{z}}$, we find

$$T = \frac{1}{2}m\mathbf{v}\cdot\mathbf{v} = \frac{1}{2}m\left(\dot{r}^2 + r^2\dot{\theta}^2 + \dot{z}^2\right) \tag{8.4.21}$$

The potential energy is found using $V = mgz$. By applying the constraints $z = ar^2$ and $\theta = \omega t$, we find:

$$\dot{\theta} = \omega \quad \text{and} \quad \dot{z} = 2ar\dot{r} \tag{8.4.22}$$

By inserting the constraints and (8.4.22) we can find the Lagrangian:

$$\begin{aligned}\mathcal{L} &= T - V \\ &= \frac{1}{2}m\left(\dot{r}^2 + r^2\dot{\theta}^2 + \dot{z}^2\right) - mgz \\ &= \frac{1}{2}m\left(\dot{r}^2 + r^2\omega^2 + 4r^2a^2\dot{r}^2\right) - mgar^2\end{aligned}$$

Note that $\mathcal{L} = \mathcal{L}(r, \dot{r})$. To get the equation of motion for the bead, we can insert the Lagrangian into the Euler-Lagrange equation,

$$\left.\begin{aligned}\frac{d}{dt}\left(\frac{\partial\mathcal{L}}{\partial\dot{r}}\right) &= \frac{d}{dt}\left(\frac{m}{2}\left[2\dot{r} + 8r^2a^2\dot{r}\right]\right) = \left(1 + 4a^2r^2\right)m\ddot{r} + 8ma^2r\dot{r}^2 \\ \frac{\partial\mathcal{L}}{\partial r} &= mr\left(r\omega + 4a^2\dot{r}^2 - 2ga\right)\end{aligned}\right\} \tag{8.4.23}$$

Equating the two terms above gives:

$$\left(1 + 4a^2r^2\right)\ddot{r} + 4a^2r\dot{r}^2 + r\left(\omega^2 - 2ga\right) = 0 \tag{8.4.24}$$

We are interested in the case when the bead is traveling in a circle of radius R. Therefore, inserting $\ddot{r} = \dot{r} = 0$ and $r = R$ into (8.4.24), we obtain

$$a = \frac{\omega^2}{2g}$$

Python Code
This code follows similar examples from Chapter 7. However, note the order of substitutions when defining the variable **find_a**. The highest derivatives must be defined first in the **subs** method.

```
from sympy import diff, symbols, Function, euler_equations, \
simplify, solve
import textwrap

print('-'*28,'CODE OUTPUT','-'*29,'\n')

wrapper = textwrap.TextWrapper(width=70)

t, m, g, a, w, R = symbols('t,m,g,a,w,R',positive=True)
r = Function('r')

L = 1/2*m*(diff(r(t),t)**2 + r(t)**2*w**2 + \
    4*r(t)**2*a**2*diff(r(t),t)**2) - \
        m*g*a*r(t)**2

euler = simplify(euler_equations(L,r(t),t)[0]).lhs

print('The Euler equation is:\n\n')
print(wrapper.fill(str(euler)), '=0')

# substitute derivatives = 0
find_a = solve(euler.subs([(diff(r(t),t,t),0),\
    (diff(r(t),t),0),(r(t),R)]),a)[0]
print('\na = ', find_a)

------------------------------- CODE OUTPUT -----------------------------
The Euler equation is:

m*(-4*a**2*r(t)**2*Derivative(r(t), (t, 2)) -
4*a**2*r(t)*Derivative(r(t), t)**2 - 2*a*g*r(t) + w**2*r(t) -
Derivative(r(t), (t, 2))) =0

a =   w**2/(2*g)
```

Mathematica Code
Finding the Euler-Lagrange equation is done the same way as finding the Euler equation from Chapter 7. Note to be careful to include the dependency of r on t as r[t] when defining the Lagrangian.

$<<$ VariationalMethods`

$L = 1/2 * m * (r'[t]\hat{\ }2 + r[t]\hat{\ }2\omega\hat{\ }2 + 4 * r[t]\hat{\ }2 * a\hat{\ }2 * r'[t]\hat{\ }2) - m * g * a * r[t]\hat{\ }2$

$-agmr[t]^2 + \frac{1}{2}m\left(\omega^2 r[t]^2 + r'[t]^2 + 4a^2 r[t]^2 r'[t]^2\right)$

euler $=$ EulerEquations$[L, r[t], t]$

$-m\left(r[t]\left(2ag - \omega^2 + 4a^2 r'[t]^2\right) + r''[t] + 4a^2 r[t]^2 r''[t]\right) == 0$

Solve[euler, a, Assumptions-$>$\{$r'[t]$==0, $r"[t]$==0\}][[1]]

$\left\{a \to \frac{\omega^2}{2g}\right\}$

Example 8.4 demonstrates several important points. First, it demonstrates the need to use the velocity in cylindrical coordinates. The velocity of a particle in polar, cylindrical, and spherical coordinates was discussed in Chapter 3. It is important when computing the kinetic energy that the velocity vector be expressed in the same coordinate system that you are using to solve the problem. Second, although this problem didn't specifically ask for the equation of motion, we found it easier to answer the problem by finding it. Once we had the general equation of motion for the bead (8.4.24), we inserted the specific conditions for the motion of interest. Finally, Example 8.4 shows how the Lagrange formulation does not need to know all of the forces acting on the system. We do not know the mathematical form of the force that keeps the bead on the wire. The aforementioned ignorance is not a problem, all we needed for the Lagrange formulation was to know how the value of the z coordinate was constrained.

The next example demonstrates the importance of calculating the Lagrangian in an appropriate coordinate system.

Example 8.5: A moving inclined plane

A frictionless inclined plane of mass M, length ℓ, and incline θ is sliding along a level horizontal frictionless surface, while a particle of mass m slides on the inclined plane, as shown in Figure 8.6. Find the acceleration of the inclined plane and the particle. Solve the problem analytically and using Mathematica or Python.

Solution:

We begin by drawing a picture of the system.

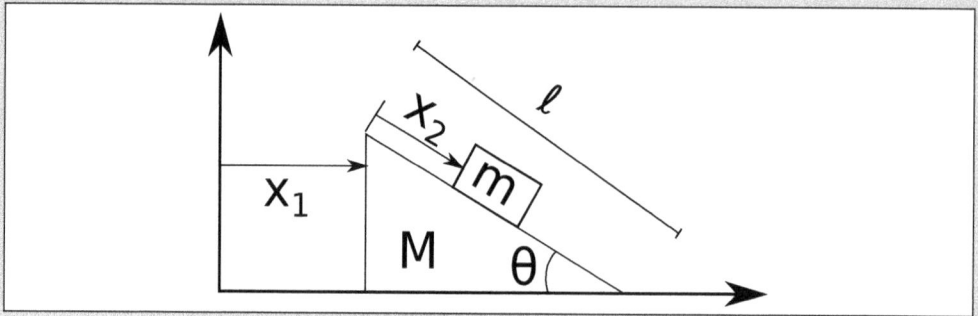

Figure 8.6: A mass m sliding down an incline plane with mass M for Example 8.5.

The figure above shows that we have created two coordinates, x_1 and x_2. The coordinate x_1 is measured from the vertical axis and describes the location of the inclined plane M. The coordinate x_2 measures the position of the particle m from the top of the inclined plane. Note that the two coordinates are not orthogonal to each other.

To calculate the Lagrangian, we need to find the kinetic energy of each object,

$$T = \frac{1}{2}Mv_1^2 + \frac{1}{2}mv_2^2$$

where $v_1^2 = \dot{x}_1$, but $v_2^2 \neq \dot{x}_2^2$. The particle's velocity $\mathbf{v}_2$ relative to the vertical axis will depend on the velocity of the inclined plane and therefore, $\mathbf{v}_2 = \dot{\mathbf{x}}_1 + \dot{\mathbf{x}}_2$, where $\dot{\mathbf{x}}_1$ and $\dot{\mathbf{x}}_2$ are *vectors* that point along the directions of x_1 and x_2 in the figure, respectively. We can find $v_2^2 = \mathbf{v}_2 \cdot \mathbf{v}_2$ by noting that the angle between $\dot{\mathbf{x}}_1$ and $\dot{\mathbf{x}}_2$ is θ,

$$v_2^2 = (\dot{\mathbf{x}}_1 + \dot{\mathbf{x}}_2) \cdot (\dot{\mathbf{x}}_1 + \dot{\mathbf{x}}_2) \tag{8.4.25}$$

$$= \dot{x}_1^2 + \dot{x}_2^2 + 2\dot{\mathbf{x}}_1 \cdot \dot{\mathbf{x}}_2 \tag{8.4.26}$$

$$= \dot{x}_1^2 + \dot{x}_2^2 + 2\dot{x}_1\dot{x}_2 \cos\theta \tag{8.4.27}$$

We now have what we need to compute the Lagrangian,

$$T = \frac{1}{2}M\dot{x}_1^2 + \frac{1}{2}m\left(\dot{x}_1^2 + \dot{x}_2^2 + 2\dot{x}_1\dot{x}_2 \cos\theta\right) \tag{8.4.28}$$

$$V = mg\left(\ell - x_2\right)\sin\theta \tag{8.4.29}$$

$$\mathcal{L} = \frac{1}{2}M\dot{x}_1^2 + \frac{1}{2}m\left(\dot{x}_1^2 + \dot{x}_2^2 + 2\dot{x}_1\dot{x}_2 \cos\theta\right) - mg\left(\ell - x_2\right)\sin\theta \tag{8.4.30}$$

Note that we used ℓ as the length of the inclined plane and the height of the particle above the horizontal axis is $(\ell - x_2)\sin\theta$. Further note that the Lagrangian is of the form $\mathcal{L} = \mathcal{L}\left(x_1, x_2, \dot{x}_1, \dot{x}_2\right)$. There are two degrees of freedom for this problem; therefore, we will have two Euler-Lagrange equations:

$$\left. \begin{aligned} \frac{d}{dt}\left(\frac{\partial \mathcal{L}}{\partial \dot{x}_1}\right) - \frac{\partial \mathcal{L}}{\partial x_1} &= 0 \\ \frac{d}{dt}\left(\frac{\partial \mathcal{L}}{\partial \dot{x}_2}\right) - \frac{\partial \mathcal{L}}{\partial x_2} &= 0 \end{aligned} \right\} \tag{8.4.31}$$

Next we will compute the necessary derivatives,

$$\frac{d}{dt}\left(\frac{\partial \mathcal{L}}{\partial \dot{x}_1}\right) = M\ddot{x}_1 + m\left(\ddot{x}_1 + 2\ddot{x}_2\cos\theta\right) \qquad \frac{\partial \mathcal{L}}{\partial x_1} = 0$$

$$\frac{d}{dt}\left(\frac{\partial \mathcal{L}}{\partial \dot{x}_2}\right) = m\left(\ddot{x}_2 + 2\ddot{x}_1\cos\theta\right) \qquad \frac{\partial \mathcal{L}}{\partial x_2} = mg\sin\theta$$

which results in two equations of motion,

$$\left.\begin{array}{r} M\ddot{x}_1 + m\left(\ddot{x}_1 + 2\ddot{x}_2\cos\theta\right) = 0 \\ \ddot{x}_2 + 2\ddot{x}_1\cos\theta = g\sin\theta \end{array}\right\} \qquad (8.4.32)$$

where the top and bottom equations of (8.4.32) result from the Euler-Lagrange equation with x_1 and x_2, respectively. The problem asks to find the acceleration of each object. Equations (8.4.32) are two coupled second-order differential equations, which we can solve algebraically for $\ddot{x}_1$ and $\ddot{x}_2$,

$$\ddot{x}_1 = \frac{-mg\cos\theta\sin\theta}{(M+m) - m\cos^2\theta} \qquad \text{and} \qquad \ddot{x}_2 = \frac{(M+m)g\sin\theta}{(M+m) - m\cos^2\theta}$$

Further simplification of the resulting accelerations can be done. However, we left them in this form so that several points can be made. First note that in the limit of $M \to \infty$, we find that the inclined plane does not accelerate and that the particle has an acceleration $\ddot{x}_2 = g\sin\theta$ which one expects from introductory physics. Likewise, when $\theta = \pi/2$, M does not accelerate and m has an acceleration of g, just as we would expect. There are other limits worth exploring, such as $M \to 0$, which we leave as an exercise for the reader.

Python Code
Notice how we needed to explicitly state the `Derivative` command in the results found in `x_solns`, where we solve for $\ddot{x}_1$ and $\ddot{x}_2$.

```python
from sympy import diff, symbols, Function, euler_equations
from sympy import  simplify, solve, sin, cos, Derivative
import textwrap

print('-'*28,'CODE OUTPUT','-'*29,'\n')

wrapper = textwrap.TextWrapper(width=74)

M, m, g, l, q,t = symbols('M,m,g,l,q,t',positive=True)
# we use q symbol for the angle theta

x1 = Function('x1')
x2 = Function('x2')

L = 1/2*M*diff(x1(t),t)**2 + 1/2*m*(diff(x1(t),t)**2 + \
    diff(x2(t),t)**2 + \
        2*diff(x1(t),t)*diff(x2(t),t)*cos(q)) - \
            m*g*(l-x2(t))*sin(q)

euler = euler_equations(L,(x1(t),x2(t)),t)
print('The Euler equations are:\n\n')
print(wrapper.fill(str(euler[0])))
print('\n' + wrapper.fill(str(euler[1])))

x_solns = simplify(\
    solve(euler,(diff(x1(t),t,t),diff(x2(t),t,t))))

print('\nThe second derivatives of x1, x2 are:\n')
print('a1 =', x_solns[Derivative(x1(t), (t, 2))])
print('a2 =', \
    simplify(x_solns[Derivative(x2(t), (t, 2))]))
```

```
--------------------------- CODE OUTPUT ---------------------------
The Euler equations are:

Eq(-1.0*M*Derivative(x1(t), (t, 2)) - m*(cos(q)*Derivative(x2(t), (t, 2))
+ Derivative(x1(t), (t, 2))), 0)

Eq(g*m*sin(q) - m*(cos(q)*Derivative(x1(t), (t, 2)) + Derivative(x2(t),
(t, 2))), 0)

The second derivatives of x1, x2 are:
a1 = g*m*sin(q)*cos(q)/(-M + m*cos(q)**2 - m)
a2 = g*(M + m)*sin(q)/(M + m*sin(q)**2)
```

> **Mathematica Code**
> Having two degrees of freedom changes little about the code compared to earlier problems. Note that the second element of the command `EulerEquations` is a list of dependent functions `{x1,x2}`.
>
> ---
> << VariationalMethods
>
> $L = 1/2 * M * \text{x1}'[t]{}^{\wedge}2 + 1/2 * m * (\text{x1}'[t]{}^{\wedge}2 + \text{x2}'[t]{}^{\wedge}2 + 2 * \text{x1}'[t] * \text{x2}'[t] * \text{Cos}[\theta]) -$
>
> $m * g * (l - \text{x2}[t]) * \text{Sin}[\theta];$
>
> euler = EulerEquations$[L, \{\text{x1}[t], \text{x2}[t]\}, t]$//FullSimplify
>
> $\left\{ (m + M)\text{x1}''[t] + m\text{Cos}[\theta]\text{x2}''[t] == 0, gm\text{Sin}[\theta] == m \left(\text{Cos}[\theta]\text{x1}''[t] + \text{x2}''[t] \right) \right\}$
>
> Solve[euler, {x1"[t], x2"[t]}][[1]]//FullSimplify
>
> $\left\{ \text{x1}''[t] \to \frac{gm\text{Sin}[2\theta]}{-m - 2M + m\text{Cos}[2\theta]}, \text{x2}''[t] \to -\frac{2g(m + M)\text{Sin}[\theta]}{-m - 2M + m\text{Cos}[2\theta]} \right\}$
> ---

One of the important issues demonstrated in Example 8.5 is about choosing the right velocity. We needed to make sure that the velocity of each particle was measured with respect to the same reference frame, in order to have a consistent kinetic energy. Also notice that $\partial \mathcal{L}/\partial x_1 = 0$. This tells us that the x_1-component of the generalized force is zero. This will have consequences which we will discuss in Section 8.7. Finally, Example 8.5 is a demonstration of a problem that has two degrees of freedom. In this case, we only needed to find the acceleration of each object. If instead we needed to find $x_1(t)$ and $x_2(t)$, we might have a difficult time solving the coupled second-order differential equations. Coupled differential equations are typically difficult to solve, and we often must rely on numerical methods.

In the next section, we will return to the Lagrange formalism and study more closely systems with constraints.

8.5 CONSTRAINT FORCES AND LAGRANGE'S EQUATION WITH UNDE-TERMINED MULTIPLIERS

In Section 8.2, we studied the example of the simple plane pendulum and we found that there was a constraint on the coordinates x and y. For a pendulum of length ℓ, there is the following relationship between x and y,

$$x^2 + y^2 = \ell^2 \tag{8.5.1}$$

Notice that (8.5.1) takes the form,

$$f(x, y) = \text{constant} \tag{8.5.2}$$

where $f(x, y) = x^2 + y^2$. We will call equations of the form (8.5.2) *constraint equations* and in this section, we will show how the Lagrange formulation allows us to directly handle constraints without the substitutions done in earlier chapters. The method is sometimes called the method of *Lagrange multipliers* or *Lagrange undetermined multipliers*. We will see that by using the constraint equation, we can also find the force causing the constraint. In the case of the pendulum, the constraint force is the tension in the pendulum's string.

Before developing the method of Lagrange multipliers, it is important to discuss types of constraints. In this chapter, we will focus on *holonomic constraints*. Holonomic constraints are constraint equations which can be written as an algebraic relationship between the coordinates. Equation (8.5.2) is an example of such a constraint. Sometimes, holonomic constraints are written in the form,

$$f(q_i; t) = 0 \tag{8.5.3}$$

where the abbreviation q_i says that f is a function of several coordinates (for example, x, y, and z in Cartesian coordinates), and the semicolon shows f's indirect dependence on time (through the coordinates q_i). For example, we can rewrite (8.5.1) in the form of (8.5.2) by writing $f(q_i; t) = f(x, y) = x^2 + y^2 - \ell^2 = 0$. Constraints that must be written as relationships involving velocities are called *nonholonomic constraints*, unless the constraint equation can be integrated to produce a relationship between coordinates. Although we will not discuss nonholonomic constraints further, the interested reader can consult the classic text by Thornton and Marion [6].

Now we will derive Lagrange's equations with undetermined multipliers. We derive the equations for a system with two degrees of freedom, x and y, and then discuss a generalization for systems with higher degrees of freedom. The system will have one holonomic constraint of the form, $f(x, y) = $ constant. We know that finding the equations of motion involves finding functions $x(t)$ and $y(t)$ that make the action integral,

$$S = \int_{t_1}^{t_2} \mathcal{L}\left(X, \dot{X}, Y, \dot{Y}\right) dt \tag{8.5.4}$$

stationary. We then write,

$$\left.\begin{aligned} X(a, t) &= x(t) + a\eta_x(t) \\ Y(a, t) &= y(t) + a\eta_y(t) \end{aligned}\right\} \tag{8.5.5}$$

where $x(t)$ and $y(t)$ are the functions that make (8.5.4) stationary, and $\eta_x(t)$ and $\eta_y(t)$ are perturbations from x and y with $\eta_x(t_1) = \eta_y(t_1) = 0$ and $\eta_x(t_2) = \eta_y(t_2) = 0$. By inserting (8.5.5) into (8.5.4) and using (7.6.3) we obtain:

$$\frac{\partial S}{\partial a} = \int_{t_1}^{t_2} \left(\frac{\partial \mathcal{L}}{\partial X} - \frac{d}{dt}\frac{\partial \mathcal{L}}{\partial \dot{X}}\right)\eta_x dt + \int_{t_1}^{t_2} \left(\frac{\partial \mathcal{L}}{\partial Y} - \frac{d}{dt}\frac{\partial \mathcal{L}}{\partial \dot{Y}}\right)\eta_y dt \tag{8.5.6}$$

However, (8.5.6) is only true for motions that satisfy the constraint $f(X, Y) = $ constant. Therefore, along the path of the particle,

$$\frac{df}{da} = \frac{\partial f}{\partial X}\eta_x + \frac{\partial f}{\partial Y}\eta_y = 0 \tag{8.5.7}$$

Recall that $dX/da = \eta_x$ and $dY/da = \eta_y$. Solving (8.5.7) for η_y/η_x and inserting the result into (8.5.6), we obtain:

$$\frac{\partial S}{\partial a} = \int_{t_1}^{t_2} \left[\left(\frac{\partial \mathcal{L}}{\partial X} - \frac{d}{dt}\frac{\partial \mathcal{L}}{\partial \dot{X}}\right) - \left(\frac{\partial \mathcal{L}}{\partial Y} - \frac{d}{dt}\frac{\partial \mathcal{L}}{\partial \dot{Y}}\right)\left(\frac{\partial f/\partial Y}{\partial f/\partial X}\right)\right]\eta_x dt \tag{8.5.8}$$

Setting $\partial S/\partial a = 0$ at $a = 0$ gives (recall that $X(a = 0, t) = x(t)$, and similar for $Y(0, t)$),

$$\left(\frac{\partial \mathcal{L}}{\partial x} - \frac{d}{dt}\frac{\partial \mathcal{L}}{\partial \dot{x}}\right)\left(\frac{\partial f}{\partial x}\right)^{-1} = \left(\frac{\partial \mathcal{L}}{\partial y} - \frac{d}{dt}\frac{\partial \mathcal{L}}{\partial \dot{y}}\right)\left(\frac{\partial f}{\partial y}\right)^{-1} \tag{8.5.9}$$

The left-hand side of (8.5.9) depends on derivatives with respect to x and $\dot{x}$, while the right-hand side depends on derivatives with respect to y and $\dot{y}$. Note that x and y are both functions of t. In order for the two sides of (8.5.9) to be equal, they must both be equal to a function of only t. We will define that function to be $-\lambda(t)$, and we call λ Lagrange's undetermined multiplier. We do not know λ, hence the name "undetermined multiplier." Setting each side of (8.5.9) equal to $-\lambda$ we obtain,

$$\left.\begin{aligned}\frac{\partial \mathcal{L}}{\partial x} - \frac{d}{dt}\frac{\partial \mathcal{L}}{\partial \dot{x}} &= -\lambda(t)\frac{\partial f}{\partial x} \\ \frac{\partial \mathcal{L}}{\partial y} - \frac{d}{dt}\frac{\partial \mathcal{L}}{\partial \dot{y}} &= -\lambda(t)\frac{\partial f}{\partial x}\end{aligned}\right\} \tag{8.5.10}$$

Equations (8.5.10) along with $f(x,y) = $ constant will produce the equations of motion for the system.

Of course, not all systems are limited to only two degrees of freedom and one constraint equation. In general, systems will have n degrees of freedom and m equations of constraint. We have already seen how to generalize the Lagrange formulation for s degrees of freedom, in which case we have s Euler-Lagrange equations to solve. However, multiple constraint equations means that there will be multiple derivatives df_j/da ($j = 1,\ldots,m$, with m equal to the number of constraints) that we will have to include in the derivation above. The result is:

Euler-Lagrange Equation with Undetermined Multipliers

$$\left.\begin{aligned}\frac{\partial \mathcal{L}}{\partial q_i} - \frac{d}{dt}\frac{\partial \mathcal{L}}{\partial \dot{q}_i} &= -\sum_{j=1}^{m}\lambda_j(t)\frac{\partial f_j}{\partial q_i} \\ f_j(q_i;t) &= 0\end{aligned}\right\} \tag{8.5.11}$$

where $i = 1,\ldots,n$, $j = 1,\ldots,m$, and we have used generalized coordinates q_i. Notice that in (8.5.11) there are m undetermined multipliers λ_j, one for each constraint equation $f_j(q_i;t) = $ constant.

Before moving on to an example, there are two important points we should make. First, you might be wondering why one would use Lagrange multipliers. Using Lagrange multipliers can save one from doing a lot of messy algebra when it comes to solving f for the relationship between coordinates. For example, in the pendulum problem, if we solved f for y, we would have $y = \pm\sqrt{\ell^2 - x^2}$. We would have to deal with both the positive and negative roots. However, as we will see in the example below, using Lagrange's multipliers will help simplify the procedure. That said, the power of modern computer algebra systems may make this argument less compelling than it once was.

The second point is more important theoretically. It turns out that the Lagrange multiplier can tell us the force causing the constraint in the system. To illustrate this, we will consider a simple system with two degrees of freedom and one constraint. The Lagrangian is,

$$\mathcal{L} = \frac{1}{2}m\left(\dot{x}^2 + \dot{y}^2\right) - V(x,y)$$

and inserting this Lagrangian into the Euler-Lagrange equation yields,

$$\left.\begin{aligned}-\frac{\partial V}{\partial x} - m\ddot{x} &= -\lambda\frac{\partial f}{\partial x} \\ -\frac{\partial V}{\partial y} - m\ddot{y} &= -\lambda\frac{\partial f}{\partial y}\end{aligned}\right\} \tag{8.5.12}$$

We will focus on the equation involving x, which can be written in the form:

$$m\ddot{x} = -\frac{\partial V}{\partial x} + \lambda\frac{\partial f}{\partial x} \tag{8.5.13}$$

The left-hand side $m\ddot{x}$ is the x-component of the net force acting on the system. In this case, there are two forces acting on the system: an external force, associated with the potential energy $V(x,y)$, and the constraint force. The first term in the right-hand side of (8.5.13) is the x-component of the external force acting on the system, gravity for example. The remaining term $\lambda\partial f/\partial x$ must therefore be the x-component of the constraint force. We can repeat a similar argument for the y-equation (8.5.12) to see that $\lambda\partial f/\partial y$ is the y-component of the constraint force. In general we have,

Constraint Force from Lagrange's Undetermined Multiplier

$$F^c_{j,q_i} = \lambda_j\frac{\partial f_j}{\partial q_i} \tag{8.5.14}$$

where F^c_{j,q_i} is the q_i-component of the constraint force associated with the constraint equation $f_j(q_i;t) = $ constant. The following example illustrates the use of the Euler-Lagrange equation to derive a constraint force.

Example 8.6: Particle on a sphere
A particle of mass m starts on top of a smooth fixed sphere of radius R. At time $t = 0$, the particle is perturbed by a very small force. The particle slides along the sphere's surface before it eventually falls off. Find the angle the particle makes as it falls off of the sphere's surface.

Solution:
Figure 8.7 illustrates the system when the particle has an angular displacement θ from its starting position.

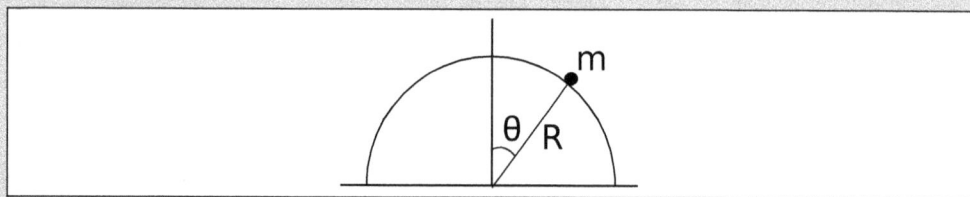

Figure 8.7: A particle of mass m on a hemisphere of radius R for Example 8.6.

To begin the problem, we start by selecting polar coordinates to solve the problem. Because the particle is moving on the surface of the sphere, our constraint equation is,

$$f(r;t) = r - R = 0 \tag{8.5.15}$$

The force of constraint is the force that keeps the particle on the sphere and $\dot{r} = 0$. When the particle leaves the sphere's surface, $\dot{r} \neq 0$; therefore, there will be an acceleration in the radial direction. We can find the relevant constraint force by finding:

$$F_r^c = \lambda \frac{\partial f}{\partial r}.$$

We know $\partial f / \partial r$, so we need to find λ. We will find λ by setting up the Euler-Lagrange equations for the system.

Using polar coordinates, we can write down the kinetic and potential energies for the system:

$$T = \frac{1}{2} m \left(\dot{r}^2 + r^2 \dot{\theta}^2 \right)$$

$$V = mgr \cos \theta$$

$$\mathcal{L} = \frac{1}{2} m \left(\dot{r}^2 + r^2 \dot{\theta}^2 \right) - mgr \cos \theta$$

where we used the horizontal line in Figure 8.7 as $y = 0$, and therefore the potential energy can be written as mgy with $y = r \cos \theta$. The Euler-Lagrange equations are:

$$\frac{\partial \mathcal{L}}{\partial r} - \frac{d}{dt} \left(\frac{\partial \mathcal{L}}{\partial \dot{r}} \right) = -\lambda \frac{\partial f}{\partial r} \tag{8.5.16}$$

$$\frac{\partial \mathcal{L}}{\partial \theta} - \frac{d}{dt} \left(\frac{\partial \mathcal{L}}{\partial \dot{\theta}} \right) = -\lambda \frac{\partial f}{\partial \theta}. \tag{8.5.17}$$

By substituting the Lagrangian $\mathcal{L}$ and the constraint equation, we obtain:

$$mr\dot{\theta}^2 - mg \cos \theta - m\ddot{r} = -\lambda \tag{8.5.18}$$

$$-mgr \sin \theta - mr^2 \ddot{\theta} - 2mr\dot{r}\dot{\theta} = 0 \tag{8.5.19}$$

Next, we apply the equation of constraint, $r = R$ and $\dot{r} = \ddot{r} = 0$, because we want to know the angle θ just before the particle leaves the sphere's surface.

$$mR\dot{\theta}^2 - mg \cos \theta = -\lambda$$
$$-mgR \sin \theta - mR^2 \ddot{\theta} = 0 \tag{8.5.20}$$

In the first equation above, we have found that $\lambda = \lambda(\theta, \dot{\theta})$. The second equation will give us a way of finding $\theta(t)$,

$$\ddot{\theta} = \frac{g}{R} \sin \theta \tag{8.5.21}$$

While we could attempt to integrate (8.5.21) and solve for θ, we actually don't need to do that. What we really need is $\dot{\theta}$, which can be found from (8.5.21) using,

$$\ddot{\theta} = \frac{d\dot{\theta}}{dt} = \frac{d\dot{\theta}}{d\theta} \frac{d\theta}{dt} = \dot{\theta} \frac{d\dot{\theta}}{d\theta} \tag{8.5.22}$$

By inserting (8.5.22) into (8.5.21), we get:

$$\int \dot{\theta} d\dot{\theta} = \int \frac{g}{R} \sin \theta d\theta$$

After integrating we find:

$$\dot{\theta}^2 = 2 \frac{g}{R} (1 - \cos \theta)$$

and inserting the result into (8.5.20) to get:

$$\lambda = mg (3 \cos \theta - 2)$$

The constraint force is then:

$$F_r^c = \lambda \frac{\partial f}{\partial r} = mg\,(3\cos\theta - 2)$$

Note that when the particle is on top of the sphere ($\theta = 0$), the constraint force is mg. Finally, the particle leaves the surface of the sphere when the constraint force is equal to zero, hence we want the angle θ_0 for which $F_r^c = 0$,

$$\theta_0 = \cos^{-1}\left(\frac{2}{3}\right)$$

Notice that this angle does not depend on the mass of the particle or the radius of the sphere!

8.6 CONSERVATION THEOREMS AND THE LAGRANGIAN

Besides being used to derive a system's equations of motion, the Lagrangian can help us identify conserved quantities in the system. In particular, we will look at conservation of momentum and energy.

8.6.1 CONSERVATION OF MOMENTUM

We already know that in Newtonian mechanics the total momentum of an isolated system of particles is conserved. We would expect to find that the total momentum is conserved for the isolated system in the Lagrange formulation as well. One important feature of isolated systems is that they are *translationally invariant*. In other words, if you move each particle in the system by the same amount and in the same direction, then there is no change in the system, other than the location of each particle. We will see that translational invariance of the Lagrangian will lead to momentum conservation in the system.

The connection between translational invariance of the Lagrangian and momentum conservation is a result of Noether's theorem, which states that whenever there is a continuous symmetry in the Lagrangian, there is a corresponding conservation law. Noether's theorem was proved by Emmy Noether (1882–1935) in 1915 and published in 1918 [7]. In this subsection, we will demonstrate the translational invariance of the Lagrangian, and we will connect the symmetry to the conservation of momentum.

Consider a system of N isolated particles of mass m_α, where $\alpha = 1, \ldots, N$. The location of each particle is denoted by $\mathbf{r}_\alpha$. Next we shift all of these particles through the same displacement $d\mathbf{r}$, which is our "translation," $\mathbf{r}_\alpha \to \mathbf{r}_\alpha + d\mathbf{r}$. We need to show that the Lagrangian will remain constant under the translation. In other words,

$$d\mathcal{L} = \mathcal{L}\,(\mathbf{r}_\alpha + d\mathbf{r}, \dot{\mathbf{r}}_\alpha; t) - \mathcal{L}\,(\mathbf{r}_\alpha, \dot{\mathbf{r}}_\alpha; t) = 0 \qquad (8.6.1)$$

Notice that there is no shift in the velocity in (8.6.1). We will discuss why below. Since the Lagrangian consists of both the kinetic and potential energies, we need to show both $dT = 0$ and $dV = 0$.

We will first demonstrate $dT = 0$. Notice that the translation does not change the velocity of the particles because $d\mathbf{r}$ is constant. The kinetic energy depends on the velocity, $T = \frac{1}{2}m\dot{\mathbf{r}} \cdot \dot{\mathbf{r}}$, and the displacements do not change the velocities; therefore, the kinetic energy does not change with the translation and therefore $dT = 0$.

The translation does affect the position of the particle. The careful reader might then think that $dV \neq 0$, because potential energies are often position dependent. Wouldn't the gravitational potential energy change with translation? Yes! However, recall that this is an *isolated* system, meaning that there are no external forces acting on the system. Furthermore, inter-particle forces are usually dependent on the distances between particles. Because all of the particles experience the same displacement, the distances between particles will not change. Hence, $dV = 0$, and therefore $d\mathcal{L} = 0$.

Now that we have demonstrated the translational invariance of the Lagrangian, we can make the connection between the symmetry and the conservation of momentum. For simplicity, we will set $N = 1$, but the following holds for any number of particles. In addition, we will use generalized coordinates with generalized unit vectors $\hat{e}_i$, again for simplicity of notation, $\mathbf{r} = \sum q_i \hat{e}_i$ and $d\mathbf{r} = \sum \hat{e}_i dq_i$. Hence, for a single particle under the translation,

$$d\mathcal{L} = \sum_{i=1}^{s} \frac{\partial \mathcal{L}}{\partial q_i} dq_i + \sum_{i=1}^{s} \frac{\partial \mathcal{L}}{\partial \dot{q}_i} d\dot{q}_i = 0 \tag{8.6.2}$$

where s is equal to the particle's number of degrees of freedom. Because $d\mathbf{r}$ is not a function of time,

$$d\dot{q}_i = d\left(\frac{dq_i}{dt}\right) = \frac{d}{dt}(dq_i) = 0$$

Therefore (8.6.2) becomes:

$$d\mathcal{L} = \sum_{i=1}^{s} \frac{\partial \mathcal{L}}{\partial q_i} dq_i = 0 \tag{8.6.3}$$

Because $dq_i \neq 0$, we must have

$$\frac{\partial \mathcal{L}}{\partial q_i} = 0 \tag{8.6.4}$$

for all i, and therefore according to the Euler-Lagrange equation,

$$\frac{d}{dt}\left(\frac{\partial \mathcal{L}}{\partial \dot{q}_i}\right) = 0 \tag{8.6.5}$$

Equation (8.6.5) then tells us that:

$$p_i = \frac{\partial \mathcal{L}}{\partial \dot{q}_i} = \text{constant} \tag{8.6.6}$$

We have already seen that p_i is the ith component of the particle's generalized momentum. The quantity p_i is sometimes also referred to as the *canonical momentum conjugate to* q_i. From (8.6.4), we see that when $\mathcal{L}$ does not explicitly depend on the coordinate q_i, then the canonical momentum conjugate to q_i is conserved. In other words, if q_i does not appear in $\mathcal{L}$, then p_i is constant. Visual inspection of the Lagrangian can tell you what momenta are conserved in the system! If q_i does not appear in $\mathcal{L}$, then changing q_i will not change $\mathcal{L}$ and $\mathcal{L}$ is said to be "invariant" under changes of q_i. Note that we can extend the work above to N-particle systems by including an additional sum over $\alpha = 1, \ldots, N$. The procedure would then follow as done above. Finally, when the Lagrangian is independent of the coordinate q_i, then q_i is called a *cyclic* coordinate. Next, we will see that time-translational symmetry in the Lagrangian will lead to conservation of energy.

8.6.2 CONSERVATION OF ENERGY

We will now see how time-translational invariance of the Lagrangian leads to conservation of energy. Time-translational invariance of the Lagrangian means that the Lagrangian does not explicitly depend on time, i.e., $\partial \mathcal{L}/\partial t = 0$.

Let us begin by considering a single-particle system with s degrees of freedom. Like the work done connecting translational invariance of the Lagrangian to the conservation of momentum, we can extend what follows to N-particle systems by including an additional sum to the equations below.

We begin by finding the total time derivative of the Lagrangian, which according to the chain-rule is

$$\frac{d\mathcal{L}}{dt} = \sum_{i=1}^{s} \left(\frac{\partial \mathcal{L}}{\partial q_i} \right) \dot{q}_i + \sum_{i=1}^{s} \left(\frac{\partial \mathcal{L}}{\partial \dot{q}_i} \right) \ddot{q}_i + \frac{\partial \mathcal{L}}{\partial t} \tag{8.6.7}$$

For now, we will not restrict ourselves to Lagrangians that are time translationally invariant. Next, we will insert the Euler-Lagrange equation

$$\frac{\partial \mathcal{L}}{\partial q_i} = \frac{d}{dt} \left(\frac{\partial \mathcal{L}}{\partial \dot{q}_i} \right) \tag{8.6.8}$$

into (8.6.7) to obtain,

$$\frac{d\mathcal{L}}{dt} = \sum_{i=1}^{s} \left(\frac{d}{dt} \frac{\partial \mathcal{L}}{\partial \dot{q}_i} \right) \dot{q}_i + \sum_{i=1}^{s} \left(\frac{\partial \mathcal{L}}{\partial \dot{q}_i} \right) \ddot{q}_i + \frac{\partial \mathcal{L}}{\partial t} \tag{8.6.9}$$

$$= \sum_{i=1}^{s} \frac{d}{dt} \left(\frac{\partial \mathcal{L}}{\partial \dot{q}_i} \dot{q}_i \right) + \frac{\partial \mathcal{L}}{\partial t} \tag{8.6.10}$$

If the Lagrangian is invariant to time translations, then $\partial \mathcal{L}/\partial t = 0$ and therefore (8.6.10) becomes,

$$\frac{d}{dt} \left[\mathcal{L} - \sum_{i=1}^{s} p_i \dot{q}_i \right] = 0 \tag{8.6.11}$$

where we used $p_i = \partial \mathcal{L}/\partial \dot{q}_i$. We see that we have a new conserved quantity in the square brackets of (8.6.11). The conserved quantity is called the *Hamiltonian* function, or simply the Hamiltonian. The Hamiltonian is defined as:

The Hamiltonian

$$H = \sum_{i=1}^{s} p_i \dot{q}_i - \mathcal{L} \tag{8.6.12}$$

where s is the number of degrees of freedom. As stated above, the Hamiltonian is an important quantity in both classical and quantum physics.

When the Lagrangian is invariant to time translations, then the Hamiltonian is constant in time. The question remains, how does any of this connect to conservation of energy? To answer that question, let us look at the simple harmonic oscillator (SHO). The Lagrangian for the SHO is

$$\mathcal{L} = \frac{1}{2} m \dot{x}^2 - \frac{1}{2} k x^2 \tag{8.6.13}$$

where we will consider the standard model of the SHO, a mass m connected to a massless spring with spring constant k displaced a distance x from its equilibrium position located at $x = 0$, allowed to slide along a horizontal frictionless surface. In this case, we have one degree of freedom and one generalized coordinate $q_1 = x$. We can find the canonical momentum conjugate to x,

$$p = \frac{\partial \mathcal{L}}{\partial \dot{x}} = m\dot{x} \tag{8.6.14}$$

Inserting $p = m\dot{x}$ and $\dot{q} = \dot{x}$ into (8.6.12), we find

$$H = m\dot{x}^2 - \mathcal{L} = 2T - (T - V) = T + V = E \tag{8.6.15}$$

We see, at least for the SHO, that the Hamiltonian is equal to the total mechanical energy. Next, we will prove that $H = T + V$ for a wider class of systems.

Under the right conditions, the Hamiltonian is the total energy of the system. Consider a system of N particles, each denoted by the index $\alpha = 1, 2, \ldots, N$ described by the generalized coordinates $q_1, \ldots, q_n$. We will now prove that if

$$\mathbf{r}_\alpha = \mathbf{r}_\alpha (q_1, \ldots, q_n) \tag{8.6.16}$$

where $\mathbf{r}_\alpha$ is the time-independent location of m_α in Cartesian coordinates, then $H = T + V$. In other words, the condition for $H = T + V$ is that the transformation between Cartesian and generalized coordinates must be time-independent. We begin by computing the kinetic energy $T = \frac{1}{2}\sum_\alpha m_\alpha \dot{\mathbf{r}}_\alpha \cdot \dot{\mathbf{r}}_\alpha$ in terms of the generalized coordinates q_i. The time derivative of (8.6.16) is

$$\dot{\mathbf{r}}_\alpha = \sum_{j=1}^{n} \frac{\partial \mathbf{r}_\alpha}{\partial q_j} \dot{q}_j \tag{8.6.17}$$

Therefore

$$\dot{\mathbf{r}}_\alpha^2 = \dot{\mathbf{r}}_\alpha \cdot \dot{\mathbf{r}}_\alpha = \left(\sum_j \frac{\partial \mathbf{r}_\alpha}{\partial q_j} \dot{q}_j \right) \cdot \left(\sum_k \frac{\partial \mathbf{r}_\alpha}{\partial q_k} \dot{q}_k \right) \tag{8.6.18}$$

The total kinetic energy is then

$$T = \frac{1}{2} \sum_\alpha m_\alpha \dot{\mathbf{r}}_\alpha^2 = \frac{1}{2} \sum_{j,k} a_{jk} \dot{q}_j \dot{q}_k \tag{8.6.19}$$

where

$$a_{jk} = \sum_\alpha m_\alpha \left(\frac{\partial \mathbf{r}_\alpha}{\partial q_j} \right) \cdot \left(\frac{\partial \mathbf{r}_\alpha}{\partial q_k} \right) \tag{8.6.20}$$

and $a_{jk} = a_{jk}(q_1, \ldots, q_n)$. Next, we will compute the canonical momentum conjugate to q_i,

$$p_i = \frac{\partial \mathcal{L}}{\partial \dot{q}_i} = \frac{\partial T}{\partial \dot{q}_i} = \sum_j a_{ij} \dot{q}_j \tag{8.6.21}$$

Proof of (8.6.21) is left as Problem 28 in the End-of-Chapters Problems. Finally, we insert (8.6.21) into (8.6.12):

$$\left. \begin{aligned} H &= \sum_i \left(\sum_j a_{ij} \dot{q}_j \right) \dot{q}_i - \mathcal{L} \\ &= 2T - (T - V) \\ &= T + V \end{aligned} \right\} \tag{8.6.22}$$

In summary, as long as $\mathcal{L}$ is invariant to translations in time (i.e., not explicitly dependent on time), the Hamiltonian is a conserved quantity. In addition, if the transformation from Cartesian coordinates to generalized coordinates is time-independent, then the Hamiltonian is equal to the total mechanical energy of the system. This is another demonstration of Noether's theorem.

Both the connections between the symmetries (translational invariance and time-translational invariance of $\mathcal{L}$), and the conservation laws (momentum and energy), result from cyclic coordinates. A wise choice for a set of generalized coordinates would be one such that there are as many cyclic coordinates as possible. If such a set could be found, then there will be many conserved quantities which will likely simplify the equations of motion for our system. Not only are simple equations of motion easier to solve, they are also easier to use to gain insight into the system of interest.

8.7 HAMILTONIAN DYNAMICS

The Hamiltonian is an important quantity in both classical and quantum physics. In quantum mechanics, the Hamiltonian is an important element of the Schrödinger equation which is used to find a particle's wave function. In classical mechanics, the Hamiltonian can be used to derive a system's equations of motion. In this section, we will show how to derive Hamilton's equations which will provide the equations of motion for a system in terms of p_i and q_i.

We begin our derivation with another look at generalized momentum. Recall that the generalized momentum can be found using the Lagrangian,

$$p_i = \frac{\partial \mathcal{L}}{\partial \dot{q}_i} \tag{8.7.1}$$

Because the Lagrangian is a function of both generalized coordinates and generalized velocities, the momentum can be written as $p_i = p_i(q_j, \dot{q}_j; t)$. The function for the momentum can in general be inverted such that,

$$\dot{q}_j = \dot{q}_j(q_k, p_k; t) \tag{8.7.2}$$

If we combine (8.7.2) with the definition of the Hamiltonian,

$$H = \sum_i p_i \dot{q}_i - \mathcal{L}(q_k, \dot{q}_k; t) \tag{8.7.3}$$

we find that the Hamiltonian is a function of generalized coordinates and generalized momenta, $H = H(q_k, p_k; t)$.

Now we are ready to derive Hamilton's equations of motion. We will derive Hamilton's equations for a system with one degree of freedom and leave the derivation for the s-degrees of freedom for Problem 31 in the End-of-Chapter Problems. We begin by calculating the total differential of $H = H(q, p; t)$,

$$dH = \frac{\partial H}{\partial q}dq + \frac{\partial H}{\partial p}dp + \frac{\partial H}{\partial t}dt \tag{8.7.4}$$

Next, we calculate dH using (8.7.3),

$$dH = \dot{q}dp + pd\dot{q} - d\mathcal{L} \tag{8.7.5}$$

$$= \dot{q}dp + pd\dot{q} - \left(\frac{\partial \mathcal{L}}{\partial q}dq + \frac{\partial \mathcal{L}}{\partial \dot{q}}d\dot{q} + \frac{\partial \mathcal{L}}{\partial t}dt\right) \tag{8.7.6}$$

$$= \dot{q}dp + pd\dot{q} - \left(\dot{p}dq + pd\dot{q} + \frac{\partial \mathcal{L}}{\partial t}dt\right) \tag{8.7.7}$$

where we used,

$$\dot{p} = \frac{d}{dt}(p) = \frac{d}{dt}\frac{\partial \mathcal{L}}{\partial \dot{q}} = \frac{\partial \mathcal{L}}{\partial q}$$

from the Euler-Lagrange equation. Collecting terms, (8.7.7) becomes,

$$dH = \dot{q}dp - \dot{p}dq - \frac{\partial \mathcal{L}}{\partial t}dt \qquad (8.7.8)$$

Finally, we equate the coefficients of dq and dp in (8.7.4) and (8.7.8) to get Hamilton's equations for a system with one degree of freedom,

$$\dot{q} = \frac{\partial H}{\partial p} \qquad \dot{p} = -\frac{\partial H}{\partial q} \qquad (8.7.9)$$

Hamilton's equations for a system of s degrees of freedom are:

Hamilton's Equations of Motion

$$\dot{q}_k = \frac{\partial H}{\partial p_k} \qquad \dot{p}_k = -\frac{\partial H}{\partial q_k} \qquad k = 1, \ldots, s \qquad (8.7.10)$$

Hamilton's equations of motion are also sometimes referred to as the canonical equations of motion, and the description of a particle's motion by these equations is referred to as *Hamiltonian dynamics*. Regardless of the number of degrees of freedom, we also have the relationship:

$$\frac{\partial H}{\partial t} = -\frac{\partial \mathcal{L}}{\partial t} \qquad (8.7.11)$$

found by equating the coefficients of dt in (8.7.4) and (8.7.8).

Finally, if we divide (8.7.4) by dt, we find that,

$$\frac{dH}{dt} = \frac{\partial H}{\partial t} \qquad (8.7.12)$$

While both the Lagrangian and the Hamiltonian can be used to find a system's equations of motion, there are some differences between the two formulations. First, notice that for a system of s degrees of freedom, Hamilton's equations result in $2s$ first-order ODEs. The Lagrange formulation for the same system results in s second-order ODEs. Hamilton's formulation uses generalized coordinates and generalized momenta, while the Lagrange formulation uses generalized coordinates and generalized velocities. These different properties are summarized in Table 8.1.

Table 8.1

Summary of the Hamilton and Lagrange formulations of a system's equations of motion for a system with s degrees of freedom and no constraints.

Formulation	Number of Equations	Order of Equations	Coordinates Used
Lagrange	s	2	q_k and $\dot{q}_k$
Hamilton	$2s$	1	q_k and p_k

You now have three different methods of formulating a system's equations of motion: Newton's second law, the Lagrangian, and the Hamiltonian. In addition, you can use conservation laws to set up and solve problems. The method you choose will depend on what information you are given, and what information you want to learn about the problem at hand. In fact, you may learn different things by solving the same problem using multiple techniques. Once students learn about Lagrangians, they sometimes ignore Newton's second law, but Newton's second law can be the easiest one to use when dealing with drag forces and (as we will see later) noninertial reference frames. Use the right tool for the job!

To end this section, we present a few examples of using the Hamilton formulation in order to find equations of motion. Notice that all of these examples will have a common theme: in order to find H, one needs to first find $\mathcal{L}$ and compute the generalized momenta.

Example 8.7: The simple pendulum revisited, again

Find Hamilton's equations for the simple plane pendulum from Figure 8.1. Using $m = 1.0$ kg, $\ell = 1.0$ m, and $g = 9.8$ m/s^2, plot $\theta(t)$ with initial conditions $\theta(0) = 3.0$ rad and $\dot{\theta}(0) = 0$.

Solution:

From our work done earlier in this chapter, we know that the Lagrangian for this system is given by (8.2.5):

$$\mathcal{L} = \frac{1}{2}m\ell^2\dot{\theta}^2 + mg\ell\cos\theta$$

where we used the coordinate θ instead of q. To find the Hamiltonian, we compute the canonical momentum conjugate to θ,

$$p_\theta = \frac{\partial\mathcal{L}}{\partial\dot{\theta}} = m\ell^2\dot{\theta} \tag{8.7.13}$$

Next, we use (8.6.12) to find the Hamiltonian,

$$
\begin{aligned}
H &= p_\theta\dot{\theta} - \mathcal{L} \\
&= p_\theta\frac{p_\theta}{m\ell^2} - \left(\frac{1}{2}m\ell^2\left(\frac{p_\theta}{m\ell^2}\right)^2 + mg\ell\cos\theta\right) \\
&= \frac{p_\theta^2}{2m\ell^2} - mg\ell\cos\theta
\end{aligned} \tag{8.7.14}
$$

Notice that we needed to substitute for $\dot{\theta}$ in order to get H in terms of θ and p_θ. Further, notice that the Hamiltonian is equal to the total mechanical energy of the system because the transformation between Cartesian coordinates and polar coordinates is not time-dependent.

Finally, we compute Hamilton's equations,

$$
\left.\begin{aligned}
\dot{\theta} &= \frac{\partial H}{\partial p_\theta} \quad\rightarrow\quad \dot{\theta} = \frac{p_\theta}{m\ell^2} \\
\dot{p}_\theta &= -\frac{\partial H}{\partial\theta} \quad\rightarrow\quad \dot{p}_\theta = -mg\ell\sin\theta
\end{aligned}\right\} \tag{8.7.15}
$$

There are a few things to notice with our solution. First, Hamilton's equation for $\dot{\theta}$ simply reproduces what was obtained by computing the canonical momentum. Second, the canonical momentum in this case is actually the angular momentum, not the linear momentum. Finally, notice that $\dot{p}_\theta$ is the torque acting on the system.

We plot a numerical solution to (8.7.15) using the following Mathematica and Python codes. Note that we use the initial conditions $\theta(0) = 3$ rad and $p_\theta(0) = 0$. The initial condition for p_θ is found using (8.7.13). We show the output for the Mathematica code in Figure 8.8.

Python Code

We use `diff` to evaluate the partial derivatives of the Hamiltonian, and `odeint` is used to integrate numerically the system of equations (8.7.15). The symbolic results are the same as the analytical ones obtained above.

```python
from sympy import diff, symbols, Function, cos

from scipy.integrate import odeint
import numpy as np
import matplotlib.pyplot as plt

print('-'*28,'CODE OUTPUT','-'*29,'\n')

ptheta, m, l, g, theta = symbols('ptheta, m, l, g, theta ',real=True)

H = ptheta**2/(2*m*l**2)-m*g*l*cos(theta)

print('theta-dot = ', diff(H,ptheta))

print('\nptheta-dot = ', -diff(H,theta))

m, l, g = 1, 1, 9.8

def deriv(y, t):
    q, p = y    #let v = dq/dt
    dydt = [p/(m*l**2), -m*g*l*np.sin(q)]
    return dydt

ics = [3,0]
t = np.linspace(0,20,100)

soln = odeint(deriv, ics, t)

plt.plot(t,soln[:,0])
plt.xlabel('time')
plt.ylabel('theta')
plt.show()

--------------------------- CODE OUTPUT ----------------------------
theta-dot =  ptheta/(l**2*m)

ptheta-dot =  -g*l*m*sin(theta)
```

Mathematica Code

We use NDSolve to solve the system of equations (8.7.15).

SetOptions[Plot, Axes->False, Frame->True, BaseStyle->{FontSize->16}];

$H = p\theta \textasciicircum 2/(2 * m * l\textasciicircum 2) - m * g * l * \mathrm{Cos}[\theta]$;

Print["θ-dot = ", $D[H, p\theta]$]

Print["pθ-dot = ", $-D[H, \theta]$]

θ-dot $= \frac{p\theta}{l^2 m}$
pθ-dot $= -glm\mathrm{Sin}[\theta]$

$l = 1.0$;

$m = 1.0$;

$g = 9.8$;

soln = NDSolve[$\{\theta'[t] == p\theta[t]/(ml\textasciicircum 2), p\theta'[t] == -mgl\mathrm{Sin}[\theta[t]], \theta[0] == 3.0,$

$p\theta[0] == 0\}, \{\theta, p\theta\}, \{t, 0, 20\}$];

Plot[$\theta[t]/$.soln, $\{t, 0, 20\}$, BaseStyle $\rightarrow$ {Black, Bold, FontSize $\rightarrow$ 22},

Frame $\rightarrow$ True, Axes->False, FrameLabel $\rightarrow$ {"time", "θ"}, ImageSize $\rightarrow$ Large]

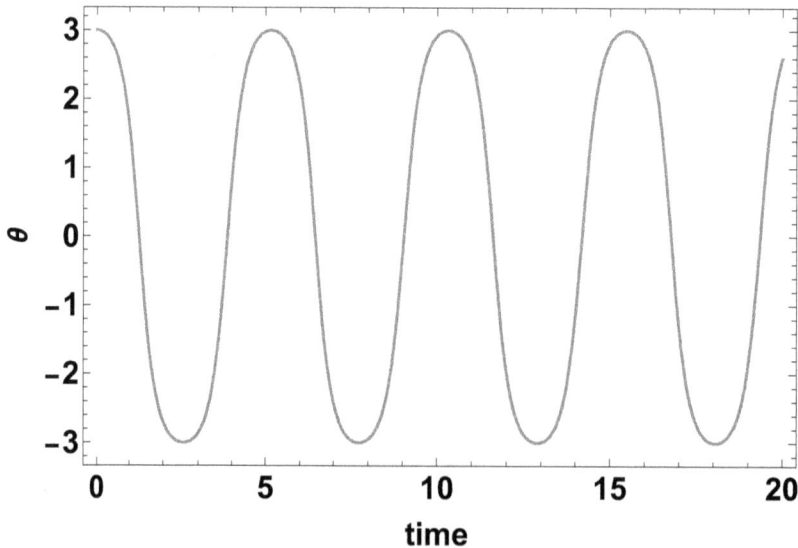

Figure 8.8: Mathematica output for Example 8.7. Notice that with the large initial amplitude, $\theta(t)$ is not sinusoidal.

Example 8.8: Mass on a cylinder

Find Hamilton's equations of motion for a particle of mass m constrained to move on the surface of a frictionless cylinder of radius R. The particle is attracted to the origin by a force $\mathbf{F} = -kr\hat{\mathbf{r}}$ where k is a positive constant, $\hat{\mathbf{r}}$ is the unit vector pointing from the origin to the particle's position, and r is the distance between the particle and the origin.

Solution:

An illustration of the system is shown in Figure 8.9.

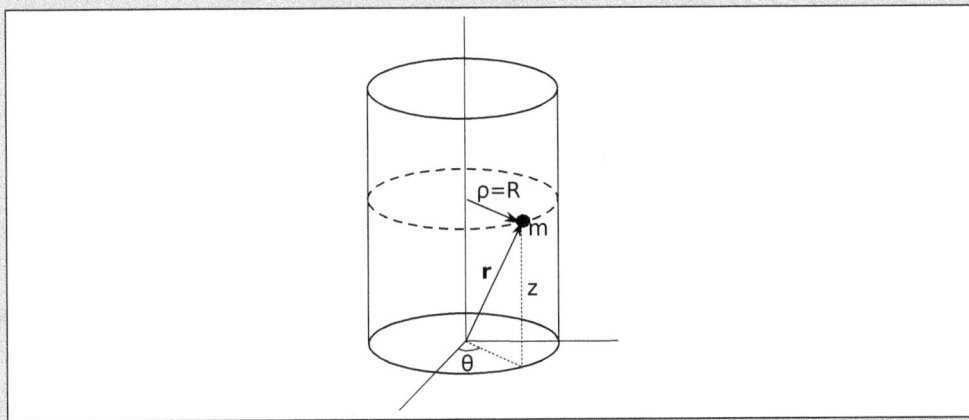

Figure 8.9: A particle of mass m moving along a cylinder of radius R, in Example 8.8.

We begin by choosing cylindrical coordinates to describe the system. Furthermore, there is one equation of constraint,

$$x^2 + y^2 = R^2 \tag{8.7.16}$$

which can also be written as $\rho^2 = R^2$. Therefore, there are two degrees of freedom in this problem.

Next, we compute the kinetic energy. In cylindrical coordinates, $v^2 = \dot{\rho}^2 + \rho^2\dot{\theta}^2 + \dot{z}^2$. However, using our constraint $\dot{\rho} = 0$, the kinetic energy of the particle becomes,

$$T = \frac{1}{2}m\left(R^2\dot{\theta}^2 + \dot{z}^2\right) \tag{8.7.17}$$

The potential energy is best found in Cartesian coordinates first. We are given that

$$\mathbf{F} = -k\mathbf{r} = -kr\hat{\mathbf{r}} = -k\left(x\hat{\mathbf{i}} + y\hat{\mathbf{j}} + z\hat{\mathbf{k}}\right) \tag{8.7.18}$$

Therefore, the potential energy is

$$\begin{aligned} V &= \frac{1}{2}kr^2 \\ &= \frac{1}{2}k\left(x^2 + y^2 + z^2\right) \end{aligned} \tag{8.7.19}$$

If we apply the constraint (8.7.16), then the potential energy becomes

$$V = \frac{1}{2}k\left(R^2 + z^2\right)$$

We next compute the Lagrangian $\mathcal{L} = T - V$,

$$\mathcal{L} = \frac{1}{2}m\left(R^2\dot{\theta}^2 + \dot{z}^2\right) - \frac{1}{2}k\left(R^2 + z^2\right) \tag{8.7.20}$$

From the Lagrangian we see that the canonical momentum conjugate to θ is conserved because θ is a cyclic coordinate. In order to find the Hamiltonian, we need to compute p_θ and p_z,

$$p_\theta = \frac{\partial\mathcal{L}}{\partial\dot{\theta}} = mR^2\dot{\theta} \quad \text{and} \quad p_z = \frac{\partial\mathcal{L}}{\partial\dot{z}} = m\dot{z} \tag{8.7.21}$$

Note that p_θ again is an angular momentum, whereas p_z is the translational momentum. Next, we use (8.7.3) to find the Hamiltonian,

$$H = p_\theta\dot{\theta} + p_z\dot{z} - \mathcal{L} \tag{8.7.22}$$

$$= p_\theta\frac{p_\theta}{mR^2} + p_z\frac{p_z}{m} - \left[\frac{1}{2}m\left(R^2\left(\frac{p_\theta}{mR^2}\right)^2 + \left(\frac{p_z}{m}\right)^2\right) - \frac{1}{2}k\left(R^2 + z^2\right)\right] \tag{8.7.23}$$

$$= \frac{p_\theta^2}{2mR^2} + \frac{p_z^2}{2m} + \frac{1}{2}k\left(R^2 + z^2\right) \tag{8.7.24}$$

Finally, we use the Hamiltonian to compute Hamilton's equations,

$$\left.\begin{aligned}
\dot{z} &= \frac{\partial H}{\partial p_z} &\rightarrow\quad \dot{z} &= \frac{p_z}{m} \\
\dot{\theta} &= \frac{\partial H}{\partial p_\theta} &\rightarrow\quad \dot{\theta} &= \frac{p_\theta}{mR^2} \\
\dot{p}_z &= -\frac{\partial H}{\partial z} &\rightarrow\quad \dot{p}_z &= -kz \\
\dot{p}_\theta &= -\frac{\partial H}{\partial\theta} &\rightarrow\quad \dot{p}_\theta &= 0
\end{aligned}\right\} \tag{8.7.25}$$

Notice that $\dot{p}_z$ is equal to the z-component of the external force acting on the system.

8.8   ADDITIONAL EXPLORATIONS INTO THE HAMILTONIAN

You might be wondering why we would ever want to use the Hamilton formulation, especially given the fact that we typically need to find the Lagrangian in order to get the Hamiltonian. Why not just use the Lagrange formulation? The reason is that Hamilton's formulation is powerful from a theoretical standpoint. For example, in the case of cyclic coordinates, the equations of motion are extremely simple. Consider Example 8.8 where we studied the mass constrained to move along the surface of a cylinder. Recall that θ was a cyclic coordinate in Example 8.8. The equations of motion relevant to θ were,

$$\dot{\theta} = \frac{p_\theta}{mR^2} \quad \text{and} \quad \dot{p}_\theta = 0.$$

Because p_θ is constant, it is easy to find $\theta(t)$,

$$\theta(t) = \int_{t_0}^{t}\frac{p_\theta}{mR^2}dt' = \frac{p_\theta}{mR^2}\left(t - t_0\right).$$

When the canonical momentum is constant, the equation of motion can be reduced to quadrature (as above). If a coordinate system can be found such that all coordinates are cyclic, then the equations of motion are trivial. In fact, it is possible to find such coordinate systems. Those transformations were developed by Carl Gustav Jacob Jacobi (1804–1851) and the work is referred to as the *Hamilton-Jacobi theory*. Hamilton-Jacobi theory is beyond the scope of this book, but the motivated reader should consult Goldstein et al. [8].

Furthermore, simply knowing the Hamiltonian alone can give us insight into all possible behaviors of the system. This can be done by using the Hamiltonian to plot the system's phase space. Unlike in Chapter 6 where we plotted x versus v, we will plot p versus q to create the phase space. In particular, we will plot the canonical momentum versus its conjugate generalized coordinate. In general, the dimension of the phase space is $2s$, where s is the system's number of degrees of freedom. Notice that $H = H(q_k, p_k)$ and will define surfaces in the phase space. To understand the value of such plots, we will revisit the simple pendulum, yet again.

Example 8.9: The phase space of the simple pendulum

Consider the simple plane pendulum. Using the Hamiltonian found in Example 8.7, plot the phase space of the simple plane pendulum using the values $m = 1.0$ kg, $\ell = 1.0$ m, and $g = 9.8$ m/s^2. Specifically, plot the contours corresponding to the initial conditions: $(q, p) = (0.5\text{ rad}, 0\text{ kg m}^2/\text{s})$, $(2.0\text{ rad}, 0\text{ kg m}^2/\text{s})$, $(\pi\text{ rad}, 0\text{ kg m}^2/\text{s})$, $(\pi\text{ rad}, \pm 4.43\text{ kg m}^2/\text{s})$ and $(\pi\text{ rad}, \pm 6.26\text{ kg m}^2/\text{s})$, and interpret the results.

Solution:
The Hamiltonian for the simple plane pendulum was found in Example 8.7,

$$H(q, p) = \frac{p^2}{2m\ell^2} - mg\ell \cos q \qquad (8.8.1)$$

where we have defined $p = p_\theta$ and $q = \theta$ for simplicity of notation.
The initial conditions correspond to the following values of H:

Initial Condition (q_0, p_0)	$H(q_0, p_0)$
$(0.5\text{ rad}, 0\text{ kg m}^2/\text{s})$	-8.60
$(2.0\text{ rad}, 0\text{ kg m}^2/\text{s})$	4.07
$(\pi\text{ rad}, 0\text{ kg m}^2/\text{s})$	9.8
$(\pi\text{ rad}, \pm 4.43\text{ kg m}^2/\text{s})$	19.61
$(\pi\text{ rad}, \pm 6.26\text{ kg m}^2/\text{s})$	29.39

Note the negative value of the first initial condition is due to the choice of y-coordinate to describe the pendulum's location. The codes to create a contour plot are shown below. The result of the Python code is shown in Figure 8.10.

Notice that we plotted q only for its possible range of $-\pi \leq q \leq \pi$. Some of the contours (contours with $H \leq 9.8$) form closed loops. Closed contours correspond to oscillatory motion, which is what we would expect for $q \leq \pi$ and an initial angular velocity of zero. The contour for $H = -8.60$, is nearly elliptical and if we were to further decrease the initial value of q, we would get a contour corresponding to a simple harmonic motion. The contour $H = 4.07$ corresponds to a larger amplitude of oscillation where the pendulum is no longer oscillating sinusoidally. The dashed line (red in the e-book) corresponds to the largest amplitude oscillation where $q = \pi$ and $p = 0$ initially. Contours with $H > 9.8$ correspond to even higher values of energy where the pendulum is now swinging in one direction continuously. Note that there are two contours for each value of $H > 9.8$, one for clockwise and the other for counterclockwise rotations.

Python Code
In Python we use the command `contour` from the Matplotlib library. Unlike in Mathematica, we must define a grid using `np.meshgrid` for which we will calculate the values of H. The value of each contour plotted is defined using the `levels` option in the `contour` command.

```python
import numpy as np
import matplotlib.pyplot as plt

m, l, g = 1, 1, 9.8

q_range = np.linspace(-np.pi,np.pi,100)
p_range = np.linspace(-10,10,100)

q, p = np.meshgrid(q_range,p_range)

H = p**2/(m*l**2) - m*g*l*np.cos(q)

contour_plot = plt.contour(q,p,H, \
        levels=[-8.6,4.07,9.8,19.61,29.39],colors=['k','k','r','k','k'],\
        linestyles=['solid','solid', 'dashed','solid','solid'])

plt.clabel(contour_plot)
plt.xlabel('theta')
plt.ylabel('p')
#plt.savefig('ex-8-9-python.eps')
plt.show()
```

Mathematica Code

We use the command `ContourPlot` to create the phase portrait for specific values of H. The option `Contours` specifies the value of the contours plotted.

```
SetOptions[ContourPlot, BaseStyle->{FontSize->16}];

l = 1.0;

m = 1.0;

g = 9.8;

H[q_, p_]:=p²/2ml² − mglCos[q];
contour1 = H[0.5, 0];

contour2 = H[2.0, 0];

contour3 = H[π, 0];

contour4 = H[π, 4.43];

contour5 = H[π, 6.26];

contourList = {contour1, contour2, contour3, contour4, contour5};

ContourPlot[H[q, p], {q, −π, π}, {p, −10, 10}, ContourShading → None,

Contours → contourList, ContourStyle → {Bold, Black, {Red, Dashed}, Black, Black},

FrameLabel → {q, p}, ContourLabels->True]
```

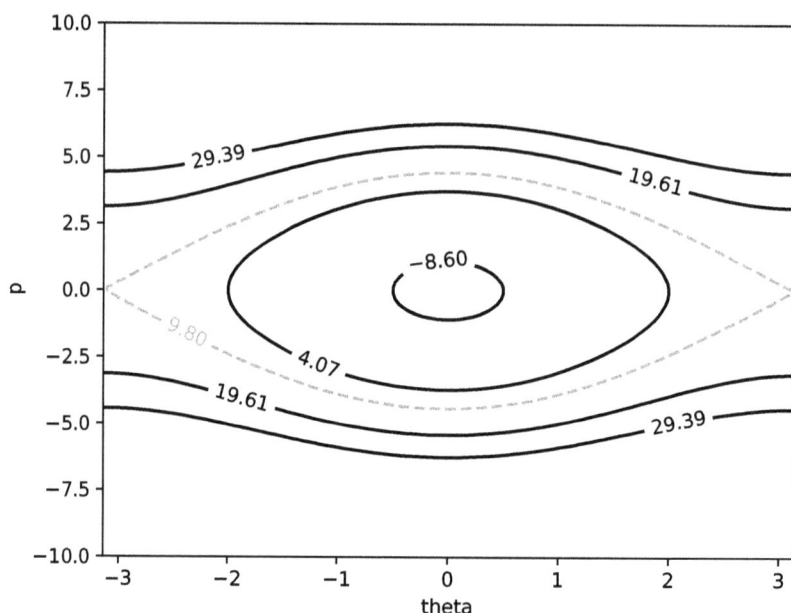

Figure 8.10: Contour plot of the simple pendulum's Hamiltonian produced by Python in Example 8.9.

The Hamilton formulation is a powerful one, especially for theoretical mechanics. Extensions of the Hamilton formulation like the Hamilton-Jacobi theory laid the groundwork for the theoretical structure of quantum mechanics. Beyond Goldstein et al. [8], the very motivated student will find Arnold [9] to be both a highly challenging and highly rewarding presentation of theoretical mechanics that goes well beyond the undergraduate-level, tying classical mechanics to advanced mathematics such as Lie algebra and symplectic geometry.

8.9 CHAPTER SUMMARY

The Lagrange formulation uses Hamilton's principle, which says that the path a particle will follow makes the action integral

$$S = \int_{t_1}^{t_2} \mathcal{L}\left(q_i \dot{q}_i; t\right) dt$$

stationary. The functions $q_j(t)$ that make the action integral stationary can be found using the Euler-Lagrange equation,

$$\frac{d}{dt}\left(\frac{\partial \mathcal{L}}{\partial \dot{q}_i}\right) - \frac{\partial \mathcal{L}}{\partial q_i} = 0$$

The Euler-Lagrange equation produces s second-order ODEs, where s is the number of degrees of freedom in the system, which can be solved for $q_i(t)$.

In the case where there are m holonomic constraints on the system, we can express the constraints as m constraint equations f_j and the Euler-Lagrange equation becomes

$$\frac{\partial \mathcal{L}}{\partial q_i} - \frac{d}{dt}\frac{\partial \mathcal{L}}{\partial \dot{q}_i} = -\sum_{j=1}^{m} \lambda_j(t)\frac{\partial f_j}{\partial q_i}$$

where λ_j are the undetermined Lagrange multipliers.

In addition to using the Lagrangian, the Hamiltonian can also be used to derive a system's equations of motion. The Hamiltonian is defined as,

$$H = \sum_{i=1}^{s} p_i \dot{q}_i - \mathcal{L}.$$

Hamilton's equations of motion are then

$$\dot{q}_i = \frac{\partial H}{\partial p_i} \quad \text{and} \quad \dot{p}_i = -\frac{\partial H}{\partial q_i}.$$

8.10 END-OF-CHAPTER PROBLEMS

The symbol ⌨ indicates a problem which requires some computer assistance, in the form of graphics, or numerical computation, or symbolic evaluation.

Section 8.1: Introduction to the Lagrangian

1. ⌨ Consider a free particle of mass m moving in three dimensions. The particle is free of any external forces. Find the Lagrangian and the equations of motion for the particle. Identify any conserved quantities. Solve this problem analytically and by using Mathematica or Python.

2. ⌨ Consider a particle of mass m moving in three dimensions near the surface of the Earth. Find the Lagrangian and the equations of motion for the particle. Identify any conserved quantities. Solve this problem analytically and by using Mathematica or Python.

3. ⌨ Consider a particle of mass $m = 1.0$ kg moving in one dimension under the influence of the force $F = -kx^3$. Find the Lagrangian and the equations of motion. Use a computer to solve for $x(t)$ in the case where $k = 0.1$ N/m^3, $x(0) = 1.0$ m, and $\dot{x}(0) = 0$ m/s.

Section 8.2: Generalized Coordinates and Degrees of Freedom

4. What is the number of degrees of freedom for a particle undergoing projectile motion without air resistance? Are there any constraints on the motion? Assume that the projectile travels a short enough distance that the Coriolis force (see Chapter 10) is negligible.

5. Consider a disk of radius R rolling without sliding down an inclined plane. What coordinates are needed to describe the state of the system? Are there constraints among those quantities?

6. A very small ball of mass m and radius a is constrained to roll along the surface of a sphere of $R \gg a$. The ball is constrained to move along geodesics on the sphere. How many degrees of freedom does the ball have?

Section 8.3: Hamilton's Principle

7. Consider a single-particle system of s degrees of freedom. The particle experiences a force which has an associated potential energy $V(q_i)$. Show that the functions $q_i(t)$ for $i = 1, \ldots, s$ that make the action integral $S = \int_{t_1}^{t_2} \mathcal{L}(q_i, \dot{q}_i)\, dt$ stationary also obey Newton's laws. In other words, extend the work done in Section 8.3 to a single-particle system with s-degrees of freedom.

8. Repeat Problem 7 for an N-particle system.

Sections 8.4 and 8.5: Examples of Lagrangian Dynamics

9. Fill in the necessary steps to calculate (8.4.19) and (8.4.20) in Example 8.3:

$$\mathcal{L} = \frac{1}{2}m\left(R^2\omega^2 + \ell^2\dot{\theta}^2 + 2R\ell\omega\dot{\theta}\sin(\theta - \omega t)\right) - mg\left(R\sin(\omega t) - \ell\cos\theta\right) \quad (8.10.1)$$

$$\ddot{\theta} + \frac{g}{\ell}\sin\theta = \frac{R\omega^2}{\ell}\cos(\theta - \omega t) \quad (8.10.2)$$

10. In Example 8.5, find the acceleration of the inclined plane and the particle in the limit of $M \to 0$. Discuss your results.

11. 🖥 A simple plane pendulum made of a mass m and massless rod of length ℓ is attached to a support that accelerates upward with an acceleration a. Write down the pendulum's x and y coordinates as a function of time t and parameter θ (defined as the angular displacement from the pendulum's stable equilibrium). Find the Lagrangian and the pendulum's equation of motion, both analytically and using a CAS. Find also the period of small oscillations around the equilibrium position.

12. 🖥 Consider the pendulum described in Problem 11. Plot the path of the pendulum bob as a function of time in the xy-plane using $\ell = 1.0$ m and $a = g/10$, $g/2$, g, and $2g$. Use $\theta(0) = 0.25$ rad and $\dot{\theta}(0) = 0$ as initial conditions.

13. 🖥 Find the equation of motion for a simple plane pendulum made of a mass m and a massless rod of length ℓ which is attached to a support that oscillates horizontally, such that $x_s = a\cos(\omega t)$, where x_s is the location of the support on the horizontal axis. The support has no vertical motion. Write down the pendulum's x and y coordinates as a function of time t and parameter θ (defined as the angular displacement from the pendulum's stable equilibrium, as in Figure 8.1). Find the Lagrangian and the pendulum's Euler-Lagrange equation of motion, both analytically and using a CAS.

14. 🖥 Consider the driven pendulum described in Problem 13. Plot the path of the pendulum bob in the xy-plane using $a = 0.25$ m, $\omega_0^2 = g/\ell = 1$ rad/s and $\omega = \omega_0/10$, $\omega = \omega_0/2$, ω_0, and $2\omega_0$. Use $\theta(0) = 0.25$ rad and $\dot{\theta}(0) = 0$ as initial conditions.

15. 🖥 A simple plane pendulum made of a mass m and massless rod of length ℓ is attached to a support that oscillates vertically, such that $y_s = a\cos(\omega t)$, where y_s is the location of the support on the vertical axis. The support has no horizontal motion. Write down the pendulum's x and y coordinates as a function of time and parameter θ (defined as the angular displacement from the pendulum's stable equilibrium, as in Figure 8.1). Find the Lagrangian and the pendulum's Euler-Lagrange equation of motion, both analytically and using a CAS.

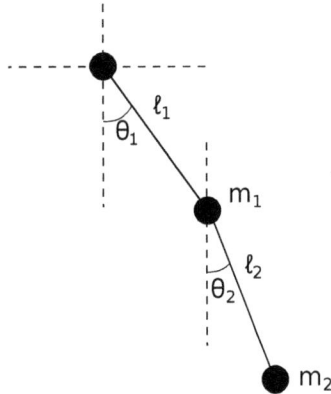

Figure 8.11: The double pendulum in Problem 8.17.

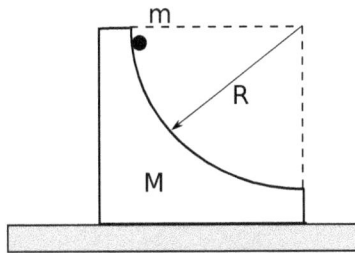

Figure 8.12: Circular wedge and mass in Problem 8.19.

16. 🖥 Consider the driven pendulum described in Problem 15. Plot the path of the pendulum bob in the xy-plane using $a = 0.25$ m, $\omega_0^2 = g/\ell = 1$ rad/s and $\omega = \omega_0/10$, $\omega_0/2$, ω_0, and $2\omega_0$. Use $\theta(0) = 0.25$ rad and $\dot{\theta}(0) = 0$ as initial conditions.

17. A double pendulum consists of two pendulums attached in series by massless rods as shown in Figure 8.11. Find the equation of motion for the double pendulum. Note that the motion of the pendulum is constrained to the plane like a simple plane pendulum.

18. 🖥 Consider a simple pendulum of mass m and initial length ℓ. After the pendulum begins to swing, its length is shortened at a rate $\dot{\ell} = -\alpha$, where α is a constant. Use the Lagrangian to compute the equations of motion of the system. Is the energy conserved in this system? Solve this problem analytically and by using Mathematica or Python.

19. A small particle with mass m slides down a smooth circular wedge with mass M and radius of curvature R as shown in Figure 8.12. The wedge is free to move horizontally along a frictionless surface. Find the equation of motion for each object.

20. Find the equations of motion for a single particle in three dimensions experiencing a central force $\boldsymbol{F}(\boldsymbol{r}) = F(r)\hat{\mathbf{r}}$, where r is the magnitude of the position vector, and $F(r)$ is a scalar function. Use spherical coordinates.

21. A particle of mass m is constrained to move along a frictionless massless circular hoop of radius a. The hoop rotates about its vertical diameter with a constant angular speed ω. Find the equilibrium positions of the particle.

22. 🖥 Consider a mass attached to a massless spring with a spring constant k and equilibrium length ℓ that is allowed to swing on a frictionless plane. This is the so-called *elastic pendulum*. Find the equation of motion for an elastic pendulum of mass m and equilibrium length ℓ. Initially, the pendulum is stretched 0.25 meters from its equilibrium length and displaced an angle of 0.5 radians from its stable equilibrium. The pendulum is then released from rest. Plot the path of the pendulum bob in the xy-plane using $\ell = 1$, $m = 1$, $k = 1$ (in SI units) for $t = 0$ to 4.4 seconds.

23. A smooth horizontal circular wire of radius R rotates with constant angular speed ω about a point on its perimeter. A particle of mass m is constrained to move along the wire as shown in Figure 8.13. Find the equation of motion for the particle. About what point does the particle oscillate?

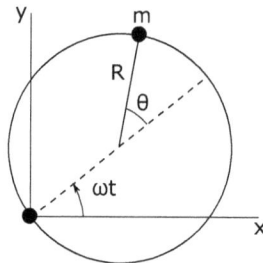

Figure 8.13: A bead restricted to move on a rotating circular wire in Problem 8.23.

24. A rope of mass M and length L is stretched out along a frictionless table such that a length z_0 is hanging over the edge of the table. Find the equation of motion of the rope if it is released from rest. Use the Lagrangian to show that the rope requires a time

$$t = \sqrt{\frac{L}{g}}\cosh^{-1}(L/z_0)$$

to completely slide off the table. You may remember this problem from Chapter 4, where we solved it using Newton's laws.

25. A particle of mass m moves in a plane under the force $F = -\alpha r^2$ where F points towards the origin, and r is the magnitude of the position vector. Find the equations of motion for the particle. Are there any conserved quantities? Is so, what are they?

26. A sphere of radius a and mass m is constrained to roll without slipping on the inside of a cylinder with a radius R. Find the equations of motion of the sphere and calculate the period of small oscillations.

27. A particle of mass m is constrained to move on a frictionless wire which is bent into the shape of a helix with $r = R$ and $z = \alpha\theta$, where R and α are positive constants. The particle is under the influence of gravity, which points in the $-z$ direction. Find the equation of motion for the particle. What is the particle's acceleration when $R \to 0$?

Section 8.6: Conservation Theorems and the Lagrangian

28. Prove (8.6.21):

$$p_i = \frac{\partial \mathcal{L}}{\partial \dot{q}_i} = \frac{\partial T}{\partial \dot{q}_i} = \sum_j a_{ij} \dot{q}_j$$

29. A particle of mass m moves within a force field that has an associated potential energy $V = -k/r$ where k is a positive constant. What are the conserved quantities for this particle?

30. Consider a system with s degrees of freedom and Lagrangian $\mathcal{L}(q_k, \dot{q}_k)$, where $k = 1, \ldots, s$. Now, let $\mathcal{F} = \mathcal{F}(q_k)$ be any function of the generalized coordinates, q_k. We can construct an additional Lagrangian,

$$\mathcal{L}' = \mathcal{L} + d\mathcal{F}/dt$$

Show that $\mathcal{L}$ and $\mathcal{L}'$ give the same equations of motion.

Section 8.7: Hamiltonian Dynamics

31. Derive Hamilton's equations of motion for a system with s degrees of freedom.

32. ⌨ A spherical pendulum is a simple pendulum that is not restricted to move in a plane. Find the equations of motion for a spherical pendulum that has a fixed support and consists of a massless rod of length ℓ attached to a mass m at the end of the rod. Using $m = 1$ and $\ell = 1$ (in SI units) plot the path of the spherical pendulum when the pendulum bob has an initial position of $\mathbf{r}_0 = \hat{\mathbf{i}}$ (where $\hat{\mathbf{i}}$ is the unit vector along the x-axis), and an initial angular velocity $\boldsymbol{\omega}_0 = \hat{\boldsymbol{\theta}} + \hat{\boldsymbol{\phi}}$ (where $\hat{\boldsymbol{\theta}}, \hat{\boldsymbol{\phi}}$ are the unit vector in spherical coordinates). All quantities are in SI units.

33. Find Hamilton's equations for a particle of mass m moving in the double-well potential, $V = -\frac{1}{2}kx^2 + \frac{1}{4}\epsilon x^4$, where k and ϵ are positive constants.

34. A particle of mass m moves within a force field that has an associated potential energy $V = -k/r$, where k is a positive constant. Find Hamilton's equations of motion.

35. ⌨ Find Hamilton's equations for the elastic pendulum described in Problem 22. Plot the path of the elastic pendulum using the same parameter values and initial conditions as in Problem 22. However, this time, plot the path of the pendulum from t = 0 to t = 10 seconds.

36. Find Hamilton's equations for a particle of mass m confined to the surface of a cone, which in cylindrical coordinates is described by $r = az$, inside a uniform gravitational field. Show that for a given energy, there are maximum and minimum values of z for which the particle is confined.

37. Consider the force

$$F = -\frac{k}{x}e^{-\lambda t} \tag{8.10.3}$$

acting on a particle of mass m moving in one dimension. Assume that the constants k and λ are both positive. We can create a "pseudo-potential energy," using the relationship, $V = -\int F dx$. Using the pseudo-potential energy, find Hamilton's equations of motion for the particle. Compute $\partial H/\partial t$ and $\partial \mathcal{L}/\partial t$ and show that they satisfy (8.7.11):

$$\frac{\partial H}{\partial t} = -\frac{\partial \mathcal{L}}{\partial t}$$

38. Recall the cycloid from the brachistochrone problem in Chapter 7,

$$y = a\left(1 - \cos\theta\right)$$
$$x = a\left(\theta - \sin\theta\right)$$

where a is a constant and θ is the angle of rotation. Find Hamilton's equations of motion for a particle moving along a cycloid.

Looking for more Hamiltonian problems? Try finding the Hamiltonian for any of the problems from Sections 8.4 and 8.5 above!

Section 8.8: Additional Explorations into the Hamiltonian

39. *Poisson brackets* are an interesting method of representing the time evolution of a quantity using the Hamiltonian. Consider two functions $g(q_k, p_k)$ and $h(q_k, p_k)$. The functions g and h are functions of the generalized coordinates and generalized momenta, similar to the Hamiltonian. The Poisson bracket is defined as:

$$[g, h] = \sum_k \left(\frac{\partial g}{\partial q_k} \frac{\partial h}{\partial p_k} - \frac{\partial g}{\partial p_k} \frac{\partial h}{\partial q_k} \right)$$

Using Poisson brackets, show:

  a. $\frac{dg}{dt} = [g, H] + \frac{\partial g}{\partial t}$, where H is the Hamiltonian.
  b. $\dot{q}_k = [q_k, H]$.
  c. $\dot{p}_k = [p_k, H]$.
  d. $[p_i, p_j] = [q_i, q_j] = 0$.
  e. $[q_i, p_j] = \delta_{ij}$, where δ_{ij} is the Kronecker delta (equal to 1 when $i = j$, and 0 otherwise).

40. If the Poisson bracket (see Problem 39) of two quantities is zero, the two quantities are said to *commute*. Show that any time-independent quantity that commutes with the Hamiltonian is conserved.

41. ⌨ Both in Chapter 5 and in Problem 33 of this chapter, we studied the double-well potential, $V = -\frac{1}{2}kx^2 + \frac{1}{4}\epsilon x^4$.

  a. Find the Hamiltonian for this potential.

  b. Using the Hamiltonian, plot the phase space for $\epsilon = 1$ and various values of k. Use $m = 1$ for the particle's mass.

  c. How does the phase space change with k? For each value of k that you try, describe the possible motions of the particle.

  d. Draw the phase space plots and identify the types of motion associated with each region of the phase space.

9 Central Forces and Planetary Motion

In this chapter we will first look at the concept of central forces and their importance in describing physical phenomena. Next we examine the problem of two objects interacting via a central force, this is the so called two-body problem. Our analysis of the two-body problem will lead to Kepler's Laws of Planetary Motion. We will obtain the solution of the two-body problem both analytically and by numerically integrating Newton's law. Finally, we study the three-body problem where three objects interact via central forces. Solving for the motion of one of the objects in the three-body problem will require us to use the computational skills developed so far in this book.

9.1 CENTRAL FORCES

In this section, we will outline some general properties of central forces.

A central force is a force on a particle that is directed along a line that joins the particle and the force center (such as the origin). When two particles interact via a central force $\mathbf{F}$, the force is always directed along the line that joins the centers of the particles, as shown in Figure 9.1. As an example, $\mathbf{F}$ may be Newton's Law of Universal Gravitation (9.1.2), in which case $\mathbf{F}_{21}$ is the force of gravitational attraction experience by m_1 due to the presence of m_2. Parallel to the force vectors in Figure 9.1 is a dashed line of length r, the distance between the centers of m_1 and m_2, and the unit vector $\hat{\mathbf{r}}$ which lies along the line joining the two masses. Note that the vector $\hat{\mathbf{r}}$ in Figure 9.1 should lie on top of the vector $\mathbf{F}_{21}$, but we placed it beside $\mathbf{F}_{21}$ for clarity in the image.

Central forces depend only on the distance r between the two interacting bodies. Therefore, spherical coordinates provide a simple way of describing central forces, especially because they do not depend on θ and ϕ. Hence, we can represent a central force generally in the form:

General Formula for Central Forces

$$\mathbf{F} = f(r)\hat{\mathbf{r}} \tag{9.1.1}$$

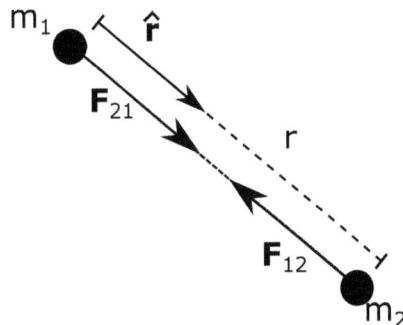

Figure 9.1: Two masses m_1 and m_2 interacting via a central force $\mathbf{F}$.

where $f(r)$ is a scalar function of the magnitude r of the position vector $\mathbf{r}$, and $\hat{\mathbf{r}}$ is a unit vector along the position vector as shown in Figure 9.1.

The best known examples of central forces in physics are Newton's Law of Universal Gravitation $\mathbf{F}_G(r)$ between two masses m_1 and m_2, and the electrostatic Coulomb force $\mathbf{F}_C(r)$ between two electric changes q_1 and q_2. The mathematical description of these two forces is very similar:

$$\mathbf{F}_G(r) = -\frac{Gm_1m_2}{r^2}\hat{\mathbf{r}} \tag{9.1.2}$$

$$\mathbf{F}_C(r) = \frac{1}{4\pi\epsilon_0}\frac{q_1q_2}{r^2}\hat{\mathbf{r}} \tag{9.1.3}$$

where $G = 6.67 \times 10^{-11}$ N m^{2}kg^2 is the universal gravitational constant and $\epsilon_0 = 8.85 \times 10^{-12}$ F/m is the corresponding constant for electrical forces, called the permittivity of free space.

9.1.1 CENTRAL FORCES AND THE CONSERVATION OF ENERGY

A very important property of central forces is that they are conservative. This can be easily seen by writing the general form of the curl of a force $\mathbf{F}$ in spherical coordinates:

$$\nabla \times \mathbf{F} = \frac{1}{r\sin\theta}\left(\frac{\partial}{\partial\theta}(F_\phi\sin\theta) - \frac{\partial F_\theta}{\partial\phi}\right)\hat{\mathbf{r}} + \frac{1}{r}\left(\frac{1}{\sin\theta}\frac{\partial F_r}{\partial\phi} - \frac{\partial}{\partial r}(rF_\phi)\right)\hat{\boldsymbol{\theta}}$$
$$+ \frac{1}{r}\left(\frac{\partial(rF_\theta)}{\partial r} - \frac{\partial F_r}{\partial\theta}\right)\hat{\boldsymbol{\phi}} \tag{9.1.4}$$

In the case of central forces we have $F_\theta = F_\phi = 0$, $\frac{\partial F_r}{\partial\phi} = 0$, $\frac{\partial F_r}{\partial\theta} = 0$ so that (9.1.4) gives:

$$\nabla \times \mathbf{F} = 0 \tag{9.1.5}$$

and therefore, $\mathbf{F}$ is a conservative force.

As we saw in Chapter 5, conservative forces can be associated with potential energy functions $V(r) = -\int \mathbf{F} \cdot d\mathbf{r}$. The corresponding scalar gravitational potential energy V_G and electrostatic potential energy V_C are:

$$V_G(r) = -\frac{Gm_1m_2}{r} \tag{9.1.6}$$

$$V_C(r) = \frac{1}{4\pi\epsilon_0}\frac{q_1q_2}{r} \tag{9.1.7}$$

Another well-known example of a central force is the intermolecular long-range Van der Waals force, which is of importance in several branches of science. In one of its simplest forms, the Van der Waals force between a pair of neutral atoms or molecules can be approximated by the Lennard-Jones potential (also termed the L-J potential or 6-12 potential). The mathematical form of the 6-12 potential $V_{LJ}(r)$ is:

$$V_{LJ}(r) = \left[\frac{c_1}{r^{12}} - \frac{c_2}{r^6}\right] \tag{9.1.8}$$

where c_1 and c_2 are constants with the appropriate SI units.

In the field of particle and atomic physics, the Yukawa short range nuclear force is of great importance, and the corresponding Yukawa potential $V_Y(r)$ is represented mathematically by:

$$V_Y(r) = c_4\frac{e^{-c_5r}}{r} \tag{9.1.9}$$

where c_4, and c_5 are constants with the appropriate SI units.

9.1.2 CENTRAL FORCES AND THE CONSERVATION OF ANGULAR MOMEN-TUM

Next we will show that central forces conserve angular momentum. Consider a system of N particles, where m_i, $\mathbf{r}_i$, and $\mathbf{p}_i$ are the mass, position, and momentum of particle i with $i = 1, \ldots, N$. We know from Chapter 4 that the total torque $\mathbf{N}$ acting on a system of particles is

$$\mathbf{N} = \sum_{i=1}^{N} \mathbf{r}_i \times \dot{\mathbf{p}}_i = \sum_{i=1}^{N} \mathbf{r}_i \times \mathbf{F}_i \qquad (9.1.10)$$

where $\mathbf{F}_i$ is the force acting on the particle with mass m_i. In Chapter 4, we further simplified this equation to show that

$$\mathbf{N} = \sum_{i=1}^{N} \left(\mathbf{r}_i \times \mathbf{F}_i^{\text{ext}} \right) + \sum_{i=1}^{N} \sum_{j>i} (\mathbf{r}_i - \mathbf{r}_j) \times \mathbf{F}_{ij} \qquad (9.1.11)$$

where $\mathbf{F}^{\text{ext}}$ is the sum of the external forces acting on m_i, and $\mathbf{F}_{ij}$ is the force of interaction between m_i and m_j. If we assume an isolated system, then $\mathbf{F}^{\text{ext}} = 0$, and the first term in (9.1.11) is equal to zero. Using a coordinate system similar to that in Figure 9.1, we can write $\mathbf{r}_i - \mathbf{r}_j = r_{ij} \hat{\mathbf{r}}_{ij}$, where $\hat{\mathbf{r}}_{ij}$ is a unit vector that lies along a line joining m_i and m_j. Furthermore, if the force of interaction between the particles is a central force, then $\mathbf{F}_{ij} = F_{ij}(r) \hat{\mathbf{r}}_{ij}$ and the second term in (9.1.11) is: $(r_{ij} \hat{\mathbf{r}}_{ij}) \times (F_{ij} \hat{\mathbf{r}}_{ij}) = 0$. Therefore, the net torque acting on the system is zero, and angular momentum is conserved.

In the next section, we will examine the so-called two-body problem in detail. This discussion will lead us to Kepler's famous laws of planetary motion.

9.2 THE TWO-BODY PROBLEM

In this section, we will consider the two-body Keplerian problem, where two masses are interacting via a central force. Such two-body systems are very common in astronomy.

We begin with the system shown in Figure 9.2 where the masses m_1 and m_2 are at locations $\mathbf{r}_1$ and $\mathbf{r}_2$ with respect to a shared origin, O. The center of mass of the two particles is located at $\mathbf{R}$ relative to the origin as shown in Figure 9.2, and it is denoted by a white circle. The two masses interact with a central force $\mathbf{F} = F(r) \hat{\mathbf{r}}$, not shown in the figure, with $\mathbf{F}$ lying along the vector $\mathbf{r}$ similar to Figure 9.1.

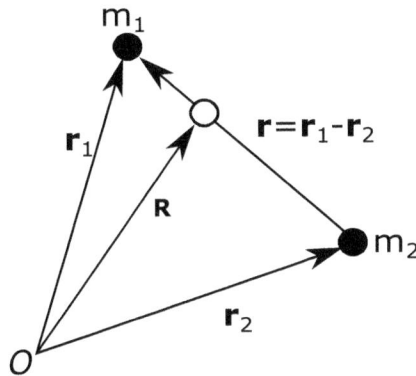

Figure 9.2: Position vectors $\mathbf{r}_1$ and $\mathbf{r}_2$ and relative distance $\mathbf{r} = \mathbf{r}_1 - \mathbf{r}_2$ of the two masses interacting with a central force $F(r) \hat{\mathbf{r}}$. The white circle at the location $\mathbf{R}$ shows the location of the system's center of mass, and O is the origin of the coordinate system.

If $\mathbf{r}_1$ and $\mathbf{r}_2$ are the position vectors of the two masses m_1 and m_2 respectively, then the vector distance between the two masses is

$$\mathbf{r} = \mathbf{r}_1 - \mathbf{r}_2 \qquad (9.2.1)$$

Newton's second law $F = ma$ for each mass can be written as:

$$m_1 \ddot{\mathbf{r}}_1 = -F(r)\hat{\mathbf{r}} \qquad (9.2.2)$$

$$m_2 \ddot{\mathbf{r}}_2 = F(r)\hat{\mathbf{r}} \qquad (9.2.3)$$

where $\hat{\mathbf{r}}$ is the unit vector in the direction of $\mathbf{r}$. Note that we are using Newton's third law to write the above equations for an attractive central force. Subtracting (9.2.2) and (9.2.3) we obtain:

$$\ddot{\mathbf{r}}_1 - \ddot{\mathbf{r}}_2 = -\left(\frac{1}{m_1} + \frac{1}{m_2}\right)F(r)\hat{\mathbf{r}} \qquad (9.2.4)$$

$$\frac{m_1 m_2}{m_1 + m_2}(\ddot{\mathbf{r}}_1 - \ddot{\mathbf{r}}_2) = -F(r)\hat{\mathbf{r}} \qquad (9.2.5)$$

Which can be rewritten as,

Newton's Second Law for the Reduced Mass μ

$$\mu\ddot{\mathbf{r}} = -F(r)\hat{\mathbf{r}} \qquad (9.2.6)$$

$$\mu = \frac{m_1 m_2}{m_1 + m_2} \qquad (9.2.7)$$

This equation tells us that the two-body system behaves just like a single mass μ with a position vector $\mathbf{r}$, moving under the influence of the central force $\mathbf{F} = F(r)\hat{\mathbf{r}}$. This single mass μ is called the *reduced mass* of the two-body system, and it represents an effective inertial mass appearing in the two-body problem. Using the reduced mass μ allows us to study the two-body problem as if it were a one-body problem. Note that the minus sign appears in (9.2.6) because we assumed an attractive force. A repulsive force would give similar results but without the minus sign in (9.2.6).

What is the reduced mass? The definition in (9.2.7) can be difficult to interpret. Consider the Earth-Sun system. In this case the mass of the Sun m_S is much greater than the mass of the Earth, m_E. If we use $m_1 = m_S$ and $m_2 = m_E$ in (9.2.7), we find that $\mu \approx m_E$. Hence, when we study the motion of the Earth-Sun system, we are essentially studying the motion of the Earth. As the masses become more comparable, we cannot associate the reduced mass with either one of the masses, as we will demonstrate later in Example 9.1.

When working with a system of multiple particles, we know that the system's center of mass can be useful in describing the behavior of the system. As we discussed in Chapter 4, the position of the center of mass of the two-body system is defined by:

$$\mathbf{R} = \frac{m_1 \mathbf{r}_1 + m_2 \mathbf{r}_2}{m_1 + m_2} \qquad (9.2.8)$$

Because the translational motion of the system as a whole is not of interest (we are focused on the particles' orbits with respect to each other), we can set $\mathbf{R} = 0$, as shown in Figure 9.3a. In other words, we can place the center of mass at the origin. This new coordinate system is different from that shown in Figure 9.2. As we will see, this new coordinate system with the center of mass at the origin, will be very useful in describing the orbits.

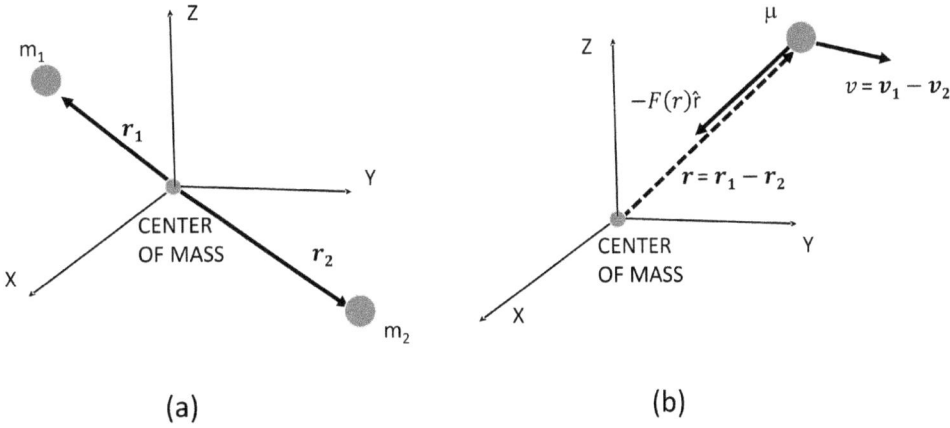

(a)	(b)

Figure 9.3: (a) The two-body system can be best described using a system of coordinates axes with the origin located at the center of mass of the system. (b) The motion can also be described by Newton's law (9.2.6), as the motion of the reduced mass μ, with position vector $\mathbf{r} = \mathbf{r}_1 - \mathbf{r}_2$, moving with the relative velocity of the two masses $\mathbf{v} = \dot{\mathbf{r}}_1 - \dot{\mathbf{r}}_2$.

Using $\mathbf{R} = 0$, we have

$$m_1\mathbf{r}_1 + m_2\mathbf{r}_2 = 0 \qquad (9.2.9)$$

By combining (9.2.9) with the definition $\mathbf{r} = \mathbf{r}_1 - \mathbf{r}_2$ for the position vectors $\mathbf{r}_1$ and $\mathbf{r}_2$, we obtain:

$$\mathbf{r}_1 = \left(\frac{m_2}{m_1 + m_2}\right)\mathbf{r} \qquad (9.2.10)$$

$$\mathbf{r}_2 = -\left(\frac{m_1}{m_1 + m_2}\right)\mathbf{r} \qquad (9.2.11)$$

Note that in the center of mass coordinate system of Figure 9.3a, the three vectors $\mathbf{r}, \mathbf{r}_1, \mathbf{r}_2$ are collinear. Equations (9.2.10) and (9.2.11) allow us to relate the particles' positions to the separation vector $\mathbf{r}$. In addition, we can write the Lagrangian for the system as:

$$\mathcal{L} = \frac{1}{2}m_1\dot{\mathbf{r}}_1^2 + \frac{1}{2}m_2\dot{\mathbf{r}}_2^2 - V(r) \qquad (9.2.12)$$

where $V(r) = -\int F(r)dr$ is the potential energy associated with the central force $\mathbf{F}$. Substituting the values of (9.2.10) and (9.2.11) in the Lagrangian (9.2.12), we find after some simple algebra that the Lagrangian of the two-body system is:

Lagrangian of the Two-Body System

$$\mathcal{L} = \frac{1}{2}\mu\,|\,\dot{\mathbf{r}}\,|^2 - V(r) \qquad (9.2.13)$$

Similar to Newton's second law, the Lagrangian shows that we are reducing the two-body problem to the motion of a single object with a mass equal to the reduced mass μ, moving with the relative velocity of the two masses $\mathbf{v} = \dot{\mathbf{r}}_1 - \dot{\mathbf{r}}_2$.

Similarly, the total angular momentum of the two-body problem is found from:

$$\mathbf{L} = m_1 \left(\mathbf{r}_1 \times \frac{d\mathbf{r}_1}{dt} \right) + m_2 \left(\mathbf{r}_2 \times \frac{d\mathbf{r}_2}{dt} \right) \tag{9.2.14}$$

Substituting (9.2.10) and (9.2.11) into (9.2.14) we can obtain the total angular momentum (see Problem 10 in the End-of-Chapter Problems):

Angular Momentum of the Two-Body System

$$\boldsymbol{\ell} = \mu \left(\mathbf{r} \times \dot{\mathbf{r}} \right) \tag{9.2.15}$$

where we have replaced $\mathbf{L}$ (for two bodies) with $\boldsymbol{\ell}$ (for one body). This equation tells us that the total angular momentum of the two-body problem is equal to the angular momentum of the reduced mass μ, located at the point $\mathbf{r} = \mathbf{r}_1 - \mathbf{r}_2$ and moving with velocity $\mathbf{v} = \dot{\mathbf{r}} = \dot{\mathbf{r}}_1 - \dot{\mathbf{r}}_2$. Again, the result is that we can reduce the two-body problem to studying the motion of a single particle of mass μ.

Furthermore, taking the time derivative of (9.2.15) results in $\dot{\boldsymbol{\ell}} = 0$; hence just as before, the angular momentum is conserved. One of the physical consequences of the conservation of the angular momentum vector $\boldsymbol{\ell}$ is that the motion of the mass μ has to remain on the same plane (which also contains the force center), in order to keep the direction of $\boldsymbol{\ell}$ constant at all times. This is shown for the special case of a circular orbit in Figure 9.4. Because the motion of the particle is restricted to a plane, polar coordinates are a useful coordinate system when dealing with central force problems. The constant nature of $\boldsymbol{\ell}$ helps explain why all of the orbits of the planets in our solar system are essentially coplanar and continue to stay that way.

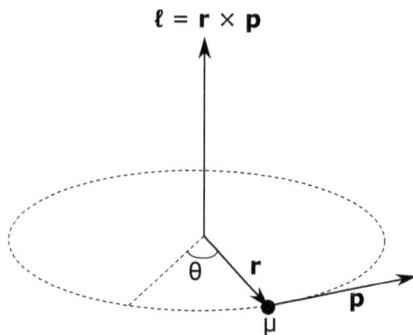

Figure 9.4: Relationship of the angular momentum vector $\boldsymbol{\ell}$, position vector r and linear momentum vector p $= \mu\mathbf{v}$.

Since the motion is restricted in to a plane, we can compute the magnitude of the angular momentum in polar coordinates. Recall that in polar coordinates $\dot{\mathbf{r}} = \dot{r}\hat{\mathbf{r}} + r\dot{\theta}\hat{\boldsymbol{\theta}}$, and substituting in (9.2.15) we obtain:

We can also compute the total energy of the two-body system using $E = T + V$,

$$E = \frac{1}{2}\mu\left(\dot{r}^2 + r^2\dot{\theta}^2\right) + V(r) \tag{9.2.17}$$

$$= \frac{1}{2}\mu\dot{r}^2 + \frac{1}{2}\mu r^2\left(\frac{\ell}{\mu r^2}\right)^2 + V(r) \tag{9.2.18}$$

This can be simplified to

Again, the physical interpretation of this equation, and the others before it, is that the two-body system can be replaced by a single reduced mass μ at a distance $r = |\mathbf{r}_1 - \mathbf{r}_2|$ from the origin (i.e., the force center and the system's center of mass), and moving with the velocity $\mathbf{v} = \dot{\mathbf{r}} = \dot{\mathbf{r}}_1 - \dot{\mathbf{r}}_2$. This is shown schematically in Figure 9.3b.

While the reduced mass is a useful tool for performing the calculations above, what we really want is to be able to calculate the motions of m_1 and m_2. In the next section, we will derive the equations of motion for the two-body problem.

9.3   EQUATIONS OF MOTION FOR THE TWO-BODY PROBLEM

When finding equations of motion, we often want to find the position of the particle as a function of time. In principle, this can be done by solving (9.2.19) for $\dot{r}$:

$$\dot{r} = \frac{dr}{dt} = \pm\sqrt{\frac{2}{\mu}\left(E - V(r)\right) - \frac{\ell^2}{\mu^2 r^2}} \tag{9.3.1}$$

which, in principle, can be integrated to get $t(r)$ and then inverted to get $r(t)$, once we have specified $V(r)$. However, $r(\theta)$ would be a more useful equation of motion because it would give us a mathematical formula for the observed motion. We can substitute,

$$\frac{dr}{dt} = \frac{dr}{d\theta}\frac{d\theta}{dt} = \frac{dr}{d\theta}\frac{\ell^2}{2\mu r^2} \tag{9.3.2}$$

into (9.3.1) and integrate to obtain

$$\theta(r) = \pm\int \frac{\ell dr}{r^2\sqrt{2\mu\left(E - V(r) - \frac{\ell^2}{r^2}\right)}} \tag{9.3.3}$$

In principle, after specifying $V(r)$, we could perform the integral to get a mathematical form for $\theta(r)$, and then invert it to get $r(\theta)$. However, in general, the integral in (9.3.3) is difficult to do and the inversion is also not easy.

Before we present the analytical solution, we demonstrate the *numerical* solution for the two-body problem. Computers can provide a means of obtaining numerical solutions, in which case it is often easier working directly with Newton's second law. In the next example, we show how numerical solutions for an object's orbit in the two-body system can be obtained by numerically solving Newton's second law.

Example 9.1: The orbits in a two-body gravitational system

Integrate Newton's second law using Newton's Law of Universal Gravitation to represent for interaction force between two masses m_1 and m_2, in order to obtain the orbits of a binary star system with masses $m_1 = 1$, $m_2 = 2$ (in arbitrary units). In order to simplify the presentation of the numerical results, use arbitrary units with $G = 1$. At time $t = 0$ the reduced mass is located at $x = 1$, $y = 0$ and has a velocity with an x-component of $vx = 0$ and a y-component of $vy = 1$.

Solution:

As discussed in this section, Newton's second law for the two masses becomes a single equation for the reduced mass:

$$\mu \ddot{\mathbf{r}} = -\frac{Gm_1m_2}{r^2}\hat{\mathbf{r}}$$

where we have used Newton's Law of Universal Gravitation as the central force. Now we need to rewrite the above formula in terms of Cartesian coordinates. The unit vector is given in Cartesian coordinates (x, y) by,

$$\hat{\mathbf{r}} = \frac{\mathbf{r}}{r} = \frac{x\hat{\mathbf{i}} + y\hat{\mathbf{j}}}{\sqrt{x^2 + y^2}}$$

By combining the previous two equations and using $r^2 = x^2 + y^2$,

$$\mu \frac{d^2\mathbf{r}}{dt^2} = -\frac{Gm_1m_2}{x^2 + y^2}\frac{x\hat{\mathbf{i}} + y\hat{\mathbf{j}}}{\sqrt{x^2 + y^2}} = -\frac{Gm_1m_2}{(x^2 + y^2)^{3/2}}\left(x\hat{\mathbf{i}} + y\hat{\mathbf{j}}\right)$$

Using $\mu = m_1m_2/(m_1 + m_2)$, we obtain the following two coupled differential equations:

$$\frac{d^2x}{dt^2} = -\frac{G(m_1 + m_2)}{(x^2 + y^2)^{3/2}}x \qquad \frac{d^2y}{dt^2} = -\frac{G(m_1 + m_2)}{(x^2 + y^2)^{3/2}}y \qquad (9.3.4)$$

Once we know the solution $\mathbf{r}(t)$ for the motion of the reduced mass μ, then we can also plot the positions of the two masses $\mathbf{r}_1$ and $\mathbf{r}_2$ by using their relationship to $\mathbf{r}(t)$:

$$\mathbf{r}_1 = \left(\frac{m_2}{m_1 + m_2}\right)\mathbf{r} \qquad (9.3.5)$$

$$\mathbf{r}_2 = -\left(\frac{m_1}{m_1 + m_2}\right)\mathbf{r} \qquad (9.3.6)$$

The positions can be calculated from $\mathbf{r}$ by using (9.2.10) and (9.2.11), by multiplying $\mathbf{r}$ with the factors $m_1/(m_1 + m_2)$ and $-m_2/(m_1 + m_2)$ to get $\mathbf{r}_1$ and $\mathbf{r}_2$, respectively.

Python Code

In this example, the first component `Y[:,0]` of the vector `Y` represents the variable $x(t)$, and the second component `Y[:,1]` represents the corresponding velocity $\dot{x}(t) = dx_1/dt$. Similarly, the third component `Y[:,2]` of `Y` represents the variable $y(t)$, and the fourth component `Y[:,3]` represents the corresponding velocity dy/dt. In Figure 9.5a we see that the reduced mass μ follows an elliptical orbit.

We also plotted the positions of the two masses in Figure 9.5b. The positions can be calculated by multiplying $\mathbf{r}$ with the factors $m_1/(m_1 + m_2)$ and $-m_2/(m_1 + m_2)$ to get $\mathbf{r}_1$ and $\mathbf{r}_2$, respectively. Notice that each mass follows an elliptical path around the center of mass, which is located at one of the foci of each ellipse.

Since the vectors $\mathbf{r}_1$, $\mathbf{r}_2$ and $\mathbf{r}$ are collinear, the line connecting the locations of the two masses at any moment passes through the origin. We also draw the line connecting the two masses m_1 and m_2 at time t=1 s.

```python
from scipy.integrate import odeint
import numpy as np
import matplotlib.pyplot as plt


# function defines the differential equations
# Y[0]=x, Y[1]=x', Y[2]=y, Y[3]=y'
def deriv(Y,t):
    return np.array([ Y[1], -(m1+m2)*\
    Y[0]/((Y[0]**2+Y[2]**2)**(3/2)),
                        Y[3], -(m1+m2)*\
    Y[2]/((Y[0]**2+Y[2]**2)**(3/2))])

t = np.linspace(0,6,600)   # t= array of times to be evaluated
m1, m2 = 1, 2              # masses

# yinit = initial conditions [x(0), x'(0), y(0), y'(0)]
yinit = np.array([1, 0, 0, 1])
Y = odeint(deriv, yinit, t)   # solve the odes using odeint

plt.subplot(2,1,1) # plot motion of reduced mass
plt.plot(Y[:,0],Y[:,2],label="Motion of reduced mass")
plt.xlim(-.25,1.1);
plt.ylim(-.5,.5);
plt.plot(np.linspace(-.25,1.1,100),[0]*100)  # draw x-axis
plt.plot([0]*100,np.linspace(-.5,.5,100))  # draw y-axis
plt.title("(a) Motion of reduced mass")

plt.subplot(2,1,2) # plot orbits of m1, m2
plt.plot(-(m2/(m1+m2))*Y[:,0],-(m2/(m1+m2))*Y[:,2])
plt.plot((m1/(m1+m2))*Y[:,0],(m1/(m1+m2))*Y[:,2],'r--')
plt.title("(b) Orbits of m1, m2")
plt.xlim(-.7,.4);
plt.ylim(-.33,.33);
plt.plot(np.linspace(-.7,.4,100),[0]*100)  # draw x-axis
plt.plot([0]*100,np.linspace(-.33,.33,100))  # draw y-axis
```

```
# draw line connecting m1 and m2 at time t=1 s
x1, y1 = np.array([[-(m2/(m1+m2))*Y[:,0][100],\
(m1/(m1+m2))*Y[:,0][100]],\
[-(m2/(m1+m2))*Y[:,2][100],(m1/(m1+m2))*Y[:,2][100]]])
for i in range(0, len(x1), 2):
    plt.plot(x1[i:i+2], y1[i:i+2], 'ro-')
plt.tight_layout()
plt.show()
```

Mathematica Code

The command `NDSolve` is used to solve the system of differential equations (9.3.4), and the `Table` command is used to construct a list of the pairs of position coordinates $\mathbf{r} = (x, y)$ from $t = 0$ to $t = 10$, in time steps of $dt = 0.01$. The coordinate pairs are stored in the variable `xyList`. Finally a graph of the orbits of the two masses is produced by using the `ListPlot` command.

The positions $\mathbf{r}_1$ and $\mathbf{r}_2$ can be calculated from $\mathbf{r}$ easily by multiplying the list `xyList` with the factors $m_1/(m_1 + m_2)$ and $-m_2/(m_1 + m_2)$ to get $\mathbf{r}_1$ and $\mathbf{r}_2$, respectively.

m1 = 1; m2 = 2; x0 = 1; y0 = 0; vx0 = 0; vy0 = 1;

sol = NDSolve[{x"$[t]$ == $-(m1 + m2) * x[t]/((x[t]^2 + y[t]^2)^{(3/2)})$,

y"$[t]$ == $-(m1 + m2) * y[t]/((x[t]^2 + y[t]^2)^{(3/2)})$, $x[0]$ == x0, $y[0]$ == y0,

$x'[0]$ == vx0, $y'[0]$ == vy0}, $\{x, y\}$, $\{t, 0, 10\}$];

x1 = x/.sol[[1]];

y1 = y/.sol[[1]];

xyList = Table[$\{x1[t], y1[t]\}$, $\{t, 0, 10, .01\}$];

gr1 = Show[ListPlot[xyList], Graphics[Disk[$\{0, 0\}$, .02]], BaseStyle->FontSize->15,

PlotLabel->"Orbit of reduced mass"];

gr2 = ListPlot[$\{-m2/(m1 + m2) * $ xyList, $m1/(m1 + m2) *$ xyList$\}$,

BaseStyle->FontSize->15, PlotLabel->"Orbits of m1,m2"];

GraphicsGrid[$\{\{$gr1, gr2$\}\}$]

(a) Motion of reduced mass

(b) Orbits of m1, m2

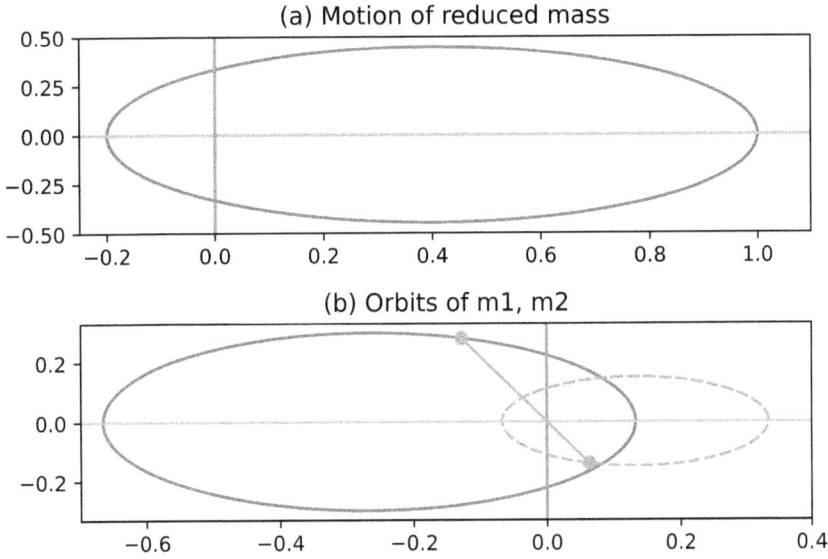

Figure 9.5: The two-body system in Example 9.1 can be described in two equivalent ways. In (a) the system is described as the motion of the reduced mass μ in an ellipse with the center of mass at the origin. In (b) the system is described by the motion of the two masses in ellipses around the center of mass which is located at the origin, and it represents one of the foci of the ellipse.

Although numerical solutions are very useful, it is very important to obtain an analytical solution, when possible. Next, we derive a general equation which provides another means of computing the *orbit* of the reduced mass for the two-body problem. This equation can also be used to find the mathematical form of the central force acting on the reduced mass, if the mathematical form of the *orbit* is known to us. In order to obtain the central force, we need to use Newton's second law in polar coordinates. Here r and θ are the usual position and angle polar coordinates. The acceleration in polar coordinates is given by:

$$\frac{d\boldsymbol{v}}{dt} = \left[\ddot{r} - r\dot{\theta}^2\right]\hat{\mathbf{r}} + \left[r\ddot{\theta} + 2\dot{r}\dot{\theta}\right]\hat{\boldsymbol{\theta}} \tag{9.3.7}$$

Newton's second law for the reduced mass μ becomes:

$$\mu\left[\ddot{r} - r\dot{\theta}^2\right]\hat{\mathbf{r}} + \mu\left[r\ddot{\theta} + 2\dot{r}\dot{\theta}\right]\hat{\boldsymbol{\theta}} = -f(r)\hat{\mathbf{r}} \tag{9.3.8}$$

where we used the reduced mass of the two-body system.

By setting the *radial* components on both sides of this equation equal to each other and the *angular* components equal to each other, we obtain:

$$\mu\left[\ddot{r} - r\dot{\theta}^2\right] = f(r) \tag{9.3.9}$$

$$r\ddot{\theta} + 2\dot{r}\dot{\theta} = 0 \tag{9.3.10}$$

Equation (9.3.10) can also be written as:

$$\mu r(r\ddot{\theta} + 2\dot{r}\dot{\theta}) = \frac{d}{dt}\left[\mu r^2 \dot{\theta}\right] = \frac{d\ell}{dt} = 0 \tag{9.3.11}$$

which is another way of showing that the magnitude of the angular momentum ℓ is constant. The radial equation (9.3.9) can be solved by making the substitution:

$$r = \frac{1}{u} \tag{9.3.12}$$

Calculating the time derivatives dr/dt and d^2r/dt^2:

$$\dot{r} = -\frac{1}{u^2}\dot{u} = -\frac{1}{u^2}\dot{\theta}\frac{du}{d\theta} = -\frac{\ell}{\mu}\frac{du}{d\theta} \tag{9.3.13}$$

$$\ddot{r} = -\frac{\ell}{\mu}\frac{d}{dt}\frac{du}{d\theta} = -\frac{\ell}{\mu}\frac{d\theta}{dt}\frac{d}{d\theta}\frac{du}{d\theta} = -\frac{\ell}{\mu}\dot{\theta}\frac{d^2u}{d\theta^2} = -\frac{\ell^2 u^2}{\mu^2}\frac{d^2u}{d\theta^2} \tag{9.3.14}$$

where we used $\dot{\theta} = \ell u^2/\mu$ and the chain rule. By substituting these derivatives dr/dt and d^2r/dt^2 into the radial equation (9.3.9), we obtain:

$$\mu\left[-\frac{\ell^2 u^2}{\mu^2}\frac{d^2u}{d\theta^2} - \frac{1}{u}\left(\frac{\ell^2 u^4}{\mu^2}\right)\right] = f\left(u^{-1}\right) \tag{9.3.15}$$

or

$$\frac{d^2u}{d\theta^2} + u = -\frac{\mu}{\ell^2 u^2}f\left(u^{-1}\right) \tag{9.3.16}$$

Finally this equation can be written in terms of r by substituting $u = 1/r$:

Two-Body Equation of Motion in Terms of $r(\theta)$ and $f(r)$

$$\frac{d^2}{d\theta^2}\left(\frac{1}{r}\right) + \frac{1}{r} = -\frac{\mu r^2}{\ell^2}f(r) \tag{9.3.17}$$

This equation is useful when we know the orbit in the form $r(\theta)$, and we want to evaluate the central force $f(r)$ which creates this orbit. In the following two examples, we use (9.3.17) to evaluate the central force $f(r)$ and the total energy E for a given orbit $r(\theta)$ for a two-body system.

Example 9.2: Finding the central force when we know the orbit

The orbit $r(\theta)$ of a mass m moving inside a central force field, is given by the expression:

$$r(\theta) = k\theta$$

(a) Find the central force $f(r)$ creating this orbit, both analytically and using a CAS.
(b) Find the orbit $\theta(t)$ with the initial condition $\theta(0) = 0$.

Solution:

(a) We use (9.3.17) to evaluate the force:

$$f(r) = -\frac{\ell^2}{\mu r^2}\left[\frac{d^2}{d\theta^2}\left(\frac{1}{r}\right) + \frac{1}{r}\right]$$

We now have

$$\frac{d^2}{d\theta^2}\left(\frac{1}{r}\right) = \frac{d^2}{d\theta^2}\left(\frac{1}{k\theta}\right) = \frac{2}{k\theta^3} = \frac{2k^2}{r^3}$$

and the $f(r)$ equation above gives:

$$f(r) = -\frac{\ell^2}{\mu r^2}\left[\frac{2k^2}{r^3} + \frac{1}{r}\right] = -\frac{\ell^2}{\mu}\left[\frac{2k^2}{r^5} + \frac{1}{r^3}\right]$$

So the central force in this system is a linear combination of the two terms $1/r^5$ and $1/r^{3.}$.

(b) We can find the orbit $\theta(t)$ by using the conservation of angular momentum:

$$\mu r^2 \dot{\theta} = \mu r^2 \frac{d\theta}{dt} = \ell$$

and substituting $r = k\theta$:

$$\mu k^2 \theta^2 \frac{d\theta}{dt} = \frac{\mu k^2}{3}\frac{d\left(\theta^3\right)}{dt} = \ell$$

This can be integrated to yield:

$$\theta^3(t) = \frac{3t\ell}{\mu k^2} + \theta^3(0)$$

and using the initial condition $\theta(0) = 0$ yields the orbit $\theta(t)$, or equivalently the orbit $r(t) = k\,\theta(t)$:

$$\theta(t) = \sqrt[3]{\frac{3t\ell}{\mu k^2}} \qquad r(t) = k\sqrt[3]{\frac{3t\ell}{\mu k^2}}$$

Python Code

We use **diff** to obtain the second derivative of $u = 1/r$, and **dsolve** in SymPy to solve the differential equation symbolically for $\theta(t)$. The output of the code agrees with the analytical solution obtained above.

```python
from sympy import  symbols,dsolve, Function, diff

print('-'*28,'CODE OUTPUT','-'*29,'\n')
r, k, theta, L, mu, t =symbols('r,k,theta,L,mu,t',real=True)

# Part (a) : find force F(r)
rr = k*theta

force = -L**2/(mu*rr**2)*(diff(1/rr,theta,theta)+1/rr)

print('The force F(r) = ',force.expand().subs({theta:r/k}))

# Part (b) : we use symbol theta2 instead of theta
# in the ODE, since theta was defined as a symbol in part (a)

theta2 = Function('theta2')

ic = {theta2(0):0}

soln = dsolve(mu*k**2*theta2(t)**2*diff(theta2(t),t)-L,\
            theta2(t),ics=ic)[0]

print('\nThe solution theta(t) = ',soln.rhs)

-------------------------- CODE OUTPUT ----------------------------
The force F(r) =   -2*L**2*k**2/(mu*r**5) - L**2/(mu*r**3)

The solution theta(t) =   3**(1/3)*(L*t/(k**2*mu))**(1/3)
```

Mathematica Code

We use **D** to obtain the second derivative of $u = 1/r$, and **DSolve** to solve the differential equation symbolically for $\theta(t)$. Note that we have to select the positive solution of the ODE.

```
rr = k ∗ θ;

force = −(l^2/(μ ∗ rr^2) ∗ (D[1/rr, {θ, 2}] + 1/rr));

Print["The force is: F(r) = ", force/.{θ->r/k}//FullSimplify];
```

The force is: $F(r) = -\dfrac{l^2\left(2k^2+r^2\right)}{r^5\mu}$

```
soln = DSolve[{θ'[t]==l/(μ ∗ k^2 ∗ θ[t]^2), θ[0]==0}, θ[t], t][[2]][[1]];


Print["The positive solution of the ODE is: "];

Print[" θ[t] = ", θ[t]/.soln];

Print["or"];

Print["θ[t]^3 = ", (θ[t]/.soln)^3]
```

The positive solution of the ODE is:
$\theta[t] = \dfrac{3^{1/3}l^{1/3}t^{1/3}}{k^{2/3}\mu^{1/3}}$
or
$\theta[t]\verb|^|3 = \dfrac{3lt}{k^2\mu}$

Example 9.3: Finding the total energy when we know the orbit

Find the total energy E for the orbit $r(\theta)$ of the mass μ in Example 9..2.

Solution:

We find the potential energy $V(r) = -\int f(r)\,dr$ by using the expression for the force $f(r)$ from the previous example:

$$V(r) = -\int f(r)\,dr = -\int \left[\frac{\ell^2}{\mu}\left[-\frac{2k^2}{r^5}+\frac{1}{r^3}\right]\right]dr = -\frac{\ell^2}{2\mu}\left[\frac{k^2}{r^4}+\frac{1}{r^2}\right]$$

The kinetic energy T is given by:

$$T = \frac{\ell^2}{2\mu r^2} + \frac{1}{2}\mu\left(\frac{dr}{dt}\right)^2$$

We can evaluate dr/dt using the chain rule, the orbit equation $r = k\theta$ and the conservation of angular momentum $\ell = \mu r^2\dot{\theta}$:

$$\frac{dr}{dt} = \frac{dr}{d\theta}\frac{d\theta}{dt} = \frac{d(k\theta)}{d\theta}\frac{d\theta}{dt} = k\dot{\theta} = k\frac{\ell}{\mu r^2}$$

Finally the total energy will be:

$$E = T + V = \frac{\ell^2}{2\mu r^2} + \frac{1}{2}\mu\left(\frac{dr}{dt}\right)^2 + V(r)$$

$$E = \frac{\ell^2}{2\mu r^2} + \frac{1}{2}\mu\left(k\frac{\ell}{\mu r^2}\right)^2 - \frac{\ell^2}{2\mu}\left[\frac{k^2}{r^4}+\frac{1}{r^2}\right] = 0$$

As discussed later in this chapter, the value of $E = 0$ corresponds to a parabolic motion of the reduced mass μ in the two-body system.

9.4 PLANETARY MOTION AND KEPLER'S FIRST LAW

In this section, we find the equation of motion for the reduced mass μ experiencing an inverse square law force:

$$\mathbf{F}(r) = -\frac{k}{r^2}\hat{\mathbf{r}} \tag{9.4.1}$$

such as Newton's Law of Universal Gravitation where $k = Gm_1m_2$, or the Coulomb force where $k = -q_1q_2/(4\pi\epsilon_0)$. In this case, we can insert $f(r) = -k/r^2$ into (9.3.17) to get

$$u'' + u = \frac{\mu k}{\ell^2} \tag{9.4.2}$$

where we used $u = 1/r$ and primes denote differentiation with respect to θ. We will solve (9.4.2) for $u(\theta)$ and then find $r(\theta)$. We begin with the substitution, $w(\theta) = u(\theta) - \mu k/\ell^2$, then (9.4.2) becomes:

$$w'' = -w \tag{9.4.3}$$

Notice that the result is a simple harmonic oscillator equation for w with a natural frequency $\omega_0 = 1$. The solution is

$$w = A\cos\theta + B\sin\theta \tag{9.4.4}$$

With a careful choice of coordinates, we can set $B = 0$. Let's choose $\theta = 0$ as the angular position of μ when it makes its closest approach $r_{\min}$ to the force center (i.e., to the center of mass). At this point, $\dot{r} = 0$ (because this is a turning point in the orbit) and due to (9.3.13), $u' = 0$. Hence, we are setting an initial condition for u to be $u'(\theta = 0) = 0$. Because $w' = u'$, we also have $w'(\theta = 0) = 0$, and this condition leads to $B = 0$. Hence we have,

$$u = A\cos\theta + \frac{\mu k}{\ell^2} \tag{9.4.5}$$

We now introduce a new constant $e = \ell^2 A/(\mu k)$ and we set $\alpha = \ell^2/(\mu k)$, to obtain:

$$\frac{1}{r} = \frac{1}{\alpha}(1 + e\cos\theta) \tag{9.4.6}$$

$$r = \frac{\alpha}{1 + e\cos\theta} \tag{9.4.7}$$

which, as we showed in Chapter 4, is the equation of a conic section in polar coordinates with one focus at the origin. The dimensionless variable e is the *eccentricity* of the conic section, and the distance α is the *semi-latus rectum*. Later in this chapter we will discuss the equation for conic sections in polar coordinates in more detail.

When studying planetary motion, Johannes Kepler (1571-1630) developed three laws of planetary motion. Kepler's first law stated that planets orbit in an ellipse and that the Sun is at one of the foci of the ellipse. An ellipse, of course, is one example of a conic section. However, starting from Newton's second law, we have found a much more general result. We have found that the motion of the reduced mass is of the form of a conic section: circle, ellipse, parabola, or hyperbola. What conditions dictate which conic section the reduced mass follows? In the next section, we will begin to answer this question.

However, before we move on, there is one piece of (9.4.6) that is still unsatisfactory, the eccentricity e is in terms of the constant of integration A. It would be better if we could rewrite the eccentricity e in terms of physical parameters, similar to the equation $\alpha = \ell^2/(\mu k)$. We begin by using our choice of coordinates where $\dot{r} = 0$ at $r = r_{\min}$ and $r_{\min}$ occurs at $\theta = 0$, hence (9.4.6) gives

$$r_{\min} = \frac{\alpha}{1+e} \qquad (9.4.8)$$

and the total energy at $r_{\min}$ is

$$E = \frac{\ell^2}{2\mu r_{\min}^2} - \frac{k}{r_{\min}} \qquad (9.4.9)$$

Substituting (9.4.8) into (9.4.9), and solving the resulting quadratic equation for the eccentricity e, we obtain

$$e = \sqrt{1 + \frac{2E\ell^2}{\mu k^2}} \qquad (9.4.10)$$

In summary, the motion of the reduced mass in the two-body system is a conic section, with one focus at the origin. In terms of the eccentricity e and the semi-latus rectum α, the equation of the conic section $r(\theta)$ in polar coordinates is:

Two-Body Equation of Motion in Terms of e and α

$$r = \frac{\alpha}{1 + e\cos\theta} \qquad (9.4.11)$$

$$e = \sqrt{1 + \frac{2E\ell^2}{\mu k^2}} \qquad (9.4.12)$$

$$\alpha = \frac{\ell^2}{\mu k} \qquad (9.4.13)$$

Note that the conic section in (9.4.11) is completely described by the two parameters e, α. These two *geometrical* parameters can be evaluated using (9.4.12) and (9.4.13), when we know the two conserved *physical* quantities in the system, i.e. when we know the total energy E and the conserved angular momentum ℓ. Alternatively, the conic section is completely described by the physical quantities E, ℓ.

9.5   ORBITS IN A CENTRAL FORCE FIELD

In this section we look at the general problem of determining the orbits of a particle with a reduced mass μ moving in a central force field $\mathbf{F}(\mathbf{r}) = -k/r^2\hat{\mathbf{r}}$. We already know that the orbits are conic sections. However, we would like to know, for example, under what conditions is a particle's orbit an ellipse, as opposed to a parabola?

We start by considering the equation for the total energy:

$$E = T + V = \frac{1}{2}\mu\dot{r}^2 + \frac{l^2}{2\mu r^2} + V(r) \qquad (9.5.1)$$

where we have returned to the generic potential energy $V(r)$. Note that for an inverse square law, $V = -k/r$. The first term in (9.5.1) can be interpreted as a kinetic energy term, because it is of the form of one half times a mass and the square of the radial velocity. It is convenient to think of the second two terms as representing potential energy.

We notice that the term

$$V_C \equiv \frac{l^2}{2\mu r^2} = \frac{1}{2}\mu r^2 \dot{\theta}^2 \qquad (9.5.2)$$

can be interpreted as a form of energy called the *centrifugal potential energy* V_C, and the corresponding force will be the *centrifugal force* F_C:

$$F_C = -\frac{\partial V_C}{\partial r} = \frac{l^2}{\mu r^3} \qquad (9.5.3)$$

note that $\mathbf{F}_C = F_C \hat{\mathbf{r}}$ points radially outward from the center of mass (the force center for the two-body problem). We can then write the *effective potential energy* $V_{\text{eff}}(r)$ as the sum of two terms, the centrifugal potential V_C and the potential energy $V(r)$ corresponding to the central force.

The Effective Potential in a Central Force Field

$$V_{\text{eff}}(r) = \frac{l^2}{2\mu r^2} + V(r) = V_C + V(r) \qquad (9.5.4)$$

Figure 9.6 shows a plot of the effective potential $V_{\text{eff}}(r)$ as a function of the distance r, and for an attractive potential $V = -k_1/r$ where $k_1 = 2.5$ in SI-units. The dashed lines in this graph indicate a repulsive positive centrifugal potential $V_C = 1/r^2$, and the attractive potential $V = -k_1/r$. The thick solid line in this figure shows the sum of these two potentials, which represents the effective potential $V_{\text{eff}}(r)$. When the distance $r \to 0$ the effective potential goes to infinity, while as $r \to \infty$ the effective potential goes to zero. The horizontal lines in this figure indicate two possible energies of the reduced mass, $E_0 = -1.55$ J and $E_1 = 1.3$ J. Note that the value of these energies are arbitrarily chosen for illustrative purposes only.

We can use the methods of Chapter 5 in order to qualitatively describe the motion of the reduced mass μ as a function of its total energy E. Recall, that we can think of the reduced mass as a particle that is rolling along the track made by the shape of the V_{eff} graph. When $E = E_0$, the particle is fixed at the minimum of V_{eff}. This means that the value of r does not change. The result is that the reduced mass has a circular orbit with $r = r_0$. When $E = E_1 < 0$ (or more generally for $E_0 < E < 0$), the particle rolls back and forth between $r = r_1$ and $r = r_2$. We know from Chapter 5 that such a "rolling back and forth" means oscillatory motion. The reduced mass gets no closer to the force center than $r = r_1$, and no farther than $r = r_2$. Hence the resulting motion is an ellipse with $r_{\text{min}} = r_1$ and $r_{\text{max}} = r_2$. Note that we are using r_{min} and r_{max} as the closest and farthest distance between the reduced mass and the force center, respectively. In general, we can see from Figure 9.6 that when $E < 0$, the motion of the reduced mass is bound to periodic orbits.

Furthermore, from Figure 9.6 we can see that when $E \geq 0$, the motion is unbound. The reduced mass comes in from infinity and gets as close to the force center as the distance r_3, which is not shown in Figure 9.6. However, one can find r_3 by solving $E = V_{\text{eff}}(r_3)$ where E is the energy of the reduced mass. The solution can be found either analytically or numerically. The unbound orbits are parabolic or hyperbolic, depending on the value of E.

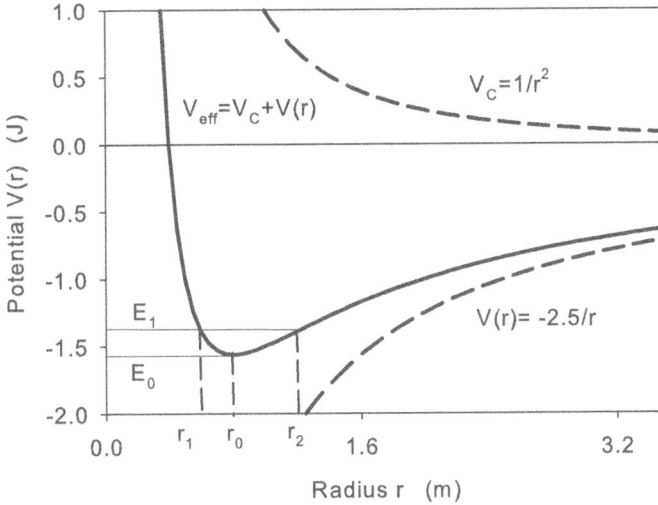

Figure 9.6: A plot of the effective potential $V_{\text{eff}}(r) = V_C + V(r)$ as a function of the distance r (thick solid line). The dashed lines indicate the repulsive positive centrifugal potential V_C and the attractive potential $V(r) = -2.5/r$. When $E = E_0$ the motion is a circle and $r = r_0$. When $E = E_1 < 0$, the motion is bound between the two circles with radii $r = r_1$ and $r = r_2$. When $E > 0$, the motion is unbound.

So far in this section, we have learned that the energy of the reduced mass determines the nature of its orbit. Now that we have a qualitative understanding of the orbit of μ, we can use our knowledge of conic sections in order to determine the conditions under which each orbit occurs. We know that conic sections have specific eccentricities, for example, $e = 0$ for a circle. Using the equation for the eccentricity (9.4.10) we can now determine the energies of each type of orbit. They are summarized in Table 9.1, and examples are shown in Figure 9.7.

Table 9.1

The energies and eccentricities associated with each type of orbit.

Eccentricity e	Energy E	Orbit shape
$e = 0$	$E = V_{\text{eff,min}}$	circle
$0 < e < 1$	$V_{\text{eff,min}} < E < 0$	ellipse
$e = 1$	$E = 0$	parabola
$e > 1$	$E > 0$	hyperbola

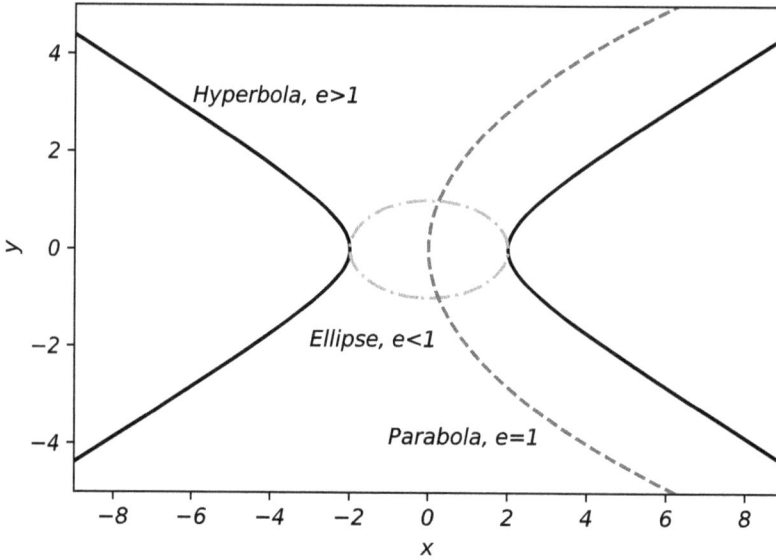

Figure 9.7: Possible conic section orbits of the reduced mass for different eccentricities e: ellipse ($e < 1$), parabola ($e = 1$) and hyperbola ($e > 1$). The orbits in this example are centered at the origin of the coordinate system.

9.6 KEPLER'S LAWS OF PLANETARY MOTION

To conclude our work on the two-body problem, we examine Kepler's Laws of Planetary Motion in detail.

9.6.1 KEPLER'S FIRST LAW

As mentioned previously, Kepler's first law states that planets orbit in an ellipse with the Sun at one focus. We have found that the general motion is that of a conic section; however, we will focus on ellipses in this section because of their importance in planetary and satellite motion.

Let us begin our discussion by recalling the path $r(\theta)$ we found in (9.4.11)-(9.4.13):

$$r = \frac{\alpha}{1 + e\cos\theta} \qquad \alpha = \frac{\ell^2}{\mu k} \qquad e = \sqrt{1 + \frac{2E\ell^2}{\mu k^2}} \tag{9.6.1}$$

and our choice of coordinates is such that $r = r_{\min}$ when $\theta = 0$, and $r = r_{\max}$ when $\theta = \pi$, as shown in Figure 9.8, where the force center is at the focus F. From (9.4.6) we can easily show,

$$r_{\min} = \frac{\alpha}{1 + e} \qquad \text{and} \qquad r_{\max} = \frac{\alpha}{1 - e} \tag{9.6.2}$$

In this section, we will assume that the reduced mass is a planet orbiting the Sun. Hence we will use the terms *perihelion* as the planet's distance of closest approach $r_{\min}$ to the Sun, and *aphelion* as the planet's greatest distance $r_{\max}$ from the Sun. We can define the *major axis* $2a$ of the ellipse as

$$r_{\min} + r_{\max} = 2a \qquad (9.6.3)$$

and the quantity a is called the *semi-major axis*.

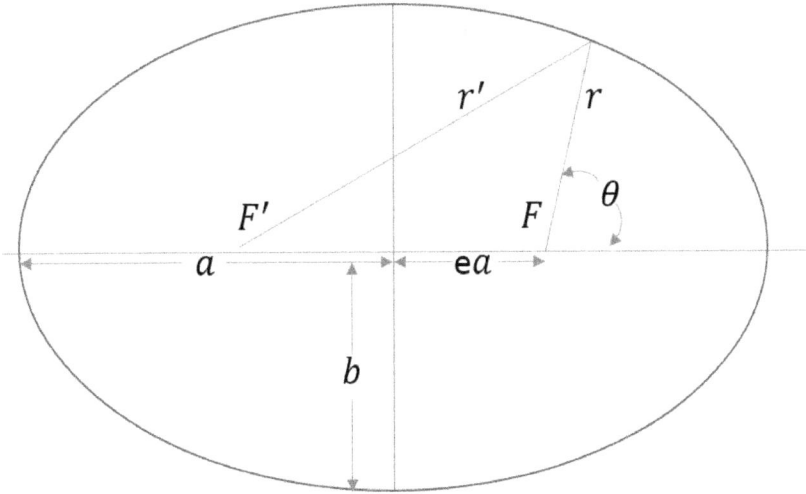

Figure 9.8: The motion of the planets around the Sun can be described by an ellipse which has the Sun at one focus, F. The second focus F' is an (empty) point in space. The ellipse can be described mathematically using (9.6.1) in polar coordinates (r, θ).

At $\theta = 90°$ and at $\theta = 270°$ the distance is equal to the semi-latus rectum distance $r = \alpha$ shown in Figure 9.9, and depends on both the semi-major axis a of the elliptical motion, and on the eccentricity e:

$$\alpha = a\left(1 - e^2\right) \qquad (9.6.4)$$

The *minor axis* $2b$ is perpendicular to the major axis $2a$, and both (a, b) are related to the energy of the orbit:

$$a = \frac{\alpha}{1 - e^2} = \frac{k}{2\,|E|} \qquad (9.6.5)$$

$$b = \frac{\alpha}{\sqrt{1 - e^2}} = \frac{\ell}{\sqrt{2\mu\,|E|}} \qquad (9.6.6)$$

Here again note the presence of the two conserved quantities E, ℓ. Furthermore, we can rewrite the perihelion and aphelion distances as:

$$r_{\min} = a(1 - e) \qquad \text{and} \qquad r_{\max} = a(1 + e)$$

The eccentricity e can also be found from $r_{\min}$ and $r_{\max}$:

$$e = \frac{r_{\max} - r_{\min}}{r_{\max} + r_{\min}} \qquad (9.6.7)$$

The area of an ellipse is $A = \pi ab$. In the special case of a circle, the eccentricity $e = 0$, resulting in $r_{\min} = r_{\max} = a = b$ and $A = \pi r^2$.

$$r_{MAX} = a(1+e) \qquad r_{MIN} = a(1-e)$$

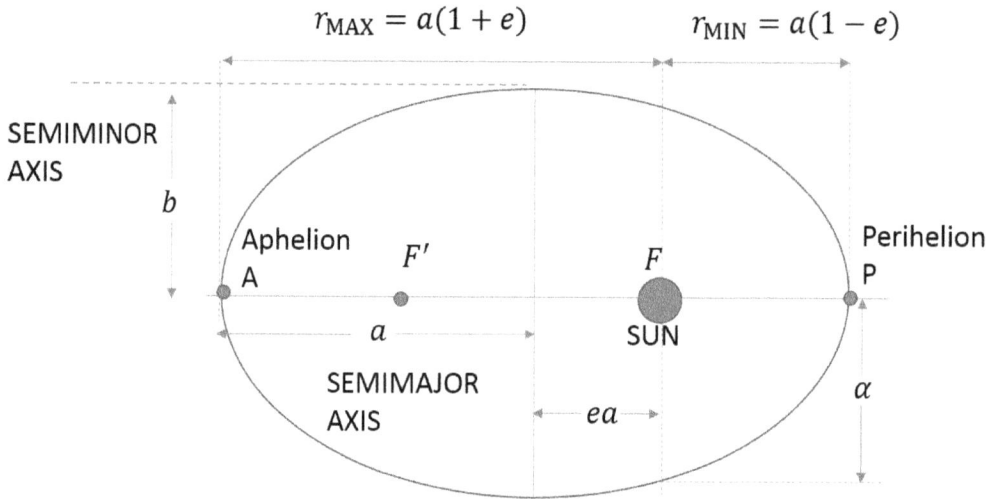

Figure 9.9: Kepler's first law places the Sun at the focus of an elliptical orbit.

The next two examples shows how to plot the orbits of planets. The first example uses (9.4.6) and data for the planets Mercury and Earth, while the second uses a numerical solution to Newton's second law.

Example 9.4: Plotting the orbits of Earth and Mercury around the sun

The Earth's distance from the Sun ranges from 147.5 million km (at perihelion), to about 152.6 million km (at aphelion), while for Mercury these distances are from 46,000,000 to 70,000,000 km. Plot the orbits of Mercury and Earth around the Sun.

Solution:

The semi-major axis a can be found directly from the distances r_{min} and r_{max} for the perihelion and aphelion by using (9.6.3), and the eccentricity e can be found by using (9.6.7). Once the geometrical properties a, e are known, we know everything about the orbits and we can plot them using the general equation (9.4.6) for the elliptical orbits.

Python Code

We evaluate α and e, and use

$$r = \frac{\alpha}{1 + e\cos\theta} \qquad (9.6.8)$$

with $x = r\cos\theta$, $y = r\sin\theta$ to create the plots.

```python
import numpy as np
import matplotlib.pyplot as plt
print('-'*28,'CODE OUTPUT','-'*29,'\n')


rminE = 147.5*10**9          # Earth data
rmaxE = 152.6*10**9

a = (rminE + rmaxE)/2
eEarth = (rmaxE - rminE)/(rminE + rmaxE)
print('The eccentricity of Earth is  e = ', round(eEarth,4))

alpha = a*(1 - eEarth**2)

theta = np.linspace(0,2*np.pi,100)
rE = alpha/(1 + eEarth*np.cos(theta))

plt.plot(rE*np.cos(theta),rE*np.sin(theta))

rminM = 46*10**9             # Mercury data
rmaxM = 70*10**9

aM = (rminM + rmaxM)/2
eMercury = (rmaxM - rminM)/(rminM + rmaxM)
print('The eccentricity of Mercury is  e = ', round(eMercury,4))

alphaM = aM*(1 - eMercury**2)
rM = alphaM/(1 + eMercury*np.cos(theta))

plt.plot(rM*np.cos(theta),rM*np.sin(theta))
plt.title('Orbits of Earth and Mecrury around the Sun')
plt.xlabel('x [m]')
plt.ylabel('y [m]')
plt.show()

--------------------------- CODE OUTPUT ---------------------------
The eccentricity of Earth is  e =  0.017
The eccentricity of Mercury is  e =  0.2069
```

Mathematica Code

We use the `PolarPlot` command to plot the two orbits, and we also print out the values of e for the two planets.

Notice the almost completely circular orbit of the Earth (because the eccentricity $e = 0.0167$ is close to zero), and the clearly elliptical orbit for Mercury (because of the larger value of $e = 0.21$).

```
rminE = 147.5 * 10^6 * 10^3;

rmaxE = 152.6 * 10^6 * 10^3;


a = (rminE + rmaxE)/2;

eEarth = (rmaxE − rminE)/(rminE + rmaxE)//N;

Print["The eccentricity of Earth is e = ", eEarth];


gr1 = PolarPlot[a * (1 − eEarth^2)/(1 + eEarth * Cos[θ]), {θ, 0, 2 * Pi},

Frame → True, FrameLabel → {"x, m", "y, m"},

PlotRange → {{−2 * 10^11, 2 * 10^11}, {−2 * 10^11, 2 * 10^11}}];


rminM = 46 * 10^6 * 10^3;

rmaxM = 70 * 10^6 * 10^3;


a = (rminM + rmaxM)/2;

eMercury = (rmaxM − rminM)/(rminM + rmaxM)//N;

Print["The eccentricity of Mercury is e = ", eMercury];


gr2 = PolarPlot[a * (1 − eMercury^2)/(1 + eMercury * Cos[θ]), {θ, 0, 2 * Pi},

Frame → True, PlotRange → {{−2 * 10^11, 2 * 10^11}, {−2 * 10^11, 2 * 10^11}}];


Show[{gr1, gr2}, BaseStyle → FontSize → 16]
```

The eccentricity of Earth is e = 0.0169943
The eccentricity of Mercury is e = 0.206897

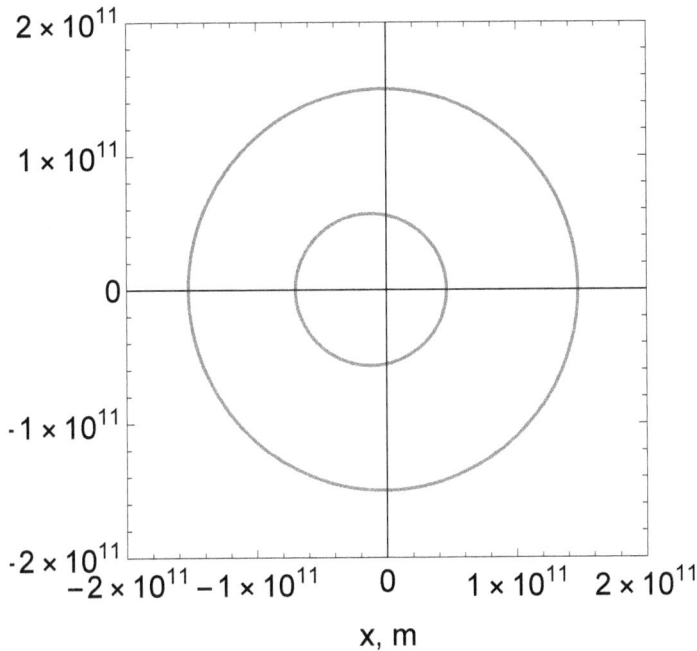

Figure 9.10: Plots of the orbits of Mercury and Earth around the Sun, from Example 9.4.

Example 9.5: The Moon's orbit around the earth

The revolution of the Moon around the Earth is a very complex process, however we can approximate the binary system Earth-Moon as a Keplerian system. Evaluate

(a) The period T of the Moon's revolution around the Earth.

(b) The eccentricity e of the Moon's orbit.

(c) Plot the Moon's orbit around the Earth.

Use the following numerical values: perigee distance $= 0.3633 \times 10^9$ m, apogee distance $= 0.4055 \times 10^9$ m, mass of the Moon $= 0.07346 \times 10^{24}$ kg, mass of the Earth $= 5.9724 \times 10^{24}$ kg, and gravitational constant $G = 6.673 \times 10^{-11}$ SI units.

Solution:

(a) We evaluate the eccentricity using

$$e = \frac{r_{\max} - r_{\min}}{r_{\max} + r_{\min}} = \frac{apogee + perigee}{apogee - perigee} = 0.0549 \qquad (9.6.9)$$

This is almost a circular orbit. The major semi-axis is evaluated from

$$a = (perigee + apogee)/2$$

(b) The period of revolution of the Moon is found from Kepler's third law:

$$P^2 = \frac{4\pi^2 a^3}{G(m_1 + m_2)} \qquad (9.6.10)$$

$$P = \sqrt{\frac{4\pi^2 a^3}{G(m_{moon} + m_{earth})}} = 27.29 \; days \qquad (9.6.11)$$

Python Code
After evaluating a and the eccentricity e, we plot the orbit using the polar coordinates expression $r = \alpha/\left(1 + e\cos\theta\right)$ and the coordinates $x = r\cos\theta$, $y = r\sin\theta$, where the angle θ ranges from 0 to 2π.

```python
import numpy as np
import matplotlib.pyplot as plt
print('-'*28,'CODE OUTPUT','-'*29,'\n')


# All quantities are in SI units
perigee = 0.3633*10**9;
apogee = 0.4055*10**9;
Mmoon = 0.07346*10**24;
Mearth = 5.9724*10**24;
G = 6.673*10**-11;


e = (apogee - perigee)/(apogee + perigee)
a = (perigee + apogee)/2;
Period = np.sqrt(4*np.pi**2*a**3/(G*(Mmoon + Mearth)));

print('\nThe semimajor axis a = ', a, 'm');
print("\nThe eccentricity of moon's orbit = ", round(e,4))
print('\nThe period of revolution T of moon = ', \
      round(Period/(3600*24),2), ' days')

theta = np.linspace(0,2*np.pi,100)
alpha = a*(1 - e**2)
r = alpha/(1 + e*np.cos(theta))

plt.plot(r*np.cos(theta),r*np.sin(theta))
plt.title("Orbit of Moon around the Earth")
plt.xlabel('x')
plt.ylabel('y')
plt.xlim(-4.2*10**8, 4.2*10**8);
plt.ylim(-4.2*10**8, 4.2*10**8);

# draw x-y-axes
plt.plot(np.linspace(-4.2*10**8,4.2*10**8,100),[0]*100)
plt.plot([0]*100,np.linspace(-4.2*10**8,4.2*10**8,100))
plt.show()

--------------------------------- CODE OUTPUT ---------------------------------

The semimajor axis a =  384400000.0 m

The eccentricity of moon's orbit =  0.0549

The period of revolution T of moon =  27.29  days
```

Mathematica Code

The code is very similar to the Python code, and we use the command `PolarPlot` to plot the Moon's orbit.

```
Print["All quantities are in SI units"];

perigee = 0.3633 * 10^9;

apogee = 0.4055 * 10^9;

e = (apogee − perigee)/(apogee + perigee);

a = (perigee + apogee)/2;


Print["semi-major axis a = ", a];

Print["Eccentricity of Moon's orbit = ", e];

Mmoon = 0.07346 * 10^24;

Mearth = 5.9724 * 10^24;

G = 6.673 * 10^ − 11;


Period = Sqrt[4 * Pi^2 * a^3/(G * (Mmoon + Mearth))];

Print["The Revolution period T of Moon = ", Period/(3600 * 24),

" days"];

All quantities are in SI units
semimajor axis a = 3.844 × 10^8
Eccentricity of moon's orbit = 0.0548907
The Revolution period T of moon = 27.2867 days

α = a * (1 − e^2);

r = α/(1 + e * Cos[θ]);

PolarPlot[r, {θ, 0, 2 * Pi}, FrameLabel->{"x [m]", "y [m]"},

Frame->True, BaseStyle->{FontSize->16},

PlotLabel->"Moon's orbit",

PlotRange->{{−4.2 * 10^8, 4.2 * 10^8}, {−4.2 * 10^8, 4.2 * 10^8}}]
```

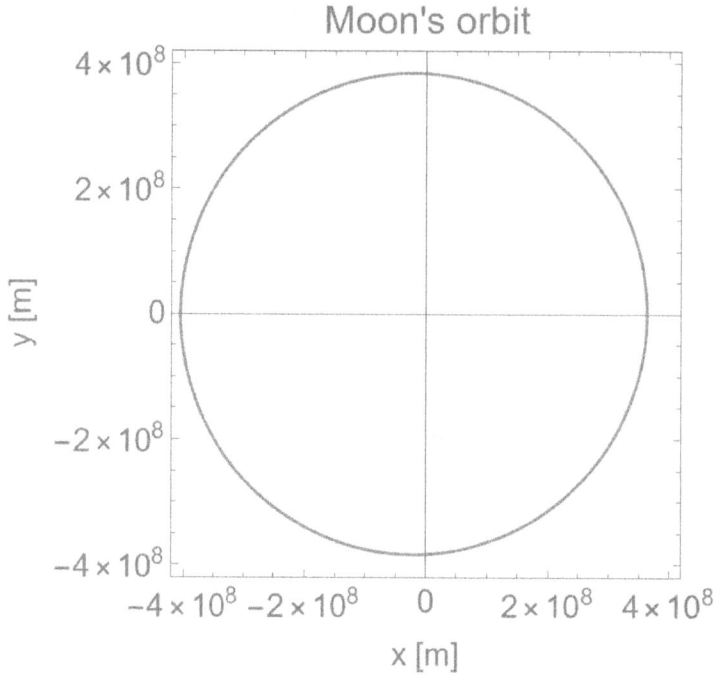

Figure 9.11: The slightly elliptical orbit of the Moon, evaluated in Example 9.5.

9.6.2 KEPLER'S SECOND LAW

Recall that the magnitude of the angular momentum vector is:

$$\ell = \mu r^2 \dot{\theta} \qquad (9.6.12)$$

From this magnitude we can now derive Kepler's second law. We begin by considering a small wedge of the planet's orbit traced out in a time dt, as shown in Figure 9.12.

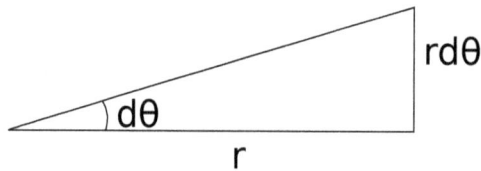

Figure 9.12: A small wedge of the planet's orbit traced out in a time dt.

The wedge shown in Figure 9.12 forms an infinitesimal triangular area element dA, which in polar coordinates can be written as:

$$dA = \frac{r^2}{2} d\theta \qquad (9.6.13)$$

By dividing with dt and using (9.6.12), we obtain Kepler's second law:

Kepler's Second Law

$$\frac{dA}{dt} = \frac{r^2}{2}\frac{d\theta}{dt} = \frac{\ell}{2\mu} = \text{constant} \qquad (9.6.14)$$

Note that Kepler's second law holds for *any* central force. Equation (9.6.14) is the mathematical form of Kepler's second law which says that a line joining a planet to the Sun sweeps out equal areas in equal time intervals. This is illustrated in Figure 9.13. The result of Kepler's second law is that the planet's speed increases as it reaches perihelion and decreases as it approaches aphelion.

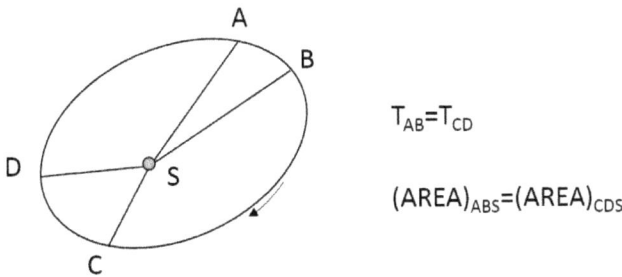

$$T_{AB} = T_{CD}$$

$$(AREA)_{ABS} = (AREA)_{CDS}$$

Figure 9.13: Kepler's second law: A line connecting a planet to the Sun (S) sweeps out equal areas in equal times. If the time T_{AB} required to travel distance AB is equal to the time T_{CD} required to travel distance CD, then the swept out area $AREA_{ABS}$ must be equal to the corresponding area $AREA_{CDS}$.

Example 9.6 shows how to evaluate the angular momentum of the Earth as it orbits around the Sun, and also how to estimate the Earth's orbital velocity.

Example 9.6: The orbital angular momentum of the Earth around the Sun

The speed of the Earth at perihelion is 30,300 m/s, and its distance from the Sun ranges from 147.3 million km (at perihelion), to about 152.6 million km (at aphelion). Evaluate:

(a) The orbital angular momentum of the Earth around the Sun.
(b) The orbital velocity of the Earth at aphelion.
(c) The speed of the Earth at the latus rectum point α of its elliptical orbit.

Solution:

(a) When the Earth is at the aphelion (or at the perihelion) of its orbit, the velocity vector $\mathbf{v} = d\mathbf{r}/dt$ is perpendicular to the position vector $\mathbf{r}$, so that the radial component of the velocity is zero $(dr/dt = 0)$; therefore, the magnitude of the angular momentum is found from:

$$|\,\boldsymbol{\ell}\,| = \left| m\left(\mathbf{r} \times \frac{d\mathbf{r}}{dt}\right) \right| = m\,|\,\mathbf{r}\,|\,|\,\frac{d\mathbf{r}}{dt}\,|\,\sin 90° = mrv$$

where v represents the magnitude of the orbital velocity at these two points and $m = 5.97 \times 10^{24}$ kg is the mass of the Earth. By using the given value of the speed at perihelion and the given distance from the Sun, we can find the magnitude of the angular momentum in SI units:

$$\ell = |\, \boldsymbol{\ell} \,| = m r_{\text{peri}} v_{\text{peri}} = \left(5.97 \times 10^{24}\ \text{kg}\right)\left(3.03 \times 10^{4}\ \text{m/s}\right)\left(147.3 \times 10^{9}\ \text{m}\right)$$
$$= 2.66 \times 10^{40}\ \text{kgm}^2/\text{s}$$

(b) Since the magnitude of the angular momentum is conserved, we have:

$$\ell = m r_{\text{peri}} v_{\text{peri}} = m r_{\text{aphe}} v_{\text{aphe}}$$

so that the orbital velocity at the aphelion is found from:

$$v_{\text{aphe}} = \ell / \left(m\, r_{\text{aphe}}\right)$$

By using this equation we find the speed at the aphelion in SI units:

$$v_{\text{aphe}} = \frac{\left(2.66 \times 10^{40}\ \text{kgm}^2/\text{s}\right)}{\left(5.97 \times 10^{24}\ \text{kg}\ \right) \times \left(152.6 \times 10^{9}\ \text{m}\right)} / = 29{,}200\ \text{m/s}$$

(c) The latus rectum point is located on the elliptical orbit of the Earth, at a distance α *vertically above the location of the Sun* in Figure 9.9. By working as in the previous example, we use the values of r_{peri} and r_{aphe} to find the values of the semi-major axis a and the eccentricity e. Once we know a and e, we calculate the distance α from (9.6.4). Finally, we use conservation of angular momentum to find the v_{latus}:

$$v_{\text{latus}} = r_{\text{peri}} v_{\text{peri}} / \alpha$$

which gives the value of $v_{\text{latus}} = 29{,}800\ m/s$. Note that the velocity vector at the latus rectum is not perpendicular to the position vector.

9.6.3 KEPLER'S THIRD LAW

Next we derive Kepler's third law in which he mathematically related the period of a planet's orbit to its semi-major axis. Using our more general result, we will derive the period-distance relationship for the reduced mass.

We begin by integrating (9.6.14), Kepler's second law,

$$\int dt = \int \frac{2\mu}{\ell} dA \tag{9.6.15}$$

$$P = \frac{2\mu}{\ell} \pi ab \tag{9.6.16}$$

where we used the area of an ellipse $A = \pi ab$. P is the period, the time needed for a line joining the reduced mass and the force center to sweep out the entire area of the ellipse. Next, we use (9.6.5) and (9.6.6) to express the period P and as a function of energy E.

$$P = \pi k \sqrt{\frac{\mu}{2}} \, |E|^{-3/2} \tag{9.6.17}$$

By substituting now $|E| = k/(2a)$ from (9.6.5), we obtain Kepler's third law:

Kepler's Third Law

$$P^2 = \frac{4\pi^2 \mu}{k} a^3 \qquad\qquad (9.6.18)$$

Kepler's original statement was that the planet's period squared was proportional to the cube of its semi-major axis. Kepler did not know the proportionality constant. However, using Newton's second law, we have found it! If we consider the central force to be Newton's Law of Universal Gravitation, the constant $k = Gm_1 m_2$ and the definition of the reduced mass $\mu = m_1 m_2 / (m_1 + m_2)$, then Kepler's third law becomes

$$P^2 = \frac{4\pi^2 a^3}{G(m_1 + m_2)} \qquad\qquad (9.6.19)$$

However, for a planet of mass m_1 orbiting the Sun of mass $m_2 = M \gg m_1$, we have

$$P^2 = \frac{4\pi^2 a^3}{GM} \qquad\qquad (9.6.20)$$

Example 9.7 illustrates Kepler's Third Law with the moons of Jupiter.

Example 9.7: The Moons of Jupiter

The following table lists the orbital properties of the four largest moons of Jupiter, discovered by Galileo Galilei. Use this data to find the mass of Jupiter.

Satellite	Semi-major axis (km)	Orbital period (days)
Io	421,700	1.7691
Europa	671,034	3.5512
Ganymede	1,070,412	7.1546
Callisto	1,882,709	16.689

Solution:

The data can be analyzed by plotting Kepler's law

$$P^2 = \frac{4\pi^2 a^3}{G(m_1 + m_2)} \qquad\qquad (9.6.21)$$

By plotting P^2 as a function of a^3 using the given table, we should obtain a a linear graph $P^2 = m\,a^3 + b$, with a slope

$$slope = \frac{4\pi^2}{G(m_1 + m_2)}$$

For Jupiter $m_1 = m_{Jupiter} \gg m_2$, so this becomes

$$slope = \frac{4\pi^2}{GM_{Jupiter}}$$

By fitting a straight line to the plot of the data, we can obtain an estimate of the mass of Jupiter:

$M_{Jupiter} = 4\pi^2 / (G\,slope)$. Note that in this problem it is important to use SI units throughout the solution, by converting the units of kilometers into meters, and the periods P into SI units by converting days into seconds.

Python Code

We use list comprehensions to obtain the variables $x = a^3$, $y = P^3$. We use SI units by converting the units of kilometers into meters, and the days into seconds. The **aValues** list is converted in SI units by converting the units of kilometers into meters, and the periods P are converted into SI units by converting days into seconds. We use the command **polyfit** from the NumPy library, to find the best straight line fit to this plot. The command **solve** from SymPy is used to obtain the value of the mass of Jupiter.

```
import numpy as np
import matplotlib.pyplot as plt
from sympy import solve, symbols
print('-'*28,'CODE OUTPUT','-'*29,'\n')

M = symbols('M')
G = 6.67408*10**-11

aValues = np.array([4.21700*10**8, 6.71034*10**8, 10.70412*10**8,\
                    18.82709*10**8])
acube = [aValues[i]**3 for i in range(4)]

periods = np.array([1.7691*3600*24, 3.5512*3600*24, 7.1546*3600*24,\
                    16.689*3600*24])
psquare = [periods[i]**2 for i in range(4)]

plt.plot(acube,psquare,'o')      # plot period^2 vs a^3
plt.xlabel('a^3 [m^3]')
plt.ylabel('P^2 [s^2]')

m,b = np.polyfit(acube, psquare, 1)
print('Best slope =', m)

mJupiter = solve(m - 4*np.pi**2/(G*M), M)
print('Mass of Jupiter = ', mJupiter[0], " kg");
x = np.linspace(0,7*10**27,100)
plt.plot(x,b + m*x)              # plot best fit line
plt.show()

-------------------------- CODE OUTPUT ------------------------------
Best slope = 3.1155753615415456e-16
Mass of Jupiter =  1.89858509859858e+27  kg
```

Mathematica Code

We set up two lists called aValues and periods, which contain the semi-major axes a and the periods P of the four moons. The aValues list is converted in SI units by converting the units of kilometers into meters, and the periods P are converted into SI units by converting days into seconds. The command Transpose is used to create a new list called list, containing the pairs (a^3, P^2) for the four moons. This is plotted using the ListPlot command, and the graph is stored in the variable $gr1$. By using the FindFit command, we fit a best line to the graph, and we obtain the best slope of this line to have a slope of 3.11×10^{-16} (SI-units).

By setting the best slope $3.11 \times 10^{-16} = 4\pi^2/(GM)$ and using the NSolve command, we obtain the mass of Jupiter as $M = 1.89 \times 10^{27}$ kg. This value is very close to the accepted value of the mass of Jupiter $M_J = 1.90 \times 10^{27}$ kg.

```
G = 6.67408 * 10^ − 11;

aValues = {421700, 671034, 1070412, 1882709} * 10^3;

periods = {1.7691, 3.5512, 7.1546, 16.689} * 3600 * 24;

list = Transpose[{aValues^3, periods^2}];

gr1 = ListPlot[list, Frame → True,

FrameLabel → {"a^3 (m^3)", "P^2 (s^2)"}, BaseStyle->FontSize->17,

PlotRange → {{0, 8 * 10^27}, {0, 2.2 * 10^12}}, PlotMarkers->Automatic];


Print["Best line fit = a+b*x with "];

bestline = FindFit[list, a + b * x, {a, b}, x];

Print["Best slope =", slope = b/.bestline];

mJupiter = NSolve[slope == 4 * Pi^2/(G * M), M];

Print["Mass of Jupiter = ", M/.mJupiter[[1]], " kg"];

Best line fit = a+b*x with
Best slope =3.1155753615415436*^-16
Mass of Jupiter = 1.89859 × 10^27 kg

gr2 = Plot[(a + b * x)/.bestline, {x, 0, 8 * 10^27},

BaseStyle->FontSize->17, Frame → True,

PlotRange → {{0, 8 * 10^27}, {0, 2.2 * 10^12}}];

Show[gr1, gr2]
```

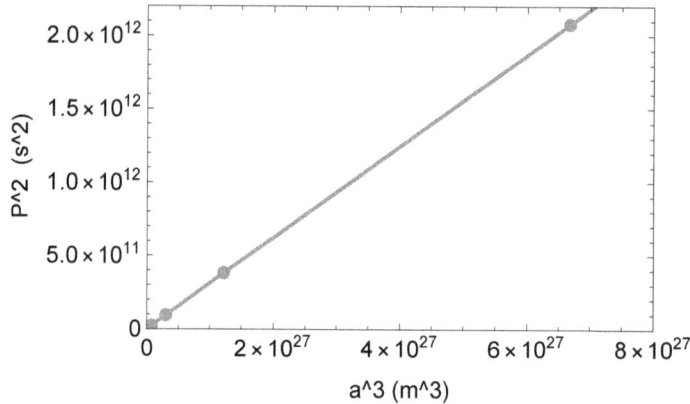

Figure 9.14: Kepler's third law applied to four of the many Moons of Jupiter, from Example 9.7.

9.7 PLANAR CIRCULAR RESTRICTED THREE-BODY PROBLEM

In this final section of Chapter 9, we examine a problem which requires the use of numerical methods, the so-called planar circular restricted three-body problem. Consider three masses, m_1, m_2, and m_3 that are interacting solely via gravitational interactions. There are no other objects exerting forces on the masses. We make the following assumptions. First, the motion of all three particles is confined to the same plane. Second, $m_3 \ll m_1$ and $m_3 \ll m_2$ and, therefore, m_3 does not affect the motion of masses m_1 and m_2. If the second assumption were the only one being made about the system, the system would be referred to as a *restricted* three-body problem.

The final assumption is that the masses m_1 and m_2 each move in a circular orbit about their shared center of mass. Because m_3 doesn't affect the motion of the other two masses, we know that, from the work done in this chapter, m_1 and m_2 will orbit their shared center of mass and that the orbit will be in the form of a conic section. In this case, we choose a specific conic section, a circle. These three assumptions together form the *planar circular restricted three-body problem*.

We use the formalism developed in Section 9.6 in order to describe the motions of m_1 and m_2. Furthermore, we will scale the variables associated with m_1 and m_2 as done in Worthington [10]:

- The sum of the masses m_1 and m_2 is unity, $m_1 + m_2 = 1$.

- The distance between m_1 and m_2 is fixed to be $r = 1$.

- The period of the orbits of m_1 and m_2 about their shared center of mass is 2π.

These scalings will simplify the resulting equations of motion. Using Kepler's third law and the scalings above, we find that $G = 1$. In addition, we can choose $m_2 = m$ and then $m_1 = 1 - m$, where $0 \le m \le 1/2$. The end result is that the equations of motion will depend on only one parameter, m.

Before deriving the equations of motion, we need to establish a coordinate system as shown in Figure 9.15.

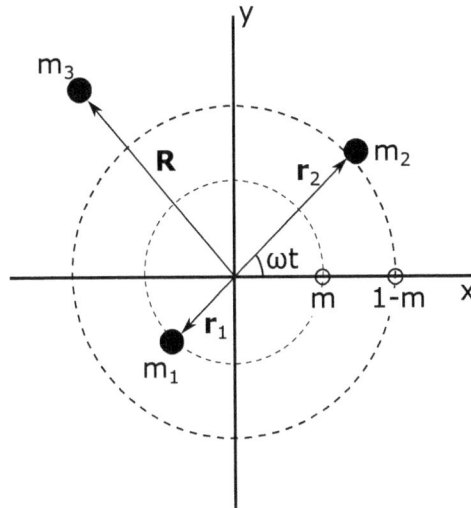

Figure 9.15: The coordinate system for the planar restricted circular three-body problem. The center of mass for m_1 and m_2 is at the origin. Mass m_1 follows a circular orbit with radius m about the center of mass, and m_2 follows a circular orbit of radius $1 - m$. The third body m_3 is located at the position $\mathbf{R}$. Not shown in this figure is a vector $\mathbf{r}$ which points from m_1 to m_2 and passes through the origin.

In order to understand Figure 9.15, we begin with the assumption that m_1 and m_2 follow circular orbits. Using (9.2.10) and (9.2.11), we can compute the radius of each orbit:

$$r_1 = \frac{m_2}{m_1 + m_2} r = m \tag{9.7.1}$$

$$r_2 = \frac{-m_1}{m_1 + m_2} r = 1 - m \tag{9.7.2}$$

Noting that $\omega = 1$ (because the period of m_1 and m_2 is 2π), we can write,

$$\boldsymbol{r}_2 = (1 - m)\cos t\,\hat{\mathbf{i}} + (1 - m)\sin t\,\hat{\mathbf{j}} \tag{9.7.3}$$

Although not shown in Figure 9.15, the separation vector $\mathbf{r} = \mathbf{r}_2 - \mathbf{r}_1$ points from m_1 to m_2 and passes through the origin. Noting that $r = 1$, we can write,

$$\mathbf{r}_1 = -m\cos t\,\hat{\mathbf{i}} - m\sin t\,\hat{\mathbf{j}} \tag{9.7.4}$$

The location of m_3 can be written as

$$\mathbf{R} = X\hat{\mathbf{i}} + Y\hat{\mathbf{j}} \tag{9.7.5}$$

We will use the Lagrangian to derive the equations of motion for the particle m_3. The kinetic energy of the mass m_3 is,

$$T_3 = \frac{1}{2}m_3\left(\dot{X}^2 + \dot{Y}^2\right) \tag{9.7.6}$$

We find the potential energy using the general formula, $V = -Gmm_2/r$ where m and m_2 are the masses of the interacting particles and r is the distance between m and m_2. There are two gravitational interactions that involve m_3, V_{13} is the potential energy between m_1 and m_3, and V_{23} is the potential energy between m_2 and m_3. They can be written as:

$$V_{13} = -\frac{G(1-m)m_3}{r_{13}} \tag{9.7.7}$$

$$V_{23} = -\frac{Gmm_3}{r_{23}} \tag{9.7.8}$$

where

$$r_{13}^2 = (X + m\cos t)^2 + (Y + m\sin t)^2 \tag{9.7.9}$$

$$r_{23}^2 = (X - (1-m)\cos t)^2 + (Y - (1-m)\sin t)^2 \tag{9.7.10}$$

are the distances between m_3 and m_1 and between m_3 and m_2, respectively.

Because we are interested only in the motion of m_3 (we already know the motion of m_1 and m_2 by our assumptions), we will write only the terms of the Lagrangian for the mass m_3. The terms for the kinetic energies of m_1 and m_2, and the term describing the gravitational potential energy between m_1 and m_2 will not contribute to the equations of motion for m_3 because they are independent of the variables X, $\dot{X}$, Y, and $\dot{Y}$. The Lagrangian of our system (including only the terms relevant to m_3) is $\mathcal{L} = T_3 - (V_{13} + V_{23})$, which, when using $G = 1$, can be written as,

$$\mathcal{L} = \frac{1}{2}m_3\left(\dot{X}^2 + \dot{Y}^2\right) + \frac{(1-m)m_3}{r_{13}} + \frac{mm_3}{r_{23}} \tag{9.7.11}$$

Note that (9.7.11) is time-dependent. We can eliminate the time-dependence by transforming our coordinate system into a frame that is rotating with the masses m_1 and m_2. In this rotating frame, m_1 and m_2 are at rest, with m_1 located at the point $(m, 0)$ and m_2 located at the point $(1 - m, 0)$. Let us use the notation (x, y) to describe the position of m_3 in the rotating frame. The transformation that connects the coordinates (X, Y) and (x, y) is the rotation matrix

$$\begin{pmatrix} X \\ Y \end{pmatrix} = \begin{pmatrix} \cos t & -\sin t \\ \sin t & \cos t \end{pmatrix} \begin{pmatrix} x \\ y \end{pmatrix} \tag{9.7.12}$$

where the angle of rotation is $\theta = \omega t = t$, since $\omega = 1$ due to our assumption about the period of revolution for masses m_1 and m_2. We can write out the matrix multiplication in (9.7.12) to find,

$$X = x\cos t - y\sin t \tag{9.7.13}$$
$$Y = x\sin t + y\cos t \tag{9.7.14}$$

Using (9.7.13) and (9.7.14) we find,

$$\dot{X} = \dot{x}\cos t - x\sin t - \dot{y}\sin t - y\cos t \qquad (9.7.15)$$

$$\dot{Y} = \dot{x}\sin t + x\cos t + \dot{y}\cos t - y\sin t \qquad (9.7.16)$$

$$r_{13}^2 = (x+m)^2 + y^2 \qquad (9.7.17)$$

$$r_{23}^2 = (x - (1-m))^2 + y^2 \qquad (9.7.18)$$

Inserting equations (9.7.15)-(9.7.18) into (9.7.11) and after some algebra, we can re-write the Lagrangian as,

$$\mathcal{L} = \frac{1}{2}m_3\left((\dot{x}-y)^2 + (\dot{y}+x)^2\right) + \frac{m_3(1-m)}{r_{13}} + \frac{mm_3}{r_{23}} \qquad (9.7.19)$$

Finally, to get the equations of motion, we use the Euler-Lagrange equations,

$$\frac{d}{dt}\frac{\partial\mathcal{L}}{\partial\dot{x}} - \frac{\partial\mathcal{L}}{\partial x} = 0 \qquad (9.7.20)$$

$$\frac{d}{dt}\frac{\partial\mathcal{L}}{\partial\dot{y}} - \frac{\partial\mathcal{L}}{\partial y} = 0 \qquad (9.7.21)$$

which result in

$$\ddot{x} - 2\dot{y} - x = -\frac{(1-m)(x+m)}{\left((x+m)^2+y^2\right)^{3/2}} - \frac{m\left[x-(1-m)\right]}{\left((x-(1-m))^2+y^2\right)^{3/2}} \qquad (9.7.22)$$

$$\ddot{y} + 2\dot{x} - y = -\frac{(1-m)\,y}{\left((x+m)^2+y^2\right)^{3/2}} - \frac{my}{\left([x-(1-m)]^2+y^2\right)^{3/2}} \qquad (9.7.23)$$

The equations of motion (9.7.22) and (9.7.23) are two coupled second-order ordinary differential equations that must be solved numerically.

Example 9.8 shows the numerical solution of this system of ODEs using Python. The corresponding code using Mathematica for this example is left for Problem 9.38.

Example 9.8: Numerical solution of the planar circular restricted three-body
problem
Solve the equations of motion (9.7.22) and (9.7.23) using Python, for various initial conditions.

Solution:
To compute the numerical solution in Python, we rewrote (9.7.22) and (9.7.23) as a system of four first-order differential equations using $u \equiv \dot{x}$ and $v \equiv \dot{y}$ as new variables. The resulting equations of motion are

$$\dot{x} = u \qquad (9.7.24)$$

$$\dot{u} = 2v + x - \frac{(1-m)(x+m)}{\left((x+m)^2+y^2\right)^{3/2}} - \frac{m\left[x-(1-m)\right]}{\left((x-(1-m))^2+y^2\right)^{3/2}} \qquad (9.7.25)$$

$$\dot{y} = v \qquad (9.7.26)$$

$$\dot{v} = -2u + y - \frac{(1-m)\,y}{\left((x+m)^2+y^2\right)^{3/2}} - \frac{my}{\left([x-(1-m)]^2+y^2\right)^{3/2}} \qquad (9.7.27)$$

The four first-order equations of motion are then solved using `odeint` for the condition $m = 0.01$. The local variables `peXterm` and `peYterm` are the right-hand side of (9.7.22) and (9.7.23), respectively. These additional variables were included to improve the clarity of the algorithm. In addition, we included the basic information needed to create a four-panel plot. The results of the code are shown in Figure 9.16.

In Figure 9.16 we see the trajectory of m_3 in the rotating coordinate frame (x, y) for four different initial conditions. In each plot, the red (in the e-book) dot near the origin is m_1 and the other dot (which is green in the e-book) is m_2. The blue (in the e-book) curve is the trajectory of m_3. Notice in the upper-left plot of Figure 9.16, m_3 is in a precessing orbit about m_1, which is obscured by the trajectory. The upper-right plot has m_3 spiraling out towards an orbit that encompasses both m_1 and m_2. The lower-right plot shows a case where m_3 orbits about m_2.

Arguably the most interesting plot in Figure 9.16 is the lower-left plot which shows an example where m_3 orbits around a so-called *Lagrange point* of the m_1-m_2 system. A Lagrange point is a point in space where the centrifugal force is balanced by the gravitational attraction. Because we transformed the coordinates of the original problem into a rotating system, we introduced a noninertial frame. We will discuss noninertial frames in the next chapter. However, for now all you need to know is that in non-inertial frames, an apparent force called the *centrifugal force* appears which is directed outward. Near two orbiting masses, such as m_1 and m_2, there are five points called Lagrange points where the centrifugal (outward) force is equal to the gravitational attraction (inward). The Lagrange points are equilibrium points, and the one represented in the lower-left panel of Figure 9.16 is called L4, which is a stable equilibrium.

For example, the *Trojan asteroids* orbit Jupiter's L4 point. In addition, Lagrange points in the Earth-Sun and Earth-Moon systems have been used by many probes and satellites. For example, the James Webb Space Telescope orbits L2, which lies along a line joining the Earth and the Sun, and would be to the right of the green dot (in the e-book) in Figure 9.16.

Python Code

```python
import numpy as np
from scipy.integrate import odeint
import matplotlib.pyplot as plt

mu = 0.01

def threeBody(vec,t):
    x, y, u, v = vec

    peXterm = -(1.0-mu)*(x+mu)/((x+mu)**2+y**2)**(1.5)-mu*\
        (x-1.0+mu)/((x-1.0+mu)**2+y**2)**(1.5)
    peYterm = -y*(1.0-mu)/((x+mu)**2+y**2)**(1.5)-mu*y/((x-1.0+\
        mu)**2+y**2)**(1.5)
    derivative = [u, v, 2*v + x + peXterm, -2*u + y + peYterm ]
    return derivative

initialCondition1 = [-0.75, 0, 0, 1]
t=np.linspace(0,20,2001)
sol1 = odeint(threeBody, initialCondition1, t)

initialCondition2 = [1.1, 0, 0, 0.3]
t=np.linspace(0,50,5001)
sol2 = odeint(threeBody, initialCondition2, t)

initialCondition3 = [0.5, 0.86, 0, 0.02]
t=np.linspace(0,50,5001)
sol3 = odeint(threeBody, initialCondition3, t)

initialCondition4 = [1.1, 0.0, 0, 0.01]
t=np.linspace(0,10,1001)
sol4 = odeint(threeBody, initialCondition4, t)

f, ((ax1, ax2), (ax3, ax4)) = plt.subplots(2,2)
plt.subplots_adjust(top = 0.99, bottom=0.01, hspace=0.75, wspace=1.0)
ax1.plot(sol1[:,0],sol1[:,1])
ax2.plot(sol2[:,0],sol2[:,1])
ax3.plot(sol3[:,0],sol3[:,1])
ax4.plot(sol4[:,0],sol4[:,1])
plt.show()
```

x0 = -0.75, y0 = 0
u0 = 0, v0 = 1.0

x0 = 1.1, y0 = 0
u0 = 0, v0 = 0.3

x0 = 0.5, y0 = 0.86
u0 = 0, v0 = 0.02

x0 = 1.1, y0 = 0
u0 = 0, v0 = 0.01

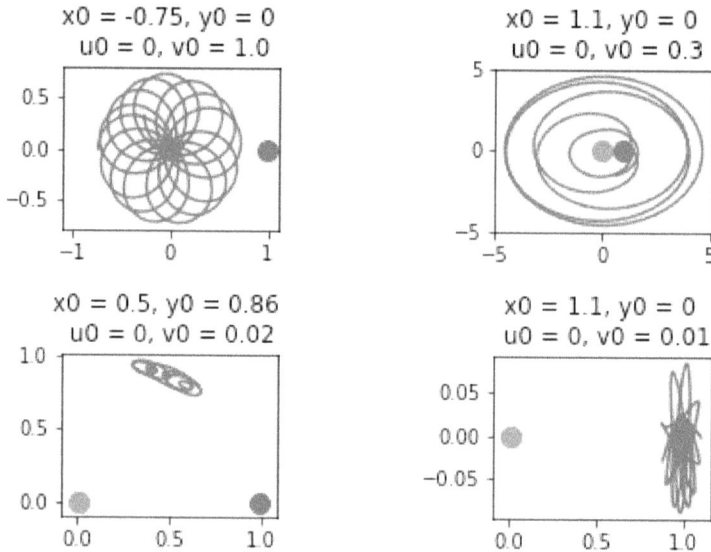

Figure 9.16: Results of the Python code for the planar restricted three-body problem, from Example 9.8. The red (in the e-book) dot near the origin is m_1, and the green (in the e-book) dot is m_2. The curve represents the path followed by m_3. The complete code for this plot can be found in this book's website.

9.8 CHAPTER SUMMARY

A central force is directed along a line that joins the particle and the force center. Central forces can be expressed in spherical coordinates as:

$$\mathbf{F} = f(r)\hat{\mathbf{r}}$$

where the vector $\mathbf{r}$ is the location of the particle relative to the force center. Central forces conserve energy and angular momentum.

In a two-body system, where two masses m_1 and m_2 interact via a central force, we can reduce the problem to that of the motion of a single particle of mass μ called the reduced mass:

$$\mu = \frac{m_1 m_2}{m_1 + m_2}$$

The reduced mass is located at $\mathbf{r} = \mathbf{r}_1 - \mathbf{r}_2$ relative to the system's center of mass. The center of mass serves as the force center.

One approach of finding the equations of motion for the reduced mass in the two-body problem, is by solving the differential equation,

$$\frac{d^2}{d\theta^2}\left(\frac{1}{r}\right) + \frac{1}{r} = -\frac{\mu r^2}{\ell^2} f(r)$$

From the equation of motion, which is derived from Newton's second law in polar coordinates, we can derive general forms of Kepler's Three Laws of Planetary Motion.

The first law is that the orbit of the reduced mass is of the form of a conic section with the force center at one focus,

$$\frac{1}{r} = \frac{1}{\alpha}\left(1 + e\cos\theta\right) \qquad \alpha = \frac{\ell^2}{\mu k} \qquad e = \sqrt{1 + \frac{2E\ell^2}{\mu k^2}}$$

For planets orbiting the Sun, their motion is an ellipse with the Sun at one focus. The shape of the ellipse can be described by

$$r_{\min} = a\left(1 - e\right) \qquad r_{\max} = a\left(1 + e\right)$$

where e is the eccentricity of the ellipse, r is the distance from the Sun to the planet, a and b are the semi-major and semi-minor axes of the ellipse. The perihelion is the point where the distance is minimum $r_{\min}$, and the aphelion is the point where the distance is maximum r_{max}. In general, the motion of the reduced mass can be any conic section, and is determined by the energy E and angular momentum ℓ which determine the eccentricity e of the orbit. If $0 < e < 1$ the motion is an ellipse, if $e = 0$ the orbit is a circle, if $e > 1$ the motion is a hyperbola, and for $e = 1$ it is a parabola.

The conservation of angular momentum leads to Kepler's second law: a line joining the reduced mass to the force center sweeps out equal areas in equal time periods.

Kepler's third law relates the period of the planet's orbit P to the semi-major axis of its orbit a.

$$P^2 = \frac{4\pi^2 a^3}{G\left(m_1 + m_2\right)}$$

The three-body problem involves interactions between three objects. We examined the circular restricted three-body problem which is a relatively simple formulation of the three-body problem. However, even the circular restricted three-body problem must be analyzed numerically. The motion of the third mass $m_3 \ll m_1, m_2$ can be rather complicated.

9.9 END-OF-CHAPTER PROBLEMS

The symbol ⌨ indicates a problem which requires some computer assistance, in the form of graphics, numerical computation, or symbolic evaluation.

Section 9.1: Central Forces

1. ⌨ A two-body system interacts with a central force $\mathbf{F}\left(\mathbf{r}\right) = f\left(r\right)\hat{\mathbf{r}}$ where $\hat{\mathbf{r}}$ is the radial unit vector. Show that (a) the angular momentum and (b) the total energy are conserved. Solve this problem both analytically and using a CAS.

2. A two-body system of masses m_1 and m_2 interacts with the central force field $F = -\left(K/r^2\right)\hat{\mathbf{r}}$, where $\hat{\mathbf{r}}$ is the radial unit vector.

   a. Find the Lagrangian of the two-body system as a function of the distance r. What are two constants of the motion?

   b. How much work is done by the force field in (a) by moving the particle from a point on the circle with radius $r = a > 0$, to another point on the circle $r = b > 0$? Does the work depend on the path? Explain.

3. A particle of mass m moves in a central force field defined by $F = -K/r^3$.

   a. Write an equation for the conservation of energy.

   b. Prove that if E is the total energy supplied to the particle, then its speed is given by

$$v = \sqrt{K/(mr^2) + 2E/m}$$

4. A particle of mass m moves in a central force field defined by $F = -k\,r^2$. It starts from rest at a point on the circle $r = a$. Prove that when it reaches the circle $r = b$ its speed will be

$$v = \sqrt{2k(a^3 - b^3)/3m}$$

5. A particle of mass m moves in a central force field $F = -K/r^n$ where K and n are constants and $K > 0$. The particle starts from rest at $r = a$ and arrives at $r = 0$ with finite speed V_0.

   a. Prove that for this physical situation, the condition $n < 1$ must be true.

   b. Prove that

$$V_0 = \sqrt{\frac{2Ka^{1-n}}{m\,(n-1)}}$$

6. Let L, M, and T represent the dimensions of length, mass and time, respectively. Find the dimensions of the universal gravitational constant G.

7. 🖥 The coordinates of a mass m are given as a function of time t, as $x = x_0 \cos(\omega_1 t)$ and $y = y_0 \sin(\omega_2 t)$, where ω_1, ω_2 are positive constants.

   a. Is the angular momentum of the mass m conserved as it moves on the xy-plane?

   b. Show that if $\omega_1 = \omega_2$, then the force acing on the mass is a central force.

   c. Evaluate the total energy when $\omega_1 = \omega_2$, and show that it is conserved.

   Solve this problem analytically and using a CAS.

8. A mass m experiences a force

$$\mathbf{F} = f\,(r)\,\hat{\mathbf{r}} - \lambda\mathbf{v}$$

   where λ is a positive constant, $f\,(r)$ is a scalar function of the radial distance r, $\mathbf{v}$ is the velocity vector, and $\hat{\mathbf{r}}$ is the unit vector along the radial direction. Show that the angular momentum varies with time as $L = L_0 e^{-\frac{\lambda}{m}t}$, where L_0 is the initial angular momentum at $t = 0$.

Sections 9.2 - 9.3: The Two-Body Problem and Equations of Motion for the Two-Body Problem

9. 🖥 Two point masses m_1, m_2 are separated by a distance r. Starting from rest, they both begin to accelerate towards each other. Show that the elapsed time t until they collide is:

$$t = \pi\sqrt{\frac{r^3}{8G(m_1 + m_2)}}$$

   You may want to evaluate the integrals with a CAS.

10. Show that the total angular momentum $\mathbf{L} = \mathbf{L}_1 + \mathbf{L}_2$ of the two-body system is equal to

$$\mathbf{L} = \mu\,(\mathbf{r} \times \mathbf{v})$$

   where μ is the reduced mass, $\mathbf{r} = \mathbf{r}_2 - \mathbf{r}_1$ is the relative position vector of the two masses, and the velocity vector $\mathbf{v} = \dot{\mathbf{r}}$. Here $\mu = \text{reduced mass} = \frac{m_1 m_2}{m_1 + m_2}$.

11. ⌨ Show that the kinetic energy of the two-body system with a reduced mass μ and relative position vector of the two masses $\mathbf{r} = \mathbf{r}_2 - \mathbf{r}_1$ is equal to:

$$T = \frac{1}{2}\mu \, |\dot{\mathbf{r}}|^2$$

where μ = reduced mass $= \frac{m_1 m_2}{m_1 + m_2}$. Solve this problem both analytically and using a CAS.

12. ⌨ In Example 9.1 we evaluated numerically and plotted the orbits in a two-body gravitational system. Use the command $\mathtt{ListAnimate}$ in Mathematica, to produce an animation of the two bodies moving around each other.

13. ⌨ The orbit of a particle of mass m in a central force located at $r = 0$ is a spiral $r = e^{-\theta}$. Prove that the force is proportional to $1/r^3$.

14. ⌨ What central force is required to produce an orbit $r^2 = a^2 \cos(2\theta)$, where a is a positive constant? It may be useful to use a CAS to evaluate and simplify the necessary derivatives.

15. Show that both $r = e^{-\theta}$ and $r = 1/\theta$ are possible orbits for a central force proportional to $1/r^3$. How is this possible?

16. A particle with mass m moves around the origin O with a speed $v = a/r$, inversely proportional to the distance from O.

   a. Show that the motion of the particle requires an inverse cube central force.

   b. What types of orbits are possible?

17. The orbit of a mass m inside a central force field is such that $r = a/\theta$, where a is a positive constant. Show that the potential must be proportional to $1/r^2$.

18. Obtain the orbit for a mass m moving in a central force field defined by $F = -K/r^3$, and describe it physically.

19. A particle of mass m moves in a central force field given in magnitude by $\boldsymbol{F} = -K\boldsymbol{r}$ where K is a positive constant and $\boldsymbol{r}$ is the position vector. If the particle starts at $r = a$, $\theta = 0$ with a speed v_0 in a direction perpendicular to the x-axis, determine and describe its orbit.

20. A particle of mass m moves in a central force field given in magnitude by $F = -k/r^3$ where k is a positive constant, and r is the distance from the force center. The particle starts at $r = a$, $\theta = 0$ with a speed v_0 in a direction making an angle α with the positive x-axis.

   a. Show that the differential equation for the orbit is given in terms of $u = 1/r$ by

$$\frac{d^2 u}{d\theta^2} + (1 - \gamma)u = 0$$

   where

$$\gamma = \frac{k}{ma^2 v_0^2 \sin^2 \alpha}$$

   b. Solve the differential equation in (a) and interpret it physically.

21. 🖳 The orbit of a mass m and angular momentum L in a central field is given by

$$r\left(\theta\right) = \frac{1}{A\cos\left(\omega_o\theta\right)}$$

where ω_o is a positive constant, and the constants A, ϕ depend on the initial conditions. This type of orbit is obtained in Problems 9.16, 9.18 and 9.20 under certain assumptions.

    a. Assuming that $\theta(0) = 0$, obtain analytical expressions for the radius $r(t)$ as a function of time.

    b. Plot $r(t)$ as a function of time t, and a polar plot $r(\theta)$ by using the numerical values $A = 1$, $L = 1$, $m = 1$, $\omega = 0.3$ (all in SI units). What is the physical meaning of these plots?

Solve this problem both analytically and also using a CAS.

Section 9.4: Planetary Motion and Kepler's First Law

22. 🖳 A planet moves around a massive Sun according to the equation

$$r = \frac{12}{3 + \cos\theta}$$

    a. Evaluate the semi-major axes a, b and plot the orbit.

    b. Find the central force for this orbit, in terms of the angular momentum and the mass m of the planet.

Solve this problem both analytically and also using a CAS.

23. 🖳 A planet moves around a massive Sun according to the equation

$$r = \frac{24}{3 + 5\cos\theta}$$

    a. Evaluate and plot the orbit.

    b. Find the central force for this orbit, in terms of the angular momentum L and the mass m of the planet.

Solve this problem both analytically and also using a CAS.

24. 🖳 A planet revolves around a massive Sun in an elliptical orbit, with one focus F at the origin, the center of the ellipse located at $(-4, 0)$ and semi-major axis $a = 10$.

    a. Find the equation of the orbit in Cartesian and polar coordinates.

    b. Find the central force for this orbit, in terms of the angular momentum L and the mass m of the planet.

Solve this problem both analytically and also using a CAS.

25. 🖳 The orbit of a mass m and angular momentum L in a central field is given by

$$r\left(\theta\right) = \frac{1}{A\cosh\left(\omega_0\theta\right)}$$

where ω_0 is a positive constant, and the constants A, ϕ depend on the initial conditions. This type of orbit is obtained in Problems 9.16, 9.18 and 9.20 under certain assumptions.

a. Assuming that $\theta(0) = 0$, obtain analytical expressions for the radius $r(t)$ as a function of time.

b. Plot $r(t)$ as a function of time t, and a polar plot $r(\theta)$ by using the numerical values $A = 1$, $L = 1$, $m = 1$, $\omega = 0.3$ (all in SI units). What is the physical meaning of these plots?

Solve this problem both analytically and also using a CAS.

26. ⌨ The distance of closest approach of Halley's Comet to the Sun is 0.57 astronomical units AU (1 AU is the mean Earth-Sun distance). The greatest distance of the comet from the Sun is 35 AU. Create a polar plot of the orbit of Halley's comet around the Sun in the vicinity of our solar system, which extends about 40 AU in diameter. Plot together the orbits of Mercury, the earth and Halley's Comet.

Section 9.5: Orbits in Central Force Fields

27. Evaluate and discuss the motion of a particle in the central force field $F = \alpha/r^2 + \beta/r^3$ for $\beta > 0$ and α, β are constants.

28. ⌨ Integrate numerically Newton's second law when the interaction force between two masses m_1 and m_2 is of the form

$$\mathbf{F}(r) = -\frac{Gm_1m_2}{r^{1.9}}\hat{\mathbf{r}} \quad \text{or} \quad \mathbf{F}(r) = -\frac{Gm_1m_2}{r^{2.1}}\hat{\mathbf{r}}$$

Evaluate numerically and discuss the orbits of a particle in these central force potentials. Use the numerical values $m_1 = 1$, $m_2 = 2$, $G = 1$ and initial conditions $x(0) = 1$, $x'(0) = 0$, $y(0) = 0$, $y'(0) = 0.5$ (all quantities in SI units). Show that the resulting orbits of the reduced mass are precessing ellipses, and discuss the differences between the two cases. Use a CAS to carry out numerical integration of Newton's law.

29. Planet A has a velocity v_A at aphelion and a velocity of v_P at perihelion during its motion around a star. A second planet B moves around the same star as planet A, in a circular orbit with a radius R and with a speed v. Show that the aphelion distance for planet A is

$$R_A = \frac{2Rv^2}{v_A(v_A + v_P)}$$

30. A comet of energy E and angular momentum L enters a region with an attractive central potential $V(r)$. Show that the distance of closest approach of the comet to the star is

$$r_{\min} = \frac{L}{\sqrt{2m(E - V(r_{\min}))}}$$

31. A comet of mass m approaches our sun with a velocity v_0 from a very large distance as shown in Figure 9.17. If there was no deflection of the comet due to the Sun, it would have passed a distance b from the Sun as shown. Show that the distance of closest approach for the comet is given by:

$$r_{\min} = \frac{k}{mv_0^2} + \sqrt{\left(\frac{k}{mv_0^2}\right)^2 + b^2}$$

where the gravitational potential is $V = -k/r$.

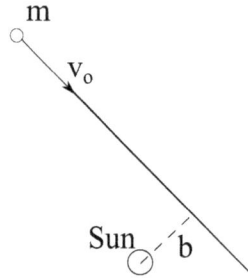

Figure 9.17: The comet of mass m approaching the Sun, from Problem 9.31.

32. A spacecraft travels from planet A to planet B in an elliptical orbit, so that the perihelion is at planet A and the aphelion is at planet B. Planets A and B move in circular orbits around a star of mass M, with radii R_A and R_B as shown in Figure 9.18. Assume that the masses of the planets have no effect on the spacecraft. Show that the spacecraft must be given a speed

$$v = \sqrt{\frac{2GMR_A}{R_B\left(R_A + R_B\right)}} - \sqrt{\frac{GM}{R_A}}$$

with respect to planet A, in order to move in the dashed elliptical orbit shown below.

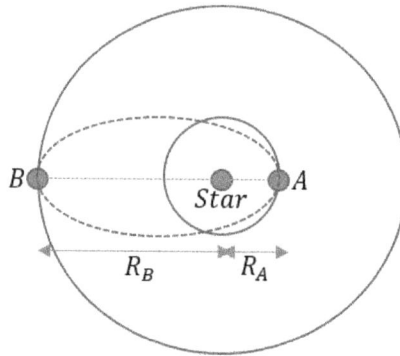

Figure 9.18: A spacecraft transiting from planet A to planet B along the dotted elliptical path, from Problem 9.32.

Section 9.6: Kepler's Laws of Planetary Motion

33. Solve the following Kepler law problems:

   a. Assuming that the planet Mars has a period about the Sun equal to approximately 687 earth days, find the mean distance of Mars from the Sun. Take the distance of the Earth from the Sun as 93 million miles.

b. Calculate the mass of the Sun using the fact that the Earth is 150 million km from it, and the Earth makes one complete revolution about the Sun in approximately 1 year.

34. Suppose that a small spherical planet has a radius of 10 km and a mean density of 5 g/cm^3.

  a. What would be the acceleration due to gravity at its surface?

  b. What would a person weigh on this planet if they weighed 80 kg on Earth?

  c. If the acceleration due to gravity on the surface of a spherical planet is g_P and the mean density and radius are given by σ and R respectively, show that $g_P = (4\pi/3)\,GR\sigma$, where G is the universal gravitational constant.

35. Solve the following Kepler law problems:

  a. Evaluate the radius of the orbit of a geostationary satellite.

  b. Evaluate the orbiting period of the space shuttle around the Earth, by assuming that its orbit is very close to the Earth's surface.

36. ⌨ Create three lists of the perihelion, aphelion and orbital periods of the planets, and use Kepler's third law to estimate the mass of the Sun, by fitting a least squares linear function. Follow a procedure similar to Example 9.7.

Section 9.7: Circular Restricted Three-Body Problem

37. Using the code provided in Section 9.7, try different initial conditions for the mass m_3 and discuss the result.

38. ⌨ Write the Mathematica code for Example 9.8, and discuss the plots for various initial conditions of the circular restricted three-body problem.

39. In this problem, we evaluate and plot the Lagrange points for the circular restricted three-body problem with $m = 0.01$. We continue with the situation where $m_1 \gg m_2$ such as the Earth-Sun-satellite system. Note that the value of m in the Earth-Sun satellite is close to 10^{-6} but we will continue with $m = 0.01$ for visualization purposes. The Lagrange points are points where $\ddot{x} = \ddot{y} = 0$ and $\dot{x} = \dot{y} = 0$ in the rotating frame discussed in Section 9.7. Using a computer, solve numerically (9.7.22) and (9.7.23) when $\ddot{x} = \ddot{y} = \dot{x} = \dot{y} = 0$. Plot the resulting points on the same graph as a circle with radius $1 - m$, which represents the orbit of m_2. You should find that there are a total of five Lagrange points and that two of the Lagrange points lie along the orbit of m_2 and three of them lie along the same line (one of those is also located on the orbit).

10 Motion in Noninertial Reference Frames

As we saw in Chapter 1, Newton's second law is valid only for inertial reference frames. However, it is sometimes the case where we need to describe the motion in a noninertial reference frame. For example, long-range missile trajectories need to be described using a noninertial reference frame, a frame fixed to the surface of the rotating (and therefore noninertial) Earth. In Chapter 1 we found that using Newton's second law in a noninertial reference frame, resulted in the appearance of inertial forces. In this chapter, we will expand on the idea of inertial forces; specifically, those arising in rotating reference frames. We will discuss how a vector in a rotating reference frame can be described in a nonrotating frame, and from that description, we will show how applying Newton's second law in a rotating frame results in inertial forces such as the Coriolis force and the centrifugal force. We will then study how the Coriolis and centrifugal forces affect the motion of a particle near the surface of the Earth. Finally, we will explore the famous problems of the Foucault pendulum and projectile motion in a noninertial frame.

10.1 MOTION IN A NONROTATING ACCELERATING REFERENCE FRAME

Although we examined the case of a nonrotating accelerating frame in Chapter 1, it is worthwhile revisiting it, this time using three dimensions. Consider a particle of mass m whose motion is being described using two reference frames as shown in Figure 10.1.

The reference frame S' shown in Figure 10.1 is an inertial frame which we will imagine to be at rest, and the frame S is accelerating away from S' with an acceleration $\mathbf{A} = \ddot{\mathbf{R}}$. In Chapter 1, we denoted the moving frame with primes. In this chapter, we will switch our notation for convenience and use primed variables to describe quantities measured in the inertial frame S' and unprimed variables for quantities measured in the noninertial frame S. We will be focusing on the motion of a particle in the noninertial frame, and we do not want to complicate the notation with additional primes.

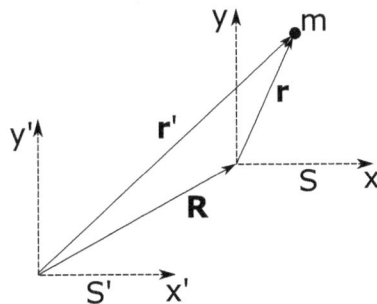

Figure 10.1: A particle of mass m located at a position $\mathbf{r}'$ relative to the origin of an inertial frame S' and at a location $\mathbf{r}$ relative to the origin of a noninertial frame S. The vector $\mathbf{R}$ measures the position of the noninertial frame's origin relative to the origin of the inertial frame.

The position of the particle measured in S' can be written as $\mathbf{r'} = \mathbf{R} + \mathbf{r}$ and the velocity of the particle, as measured in the frame S' is therefore,

$$\dot{\mathbf{r}}' = \dot{\mathbf{r}} + \mathbf{V} \tag{10.1.1}$$

where $\mathbf{V} = \dot{\mathbf{R}}$ is the instantaneous velocity of the frame S relative to S'. For example, if S' is a stationary reference frame on the ground and S is fixed in an accelerating car, then the particle's velocity relative to the ground ($\dot{\mathbf{r}}'$) is the velocity of the particle relative to the car ($\dot{\mathbf{r}}$) plus the velocity of the car ($\mathbf{V}$). It should be noted that (10.1.1) only holds for non-relativistic cases.

Next, we differentiate (10.1.1) to get the acceleration of the particle as measured in the noninertial frame,

$$\ddot{\mathbf{r}} = \ddot{\mathbf{r}}' - \mathbf{A} \tag{10.1.2}$$

where $\mathbf{A} = \dot{\mathbf{V}}$. Notice that similar to what was found in Chapter 1, an additional acceleration is measured in S', which is not measured in S. Multiplying (10.1.2) by the mass of the particle m,

$$m\ddot{\mathbf{r}} = m\ddot{\mathbf{r}}' - m\mathbf{A} \tag{10.1.3}$$
$$\mathbf{F} = \mathbf{F}' - m\mathbf{A} \tag{10.1.4}$$

where $\mathbf{F} = m\ddot{\mathbf{r}}$ are the forces acting on the particle as measured in the noninertial frame, and $\mathbf{F}' = m\ddot{\mathbf{r}}'$ are the forces acting on the system as measured in the inertial frame. The forces acting on the particle such as gravity, air resistance, friction, etc. These forces are measured in both frames. However, there is an additional force measured in the noninertial frame, $-m\mathbf{A}$. This additional force is sometimes called an *inertial force*, a fictitious force, or a pseudo-force. In order to use Newton's second law to describe motion in a noninertial frame, we need to include the additional inertial force, which is not due to the particle's interaction with another body, but rather arises from the acceleration of the noninertial frame.

Although the inertial force is sometimes called the fictitious force, inertial forces are real to those moving in a noninertial reference frame. You have experienced noninertial forces in an accelerating car as you are being pushed back on the seat, or on a rotating carnival ride, where you are pushed against the side of a carriage as the ride spins. In each case, the force is due to your inertia as opposed to the interaction with another body, hence the name inertial force, since it is introduced in order to account for all of the accelerations you are experiencing. By including inertial forces, we are able to use Newton's second law to describe motion in noninertial reference frames.

Example 10.1: A pendulum with an accelerating support

Consider a simple plane pendulum of mass m and a massless rod of length ℓ in a downward uniform gravitational field $\mathbf{g}$, and whose support has a vertical upward acceleration of $\mathbf{a}$ as shown in Figure 10.2. Find the frequency of small oscillations of the pendulum.

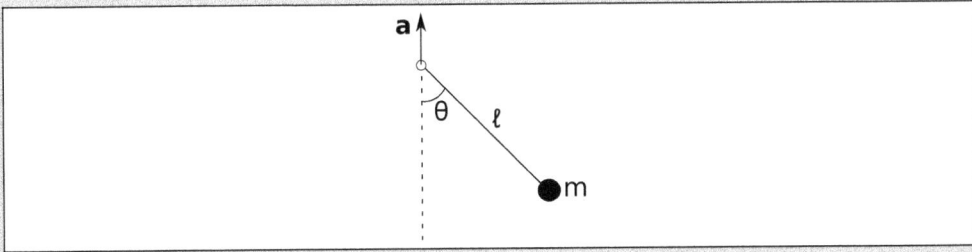

Figure 10.2: An upward accelerating pendulum with mass m and length ℓ from Example 10.1.

Solution:
According to (10.1.3), the force in the frame S which is accelerating with the pendulum is,

$$\mathbf{F} = (\mathbf{T} + m\mathbf{g}) - m\mathbf{a} \qquad (10.1.5)$$

where $\mathbf{T}$ is the tension in the pendulum, $\mathbf{g}$ is the acceleration due to gravity, and $\mathbf{a}$ is a vector that has a magnitude of a and points in the direction opposite of $\mathbf{g}$. The term in the parentheses of (10.1.5) is $\mathbf{F}'$, the sum of the forces acting on the system according to Newton's second law in the inertial frame. We can rewrite (10.1.5) as

$$\mathbf{F} = \mathbf{T} + m\mathbf{g}_{\mathrm{eff}} \qquad (10.1.6)$$

where $\mathbf{g}_{\mathrm{eff}} = \mathbf{g} - \mathbf{a}$ is the effective gravitational field which includes the gravitational attraction and the acceleration of the pendulum's support. It then follows that in the frame S, the equation of motion of the pendulum for small oscillations is,

$$\ddot{\theta} + \frac{g_{\mathrm{eff}}}{\ell}\theta = 0 \qquad (10.1.7)$$

where θ is the angle that the pendulum makes with the vertical (as usual). From Chapter 6 we know that the coefficient in front of θ in (10.1.7) is the square of ω_0, the frequency of small oscillations. Therefore, we find

$$\omega_0 = \sqrt{\frac{|\mathbf{g} - \mathbf{a}|}{\ell}} \qquad (10.1.8)$$

We see that the upward acceleration effectively reduces the gravitational field. The result is that the accelerated frame is the same as having an additional gravitational field (in this case, whose acceleration is in the opposite direction of $\mathbf{g}$). The fact that accelerating frames are indistinguishable from gravitational fields is an important element in the general theory of relativity.

10.2 ANGULAR VELOCITY AS A VECTOR

Before we move on to rotating frames, we need to first reexamine angular velocity, and in particular, the vector nature of angular velocity. Euler's rotation theorem says that any infinitesimal displacement of a rigid object such that a point on the body remains fixed, is equivalent to a rotation about an axis that runs through the fixed point. This theorem is difficult to prove, but we don't need to prove it here. However, as an illustrative example, consider a wheel rolling along a road. Euler's theorem says that an infinitesimal displacement of the wheel can be described as a rotation about the contact point between the wheel and the road.

Euler's theorem tells us that in order to specify the rotation about a point, we need only the direction of the axis and the amount rotated about the axis. Of course, if we are interested in the rate of rotation, the *angular velocity*, then we would need the direction of the axis of rotation *and* the rate of the rotation. This means we can write the angular velocity as a vector ω that lies along the axis of rotation and whose magnitude is the rate of rotation. A stationary spinning top may have an angular velocity of 2π rad/sec with an axis that is vertically oriented. However, does ω point up or down? The answer to that is determined by the right-hand rule,

Right-Hand Rule for Rotation

Curl the fingers of your right hand in the direction of the rotation. Your thumb points in the direction of ω.

The right-hand rule is illustrated in Figure 10.3. The circle with arrows in Figure 10.3 represents a wheel rotating in a counterclockwise direction. As you curl the fingers of your right hand in the direction of the arrows on the wheel, your thumb should point up, away from the page. Your thumb is pointing in the direction of the wheel's angular velocity.

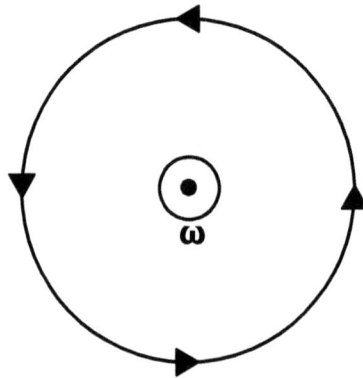

Figure 10.3: The right-hand rule. The large circle represents a wheel and the arrows on the circle represent a counter clockwise rotation of the wheel. Curl the fingers of your right hand in the direction of the arrows on the circle. Your thumb should point in the direction of the vector ω.

If an object's angular velocity is changing with time, then both the rate of rotation and/or the orientation of the rotation axis is changing in time. Situations that involve changes in the orientation of the rotation axis can be difficult to describe mathematically. In this chapter, we will focus primarily on systems with fixed angular velocities, i.e., both the rate of rotation and the direction of the rotation axis will remain constant.

In addition to a formal definition of angular velocity, we will also need a relationship between the velocity of a particle and its angular velocity. Recall that the tangential velocity v of a particle moving along a circle of radius r with angular velocity ω is $v = \omega r$. In this particular case, the axis of rotation went through the center of the circle and was perpendicular to the plane of the circle. This simple relationship between v and ω results from the fact that all of the motion is restricted to a plane. That is not generally true.

Consider a particle, represented by the black dot in Figure 10.4 fixed to a location on the Earth's surface in the Northern Hemisphere. The angular velocity of the Earth is essentially constant and points from the South Pole to the North Pole. While every point on the Earth has the same angular velocity, the particle's latitude will affect the particle's tangential velocity, $\mathbf{v}$.

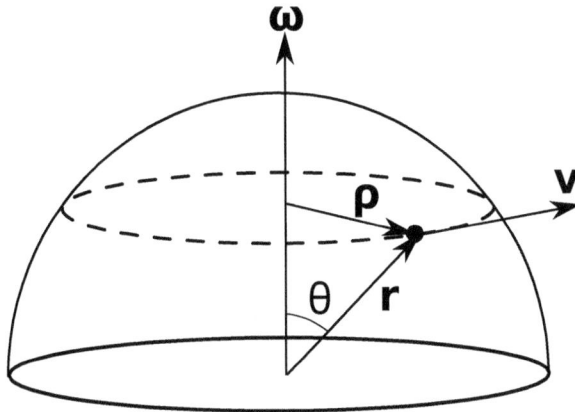

Figure 10.4: The Northern Hemisphere of the Earth showing the location of a point particle (black dot) and its velocity, $\mathbf{v}$, which is tangent to the dashed circle.

In Figure 10.4, the particle travels in a circle of radius $\rho = |\boldsymbol{\rho}|$ as the Earth rotates. The latitude of the particle is $\pi/2 - \theta$. As θ increases, the particle needs to travel a greater distance in the same amount of time, while the angular velocity is the same at all latitudes. Therefore, $v = |\mathbf{v}|$ must increase with an increasing θ. We recall that the tangential velocity of the particle is $v = \rho\omega$ where $\rho = r\sin\theta$. Therefore, $v = r\omega\sin\theta$ and we can write the tangential velocity as,

$$\mathbf{v} = \boldsymbol{\omega} \times \mathbf{r} \qquad (10.2.1)$$

Notice that (10.2.1) is the general form for the familiar equation $v = r\omega$, where $\theta = \pi/2$. We need to use (10.2.1) in this case because the origin of the inertial frame is the center of the Earth, and the center of the Earth is not the center of the circle along which the particle travels. Furthermore, the center of the Earth will serve as the origin of the inertial frame that we will use in future problems.

Equation (10.2.1) is not unique to the velocity vector. Note that we can rewrite (10.2.1) as,

$$\frac{d\mathbf{r}}{dt} = \boldsymbol{\omega} \times \mathbf{r} \qquad (10.2.2)$$

which is not a formula unique to the vector $\mathbf{r}$. In fact, for any vector $\mathbf{Q}$ that is constant in the rotating reference frame, its time derivative as measured in the nonrotating frame is,

$$\frac{d\mathbf{Q}}{dt} = \boldsymbol{\omega} \times \mathbf{Q} \qquad (10.2.3)$$

In the next section, we will provide a derivation of a more general form of (10.2.3), which will include vectors that are not constant in the rotating reference frame.

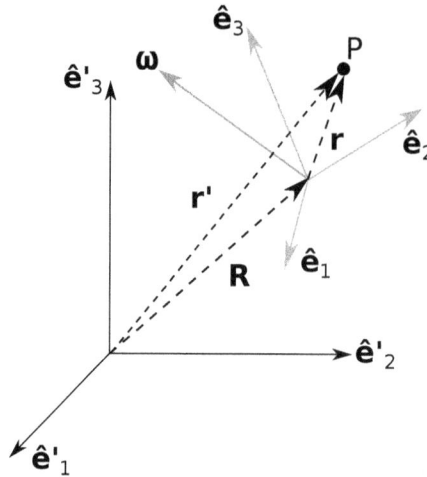

Figure 10.5: The inertial frame S' (primed unit vectors) and the noninertial rotating frame S (unprimed unit vectors, red in the e-book). The vector $\mathbf{R}$ represents the location of the origin of the noninertial frame as measured in the inertial system. The vector $\boldsymbol{\omega}$ represents the angular velocity of the noninertial frame.

10.3 TIME DERIVATIVES OF VECTORS IN ROTATING COORDINATE FRAMES

Recall that in Section 10.1, we related quantities such as velocity and acceleration between inertial and noninertial frames. This was easily done because neither frame was rotating. However, once we allow for the noninertial frame to rotate with respect to the inertial frame, things become a bit more difficult. We will still be able to relate quantities such as the acceleration measured in the rotating frame to the acceleration measured in the inertial frame, but to do so, we will need to include the angular velocity of the rotating frame.

We begin by setting up two specific reference frames. As in Section 10.1, we will use primed coordinates to denote the fixed inertial frame as S', and unprimed coordinates for the noninertial frame S. The origin of the noninertial frame S is located at a location $\mathbf{R}$ from the origin of S' and the noninertial frame is rotating with an angular velocity $\boldsymbol{\omega}$ with respect to S'. The two coordinate systems are shown in Figure 10.5.

In Figure 10.5, the inertial frame S' is represented by the primed unit vectors $\hat{\mathbf{e}}'_i$, where we are using generalized unit vectors in order to simplify an upcoming derivation. The noninertial frame S is represented by the unprimed unit vectors $\hat{\mathbf{e}}_i$ and are colored red in the e-book. The unit vectors $\hat{\mathbf{e}}_i$ are fixed in the rotating frame. In other words, $\hat{\mathbf{e}}_i$ rotate with the frame and, therefore, an observer in the rotating frame will measure $\hat{\mathbf{e}}_i$ to be constant in both magnitude and direction. The vector $\boldsymbol{\omega}$ (green in the e-book) represents the angular velocity of the noninertial frame. The point P can be located using either coordinate system. As measured in the inertial frame S', the point P is located at, $\mathbf{r}' = \mathbf{R} + \mathbf{r}$.

Now let us consider an arbitrary vector $\mathbf{Q}$ which is measured in the noninertial frame. The vector $\mathbf{Q}$ could be the position of a particle at the point P, the velocity of the particle, or any other vector quantity. We will compute the time derivative of $\mathbf{Q}$ as measured in the inertial (fixed) frame and the noninertial (rotating) frame using the following notation,

$$\left(\frac{d\mathbf{Q}}{dt}\right)_f = \text{the time derivative of } \mathbf{Q} \text{ relative to the inertial (fixed) frame } S'$$

$$\left(\frac{d\mathbf{Q}}{dt}\right)_r = \text{the time derivative of } \mathbf{Q} \text{ relative to the noninertial (rotating)}$$

$$\text{frame } S.$$

We begin by writing $\mathbf{Q}$ in terms of the unit vectors that are fixed in the noninertial frame,

$$\mathbf{Q} = \sum_{i=1}^{3} Q_i \hat{\mathbf{e}}_i \tag{10.3.1}$$

The coefficients Q_i are the same in each frame. Observers in the inertial frame, would see $\hat{\mathbf{e}}_i$ vary with time, but that would be the only difference between what is observed in the two frames. If we calculate the time derivative of $\mathbf{Q}$ relative to the noninertial frame we find,

$$\left(\frac{d\mathbf{Q}}{dt}\right)_r = \sum_{i=1}^{3} \dot{Q}_i \hat{\mathbf{e}}_i \tag{10.3.2}$$

Because the scalar coefficients Q_i are the same in each frame, we do not need to distinguish whether $\dot{Q}_i$ is relative to the inertial or noninertial frame. Note that the derivative does not affect the unit vectors because they are constant relative to the noninertial frame. Next, we compute the time derivative of $\mathbf{Q}$ relative to the fixed frame,

$$\left(\frac{d\mathbf{Q}}{dt}\right)_f = \sum_{i=1}^{3} \dot{Q}_i \hat{\mathbf{e}}_i + \sum_{i=1}^{3} Q_i \left(\frac{d\hat{\mathbf{e}}_i}{dt}\right)_f \tag{10.3.3}$$

The first term on the right-hand side of (10.3.3) is what we found when computing the time derivative relative to the noninertial frame. Notice that the second term in the right-hand side of (10.3.3) is the time derivative of the vector $\hat{\mathbf{e}}_i$, which is constant in the rotating frame. We can use the general equation (10.2.3) to write,

$$\left(\frac{d\hat{\mathbf{e}}_i}{dt}\right)_f = \boldsymbol{\omega} \times \hat{\mathbf{e}}_i \tag{10.3.4}$$

Therefore, (10.3.3) becomes,

$$\left(\frac{d\mathbf{Q}}{dt}\right)_f = \sum_{i=1}^{3} \dot{Q}_i \hat{\mathbf{e}}_i + \sum_{i=1}^{3} \boldsymbol{\omega} \times (Q_i \hat{\mathbf{e}}_i) \tag{10.3.5}$$

or using (10.3.2):

Time Derivative of a Vector Q Relative to an Inertial Frame

$$\left(\frac{d\mathbf{Q}}{dt}\right)_f = \left(\frac{d\mathbf{Q}}{dt}\right)_r + \boldsymbol{\omega} \times \mathbf{Q} \tag{10.3.6}$$

Notice that (10.3.6) is what we would expect from our studies of translationally moving reference frames. To better understand this, let $\mathbf{Q} = \mathbf{r}$, the position of a particle as measured in the noninertial (rotating) frame. The first term in the right-hand side of (10.3.6) is the velocity measured in the rotating frame. The second term is the translational velocity of the particle at $\mathbf{r}$ due to the rotation of the noninertial frame. Hence, the velocity measured by an observer in the fixed frame is the velocity of the particle in the moving frame plus the velocity of the frame itself. Note that one result of (10.3.6) is that the angular acceleration $\dot{\boldsymbol{\omega}}$ is the same in each frame because $\boldsymbol{\omega} \times \boldsymbol{\omega} = 0$.

Equation (10.3.6) tells us that a vector's rate of change can be related between inertial and noninertial frames. Newton's second law provides a particle's equations of motion by stating that the net force acting on a particle is proportional to the particle's change in velocity as measured in an inertial frame. Thus, (10.3.6) allows us to relate the acceleration of a particle in an inertial frame, found from Newton's second law, to situations where the observer is in a noninertial frame. In the next section, we will use (10.3.6) to find Newton's second law in a rotating frame.

10.4 NEWTON'S SECOND LAW IN A ROTATING FRAME

In order to find an equation which describes the motion of a particle in a rotating frame, we will need to modify Newton's second law for use in noninertial frames. To find the form of Newton's second law in a rotating frame S, we will use as a reference Figure 10.5. Suppose that there is a particle with mass m located at the point P in Figure 10.5. We will allow the noninertial frame to be both rotating and have a linear acceleration relative to S'. It will be helpful to think of the inertial frame as fixed. Let $\boldsymbol{\omega}$ be the angular velocity of the noninertial frame S.

In the inertial frame S' we can write Newton's second law as,

$$\mathbf{F}' = m \left(\frac{d\mathbf{v}'}{dt} \right)_f \tag{10.4.1}$$

where $\mathbf{v}'$ is the velocity of the particle relative to the inertial frame S', and $\mathbf{F}'$ is the net force acting on the particle. The net force $\mathbf{F}'$ includes the interaction forces between the mass m and other bodies such as gravity, air resistance, friction, etc. In order to find the form of Newton's second law for a *noninertial* frame, we will need to rewrite the acceleration as measured in the inertial frame $(d\mathbf{v}'/dt)_f$ in terms of the acceleration as measured in the noninertial frame.

We begin by finding $\mathbf{v}'$. Figure 10.5 shows that,

$$\mathbf{r}' = \mathbf{R} + \mathbf{r} \tag{10.4.2}$$

The velocity $\mathbf{v}'$ is found by taking the time derivative of $\mathbf{r}'$ relative to the inertial frame.

$$\mathbf{v}' = \left(\frac{d\mathbf{r}'}{dt} \right)_f = \left(\frac{d\mathbf{R}}{dt} \right)_f + \left(\frac{d\mathbf{r}}{dt} \right)_f \tag{10.4.3}$$

We use (10.3.6) on the second term in the right-hand side of (10.4.3) in order to find how an observer in the inertial frame S' measures the velocity of the particle moving in the noninertial frame S,

$$\left(\frac{d\mathbf{r}}{dt} \right)_f = \left(\frac{d\mathbf{r}}{dt} \right)_r + \boldsymbol{\omega} \times \mathbf{r} \tag{10.4.4}$$

where $\boldsymbol{\omega}$ is the angular velocity of S. By using (10.4.4) and defining the following:

$$\mathbf{v}' \equiv \left(\frac{d\mathbf{r}'}{dt}\right)_f \tag{10.4.5}$$

$$\mathbf{V} \equiv \left(\frac{d\mathbf{R}}{dt}\right)_f \tag{10.4.6}$$

$$\mathbf{v} \equiv \left(\frac{d\mathbf{r}}{dt}\right)_r \tag{10.4.7}$$

we can write:

$$\mathbf{v}' = \mathbf{V} + \mathbf{v} + (\boldsymbol{\omega} \times \mathbf{r}) \tag{10.4.8}$$

where:

$\mathbf{v}'$ = the velocity of the particle relative to the inertial frame S'
$\mathbf{V}$ = the linear velocity of the noninertial frame relative to S'
$\mathbf{v}$ = the velocity of the particle relative to the noninertial frame S
$\boldsymbol{\omega}$ = the angular velocity of the noninertial frame S relative to S'
$\boldsymbol{\omega} \times \mathbf{r}$ = the velocity of the particle due to the rotation of S relative to S'

Next, we return to (10.4.1) and insert (10.4.8) to obtain:

$$\mathbf{F}' = m\left[\left(\frac{d\mathbf{V}}{dt}\right)_f + \left(\frac{d\mathbf{v}}{dt}\right)_f + (\dot{\boldsymbol{\omega}} \times \mathbf{r}) + \boldsymbol{\omega} \times \left(\frac{d\mathbf{r}}{dt}\right)_f\right] \tag{10.4.9}$$

Now, we will go through the right-hand side of (10.4.9) term by term. The first term in the right-hand side of (10.4.9) is the linear acceleration of the noninertial frame,

$$\mathbf{A} = \left(\frac{d\mathbf{V}}{dt}\right)_f \tag{10.4.10}$$

The second term in the right-hand side of (10.4.9) is rewritten using (10.3.6):

$$\left(\frac{d\mathbf{v}}{dt}\right)_f = \left(\frac{d\mathbf{v}}{dt}\right)_r + (\boldsymbol{\omega} \times \mathbf{v}) \tag{10.4.11}$$

$$= \mathbf{a} + (\boldsymbol{\omega} \times \mathbf{v}) \tag{10.4.12}$$

where $\mathbf{a} = (d\mathbf{v}/dt)_r$ is the acceleration of the particle as measured in the noninertial frame. The third term in the right-hand side of (10.4.9) does not need simplification because $\dot{\boldsymbol{\omega}}$ is the same in each reference frame. Finally, the fourth term in the right-hand side of (10.4.9) is rewritten using (10.3.6),

$$\boldsymbol{\omega} \times \left(\frac{d\mathbf{r}}{dt}\right)_f = \boldsymbol{\omega} \times \left(\frac{d\mathbf{r}}{dt}\right)_r + \boldsymbol{\omega} \times (\boldsymbol{\omega} \times \mathbf{r}) \tag{10.4.13}$$

$$= (\boldsymbol{\omega} \times \mathbf{v}) + [\boldsymbol{\omega} \times (\boldsymbol{\omega} \times \mathbf{r})] \tag{10.4.14}$$

By inserting (10.4.10)-(10.4.14) into (10.4.9), we obtain,

$$\mathbf{F}' = m\mathbf{A} + m\mathbf{a} + 2m\left(\boldsymbol{\omega} \times \mathbf{v}\right) + m\left(\dot{\boldsymbol{\omega}} \times \mathbf{r}\right) + m\left[\boldsymbol{\omega} \times (\boldsymbol{\omega} \times \mathbf{r})\right] \tag{10.4.15}$$

If we define

$$\mathbf{F} \equiv \mathbf{F}' - m\mathbf{A} - 2m\left(\boldsymbol{\omega} \times \mathbf{v}\right) - m\left(\dot{\boldsymbol{\omega}} \times \mathbf{r}\right) - m\left[\boldsymbol{\omega} \times (\boldsymbol{\omega} \times \mathbf{r})\right] \tag{10.4.16}$$

as the force experienced by the particle in the noninertial frame, then for noninertial frames, Newton's second law takes the form

$$\mathbf{F} = m\mathbf{a} \tag{10.4.17}$$

Recall that $\mathbf{a}$ is the acceleration of the particle as measured in the noninertial frame. Although the above equation looks like how we have written Newton's second law from Chapter 1 and onwards, it is important to remember all of the terms that are packed into $\mathbf{F}$ as defined in (10.4.16). The force $\mathbf{F}$ is sometimes called the *effective force* acting on the particle because it includes both interaction forces and inertial forces.

Notice that there are four inertial forces introduced in (10.4.16). The inertial force $-m\mathbf{A}$ is the same one we found in Section 10.1, and it is due to the linear acceleration of S. The second term in (10.4.16), $-2m(\boldsymbol{\omega} \times \mathbf{v})$, is the so-called *Coriolis force* which we will discuss in detail later. The term, $-m(\dot{\boldsymbol{\omega}} \times \mathbf{r})$, is the inertial force associated with the angular acceleration of S'. For most of the problems in this chapter, the frame S will be attached to the surface of the Earth, and to a good approximation, $\dot{\boldsymbol{\omega}} = 0$. This force will not be present in our discussion. The final term, $-m[\boldsymbol{\omega} \times (\boldsymbol{\omega} \times \mathbf{r})]$, is the *centrifugal force* which we will also discuss later in detail.

We have learned that Newton's second law can be used, with modification, for problems involving noninertial frames. The trick is that we have to include additional inertial forces and not just the forces of interactions between bodies in the system. Just like in Section 10.1, these inertial forces are real to the observer in the noninertial frames.

Next, we will discuss further the Coriolis and centrifugal forces. These two inertial forces play an important role when describing the motion of an object near the Earth's surface.

For the remainder of this chapter, the inertial frame S' has its origin at the center of the Earth and $\hat{\mathbf{e}}_3'$ points along the Earth's rotation axis from the Earth's center to its North Pole. Hence $\boldsymbol{\omega} = \omega \hat{\mathbf{e}}_3'$ where $\omega = 2\pi$ rad/day. Furthermore, the noninertial frame S has its origin fixed at one point on the Earth's surface, the $\hat{\mathbf{e}}_1\hat{\mathbf{e}}_2$-plane is tangent to the Earth's surface at the origin of S, and $\hat{\mathbf{e}}_3$ is pointing locally up (away from the center of the Earth).

10.4.1 CENTRIFUGAL FORCE

As we saw in the previous section, the term

$$\mathbf{F}_{\text{cent}} = -m[\boldsymbol{\omega} \times (\boldsymbol{\omega} \times \mathbf{r})] \tag{10.4.18}$$

in (10.4.16) is called the centrifugal force and is present due to the rotation of the noninertial frame. We can understand the direction of the centrifugal force as follows.

In Figure 10.6, a particle of mass m (not shown) is located at $\mathbf{r}$ in a reference frame that is rotating relative to an inertial frame. In the inertial frame, an observer would see the particle moving in a circle of radius $\rho = r\sin\theta$. In order to find the direction of the centrifugal force $-m[\boldsymbol{\omega} \times (\boldsymbol{\omega} \times \mathbf{r})]$, we first need to find the direction of the vector $\boldsymbol{\omega} \times \mathbf{r}$. The vector $\boldsymbol{\omega} \times \mathbf{r}$ is perpendicular to both $\boldsymbol{\omega}$ and $\mathbf{r}$ and therefore lies in the plane of the circle as shown most clearly in the bird's-eye view in Figure 10.6(b). Next, the direction of $\boldsymbol{\omega} \times (\boldsymbol{\omega} \times \mathbf{r})$ is perpendicular to both $\boldsymbol{\omega}$ and $\boldsymbol{\omega} \times \mathbf{r}$, and therefore points radially in towards the center of the circle. Finally, the centrifugal force is $-m[\boldsymbol{\omega} \times (\boldsymbol{\omega} \times \mathbf{r})]$, which points in the opposite direction of $\boldsymbol{\omega} \times (\boldsymbol{\omega} \times \mathbf{r})$ and therefore points radially outward from the center of the circle.

We can compute the magnitude of the centrifugal force from (10.4.18). Let us focus on describing the motion along the circle in cylindrical coordinates ρ, ϕ, and z. Note that in this case, we want an azimuthal angle to describe the motion *around* the circle of radius ρ, as shown in Figure 10.6a. Hence we will use ϕ to denote that angle consistent with spherical coordinates. Using cylindrical polar coordinates, we find $\boldsymbol{\omega} \times \mathbf{r} = \omega r\sin\theta\hat{\boldsymbol{\phi}}$. Furthermore,

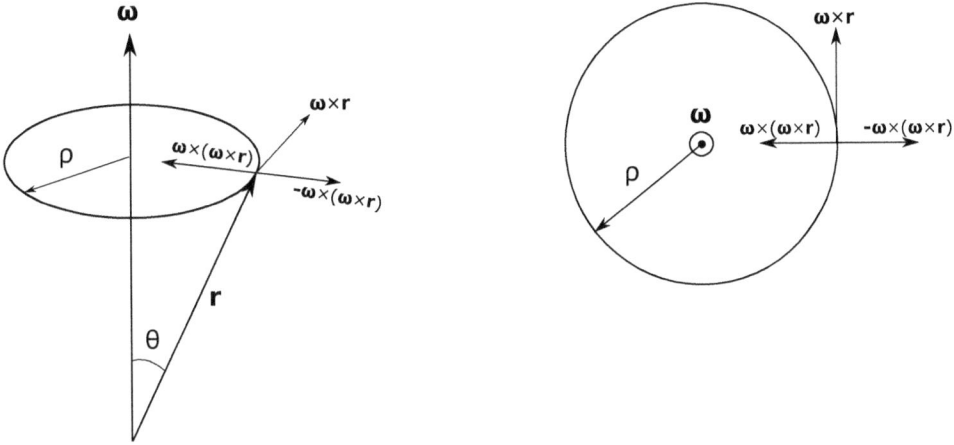

(a) The perspective from a distant observer.

(b) A bird's-eye view of (a) looking down on the vector $\boldsymbol{\omega}$.

Figure 10.6: Centrifugal force in two perspectives. A particle is located at $\mathbf{r}$ in a rotating reference frame which has an angular velocity ω. An observer in a noninertial frame would see the particle as moving in a circle with a radius $\rho = r\sin\theta$. The particle experiences a centrifugal force $-m\left[\boldsymbol{\omega}\times(\boldsymbol{\omega}\times\mathbf{r})\right]$ that points radially outward from the particle's center of revolution.

$\boldsymbol{\omega}\times(\boldsymbol{\omega}\times\mathbf{r}) = -\omega^2 r\sin\theta\,\hat{\boldsymbol{\rho}}$ (note that $\boldsymbol{\omega} = \omega\hat{\mathbf{z}}$ in cylindrical polar coordinates with $\hat{\mathbf{z}}\times\hat{\boldsymbol{\phi}} = -\hat{\boldsymbol{\rho}}$). Therefore, we find that

$$\mathbf{F}_{\text{cent}} = m\rho\omega^2\hat{\boldsymbol{\rho}} \tag{10.4.19}$$

Hence, a particle moving along a circle would experience an outward force with a magnitude $m\rho\omega^2$. Notice that this outward force is consistent with the familiar centripetal force $\mathbf{F}_c = -m r\omega^2\hat{\boldsymbol{\rho}}$. Recall that the centripetal force is the net force that keeps a particle moving along a circular path with a constant speed. Now that we have an understanding of noninertial frames, we have another means of describing uniform circular motion. From the perspective of the particle, which is in a noninertial frame, there is a centrifugal force acting on it, such that the net force is zero in the radial direction.

Because the Earth is rotating, objects near the surface of the Earth experience the centrifugal force. The outward-pointing centrifugal force reduces the effective acceleration due to gravity by an amount depending on the particle's latitude. The next example illustrates this.

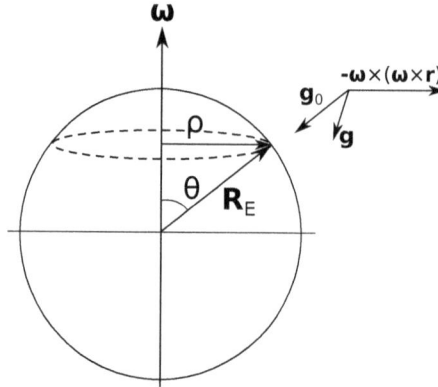

Figure 10.7: A particle located at $\mathbf{R}_E$ on the Earth's surface (solid circle) moves in a circle (dotted circle) as the Earth rotates with an angular velocity $\boldsymbol{\omega}$, in Example 10.2. Vectors for the effective gravitational acceleration $\mathbf{g}$, the centrifugal acceleration $\boldsymbol{\omega} \times (\boldsymbol{\omega} \times \mathbf{r})$, and the acceleration due to gravity $\mathbf{g}_0$ are also shown.

Example 10.2: Gravity and the centrifugal force

Consider a particle of mass m at rest on the surface of the Earth. Find the effective force acting on the particle and compute the difference in the effective acceleration due to gravity between the North Pole and a point on the equator.

Solution:

It is helpful to include a diagram of the particle's location before we perform any calculations. Figure 10.7 shows a particle's position $\mathbf{R}_E$ on the Earth's surface. The position vector $\mathbf{R}_E$ has a magnitude of 6.371 km and points in a direction from the center of the Earth to the location of the particle on the Earth's surface. The angle θ is the same angle from Figure 10.6a and is related to the particle's latitude λ by $\lambda = \pi/2 - \theta$. To a stationary observer above the Earth, the particle appears to move in a circle of radius $\rho = R_E \sin \theta$.

According to (10.4.16), the effective force acting on the particle is,

$$\mathbf{F} = \mathbf{F}' + m\mathbf{g}_0 - m\left[\boldsymbol{\omega} \times (\boldsymbol{\omega} \times \mathbf{R}_E)\right] \qquad (10.4.20)$$

where

$$\mathbf{g}_0 = -\frac{GM_E}{R_E^2}\,\hat{\mathbf{R}}_E$$

is the acceleration due to the gravitational attraction, found from Newton's universal law of gravitation, the Earth's mass is $M_E = 5.972 \times 10^{24}$ kg and $\mathbf{F}'$ is the sum of all of the forces other than gravity acting on the particle. The only inertial force acting on the particle is the centrifugal force. The Coriolis force is zero because the particle is not moving relative to the Earth's surface and, to good approximation, $\boldsymbol{\omega}$ is a constant and therefore $\dot{\boldsymbol{\omega}} = 0$.

We can collect the second and third terms in (10.4.20) and define a new acceleration due to gravity called an *effective acceleration due to gravity* $\mathbf{g}$ where,

$$\mathbf{g} = \mathbf{g}_0 - \boldsymbol{\omega} \times (\boldsymbol{\omega} \times \mathbf{r}) \qquad (10.4.21)$$

and it is this effective acceleration **g** that determines the acceleration due to free fall and the period of a pendulum at a point on Earth. Note that the centrifugal acceleration effectively reduces the acceleration due to gravity. In addition, a plumb line will lie along the direction of **g** and will slightly deviate from true vertical because of the outward-directed centrifugal force shown in the diagram above. Furthermore, the surfaces of the Earth's oceans are perpendicular to **g**, not **g**$_0$.

We can compute the effective acceleration due to gravity at the North Pole and equator by first finding the magnitude of **g**$_0$,

$$g_0 = \frac{GM_E}{R_E^2} = \frac{\left(6.674 \times 10^{-11} \text{ Nm}^2/\text{kg}^2\right)\left(5.972 \times 10^{24} \text{ kg}\right)}{\left(6.371 \times 10^6 \text{ m}\right)^2} = 9.820 \text{ m}/s^2$$

Next, we compute the magnitude of the centripetal acceleration at the North Pole and equator. We will use $\mathbf{a}_{\text{cent}} = R_E\,\omega^2 \sin\theta\,\hat{\rho}$ which comes from dividing (10.4.19) by the mass m. Note that we want the component $a_{\text{cent}} \sin\theta$ of $\mathbf{a}_{\text{cent}}$ which is parallel to $\mathbf{R}_E$. Note that for the North Pole, $\theta = 0$ and for the equator $\theta = \pi/2$. In each case $\omega = 2\pi \text{ rad/day} = 7.272 \times 10^{-5} \text{ rad/s}$,

$$a_{\text{cent}} = R_E\omega^2 \sin^2 0 = 0 \text{ m}/s^2 \qquad\qquad \text{North Pole}$$

$$a_{\text{cent}} = R_E\omega^2 \sin^2 \frac{\pi}{2} = 0.0337 \text{ m}/s^2 \qquad\qquad \text{Equator}$$

The effective gravitational acceleration is therefore, $g = g_0 - a_{\text{cent}}$, and we find that the effective acceleration due to gravity at the North Pole is reduced by 0.0337 m/s^2, or about 0.3%, greater than at the equator. The centrifugal force is responsible for the oblateness of the Earth. The Earth's polar radius is 21.4 km greater than its equatorial radius! The additional bulge at the Earth's equator actually means that the gravitational acceleration at the poles is actually 0.5% greater than that at the equator.

A result of Example 10.2 is that in general, we can rewrite the effective force (10.4.16) acting on a particle near the surface of the Earth as,

$$\mathbf{F} = \mathbf{F}' + m\mathbf{g} - 2m\left(\boldsymbol{\omega} \times \mathbf{v}\right) \tag{10.4.22}$$

where $\mathbf{F}'$ is the sum of all of the interaction forces experienced by the particle excluding its gravitational attraction to the Earth. We then combined $m\mathbf{g}_0$ with the centrifugal force into a new force $m\mathbf{g}$. The final term in (10.4.22) is the Coriolis force, which will be the topic of the next section.

10.4.2 CORIOLIS FORCE

When a particle is moving with a velocity **v** relative to a rotating reference frame, an additional inertial force called the Coriolis force will be present. The Coriolis force is,

$$\mathbf{F}_{\text{cor}} = -2m\left(\boldsymbol{\omega} \times \mathbf{v}\right) \tag{10.4.23}$$

It is clear from (10.4.23) that the Coriolis force depends on the velocity of a particle. Particles that are not moving will not experience a Coriolis force. We can estimate the size of the Coriolis force using (10.4.23). Using $\omega \approx 7.3 \times 10^{-5}$ rad/sec as in Example 10.2, we find that a golf ball at a latitude of 45 degrees North traveling at 70 m/s will experience a Coriolis force of 0.007 N, a very small effect considering its time of flight. An intercontinental ballistic missile traveling at 7 km/s will experience a Coriolis force of 0.7 N which could have a more considerable effect on the missile. However, missiles are also affected by the weather and their courses are corrected by computer guidance systems.

The direction of the Coriolis acceleration (and therefore of the Coriolis force) is determined by the $-\boldsymbol{\omega} \times \mathbf{v}$ term. As shown in Figure 10.8, the Coriolis acceleration is to the right of the velocity when the particle is moving in the Earth's Northern Hemisphere. In the Southern Hemisphere, the Coriolis acceleration will be to the left of the particle's velocity.

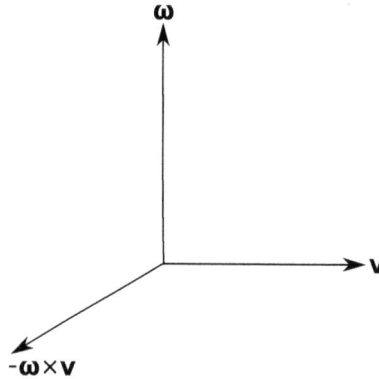

Figure 10.8: The direction of the Coriolis acceleration $\mathbf{a}_{\mathrm{cor}} = -2\,(\boldsymbol{\omega} \times \mathbf{v})$ of a particle moving with a velocity $\mathbf{v}$ relative to the angular velocity $\boldsymbol{\omega}$ of a rotating reference frame. The acceleration deflects the particle towards its right in the Northern Hemisphere, and towards its left in the Southern Hemisphere.

Besides missile trajectories, the Coriolis force has an important impact on Earth's weather, deflecting air currents in the atmosphere. In the Northern Hemisphere, air is deflected toward the right, which leads to high pressure to the right of the airflow, and low pressure to the left (opposite in Southern Hemisphere), resulting in the counterclockwise circulation of air in the Northern Hemisphere and the clockwise circulation in the Southern.

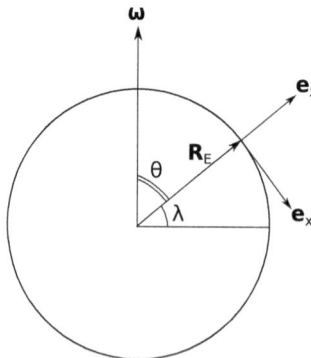

Figure 10.9: Rotating coordinate system $(\mathbf{e}_x, \mathbf{e}_y, \mathbf{e}_z)$ is attached to the Earth's surface (circle), in Example 10.3. Note that $\mathbf{e}_y$ is not shown, but points into the page.

Example 10.3: The horizontal deflection of a plumb line

Consider a particle of mass m falling freely in Earth's gravitational field. The particle is released from rest at a height H above the Earth's surface and is at a latitude of λ. Find the particle's horizontal deflection from the plumb line due to the Coriolis force. Using $H = 100$ meters, plot the horizontal deflection in centimeters as a function of latitude.

Solution:

A diagram showing the coordinate system appears in Figure 10.9, which depicts the Earth as the circle. The origin of the rotating frame S is located at a latitude λ, and is fixed on the Earth's surface. We have chosen our coordinates in the following way. The $\hat{\mathbf{e}}_z$-direction is locally upward, opposing the direction of the effective acceleration due to gravity. The direction, $\hat{\mathbf{e}}_x$, points towards the South. Not shown is the direction $\hat{\mathbf{e}}_y$ that points towards the East in order to preserve a right-handed coordinate system. Note that the angular velocity vector $\boldsymbol{\omega} = \omega \hat{\mathbf{z}}'$, where $\hat{\mathbf{z}}'$ is a unit vector in the inertial frame (fixed to the center of the Earth), and points from the Earth's center to the North Pole.

In order to find the horizontal displacement of the falling particle, we need to solve Newton's second law for rotation using our established coordinate system. Newton's second law is,

$$\mathbf{F} = m\ddot{\mathbf{r}} = m\mathbf{g} - 2m\left(\boldsymbol{\omega} \times \mathbf{v}\right) \tag{10.4.24}$$

Hence, we will need to integrate Newton's second law two times in order to get the displacement we are seeking. Note that the vectors $\mathbf{g}$ and $\mathbf{v}$ are already in the unprimed noninertial coordinate system. However, ω is in the primed or inertial coordinate system. We can transform ω to the noninertial coordinate system, by thinking of the unprimed coordinates as a rotation of the primed coordinates through an angle θ about the $\hat{\mathbf{e}}_y$ direction. In order to obtain the form of $\boldsymbol{\omega}$ in the noninertial coordinates, we can apply the rotation matrix about the y-axis to the vector $\boldsymbol{\omega}$ in the inertial coordinates,

$$\begin{pmatrix} \omega_x \\ \omega_y \\ \omega_z \end{pmatrix} = \begin{pmatrix} \cos\theta & 0 & -\sin\theta \\ 0 & 1 & 0 \\ \sin\theta & 0 & \cos\theta \end{pmatrix} \begin{pmatrix} 0 \\ 0 \\ \omega \end{pmatrix} = \begin{pmatrix} -\omega\sin\theta \\ 0 \\ \omega\cos\theta \end{pmatrix} = \begin{pmatrix} -\omega\cos\lambda \\ 0 \\ \omega\sin\lambda \end{pmatrix} \tag{10.4.25}$$

where we used $\lambda = \pi/2 - \theta$. If we assume that the x and y components of the falling particle's velocity are negligible and use $v_z = -gt$, we can compute the Coriolis acceleration $\omega \times \mathbf{v}$ using (10.4.25),

$$\boldsymbol{\omega} \times \mathbf{v} = \begin{vmatrix} \hat{\mathbf{e}}_x & \hat{\mathbf{e}}_y & \hat{\mathbf{e}}_z \\ -\omega\cos\lambda & 0 & \omega\sin\lambda \\ 0 & 0 & -gt \end{vmatrix} = -\omega\, g\, t\cos\lambda\, \hat{\mathbf{e}}_y \tag{10.4.26}$$

Using $\mathbf{g} = -g\hat{\mathbf{e}}_z$ and (10.4.26) we find that (10.4.24) becomes three equations,

$$\ddot{x} = 0 \tag{10.4.27}$$
$$\ddot{y} = 2\omega\, g\, t\cos\lambda \tag{10.4.28}$$
$$\ddot{z} = -g \tag{10.4.29}$$

which shows us that there is no deflection to the South (x-direction) and also shows that the vertical motion is a free fall. The deflection is towards the East (y-direction) and is found by integrating (10.4.28) twice, once with with initial condition $\dot{y}(0) = 0$ and then a second time with initial condition $y(0) = 0$. The result is,

$$y(t) = \frac{1}{3}\omega\,g\,t^3 \cos\lambda \qquad (10.4.30)$$

We can find the time of flight t by integrating (10.4.29) twice to get $z(t) = z(0) - \frac{1}{2}gt^2$ (note $\dot{z}(0) = 0$) with $z(0) = H$. Solving for t, we find $t = \sqrt{2H/g}$ and therefore the displacement from the plumb line of the falling particle is:

$$y = \frac{1}{3}\omega\sqrt{\frac{8H^3}{g}}\cos\lambda \qquad (10.4.31)$$

How large is this? An object dropped from a height of 100 m at a latitude of 45° North is deflected by 1.55 cm, if we neglect the effects of air resistance.

Python Code

We convert the latitude from radians to degrees to create the plot. Notice that we multiply the variable `deflection` by 100 to get the deflection in centimeters.

```
import numpy as np
import matplotlib.pyplot as plt


w = 7.272*10**(-5)   # angular speed of Earth
g = 9.8
H = 100


l = np.linspace(0,np.pi/2) # latitude, lambda

deflection = 1/3*w*np.sqrt(8*H**3/g)*np.cos(l)

plt.plot(l*180/np.pi,deflection*100)
plt.ylabel("deflection (cm)")
plt.xlabel("latitude")
plt.show()
```

Mathematica Code

Note that we plotted latitude in degrees and needed to adjust the argument of `Cos` accordingly. We multiplied the function `Deflection` by 100 in `Plot` to obtain centimeters.

```
SetOptions[Plot, Axes->False, Frame->True, BaseStyle->{FontSize->16}];

H = 100;

w = 7.272 * 10^(-5);  (* angular velocity of Earth *)

g = 9.8;

Deflection[λ_]:=1/3 * w * Sqrt[8 * H^3/g] * Cos[λ * π/180];

Plot[100 * Deflection[λ], {λ, 0, 90},

FrameLabel->{"latitude (degrees)", "deflection (cm)"}]
```

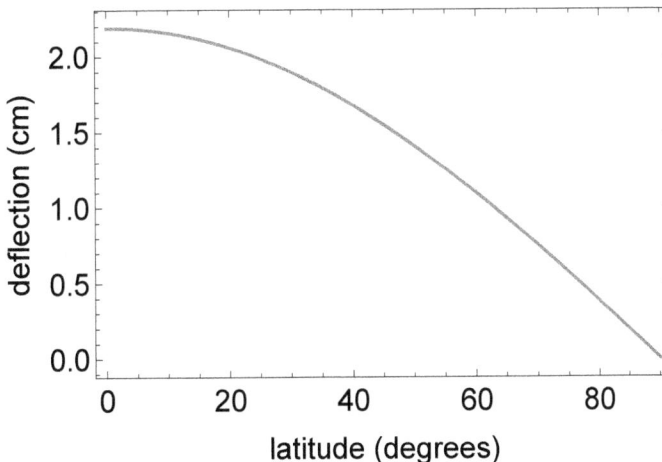

Figure 10.10: Output of the code for Example 10.3. The deflection is highest (2.2 cm) at the equator ($\lambda = 0$) and falls to zero at the North Pole.

10.5 FOUCAULT PENDULUM

One of the most famous problems involving noninertial reference frames is the Foucault pendulum, which was first devised by French physicist Jean-Bernard-Léon Foucault (1819–1868). The pendulum consists of a heavy mass m suspended from a tall ceiling by a long and light wire. As the pendulum oscillates, it precesses in the horizontal plane. In some cases, the pendulum is set up such that it knocks over pegs as it precesses. You might have seen a Foucault pendulum in-person at a science museum or university. A quick search for "Foucault pendulum" online will result in many videos of Foucault pendulums in action. Foucault originally conceived the pendula as a demonstration of the Earth's rotation. In 1851 he used a 28-kg mass suspended from a 67-m wire from the dome of the Pantheon in Paris, France. In this section, we will derive the equations of motion for the Foucault pendulum and show how its precession period is proportional to the pendulum's latitude.

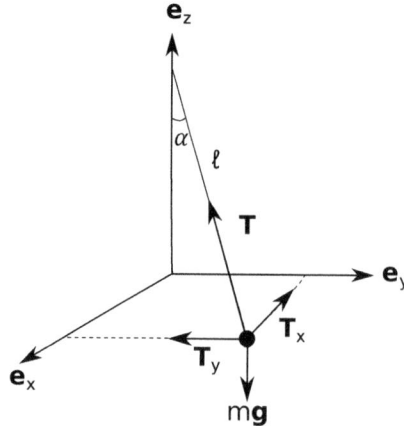

Figure 10.11: A Foucault pendulum of mass m and length ℓ suspended from a large vertical height at a point on the z-axis.

We begin with a schematic diagram of the pendulum. Figure 10.11 shows a Foucault pendulum of mass m connected to a long light wire of length ℓ. The coordinate frame illustrated is a noninertial frame whose xy-plane is tangent to the Earth's surface and whose z-axis is oriented along the local vertical direction. The angle α is small. There are two forces acting on the Foucault pendulum, gravity, and tension. The force of gravity $m\mathbf{g}$ uses the effective gravitational acceleration $\mathbf{g}$ which includes the centrifugal force as previously discussed. The tension in the wire $\mathbf{T}$ is illustrated in Figure 10.11, which also includes the x- and y-components of $\mathbf{T}$. Note that the z-component of $\mathbf{T}$ will be discussed later in the problem. Finally, we assume that the Foucault pendulum's motion is largely confined to the xy-plane. In other words, $\dot{z} \ll \dot{x}$ and $\dot{z} \ll \dot{y}$.

Assuming that the Earth's angular acceleration is constant, which over the period of the pendulum is an excellent approximation, we can use (10.4.22) to find the acceleration of the mass m by dividing (10.4.22) by m and using $\mathbf{F}' = \mathbf{T}$,

$$\mathbf{a} = \frac{\mathbf{T}}{m} + \mathbf{g} - 2\left(\boldsymbol{\omega} \times \mathbf{v}\right) \tag{10.5.1}$$

To get $x(t)$ and $y(t)$ (note that with our assumptions $z(t) = 0$), we need to find the components of the vector equation (10.5.1), and integrate and solve the resulting second-order ODE.

To get the ODE, we write the components of each of the vectors on the right-hand side of (10.5.1), noting that $\mathbf{a} = \ddot{x}\hat{\mathbf{e}}_x + \ddot{y}\hat{\mathbf{e}}_y + \ddot{z}\hat{\mathbf{e}}_z$. We can see in Figure 10.11, that $\mathbf{g} = -g\hat{\mathbf{e}}_z$ where g is the magnitude of the local effective acceleration due to gravity. Next, the components of the tension can be found using the fact that α is a small angle and ℓ is large,

$$T_x \approx -T\frac{x}{\ell} \qquad T_y \approx -T\frac{y}{\ell} \qquad T_z \approx T \tag{10.5.2}$$

where we used $\sin\alpha \approx x/\ell$ and y/ℓ for T_x and T_y, respectively. The Coriolis acceleration $\boldsymbol{\omega} \times \mathbf{v}$ can be found using the method outlined in Example 10.3,

$$\boldsymbol{\omega} \times \mathbf{v} = \begin{vmatrix} \hat{\mathbf{e}}_x & \hat{\mathbf{e}}_y & \hat{\mathbf{e}}_z \\ -\omega\cos\lambda & 0 & \omega\sin\lambda \\ \dot{x} & \dot{y} & 0 \end{vmatrix} \tag{10.5.3}$$

$$= \left(-\dot{y}\,\omega\sin\lambda\right)\hat{\mathbf{e}}_x + \left(\dot{x}\,\omega\sin\lambda\right)\hat{\mathbf{e}}_y + \left(-\dot{y}\,\omega\cos\lambda\right)\hat{\mathbf{e}}_z \tag{10.5.4}$$

where we used $\dot{z} \simeq 0$, since the pendulum's motion is confined to the xy-plane. Therefore, inserting the vector components into (10.5.1) we obtain,

$$
\left.
\begin{aligned}
\ddot{x} &= -\frac{T}{m}\frac{x}{\ell} + 2\dot{y}\,\omega\sin\lambda \\[2mm]
\ddot{y} &= -\frac{T}{m}\frac{y}{\ell} - 2\dot{x}\,\omega\sin\lambda \\[2mm]
\ddot{z} &= \frac{T}{m} - g + 2\dot{y}\,\omega\cos\lambda
\end{aligned}
\right\}
\tag{10.5.5}
$$

The pendulum stays in the xy-plane; therefore, $\ddot{z} \approx 0$. Furthermore, $\dot{y}\omega$ is small compared to the acceleration due to the gravity and the tension. Therefore, we can write $T = mg$ using the equation for $\ddot{z}$ in (10.5.5) and we can rewrite (10.5.5) as,

$$
\left.
\begin{aligned}
\ddot{x} + \beta^2 x &= 2\omega_z \dot{y} \\[1mm]
\ddot{y} + \beta^2 y &= -2\omega_z \dot{x}
\end{aligned}
\right\}
\tag{10.5.6}
$$

where $\beta^2 = T/(m\,\ell) \simeq g/\ell$, and $\omega_z = \omega\sin\lambda$ is the $\hat{\mathbf{e}}_z$ component of the angular velocity. Notice that (10.5.6) consists of a pair of *coupled* second-order differential equations. The equations are called coupled because the equation for $\ddot{x}$ contains a term with $\dot{y}$ and vice-versa. A common means of solving coupled ODEs analytically is by multiplying one of the equations by $i = \sqrt{-1}$ and adding the two equations together. The result of multiplying the $\ddot{y}$ equation by i and adding it to the $\ddot{x}$ equation is,

$$
(\ddot{x} + i\ddot{y}) + \beta^2(x + iy) = -2\omega_z i\,(\dot{x} + i\dot{y})
\tag{10.5.7}
$$

We see that (10.5.7) is a second-order ODE in the variable $q = x + iy$,

$$
\ddot{q} + 2i\omega_z \dot{q} + \beta^2 q = 0
\tag{10.5.8}
$$

The form of (10.5.8) should be familiar; it is similar to a damped harmonic oscillator. The solution to (10.5.8) is,

$$
q(t) = e^{-i\omega_z t}\left[A\exp\left(it\sqrt{\omega_z^2 + \beta^2}\right) + B\exp\left(-it\sqrt{\omega_z^2 + \beta^2}\right) \right]
\tag{10.5.9}
$$

Notice that if the Earth were not rotating, $\omega_z = 0$ and (10.5.8) would become,

$$
\ddot{q}' + \beta^2 q' = 0
\tag{10.5.10}
$$

which is the equation for simple harmonic motion, where we used the notation of $q' = x' + iy'$ for the value of q in a nonrotating frame. From our work in Chapter 6, we know that

$$
q'(t) = A e^{i\beta t} + B e^{-i\beta t}
\tag{10.5.11}
$$

Furthermore, from (10.5.10) we see that $\beta = \sqrt{g/\ell}$ is the Foucault pendulum's frequency of small oscillation when the Earth is not rotating. Because $\beta \gg \omega_z$, we can rewrite (10.5.9) as

$$
q(t) = e^{-i\omega_z t}\left[A e^{i\beta t} + B e^{-i\beta t} \right]
\tag{10.5.12}
$$

or

$$
q(t) = e^{-i\omega_z t} q'(t)
\tag{10.5.13}
$$

We can get a better physical insight to the motion of the pendulum if we write (10.5.13) as

$$
\begin{aligned}
x + iy &= (x' + iy')e^{-i\omega_z t} \\
&= (x' + iy')(\cos(\omega_z t) - i\sin(\omega_z t)) \\
&= (x'\cos(\omega_z t) + y'\sin(\omega_z t)) + i\,(-x'\sin(\omega_z t) + y'\cos(\omega_z t))
\end{aligned}
\tag{10.5.14}
$$

Next, we equate the real and imaginary parts to get the solution we are after, $x(t)$ and $y(t)$,

$$\left. \begin{array}{l} x(t) = x' \cos(\omega_z t) + y' \sin(\omega_z t) \\ y(t) = - x' \sin(\omega_z t) + y' \cos(\omega_z t) \end{array} \right\} \tag{10.5.15}$$

It is easier to get some physical insight into these equations if we rewrite (10.5.15) as a matrix equation,

$$\left(\begin{array}{c} x(t) \\ y(t) \end{array} \right) = \left(\begin{array}{cc} \cos(\omega_z t) & \sin(\omega_z t) \\ -\sin(\omega_z t) & \cos(\omega_z t) \end{array} \right) \left(\begin{array}{c} x'(t) \\ y'(t) \end{array} \right) \tag{10.5.16}$$

The appearance of the rotation matrix in (10.5.16) tells us that the Foucault pendulum's plane of oscillation is rotating through an angle $\omega_z t$ with a rotation frequency of $\omega_z = \omega \sin \lambda$. Therefore, the precession rate ω_z of the pendulum depends on the latitude of the pendulum. We also see how the Foucault pendulum gives a demonstration of the Earth's rotation, as its rotation rate is dependent on the Earth's angular velocity.

Example 10.4: Motion of Foucault pendulum in the plane

In 1851 Jean-Bernard-Léon Foucault demonstrated a Foucault pendulum with length ℓ of 67 meters in Paris, France. Plot the solution to (10.5.6) with the initial conditions $x(0) = 0.1$, $y(0) = 0$, $\dot{x}(0) = 0$, and $\dot{y}(0) = 0$ and interpret the result.

Solution:

Paris, France has a latitude of 48.8°N. The codes below create the plot. The Mathematica output appears in Figure 10.12. Notice that the pendulum begins at the point $(0.1, 0)$ and oscillates to the point $(-0.1, 0)$. As time progresses, the pendulum's plane rotates. We integrated (10.5.6) for a quarter of the pendulum's plane of oscillation's period which is approximately:

$$\tau = \frac{\omega_z}{2\pi} = \frac{\omega \sin \lambda}{2\pi} = 108,072 \text{ s} = 30 \text{ hours} \tag{10.5.17}$$

Python Code

To solve the problem in Python, we need to first rewrite (10.5.6) as a system of first-order equations:

$$\left. \begin{array}{l} \dot{x} = r \\ \dot{y} = s \\ \dot{r} = -\beta^2 x + 2\omega_z s \\ \dot{s} = -\beta^2 y - 2\omega_z r \end{array} \right\} \tag{10.5.18}$$

We then use `odeint` to solve (10.5.18). The solutions x, y are stored in `x_coord` and `y_coord`, respectively.

```
from scipy.integrate import odeint
import numpy as np

w = 7.272*10**(-5)      #angular velocity of Earth
lam = 48.8*np.pi/180    #lambda
l, g = 67, 9.8          #length of pendulum, g
wz = w*np.sin(lam)
b = g/l                 #beta-squared

def deriv(u,t):
    x, y, r, s = u
    dydt = [r, s,   -b**2*x + 2*wz*s, -b**2*y - 2*wz*r ]
    return dydt

t = np.linspace(0,27000,10000)
ics = [0.1,0,0,0]

soln = odeint(deriv, ics, t)

x_coord = soln[:,0]
y_coord = soln[:,1]

plt.plot(x_coord,y_coord)
plt.ylabel('y')
plt.xlabel('x')
plt.show()
```

Mathematica code

We use **NDSolve** to integrate (10.5.6) and plotted the solution using **ParametricPlot** for one quarter of the period of the pendulum's plane of oscillation.

$l = 67$;

$\lambda = 48.8 * \text{Pi}/180$;

$g = 9.8$;

$w = 7.727 * 10^{\wedge}(-5)$;

$\text{wz} = w * \text{Sin}[\lambda]$;

$\beta = g/l$;

$\text{soln} = \text{NDSolve}[\{x"[t] + \beta^{\wedge}2 * x[t] == 2 * \text{wz} * y'[t],$

$y"[t] + \beta^{\wedge}2 * y[t] == - 2 * \text{wz} * x'[t],$

$x[0] == 0.1, y[0] == 0, x'[0] == 0, y'[0] == 0\}, \{x[t], y[t]\}, \{t, 0, 27000\}]$;

$\text{ParametricPlot}[\{x[t], y[t]\}/.\text{soln}, \{t, 0, 27000\}, \text{FrameLabel->}\{x, y\}]$

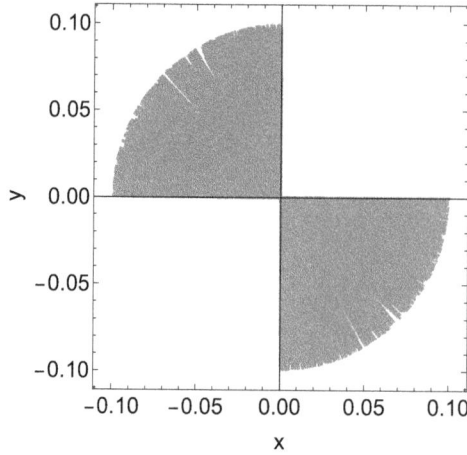

Figure 10.12: Mathematica output for Example 10.4. The pendulum sweeps out a quarter of a period over the integration time.

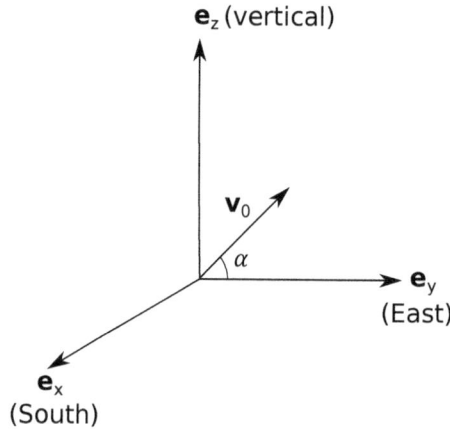

Figure 10.13: Coordinate system for a projectile launched in the yz-plane towards the East, at an angle α and with a speed v_0.

10.6 PROJECTILE MOTION IN A NONINERTIAL FRAME

As a final section to this chapter, we will explore the problem of projectile motion in a noninertial reference frame under the influence of Earth's gravitation. We will compute the Coriolis deflection of a projectile launched at a latitude $\lambda = 50°$ North at an angle $\alpha = 45°$, with a speed of $v_0 = 200$ m/s towards the East.

To address the problem of projectile motion in a noninertial frame, we will need to set up a coordinate system as shown in Figure 10.13. We have shown the noninertial frame, the frame in which observers are watching the projectile's flight. This frame is located in the Earth's Northern Hemisphere. The $\hat{\mathbf{e}}_y$ and $\hat{\mathbf{e}}_x$ directions correspond to East and South, respectively. The vector $\hat{\mathbf{e}}_z$ is oriented to point along the vertical direction. The projectile is launched in the yz-plane, such that if the Earth were not rotating, the projectile would land at a location $\mathbf{R} = R\hat{\mathbf{e}}_y$, and it would land a distance R from the origin along the y-axis.

By including the effect of the Earth's rotation, we expect that the projectile will land somewhere to the South of the y-axis. To find where the projectile lands, we will follow a process similar to finding the rotation rate of the Foucault pendulum in the previous section. We begin by writing the acceleration of the projectile using (10.4.22),

$$\mathbf{a} = \mathbf{g} - 2\left(\boldsymbol{\omega} \times \mathbf{v}\right) \tag{10.6.1}$$

Other than gravity, there are no external forces acting on the system. Therefore, $\mathbf{F}' = 0$. As with the Foucault pendulum, the effective acceleration due to gravity is $\mathbf{g} = -g\hat{\mathbf{e}}_z$. The Coriolis acceleration is,

$$\boldsymbol{\omega} \times \mathbf{v} = \begin{vmatrix} \hat{\mathbf{e}}_x & \hat{\mathbf{e}}_y & \hat{\mathbf{e}}_z \\ -\omega\cos\lambda & 0 & \omega\sin\lambda \\ \dot{x} & \dot{y} & \dot{z} \end{vmatrix} \tag{10.6.2}$$

$$\boldsymbol{\omega} \times \mathbf{v} = \left(-\dot{y}\omega\sin\lambda\right)\hat{\mathbf{e}}_x + \left(\dot{z}\omega\cos\lambda + \dot{x}\omega\sin\lambda\right)\hat{\mathbf{e}}_y + \left(-\dot{y}\omega\cos\lambda\right)\hat{\mathbf{e}}_z \tag{10.6.3}$$

Notice that we made no assumptions about the components of the projectile's velocity. Inserting the accelerations into (10.6.1), we obtain the following equations,

$$\left.\begin{aligned} \ddot{x} &= 2\dot{y}\omega\sin\lambda \\ \ddot{y} &= -2\omega\left(\dot{z}\cos\lambda + \dot{x}\sin\lambda\right) \\ \ddot{z} &= -g + 2\dot{y}\omega\cos\lambda \end{aligned}\right\} \tag{10.6.4}$$

The result is a system of coupled ODEs. While it is possible to argue some approximations to simplify (10.6.4) (such as $2\dot{y}\omega\cos\lambda \ll g$), we will seek a numerical solution of (10.6.4) in Example 10.4.

> ### Example 10.5: Projectile motion in noninertial frames
>
> Using numerical methods, solve (10.6.4) to find the range and the Coriolis deflection of a projectile launched at a latitude $\lambda = 50°$ North at an angle $\alpha = 45°$ with a speed of $v_0 = 200$ m/s towards the East.
>
> *Solution:*
> The Mathematica and Python codes are shown below.

> **Python Code**
> To solve the problem in Python, we need to first rewrite (10.6.4) as a system of first-order equations:
>
> $$\left.\begin{aligned} \dot{x} &= q \\ \dot{y} &= r \\ \dot{z} &= s \\ \dot{q} &= 2r\omega\sin\lambda \\ \dot{r} &= -2\omega\left(s\cos\lambda + q\sin\lambda\right) \\ \dot{s} &= -g + 2r\omega\cos\lambda \end{aligned}\right\} \tag{10.6.5}$$
>
> We then use `odeint` and the solutions x, y, z are stored in `x_coord`, `y_coord`, and `z_coord`, respectively. To get the time of flight, we need to know when the $z(t)$ crosses zero (for the numerical model, the projectile can fall through the Earth's surface). Using `np.where` we can find the index values where `z_coord` changes sign. Using `np.diff`, we get the index before the sign change (hence the addition of 1 in `tof_index`). Finally, we can find the time of flight, range, and Coriolis deflection by inserting `tof_index` into the respective arrays.

```
from scipy.integrate import odeint
import numpy as np
import matplotlib.pyplot as plt
print('-'*28,'CODE OUTPUT','-'*29,'\n')


w = 7.272*10**(-5)                     # angular velocity of Earth
a, l = 45*np.pi/180, 50*np.pi/180 #launch angle alpha. latitude
v0, g = 200, 9.8                       # initial velocity, g

z0_dot = v0*np.sin(a)
y0_dot = v0*np.cos(a)

def deriv(u,t):
    x, y, z, q, r, s = u
    dydt = [q, r, s, 2*r*w*np.sin(l), \
        -2*w*(s*np.cos(l) + q*np.sin(l)),\
            -g + 2*r*w*np.cos(l)]
    return dydt

t = np.linspace(0,1000,10000)
ics = [0, 0, 0, 0, y0_dot, z0_dot]
soln = odeint(deriv, ics, t)

x_coord = soln[:,0]
y_coord = soln[:,1]
z_coord = soln[:,2]

z_zero_crossings = np.where(np.diff(np.sign(z_coord)))[0]
print('The indices of the zero crossing are:',z_zero_crossings)

tof_index = z_zero_crossings[1]+1 #index of time of flight

time_of_flight = t[tof_index]
print('\nThe time of flight is: ', round(time_of_flight,2),' s')

print('\nThe Coriolis deflection = ',\
round(x_coord[tof_index],2),' m')

print('\nThe range = ', round(y_coord[tof_index],2),' m')

--------------------------- CODE OUTPUT ----------------------------
The indices of the zero crossing are: [  0 288]

The time of flight is: 28.9  s

The Coriolis deflection =  6.58  m

The range =  4085.64  m
```

Mathematica code

The code begins by defining the necessary parameters, including the components of the initial velocity which are stored as zODot and yODot. We solve (10.6.4) using NDSolve, as usual. The time of flight T is obtained by finding when the altitude of the projectile z equals zero, by using FindRoot. The index [[1,2]] is used to remove the replacement rule that Mathematica outputs for the FindRoot command. Finally, we insert the time of flight into x[t] to get the Coriolis deflection and y[t] to get the projectile's range.

$\omega = 7.27210^{\wedge}(-5);$

$\alpha = 45 * \text{Pi}/180;$

$\lambda = 50 * \text{Pi}/180;$

$\text{v0} = 200;$

$g = 9.8;$

$\text{zODot} = \text{v0Sin}[\alpha];$

$\text{yODot} = \text{v0Cos}[\alpha];$

$\text{soln} = \text{NDSolve}[\{x"[t] == 2y'[t]\omega\text{Sin}[\lambda],$

$y"[t] == -2z'[t]\omega\text{Cos}[\lambda] - 2x'[t]\omega\text{Sin}[\lambda],$

$z"[t] == -g + 2y'[t]\omega\text{Cos}[\lambda], x'[0] == 0, y'[0] == \text{yODot},$

$z'[0] == \text{zODot}, x[0] == 0, y[0] == 0, z[0] == 0\}, \{x, y, z\},$

$\{t, 0, 2000\}];$

$T = \text{FindRoot}[(z[t]/.\text{soln}) == 0, \{t, 10, 2000\}][[1, 2]];$

$\text{Print}[\text{"The time of flight T = "}, \text{NumberForm}[T, 4],$

$\text{" s"}]$

The time of flight T = 28.9 s

$\text{Print}[\text{"The Coriolis deflection is: "},$

$\text{NumberForm}[x[T]/.\text{soln}[[1]], 3], \text{" m"}]$

The Coriolis deflection is: 6.58 m

$\text{range} = y[T]/.\text{soln}[[1]];$

$\text{Print}[\text{"The range is R = "}, \text{NumberForm}[\text{range}], \text{" m"}]$

The range is R = 4085.3 m

10.7 CHAPTER SUMMARY

It is sometimes necessary to describe the motion of a particle in a noninertial reference frame. Although Newton's second law is valid only in inertial frames, we can modify Newton's second law for use in a noninertial frame. The time derivative of any vector $\mathbf{Q}$ relative to a rotating noninertial frame, is related to that of an inertial frame by the relationship:

$$\left(\frac{d\mathbf{Q}}{dt}\right)_f = \left(\frac{d\mathbf{Q}}{dt}\right)_r + \boldsymbol{\omega} \times \mathbf{Q}$$

where $\boldsymbol{\omega}$ is the angular velocity of the noninertial frame with respect to the inertial frame.

We can rewrite Newton's second law for rotating frames as,

$$m\mathbf{a} = \mathbf{F}' - m\mathbf{A} - 2m\left(\boldsymbol{\omega} \times \mathbf{v}\right) - m\left(\dot{\boldsymbol{\omega}} \times \mathbf{r}\right) - m\left[\boldsymbol{\omega} \times \left(\boldsymbol{\omega} \times \mathbf{r}\right)\right]$$

where:

$\mathbf{a}$ = the acceleration relative to the noninertial frame

$\mathbf{F}'$ = the sum of the "interaction" forces acting on the particle (gravity, friction)

$\mathbf{A}$ = the linear acceleration of the noninertial frame relative to an inertial frame

$-2\left(\boldsymbol{\omega} \times \mathbf{v}\right)$ =the Coriolis acceleration

$-\dot{\boldsymbol{\omega}} \times \mathbf{r}$ = an acceleration due to the angular acceleration of the noninertial frame

$-\boldsymbol{\omega} \times \left(\boldsymbol{\omega} \times \mathbf{r}\right)$ = the centrifugal acceleration.

The last four terms in the equation above are referred to as "inertial forces" or fictitious forces, because they arise from the acceleration of the noninertial frame and are not due to the interaction between the particle and another body.

10.8 END-OF-CHAPTER PROBLEMS

The symbol ⌨ indicates a problem which requires some computer assistance, in the form of graphics, numerical computation, or symbolic evaluation.

Section 10.1: Motion in a Nonrotating Accelerating Reference Frame

1. A pendulum with mass m hangs from the ceiling of an accelerating car by a massless rod. As the car accelerates forward with a constant acceleration $\mathbf{A}$, the pendulum tilts backwards at an angle ϕ_0. Find the pendulum's equilibrium angle ϕ_0.

Section 10.2: Angular Velocity as a Vector

2. Show that relative angular velocities add in the same way as translational velocities. For example, we know that if a particle is moving relative to a frame S with a velocity $\mathbf{v}$ and that the frame S is moving with a velocity $\mathbf{V}$ relative to the inertial frame S', then the velocity of the particle relative to S' is $\mathbf{v}' = \mathbf{V} + \mathbf{v}$. Repeat this line of reasoning with a body rotating with an angular velocity $\boldsymbol{\omega}$ relative to a rotating frame S which rotates with an angular velocity $\boldsymbol{\Omega}$ relative to an inertial frame S'.

Show that the angular velocity of the body relative to S' is $\boldsymbol{\omega}' = \boldsymbol{\Omega} + \boldsymbol{\omega}$. Assume that the reference frames S and S' have the same origin.

Section 10.4: Newton's Second Law in a Rotating Frame

3. 🖥 Consider a bullet fired horizontally towards the South with a velocity $v_0 = 500$ m/s at a latitude $\lambda = 50°$ North. How does the Coriolis force acting on the bullet compare to is weight? Solve this problem analytically and by using either Python or Mathematica.

4. A bucket of water is rotating about its vertical symmetry axis with an angular velocity ω. What is the shape of the water's surface in the bucket?

5. 🖥 Atomic clocks show that a modern day is longer by about 1.7 milliseconds than a century ago. Hence although small, the angular acceleration of the Earth is not zero $\dot{\omega} \neq 0$, where ω is the angular velocity of the Earth. Compute the acceleration $\dot{\boldsymbol{\omega}} \times \mathbf{r}$ due to the changing angular velocity of the Earth, on a particle fixed to the Earth's surface located in Williamsburg, Virginia, USA (latitude $37.3°N$ and a longitude of $76.6°W$). Solve this problem analytically and using Python or Mathematica.

6. Let S be a frame rotating with a constant angular velocity $\boldsymbol{\omega}$ relative to an inertial frame S' , with the origins of both frames being the same. The particle has a potential energy V. Find the Lagrangian in terms of the position and velocity coordinates $\mathbf{r}$ and $\dot{\mathbf{r}}$ in S. Show that the Euler-Lagrange equations yield

$$\mathbf{F}' = m\mathbf{a} + 2m\left(\boldsymbol{\omega} \times \mathbf{v}\right) + m\boldsymbol{\omega} \times \left(\boldsymbol{\omega} \times \mathbf{r}\right)$$

7. 🖥 Using the effective gravitational acceleration, plot the time it takes for a particle of mass m to fall a distance of 100 meters, as a function of latitude. Where is this time the greatest? Ignore air resistance and assume the particle is released from rest.

8. A baseball of mass m is launched vertically upward from the ground with an initial velocity v_0 at a latitude λ. Find the Coriolis deflection of the ball when it returns to the ground.

9. 🖥 Find the deflection from the plumb line for a 1.0 kg particle falling from a height $H = 100$ m under the Earth's gravity, experiencing linear air resistance. Assume that the horizontal component of the particle's velocity is small, so that the drag force acts only in the vertical direction. Thus the drag force is $F_d = -b\dot{z}\hat{z}$, where $b = 0.1$ Ns/m. Compare the answer to the result if the particle fell from the same height in vacuum at a latitude of 45 degrees.

10. 🖥 Consider a particle of mass m confined to move in a vertical plane rotating with a constant angular velocity $\boldsymbol{\omega}$ about the vertical. Let x be the horizontal coordinate on the plane and z be the vertical coordinate. Using analytical methods, solve the equations of motion for the particle and describe the particle's motion. Using Python or Mathematica, solve the equations of motion using initial conditions $\mathbf{r}_0 = 1\hat{\mathbf{i}} - 1\hat{\mathbf{j}}$ and $\mathbf{v}_0 = 1\hat{\mathbf{i}} + 1\hat{\mathbf{j}}$ (in SI units). Plot the motion of the particle on the plane using $\omega = 1$ rad/s.

11. 🖥 Let us consider again Problem 10, but now assume the plane's angular velocity is increasing in time, such that its angular acceleration is $\boldsymbol{\alpha} = \alpha\hat{\mathbf{e}}_z$ where α is a positive

constant. However, now solve the problem using only Python or Mathematica. Plot the motion of the particle for $\alpha = 0.5$, and $\alpha = 1.0$ rad/s^2. Assume that the initial angular velocity of the plane is zero.

12. A baseball is dropped from a height of 30 m at a latitude of 32° South. Ignoring air resistance, in what direction and how far did the ball land from the vertical? How does the direction compare to the result of Example 10.3? Explain any similarities or differences.

Section 10.5 Foucault Pendulum

13. Show that (10.5.9)

$$q(t) = e^{-i\omega_z t}\left[A\exp\left(it\sqrt{\omega_z^2 + \beta^2}\right) + B\exp\left(-it\sqrt{\omega_z^2 + \beta^2}\right)\right]$$

is the solution to the second-order ODE (10.5.8) :

$$\ddot{q} + 2i\,\omega_z\dot{q} + \beta^2 q = 0$$

14. 🖳 What is the rate of rotation of a Foucault pendulum's plane of oscillation if the pendulum is located in Paris, France? How about Paris, Kentucky, USA. Plot the Foucault pendulum's rotation rate as a function of latitude. Where is its maximum? What is the minimum value, and where does this minimum value occur?

15. Consider a Foucault pendulum experiencing a damping force $\mathbf{F}' = -b\,v\,\hat{\mathbf{v}}$. Is the precession frequency changed?

16. Is the Foucault pendulum's precession frequency significantly changed if $\dot{\boldsymbol{\omega}} \neq 0$, where $\boldsymbol{\omega}$ is the angular velocity of the Earth about its rotation axis? To answer this question, assume that the magnitude of the Earth's angular velocity changes at a constant rate, but its direction does not change (this is not true over long time periods). Furthermore, assume that the change in the angular velocity is very small. For example, consider the situation described in Problem 26 from Chapter 4 which states that the Earth's angular velocity changed from 7.6×10^{-5} rad/s to 7.3×10^{-5} rad/s during a time period of 350 million years. Let the pendulum have a mass m and a length ℓ (assumed to be very long) and to be located at a latitude λ in the Northern Hemisphere.

Section 10.6: Projectile Motion in a Noninertial Frame

17. 🖳 In our study of the projectile motion in a noninertial frame, we derived the equations

$$\ddot{x} = 2\dot{y}\omega \sin\lambda$$
$$\ddot{y} = -2\omega(\dot{z}\cos\lambda + \dot{x}\sin\lambda) \qquad (1)$$
$$\ddot{z} = -g + 2\dot{y}\omega\cos\lambda$$

Assuming that the approximation $2\dot{y}\omega\cos\lambda \ll g$ holds, solve the resulting system of two coupled ODEs in closed form. You will want to use a symbolic ODE solver. After finding the solutions, perform a Taylor series expansion for $x(t)$ and $y(t)$, and interpret your results.

18. 🖳 Use a CAS to find the range and Coriolis deflection of a 1.0 kg projectile experiencing linear air resistance of the form $\mathbf{F}' = -bv\hat{\mathbf{v}}$ where $b = 0.3$ SI units. The initial velocity $v_0 = 200$ m/s, the latitude $\lambda = 50°$N and the angle of launching is $\alpha = 45°$.

19. 🖵 Repeat Problem 18, but this time, the projectile experiences quadratic air resistance of the form $\mathbf{F}' = -bv^2\hat{\mathbf{v}}$ where $b = 0.01$ SI units.

20. A projectile of mass m is fired due East from a point on the Earth's surface with a latitude of λ (in the Northern Hemisphere). The projectile's initial speed is v_0 and is launched at an angle α with respect to the horizontal. Calculate the lateral deflection of the projectile as it hits the ground. Do this problem without the aid of a computer.

21. There is a legend that during World War I, the British navy consistently missed German ships when fighting near the Falkland Islands because their ships did not properly account for the Coriolis force. We will not discuss the validity of this legend here, but for a moment, let's suppose that it is true and that the Coriolis force was known, but the ship's guns were set up to hit their targets for battles in the Northern Hemisphere. The Falkland Islands are near 50° South latitude. Suppose the guns were set to accurately hit targets at 50° *North* latitude, by how much did the British ships miss their target during the Falkland Islands engagement? Assume that the German ships were due East of the British ships, that the British ships' guns had a muzzle velocity of 500 m/s and the shells were shot at an angle of 20° with respect to the horizontal.

11 Rigid Body Motion

We begin this chapter by reviewing the rotational motion of a single particle around an arbitrary axis and the concepts of the moment of inertia and the center of mass. In particular we focus on how the center of mass simplifies the description of the translational and rotational motion of a system of particles. After the review, we explore generalized definitions of the moment of inertia, including products of inertia and the inertia tensor, and we demonstrate how to calculate these quantities for a variety of solids. We will see that the moment of inertia tensor of a solid depends on the choice of the coordinate system and discuss the parallel axis theorem for rigid bodies. This is followed by a discussion of eigenvalues and eigenvectors of matrices, and how they can be used to describe the principal axes of a rigid body. The chapter will conclude with a discussion of the Euler equations and a description of the precessional motion of spinning tops and gyroscopes.

11.1 ROTATIONAL MOTION OF PARTICLES AROUND A FIXED AXIS

In this section, we review rotational motion concepts with a focus on the rotational motion of a particle that is confined to move in a circle of radius r in the xy-plane with an angular velocity $\boldsymbol{\omega} = \omega_z \hat{\mathbf{k}}$, as illustrated in Figure 11.1. We later develop equations to describe more general rotational motion.

As discussed in Chapter 10, the instantaneous velocity $\mathbf{v}$ of mass m is given by the cross product of the angular velocity vector, which, in general, takes the form $\boldsymbol{\omega} = \omega_x \hat{\mathbf{i}} + \omega_y \hat{\mathbf{j}} + \omega_z \hat{\mathbf{k}}$ and of the position vector $\mathbf{r}$:

$$\mathbf{v} = \boldsymbol{\omega} \times \mathbf{r} \tag{11.1.1}$$

Recall from Chapter 4 that the angular momentum is a useful quantity for describing rotational motion. The angular momentum vector $\boldsymbol{\ell}$ is given by the cross product of the position vector $\mathbf{r}$ and the momentum vector $\mathbf{p} = m\mathbf{v}$:

$$\boldsymbol{\ell} = \mathbf{r} \times \mathbf{p} = m\left(\mathbf{r} \times \mathbf{v}\right) \tag{11.1.2}$$

The position, velocity, angular momentum, and angular velocity vectors are shown in Figure 11.1, where in the case of $\mathbf{r} = x\hat{\mathbf{i}} + y\hat{\mathbf{j}}$ and $\boldsymbol{\omega} = \omega_z \hat{\mathbf{k}}$, the angular momentum vector and the angular velocity vectors point in the same direction, the z-axis.

Since the position and velocity vectors are at right angles to each other, we can write $v = \omega r$, and the last equation becomes:

$$\ell = mrv = mr^2\omega \tag{11.1.3}$$

As usual, unbolded variables represent magnitudes of vectors, e.g., $\omega = |\boldsymbol{\omega}|$. Notice that in this equation, the angular momentum is written as a quantity mr^2 multiplied by the velocity v. If we relate this to the definition of linear momentum $p = mv$, we can think of (11.1.3) as a "rotational inertia" (mr^2) multiplied by the angular velocity ω. This "rotational inertia" is called the moment of inertia, I, and is defined by:

$$I = mr^2 \tag{11.1.4}$$

where r is the distance of the mass m from the rotational axis.

The kinetic energy of the rotating mass m is found from $T = \frac{1}{2}mv^2$, and by substituting $v = \omega r$, we obtain:

$$T = \frac{1}{2}I\omega^2 \tag{11.1.5}$$

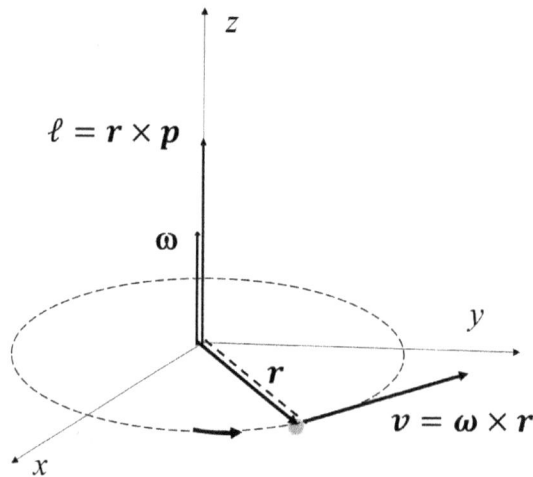

Figure 11.1: Rotational motion of a particle in the xy-plane, showing the position, velocity, angular momentum, and angular velocity vectors.

Recall from Chapter 4 that the torque acting on the particle is $\mathbf{N} = \mathbf{r} \times \mathbf{F}$. We can rewrite the torque as the time derivative of the angular momentum, by using Newton's second law for rotational motion. Using (11.1.3) we can write $\ell = I\omega$, and Newton's second law of rotation becomes:

$$N = \frac{d}{dt}\left(I\omega\right) = I\dot{\omega} = I\alpha \qquad (11.1.6)$$

where $\alpha = \dot{\omega}$ is the angular acceleration.

Let's reconsider (11.1.3) as a vector relationship:

$$\boldsymbol{\ell} = I\boldsymbol{\omega} \qquad (11.1.7)$$

Since the moment of inertia is defined as a scalar quantity in (11.1.4), this vector equation states that in the example of Figure 11.1, the angular momentum vector $\boldsymbol{\ell}$ and the angular velocity vector $\boldsymbol{\omega}$ point in the same direction along the rotational axis. In addition, we can rewrite (11.1.6) as,

$$\mathbf{N} = I\boldsymbol{\alpha} \qquad (11.1.8)$$

Note that the torque vector $\mathbf{N}$ and the angular acceleration vector $\boldsymbol{\alpha}$ point in this same direction in this case.

Example 11.1 shows an evaluation of the angular momentum vector using Python and Mathematica.

Example 11.1: Use Python and Mathematica to evaluate the angular momentum and torque $\mathbf{N}$ acting on the mass m in the situation shown in Figure 11.1. Assume that the mass m moves with a constant angular velocity ω around the z-axis in a circle of radius R.

Solution:
We use cylindrical coordinates with the angle increasing with time according to $\theta = \omega t$.

The position vector

$$\mathbf{r} = R\cos(\omega t)\,\hat{\mathbf{i}} + R\sin(\omega t)\,\hat{\mathbf{j}}$$

The codes calculate the cross product $\boldsymbol{\omega} \times \mathbf{r}$ and verify that this is indeed equal to the velocity vector $(dx/dt, dy/dt)$.

In this example, the angular momentum vector points always along the z-axis and its magnitude is equal to $\ell_z = mR^2\omega$.

The torque $\mathbf{N} = I\dot{\boldsymbol{\omega}} = 0$, since the mass moves with a constant angular velocity ω around the z-axis.

Python Code
We define the position vector as a list, and use the **cross** function in NumPy to evaluate the cross products.

```
from numpy import     cross
from sympy import    sin, cos, symbols, simplify
print('-'*28,'CODE OUTPUT','-'*29,'\n')

R, omega, t, m , theta = symbols('R, omega, t, m, theta',real=True)

r_vector =  [R*cos(omega*t),R*sin(omega*t),0]

omega_vector =  [0,0,omega]

v_vector = simplify(cross(omega_vector, r_vector) )
print('\nv_vector = ',v_vector)

L_vector = m* cross(r_vector,v_vector )

print('\nL = ',simplify(L_vector))

---------------------------- CODE OUTPUT ----------------------------

v_vector =  [-R*omega*sin(omega*t), R*omega*cos(omega*t), 0]

L =  [0, 0, R**2*m*omega]
```

Mathematica Code
The code uses the `Cross` command to evaluate the cross products.

rvector $= \{R * \text{Cos}[\omega * t], R * \text{Sin}[\omega * t], 0\}$;

ωvector $= \{0, 0, \omega\}$;

vvector $= \text{Cross}[\omega\text{vector}, \text{rvector}]$;

Print["v = ωvector x r = ", vvector];

v $= \omega$vector x r $= \{-R\omega\text{Sin}[t\omega], R\omega\text{Cos}[t\omega], 0\}$

$L = m * \text{Cross}[\text{rvector}, \text{vvector}]$;

Print["Angular momentum L = ", $L//$Simplify]

Angular momentum L $= \{0, 0, mR^2\omega\}$

Let us now consider a particle of mass m which rotates around the z-axis at a constant angle θ and with an instantaneous angular velocity ω, as shown in Figure 11.2. In this situation the angular momentum vector is perpendicular to the shaded plane in the figure, which is defined by the position and velocity vectors. As a result, in the example of Figure 11.2 the angular momentum vector $\boldsymbol{\ell}$ and the angular velocity vector $\boldsymbol{\omega}$ do *not* point in the same direction, and the direction of $\boldsymbol{\ell}$ changes continuously in space during the rotational motion.

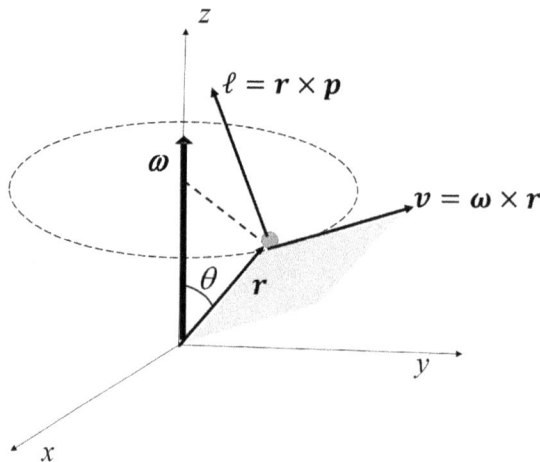

Figure 11.2: A particle of mass m revolves around the z-axis on a plane parallel to the xy-plane with angular velocity vector $\boldsymbol{\omega}$. In this example the angular momentum vector $\boldsymbol{\ell}$ and angular velocity vectors $\boldsymbol{\omega}$ are not pointing in the same direction. Compare this figure with the situation in Figure 11.1.

In this situation the equation $\boldsymbol{\ell} = I\boldsymbol{\omega}$ does not apply because $\boldsymbol{\ell}$ and $\boldsymbol{\omega}$ do not point in the same direction as in Figure 11.1. We must redefine the moment of inertia concept for the mass m. As we will see later in this chapter, we need to replace the scalar quantity I in this equation with a more general physical quantity, the *moment of inertia tensor* **I**.

Example 11.2: Use Python and Mathematica to evaluate the angular momentum and torque for the rotating mass m shown in Figure 11.2, assuming that $\omega = |\boldsymbol{\omega}|$ is constant.

Solution:
We follow the same method as in the previous example, using cylindrical coordinates. The position vector

$$\mathbf{r} = R\sin\theta\cos(\omega t)\,\hat{\mathbf{i}} + R\sin\theta\sin(\omega t)\,\hat{\mathbf{j}} + R\cos\theta\hat{\mathbf{k}}$$

The length $R\sin\theta$ appearing in the expressions for the coordinates x, y represents the radius of rotation of the particle around the z-axis.

The components of the angular momentum are:

$$\ell_x = -mR^2\omega\sin\theta\cos\theta\cos(\omega t)$$

$$\ell_y = -mR^2\omega\sin\theta\cos\theta\sin(\omega t)$$

$$\ell_z = -mR^2\omega\sin^2\theta$$

Clearly in this situation the angular momentum vector $\boldsymbol{\ell}$ has nonzero components along the $x, y,$ and z axes. By inspection of the calculated components ℓ_x and ℓ_y, we see that the vector $\boldsymbol{\ell}$ rotates around the z-axis with a constant angular speed ω, and that the z-component $\ell_z = m R^2\omega\sin^2\theta$ stays constant with time. The codes for Figure 11.2 follow.

Python Code
The code is very similar to the previous example, using a different position vector.

```
from numpy import     cross
from sympy import   sin, cos, symbols, simplify
print('-'*28,'CODE OUTPUT','-'*29,'\n')


R, omega, t, m , theta = symbols('R, omega, t, m, theta',real=True)

r_vector =  [R*sin(theta)*cos(omega*t),R*sin(theta)*sin(omega*t),\
                    R*cos(theta)]

omega_vector =   [0,0,omega]
v_vector = simplify(cross(omega_vector, r_vector) )
L_vector = m* cross(r_vector,v_vector )

print('Lx_component = ',L_vector[0])
print('Ly_component = ',L_vector[1])
print('Lz_component = ',simplify(L_vector[2]))

--------------------------- CODE OUTPUT ---------------------------
Lx_component =   -R**2*m*omega*sin(theta)*cos(theta)*cos(omega*t)
Ly_component =   -R**2*m*omega*sin(theta)*sin(omega*t)*cos(theta)
Lz_component =   R**2*m*omega*sin(theta)**2
```

Mathematica Code

rvector $= \{R * \text{Sin}[\theta] * \text{Cos}[\omega * t], R * \text{Sin}[\theta] * \text{Sin}[\omega * t], R * \text{Cos}[\theta]\};$

ωvector $= \{0, 0, \omega\};$

vvector $= \text{Cross}[\omega\text{vector}, \text{rvector}];$

Print["v $= \omega$vector x r $= $ ", vvector];

v $= \omega$vector x r $= \{-R\omega\text{Sin}[\theta]\text{Sin}[t\omega], R\omega\text{Cos}[t\omega]\text{Sin}[\theta], 0\}$

$L = m * \text{Cross}[\text{rvector}, \text{vvector}];$

Print["The angular momentum L is: "];

Print["L $= $ ", L//Simplify]

The angular momentum L is:
L $= \{-mR^2\omega\text{Cos}[\theta]\text{Cos}[t\omega]\text{Sin}[\theta], -mR^2\omega\text{Cos}[\theta]\text{Sin}[\theta]\text{Sin}[t\omega], mR^2\omega\text{Sin}[\theta]^2\}$

Before we discuss the rotational properties of solid bodies, let us review the center of mass and how it is used to describe the rotation of a system of particles.

11.2 REVIEW OF ROTATIONAL PROPERTIES FOR A SYSTEM OF PARTICLES

In this section we summarize some of the results we already saw in Chapters 4 and 5 about the momentum, angular momentum, and energy of a system of particles. In particular, we look again at the importance of the center of mass, and how it can help us simplify the description of the motion of rigid bodies.

11.2.1 CENTER OF MASS

Let us consider a collection of $i = 1, 2, \ldots, N$ discrete particles as in Figure 11.3, each with mass m_i and located at a position $\mathbf{r}_i$ relative to the origin.

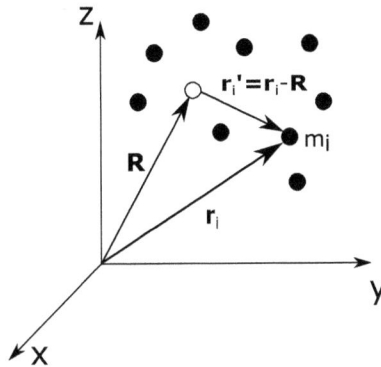

Figure 11.3: Collection of discrete particles (filled circles), with the empty circle representing their center of mass located at $\mathbf{R}$.

Let $\mathbf{R}$ represent the location of the center of mass relative to the origin, shown as a white circle in Figure 11.3, and $\mathbf{r}'_i$ denote the positions of the particles relative to the center of mass. Then $\mathbf{r}_i = \mathbf{r}'_i + \mathbf{R}$ and the position of the center of mass can be found from:

> **The Position of the Center of Mass for N Discrete Particles**
>
> $$\mathbf{R} = \frac{1}{M} \sum_{i=1}^{N} m_i \mathbf{r}_i \qquad M = \sum_{i=1}^{N} m_i \qquad \sum_{i=1}^{N} m_i \mathbf{r}'_i = 0 \qquad (11.2.1)$$

In the case of a continuous distribution of mass M, the summations are replaced by integrals:

> **The Position of the Center of Mass for a Continuous Mass Distribution**
>
> $$\mathbf{R} = \frac{1}{M} \int \mathbf{r}\, dm \qquad M = \int dm \qquad \int \mathbf{r}'\, dm = 0 \qquad (11.2.2)$$

11.2.2 MOMENTUM OF A SYSTEM OF PARTICLES

From Chapter 4 we recall that *the total momentum of a system of particles is equal to the product of the total mass M of the system and the velocity of the center of mass.* We can think of the system of particles as a single particle of mass M located at the system's center of mass, and of the net external forces $\mathbf{F}_{\text{ext}}$ as acting on the center of mass of the system. Newton's second law can be written in terms of the motion of the center of mass as follows:

> ### Total Momentum of a System of Particles or Continuous Mass Distribution
>
> $$\mathbf{P} = \sum_{i=1}^{N} \mathbf{p}_i = \sum_{i=1}^{N} m_i \dot{\mathbf{r}}_i = M\dot{\mathbf{R}} \qquad (11.2.3)$$
>
> $$\dot{\mathbf{P}} = M\ddot{\mathbf{R}} = \mathbf{F}_{\text{ext}} \qquad (11.2.4)$$

As a consequence, if the total external force acting on a system of particles is zero, then the center of mass is either at rest, or it moves with constant velocity and the total linear momentum of the system is conserved.

11.2.3 ANGULAR MOMENTUM OF A SYSTEM OF PARTICLES

From Chapter 4 we also recall that the total angular momentum of the system is the sum of two terms:

> ### Total Angular Momentum of a System of Particles
>
> $$\mathbf{L} = \sum_{i=1}^{N} (\mathbf{r}_i' \times \mathbf{p}_i') + \mathbf{R} \times \mathbf{P} \qquad (11.2.5)$$
>
> $$\dot{\mathbf{L}} = \mathbf{N} \qquad (11.2.6)$$

The first term $\sum (\mathbf{r}_i' \times \mathbf{p}_i')$ represents the angular momentum of the particles about the center of mass, and the second term $(\mathbf{R} \times \mathbf{P})$, is the angular momentum of the center of mass about the origin. In this situation Newton's second law can be stated as follows: if the internal forces between the particles in a system of particles are central forces, then the total external torque $\mathbf{N}$ on this system is equal to the time rate of change of the total angular momentum $\dot{\mathbf{L}}$ of the system. As a consequence, if the total external torque acting on a system of particles is zero, then the total angular momentum of the system is conserved.

11.2.4 WORK AND KINETIC ENERGY FOR A SYSTEM OF PARTICLES

In Chapter 5 we saw that the total kinetic energy of a system of particles is also the sum of two terms.

Kinetic Energy of a System of Particles

$$T = \frac{1}{2}M\dot{\mathbf{R}}^2 + \sum_{i=1}^{N}\left[\frac{1}{2}m_i\left(\dot{\mathbf{r}}_i'\right)^2\right] \tag{11.2.7}$$

$$W_{12} = T_2 - T_1 \tag{11.2.8}$$

The first term is the translational kinetic energy of the center of mass $\frac{1}{2}M\dot{\mathbf{R}}^2$, and the second term $\sum \frac{1}{2}m_i\left(\dot{\mathbf{r}}_i'\right)^2$, represents the kinetic energy of each particle relative to the center of mass. We also saw that the total work W_{12} done in moving a system of particles from a state of kinetic energy T_1 to a state of kinetic energy T_2 is equal to the change in kinetic energy. Furthermore, for a rigid body, the only motion relative to the center of mass is rotation. Hence, the second term in (11.2.7) is the kinetic energy of rotation about the center of mass. We can then think of equation (11.2.7) as stating that the total kinetic energy of a rigid body is equal to the translational kinetic energy of the center of mass plus the kinetic energy of rotation about the rigid body's center of mass.

For the potential energy, we need to consider both external and internal forces, and their corresponding potential energies. The potential energy of the ith mass is:

$$V_i = V_i^{\text{ext}} + \sum_{j \neq i} V_{i,j}^{\text{int}} \tag{11.2.9}$$

where V_i^{ext} is the potential energy due to the external forces acting on m_i, and $V_{i,j}^{\text{int}}$ is the potential energy due to the internal interaction force between m_i and m_j. An example of such an interaction force could be gravitational, electrostatic, or any other central force. In general for central forces, $V_{i,j}^{\text{int}} = V_{i,j}^{\text{int}}\left(|\mathbf{r}_i - \mathbf{r_j}|\right)$, i.e., the potential energy depends only on the distance between the two particles. Therefore, the total potential energy of the system is:

$$V = \sum_{i=1}^{N} V_i^{\text{ext}} + \sum_{i=1}^{N}\sum_{j \neq i} V_{i,j}^{\text{int}}\left(|\mathbf{r}_i - \mathbf{r_j}|\right) \tag{11.2.10}$$

Rigid bodies are defined as solids in which the particles that make up the solid are at fixed distances from each other. Therefore, the internal potential energy can be ignored since it will be constant for each particle. Hence, for a rigid body we will only have to worry about external forces; the internal central forces are irrelevant. When all internal and external forces acting on a system of particles are conservative, the total mechanical energy $E = T + V$ is conserved.

In Chapters 4 and 5 we saw that the above results for linear momentum, angular momentum, and kinetic energy holds also for continuous mass distributions, with the summations replaced by appropriate integrals.

11.3 MOMENT OF INERTIA TENSOR

In this section, we develop a more general definition of the moment of inertia. Figure 11.4 shows a rigid body (shaded region) rotating around a fixed axis AB, with an angular velocity vector $\boldsymbol{\omega}$. We describe this rigid body as a collection of finite masses m_i located at positions $\mathbf{r}_i$, with respect to a coordinate system xyz which is fixed on the rotating body. The instantaneous velocity $\mathbf{v}_i$ of mass m_i is given by the cross product of the angular velocity vector $\boldsymbol{\omega} = \omega_x\hat{\mathbf{i}} + \omega_y\hat{\mathbf{j}} + \omega_z\hat{\mathbf{k}}$ and the positions $\mathbf{r}_i = x_i\hat{\mathbf{i}} + y_i\hat{\mathbf{j}} + z_i\hat{\mathbf{k}}$:

$$\mathbf{v}_i = \boldsymbol{\omega} \times \mathbf{r}_i \tag{11.3.1}$$

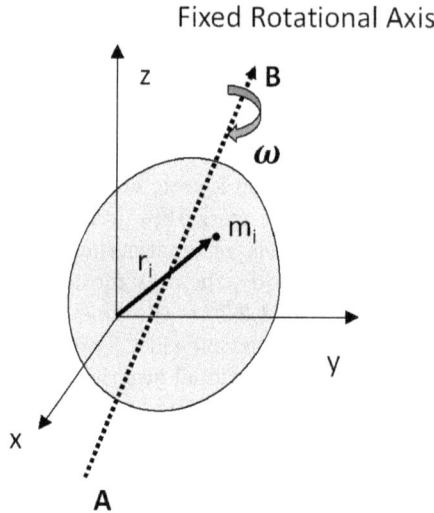

Figure 11.4: Rigid body with a fixed axis of rotation AB. The coordinate system shown is fixed on the rotating body.

The total angular momentum vector $\mathbf{L}$ relative to the center of the coordinate system is given by the vector sum of the angular momenta of the masses m_i:

$$\mathbf{L} = \sum m_i \left(\mathbf{r}_i \times \mathbf{v}_i\right) = \sum m_i \left[\mathbf{r}_i \times \left(\boldsymbol{\omega} \times \mathbf{r}_i\right)\right] \tag{11.3.2}$$

where the summation $i = 1 \ldots n$ runs over the masses m_i which make up the rigid body.

We now use the identity of the triple cross product $\mathbf{A} \times (\mathbf{B} \times \mathbf{A}) = A^2\mathbf{B} - (\mathbf{A} \cdot \mathbf{B})\mathbf{A}$ to evaluate $\mathbf{r}_i \times (\boldsymbol{\omega} \times \mathbf{r}_i)$:

$$\mathbf{r}_i \times \left(\boldsymbol{\omega} \times \mathbf{r}_i\right) = \mathrm{r}_i^2\boldsymbol{\omega} - \mathbf{r}_i\left(\mathbf{r}_i \cdot \boldsymbol{\omega}\right) \tag{11.3.3}$$

$$\mathbf{r}_i \times \left(\boldsymbol{\omega} \times \mathbf{r}_i\right) = \left(x_i^2 + y_i^2 + z_i^2\right)\left(\omega_x\hat{\mathbf{i}} + \omega_y\hat{\mathbf{j}} + \omega_z\hat{\mathbf{k}}\right) - \left(x_i\hat{\mathbf{i}} + y_i\hat{\mathbf{j}} + z_i\hat{\mathbf{k}}\right)\left(x_i\omega_x + y_i\omega_y + z_i\omega_z\right) \tag{11.3.4}$$

Substituting (11.3.4) into (11.3.2) and collecting the terms with $\hat{\mathbf{i}}$, $\hat{\mathbf{j}}$, and $\hat{\mathbf{k}}$:

$$\mathbf{L} = L_x\hat{\mathbf{i}} + L_y\hat{\mathbf{j}} + L_z\hat{\mathbf{k}} \tag{11.3.5}$$

$$\mathbf{L} = \left(I_{xx}\omega_\mathrm{x} + I_{xy}\omega_y + I_{xz}\omega_z\right)\hat{\mathbf{i}} + \left(I_{yx}\omega_\mathrm{x} + I_{yy}\omega_y + I_{yz}\omega_z\right)\hat{\mathbf{j}} + \left(I_{zx}\omega_x + I_{zy}\omega_y + I_{zz}\omega_z\right)\hat{\mathbf{k}} \tag{11.3.6}$$

where the scalar quantities I_{kl} ($k, l = x$, y, or z) are the components of the *moment of inertia tensor*, $\mathbf{I}$, and are defined by:

Moment of Inertia Tensor for Discrete Masses m_i

$$
\begin{aligned}
I_{xx} &= \sum_{i=1}^{N} m_i \left(y_i^2 + z_i^2 \right) = \sum_{i=1}^{N} m_i \left(r_i^2 - x_i^2 \right) \\
I_{yy} &= \sum_{i=1}^{N} m_i \left(x_i^2 + z_i^2 \right) = \sum_{i=1}^{N} m_i \left(r_i^2 - y_i^2 \right) \\
I_{zz} &= \sum_{i=1}^{N} m_i \left(x_i^2 + y_i^2 \right) = \sum_{i=1}^{N} m_i \left(r_i^2 - z_i^2 \right)
\end{aligned} \right\} \tag{11.3.7}
$$

$$
I_{xy} = I_{yx} = -\sum_{i=1}^{N} m_i x_i y_i \qquad I_{xz} = I_{zx} = -\sum_{i=1}^{N} m_i x_i z_i \qquad I_{yz} = I_{zx} = -\sum_{i=1}^{N} m_i y_i z_i \tag{11.3.8}
$$

From these equations it is clear that the moment of inertia matrix $\mathbf{I}$ is symmetric, i.e. $I_{xy} = I_{yx}$, $I_{xz} = I_{zx}$, and $I_{yz} = I_{zy}$.

Equations (11.3.6) and (11.3.8) can be written as the product of two matrices $\mathbf{I}$ and $\boldsymbol{\omega}$:

Angular Momentum and Moment of Inertia Tensor $\mathbf{I}$

$$
\mathbf{L} = \mathbf{I} \cdot \boldsymbol{\omega} \tag{11.3.9}
$$

$$
\mathbf{I} = \begin{bmatrix} I_{xx} & I_{xy} & I_{xz} \\ I_{yx} & I_{yy} & I_{yz} \\ I_{zx} & I_{zy} & I_{zz} \end{bmatrix} \qquad \text{and} \qquad \boldsymbol{\omega} = \begin{bmatrix} \omega_x \\ \omega_y \\ \omega_z \end{bmatrix} \tag{11.3.10}
$$

The terms I_{xx}, I_{yy}, and I_{zz} are referred to as *moments of inertia*. These are the moments of inertia of the rigid body about each axis. The terms I_{xy}, I_{yz}, I_{xz}, ... are called *products of inertia*. As we will see later, the products of inertia are zero if the x, y, and z axes correspond to the rigid body's axes of symmetry. The products of inertia measure the symmetry of the rigid body's mass distribution about the x, y and z axes.

To understand the importance of the product of inertia, consider the following. Suppose you are interested in balancing an automobile tire. In the case of the tire, the axis of symmetry is the axle passing through the center of the tire and perpendicular to the plane of the tire. A proper tire rotates about the axle, or in other words, its angular momentum vector points along the direction of the axle. Suppose the axle points in the z-direction. We want $\mathbf{L} = I_{zz}\omega\hat{\mathbf{k}}$ to be the angular momentum of the wheel. However, suppose the mass of the wheel is not evenly distributed about the axle, then the axle is no longer an axis of symmetry, and there will be at least two nonzero products of inertia. For simplicity, suppose that $I_{zy} \neq 0$ and all other products of inertia are equal to zero. Then in this case, the wheel will precess (wobble), because $\mathbf{L}$ is no longer parallel to the axle (which points in the z-direction). Hence automotive technicians will add small weights to a tire to balance it, such that the wheel's angular momentum will be parallel to its axis of rotation.

Equations (11.3.7) can also be written as a single equation, as follows:

$$
I_{kl} = \sum_{i=1}^{N} m_i \left(\delta_{kl} \sum_{s=1}^{3} u_{is}^2 - u_{ik} u_{il} \right) \tag{11.3.11}
$$

where we use the compact notation u_{is} ($s = 1, 2, 3$) to denote the three components of the position vector $\mathbf{r}_i = (x_i, y_i, z_i)$ of the mass m_i. For example, u_{32} is the y-coordinate of m_3. The symbol δ_{kl} in (11.3.11) is the Kronecker delta which equals 1 when $k = l$, and 0 otherwise.

It is important to remember that the elements of the inertia tensor depend on the choice of origin for the coordinate system. Example 11.3 shows how to calculate the elements of the moment of inertia tensor for a single particle.

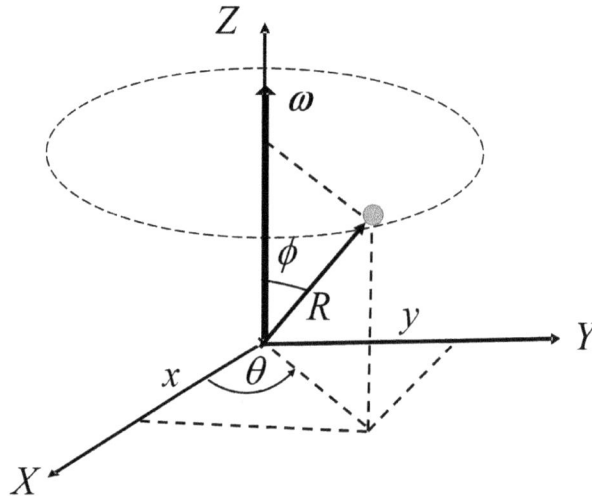

Figure 11.5: Single particle attached to a rod, from Example 11.3.

Example 11.3: Moment of inertia tensor for a single particle

A particle of mass m is attached to a thin massless rod of length R and rotates around the z-axis at a constant angle ϕ as shown in Figure 11.5, with a constant angular velocity ω. Calculate the moment of inertia tensor with respect to the XYZ coordinate system. In which direction is the angular momentum vector $\boldsymbol{\ell}$ pointing?

Solution

The instantaneous position of the mass m is best described by cylindrical coordinates, with the constant radius $r = R \sin\phi$, i.e.,

$$x = r\cos\theta = R\sin\phi\cos\theta \qquad y = r\sin\theta = R\sin\phi\sin\theta \qquad z = R\cos\phi$$

For a constant angular velocity, the angle $\theta = \omega t$. Substituting these values for x, y, z in the expressions for inertia tensor. we obtain:

$$I_{xx} = m\left(y^2 + z^2\right) = mR^2\left(\sin^2\phi\sin^2\theta + \cos^2\phi\right)$$

Similarly
$$I_{yy} = m\left(x^2 + z^2\right) = mR^2\left(\sin^2\phi\cos^2\theta + \cos^2\phi\right)$$
$$I_{zz} = m\left(x^2 + y^2\right) = mR^2\left(\sin^2\phi\cos^2\theta + \sin^2\phi\sin^2\theta\right) = mR^2\sin^2\phi$$
$$I_{xy} = -mxy = -mR^2\sin^2\phi\sin\theta\cos\theta \qquad I_{xz} = -mxz = -mR^2\sin\phi\cos\phi\cos\theta$$
$$I_{yz} = -myz = -mR^2\sin\phi\cos\phi\sin\theta$$

The complete inertia tensor is then:

$$\mathbf{I} = mR^2 \begin{bmatrix} \sin^2\phi\sin^2\theta + \cos^2\phi & -\sin^2\phi\sin\theta\cos\theta & -\sin\phi\cos\phi\cos\theta \\ -\sin^2\phi\sin\theta\cos\theta & \sin^2\phi\cos^2\theta + \cos^2\phi & -\sin\phi\cos\phi\sin\theta \\ -\sin\phi\cos\phi\cos\theta & -\sin\phi\cos\phi\sin\theta & \sin^2\phi \end{bmatrix}$$

The angular momentum vector can now be found by multiplying this matrix $\mathbf{I}$ with the angular momentum matrix $\boldsymbol{\omega} = (0, 0, \omega)$.

$$\boldsymbol{\ell} = \mathbf{I}\boldsymbol{\omega}$$

Since $\boldsymbol{\omega} = (0, 0, \omega)$, the matrix multiplication yields the third column of the matrix $\mathbf{I}$ multiplied by ω, i.e.,

$$\ell_x = -mR^2\omega\sin\phi\cos\phi\cos\theta = -mR^2\omega\sin(2\phi)\cos(\omega t)$$

$$\ell_y = -mR^2\omega\sin\phi\cos\phi\sin\theta = -mR^2\omega\sin(2\phi)\sin(\omega t)$$

$$\ell_z = mR^2\omega\sin^2\phi$$

where we used the trig identity $2\cos\phi\sin\phi = \sin(2\phi)$, and substituted $\theta = \omega t$.

The physical interpretation of these equations is that as the mass m rotates around the z-axis at the fixed angle ϕ, the angular momentum vector $\boldsymbol{\ell}$ rotates around the z-axis also, but with a different angle equal to 2ϕ. The z-component of the angular momentum vector stays constant in time, and is equal to $\ell_z = mR^2\omega\sin^2\phi$.

Python Code

We use the `Matrix` command from SymPy to define the matrices $\mathbf{I}, \boldsymbol{\omega}$ and the `dot` command from NumPy to evaluate the product $\boldsymbol{\ell} = \mathbf{I}\boldsymbol{\omega}$ of the two matrices.

```python
from numpy import dot
from sympy import  Matrix,sin, cos, symbols, simplify
print('-'*28,'CODE OUTPUT','-'*29,'\n')

R, omega, t, m , theta, phi = symbols('R,omega,t,m,theta, phi',real=True)

x = R*sin(phi)*cos(theta)
y = R*sin(phi)*sin(theta)
z = R*cos(phi)

rsq = x**2+y**2+z**2

Ixx, Iyy, Izz = m*(rsq-x**2), m*(rsq-y**2), m*(rsq-z**2)
Ixy, Ixz, Iyz = -m*x*y, -m*x*z, -m*y*z

I = Matrix([[Ixx,Ixy,Ixz],[Ixy,Iyy,Iyz],[Ixz,Iyz,Izz]])

omega_matrix=Matrix([0,0,omega])

print('Lx-component = ', dot(I,omega_matrix)[0][0])

print('\nLy-component = ', dot(I,omega_matrix)[1][0])

print('\nLz-component = ', simplify(dot(I,omega_matrix)[2][0]))

-------------------------- CODE OUTPUT --------------------------
Lx-component =   -R**2*m*omega*sin(phi)*cos(phi)*cos(theta)

Ly-component =   -R**2*m*omega*sin(phi)*sin(theta)*cos(phi)

Lz-component =   R**2*m*omega*sin(phi)**2
```

Mathematica Code

We define the matrices $\mathbf{I}, \boldsymbol{\omega}$ as lists, and use the Dot command in Mathematica to evaluate the product $\boldsymbol{\ell} = \mathbf{I}\boldsymbol{\omega}$ of the two matrices.

```
x = R * Sin[ϕ] * Cos[θ];

y = R * Sin[ϕ] * Sin[θ];

z = R * Cos[ϕ];

ωvector = {0, 0, ω};

Ixx = m * (y^2 + z^2);

Iyy = m * (x^2 + z^2);

Izz = m * (x^2 + y^2);

Ixy = Iyx = −m * x * y;

Iyz = Izy = −m * y * z;

Ixz = Izx = −m * x * z;

Inertia = {{Ixx, Ixy, Ixz}, {Iyx, Iyy, Iyz}, {Izx, Izy, Izz}};

L = Dot[Inertia, ωvector];

Print["Angular momentum L = ", L//Simplify];

L = {−mR²ωCos[θ]Cos[ϕ]Sin[ϕ], −mR²ωCos[ϕ]Sin[θ]Sin[ϕ], mR²ωSin[ϕ]²}
```

If we are dealing with a continuous uniform distribution of particles instead of a collection of discrete masses m_i, the summations in the above equations become integrals over the continuous variables (x, y, z), and the density of the material $\rho(x, y, z)$ must be included inside the integrals. For solids in three dimensions $dm = \rho\, dV$, and the components of the inertia tensor become:

Inertia Tensor for a Continuous Mass Distribution

$$\left. \begin{aligned} I_{xx} &= \iiint\limits_{V} \left(y^2 + z^2\right) \rho\, \mathrm{d}x\mathrm{d}y\mathrm{d}z \\[2mm] I_{yy} &= \iiint\limits_{V} \left(x^2 + z^2\right) \rho\, \mathrm{d}x\mathrm{d}y\mathrm{d}z \\[2mm] I_{zz} &= \iiint\limits_{V} \left(x^2 + y^2\right) \rho\, \mathrm{d}x\mathrm{d}y\mathrm{d}z \end{aligned} \right\} \qquad (11.3.12)$$

$$I_{xy} = -\iiint\limits_{V} xy\rho\, \mathrm{d}x\mathrm{d}y\mathrm{d}z \qquad I_{xz} = -\iiint\limits_{V} xz\rho\, \mathrm{d}x\mathrm{d}y\mathrm{d}z \qquad I_{yz} = -\iiint\limits_{V} yz\rho\, \mathrm{d}x\mathrm{d}y\mathrm{d}z$$

$$(11.3.13)$$

Example 11.4 shows how to evaluate the elements of the inertia tensor for a uniform cube by using *symbolic* integration.

Example 11.4: The moment of inertia of a cube around its center of mass

Consider a uniform cube of side a, evaluate the moment of inertia tensor with respect to the Cartesian coordinate system with the origin $(0, 0, 0)$ located at the center of mass, and the x, y, and z axes parallel to the edges of the cube.

Solution:

We use Cartesian coordinates for the calculation, with the limits of (x, y, z) varying from $-a/2$ to $a/2$. The mass element is $dm = (M/a^3)\, dV = (M/a^3)\, dx\, dy\, dz$ and the moments of inertia elements are:

$$I_{xx} = \iiint_V \left(y^2 + z^2\right) dm = \frac{M}{a^3} \int_{-a/2}^{a/2} \int_{-a/2}^{a/2} \int_{-a/2}^{a/2} \left(y^2 + z^2\right) dx\, dy\, dz$$

$$I_{xx} = \frac{M}{a^3} \int_{-a/2}^{a/2} dx \int_{-a/2}^{a/2} dy \left[y^2 z + \frac{z^3}{3}\right]_{z=-a/2}^{z=a/2}$$

$$I_{xx} = \frac{M}{a^3} \int_{-a/2}^{a/2} dx \left[\frac{y^3}{3} a + \frac{a^3}{12} y\right]_{y=-a/2}^{y=a/2} = \frac{M}{a^3} \int_{-a/2}^{a/2} dx \left[\frac{a^3}{12} a + \frac{a^3}{12} a\right] = \frac{1}{6} M a^2$$

The products of inertia are:

$$I_{xy} = -\iiint_V x\, y\, dm = -\frac{M}{a^3} \int_{-a/2}^{a/2} \int_{-a/2}^{a/2} \int_{-a/2}^{a/2} x\, y\, dx\, dy\, dz$$

$$I_{xy} = -\frac{M}{a^3} \int_{-a/2}^{a/2} x\, dx \int_{-a/2}^{a/2} y\, dy \int_{-a/2}^{a/2} dz = -\frac{M}{a^3}\, 0\, a = 0$$

By symmetry, $I_{xx} = I_{yy} = I_{zz} = Ma^2/6$ and $I_{xy} = I_{xz} = I_{yz} = 0$.

$$I = \begin{pmatrix} Ma^2/6 & 0 & 0 \\ 0 & Ma^2/6 & 0 \\ 0 & 0 & Ma^2/6 \end{pmatrix}$$

Python Code

The `integrate` command in SymPy is used to carry out the triple integrals, and the `Matrix` command is used to create the 3×3 inertia tensor.

```python
from sympy import symbols, integrate, Matrix
print('-'*28,'CODE OUTPUT','-'*29,'\n')


x, y, z, a, m = symbols('x, y, z, a, m',real=True)   #  symbols


# evaluate the integrals using triple itegration in SymPy
Ixx = Iyy =Izz = integrate(y**2+x**2, (x,-a/2,a/2),\
    (y,-a/2,a/2),( z,-a/2,a/2))*m/a**3


Ixy = Ixz = Iyz= -integrate(x*y,(x,-a/2,a/2),\
    (y,-a/2,a/2), (z,-a/2,a/2))*m/a**3


I = Matrix([[Ixx,Ixy,Ixz], [Ixy,Iyy,Iyz], [Ixz,Iyz,Izz]])
print('The inertia tensor I is:\n')
I

------------------------------- CODE OUTPUT -------------------------------
The inertia tensor I is:
Matrix([
[a**2*m/6,        0,        0],
[       0, a**2*m/6,        0],
[       0,        0, a**2*m/6]])
```

Mathematica Code

The `Integrate` command is used to carry out the symbolic triple integrals, and the `MatrixForm` command is used to print the 3×3 inertia tensor as a matrix.

Ixx =

Iyy = Izz = Integrate$[y^2 + z^2, \{x, -a/2, a/2\}, \{y, -a/2, a/2\}, \{z, -a/2, a/2\}]$;

Ixy = Iyx = Integrate$[x * y, \{x, -a/2, a/2\}, \{y, -a/2, a/2\}, \{z, -a/2, a/2\}]$;

Iyz = Izy = Integrate$[y * z, \{x, -a/2, a/2\}, \{y, -a/2, a/2\}, \{z, -a/2, a/2\}]$;

Ixz = Izx = Integrate$[x * z, \{x, -a/2, a/2\}, \{y, -a/2, a/2\}, \{z, -a/2, a/2\}]$;

Inertia $= m/a^3 * \{\{Ixx, Ixy, Ixz\}, \{Iyx, Iyy, Iyz\}, \{Izx, Izy, Izz\}\}$;

Print["Moment of inertia tensor = ", Inertia//MatrixForm];

Moment of inertia tensor $= \begin{pmatrix} \frac{a^2 m}{6} & 0 & 0 \\ 0 & \frac{a^2 m}{6} & 0 \\ 0 & 0 & \frac{a^2 m}{6} \end{pmatrix}$

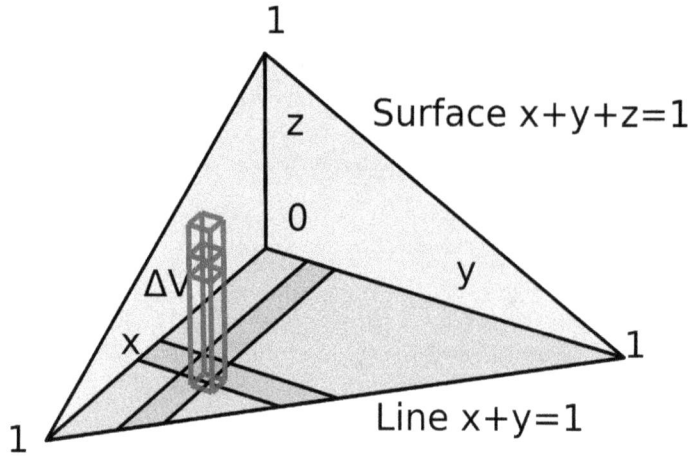

Figure 11.6: Evaluating the moment of inertia tensor of a uniform triangular pyramid in Example 1.5.

Example 11.5 shows how to evaluate the elements of the inertia tensor for a triangular pyramid with a uniform density by using *numerical* integration.

Example 11.5: The moment of inertia of a uniform triangular pyramid
Consider the triangular pyramid with vertices at the origin, $(1,0,0)$, $(0,1,0)$, and $(0,0,1)$ shown in Figure 11.6. The pyramid has a density $\rho = 1$ (in SI units). Use numerical integration to evaluate the moment of inertia tensor with respect to the Cartesian coordinate system with the origin at $(0,0,0)$.

Solution:
In order to find the moment of inertia tensor, we first need to choose a coordinate system and identify the limits of integration. We use Cartesian coordinates for the calculation, and we need to figure out the limits of integration to evaluate (11.3.12) and (11.3.13).

The volume element dV appears in Figure 11.6. Notice that its height is determined by its x and y coordinates. The "front" face of the pyramid in the figure is described by the equation $x + y + z = 1$. Therefore, the limits for the z integral will range from 0 to $1 - x - y$. Next, the distance the volume element can be translated along the y-axis in the xy-plane is determined by its x-coordinate. Therefore, the limits for the y integral will be from 0 to $1 - x$ (the equation of the line that forms the base of the pyramid in the xy-plane). Finally, x can range from 0 to 1.

Now that we have all of the pieces, we can perform the integrals. The mass element with $\rho = 1$ is $dm = \rho\, dV = dx\, dy\, dz$, and the moments of inertia elements are:

$$I_{xx} = \iiint_V \left(y^2 + z^2 \right) dm = \iiint_V \left(y^2 + z^2 \right) dx\, dy\, dz$$

$$I_{xx} = \int_0^1 dy \int_0^{1-y} dx \left[y^2 z + \frac{z^3}{3} \right]_{z=0}^{z=1-x-y}$$

$$I_{xx} = \int_0^1 dy \int_0^{1-y} dx \left[y^2 (1 - x - y) + \frac{(1 - x - y)^3}{3} \right]$$

The integrals are rather tedious to evaluate by hand, so we use the alternative numerical integration routines available in a CAS.

Python Code

The Python code uses the `scipy.integrate.tplquad()` function to evaluate the triple integrals. In order to simplify the code, we define a function `f` which evaluates the command `scipy.integrate.tplquad()` for the triple integrals, and is called for each of the components of the inertia tensor I_{xx}, I_{xy}, etc.

The important feature of the double integration here is the implementation of the integration limits in the line of code `tplquad(i, 0, 1, 0, yupper)`. Here `yupper` is the function $1 - x$ representing the upper limit in the y-integration, and `i` is the function to be integrated.

In this example we use a NumPy array to represent the moment of inertia matrix.

Note that the order of the integrations `tplquad(i, 0, 1, 0, x2, 0, z2)` is again important, and we define the upper limits of integration as two functions: x2 and z2.

```python
from scipy.integrate import tplquad
import numpy as np
from sympy import symbols
print('-'*28,'CODE OUTPUT','-'*29,'\n')


z2 = lambda y, z: 1-y-z #upper limit for z integration
x2 = lambda z: 1-z      #upper limit for x integration


def f(i):
    return tplquad(i, 0, 1,   0, x2,   0, z2)[0]

Ixx = lambda z, y, x: (y**2+z**2)
Iyy = lambda z, y, x: (x**2+z**2)
Izz = lambda z, y, x: (y**2+x**2)


Ixy = lambda z, y, x: -x*y
Ixz = lambda z, y, x: -x*z
Iyz = lambda z, y, x: -y*z


I = np.array([[f(Ixx),f(Ixy),f(Ixz)],[f(Ixy),f(Iyy),f(Iyz)],\
[f(Ixz),f(Iyz),f(Izz)]])
print('Moment of inertia tensor I=')
I


-------------------------- CODE OUTPUT ----------------------------
Moment of inertia tensor I=
array([[ 0.03333333, -0.00833333, -0.00833333],
       [-0.00833333,  0.03333333, -0.00833333],
       [-0.00833333, -0.00833333,  0.03333333]])
```

Mathematica Code
We evaluate the triple integrals in Mathematica using the `NIntegrate` command. Notice that the limits of the integral that we would do last by hand are the first limits in `NIntegrate`.

```
Ixx = NIntegrate[y^2 + z^2, {x, 0, 1}, {y, 0, 1 − x}, {z, 0, 1 − x − y}];

Iyy = NIntegrate[x^2 + z^2, {x, 0, 1}, {y, 0, 1 − x}, {z, 0, 1 − x − y}];

Izz = NIntegrate[y^2 + z^2, {x, 0, 1}, {y, 0, 1 − x}, {z, 0, 1 − x − y}];

Ixy = Iyx = −NIntegrate[x * y, {x, 0, 1}, {y, 0, 1 − x}, {z, 0, 1 − x − y}];

Iyz = Izy = −NIntegrate[y * z, {x, 0, 1}, {y, 0, 1 − x}, {z, 0, 1 − x − y}];

Ixz = Izx = −NIntegrate[x * z, {x, 0, 1}, {y, 0, 1 − x}, {z, 0, 1 − x − y}];


Inertia = {{Ixx, Ixy, Ixz}, {Iyx, Iyy, Iyz}, {Izx, Izy, Izz}};

Print["Moment of inertia tensor = ", Inertia//MatrixForm];
```

$$\text{Moment of inertia tensor} = \begin{pmatrix} 0.0333333 & -0.00833333 & -0.00833333 \\ -0.00833333 & 0.0333333 & -0.00833333 \\ -0.00833333 & -0.00833333 & 0.0333333 \end{pmatrix}$$

11.4 KINETIC ENERGY AND THE INERTIA TENSOR

We now proceed to evaluate the kinetic energy of a rigid body and demonstrate how it relates to the moment of inertia tensor. The total kinetic energy is the sum of the kinetic energies for the masses m_i moving with velocities $\mathbf{v}_i$:

$$T = \sum \left(\frac{1}{2} m_i \mathbf{v}_i \cdot \mathbf{v}_i \right) = \frac{1}{2} \sum \left(m_i \left[(\boldsymbol{\omega} \times \mathbf{r}_i) \cdot \mathbf{v}_i \right] \right) \tag{11.4.1}$$

where we used $\mathbf{v}_i = \boldsymbol{\omega} \times \mathbf{r}_i$, and we have dropped the limits of the summation for simplicity. Note that we used $\boldsymbol{\omega}_i = \boldsymbol{\omega}$ for all values of i, because for a rigid body, each particle has the same angular velocity. The fact that $\boldsymbol{\omega}$ is the same for each particle is due to the fact that the distances between particles are fixed in a rigid body. We now use the property of the scalar triple product $(\mathbf{A} \times \mathbf{B}) \cdot \mathbf{C} = \mathbf{A} \cdot (\mathbf{B} \times \mathbf{C})$ in order to simplify T:

$$T = \frac{1}{2} \sum (m_i \boldsymbol{\omega} \cdot [(\mathbf{r}_i \times \mathbf{v}_i)]) = \frac{1}{2} \boldsymbol{\omega} \cdot \left(\sum m_i [(\mathbf{r}_i \times \mathbf{v}_i)] \right) = \frac{1}{2} \boldsymbol{\omega} \cdot \mathbf{L} \tag{11.4.2}$$

where we used the definition of the angular momentum of a system of particles $\mathbf{L} = \sum m_i [(\mathbf{r}_i \times \mathbf{v}_i)]$ from (11.3.2). We now use (11.3.6) for the angular momentum $\mathbf{L}$. By writing out the components we obtain:

$$T = \frac{1}{2} \boldsymbol{\omega} \cdot \mathbf{L} = \frac{1}{2} I_{xx}\omega_x^2 + \frac{1}{2} I_{yy}\omega_y^2 + \frac{1}{2} I_{zz}\omega_z^2 + I_{xy}\omega_x\omega_y + I_{yz}\omega_y\omega_z + I_{zx}\omega_z\omega_x \tag{11.4.3}$$

By using (11.3.9), this equation can be written in a compact form as a matrix equation:

Kinetic Energy of a Rigid Body

$$T = \frac{1}{2} (\boldsymbol{\omega})^{\mathrm{T}} \cdot \mathbf{L} = \frac{1}{2} (\boldsymbol{\omega})^{\mathrm{T}} \cdot \mathbf{I} \cdot \boldsymbol{\omega} \qquad (11.4.4)$$

Here the dot product indicates the multiplication of the three matrices, and $(\boldsymbol{\omega})^{\mathrm{T}}$ indicates the transpose of the column matrix representing the angular velocity vector $\boldsymbol{\omega}$ (i.e., a row vector).

If the angular velocity vector $\boldsymbol{\omega}$ points in the direction of the unit vector $\hat{\mathbf{n}}$, we can write $\boldsymbol{\omega} = \omega \hat{\mathbf{n}}$, where ω is the magnitude of the angular frequency. Therefore, the above matrix equation becomes:

$$T = \frac{1}{2}\omega^2 (\hat{\mathbf{n}})^{\mathrm{T}} \cdot \mathbf{I} \cdot \hat{\mathbf{n}} \qquad (11.4.5)$$

From introductory physics we recall that the kinetic energy for rotation around a fixed axis is:

$$T = \frac{1}{2} I_n \omega^2 \qquad (11.4.6)$$

where I_n is the *scalar* moment of inertia around this axis. By comparing the last two equations, we obtain the general expression:

Moment of Inertia for Rotation Around an Axis in the Unit Vector Direction $\hat{\mathbf{n}}$

$$I_n = (\hat{\mathbf{n}})^{\mathrm{T}} \cdot \mathbf{I} \cdot \hat{\mathbf{n}} \qquad (11.4.7)$$

The products of inertia can also be found using an expression similar to (11.4.7). For example, I_{xy} is obtained by calculating the quantity:

$$I_{xy} = \left(\hat{\mathbf{i}}\right)^{\mathrm{T}} \cdot \mathbf{I} \cdot \hat{\mathbf{j}} \qquad (11.4.8)$$

Example 11.6:
Use SymPy and Mathematica to find the scalar moments of inertia I_n of a cube when it is rotated around
(a) the main diagonal,
(b) a face diagonal on the xy-plane,
(c) the edge that runs along the x-axis.

Solution
We first evaluate the inertia tensor with respect to a coordinate system centered at one of the vertices, as in Example 11.4. Once the tensor is evaluated, the scalar moment of inertia I is obtained using (11.4.7) and the appropriate unit vector $\hat{\mathbf{n}}$ pointing along the direction of rotation.

For the main diagonal we have $\hat{\mathbf{n}} = \frac{1}{\sqrt{3}}\left(\hat{\mathbf{i}} + \hat{\mathbf{j}} + \hat{\mathbf{k}}\right)$, for the face diagonal on the xy-plane $\hat{\mathbf{n}} = \frac{1}{\sqrt{2}}\left(\hat{\mathbf{i}} + \hat{\mathbf{j}}\right)$, and for the edge that runs along the x-axis, $\hat{\mathbf{n}} = \hat{\mathbf{i}}$.

The codes below give $I = ma^2/6$, $I = 2ma^2/3$ and $I = 5ma^2/12$, correspondingly. As we may have expected from geometrical considerations, the largest value of I is obtained for the face diagonal of the cube.

Python Code

The tensor is evaluated using SymPy as in previous examples. We use the command **transpose** from Numpy to find the transpose of the matrix $\boldsymbol{\omega} = \omega\hat{\mathbf{n}}$, and the command @ to multiply the three matrices.

```python
from sympy import symbols, integrate, Matrix, sqrt
import numpy as np
print('-'*28,'CODE OUTPUT','-'*29,'\n')

x, y, z, a, m = symbols('x, y, z, a, m',real=True)   #  symbols

# evaluate the integrals using triple itegration in SymPy
Ixx = integrate(y**2+x**2, (x,0,a), (y,0,a),(z,0,a))
Iyy = integrate(y**2+x**2, (x,0,a), (y,0,a),(z,0,a))
Izz = integrate(y**2+x**2, (x,0,a), (y,0,a),(z,0,a))

Ixy = -integrate(x*y,(x,0,a), (y,0,a), (z,0,a))
Ixz = -integrate(x*y,(x,0,a), (y,0,a), (z,0,a))
Iyz= -integrate(x*y,(x,0,a), (y,0,a), (z,0,a))

I = np.array([[Ixx,Ixy,Ixz],[Ixy,Iyy,Iyz],[Ixz,Iyz,Izz]])*m/a**3

omega = np.array([1/sqrt(3),1/sqrt(3),1/sqrt(3)])
omega_trnsp = np.transpose(omega)

print('(a) When rotational axis is along diagonal,\
I = ',omega_trnsp @ (I @ omega))

omega = np.array([1/sqrt(2),1/sqrt(2),0])
omega_trnsp = np.transpose(omega)

print('\n(b) When rotational axis is along face diagonal,\
I = ',omega_trnsp @ (I @ omega))

omega = np.array([1,0,0])
omega_trnsp = np.transpose(omega)

print('\n(c) When rotational axis is along edge of cube,\
I = ',omega_trnsp @ (I @ omega))

--------------------------- CODE OUTPUT ----------------------------
(a) When rotational axis is along diagonal,I =   a**2*m/6

(b) When rotational axis is along face diagonal,I =   5*a**2*m/12

(c) When rotational axis is along edge of cube,I =   2*a**2*m/3
```

Mathematica Code

The tensor is evaluated using `Integrate` as in previous examples. We use the period command to multiply the three matrices.

```
ρ = m/a^3;

Ixx = Iyy = Izz = Integrate[ρ(y^2 + z^2), {x,0,a}, {y,0,a}, {z,0,a}];

Ixy = Iyz = Ixz = −Integrate[ρ(x * y), {x,0,a}, {y,0,a}, {z,0,a}];
```

$$iCube = \begin{pmatrix} Ixx & Ixy & Ixz \\ Ixy & Iyy & Iyz \\ Ixz & Iyz & Izz \end{pmatrix} ;$$

```
ω = 1/Sqrt[3] * {1,1,1};

Print["(a) When the rotational axis is along cube diagonal, I=", ω.iCube.ω];
```

(a) When the rotational axis is along cube diagonal, I=$\frac{a^2 m}{6}$

```
ω = 1/Sqrt[2] * {1,1,0};

Print["(b) When the rotational axis is along face diagonal, I=",

ω.iCube.ω//Simplify];
```

(b) When the rotational axis is along face diagonal, I=$\frac{5a^2 m}{12}$

```
ω = {1,0,0};

Print["(c) When the rotational axis is along edge of cube, I=", ω.iCube.ω];
```

(c) When the rotational axis is along edge of cube, I=$\frac{2a^2 m}{3}$

11.5   INERTIA TENSOR IN DIFFERENT COORDINATE SYSTEMS – THE PARALLEL AXIS THEOREM

Two important theorems which simplify the calculations of the moments of inertia of a rigid body, are the *parallel axis theorem* and the *perpendicular axis theorem*.

In its simplest form, the *parallel axis theorem* states that if I is the scalar moment of inertia of a body around an axis AB, and I_{CM} is the corresponding moment of inertia about a second axis parallel to AB and passing through the center of mass, then I and I_{CM} are related by:

A Simple Form of the Parallel Axis Theorem

$$I = I_{CM} + md^2 \tag{11.5.1}$$

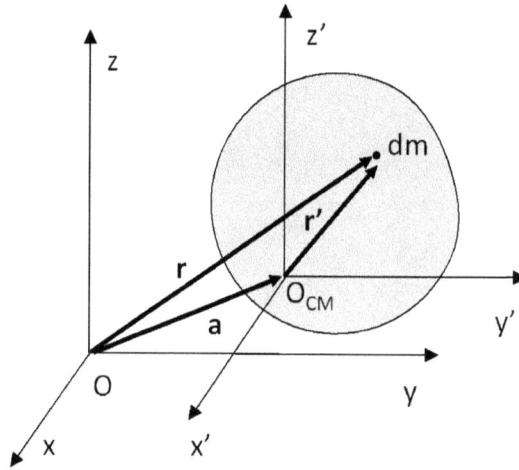

Figure 11.7: Two coordinate systems with their axes parallel to each other, are shifted in space relative to each other by a vector **a**.

where d is the shortest distance between the two parallel axes.

The *perpendicular axis theorem* refers to a thin lamina mass which is distributed on the xy-plane of the Cartesian coordinate system. If I_x, I_y, and I_z are the scalar moments of inertia of this distribution around the x, y, and z axes respectively, then these are related by:

The Perpendicular Axis Theorem for a Lamina

$$I_z = I_x + I_y \qquad (11.5.2)$$

More generally, it is convenient and useful to know the relationship between inertia tensors expressed in different coordinate systems.

We now prove the parallel axis theorem. Figure 11.7 shows two Cartesian coordinate systems with origins O and O_{CM} and with their axes parallel to each other. Note that O_{CM} is located at the object's center of mass. If **a** is the vector connecting the two origins, the relationship of any position vectors $\mathbf{r}_i$ and $\mathbf{r}'_i$ in the two coordinate system is:

$$\mathbf{r}_i = \mathbf{r}'_i + \mathbf{a} \qquad (11.5.3)$$

Let **I** be the inertia tensor defined in the coordinate system (x, y, z) with the origin fixed at point O, and $\mathbf{I}'$ be the inertia tensor for the center of mass coordinate system (x', y', z') with its origin at O_{CM}. We wish to find the relation between **I** and $\mathbf{I}'$. As we saw previously in (11.3.11), the components of the inertia tensor **I** with respect to O can be written in compact form as:

$$I_{kl} = \sum_{i=1}^{N} m_i \left(\delta_{kl} \sum_{s=1}^{3} u_{is}^2 - u_{ik} u_{il} \right) \qquad (11.5.4)$$

where we again use the compact notation u_{is} $(s = 1, 2, 3)$ to denote the three components of the mass m_i's position vector $\mathbf{r}_i = (x_i, y_i, z_i)$, and δ_{kl} is the Kronecker delta.

Similarly, the components for the tensor with respect to the second coordinate system O_{CM} are:

$$I'_{kl} = \sum_{i=1}^{N} m_i \left(\delta_{kl} \sum_{s=1}^{3} (u'_{is})^2 - u'_{ik} u'_{il} \right) \tag{11.5.5}$$

Equation (11.5.3) can be written in terms of these components as:

$$u_{is} = u'_{is} + a_s \qquad s = 1, 2, 3 \tag{11.5.6}$$

Substituting (11.5.6) into (11.5.4), by expanding the terms and rearranging, we obtain:

$$I_{kl} = \sum_{i=1}^{N} m_i \left(\delta_{kl} \sum_{s=1}^{3} (u'_{is} + a_s)^2 - (u'_{ik} + a_k)(u'_{il} + a_l) \right)$$

$$= \sum_{i=1}^{N} m_i \left[\delta_{kl} \sum_{s=1}^{3} (u'_{is})^2 - u'_{ik} u'_{il} \right] + \sum_{i=1}^{N} m_i \left[\delta_{kl} \sum_{s=1}^{3} a_s^2 - a_k a_l \right]$$

$$+ 2 \sum_{i=1}^{N} m_i \left[\delta_{kl} (u'_{is})^2 a_s \right] - \sum_{i=1}^{N} m_i u'_{ik} a_l - \sum_{i=1}^{N} m_i u'_{il} a_k \tag{11.5.7}$$

The last three terms are zero because of the definition of the center of mass:

$$\sum_{i=1}^{N} m_i \mathbf{r}'_i = 0 \implies \sum_{i=1}^{N} m_i u'_{is} = 0 \tag{11.5.8}$$

The coefficients a_k and a_l factor out of the sums, and (11.5.7) becomes:

$$I_{kl} = \sum_{i=1}^{N} m_i \left[\delta_{kl} \sum_{s=1}^{3} (u'_{is})^2 - u'_{ik} u'_{il} \right] + \sum_{i=1}^{N} m_i \left[\delta_{kl} \sum_{s=1}^{3} a_s^2 - a_k a_l \right] \tag{11.5.9}$$

$$I_{kl} = I'_{kl} + \sum_{i=1}^{N} m_i \left[\delta_{kl} \sum_{s=1}^{3} a_s^2 - a_k a_l \right] \tag{11.5.10}$$

The total mass is $M = \sum m_i$, so (11.5.10) gives the generalized parallel axes theorem, also known as the *Steiner theorem*, for the elements of the inertia tensor in any coordinate system:

The Parallel Axis Theorem

$$I_{kl} = I'_{kl} + M \left[a^2 \delta_{kl} - a_k a_l \right] \tag{11.5.11}$$

Note that $\mathbf{I}'$ is the object's moment of inertia for a coordinate system whose origin is located at the object's center of mass.

By applying the parallel axis theorem for the diagonal elements of the tensor so that $\delta_{kl} = 1$, we recover (11.5.1), the simple form of the parallel axis theorem:

$$I_{kk} = I'_{kk} + M \left[a^2 - a_k^2 \right] = I'_{kk} + M d_k^2 \tag{11.5.12}$$

where d_k is the shortest distance from the axis of rotation to the center of mass.

11.6 PRINCIPAL AXES OF ROTATION

A very useful tool for describing the motion of rigid bodies is the concept of *principal axes of rotation*. This is best described within the general theory of linear algebra, and is closely related to the concepts of eigenvalues and eigenvectors of matrices.

Our discussion of principal axes of rotation begins with a review of the inertia tensor itself. First, it is important to remember that the elements of the inertia tensor depend on the axes and choice of origin for the system. Second, recall that the inertia tensor is a real symmetric matrix, i.e., $I_{kl} = I_{lk}$ ($k, l = 1, 2, 3$). There is a theorem in Linear Algebra that states that any real symmetric matrix can be diagonalized. In other words, the matrix can be rewritten in a form which has zeros in all of the off-diagonal elements. Physically, this means that for *any* choice of the origin for any rigid body, there always exists a set of axes in which the inertia tensor is diagonal, i.e., it is of the form:

$$\mathbf{I} = \begin{bmatrix} \lambda_1 & 0 & 0 \\ 0 & \lambda_2 & 0 \\ 0 & 0 & \lambda_3 \end{bmatrix} \tag{11.6.1}$$

This special set of axes is known as *the principal axes of the rigid body*. According to the same theorem from Linear Algebra, there are three principal axes of rotation, and they are mutually orthogonal. In addition, any axis of symmetry through the origin is a principal axis.

The main property of the principal axes is that when rotation takes place around one of these three axes, the angular momentum $\mathbf{L}$ and angular velocity vectors $\boldsymbol{\omega}$ are parallel to each other. Therefore, if $\boldsymbol{\omega}$ points along the direction of a principal axis, then $\mathbf{L} = \lambda \boldsymbol{\omega}$ (where λ is a scalar), and $\mathbf{L}$ and $\boldsymbol{\omega}$ to point in the same direction. However, recall that in general, $\mathbf{L} = \mathbf{I} \cdot \boldsymbol{\omega}$. Hence:

$$\mathbf{I} \cdot \boldsymbol{\omega} = \lambda \boldsymbol{\omega} \tag{11.6.2}$$

The scalar λ is called a *principal moment of inertia*, and it is the moment of inertia of the object when it is rotating about the principal axis $\boldsymbol{\omega}$. Before discussing principal moments and principal axes further, it is important to pause and discuss some general properties of (11.6.2).

The equation $\mathbf{I} \cdot \boldsymbol{\omega} = \lambda \boldsymbol{\omega}$ is an *eigenvalue problem* in the theory of Linear Algebra. In general, in an eigenvalue problem we are given a square matrix $\mathbf{A}$, and we are looking for two quantities, a vector $\mathbf{a}$ and a scalar quantity λ, such that:

$$\mathbf{A} \, \mathbf{a} = \lambda \, \mathbf{a} \tag{11.6.3}$$

The vector $\mathbf{a}$ is called an *eigenvector of the square matrix* $\mathbf{A}$, *corresponding to the eigenvalue* λ. Hence, according to (11.6.2), the principal axes are the eigenvectors of the inertia tensor and principal moments are the eigenvalues associated with each principal axis. We see from (11.6.3), that the matrix $\mathbf{A}$ scales the vector $\mathbf{a}$ by a value λ.

Physically, from (11.6.3) we see that for $\lambda > 0$, the matrix $\mathbf{A}$ elongates or shortens the vector $\mathbf{a}$, but it does not change the direction of $\mathbf{a}$ (the vector $\mathbf{a}$ stays along the same line after multiplication). If $\lambda = 1$, then $\mathbf{a}$ remains unchanged after multiplication by $\mathbf{A}$. When $\lambda < 0$, the vector $\mathbf{a}$ is still scaled, but now it points in the direction opposite of its original orientation. The vector $\mathbf{a}$ stays along the same line after multiplication by A. In other words, the direction of $\mathbf{a}$ does not change, insofar as it lies along the same line it was on before the multiplication.

According to another theorem in Linear Algebra, in order for the eigenvalue equation (11.6.3) to have a nontrivial solution $\mathbf{a} \neq \mathbf{0}$, the determinant of the matrix $(\mathbf{A} - \lambda \mathcal{I})$ must be zero, i.e.,

$$\det \left(\mathbf{A} - \lambda \, \mathcal{I} \right) = 0 \tag{11.6.4}$$

where $\mathcal{I}$ is the square identity matrix, consisting of ones along the diagonal and zeros everywhere else. Equation (11.6.4) is called the *characteristic equation* of our eigenvalue problem. In Chapter 13, we will revisit the eigenvalue problem in more detail.

There are two steps in finding the eigenvalues and eigenvectors of a given moment of inertia matrix:

1. Solve the characteristic equation (11.6.4), in order to find the three eigenvalues λ_1, λ_2, and λ_3. These eigenvalues will depend of course on the physical parameters of the rigid body.

2. Substitute the first eigenvalue λ_1 into $\mathbf{I}\mathbf{a}_1 = \lambda_1\mathbf{a}_1$, in order to find the corresponding eigenvector $\mathbf{a}_1$. Next, we repeat this step for each of the remaining eigenvalues in order to obtain their eigenvector.

At the end of this process, we will have found three eigenvectors $\mathbf{a}_1$, $\mathbf{a}_2$, and $\mathbf{a}_3$ and the corresponding eigenvalues λ_1, λ_2, and λ_3, respectively. The resulting eigenvalues are the *principal moments of inertia* and the resulting eigenvectors are the *principal axes*. The principal moment of inertia λ_i is the rigid body's moment of inertia for a rotation about the principal axis $\mathbf{a}_i$.

The Mathematica and Python codes in Examples 11.7 and 11.8 shows how to evaluate the principal moments of inertia and principal axes of a cube, with respect to two different coordinate systems.

Example 11.7:
Find the inertia tensor, principal moments of inertia and principal axes for a cube with uniform density, mass m, and side a, with respect to a coordinate system defined by its edges and the origin located at one of the corners.

Solution:
The codes below calculate the elements of the symmetric inertia tensor, by carrying out the triple integrals from $x = 0$ to $x = a$, and similarly for the y and z axes. Note that we use $a = 1$ for plotting purposes.

Both codes yield the eigenvalues $I_1 = I_2 = 11\,m\,a^2/12$ and $I_3 = m\,a^2/6$, and the eigenvectors $(-1,0,1)$, $(-1,1,0)$ and $(1,1,1)$. It is easily verified that these three vectors are perpendicular to each other.

Python Code
The commands `Eigenvects` and `Eigenvalues` from SymPy are used to find the three eigenvectors and eigenvalues of the inertia tensor. These two commands return a Python dictionary, together with the multiplicity of each eigenvalue. The eigenvectors are stored in the parameters $vec1$, $vec2$, and $vec3$ and the commands `Poly3DCollection` and `quiver` produce a 3D plot of the cube and of the three eigenvectors $(1,1,1)$, $(-1,1,0)$ and $(-1,0,1)$.

```
from sympy import Matrix, symbols, integrate
import matplotlib.pyplot as plt
from mpl_toolkits.mplot3d.art3d import Poly3DCollection

x, y, z, a, m = symbols('x, y, z, a, m',real=True)  # symbols

# evaluate the integrals using triple integration in SymPy
Ixx = integrate(y**2+x**2, (x,0,a), (y,0,a),\
                ( z,0,a))*m/a**3
Iyy = integrate(y**2+x**2, (x,0,a), (y,0,a),\
                ( z,0,a))*m/a**3
Izz = integrate(y**2+x**2, (x,0,a), (y,0,a),\
                ( z,0,a))*m/a**3

Ixy = -integrate(x*y,(x,0,a), (y,0,a), (z,0,a))*m/a**3
Ixz = -integrate(x*y,(x,0,a), (y,0,a), (z,0,a))*m/a**3
Iyz= -integrate(x*y,(x,0,a), (y,0,a), (z,0,a))*m/a**3

I = Matrix([[Ixx,Ixy,Ixz], [Ixy,Iyy,Iyz], [Ixz,Iyz,Izz]])
print('-'*28,'CODE OUTPUT','-'*29,'\n')

print('The eigenvalues are: ')
print(I.eigenvals())

print('\n The eigenvectors are:')
vec1 = I.eigenvects()[0][2][0]
vec2 = I.eigenvects()[1][2][0]
vec3 = I.eigenvects()[1][2][1]
print(vec1)
print(vec2)
print(vec3)

--------------------------- CODE OUTPUT ----------------------------
The eigenvalues are:
{a**2*m/6: 1, 11*a**2*m/12: 2}

 The eigenvectors are:
Matrix([[1], [1], [1]])
Matrix([[-1], [1], [0]])
Matrix([[-1], [0], [1]])
```

```
fig = plt.figure()
ax = fig.add_subplot(projection='3d')    # set up 3D plot
ax.quiver(0,0,0,vec1[0],vec1[1],vec1[2],color='r',\
length=1,arrow_length_ratio=.3)
ax.quiver(0,0,0,vec2[0],vec2[1],vec2[2],color='b',\
length=1,arrow_length_ratio=.3)
ax.quiver(0,0,0,vec3[0],vec3[1],vec3[2],color='g',\
length=1,arrow_length_ratio=.3)

ax.set_xlabel('X')      # label,set limits on x,y,z axes
ax.set_ylabel('Y')
ax.set_zlabel('Z')
ax.set_xlim(-1.2,1.2);
ax.set_ylim(-1.2,1.2);
ax.set_zlim(-1.2,1.2);

shape=[((0,0,0),(1,0,0),(1,1,0),(0,1,0),\
       (0,0,0),(0,0,1),(0,1,1),(0,1,0),\
       (0,0,0),(0,0,1),(1,0,1),(1,1,1),(0,1,1),\
    (0,1,0),(1,1,0),(1,1,1),(1,1,0),(1,0,0),(1,0,1),(0,0,1) )]
cube =Poly3DCollection(shape,color='k',alpha=0.1)  # draw the cube
ax.add_collection3d(cube)
plt.show()
```

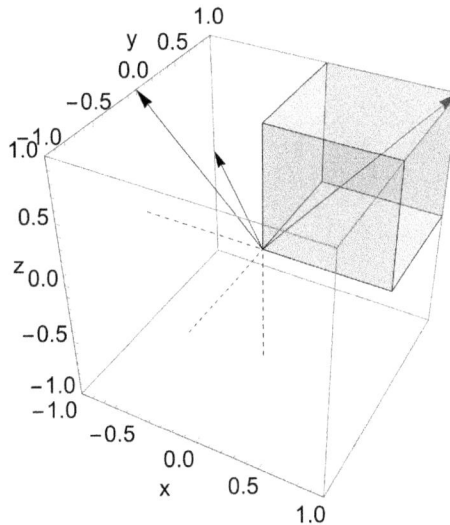

Figure 11.8: Eigenvectors of a uniform cube, with respect to a coordinate system defined by its edges and the origin located at one of the corners, from Example 11.7.

Mathematica Code

The commands Eigenvectors and Eigenvalues are used to find the three eigenvectors and eigenvalues of the inertia tensor. The eigenvectors are stored in the parameters *vec1*, *vec2*, and *vec3* and the commands Graphics3D and Arrow are then combined to produce a 3D plot of the cube and of the 3 eigenvectors in Figure 11.8.

```
ρ = m/a^3;

Ixx = Iyy = Izz = Integrate[ρ * y^2 + z^2), {x, 0, a}, {y, 0, a},
{z, 0, a}];

Ixy = Iyz = Ixz = −Integrate[ρ * (x * y), {x, 0, a}, {y, 0, a}, {z, 0, a}];

           ⎛ Ixx  Ixy  Ixz ⎞
iCube =    ⎜ Ixy  Iyy  Iyz ⎟ ;
           ⎝ Ixz  Iyz  Izz ⎠
Print["The tensor I = ", iCube//MatrixForm];

Print["The Eigenvalues of I are: ", Eigenvalues[iCube]];

Print["The Eigenvectors of I are: ", Eigenvectors[iCube]];


vec1 = Eigenvectors[iCube][[1]];

vec2 = Eigenvectors[iCube][[2]];

vec3 = Eigenvectors[iCube][[3]];
```

The Eigenvalues of I are: $\left\{ \frac{11a^2m}{12}, \frac{11a^2m}{12}, \frac{a^2m}{6} \right\}$

The Eigenvectors of I are: $\{\{-1, 0, 1\}, \{-1, 1, 0\}, \{1, 1, 1\}\}$

```
Graphics3D[{Arrow[{{0, 0, 0}, vec1}], Arrow[{{0, 0, 0}, vec2}],

Arrow[{{0, 0, 0}, vec3}], FaceForm[{Opacity[.2]}], Cuboid[{0, 0, 0},

{1, 1, 1}], Dashed, Line[{{-1, 0, 0}, {1, 0, 0}}],

Line[{{0, -1, 0}, {0, 1, 0}}], Line[{{0, 0, -1}, {0, 0, 1}}]},

Axes → True, BaseStyle → FontSize → 18, AxesLabel → {x, y, z},

PlotRange → {{-1, 1}, {-1, 1}, {-1, 1}}]
```

Example 11.8:
Find the inertia tensor, principal moments and principal axes for a cube with uniform density, mass m, and side a, with respect to a coordinate system with origin at the center of mass of the cube, and axes parallel to the edges of the cube.

Solution
The codes are very similar to Example 11.7, with the difference being that the integrations are carried out from $x = -a/2$ to $x = a/2$, instead of from $x = 0$ to $x = a$ used in Example 11.7.

The codes yield $I_1 = I_2 = I_3 = ma^2/6$ and the three eigenvectors are the unit vectors $(1,0,0)$, $(0,1,0)$ and $(0,0,1)$. The principal moments of inertia with respect to the center of the cube are equal to each other, and the corresponding moment of inertia matrix is diagonal. This means that if the cube is rotated around *any* axis through its center, the scalar moment of inertia will always be the same and equal to $I = ma^2/6$. The results are shown in Figure 11.9.

Python Code

```
from sympy import Matrix, symbols, integrate
import matplotlib.pyplot as plt
from mpl_toolkits.mplot3d.art3d import Poly3DCollection

x, y, z, a, m = symbols('x, y, z, a, m',real=True)  # symbols

# evaluate the integrals using triple integration in SymPy
Ixx = integrate(y**2+x**2, (x,-a/2,a/2), (y,-a/2,a/2),\
               ( z,-a/2,a/2))

Iyy = integrate(y**2+x**2, (x,-a/2,a/2), (y,-a/2,a/2),\
               ( z,-a/2,a/2))

Izz = integrate(y**2+x**2, (x,-a/2,a/2), (y,-a/2,a/2),\
               ( z,-a/2,a/2))

Ixy = -integrate(x*y,(x,-a/2,a/2), (y,-a/2,a/2), (z,-a/2,a/2))

Ixz = -integrate(x*y,(x,-a/2,a/2), (y,-a/2,a/2), (z,-a/2,a/2))

Iyz= -integrate(x*y,(x,-a/2,a/2), (y,-a/2,a/2), (z,-a/2,a/2))

I = m/a**3*Matrix([[Ixx,Ixy,Ixz], [Ixy,Iyy,Iyz], [Ixz,Iyz,Izz]])
```

```
print('-'*28,'CODE OUTPUT','-'*29,'\n')

print('The eigenvalues are: ')
print(I.eigenvals())

print('\n The eigenvectors are:')
vec1 = I.eigenvects()[0][2][0]
vec2 = I.eigenvects()[0][2][1]
vec3 = I.eigenvects()[0][2][2]
print(vec1)
print(vec2)
print(vec3)


fig = plt.figure()
ax = fig.add_subplot(projection='3d')    # set up 3D plot
ax.quiver(0,0,0,vec1[0],vec1[1],vec1[2],color='r',\
length=1,arrow_length_ratio=.3)
ax.quiver(0,0,0,vec2[0],vec2[1],vec2[2],color='b',\
length=1,arrow_length_ratio=.3)
ax.quiver(0,0,0,vec3[0],vec3[1],vec3[2],color='g',\
length=1,arrow_length_ratio=.3)

ax.set_xlabel('X')       # label,set limits on x,y,z axes
ax.set_ylabel('Y')
ax.set_zlabel('Z')
ax.set_xlim(-1.2,1.2);
ax.set_ylim(-1.2,1.2);
ax.set_zlim(-1.2,1.2);

shape=[((-0.5,-0.5,-0.5),(0.5,-.5,-.5),(.5,.5,-.5),(-.5,0.5,-.5),\
(-0.5,-0.5,-0.5),(-.5,-.5,.5),(-.5,.5,.5),(-.5,.5,-.5)),\
(-0.5,-0.5,-0.5),(-.5,-.5,.5),(.5,-.5,.5),(.5,.5,.5),(-.5,.5,.5),\
(-.5,.5,-.5),(.5,.5,-.5),(.5,.5,.5),(.5,.5,-.5),(.5,-.5,-.5),\
(.5,-.5,.5),(-.5,-.5,.5)) ]

cube =Poly3DCollection(shape,color='k',alpha=0.1)  # draw the cube
ax.add_collection3d(cube)
plt.show()

--------------------------- CODE OUTPUT ----------------------------------
The eigenvalues are:
{a**2*m/6: 3}

 The eigenvectors are:
Matrix([[1], [0], [0]])
Matrix([[0], [1], [0]])
Matrix([[0], [0], [1]])
```

Mathematica Code

```
ρ = m/a^3;

Ixx = Integrate[(y^2 + z^2), {x, -a/2, a/2}, {y, -a/2, a/2}, {z, -a/2, a/2}];

Iyy = Izz = Ixx;

Ixy = Iyz = Ixz = -Integrate[(x * y), {x, -a/2, a/2}, {y, -a/2, a/2}, {z, -a/2, a/2}];
```

$$\text{iCube} = \rho * \begin{pmatrix} \text{Ixx} & \text{Ixy} & \text{Ixz} \\ \text{Ixy} & \text{Iyy} & \text{Iyz} \\ \text{Ixz} & \text{Iyz} & \text{Izz} \end{pmatrix};$$

```
Print["The tensor I = ", iCube//MatrixForm];
```

$$\text{The tensor I} = \begin{pmatrix} \frac{a^2 m}{6} & 0 & 0 \\ 0 & \frac{a^2 m}{6} & 0 \\ 0 & 0 & \frac{a^2 m}{6} \end{pmatrix}$$

```
Print["The Eigenvalues of I are: ", Eigenvalues[iCube]];

Print["The Eigenvectors of I are: ", Eigenvectors[iCube]];


vec1 = Eigenvectors[iCube][[1]];

vec2 = Eigenvectors[iCube][[2]];

vec3 = Eigenvectors[iCube][[3]];
```

The Eigenvalues of I are: $\left\{ \frac{a^2 m}{6}, \frac{a^2 m}{6}, \frac{a^2 m}{6} \right\}$

The Eigenvectors of I are: $\{\{0, 0, 1\}, \{0, 1, 0\}, \{1, 0, 0\}\}$

```
Graphics3D[{Arrow[{{0, 0, 0}, vec1}], Arrow[{{0, 0, 0}, vec2}], Arrow[{{0, 0, 0}, vec3}],

FaceForm[{Opacity[.2]}], Cuboid[{-.5, -.5, -.5},

{.5, .5, .5}], Dashed, Line[{{-1, 0, 0}, {1, 0, 0}}], Line[{{0, -1, 0}, {0, 1, 0}}],

Line[{{0, 0, -1}, {0, 0, 1}}]}, Axes → True, BaseStyle → FontSize → 18,

AxesLabel → {x, y, z}, PlotRange → {{-1, 1}, {-1, 1}, {-1, 1}}]
```

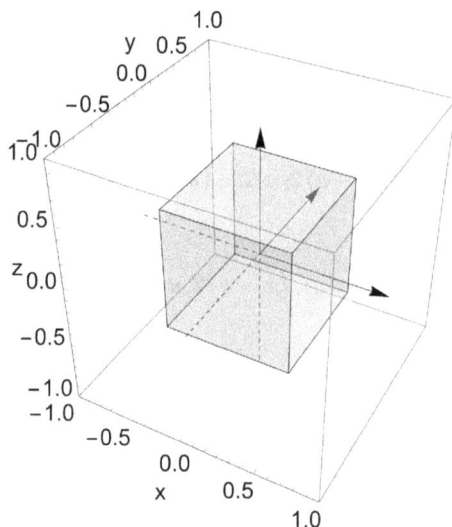

Figure 11.9: Eigenvectors of a uniform cube, with respect to a coordinate system located at the center of mass and with coordinate axes parallel to the edges of the cube, from Example 11.8.

11.7 PRECESSION OF A SYMMETRIC SPINNING TOP WITH ONE POINT FIXED AND EXPERIENCING A WEAK TORQUE

The motion of spinning tops is a fascinating example of the rotational motion of rigid bodies around an axis, and it has important practical applications in gyroscopes. It will also serve as an example of how the concept of principal axes can be applied to solve problems in rotational mechanics.

Figure 11.10 shows a symmetric spinning top of mass m, whose vertex is fixed at the origin, placed on a frictionless surface, with its rotational axis at an angle θ with respect to the vertical. If the top is spinning rapidly, it will not fall as we might expect due to the effect of gravity, but its symmetry axis describes a cone in space at the constant angle θ, as shown in Figure 11.10. This type of motion is called *precession* and is due to the torque acting on the fast spinning object.

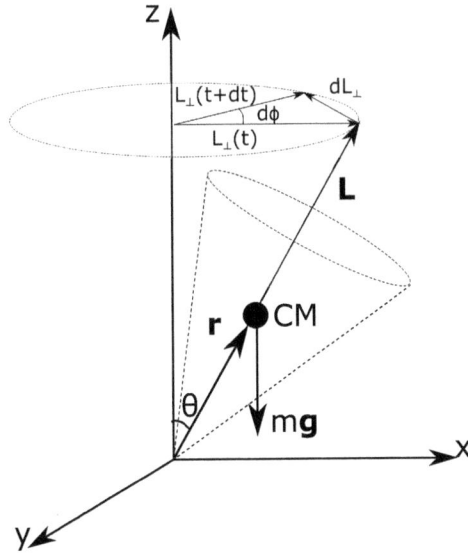

Figure 11.10: Spinning top precesses at a fixed angle θ due to the gravitational torque. The top is rotating about its symmetry axis which points in the same direction as **L**.

Let us take the contact point to be the origin O of the coordinate axes. The total angular momentum of the spinning top with respect to this origin is the sum of the angular momentum $\mathbf{L}_{\text{spin}}$ (due to the spinning motion) and the angular momentum $\mathbf{L}_{\text{CM}}$ (due to the motion of the center of mass about the pivot point):

$$\mathbf{L} = \mathbf{L}_{\text{spin}} + \mathbf{L}_{\text{CM}} \tag{11.7.1}$$

If the top spins very fast around its symmetry axis, we can assume that $\mathbf{L}_{\text{CM}} \ll \mathbf{L}_{\text{spin}}$, and therefore we can neglect the part of the angular momentum due to the motion of the center of mass, so that $\mathbf{L} = \mathbf{L}_{\text{spin}}$. With this assumption, the vector **L** points along the direction of the position vector **r** of the center of mass with respect to the origin, as shown in Figure 11.10.

The forces acting on the spinning top are the weight $\mathbf{W} = m\mathbf{g}$ and the normal force $\mathbf{F}_N$ pointing upwards at the point of contact with the horizontal surface. This torque with respect to the pivot point is produced by the gravitational force, and it is given by:

$$\mathbf{N} = \mathbf{r} \times m\mathbf{g} \tag{11.7.2}$$

where **r** is the location of the center of mass relative to the origin. The magnitude of this torque is equal to:

$$|\mathbf{N}| = |\mathbf{r} \times m\mathbf{g}| = rmg\sin\theta \tag{11.7.3}$$

where θ is also the angle between the two vectors **r** and $m\mathbf{g}$.

The important point about the motion of the spinning top is that this torque will cause a change in the *direction* of the angular momentum vector **L**, while the *magnitude* of the vector **L** will remain constant and equal to:

$$L = |\mathbf{L}| = I\omega \tag{11.7.4}$$

where I is the principal moment of inertia associated with the top's principal axis that points in the direction of **L**, and ω is the magnitude of the angular velocity about the top's axis of rotation. The torque **N** with respect to the pivot point is found from Newton's second law of rotation:

$$\mathbf{N} = \frac{d\mathbf{L}}{dt} \tag{11.7.5}$$

Since the magnitude of **L** remains constant, this equation can be written as $|\mathbf{N}| = d|\mathbf{L}|/dt$ and by combining (11.7.5) and (11.7.3) we obtain:

$$rmg \sin \theta = \frac{dL}{dt} \tag{11.7.6}$$

During the precession, the component of **L** perpendicular to the z-axis, labeled $\mathbf{L}_\perp$ in Figure 11.10, changes direction. Figure 11.10 shows the location of $\mathbf{L}_\perp$ at a time t and a later time $t + dt$, with $d\phi$ representing the angle between the initial and final angular momentum vectors. Since the magnitude L remains constant and the angle $d\phi$ is infinitesimally small, we can write:

$$\sin (d\phi) \simeq d\phi = \frac{dL}{L_\perp} = \frac{dL}{L \sin \theta} = \frac{rmg dt}{I\omega} \tag{11.7.7}$$

where we used (11.7.6) and (11.7.4).

The angular frequency Ω_P of the precession motion is given by the time rate of change of the angle ϕ:

$$\Omega_P = \frac{d\phi}{dt} \tag{11.7.8}$$

Finally by combining (11.7.8) and (11.7.7), we obtain:

$$\Omega_P = \frac{rmg}{I\omega} \tag{11.7.9}$$

According to this expression, the precessional frequency for a fast spinning top precessing at a constant angle θ will depend on these factors: the weight of the top mg, the distance of the center of mass r from the contact point, the spinning angular frequency ω and the moment of inertia I around the symmetry axis.

As discussed in Chapter 10, the Earth has an equatorial bulge; therefore, it experiences a gravitational torque due to the Sun and the Moon. These torques cause the Earth's axis of rotation to precess slowly, completing one revolution every 26,000 years. The phenomenon is called the *precession of the equinoxes*. One of the results is that Polaris is not always the North Star, and the star that the Earth's North Pole points towards changes as the Earth precesses.

11.8 RIGID BODY MOTION IN 3D AND EULER'S EQUATIONS

In this section we derive the equations of motion for a rotating rigid body in the form known as *Euler's equations*.

Let us consider a rigid body which rotates about an axis that passes through a fixed point. An example of this could be the spinning top shown in Figure 11.10. In order to specify the motion of the rigid body in space, we need six coordinates, so there are six degrees of freedom for this system. These coordinates are usually taken to be the three coordinates of the center of mass of the body, plus three more coordinates which are usually taken to be angular velocities expressing the rotation around the center of mass. If there is a fixed point on the rigid body, then we need only three coordinates to describe the rotational motion.

Since the rigid body is rotating, the principal axes are fixed on the body and therefore also rotate in space, so they are not stationary with respect to the inertial (fixed) $X'Y'Z'$-frame shown in Figure 11.11. This means that when we use the principal axes to describe the motion, we are by necessity dealing with noninertial rotating reference frames, and we can use the techniques from Chapter 10 to describe the motion of the rotating body.

The specific non-inertial frame defined by the principal axes of the rotating body is called the *body coordinate frame*, which is the XYZ coordinate system in Figure 11.11, and it is usually taken to be the principal axes of the object. The inertial $X'Y'Z'$-frame is called the *space coordinate frame*.

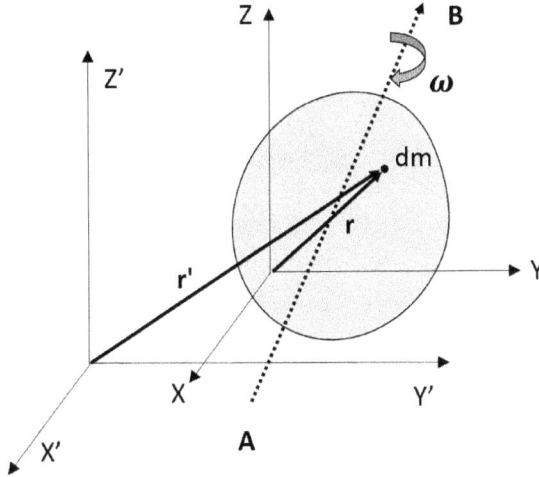

Figure 11.11: Motion of a rigid body rotating about the axis AB can be described in two coordinate systems. The space coordinate frame $X'Y'Z'$ is an inertial coordinate system. The body coordinate frame XYZ is fixed on, and rotates with the rigid body, and is therefore a noninertial frame. The body coordinate frame is usually taken to correspond to the principal axes of the rotating rigid body.

Newton's second law in an inertial frame of reference says that the time derivative of the angular momentum $\mathbf{L}$ in the inertial frame equals the applied torque $\mathbf{N}$:

$$\left(\frac{d\mathbf{L}}{dt}\right)_{\text{space}} = \frac{d}{dt}\left(\mathbf{I}_{\text{space}}\boldsymbol{\omega}\right) = \mathbf{N} \tag{11.8.1}$$

where $\mathbf{I}_{\text{space}}$ is the moment of inertia tensor calculated in the space (inertial) frame, and $\boldsymbol{\omega} = (\omega_1, \omega_2, \omega_3)$ is the angular velocity. This equation does not lend itself easily to calculations, because both $\mathbf{I}_{\text{space}}$ and $\boldsymbol{\omega}$ change during the motion. However, the calculation can be simplified using the body coordinate frame.

Let $\hat{\mathbf{e}}_1$, $\hat{\mathbf{e}}_2$, and $\hat{\mathbf{e}}_3$ be the unit vectors of the body frame pointing along the principal axes of the rotating body, the axes $X, Y,$ and Z in Figure 11.11. In the body coordinate frame, the moment of inertia tensor is constant and diagonal:

$$\mathbf{I} = \begin{bmatrix} I_1 & 0 & 0 \\ 0 & I_2 & 0 \\ 0 & 0 & I_3 \end{bmatrix} \tag{11.8.2}$$

where I_1, I_2 and I_3 are the object's principal moments of inertia. The angular momentum in the body frame $\mathbf{L}$ can be written as:

$$\mathbf{L} = L_1\hat{\mathbf{e}}_1 + L_2\hat{\mathbf{e}}_2 + L_3\hat{\mathbf{e}}_3 = I_1\omega_1\hat{\mathbf{e}}_1 + I_2\omega_2\hat{\mathbf{e}}_2 + I_3\omega_3\hat{\mathbf{e}}_3 \tag{11.8.3}$$

where ω_k are the components of the angular velocity.

In Chapter 10 we saw that in a rotating reference frame, the time derivative in (11.8.1) must be replaced with the following time derivative in the body frame:

$$\left(\frac{d\mathbf{L}}{dt}\right)_{\text{space}} = \left(\frac{d\mathbf{L}}{dt}\right)_{\text{body}} + \boldsymbol{\omega} \times \mathbf{L} \tag{11.8.4}$$

By substituting $\mathbf{L} = \mathbf{I}\boldsymbol{\omega}$ in this equation, and since the inertia tensor and the principal moments I_k do not depend on time, we have:

$$\left(\frac{d\mathbf{L}}{dt}\right)_{\text{body}} = \mathbf{I}\dot{\boldsymbol{\omega}} \tag{11.8.5}$$

By substituting (11.8.1) and (11.8.5) into (11.8.4), we arrive at the following vector form of Euler's equations:

$$\mathbf{I}\dot{\boldsymbol{\omega}} + \boldsymbol{\omega} \times (\mathbf{I}\boldsymbol{\omega}) = \mathbf{N} \tag{11.8.6}$$

For the principal axes coordinates, we have $L_k = I_k\omega_k$. Finally, by writing out the components of the cross product in (11.8.6), the components of this vector equation become:

Euler's Equations for a Rigid Body

$$I_1\dot{\omega}_1 + (I_3 - I_2)\omega_2\omega_3 = N_1$$
$$I_2\dot{\omega}_2 + (I_1 - I_3)\omega_3\omega_1 = N_2 \tag{11.8.7}$$
$$I_3\dot{\omega}_3 + (I_2 - I_1)\omega_1\omega_2 = N_3$$

where N_k ($k = 1, 2, 3$) are the components of the applied torque. This system of three coupled differential equations for $(\omega_1, \omega_2, \omega_3)$ is known as Euler's equations for a rigid body. In special cases they can be solved analytically; however, in most cases they must be integrated numerically, in order to obtain the time dependent functions $(\omega_1(t), \omega_2(t), \omega_3(t))$.

11.9 FORCE-FREE SYMMETRIC TOP

In this section, we will examine an application of Euler's equations for a symmetric top with $I_1 = I_2 \neq I_3$ which experiences no net external torque, hence $N_1 = N_2 = N_3 = 0$. In this case, Euler's equations become

$$\left. \begin{array}{r} I\dot{\omega}_1 + (I_3 - I)\omega_2\omega_3 = 0 \\ I\dot{\omega}_2 + (I - I_3)\omega_3\omega_1 = 0 \\ I_3\dot{\omega}_3 = 0 \end{array} \right\} \tag{11.9.1}$$

where we set $I_1 = I_2 = I$. We will solve (11.9.1) for $\boldsymbol{\omega} = (\omega_1, \omega_2, \omega_3)$ in order to understand the behavior of the top's axis of rotation, which points in the direction of $\boldsymbol{\omega}$. Because there are no forces acting on the top, the top's center of mass is either at rest or moving with a constant velocity. Without loss of generality, we can therefore choose a coordinate system for the space frame such that the top's center of mass is at rest at the space frame's origin. Finally, we also assume that $\boldsymbol{\omega}$ does not lie along a principal axis of the top; otherwise, the motion would be trivial.

Note that according to (11.9.1), ω_3 is constant ($\dot{\omega}_3 = 0$). The remaining two equations can then be written as

$$\left.\begin{aligned}\dot{\omega}_1 &= -\left(\frac{I_3 - I}{I}\omega_3\right)\omega_2 \\ \dot{\omega}_2 &= \left(\frac{I_3 - I}{I}\omega_3\right)\omega_1\end{aligned}\right\} \qquad (11.9.2)$$

We can define a constant angular frequency Ω such that:

$$\Omega = \frac{I_3 - I}{I}\omega_3 \qquad (11.9.3)$$

which is constant. Equations (11.9.2) then become

$$\left.\begin{aligned}\dot{\omega}_1 &= -\Omega\omega_2 \\ \dot{\omega}_2 &= \Omega\omega_1\end{aligned}\right\} \qquad (11.9.4)$$

which are a pair of coupled second-order differential equations. In other words, there are ω_2 terms in the equation for $\dot{\omega}_1$, and vice-versa. The result is that the value of ω_1 depends on ω_2, and the value of ω_2 depends on ω_1. To solve this system in closed form, we need to decouple the equations (i.e., write each of the two equations as an equation of a single variable) or we need to rewrite the equations as one single differential equation of a single variable. We will take the latter approach by extending the problem to the complex plane. If we multiply the $\dot{\omega}_2$ equation in (11.9.2) by $i = \sqrt{-1}$ and add the two equations, we obtain

$$\dot{\eta} - i\Omega\eta = 0 \qquad (11.9.5)$$

where $\eta = \omega_1 + i\omega_2$. The solution to (11.9.5) is

$$\eta = Ae^{i\Omega t} \qquad (11.9.6)$$

Note that once we introduce complex numbers, the constant in the solution can also be complex. In that case, we write, $A = \alpha e^{i\beta}$ and (11.9.6) becomes

$$\eta = \alpha e^{i(\Omega t + \beta)} \qquad (11.9.7)$$

where α and β are real constants. The term β is a phase term, and it can be set equal to zero by choosing an appropriate initial time for the problem. This is similar to observing a sinusoidal wave as it passes by. By setting the initial time $t_0 = 0$ to be the time when a crest passes your position, the resulting wave motion can then be described as a cosine function with the phase equal to zero. In what follows, we continue to use (11.9.6) as the solution to (11.9.5) with A as a real constant.

Using the Euler relationship for exponentials, we can rewrite (11.9.6) as

$$\omega_1 + i\omega_2 = A\left(\cos(\Omega t) + i\sin(\Omega t)\right) \qquad (11.9.8)$$

or

$$\left.\begin{aligned}\omega_1 &= A\cos(\Omega t) \\ \omega_2 &= A\sin(\Omega t)\end{aligned}\right\} \qquad (11.9.9)$$

which describe circular motion of radius A and angular frequency Ω.

Recall that ω_3 is constant. Therefore, the magnitude of $\boldsymbol{\omega}$ is also constant, since

$$\omega = \sqrt{\omega_1^2 + \omega_2^2 + \omega_3^2} = \sqrt{A^2 + \omega_3^2} \qquad (11.9.10)$$

Equation (11.9.9) tell us that the projection of $\boldsymbol{\omega}$ onto the x_1x_2-plane in the body frame describes a circle. Hence, $\boldsymbol{\omega}$ precesses about the symmetry axis x_3 with a precession frequency Ω. An observer in the body frame will therefore observe $\boldsymbol{\omega}$ as tracing out a cone as shown in Figure 11.12(a). The resulting cone is sometimes called a *body cone*.

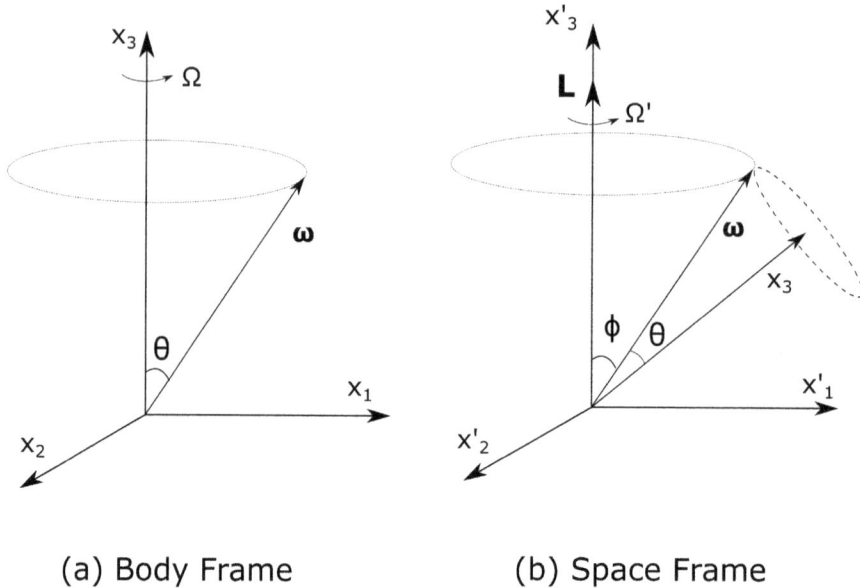

(a) Body Frame   (b) Space Frame

Figure 11.12: Precession of $\boldsymbol{\omega}$ about the symmetry axis of the force-free symmetric top as observed in the (a) body frame and (b) space frame.

Because we are examining the force-free case, the angular momentum $\mathbf{L}$ of the top is constant. Hence an observer in the space frame will see the plane containing $\boldsymbol{\omega}$ and $\mathbf{e}_3$ (the top's symmetry axis) precess about $\mathbf{L}$ with a precession frequency Ω'. It can be shown that $\Omega' = L/I$. The cone swept out by the plane containing $\boldsymbol{\omega}$ and $\mathbf{e}_3$ is sometimes called the *space cone*. The space cone is shown in Figure 11.12(b).

In addition to the precession due to the gravitational torques provided by the Sun and the Moon, the Earth experiences another precession due to the Earth's oblateness. The Earth is slightly flattened at the poles due to its rotation (see Chapter 10). The result of the flattening is that the Earth's principal moments of inertia are not all identical. The Earth's principal moment of inertia about its polar axis is slightly larger than its other two moments (by about 1 part in 300). Hence (11.9.9) tells us that there should be a precession with a frequency of $\omega_3/300$ where ω_3 is equal to one rotation per day. The result is a small precession (or wobble) of the Earth's polar axis with a period of 300 days. The wobble was discovered by an amateur astronomer Seth Chandler (1846-1913). However, the period is closer to 400 days because the Earth is not a perfectly rigid rotator.

11.10 CHAPTER SUMMARY

The tangential velocity $\mathbf{v}$ and angular momentum vector $\boldsymbol{\ell}$ of a mass m are given by the cross products:

$$\mathbf{v} = \boldsymbol{\omega} \times \mathbf{r} \qquad \boldsymbol{\ell} = \boldsymbol{r} \times \mathbf{p} = m\left(\boldsymbol{r} \times \mathbf{v}\right)$$

where $\mathbf{r}$ is the position vector and $\mathbf{p} = m\mathbf{v}$ is the momentum vector.

When the position and velocity vectors are at right angles to each other, we can write these equations as:

$$v = \omega r \qquad \ell = mrv = mr^2\omega = I\omega$$

where $I = mr^2$ is the scalar moment of inertia around the rotational axis. The kinetic energy of the rotating mass m is found from:

$$T = \frac{1}{2}I\omega^2$$

A rigid body is a collection of finite masses m_i located at positions $\mathbf{r}_i$, with respect to a coordinate system xyz which is fixed in the body, rotating around a fixed axis AB with an angular velocity vector $\boldsymbol{\omega} = (\omega_x,\ \omega_y,\ \omega_z)$. The tangential velocity $\mathbf{v}_i$ of mass m_i is given by the cross product:

$$\mathbf{v}_i = \boldsymbol{\omega} \times \mathbf{r}_i$$

The angular momentum vector $\mathbf{L}$ of a system of particles with masses m_i relative to the center of the mass coordinate system is given by the vector sum of the angular momenta of the masses:

$$\mathbf{L} = \sum m_i\left(\mathbf{r}_i \times \mathbf{v}_i\right) = \sum m_i\left[\mathbf{r}_i \times \left(\boldsymbol{\omega} \times \mathbf{r}_i\right)\right]$$

In cases where the vectors $\mathbf{L}$ and $\boldsymbol{\omega}$ do not point in the same direction, the equation $L = I\omega$ does not apply, and we replace the scalar quantity I with the moment of the symmetric *inertia tensor* $\mathbf{I}$ defined by:

$$\mathbf{L} = \mathbf{I} \cdot \boldsymbol{\omega}$$

$$\mathbf{I} = \begin{bmatrix} I_{xx} & I_{xy} & I_{xz} \\ I_{yx} & I_{yy} & I_{yz} \\ I_{zx} & I_{zy} & I_{zz} \end{bmatrix} \qquad \text{and} \qquad \boldsymbol{\omega} = \begin{bmatrix} \omega_x \\ \omega_y \\ \omega_z \end{bmatrix}$$

where the scalars I_{kl} $(k,l = 1,2,3)$ are the components of the inertia tensor $\mathbf{I}$, defined by:

$$I_{xx} = \sum m_i\left(y_i^2 + z_i^2\right) \qquad I_{xy} = I_{yx} = -\sum m_i x_i y_i \qquad \text{, etc.}$$

The elements of the inertia tensor depend on the choice of origin for the system. For a mass m with density ρ these elements are given by:

$$I_{xx} = \iiint_V \rho\left(y^2 + z^2\right)\mathrm{d}x\mathrm{d}y\mathrm{d}z \qquad I_{xy} = I_{yx} = -\iiint_V \rho xy\mathrm{d}x\mathrm{d}y\mathrm{d}z \qquad \text{etc.}$$

The kinetic energy and moment of inertia tensor are related by the matrix multiplication equation:

$$T = \frac{1}{2}\boldsymbol{\omega} \cdot \mathbf{L} = \frac{1}{2}\left(\boldsymbol{\omega}\right)^{\mathrm{T}} \cdot \mathbf{I} \cdot \boldsymbol{\omega}$$

where $(\boldsymbol{\omega})^{\mathrm{T}}$ indicates the transpose of the column matrix $\boldsymbol{\omega}$.

The scalar moment of inertia I for rotation around an axis in the direction $\hat{\mathbf{n}}$ of any unit vector, is found from the inertia tensor using the matrix multiplication equation:

$$I_n = (\hat{\mathbf{n}})^{\mathrm{T}} \cdot \mathbf{I} \cdot \hat{\mathbf{n}}$$

The parallel axis theorem for scalar moments of inertia states: if I is the scalar moment of inertia of a body around an axis AB, and I_{CM} is the corresponding moment of inertia about a second axis parallel to AB and passing through the center of mass at a distance d, then I and I_{CM} are related by:

$$I = I_{CM} + md^2$$

The generalized *parallel axis theorem between moment of inertia tensors* is as follows: If $\mathbf{I}$ is the inertia tensor defined in a coordinate system with the origin fixed at point O, and $\mathbf{I}'$ is the inertia tensor defined in a center-of mass coordinate system with its origin at the center-of-mass O_{CM} and whose axes are parallel to the previous coordinate system, then the elements of the tensors $\mathbf{I}$ and $\mathbf{I}'$ are related by:

$$I_{kl} = I'_{kl} + M \left[a^2 \delta_{kl} - a_k a_l \right]$$

where $\mathbf{a}$ is the vector connecting the two origins O and O_{CM}.

The *principal axes of the rigid body* are a special orientation of the axes, and all rigid bodies have three principal axes. The main property of the principal axes is that when rotation takes place around one of these three axes, the angular momentum $\mathbf{L}$ and angular velocity vectors $\boldsymbol{\omega}$ are parallel to each other and:

$$\mathbf{L} = \mathbf{I} \cdot \boldsymbol{\omega} = \lambda \boldsymbol{\omega}$$

where λ is a scalar constant. This equation is an *eigenvalue problem* in which we are looking for the eigenvalues and eigenvectors of the moment of the inertia matrix $\mathbf{I}$.

The principal axes is a noninertial frame fixed on the body, called the *body coordinate frame*, while the inertial xyz-frame for a stationary observer is called the *space coordinate frame*.

When using the principal axes of coordinates, we have $L_k = I_k \omega_k$, and the motion of a rigid body can be described by the *Euler equations*:

$$I_1 \dot{\omega}_1 + (I_3 - I_2)\omega_2 \omega_3 = N_1$$
$$I_2 \dot{\omega}_2 + (I_1 - I_3)\omega_3 \omega_1 = N_2$$
$$I_3 \dot{\omega}_3 + (I_2 - I_1)\omega_1 \omega_2 = N_3$$

where (N_1, N_2, N_3) is the external torque acting on the rigid body.

11.11   END-OF-CHAPTER PROBLEMS

The symbol ⌨ indicates a problem which requires some computer assistance, in the form of graphics, numerical computation, or symbolic evaluation.

Section 11.1: Rotational Motion of Particles Around a Fixed Axis

1. Human walking can be modeled as a swinging simple pendulum pivoting at the hip joint. Estimate the period of this simple pendulum. Make reasonable numerical estimates for the model, and neglect the effect of knee joints.

2. Find the principal axes and principal moments of inertia for a system of two masses m_1, m_2 connected by a massless rigid rod of length l.

3. 🖵 A cylinder has mass m and radius R, and is attached to a hanging mass m. The cylinder rolls without slipping upwards on an inclined plane, as shown in Figure 11.13. A massless string is wrapped around the cylinder.

   a. Find the magnitude and direction of the acceleration of the hanging mass.

   b. What are the magnitude and direction of the force of static friction at the contact point between the cylinder and the inclined plane?

   Use a CAS to solve the system of equations.

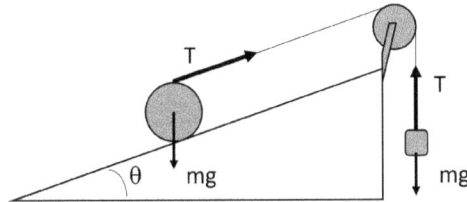

Figure 11.13: Cylinder on an inclined plane attached to a hanging mass, in Problem 11.3.

4. A uniform hoop of mass m and radius R hangs in a vertical plane and is supported by a nail at a point on the circumference, as shown in Figure 11.14. Calculate the natural frequency of small oscillations.

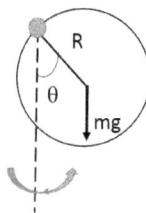

Figure 11.14: Uniform hoop supported by a nail, located at the blue circle (in the e-book), in Problem 11.4.

5. Baseball players always try to hit what is known as the *sweet spot* on the baseball bat. The sweet spot is where the hit delivers the most energy to the ball, and at the same time minimizes the force on the hands of the player. To describe the physics of the sweet spot, the baseball bat can be assumed to be rotating in space around a pivot point at its end without linear displacement, as shown in Figure 11.15. At what distance D along the baseball bat is the sweet spot located?

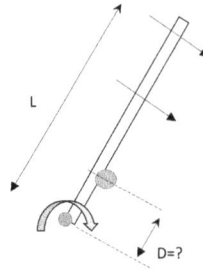

Figure 11.15: Baseball bat hitting a baseball at the sweet spot D away from the pivot point, in Problem 11.5.

6. A torsion pendulum consists in general of a mass m hanging from a wire that can be twisted, producing a torque $N = I\alpha$ proportional to the angle θ describing the twisting. A torsion pendulum satisfies an equation similar to Hooke's Law:

$$I_1\ddot{\theta} + A\theta = 0$$

where I_1 is the moment of inertia of the attached object around the rotational axis, and A is the torsion coefficient describing the stiffness of the wire. Consider the torsion pendulum shown in Figure 11.16, which consists of a thin disk of radius R and mass M, with a cylindrical mass M of radius $R/4$ placed on top of the disk. Find the ratio of the periods of this torsion pendulum, with and without the presence of the cylindrical mass on top of the thin disk.

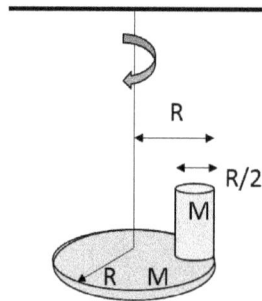

Figure 11.16: A Torsion pendulum in Problem 11.6.

7. A bowling ball of uniform density is thrown along a horizontal alley with initial velocity v_0 in such a way that it initially slides without rolling. The ball has mass m, the coefficient of static friction with the floor is μ_s, and the coefficient of sliding friction with the floor is μ_k. Compute how far the ball will slide before it starts rolling. You can ignore the effect of air resistance.

8. A coin with radius R and mass m is spinning about its axis of symmetry through its center with angular frequency ω_0 as shown in Figure 11.17a. The coin is placed down on a horizontal surface as shown in Figure 11.17b. The coin stops slipping and starts rolling away. What is the velocity of the coin when it rolls away?

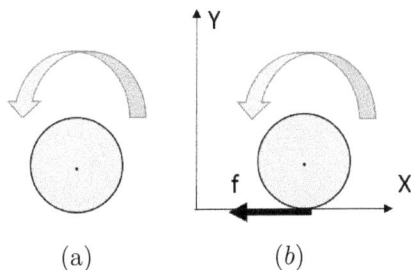

Figure 11.17: Spinning coin, in Problem 11.8.

9. A wheel of mass M and radius R is projected along a horizontal surface with an initial linear velocity v_0 and an initial angular velocity ω_0, as shown in Figure 11.18. The wheel starts sliding along the surface. Let the coefficient of friction between the wheel and the surface be μ.

    a. How long does it take for the wheel to stop sliding?

    b. What is the velocity of the center of mass of the wheel when the sliding stops?

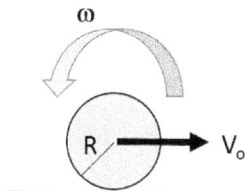

Figure 11.18: Wheel projected along a horizontal surface, in Problem 11.9.

10. A solid uniform cylinder of mass m and radius R is placed on a plane inclined at angle θ with respect to the horizontal, as shown in Figure 11.19. Let a be the acceleration of the axis of the cylinder along the incline. The coefficient of friction between the cylinder and plane is μ. For θ less than some critical angle θ_c, the cylinder will roll down the incline without slipping.

    a. What is the critical angle θ_c?

    b. What is the acceleration a for angles less than the critical angle?

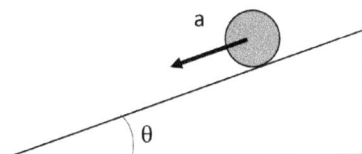

Figure 11.19: Cylinder rolling down an inclined plane, in Problem 11.10.

11. A wheel of radius R and moment of inertia I is mounted on a frictionless axle through its center. A flexible, weightless cord is wrapped around the rim of the wheel and carries a body of mass M which begins descending as shown in Figure 11.20. What is the tension in the cord?

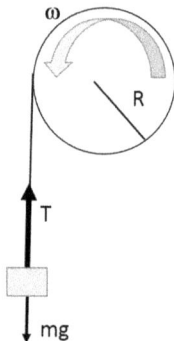

Figure 11.20: Mass hanging from a wheel, in Problem 11.11.

12. A thin uniform stick of mass m is resting with its bottom end on a frictionless table. The stick is released from rest at an angle θ_0 with respect to the vertical, as shown in Figure 11.21. Find the force exerted by the table upon the stick at an infinitesimally small time after its release.

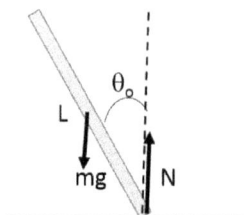

Figure 11.21: Thin uniform stick in the process of falling over, in Problem 11.12.

Section 11.2: Review of Rotational Properties for a System of Particles

13. A rigid body consists of three masses $m_1 = 2$, $m_2 = 1$, $m_3 = 4$ located at positions $\mathbf{r}_1 = (1, -1, 1)$, $\mathbf{r}_2 = (2, 0, 2)$ and $\mathbf{r}_3 = (-1, 1, 0)$, with all units in the SI system.

   a. Calculate the angular momentum of the system when the angular velocity is $\boldsymbol{\omega} = (3, -2, 4)$.

   b. Find the principal axes and principal moments of inertia of this system.

14. A billiard ball of radius R and mass M is struck with a horizontal cue stick at a height h above the billiard table, as shown in Figure 11.22. Find the value of h for which the ball will roll without slipping.

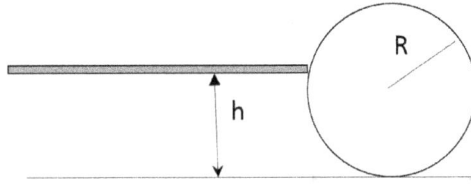

Figure 11.22: Billiard ball being struck by a horizontal cue, in Problem 11.14.

Section 11.3: Moment of Inertia Tensor

15. A square plate with mass m and side a, has a uniform mass density and lies on the xy-plane of a xyz coordinate system, with the origin located at one of its corners. The square plate has a thickness b.

    a. Find the moment of inertia tensor of the square plate with respect to this coordinate system.

    b. Find the principal moments of inertia and principal axes for this plate, when the thickness $b << a$, i.e., when the plate is very thin.

16. Two equal point masses M are connected by a massless rigid rod of length $2b$, to form a dumbbell. Choose a coordinate system in which the dumbbell is on the xy-plane at an angle ϕ with respect to the x-axis, and with the origin located at the center of mass of the dumbbell, as shown in Figure 11.23. The angular velocity ω of rotation around the z-axis is a constant in time.

    a. Calculate the elements of the inertia tensor for this coordinate system.

    b. Find and discuss the principal axes and principal moments of inertia.

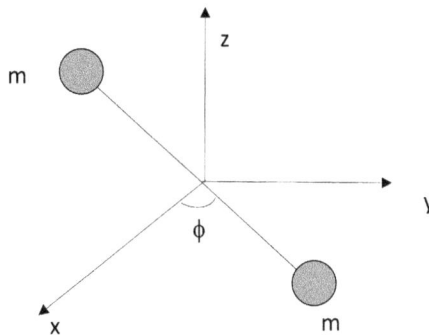

Figure 11.23: Two point masses connected by a massless rod, in Problem 11.16.

17. Find the principal moments of inertia and principal axes for a uniform cylinder of height H and radius R, with respect to a coordinate system located at the center of mass of the cylinder.

Section 11.4: Kinetic Energy and the Inertia Tensor

18. 🖥 Four masses all with mass m, lie in the xy-plane at positions $(x, y) = (a, 0)$, $(-a, 0)$, $(0, +2a)$, and $(0, -2a)$. These are joined by massless rods to form a rigid body.

    a. Find the moment of inertia tensor using the xyz-axes as a reference system.

    b. Consider a direction given by a unit vector $\hat{n}$ that makes equal angles with the positive xyz-axes. Find the moment of inertia for rotation about this axis.

    c. Find the angular momentum with respect to the unit vector $\hat{n}$.

19. A compound pendulum has mass m and principal moments of inertia I_1, I_2, and I_3. The pendulum oscillates about a horizontal axis which makes angles α, β, and γ with respect to the principal axes of inertia. Show that the period of small oscillations is

$$T = 2\pi \sqrt{\frac{mgd}{I}}$$

where d is the distance from the center of mass to the axis of rotation, and

$$I = md^2 + I_1 \cos^2 \alpha + I_2 \cos^2 \beta + I_3 \cos^2 \gamma$$

20. A spherical ball of mass m and radius r rolls without slipping on a track as shown in Figure 11.24. Find the minimum height h above the top position in the loop that will permit the ball to maintain constant contact with the rail of the loop.

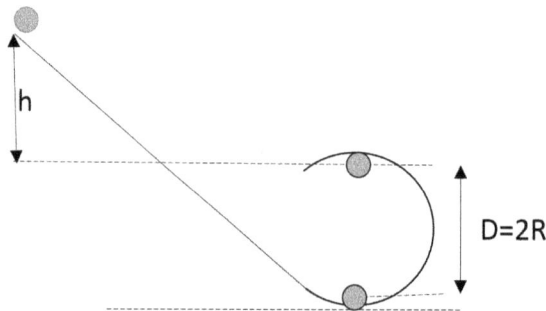

Figure 11.24: Ball rolling without slipping on a track, in Problem 11.20.

21. A uniform plank of length $2a$ is held temporarily so that one end leans against a frictionless vertical wall, and the other end rests on a frictionless floor making an angle θ_0 with the floor, as shown in Figure 11.25. When the plank is released, it will slide down under the influence of gravity. Show that the time t that it will take the plank to reach a new angle θ is:

$$t = \int_0^t \frac{d\theta}{\sqrt{\frac{3g}{2a} (\sin \theta_0 - \sin \theta)}}$$

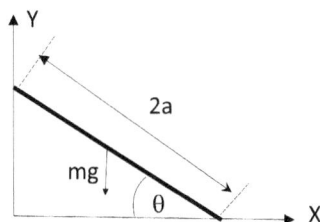

Figure 11.25: Uniform plank temporarily leaning against a frictionless vertical wall, in Problem 11.21.

22. A uniform solid ball of radius a rolls with velocity v on a level surface, and it collides inelastically with a step of height $h < a$, as shown in Figure 11.26. Find the minimum velocity for which the ball will "trip" up over the step. Assume that no slipping occurs at the impact point.

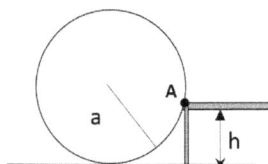

Figure 11.26: Solid ball tripping over a ledge, in Problem 11.22.

23. A particle of mass m and speed v collides elastically with the end of a uniform thin rod of mass M and length L as shown in Figure 11.27. The collision takes place on a horizontal surface, so gravity can be ignored. Show that if mass m is stationary after the collision, then $M = 4m$.

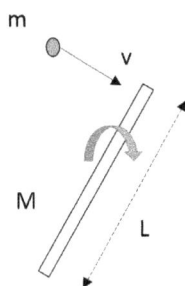

Figure 11.27: Particle colliding with the end of a uniform thin rod, in Problem 11.23.

24. Solve the following problems involving different physical pendulums. A physical pendulum is a rigid body that swings freely about some pivot point.

a. A rectangular thin plate with sides (a, b) is hung vertically from its edge of length a. Show that the period of small oscillations is $T = 2\pi\sqrt{2b/(3g)}$.

b. A uniform solid sphere with radius R is hung vertically from a point on its surface. Show that the period of small oscillations is $T = 2\pi\sqrt{7R/(5g)}$.

c. In a simple pendulum, the string is replaced with a solid thin rod of mass M and length L, and a mass m is attached at the end of this rod. Show that the period of small oscillations is $T = 2\pi\sqrt{2(M + 3m)L/[3(M + 2m)g]}$.

25. Given that the moment of inertia of a cube about an axis that passes through the center of mass and the center of one face is I_0, find the moment of inertia about an axis passing through the center of mass and one corner of the cube.

Section 11.6: Principal Axes of Rotation

26. ⌨ A thin disk of radius R and mass M lying in the xy-plane has a point mass $m = 5M/4$ attached on its edge as shown in Figure 11.28. We ignore gravity.

   a. Show that the moment of inertia tensor of the disk about its center of mass is:

   $$\mathbf{I} = \frac{MR^2}{4}\begin{bmatrix} 1 & 0 & 0 \\ 0 & 1 & 0 \\ 0 & 0 & 2 \end{bmatrix}$$

   b. Find the moment of inertia tensor of the combination of disk and point mass when the system rotates around the pivot point A at the origin, in the coordinate system shown in Figure 11.28.

   c. Find the principal moments and the principal axes.

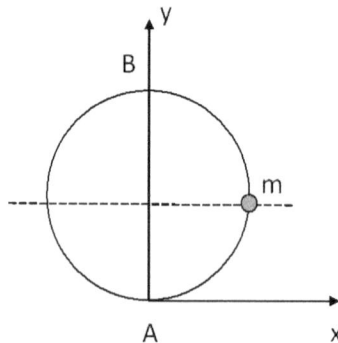

Figure 11.28: Thin disk with a point mass (solid dot) attached at the edge, in Problem 11.26.

27. A torsion pendulum consists of a vertical wire attached to a cube which may rotate about the vertical. The cube is hung from a corner, then it is hung from midway along an edge, and finally it is hung from the middle of a face. In each case, the cube is allowed to oscillate around the pivot point. Which one of the three periods of oscillation of the cube is largest?

Sections 11.7-11.8: Motion of a Spinning Top- Precession - Rigid Body Motion in 3D and Euler's Equations

28. 🖵 Consider a rotating object which is not experiencing any external torques. Using Euler's equations, show that the magnitude of the angular momentum is constant. Show also that the rotational kinetic energy is constant. Use a CAS to carry out the algebra.

29. 🖵 A thin rectangular plate has dimensions a and b. Find the torque required to rotate this plate with constant angular velocity ω around one of its diagonals.

30. 🖵 A uniform rigid wheel is located on the xy-plane, and has principal moments of inertia $I_1 = I_2 \neq I_3$ about its body-fixed principal axes $\hat{x}_1$, $\hat{x}_2$ and $\hat{x}_3$ as shown in Figure 11.29. The wheel is attached at its center of mass to a bearing which allows frictionless rotation about one space-fixed axis. The wheel is "dynamically balanced", i.e., it can rotate at constant ω, so that it can exert no torque on its bearing.

   a. Examine and integrate the Euler equations, first without any given initial conditions, and secondly by using the initial conditions $\omega_1(0) = 1$ and $\omega_2(0) = 0$.

   b. What conditions must the components of ω satisfy for this dynamically balanced system? Discuss the motion.

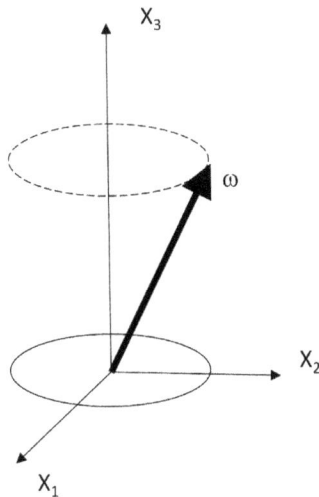

Figure 11.29: Uniform rigid wheel rotating around the axis $\boldsymbol{\omega}$, in Problem 11.30.

12 Coupled Oscillations

In this chapter, we explore the properties of *coupled harmonic oscillators*. These systems can be analyzed by using either the Lagrangian approach of Chapter 8, or alternatively using Newton's second law. The simplest form of these systems in mechanics contains two masses connected by springs to each other. A second simple example of coupled mechanical oscillators is the double pendulum, which also exhibits a wide range of interesting behaviors. We will see that these simple oscillating systems can exhibit normal modes of oscillation, which are patterns of motion in which all parts of the system move sinusoidally with the same frequency. The frequencies of the normal modes of a system are known as its natural frequencies of oscillation. We will find that any motion exhibited by the system can be expressed as a linear combination of these normal modes.

The discussion of the two-mass system will lead us to a more general description of linearly coupled harmonic systems, and how their equations of motion can be written in matrix form. The best way to obtain solutions to the equations of motions for coupled oscillations is by using standard techniques from Linear Algebra, in order to find the eigenvalues and eigenvectors of a matrix. The eigenvectors and eigenvalues of the matrix characterizing the oscillating system are closely related to its normal modes.

This chapter will conclude with a general treatment of coupled oscillations, and a discussion of normal coordinates.

12.1 COUPLED OSCILLATIONS OF A TWO-MASS, THREE-SPRING SYSTEM

In this section we introduce a simple system exhibiting coupled oscillations, consisting of two masses connected by springs to each other and to two supporting walls.

12.1.1 EQUATIONS OF MOTION

Consider two masses m_1 and m_2 attached to three springs with spring constants k_1, k_2, and k_3 and to two fixed walls, as shown in Figure 12.1.

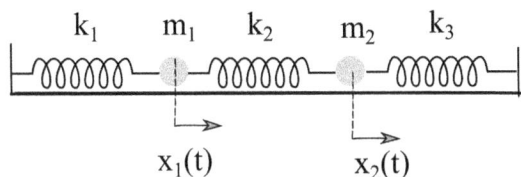

Figure 12.1: System of two coupled harmonic oscillators consisting of two masses m_1 and m_2 connected with three springs with constants k_1, k_2, and k_3.

Let us denote by $x_1(t)$ and $x_2(t)$ the horizontal displacements of the two masses from their respective equilibrium points. The force on the first mass due to the first spring is $-k_1 x_1$. The middle spring will be stretched by a distance $(x_1 - x_2)$, and the force on the first mass due to this middle spring will be $-k_2(x_1 - x_2)$. The total force on the first mass must then be $F_1 = -k_1 x_1 - k_2(x_1 - x_2)$. Similarly, the force on the second mass due to the middle spring is $-k_2(x_2 - x_1)$, and the force on the second mass due to the third spring will be $-k_3 x_2$.

Using the notation $\ddot{x}$ for acceleration, the equations of motion from Newton's second law $F = ma$ for the two masses are:

$$\left.\begin{array}{l} m_1\ddot{x}_1 = -k_1x_1 - k_2(x_1 - x_2) \\ m_2\ddot{x}_2 = -k_2(x_2 - x_1) - k_3x_2 \end{array}\right\} \qquad (12.1.1)$$

In a more general way, we can obtain the same equations by starting with the Lagrangian formulation of Chapter 8. The potential energies of the two end springs are $V_1 = k_1x_1^2/2$ and $V_3 = k_3x_2^2/2$, while the potential energy for the middle spring is $V_2 = k_2(x_1 - x_2)^2/2$, so that the Lagrangian is equal to:

$$\mathcal{L} = T - V_{Total} = \frac{1}{2}m_1\dot{x}_1{}^2 + \frac{1}{2}m_2\dot{x}_2{}^2 - \frac{1}{2}k_1x_1^2 - \frac{1}{2}k_3x_2^2 - \frac{1}{2}k_2(x_1 - x_2)^2 \qquad (12.1.2)$$

The Euler-Lagrange equations are:

$$\frac{d}{dt}\left(\frac{\partial \mathcal{L}}{\partial \dot{x}_1}\right) - \frac{\partial \mathcal{L}}{\partial x_1} = 0 \quad \rightarrow \quad m_1\ddot{x}_1 = -k_1x_1 - k_2(x_1 - x_2) \qquad (12.1.3)$$

$$\frac{d}{dt}\left(\frac{\partial \mathcal{L}}{\partial \dot{x}_2}\right) - \frac{\partial \mathcal{L}}{\partial x_2} = 0 \quad \rightarrow \quad m_2\ddot{x}_2 = -k_2(x_2 - x_1) - k_3x_2 \qquad (12.1.4)$$

These are of course the same equations as in (12.1.1).

12.1.2 NUMERICAL SOLUTION OF THE EQUATIONS OF MOTION

In general, it is not possible to obtain the solutions $x_1(t)$ and $x_2(t)$ of the system of equations (12.1.1) analytically, and they must be obtained by numerically integrating the equations for given initial conditions of the system. The initial conditions are usually given as the initial positions and initial speeds of the two masses.

Example 12.1 shows how to obtain and plot the numerical solutions $x_1(t)$ and $x_2(t)$, with the initial conditions $x_1(0) = 0$, $x_2(0) = a$, $\dot{x}_1(0) = a$, and $\dot{x}_2(0) = 0$.

Example 12.1: Numerical solution for the general case of two coupled oscillating masses

Integrate (12.1.1) for $k_1 = 4\,\mathrm{N/m}$, $k_2 = 2\,\mathrm{N/m}$, $k_3 = 3\,\mathrm{N/m}$, $m_1 = 1\,\mathrm{kg}$, $m_2 = 2\,\mathrm{kg}$ and plot the numerical solutions $x_1(t)$ and $x_2(t)$, with the initial conditions $x_1(0) = 0$, $x_2(0) = 1\,\mathrm{m}$, $\dot{x}_1(0) = 0$ and $\dot{x}_2(0) = 0$. This situation corresponds to the case where the first mass m_1 is initially at rest at its equilibrium position ($x_1(0) = 0$ and $\dot{x}_1(0) = 0$), and the second mass is pulled a distance $a = 1$ meter from its equilibrium and released from rest ($\dot{x}_2(0) = 0$).

Solution:

In this example we use the `NDSolve` command in Mathematica, and the `odeint` library in Python to numerically solve the system of differential equations, and to obtain the numerical solutions $x_1(t)$ and $x_2(t)$.

The plots of $x_1(t)$ and $x_2(t)$ in Figure 12.2 are obviously complex, and it is not possible to give a simple physical description of the motion of the two masses. The key physical component which creates this complex behavior is the middle spring in Figure 12.1, since this is the component that couples the motion of the two masses.

Python Code

As usual for solving second-order ODEs with Python, we define a vector X = [x1,y1,x2,y2] where y1,y2 represent the speeds $v_1 = dx_1/dt$ and $v_2 = dx_2/dt$. We use the command `odeint` and define a function `solvODE` whose input are X and the time array t, and the output is an array `soln` with columns representing $x_1(t), dx_1/dt, x_2(t), dx_2/dt$. The plots of $x_1(t)$ and $x_2(t)$ are shown in Figure 12.2.

```python
import numpy as np
from scipy.integrate import odeint
import matplotlib.pyplot as plt

def solveODE(X, t):
    # X :   X = [x1,y1,x2,y2]
    # t :   time
    x1, y1, x2, y2 = X

    f = [y1, k2 * (x2 - x1 ) / m1 -k1*x1/m1,
         y2, k2 * (x1 - x2 ) / m2 -k3*x2/m2]
    return f

k1, k2, k3, m1, m2 = 4, 2, 3, 1, 2   # Parameter values

# Initial conditions (x1,y1 = dx1/dt, x2, y2 = dx2/dt)
# x1, x2= initial positions, y1, y2 = initial velocities
x1, y1, x2, y2 = 0, 0, 1, 0

# times for evaluating x(t)
t = np.linspace(0,20,100)

# Pack up the parameters and initial conditions:
X = [x1, y1, x2, y2]

# Call the ODE solver.
soln = odeint(solveODE, X, t)

plt.subplot(1,2,1)
plt.plot(t, soln[:,0])
plt.title('(a) x1(t)')

plt.subplot(1,2,2)
plt.plot(t, soln[:,2])
plt.title('(b) x2(t)')
plt.tight_layout()
plt.show()
```

Mathematica Code

We use NDSolve to solve the system of ODEs from $t = 0$ to $t = 20$ s.

The numerical values for the parameters m_1, m_2, k_1, k_2, and k_3 are needed in order to plot the solutions using the Plot and GraphicsGrid commands.

```
SetOptions[Plot, BaseStyle->{FontSize->14}];

k1 = 4; k2 = 2; k3 = 3; m1 = 1; m2 = 2;

sol = NDSolve[{m1 * x1"[t] == −k1 * x1[t] + k2 * (x2[t] − x1[t]),

m2 * x2"[t] == −k3 * x2[t] + k2 * (x1[t] − x2[t]), x1[0] == 0, x1'[0] == 0,

x2[0] == 1, x2'[0] == 0}, {x1, x2}, {t, 0, 20}];


gr1 = Plot[x1[t]/.sol, {t, 0, 20}, PlotLabel → "(a) x1[t]"];

gr2 = Plot[x2[t]/.sol, {t, 0, 20}, PlotLabel → "(b) x2[t]"];


GraphicsGrid[{{gr1, gr2}}, Frame → True]
```

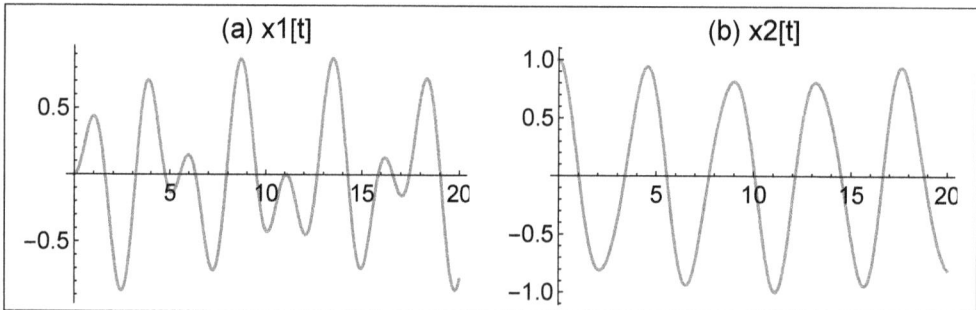

Figure 12.2: Plots of the positions $x_1(t)$, $x_2(t)$ of the two masses for the two-mass three-spring system, from Example 12.1.

12.1.3   EQUAL MASSES AND IDENTICAL SPRINGS: THE NORMAL MODES

In Example 12.2, we look at the special case of equal masses ($m_1 = m_2$), and springs with equal spring constants ($k_1 = k_2 = k_3$). This is an interesting physical situation, in which the analytical solutions are simple, and it may be easy to understand the physics of the situation. This example introduces the concept of normal modes in a simple and clear manner.

Example 12.2: Equal masses and identical springs: the antisymmetric oscillation

Solve analytically the two-mass three-spring system and plot the numerical solutions $x_1(t)$ and $x_2(t)$ for the special case of identical masses and identical springs, $m_1 = m_2$ and $k_1 = k_2 = k_3$. Use the numerical values $a = 1.0$ m, $k = 1.0$ N/m, $m = 1.0$ kg.
  (a) Use the initial conditions $x_1(0) = a$, $x_2(0) = -a$, $\dot{x}_1(0) = 0$ and $\dot{x}_2(0) = 0$.
  (b) Use the initial conditions $x_1(0) = a$, $x_2(0) = a$, $\dot{x}_1(0) = 0$ and $\dot{x}_2(0) = 0$.

Solution:

(a) In this situation the two masses are initially displaced from their equilibrium positions by equal and *opposite* distances a, and they are then released from rest.

The analytical solutions in this case from CAS are

$$x_1(t) = a \cos\left(\sqrt{3k/m}\, t\right)$$

$$x_2(t) = -a \cos\left(\sqrt{3k/m}\, t\right)$$

The minus sign in $x_2(t)$ tells us that the two masses will move together with the same speed, but they will be completely out of phase as shown in Figure 12.3ab.

The frequency of oscillation for both masses in this situation is $\omega_2 = \sqrt{3k/m}$.

This situation is referred to as the *antisymmetric oscillation* of the system.

(b) In this situation the two masses are pulled the same distance a from their corresponding equilibrium, and they are released from rest ($\dot{x}_1(0) = 0$ and $\dot{x}_2(0) = 0$).

The analytical solutions are

$$x_1(t) = x_2(t) = a \cos\left(\sqrt{k/m}\, t\right)$$

This tells us that if the two masses are initially displaced from equilibrium by the same distance and released from rest, the two masses will move together with the same speed and in phase, as if the middle spring was not present. This makes physical sense, since in this situation the middle spring will be unstretched from its natural length, and will remain unstretched during the motion of the two masses.

The frequency of oscillation for both masses in this situation is $\omega_1 = \sqrt{k/m}$, i.e., the same frequency as if only one of the two masses were attached to a single spring with a spring constant k.

This situation is referred to as the *symmetric oscillation* of the system.

Python Code

(a) We use the command `dsolve_system` in SymPy to solve analytically the system of ODEs with the given initial conditions. The command `lambdify` converts the symbolic answer $x_1(t)$ and $x_2(t)$ into NumPy arrays which are plotted in Figures 12.3a and 12.3b.

(b) We use the same code (not shown here) and change only the initial conditions $x_1(0) = a$, $x_2(0) = a$, $\dot{x}_1(0) = 0$ and $\dot{x}_2(0) = 0$. The positions $x_1(t)$ and $x_2(t)$ are plotted in Figures 12.3c and 12.3d.

```
from sympy import Function,  Eq, Derivative as D, symbols,\
lambdify, simplify
from sympy.solvers.ode.systems import dsolve_system
import numpy as np
import matplotlib.pyplot as plt
print('-'*28,'CODE OUTPUT','-'*29,'\n')


k, m, t, x0, v0, a = symbols('k, m, t, x0, v0, a',real=True,\
positive=True)
x1, x2 = symbols('x1, x2', cls=Function)

eq1 = Eq(D(x1(t),t,t), -(k/m)* x1(t) -(k/m)*(x1(t)-x2(t)))
eq2 = Eq(D(x2(t),t,t), -(k/m)* x2(t) -(k/m)*(x2(t)-x1(t)))

# initial conditions x1(0)=a, v1(0)=0, x2(0)=-a and v2(0)=1
initCondits = {x1(0): a, D(x1(t),t).subs(t, 0): 0,\
               x2(0): -a, D(x2(t),t).subs(t, 0): 0 }
# solve the system of two differential equations symbolically
soln = dsolve_system((eq1, eq2), [x1(t),x2(t)], t, initCondits)

X1 = soln[0][0].rhs      # extract the solutions x1(t) and x2(t)
X2 = soln[0][1].rhs
print('x1(t) =',X1)
print('\nx2(t) =',X2)


# make x1(t) and x2(t) functions using lambdify

x1soln = lambdify(t, X1.subs({k:1,m:1,a:1}),'numpy')
x2soln = lambdify(t, X2.subs({k:1,m:1,a:1}),'numpy')

# evaluate x(t) values from t=0 to t=10 s
xvals = np.arange(0,5.5,.1)

plt.plot(x1soln(xvals),label='(a) x1(t)')
plt.plot(x2soln(xvals),'r--',label='(b) x2(t)')
plt.xlabel('Time t [s]')
plt.ylabel('x1(t), x2(t)')
plt.legend()
plt.show()

-------------------------- CODE OUTPUT ----------------------------
x1(t) = a*cos(sqrt(3)*sqrt(k)*t/sqrt(m))

x2(t) = -a*cos(sqrt(3)*sqrt(k)*t/sqrt(m))
```

Mathematica Code
We use `DSolve` to solve symbolically the system of differential equations, and `ExpToTrig` is used to convert the solutions into trig functions. Note also the use of the `Simplify` command.

```
SetOptions[Plot, BaseStyle->{FontSize->14}];

sol =
ExpToTrig[DSolve[{m * x1''[t] == -k * x1[t] + k * (x2[t] - x1[t]),
m * x2''[t] == -k * x2[t] + k * (x1[t] - x2[t]), x1[0] == a, x1'[0] == 0,
x2[0] == -a, x2'[0] == 0}, {x1[t], x2[t]}, t]]//Simplify
numValues = {a → 1, k → 1, m → 1};
```

$$\left\{\left\{x1[t] \to a\mathrm{Cos}\left[\frac{\sqrt{3}\sqrt{k}t}{\sqrt{m}}\right], x2[t] \to -a\mathrm{Cos}\left[\frac{\sqrt{3}\sqrt{k}t}{\sqrt{m}}\right]\right\}\right\}$$

```
gr1 = Plot[x1[t]/.sol/.numValues, {t, 0, 10}, PlotLabel → "(a) x1[t]"];
gr2 = Plot[x2[t]/.sol/.numValues, {t, 0, 10}, PlotLabel → "(b) x2[t]",
PlotStyle->Dashed];


GraphicsGrid[{{gr1, gr2}}, Frame → True]

sol =
ExpToTrig[DSolve[{m * x1''[t] == -k * x1[t] + k * (x2[t] - x1[t]),
m * x2''[t] == -k * x2[t] + k * (x1[t] - x2[t]), x1[0] == a, x1'[0] == 0, x2[0] == a,
x2'[0] == 0}, {x1[t], x2[t]}, t]]//Simplify
numValues = {a → 1, k → 1, m → 1};
```

$$\left\{\left\{x1[t] \to a\mathrm{Cos}\left[\frac{\sqrt{k}t}{\sqrt{m}}\right], x2[t] \to a\mathrm{Cos}\left[\frac{\sqrt{k}t}{\sqrt{m}}\right]\right\}\right\}$$

```
gr1 = Plot[x1[t]/.sol/.numValues, {t, 0, 10}, PlotLabel → "(c) x1[t]"];
gr2 = Plot[x2[t]/.sol/.numValues, {t, 0, 10}, PlotLabel → "(d) x2[t]",
PlotStyle->Dashed];


GraphicsGrid[{{gr1, gr2}}, Frame → True]
```

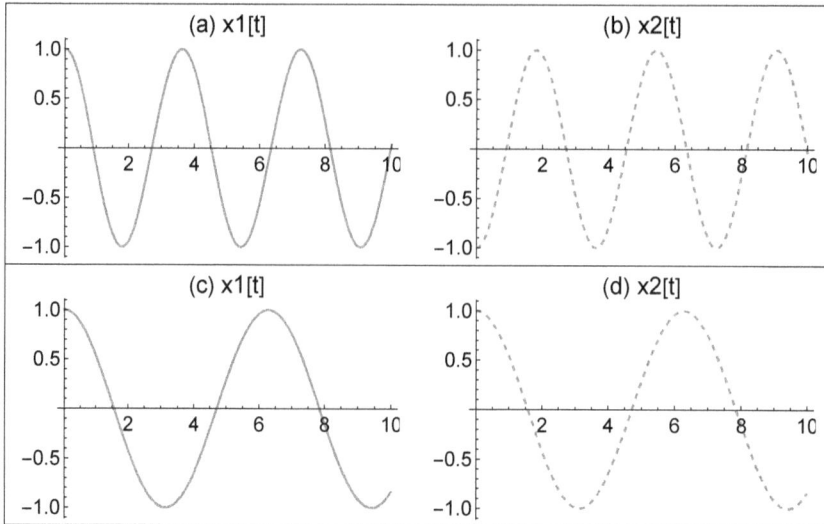

Figure 12.3: (a),(b) Plots of the *antisymmetric* normal mode of the two-mass three-spring system, from Example 12.2. The two masses in this case are completely out of phase and $x_1(t) = -x_2(t)$. (c),(d) Plots of the *symmetric* normal mode of the system, in which the two masses are completely in phase and $x_1(t) = x_2(t)$.

Example 12.2 showed that the system of two equal masses and three identical springs in Figure 12.1 has two *natural frequencies* given by $\omega_1 = \sqrt{k/m}$ and $\omega_2 = \sqrt{3k/m}$. By properly choosing the initial conditions in the system as in the two examples, we can force both masses to oscillate with a single frequency, either ω_1 or ω_2. In these special situations, the two natural frequencies are uncoupled from each other, and we say that these are the *normal modes* of the oscillating system.

Figure 12.4 shows schematically the motion of the two masses in either the *symmetric* oscillation pattern with frequency $\omega_1 = \sqrt{k/m}$ (left panel), or an *antisymmetric* oscillation with frequency $\omega_2 = \sqrt{3k/m}$ (right panel).

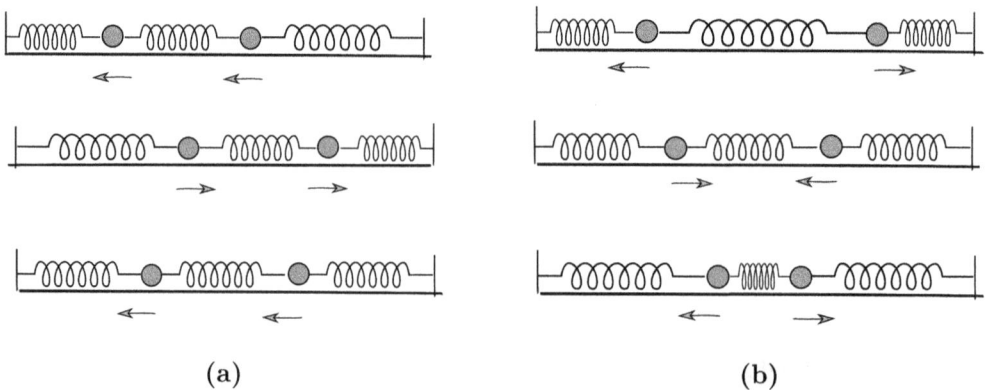

(a)   (b)

Figure 12.4: The normal modes of the two-mass, three-spring system in Figure 12.1, corresponding to the natural frequencies (a) $\omega_1 = \sqrt{k/m}$ and (b) $\omega_2 = \sqrt{3k/m}$. These normal modes represent *symmetric* and *antisymmetric* oscillations, respectively.

12.1.4 GENERAL CASE: LINEAR COMBINATION OF NORMAL MODES

If the two masses are displaced at some random distances and are released, they will oscillate in a complex manner, which can be described as the linear combination of oscillations with frequencies ω_1 and ω_2.

The Python and Mathematica codes in Example 12.3 illustrate the general case, where the solutions $x_1(t)$ and $x_2(t)$ are indeed linear combinations of trigonometric functions involving the two frequencies $\omega_1 = \sqrt{k/m}$ and $\omega_2 = \sqrt{3k/m}$. This example shows how occasionally we can numerically decouple the motions corresponding to these two frequencies.

Example 12.3: Analytical solutions for equal masses and identical springs; decoupling of the two frequencies

Integrate (12.1.1) for identical springs and identical masses, with the initial conditions $x_1(0) = 0$, $x_2(0) = a$, $\dot{x}_1(0) = 0$ and $\dot{x}_2(0) = 0$.

Solution:

In this case, the second mass in Figure 12.1 is moved from equilibrium by a distance a and is released from rest. The first mass is initially at rest at its equilibrium position. The analytical solutions obtained from the CAS are

$$x_1(t) = (a/2) \left[\cos\left(\sqrt{k/m}\, t \right) - \cos\left(\sqrt{3k/m}\, t \right) \right]$$

$$x_2(t) = (a/2) \left[\cos\left(\sqrt{k/m}\, t \right) + \cos\left(\sqrt{3k/m}\, t \right) \right]$$

and these are shown in Figures 12.5 (a) and (b) below. Once more, the functions $x_1(t)$ and $x_2(t)$ are complicated, and it is difficult to describe how exactly the two masses are moving. This is because mathematically both $x_1(t)$ and $x_2(t)$ contain the frequencies ω_1 and ω_2.

Figures 12.5 (c) and (d) show plots of the function

$$x_1(t) + x_2(t) = a \cos\left(\sqrt{k/m}\, t \right)$$

$$x_1(t) - x_2(t) = -a \cos\left(\sqrt{3k/m}\, t \right)$$

By using these linear combinations, it is now possible to *decouple* the two normal modes, so that the motions shown in Figures 12.5 (c) and (d) are simple cosine functions with frequencies $\omega_1 = \sqrt{k/m}$ and $\omega_2 = \sqrt{3k/m}$.

Python Code

We use the command `dsolve_system` in SymPy to solve analytically the system of ODEs with the given initial conditions. The command `lambdify` converts the symbolic answer $x_1(t)$ and $x_2(t)$ into NumPy arrays which are plotted in four panels.

```python
from sympy import Function,  Eq, Derivative as D, symbols,\
lambdify, simplify
from sympy.solvers.ode.systems import dsolve_system
import numpy as np
import matplotlib.pyplot as plt
print('-'*28,'CODE OUTPUT','-'*29,'\n')

k, m, t, x0, v0, a = symbols('k, m, t, x0, v0, a',real=True,\
positive=True)
x1, x2 = symbols('x1, x2', cls=Function)

def plotx(y,ttle):              # function to plot x1,x2 etc
    plt.plot(xvals,y,'b');
    plt.title(ttle);


eq1 = Eq(D(x1(t),t,t), -(k/m)* x1(t) -(k/m)*(x1(t)-x2(t)))
eq2 = Eq(D(x2(t),t,t), -(k/m)* x2(t) -(k/m)*(x2(t)-x1(t)))

# initial conditions x1(0)=0, v1(0)=0, x2(0)=0  and v2(0)=1
initCondits = {x1(0): 0, D(x1(t),t).subs(t, 0): 0,\
            x2(0): a, D(x2(t),t).subs(t, 0): 0 }
# solve the system of two differential equations symbolically
soln = dsolve_system((eq1, eq2), [x1(t),x2(t)], t, initCondits)

X1 = soln[0][0].rhs       # extract the solutions x1(t) and x2(t)
X2 = soln[0][1].rhs
print('\nx1(t) =\n',simplify(X1))
print('\nx2(t) =\n',simplify(X2))

---------------------------- CODE OUTPUT ----------------------------

x1(t) =
 a*(cos(sqrt(k)*t/sqrt(m)) - cos(sqrt(3)*sqrt(k)*t/sqrt(m)))/2

x2(t) =
 a*(cos(sqrt(k)*t/sqrt(m)) + cos(sqrt(3)*sqrt(k)*t/sqrt(m)))/2
```

```
# make x1(t) and x2(t) functions using lambdify
x1soln = lambdify(t, X1.subs({k:1,m:1,a:1}),'numpy')
x2soln = lambdify(t, X2.subs({k:1,m:1,a:1}),'numpy')

xvals = np.arange(0,10,.2)  # evaluate x(t) from t=0 to t=10 s

plt.subplot(2,2,1)   # plot panels for x1, x2, x1+x2 and x1-x2
plotx(x1soln(xvals),'(a) x1(t)')
plt.subplot(2,2,2)
plotx(x2soln(xvals),'(b) x2(t)')
plt.subplot(2,2,3)
plotx(x1soln(xvals)+x2soln(xvals),'(c) x1(t)+x2(t)')
plt.subplot(2,2,4)
plotx(x1soln(xvals)-x2soln(xvals),'(d) x1(t)-x2(t)')
plt.tight_layout()
plt.show()
```

Mathematica Code

We use `DSolve` to solve symbolically the system of differential equations, and `ExpToTrig` is used to convert the solutions into trig functions.

sol =

ExpToTrig[DSolve[$\{m * \text{x1}''[t] == -k * \text{x1}[t] + k * (\text{x2}[t] - \text{x1}[t])$,

$m * \text{x2}''[t] == -k * \text{x2}[t] + k * (\text{x1}[t] - \text{x2}[t]), \text{x1}[0] == 0, \text{x1}'[0] == 0$,

$\text{x2}[0] == a, \text{x2}'[0] == 0\}, \{\text{x1}[t], \text{x2}[t]\}, t]][[1]]$//Simplify

numValues = $\{a \to 1, k \to 1, m \to 1\}$;

$\left\{\text{x1}[t] \to \frac{1}{2}a\left(\text{Cos}\left[\frac{\sqrt{k}t}{\sqrt{m}}\right] - \text{Cos}\left[\frac{\sqrt{3}\sqrt{k}t}{\sqrt{m}}\right]\right), \text{x2}[t] \to \frac{1}{2}a\left(\text{Cos}\left[\frac{\sqrt{k}t}{\sqrt{m}}\right] + \text{Cos}\left[\frac{\sqrt{3}\sqrt{k}t}{\sqrt{m}}\right]\right)\right\}$

gr1 = Plot[x1[t]/.sol/.numValues, $\{t, 0, 10\}$, PlotLabel → "(a) x1[t]"];

gr2 = Plot[x2[t]/.sol/.numValues, $\{t, 0, 10\}$, PlotLabel → "(b) x2[t]"];

gr3 = Plot[(x1[t] + x2[t])/.sol/.numValues, $\{t, 0, 10\}$,

PlotLabel → "(c) x1[t]+x2[t]"];

gr4 = Plot[(x1[t] − x2[t])/.sol/.numValues, $\{t, 0, 10\}$,

PlotLabel → "(c) x1[t]-x2[t]"];

GraphicsGrid[$\{\{$gr1, gr2$\}, \{$gr3, gr4$\}\}$, Frame → True]

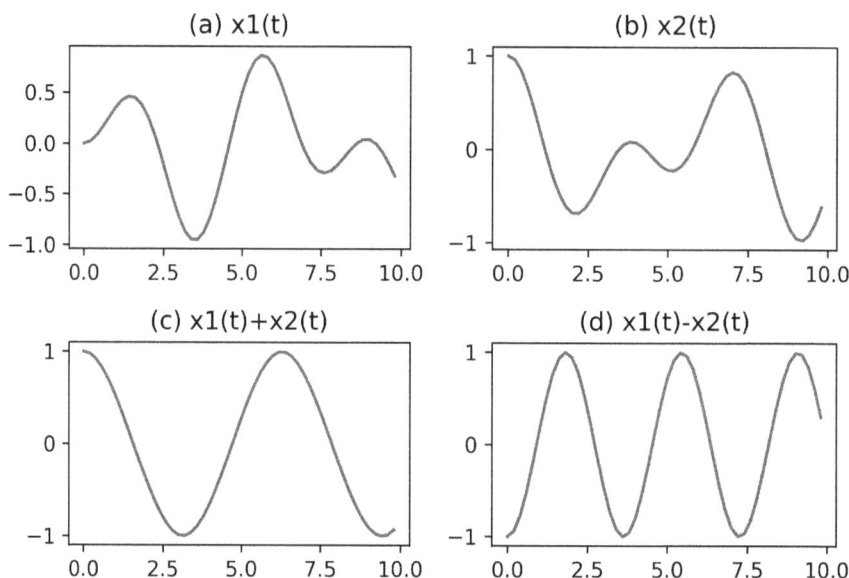

Figure 12.5: Plots of the positions (a) $x_1(t)$, (b) $x_2(t)$, (c) $x_1(t)+x_2(t)$ and (d) $x_1(t)-x_2(t)$ of the two-mass, three-spring system from Example 12.3.

In Example 12.4 we examine one more interesting behavior of the two-mass, three-spring system, the case of *weakly coupled oscillators*. In this example the weak coupling is established by choosing a middle spring with smaller spring constant than the two end springs ($k_1 = k_3 = k$ and $k_2 << k$).

Example 12.4: Weakly coupled oscillators
Integrate (12.1.1) and plot the numerical solutions $x_1(t)$ and $x_2(t)$ for the special case of identical masses ($m_1 = m_2 = 1\,\text{kg}$), identical end springs ($k_1 = k_3 = 1\,\text{N/m}$), and a smaller middle spring constant ($k_2 = 0.2\,\text{N/m}$). Consider the physical situation with the initial conditions $x_1(0) = 1\,\text{m}$, $x_2(0) = 0$, $\dot{x}_1(0) = 0$, and $\dot{x}_2(0) = 0$.

Solution:
The codes for Python and Mathematica are the same as Example 12.1, with different numerical values of the parameters. The output of the Mathematica code in Figures 12.6 (a) and (b) of Figure 12.6 shows the solutions $x_1(t)$ and $x_2(t)$. The motion of the two masses shows a clear beat pattern, in which mass m_1 reaches a maximum amplitude of oscillation at the same time that m_2 reaches a minimum amplitude, and vice versa.

In this example it is again possible to *uncouple* the motions of the two masses, by plotting the sum $x_1(t)+x_2(t)$ and the difference $x_1(t)-x_2(t)$ of the two functions $x_1(t)$ and $x_2(t)$. This is shown in Figures 12.6 (c) and (d), where the two linear combinations $x_1(t) + x_2(t)$ and $x_1(t) - x_2(t)$ can be seen to have pure harmonic oscillations with different frequencies, corresponding to the two normal modes of the system. We will see a more general method of uncoupling the normal modes of the system later in this chapter, when we discuss the concept of normal coordinates.

In the next two sections, we develop a more formal mathematical analysis of the normal modes for the system in Figure 12.1, by using the techniques of Linear Algebra.

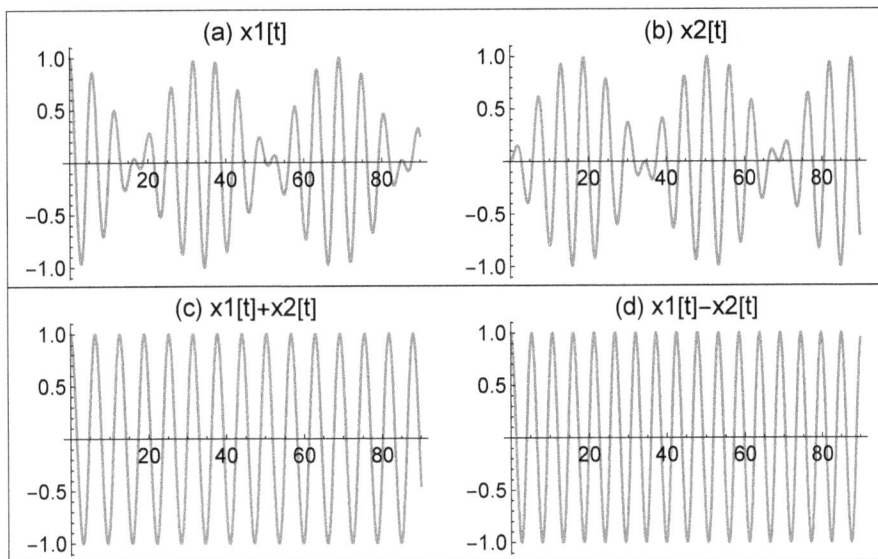

Figure 12.6: Plots of the positions (a) $x_1(t)$ and (b) $x_2(t)$, (c) $x_1(t) + x_2(t)$ and (d) $x_1(t) - x_2(t)$ of the weakly coupled two-mass, three-spring system, from Example 12.4.

12.2 NORMAL MODE ANALYSIS OF THE TWO-MASS, THREE-SPRING SYSTEM

We now proceed to analyze the system of equations (12.1.1) using two different methods. In Subsection 12.2.1, we show how to obtain the analytical solution for the two-mass, three-spring system by using the standard matrix techniques of Linear Algebra. In Subsection 12.2.2, we show how to solve the same problem by turning it into an eigenvalue/eigenvector type of problem, which can be easily analyzed using the commands available in Mathematica and Python.

12.2.1 EQUAL MASSES AND IDENTICAL SPRINGS – ANALYTICAL SOLUTION

In the case of equal masses $m_1 = m_2 = m$, and identical spring constants $k_1 = k_2 = k_3 = k$, the equations of motion (12.1.1) become:

$$m\ddot{x}_1 = -kx_1 - k(x_1 - x_2) \qquad m\ddot{x}_2 = -k(x_2 - x_1) - kx_2 \qquad (12.2.1)$$

This system of equations can be written in a compact matrix form:

$$\begin{pmatrix} m & 0 \\ 0 & m \end{pmatrix} \begin{pmatrix} \ddot{x}_1 \\ \ddot{x}_2 \end{pmatrix} = \begin{pmatrix} -2k & k \\ k & -2k \end{pmatrix} \begin{pmatrix} x_1 \\ x_2 \end{pmatrix} \qquad (12.2.2)$$

We can now solve this matrix equation by using the standard methods of Linear Algebra, and by following the same eigenvalue problem procedure we used in Chapter 11 for the principal moments of inertia.

We proceed in two steps, first we find the natural frequencies ω of the system, and second we find the positions $x_1(t)$ and $x_2(t)$, as follows.

Since we expect oscillatory motion, we try solutions of the form:

$$x_1(t) = A_1 e^{i\omega t} \qquad x_2(t) = A_2 e^{i\omega t} \qquad (12.2.3)$$

where A_1 and A_2 are the unknown amplitudes of oscillation for the two masses, and ω is the unknown frequency of oscillation. Substituting these into the matrix equation (12.2.2):

$$\begin{pmatrix} m & 0 \\ 0 & m \end{pmatrix} \begin{pmatrix} -\omega^2 A_1 e^{i\omega t} \\ -\omega^2 A_2 e^{i\omega t} \end{pmatrix} = \begin{pmatrix} -2k & k \\ k & -2k \end{pmatrix} \begin{pmatrix} A_1 e^{i\omega t} \\ A_2 e^{i\omega t} \end{pmatrix} \tag{12.2.4}$$

By canceling the exponential factor $e^{i\omega t}$ which is common to all terms, and combining the matrices, we obtain:

$$\begin{pmatrix} -\omega^2 m + 2k & -k \\ -k & -\omega^2 m + 2k \end{pmatrix} \begin{pmatrix} A_1 \\ A_2 \end{pmatrix} = \begin{pmatrix} 0 \\ 0 \end{pmatrix} \tag{12.2.5}$$

This matrix equation represents a system of equations. A theorem from Linear Algebra says that if the determinant of the matrix is non-zero, then there is a unique solution, which in this case is the trivial solution $A_1 = A_2 = 0$. However, in order for multiple solutions to exist, the determinant of the matrix must be zero. We are interested in a nontrivial solution $A_1, A_2 \neq 0$, so we solve for the values of ω which cause the determinant to be zero.

We set the determinant of the matrix equal to zero:

$$\det \begin{pmatrix} -\omega^2 m + 2k & -k \\ -k & -\omega^2 m + 2k \end{pmatrix} = 0 \tag{12.2.6}$$

$$(\omega^2 m - 2k)(\omega^2 m - 2k) - k^2 = 0 \tag{12.2.7}$$

Solving for ω, we obtain four possible solutions, only two of which are positive:

$$\omega_1 = \sqrt{\frac{k}{m}}, \quad \omega_2 = \sqrt{\frac{3k}{m}} \tag{12.2.8}$$

This completes the first part of the analysis, where we determined the two natural frequencies ω_1 and ω_2. In the previous section, we found that oscillatory solutions to our system of equations can contain one or both of these two frequencies.

In order to complete the description of the system, we must also find the two unknown amplitudes of oscillation A_1 and A_2. If we substitute $\omega_1 = \sqrt{\frac{k}{m}}$ into the matrix equation (12.2.6), we obtain:

$$\left[\begin{array}{cc} \left(\sqrt{\frac{k}{m}}\right)^2 m - 2k & k \\ k & \left(\sqrt{\frac{k}{m}}\right)^2 m - 2k \end{array} \right] \begin{pmatrix} A_1 \\ A_2 \end{pmatrix} = \begin{bmatrix} -k & k \\ k & -k \end{bmatrix} \begin{pmatrix} A_1 \\ A_2 \end{pmatrix} = 0 \tag{12.2.9}$$

By multiplying the matrices, we obtain these two equations:

$$-kA_1 + kA_2 = 0 \tag{12.2.10}$$

$$kA_1 - kA_2 = 0 \tag{12.2.11}$$

It is clear that these two equations are identical, and that $A_1 = A_2$. Note that this always happens when we are finding the eigenvectors of a 2×2 matrix, one of the equations will be redundant and can just be ignored. We conclude that when $\omega_1 = \sqrt{k/m}$, the two amplitudes are equal $A_1 = A_2$, and therefore $x_1(t) = x_2(t) = A_1 e^{i\omega t}$. We can now write the first solution for the motion of the two masses, which corresponds to the first normal model $\omega_1 = \sqrt{k/m}$:

$$x_1(t) = x_2(t) = A_1 e^{i\omega_1 t} \tag{12.2.12}$$

Since $A_1 = A_2$ and $x_1(t) = x_2(t)$, this type of motion corresponds to both masses moving in the same direction and in phase at all times, as we saw previously in Figure 12.4a. This type of motion is known as the first *normal mode* or the *symmetric mode of oscillation*, and the general motion of the system in this mode can be written in terms of trigonometric functions:

$$x_1(t) = x_2(t) = D_1 \cos(\omega_1 t - \phi_1) \tag{12.2.13}$$

In matrix notation, the first normal mode can be written as:

$$\begin{pmatrix} x_1(t) \\ x_2(t) \end{pmatrix} = D_1 \begin{pmatrix} 1 \\ 1 \end{pmatrix} \cos(\omega_1 t - \phi_1) \tag{12.2.14}$$

By working in a similar fashion for the second natural frequency of the system, we substitute $\omega_2 = \sqrt{3k/m}$ into (12.1.1), and obtain $A_1 = -A_2$. Since $A_1 = -A_2$, this type of motion corresponds to the two masses moving in opposite directions, while the center of mass remains stationary, as shown in Figure 12.4b. This type of motion is known as the second *normal mode* or the *antisymmetric mode of oscillation*.

We can then write the second possible solution corresponding to $\omega_2 = \sqrt{3k/m}$ as:

$$\begin{pmatrix} x_1(t) \\ x_2(t) \end{pmatrix} = D_2 \begin{pmatrix} 1 \\ -1 \end{pmatrix} \cos(\omega_2 t - \phi_2) \tag{12.2.15}$$

In general, the motion of the system will be a linear combination of the two normal modes, corresponding to the frequencies $\omega_1 = \sqrt{k/m}$ and $\omega_2 = \sqrt{3k/m}$.

A faster method of obtaining the normal mode frequencies ω_1 and ω_2 and the amplitudes A_1 and A_2, is by solving the matrix equation (12.2.5) using the symbolic capabilities of Python and Mathematica. Example 12.5 shows how to implement the above procedure in a CAS, to obtain the general solution for the two-mass, three-spring system.

Example 12.5: Solving the two-mass, three-spring system by solving the matrix equation

Use a CAS to solve the two-mass, three-spring system in Figure 12.1 as a matrix equation problem. Consider specifically the case of equal masses and identical springs, and follow the steps above: in the symbolic codes, substitute $x_1(t) = A_1 e^{i\omega t}$ and $x_2(t) = A_2 e^{i\omega t}$ in the differential equations, cancel out the $e^{i\omega t}$ terms, and setup the matrix A. Find the normal mode frequencies ω_1 and ω_2 and the general relationship between the amplitudes A_1 and A_2 by solving the characteristic equation.

Solution:

The codes produce the first normal mode frequency $\omega_1 = \sqrt{k/m}$, and the corresponding relationship between the amplitudes is $A_1 = A_2$. This is of course the symmetric mode of oscillation for the system that we saw previously.

The codes also produce the second normal mode frequency $\omega_2 = \sqrt{3k/m}$, and the corresponding relationship between the amplitudes $A_1 = -A_2$ for the antisymmetric mode of oscillation.

Python Code

We enter the ODEs for the variables $x_1(t)$, $x_2(t)$ to obtain the symbolic equations eq1 and eq2. The matrix A is formed from the coefficients A_1, A_2 of eq1 and eq2, using the `.coeff()` method.

The natural frequencies are found using the `solve` command in SymPy to solve the characteristic equation `det(A)=0`.

```
from sympy import symbols, exp, I, diff, solve, expand,  Matrix

k, m, omega, x1, x2, A1, A2, t = \
symbols("k, m, omega, x1, x2, A1, A2, t")
#  define all symbols for variables

print('-'*28,'CODE OUTPUT','-'*29,'\n')

# define x1 and x2 as complex exponentials
x1 = A1*exp(I*omega*t)
x2 = A2*exp(I*omega*t)

# differential equations for x1, x2 divided by exp(I*omega*t)
eq1 = expand((m*diff(x1,t,t)+k*x1+k*(x1-x2))/exp(I*omega*t))
eq2 = expand((m*diff(x2,t,t)+k*x2+k*(x2-x1))/exp(I*omega*t))

# Matrix A has coeffcients of A1, A2 in the system of equations
A = Matrix([[eq1.coeff(A1),eq1.coeff(A2)],[eq2.coeff(A1),eq2.coeff(A2)]])

print('The matrix of coefficients for A1, A2 is:\n')
print(A)

sol = solve(A.det(),omega)  # set det(A)=0 and solve for omega

print('\nNatural frequency omega1 = ',sol[0])
print('Natural frequency omega2 = ',sol[1])
print('\nNatural frequency omega3 = ',sol[2])
print('Natural frequency omega4 = ',sol[3])

-------------------------- CODE OUTPUT ----------------------------
The matrix of coefficients for A1, A2 is:
Matrix([[2*k - m*omega**2, -k], [-k, 2*k - m*omega**2]])

Natural frequency omega1 =  -sqrt(3)*sqrt(k/m)
Natural frequency omega2 =  sqrt(3)*sqrt(k/m)

Natural frequency omega3 =  -sqrt(k/m)
Natural frequency omega4 =  sqrt(k/m)
```

Mathematica Code

The code follows the same steps in the analytical solution and in the above Python code, by evaluating the elements of the matrix and solving the characteristic equation $\det(A)=0$ to obtain the frequencies $\omega_1 = \sqrt{k/m}$ and $\omega_2 = \sqrt{3k/m}$. The command Coefficient is used to find the elements of the matrix.

$x1 = A1 * \text{Exp}[I * \omega * t]; x2 = A2 * \text{Exp}[I * \omega * t];$

$eq1 = m * D[x1, \{t, 2\}] + k * x1 + k * (x1 - x2);$

$expr1 = (eq1)/\text{Exp}[I * \omega * t]//\text{Simplify};$

Print["The first ODE yields the equation : ", expr1]

The first ODE yields the equation : $2A1k - A2k - A1m\omega^2$

$eq2 = m * D[x2, \{t, 2\}] + k * x2 + k * (x2 - x1);$

$expr2 = (eq2)/\text{Exp}[I * \omega * t]//\text{Simplify};$

Print["The second ODE yields the equation : ", expr2]

The second ODE yields the equation : $-A1k + 2A2k - A2m\omega^2$

$A = \{\{\text{Coefficient}[expr1, A1], \text{Coefficient}[expr1, A2]\},$

$\{\text{Coefficient}[expr2, A1], \text{Coefficient}[expr2, A2]\}\};$

Print["The matrix A is: A = ", A]

The matrix A is: A = $\{\{2k - m\omega^2, -k\}, \{-k, 2k - m\omega^2\}\}$

Print["The natural frequencies are: "]

Print[Solve[Det[A]==0, \omega]]

The natural frequencies are:
$\left\{\left\{\omega \to -\frac{\sqrt{k}}{\sqrt{m}}\right\}, \left\{\omega \to \frac{\sqrt{k}}{\sqrt{m}}\right\}, \left\{\omega \to -\frac{\sqrt{3}\sqrt{k}}{\sqrt{m}}\right\}, \left\{\omega \to \frac{\sqrt{3}\sqrt{k}}{\sqrt{m}}\right\}\right\}$

12.2.2 SOLVING THE TWO-MASS AND THREE-SPRING SYSTEM AS AN EIGEN-VALUE PROBLEM

In this section, we show how to solve the general two-mass, three-spring system as an eigenvalue problem, using the commands available in Mathematica and Python. Let us consider again the general equations of motion (12.1.1), for the system of two masses in

Figure 12.1:

$$m_1\ddot{x}_1 = -k_1 x_1 - k_2(x_1 - x_2) \qquad m_2\ddot{x}_2 = -k_2(x_2 - x_1) - k_3 x_2 \tag{12.2.16}$$

or after dividing the first equation by m_1, and the second equation by m_2:

$$\ddot{x}_1 = -\frac{k_1}{m_1}x_1 - \frac{k_2}{m_1}(x_1 - x_2) \qquad \ddot{x}_2 = -\frac{k_2}{m_2}(x_2 - x_1) - \frac{k_3}{m_2}x_2 \tag{12.2.17}$$

By substituting a trial solution of the form $x_1(t) = A_1 e^{i\omega t}$ and $x_2(t) = A_2 e^{i\omega t}$ and canceling the common factor $e^{i\omega t}$, these equations yield:

$$-A_1\omega^2 = -\frac{k_1}{m_1}A_1 - \frac{k_2}{m_1}(A_1 - A_2) \qquad -A_2\omega^2 = -\frac{k_2}{m_2}(A_2 - A_1) - \frac{k_3}{m_2}A_2 \tag{12.2.18}$$

These can be written in compact matrix form as:

$$\begin{pmatrix} \frac{k_1+k_2}{m_1} & \frac{-k_2}{m_1} \\[2mm] \frac{-k_2}{m_2} & \frac{k_2+k_3}{m_2} \end{pmatrix} \begin{pmatrix} A_1 \\ A_2 \end{pmatrix} = \omega^2 \begin{pmatrix} A_1 \\ A_2 \end{pmatrix} \tag{12.2.19}$$

You will recognize that this equation is an *eigenvalue problem* in the theory of Linear Algebra, similar to the eigenvalue problems we encountered for the moment of inertia matrix in Chapter 11.

For the two-mass system of Figure 12.1, the eigenvalue problem to be solved becomes clear by writing (12.2.19) in this matrix form:

The Two-Mass System as an Eigenvalue Problem

$$\mathbf{GA} = \omega^2 \mathbf{A} \tag{12.2.20}$$

$$\mathbf{A} = \begin{pmatrix} A_1 \\ A_2 \end{pmatrix} \qquad \mathbf{G} = \begin{pmatrix} \frac{k_1+k_2}{m_1} & \frac{-k_2}{m_1} \\[2mm] \frac{-k_2}{m_2} & \frac{k_2+k_3}{m_2} \end{pmatrix}$$

We are looking for the eigenvalues $\lambda = \omega^2$ of the square matrix $\mathbf{G}$, which will give us the natural frequencies of oscillation. We are also looking for the corresponding eigenvectors $\mathbf{A}$, which will give us the normal modes of oscillation corresponding to each natural frequency. In order for the eigenvalue equation (12.2.20) to have a nontrivial solution ($\mathbf{A} \neq \mathbf{0}$), we must have:

$$\det \begin{bmatrix} \frac{k_1+k_2}{m_1} - \omega^2 & \frac{-k_2}{m_1} \\[2mm] \frac{-k_2}{m_2} & \frac{k_2+k_3}{m_2} - \omega^2 \end{bmatrix} = 0 \tag{12.2.21}$$

which is the characteristic equation of our eigenvalue problem. From this point on, we proceed by following the same two-step method used in the previous section. First, we must solve the characteristic equation (12.2.21) in order to find the frequencies ω_1 and ω_2. These frequencies will depend on k_1, k_2, k_3, m_1, and m_2. In the next step, we substitute the first natural frequency ω_1 into (12.2.19), in order to find (A_1, A_2), the first normal mode. Finally, we repeat the previous steps using the second natural frequency ω_2, in order to find (A_1, A_2) for the second normal mode.

The Mathematica and Python codes in Example 12.6 show how to find the eigenvectors and eigenvalues for two different cases of the two-mass system in Figure 12.1.

Example 12.6: The two-mass system as an eigenvalue/eigenvector problem

Solve the two oscillating mass systems in Figure 12.1 as an eigenvalue/eigenvector problem, in these two cases:
(a) Identical springs and identical masses.
(b) Identical springs and unequal masses.

Solution:
(a) In this case the matrix is

$$\mathbf{G} = \begin{pmatrix} \frac{2k}{m} & \frac{-k}{m} \\ \frac{-k}{m} & \frac{2k}{m} \end{pmatrix}$$

The codes produce the first eigenvalue $\omega_1^2 = k/m$ or $\omega_1 = \sqrt{k/m}$, and the corresponding eigenvector,

$$\begin{pmatrix} A_1 \\ A_2 \end{pmatrix} = \begin{pmatrix} 1 \\ 1 \end{pmatrix}$$

This is what we also obtained in Section 12.2 for the first normal mode of the oscillating two-mass system. Similarly, the second eigenvalue $\omega_2 = \sqrt{3k/m}$, and the corresponding eigenvector

$$\begin{pmatrix} A_1 \\ A_2 \end{pmatrix} = \begin{pmatrix} -1 \\ 1 \end{pmatrix}$$

This is again the same result we obtained in Section 12.2 for the second normal mode.

(b) In this case the matrix is

$$\mathbf{G} = \begin{pmatrix} \frac{2k}{m_1} & \frac{-k}{m_1} \\ \frac{-k}{m_2} & \frac{2k}{m_2} \end{pmatrix}$$

and the CAS produces the two eigenvalues and eigenvectors:

$$\omega_1 = \sqrt{\frac{k(m_1 + m_2 - z)}{m_1 m_2}} \qquad \begin{pmatrix} A_1 \\ A_2 \end{pmatrix} = \begin{pmatrix} \frac{m_1 - m_2 + z}{m_1} \\ 1 \end{pmatrix}$$

$$\omega_2 = \sqrt{\frac{k(m_1 + m_2 + z)}{m_1 m_2}} \qquad \begin{pmatrix} A_1 \\ A_2 \end{pmatrix} = \begin{pmatrix} \frac{-m_1 + m_2 + z}{m_1} \\ 1 \end{pmatrix}$$

where $z = \sqrt{m_1^2 - m_1 m_2 + m_2^2}$. In this two-mass, three-spring system with unequal masses, the ratio of the amplitudes A_1/A_2 depends in a complicated manner on the masses (m_1, m_2), but it does not depend on the spring constant k.

Python Code – Part (a)

In the Python code we use the command `eigenvals` and `eigenvects`, which produces the same eigenvalues and eigenvectors as the analytical solution, each eigenvalues with a multiplicity of 1. For example, the output `(k/m,1,[Matrix([[1],[1]])` is interpreted as the frequency $\omega_1 = \sqrt{k/m}$ with a multiplicity of 1, and the `Matrix([[1],[1]])` result indicates the corresponding eigenvector

$$\begin{pmatrix} A_1 \\ A_2 \end{pmatrix} = \begin{pmatrix} 1 \\ 1 \end{pmatrix}$$

```
from sympy import Matrix, symbols

print('-'*28,'CODE OUTPUT','-'*29,'\n')

k, m = symbols('k, m')

A = Matrix([[2*k/m,-k/m],[-k/m,2*k/m]])

omega_sq = list(A.eigenvals().keys())

eigenvects = list(A.eigenvects())

print('The square of the first frequency is: ',omega_sq[0])

print('\nThe first eigenvector is: ',eigenvects[1][2][0])

print('\nThe square of the second frequency is: ',omega_sq[1])

print('\nThe second eigenvector is: ',eigenvects[0][2][0])


---------------------------- CODE OUTPUT ----------------------------
The square of the first frequency is:  3*k/m

The first eigenvector is:  Matrix([[-1], [1]])

The square of the second frequency is:  k/m

The second eigenvector is:  Matrix([[1], [1]])
```

Python Code – Part (b)

This code is essentially the same as part (a) with a different matrix A. We use the `.subs` method to make the replacement $z = \sqrt{m_1^2 - m_1 m_2 + m_2^2}$, as explained above.

```
from sympy import Matrix, symbols, sqrt, simplify

print('-'*28,'CODE OUTPUT','-'*29,'\n')

k, m1, m2, z = symbols('k, m1, m2, z')

A = Matrix([[2*k/m1,-k/m1],[-k/m2,2*k/m2]])

omega_sq = list(A.eigenvals())

eigenvects = list(A.eigenvects())

print('The square of the first frequency is:\n',\
    simplify(omega_sq[0].subs(sqrt(m1**2 - m1*m2 + m2**2),z)))

print('\nThe square of the second frequency is:\n',\
    simplify(omega_sq[1].subs(sqrt(m1**2 - m1*m2 + m2**2),z)))

print('\nAfter the z-substitution, the eigenvector #1 is:')
u = eigenvects[0][2][0]
print(simplify(u.subs(sqrt(m1**2 - m1*m2 + m2**2),z)))

print('\nAfter the z-substitution, the eigenvector #2 is:')
u = eigenvects[1][2][0]
print(simplify(u.subs(sqrt(m1**2 - m1*m2 + m2**2),z)))


--------------------------- CODE OUTPUT ----------------------------
The square of the first frequency is:
 k*(m1 + m2 - z)/(m1*m2)

The square of the second frequency is:
 k*(m1 + m2 + z)/(m1*m2)

After the z-substitution, the eigenvector #1 is:
Matrix([[(m1 - m2 + z)/m1], [1]])

After the z-substitution, the eigenvector #2 is:
Matrix([[(m1 - m2 - z)/m1], [1]])
```

Mathematica Code – Part (a)
The code below uses the commands Eigenvalues and Eigenvectors . We use the
replacement command /. to make the replacement $z = \sqrt{m_1^2 - m_1 m_2 + m_2^2}$.

$$A = \begin{pmatrix} 2*k/m & -k/m \\ -k/m & 2*k/m \end{pmatrix};$$

Print["The frequencies are: ", Sqrt[Eigenvalues[A]]]

The frequencies are: $\left\{ \sqrt{3}\sqrt{\frac{k}{m}}, \sqrt{\frac{k}{m}} \right\}$

Print["The eigenvectors are: ", Eigenvectors[A]]

The eigenvectors are: $\{\{-1,1\},\{1,1\}\}$

Mathematica Code – Part (b)

$$A = \begin{pmatrix} 2*k/m1 & -k/m1 \\ -k/m2 & 2*k/m2 \end{pmatrix};$$

rule $= \left\{ \sqrt{m1^2 - m1m2 + m2^2} \text{->} z \right\};$

Print["The frequencies after the z-substitution are:",

Sqrt[Eigenvalues[A]]/.rule]

The frequencies after the z-substitution are: $\left\{ \sqrt{\frac{k(m1+m2-z)}{m1m2}}, \sqrt{\frac{k(m1+m2+z)}{m1m2}} \right\}$

Print["The eigenvectors are:",

Eigenvectors[A]/.rule//Simplify]

The eigenvectors are: $\left\{ \left\{ \frac{m1-m2+z}{m1}, 1 \right\}, \left\{ -\frac{-m1+m2+z}{m1}, 1 \right\} \right\}$

12.3 DOUBLE PENDULUM

Figure 12.7 shows a double pendulum, consisting of two masses m_1 and m_2 attached to
massless rigid rods of lengths L_1 and L_2. We can treat this problem using the same methods
as for the two-mass three-spring oscillating system, by developing the equations of motion

and evaluating the natural frequencies. In Subsection 12.3.1, we will develop the Lagrangian and the equations of motion. In Subsection 12.3.2, we will find the analytical solution for the special case of two identical coupled pendula. In Subsection 12.3.3, we will treat the double pendulum as an eigenvalue problem, and show how to obtain the natural frequencies and the amplitudes of the normal modes.

12.3.1 LAGRANGIAN AND EQUATIONS OF MOTION – NUMERICAL SOLUTIONS

The position (x_1, y_1) of the first mass is $(x_1, y_1) = (L_1 \sin\theta_1, L_1 \cos\theta_1)$, so that the kinetic energy of mass m_1 is:

$$T_1 = \frac{1}{2}m_1\left(\dot{x}_1^2 + \dot{y}_1^2\right) = \frac{1}{2}m_1\left(L_1^2\cos^2\theta_1\dot{\theta}_1^2 + L_1^2\sin^2\theta_1\dot{\theta}_1^2\right) = \frac{1}{2}m_1 L_1^2\dot{\theta}_1^2 \tag{12.3.1}$$

The location (x_2, y_2) of m_2 is shifted with respect to the first mass by (x_1, y_1), so that $(x_2, y_2) = (L_1\sin\theta_1 + L_2\sin\theta_2, L_1\cos\theta_1 + L_2\cos\theta_2)$. Therefore, the kinetic energy of the second mass is:

$$T_2 = \frac{1}{2}m_2\left(\dot{x}_2^2 + \dot{y}_2^2\right) \tag{12.3.2}$$

$$= \frac{1}{2}m_2\left\{\frac{d}{dt}\left(L_1\sin\theta_1 + L_2\sin\theta_2\right)^2 + \frac{d}{dt}\left(L_1\cos\theta_1 + L_2\cos\theta_2\right)^2\right\} \tag{12.3.3}$$

After differentiating and collecting terms, we find:

$$T_2 = \frac{1}{2}m_2\left(L_1^2\dot{\theta}_1^2 + 2L_1 L_2\dot{\theta}_1\dot{\theta}_2\cos\left(\theta_1 - \theta_2\right) + L_2^2\dot{\theta}_2^2\right) \tag{12.3.4}$$

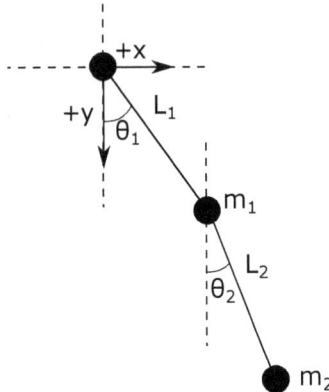

Figure 12.7: The double pendulum oscillator is characterized by the two angles $(\theta_1(t), \theta_2(t))$.

The total kinetic energy is then:

$$T = \frac{1}{2}\left(m_1 + m_2\right)L_1^2\dot{\theta}_1^2 + m_2 L_1 L_2\dot{\theta}_1\dot{\theta}_2\cos\left(\theta_1 - \theta_2\right) + \frac{1}{2}m_2 L_2^2\dot{\theta}_2^2 \tag{12.3.5}$$

For amplitudes of small oscillations we use the approximation $\cos\left(\theta_1 - \theta_2\right) \simeq 1$, so that:

$$T = \frac{1}{2}\left(m_1 + m_2\right) L_1^2 \dot{\theta}_1^2 + m_2 L_1 L_2 \dot{\theta}_1 \dot{\theta}_2 + \frac{1}{2} m_2 L_2^2 \dot{\theta}_2^2 \qquad (12.3.6)$$

The total potential energy V is the sum of potential energies for each pendulum:

$$V = m_1 g y_1 + m_2 g y_2 = \left(m_1 + m_2\right) g L_1 \cos\theta_1 + m_2 g L_2 \cos\theta_2 \qquad (12.3.7)$$

For amplitudes of small oscillations, we use the approximation $\cos\theta_1 \simeq 1 - \frac{\theta_1^2}{2}$ and $\cos\theta_2 \simeq 1 - \frac{\theta_2^2}{2}$ so that:

$$V = \left(m_1 + m_2\right) g L_1 \left(1 - \frac{\theta_1^2}{2}\right) + m_2 g L_2 \left(1 - \frac{\theta_2^2}{2}\right) \qquad (12.3.8)$$

Using these small angle approximations, the Lagrangian $L = T - V$ of the system is:

$$\mathcal{L} = \frac{1}{2}\left(m_1 + m_2\right) L_1^2 \dot{\theta}_1^2 + m_2 L_1 L_2 \dot{\theta}_1 \dot{\theta}_2 + \frac{1}{2} m_2 L_2^2 \dot{\theta}_2^2$$
$$- \left(m_1 + m_2\right) g L_1 \left(1 - \frac{\theta_1^2}{2}\right) - m_2 g L_2 \left(1 - \frac{\theta_2^2}{2}\right) \qquad (12.3.9)$$

We can now find the Euler-Lagrange equations:

$$\frac{d}{dt}\left(\frac{\partial \mathcal{L}}{\partial \dot{\theta}_1}\right) - \frac{\partial \mathcal{L}}{\partial \theta_1} = 0 \qquad\qquad \frac{d}{dt}\left(\frac{\partial \mathcal{L}}{\partial \dot{\theta}_2}\right) - \frac{\partial \mathcal{L}}{\partial \theta_2} = 0 \qquad (12.3.10)$$

By evaluating the derivatives and simplifying, we obtain:

$$\left(m_1 + m_2\right) L_1 \ddot{\theta}_1 + m_2 L_2 \ddot{\theta}_2 = - \left(m_1 + m_2\right) g \theta_1 \qquad (12.3.11)$$

$$m_2 L_1 \ddot{\theta}_1 + m_2 L_2 \ddot{\theta}_2 = -m_2 g \theta_2 \qquad (12.3.12)$$

If we desire to do so, we can use the commands in Python and Mathematica to carry out a numerical integration of this system of differential equations, and plot the functions $\theta_1(t)$ and $\theta_2(t)$. This is left as a problem at the end of the chapter. Next, we will find analytical solutions for $\theta_1(t)$ and $\theta_2(t)$, for the special case of two coupled identical pendula.

12.3.2 IDENTICAL MASSES AND LENGTHS - ANALYTICAL SOLUTIONS

We now use a trial solution in the form of exponential functions $\theta_1(t) = A_1 e^{i\omega t}$ and $\theta_2(t) = A_2 e^{i\omega t}$. Substituting and canceling the common exponential term $e^{i\omega t}$, we find:

$$- \left(m_1 + m_2\right) L_1 A_1 \omega^2 - m_2 L_2 A_2 \omega^2 = - \left(m_1 + m_2\right) g A_1 \qquad (12.3.13)$$

$$- m_2 L_1 A_1 \omega^2 - m_2 L_2 A_2 \omega^2 = -m_2 g A_2 \qquad (12.3.14)$$

After collecting the terms:

$$\left(m_1 + m_2\right)\left\{-L_1 \omega^2 + g\right\} A_1 - m_2 L_2 \omega^2 A_2 = 0 \qquad (12.3.15)$$

$$-m_2 L_1 \omega^2 A_1 - m_2 \left\{L_2 \omega^2 - g\right\} A_2 = 0 \qquad (12.3.16)$$

This system of linear equations for A_1 and A_2 will have a nontrivial solution, only if the determinant is zero:

$$\det \begin{pmatrix} \left(m_1 + m_2\right)\left\{-L_1 \omega^2 + g\right\} & -m_2 L_2 \omega^2 \\ -m_2 L_1 \omega^2 & -m_2 \left\{L_2 \omega^2 - g\right\} \end{pmatrix} = 0 \qquad (12.3.17)$$

In principle, we can obtain the solution to this characteristic equation; however, the resulting algebraic expressions are very complicated. So instead of looking at the completely general solution, let us look at the special case of equal lengths ($L_1 = L_2 = L$), and identical masses ($m_1 = m_2 = m$). In this special case, (12.3.17) becomes:

$$\det \begin{pmatrix} 2m\{-L\omega^2 + g\} & -mL\omega^2 \\ -mL\omega^2 & -m\{L\omega^2 - g\} \end{pmatrix} = 0 \qquad (12.3.18)$$

which gives the two possible solutions:

$$\omega_1 = \sqrt{\left(2+\sqrt{2}\right)\frac{g}{L}} \qquad \omega_2 = \sqrt{\left(2-\sqrt{2}\right)\frac{g}{L}} \qquad (12.3.19)$$

As in the case of the two-mass, three-spring oscillators, this completes the first part of the analysis, where we determined the frequencies ω_1 and ω_2 of the normal modes. In order to complete the description of the system, we must also find the two unknown amplitudes of oscillation A_1 and A_2. Substituting the value of $\omega_1 = \sqrt{\left(2-\sqrt{2}\right)g/L}$ into (12.3.15), we find:

$$2m\left\{-L\left(2-\sqrt{2}\right)\frac{g}{L}+g\right\}A_1 - mL\left(2-\sqrt{2}\right)\frac{g}{L}A_2 = 0 \qquad (12.3.20)$$

Simplifying and canceling the common factors:

$$\frac{A_1}{A_2} = \frac{\left(2-\sqrt{2}\right)}{2\left(-1+\sqrt{2}\right)} = -\frac{1}{\sqrt{2}} \qquad (12.3.21)$$

The solution for $\omega_1 = \sqrt{\left(2-\sqrt{2}\right)g/L}$ is then $A_2 = \sqrt{2}A_1$, and we can write the first possible solution corresponding to the first normal mode:

$$\theta_1(t) = A_1 e^{i\omega_1 t} \qquad \theta_2(t) = \sqrt{2}\theta_1(t) = \sqrt{2}A_1 e^{i\omega_1 t} \qquad (12.3.22)$$

Since $A_2 = \sqrt{2}A_1$ and $\theta_2(t) = \sqrt{2}\theta_1(t)$, this type of motion corresponds to both masses moving in the same direction and in phase at all times, as shown in Figure 12.8a. The amplitude of the second pendulum must be equal to $\sqrt{2}$ times larger than the amplitude of the first pendulum. This is the first normal mode or the symmetric mode of oscillation, and the general motion of the system in this mode can be written as a linear combination of trigonometric functions:

$$\theta_2(t) = \sqrt{2}\theta_1(t)\cos\left(\omega_1 t - \phi_1\right) \qquad (12.3.23)$$

In matrix notation, the first normal mode can be written as:

$$\begin{pmatrix} \theta_1(t) \\ \theta_2(t) \end{pmatrix} = D_1 \begin{pmatrix} 1 \\ \sqrt{2} \end{pmatrix} \cos\left(\omega_1 t - \phi_1\right) \qquad (12.3.24)$$

By working in a similar manner with the second frequency $\omega_2 = \sqrt{\left(2+\sqrt{2}\right)g/L}$, we find:

$$\begin{pmatrix} \theta_1(t) \\ \theta_2(t) \end{pmatrix} = D_2 \begin{pmatrix} -1 \\ \sqrt{2} \end{pmatrix} \cos\left(\omega_2 t - \phi_1\right) \qquad (12.3.25)$$

This is the second normal mode or the antisymmetric mode of oscillation, in which both masses moving in the same direction completely out of phase at all times, as shown in Figure 12.8b. The amplitude of the second pendulum is again $\sqrt{2}$ times larger than the amplitude of the first pendulum.

(a)   (b)

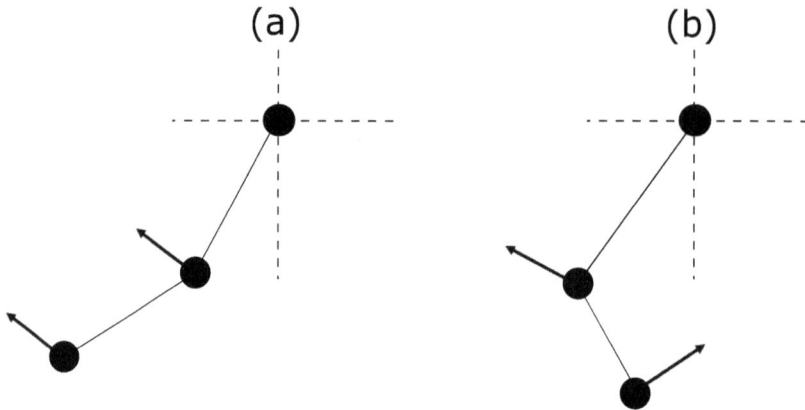

Figure 12.8: Symmetric and antisymmetric normal modes of small angle oscillations for the double pendulum with equal masses and equal lengths. The two natural frequencies are $\omega = \sqrt{\left(2 \pm \sqrt{2}\right) g/L}$, and the amplitudes are related by $A_2 = \pm\sqrt{2}A_1$.

The above results can be obtained using a CAS, as shown in Example 12.7.

Example 12.7: The natural frequencies and normal modes of the double pendulum

Find the natural frequencies of the double pendulum with equal masses and equal lengths.

Solution:

When $m_1 = m_2 = m$ and $L_1 = L_2 = L$, the system of equations (12.3.15) and (12.3.16) can be written in matrix form as:

$$\begin{pmatrix} -2m\left(L\omega^2 - g\right) & -mL\omega^2 \\ -mL\omega^2 & -m\left(L\omega^2 - g\right) \end{pmatrix} \begin{pmatrix} A_1 \\ A_2 \end{pmatrix} = \begin{pmatrix} 0 \\ 0 \end{pmatrix}$$

Once more, we accept only the two positive values of the natural frequency ω. The solutions are

$$\omega_1 = \sqrt{\left(2 - \sqrt{2}\right) g/L}$$

and

$$\omega_2 = \sqrt{\left(2 + \sqrt{2}\right) g/L}$$

Python Code

The code uses the command `Matrix` is Sympy to define the matrix, and `Solve` is used to solve the characteristic equation for the natural frequency ω.

```python
from sympy import Matrix, symbols, solve, det

print('-'*28,'CODE OUTPUT','-'*29,'\n')

k, m, L, omega, g, A1, A2 = symbols('k, m,  L, omega, g, A1, A2')

A = Matrix([[2*m*(-L*omega**2+g),-m*L*omega**2],\
            [-m*L*omega**2,-m*(L*omega**2 - g)]])

omega_vals = solve(det(A),omega)

print('Eigenvalues #1 is: ',omega_vals[0])
print('\nEigenvalues #2 is: ',omega_vals[1])
print('\nEigenvalues #3 is: ',omega_vals[2])
print('\nEigenvalues #4 is: ',omega_vals[3])

---------------------------------- CODE OUTPUT ----------------------------------
Eigenvalues #1 is:   -sqrt(-sqrt(2)*g/L + 2*g/L)

Eigenvalues #2 is:   sqrt(-sqrt(2)*g/L + 2*g/L)

Eigenvalues #3 is:   -sqrt(sqrt(2)*g/L + 2*g/L)

Eigenvalues #4 is:   sqrt(sqrt(2)*g/L + 2*g/L)
```

Mathematica Code

The code uses the commands `Det` and `Solve` to find the four roots of the characteristic equation for the natural frequency ω.

$$A = \begin{pmatrix} 2*m*(-L*\omega\wedge 2 + g) & -m*L*\omega\wedge 2 \\ -m*L*\omega\wedge 2 & -m*(L*\omega\wedge 2 - g) \end{pmatrix};$$

$$\text{Do}\left[\text{Print}\left[\text{Solve}\left[\text{Det}[A] == \begin{pmatrix} 0 \\ 0 \end{pmatrix}, \omega\right][[i]]\right],\right.$$

$$\{i, 1, 4\}]$$

$$\left\{\omega \to -\sqrt{\frac{2g}{L} - \frac{\sqrt{2}g}{L}}\right\}$$

$$\left\{\omega \to \sqrt{\frac{2g}{L} - \frac{\sqrt{2}g}{L}}\right\}$$

$$\left\{\omega \to -\sqrt{\frac{2g}{L} + \frac{\sqrt{2}g}{L}}\right\}$$

$$\left\{\omega \to \sqrt{\frac{2g}{L} + \frac{\sqrt{2}g}{L}}\right\}$$

12.3.3 DOUBLE PENDULUM AS AN EIGENVECTOR/EIGENVALUE PROBLEM

We already saw that it is possible to solve the two-mass, three-spring system as an eigenvalue/eigenvector problem. Let's see how we can do that for the double pendulum problem. One problem we encounter is that the equations of motion (12.3.11) and (12.3.12) have terms containing both $\ddot{\theta}_1$ and $\ddot{\theta}_2$. Let us then first isolate the $\ddot{\theta}_1$ and $\ddot{\theta}_2$ terms into separate equations, so that the eigenvalue problem becomes more clear. By subtracting (12.3.11) and (12.3.12), we obtain the equation for $\ddot{\theta}_1$:

$$m_1 L_1 \ddot{\theta}_1 = -(m_1 + m_2) g \theta_1 + m_2 g \theta_2 \qquad (12.3.26)$$

By substituting $\ddot{\theta}_1$ from this expression in (12.3.11), we obtain the equation for $\ddot{\theta}_2$:

$$m_1 L_2 \ddot{\theta}_2 = -(m_1 + m_2) g \theta_1 + (m_1 + m_2) g \theta_2 \qquad (12.3.27)$$

As previously, we substitute the exponential solutions $\theta_1(t) = A_1 e^{i\omega t}$ and $\theta_2(t) = A_2 e^{i\omega t}$ into these two equations to obtain:

$$-A_1 \omega^2 = -\frac{(m_1 + m_2) g}{m_1 L_1} A_1 + \frac{m_2 g}{m_1 L_1} A_2 \qquad (12.3.28)$$

$$-A_2 \omega^2 = -\frac{(m_1 + m_2) g}{m_1 L_2} A_1 - \frac{g (m_1 + m_2)}{m_1 L_2} A_2 \qquad (12.3.29)$$

In matrix form:

$$\begin{pmatrix} \frac{(m_1 + m_2)g}{m_1 L_1} & -\frac{m_2 g}{m_1 L_1} \\ \frac{(m_1 + m_2)g}{m_1 L_2} & -\frac{g(m_1 + m_2)}{m_1 L_2} \end{pmatrix} \begin{pmatrix} A_1 \\ A_2 \end{pmatrix} = \omega^2 \begin{pmatrix} A_1 \\ A_2 \end{pmatrix} \qquad (12.3.30)$$

This is once more the standard form of the eigenvalue problem, and we can proceed by finding the eigenvalues and eigenvectors of the 2×2 matrix on the left-hand side of this matrix equation. Finding the natural frequencies and the amplitudes of oscillation A_1 and A_2 in Python and Mathematica is left as an exercise, see the End-of-Chapter Problems.

12.4 GENERAL THEORY OF SMALL OSCILLATIONS AND NORMAL CO-ORDINATES

In this section, we develop the general mathematical formalism for the normal modes of coupled oscillators. The first subsection shows how the Lagrangian can be reduced into a quadratic function of the generalized coordinates and generalized velocities of the system. In the second subsection, the equations of motion are reduced into a matrix equation, and the solutions of this matrix equation are shown to be closely related to the concepts of normal modes and normal coordinates.

12.4.1 LAGRANGIAN FOR SMALL OSCILLATIONS AROUND AN EQUILIBRIUM POSITION

In Section 8.2, we introduced the concept of generalized coordinates $(q_1, q_2, \ldots, q_s)$ or q_j, with $j = 1, 2, \ldots, s$ where s represents the number of degrees of freedom in the system. The generalized coordinates describe any coordinate system that completely specifies the state of a system. We also introduced the corresponding generalized velocities $\dot{q}_j$, in order to describe the first derivative of the generalized coordinate with respect to time. These generalized quantities were useful in describing the Euler-Lagrange equations of the system in the compact form:

$$\frac{d}{dt}\left(\frac{\partial \mathcal{L}}{\partial \dot{q}_j}\right) - \frac{\partial \mathcal{L}}{\partial q_j} = 0 \tag{12.4.1}$$

We also considered a system of N particles, denoted by the index $\alpha = 1,\ 2,\ldots,N$, and assumed that the Cartesian coordinates $\mathbf{r}_\alpha$, describing the location of the particle with mass, m_α, are given in terms of the generalized coordinates by:

$$\mathbf{r}_\alpha = \mathbf{r}_\alpha\left(q_1,\ldots,q_n\right), \tag{12.4.2}$$

We then showed that the kinetic energy which is written in Cartesian coordinates as

$$T = \frac{1}{2}\sum_\alpha m_\alpha \dot{\mathbf{r}}_\alpha \cdot \dot{\mathbf{r}}_\alpha \tag{12.4.3}$$

becomes the following expression in generalized coordinates:

$$T = \frac{1}{2}\sum_{j,k} a_{jk}\dot{q}_j\dot{q}_k \tag{12.4.4}$$

where

$$a_{jk} = \sum_\alpha m_\alpha \left(\frac{\partial \mathbf{r}_\alpha}{\partial q_j}\right)\cdot\left(\frac{\partial \mathbf{r}_\alpha}{\partial q_k}\right) \tag{12.4.5}$$

and the coefficients $a_{jk} = a_{jk}\left(q_1,\ldots,q_s\right)$ depend on the coordinates q_j.

Next, we assume that the potential energy of the system depends only on the coordinates q_j:

$$V = V\left(q_1,\ldots,q_s\right) \tag{12.4.6}$$

Under the above assumptions, our Lagrangian has the form:

$$\mathcal{L} = T - V = \frac{1}{2}\sum_{j,k} a_{jk}\dot{q}_j\dot{q}_k - V\left(q_1,\ldots,q_s\right) \tag{12.4.7}$$

Let us now assume that the system performs small oscillations around a point of stable equilibrium $\mathbf{q}_{EQ} = \left(q_1,\ q_2,\ \ldots,q_s\right)_{EQ}$. Without loss of generality, we can redefine our coordinates so that this equilibrium point is the origin of our generalized coordinate system, i.e. ,$\mathbf{q}_{EQ} = 0$. Next, we expand the potential energy around this point:

$$V = V(0) + \sum_{i=1}^{s}\frac{\partial V}{\partial q_i}q_i + \frac{1}{2}\sum_{j,k=1}^{s}\frac{\partial^2 V}{\partial q_j \partial q_k}q_j q_k + \ldots \tag{12.4.8}$$

where all derivatives are evaluated at the equilibrium point located at the origin of our coordinate system, $\mathbf{q}_{EQ} = 0$. At the equilibrium points we have $\partial V/\partial q_i = 0$, and we can further simplify this expression by setting $V(0) = 0$ to obtain:

$$V = \frac{1}{2}\sum_{j,k=1}^{s} K_{jk}q_j q_k \tag{12.4.9}$$

where we defined the constants

$$K_{jk} = K_{kj} = \frac{\partial^2 V}{\partial q_j \partial q_k} \tag{12.4.10}$$

We can also Taylor expand the coefficients a_{jk} in (12.4.5) around the equilibrium point, and keep only the constant term in the expansion, so that the a_{jk} are now constants. By combining (12.4.4) and (12.4.9), we obtain the following expression for the kinetic energy T, for the potential energy V and the Lagrangian $\mathcal{L}$:

Generalized Lagrangian for Small Oscillations about Equilibrium

$$T = \frac{1}{2} \sum_{j,k=1}^{s} a_{jk} \dot{q}_j \dot{q}_k \qquad a_{jk} = \text{constants} \qquad (12.4.11)$$

$$V = \frac{1}{2} \sum_{j,k=1}^{s} K_{jk} q_j q_k \qquad K_{jk} = \text{constants} \qquad (12.4.12)$$

$$\mathcal{L} = T - V = \frac{1}{2} \sum_{j,k=1}^{s} a_{jk} \dot{q}_j \dot{q}_k - \frac{1}{2} \sum_{j,k=1}^{s} K_{jk} q_j q_k \qquad (12.4.13)$$

Example 12.8 shows how to calculate the coefficients a_{jk} and K_{jk} for the double pendulum system we studied in the previous section.

Example 12.8: Finding the generalized Lagrangian for the double-pendulum

Calculate the coefficients a_{jk} and K_{jk} for the double pendulum.

Solution:

The kinetic energy of the double pendulum for small oscillations was found in the previous section to be:

$$T = \frac{1}{2}(m_1 + m_2) L_1^2 \dot{\theta}_1^2 + m_2 L_1 L_2 \dot{\theta}_1 \dot{\theta}_2 + \frac{1}{2} m_2 L_2^2 \dot{\theta}_2^2$$

$$T = \frac{1}{2}(m_1 + m_2) L_1^2 \dot{\theta}_1^2 + \frac{1}{2} m_2 L_1 L_2 \dot{\theta}_1 \dot{\theta}_2 + \frac{1}{2} m_2 L_1 L_2 \dot{\theta}_1 \dot{\theta}_2 + \frac{1}{2} m_2 L_2^2 \dot{\theta}_2^2$$

The coefficients a_{jk} are the coefficients appearing in (12.4.4) for the kinetic energy T. Since our generalized coordinates are $q_1 = \theta_1$ and $q_2 = \theta_2$, then a_{jk} are the coefficients of the $\dot{\theta}_1^2/2$, $\dot{\theta}_2^2/2$, and $(\dot{\theta}_1 \dot{\theta}_2)/2$ terms in this expression for T. Therefore,

$$a_{11} = \text{coefficient of } (\dot{\theta}_1^2)/2 = (m_1 + m_2) L_1^2$$

$$a_{12} = a_{21} = \text{coefficient of } (\dot{\theta}_1 \dot{\theta}_2)/2 = m_2 L_1 L_2$$

$$a_{22} = \text{coefficient of } (\dot{\theta}_2^2)/2 = m_2 L_2^2$$

The potential energy of the double pendulum was found in the previous section to be:

$$V = (m_1 + m_2) g L_1 \left(1 - \frac{\theta_1^2}{2}\right) + m_2 g L_2 \left(1 - \frac{\theta_2^2}{2}\right)$$

The coefficients K_{jk} are the coefficients appearing in (12.4.9) for the potential energy V, for the θ_1^2, θ_2^2, and $\theta_1 \theta_2$ terms. Therefore,

$$K_{11} = \text{coefficient of } (\theta_1^2) = (m_1 + m_2) g L_1$$

$$K_{12} = K_{21} = \text{coefficient of } (\theta_1 \theta_2) = 0$$

$$K_{22} = \text{coefficient of } (\theta_2^2) = m_2 g L_2$$

We can then use the constant coefficients a_{jk} and K_{jk} as two 2×2 matrices. The coefficients a_{jk} are used to create the following matrix $\mathbf{M}$, and the coefficients K_{jk} are used to create the matrix $\mathbf{K}$:

$$\mathbf{M} = \begin{pmatrix} (m_1 + m_2) L_1^2 & m_2 L_1 L_2 \\ m_2 L_1 L_2 & m_2 L_2^2 \end{pmatrix} \quad \mathbf{K} = \begin{pmatrix} (m_1 + m_2) g L_1 & 0 \\ 0 & m_2 g L_2 \end{pmatrix}$$

As we will see next, the matrix $\mathbf{M}$ can be thought of as a generalized "mass matrix" and the matrix $\mathbf{K}$ as a generalized "spring constant" matrix $\mathbf{K}$.

12.4.2 EQUATIONS OF MOTION FOR SMALL OSCILLATIONS AROUND AN EQUILIBRIUM POINT

We can now proceed to derive the general equations of motion, by evaluating the Euler-Lagrange expressions for the coordinate q_j:

$$\frac{\partial \mathcal{L}}{\partial q_j} = \frac{\partial V}{\partial q_j} = \sum_{k=1}^{s} K_{jk} q_k \tag{12.4.14}$$

$$\frac{d}{dt} \left(\frac{\partial \mathcal{L}}{\partial \dot{q}_j} \right) = \frac{d}{dt} \left(\frac{\partial T}{\partial \dot{q}_j} \right) = \frac{d}{dt} \left(\sum_{k=1}^{s} a_{jk} \dot{q}_k \right) = \sum_{k=1}^{n} a_{jk} \ddot{q}_k \tag{12.4.15}$$

The Euler-Lagrange equations for q_j then become:

$$\sum_{k=1}^{s} a_{jk} \ddot{q}_k = -\sum_{k=1}^{s} K_{jk} q_k \tag{12.4.16}$$

We will have one such equation for each coordinate q_j. Finally, we can write the equations of motion in compact form as a single matrix equation:

General Equations of Motion for Small Oscillations around Equilibrium

$$\mathbf{M} \ddot{\mathbf{q}} = -\mathbf{K} \mathbf{q} \tag{12.4.17}$$

$$\begin{pmatrix} a_{11} & a_{12} & \cdots & a_{1s} \\ a_{21} & a_{22} & \cdots & a_{2s} \\ \vdots & \vdots & \ddots & \vdots \\ a_{s1} & a_{s2} & \cdots & a_{ss} \end{pmatrix} \begin{pmatrix} \ddot{q}_1 \\ \ddot{q}_2 \\ \vdots \\ \ddot{q}_s \end{pmatrix} = - \begin{pmatrix} K_{11} & K_{12} & \cdots & K_{1s} \\ K_{21} & K_{22} & \cdots & K_{2s} \\ \vdots & \vdots & \ddots & \vdots \\ K_{s1} & K_{s2} & \cdots & K_{ss} \end{pmatrix} \begin{pmatrix} q_1 \\ q_2 \\ \vdots \\ q_s \end{pmatrix}$$
$$\tag{12.4.18}$$

with the $s \times 1$ column matrix $\mathbf{q}$ denoting the generalized coordinates of the system, $\mathbf{M}$ representing the symmetric $s \times s$ generalized "mass matrix", and $\mathbf{K}$ the symmetric generalized "spring constant" matrix.

$$\mathbf{q} = \begin{pmatrix} q_1 \\ \vdots \\ q_s \end{pmatrix} \quad \mathbf{M} = \begin{pmatrix} a_{11} & \cdots & a_{1s} \\ \vdots & \ddots & \vdots \\ a_{s1} & \cdots & a_{ss} \end{pmatrix} \quad \mathbf{K} = \begin{pmatrix} K_{11} & \cdots & K_{1s} \\ \vdots & \ddots & \vdots \\ K_{s1} & \cdots & K_{ss} \end{pmatrix} \quad (12.4.19)$$

Since we expect oscillatory motion, we try solutions of the form $q_j(t) = A_j e^{i\omega t}$ where A_j are the amplitudes of oscillation and ω is the natural frequency of the system. By substituting into (12.4.17) and canceling the exponential terms as usual, we obtain:

Generalized Equations of Motion for Small Oscillations around Equilibrium

$$\omega^2 \mathbf{M} \mathbf{A} = \mathbf{K} \mathbf{A} \qquad (12.4.20)$$

with the $s \times 1$ column matrix $\mathbf{A}$ (whose elements are the A_j from the trial solution q_j) denoting the unknown amplitude matrix.

Equation (12.4.20) is the generalized matrix form of the familiar equation $m\omega^2 = k$, which describes the oscillation frequency ω for a mass m attached to a spring with spring constant k. This equation can be solved in principle by using any of the Linear Algebra techniques we saw previously in this book.

Example 12.9 shows one method of solving (12.4.20) for the double pendulum.

Example 12.9: Finding the natural frequencies and normal modes of the double pendulum, again

Solve (12.4.20) for the double pendulum with equal masses and equal lengths, and find the natural frequencies and the normal modes.

Solution:

In Example 12.9, we calculated the coefficients a_{jk} and K_{jk} of the matrices $\mathbf{M}$ and $\mathbf{K}$ for the double pendulum. Equation (12.4.20) then becomes:

$$\omega^2 \begin{pmatrix} (m_1 + m_2) L_1^2 & m_2 L_1 L_2 \\ m_2 L_1 L_2 & m_2 L_2^2 \end{pmatrix} \begin{pmatrix} A_1 \\ A_2 \end{pmatrix} = \begin{pmatrix} (m_1 + m_2) g L_1 & 0 \\ 0 & m_2 g L_2 \end{pmatrix} \begin{pmatrix} A_1 \\ A_2 \end{pmatrix}$$

By collecting terms, i.e., writing (12.4.20) as $(\omega^2 \mathbf{M} - \mathbf{K}) \mathbf{A} = 0$, we obtain:

$$\begin{pmatrix} (m_1 + m_2)(L_1^2 \omega^2 - g L_1) & m_2 L_1 L_2 \omega^2 \\ m_2 L_1 L_2 \omega^2 & m_2 (L_2^2 \omega^2 - g L_2) \end{pmatrix} \begin{pmatrix} A_1 \\ A_2 \end{pmatrix} = \begin{pmatrix} 0 \\ 0 \end{pmatrix}$$

Simplifying and setting $L_1 = L_2 = L$ and $m_1 = m_2 = m$:

$$\begin{pmatrix} 2mL(L\omega^2 - g) & mL^2 \omega^2 \\ mL^2 \omega^2 & mL(L\omega^2 - g) \end{pmatrix} \begin{pmatrix} A_1 \\ A_2 \end{pmatrix} = \begin{pmatrix} 0 \\ 0 \end{pmatrix}$$

This, of course, is the same equation we obtained and solved in Example 12.7, and we can reuse the code for Example 12.7 to find ω and $\mathbf{A}$.

In the next subsection, we proceed to show how the above general matrix formalism of the equations of motion leads to the concept of normal coordinates.

12.4.3 NORMAL COORDINATES

Let us consider the general matrix equation (12.4.20), and rewrite it in the form:

$$\left(\mathbf{K} - \omega_j^2 \mathbf{M}\right)\mathbf{A}_j = 0 \quad \text{where} \quad \mathbf{A}_j = \begin{pmatrix} A_1 \\ A_2 \\ \cdots \\ A_s \end{pmatrix}_j \quad (j = 1 \ldots s) \tag{12.4.21}$$

$$\mathbf{K}\mathbf{A}_j = \omega_j^2 \mathbf{M}\mathbf{A}_j \tag{12.4.22}$$

where s is equal to the number of degrees of freedom of the coupled oscillator. Equation (12.4.22) tells us that the column vector $\mathbf{A}_j$ can be characterized as a generalized eigenvector of the real-valued symmetric square matrix $\left(\mathbf{K} - \omega_j^2 \mathbf{M}\right)$.

A general theorem of Linear Algebra for real-valued symmetric square matrices, proves that these s generalized eigenvectors $\mathbf{A}_j$ are linearly independent $(s \times 1)$ column vectors, and that they form a complete set of basis vectors for the space of all $(s \times 1)$ column vectors. Therefore, we can write *any* solution $\mathbf{q}$ of the equations of motion (12.4.17) as a linear combination of these eigenvectors:

$$\mathbf{q} = \sum_{j=1}^{s} c_j(t)\mathbf{A}_j \tag{12.4.23}$$

where $c_j(t)$ are time-dependent coefficients. Our goal is to show that these coefficients $c_j(t)$ oscillate independently of each other, i.e., that each coefficient corresponds to a different frequency ω_j.

Substituting (12.4.23) into (12.4.17) and using (12.4.22), we obtain:

$$\mathbf{M}\ddot{\mathbf{q}} = -\mathbf{K}\mathbf{q} \tag{12.4.24}$$

$$\sum \ddot{c}_j(t)\mathbf{M}\mathbf{A}_j = -\sum c_j(t)\mathbf{K}\mathbf{A}_j \tag{12.4.25}$$

$$\sum \ddot{c}_j(t)\mathbf{M}\mathbf{A}_j = -\sum c_j(t)\omega_j^2\mathbf{M}\mathbf{A}_j \tag{12.4.26}$$

Since the vectors are linearly independent, the last equation can hold only if the coefficients of the vector $\mathbf{M}\mathbf{A}_j$ are equal:

$$\ddot{c}_j(t) = -c_j(t)\omega_j^2 \tag{12.4.27}$$

The solutions of these s equations are the sinusoidal functions $c_j(t)$ given by:

$$c_j(t) = b_j \sin\left(\omega_j t - \phi_j\right) \tag{12.4.28}$$

where ϕ_j is a phase angle, and b_j is the amplitude; both parameters (ϕ_j, b_j) are determined by the initial conditions of the system. This equation establishes that each coefficient $c_j(t)$ oscillates with a different frequency ω_j. By substituting (12.4.28) into (12.4.23), we obtain the desired general result:

$$\mathbf{q} = \sum_{j=1}^{s} b_j \sin\left(\omega_j t - \phi_j\right)\mathbf{A}_j \tag{12.4.29}$$

In words, this equation tells that any solution $\mathbf{q}$ of the equations of motion can be written as the linear combination of s independent oscillating terms, each of which has its own characteristic natural frequency ω_j. These oscillating terms are known as the *normal coordinates* $c_j(t)$ of the system. Hence all motions of a linear coupled oscillator can be expressed as a linear combination of normal coordinates.

The material in this section is a formal way of demonstrating the result of Example 12.4. In that example we found that the analytical solution for equal masses and equal spring constants, could be written as the sum of two terms, $\cos\left(\sqrt{k/m}t\right)$ and $\cos\left(\sqrt{3k/m}t\right)$. For example, the solution $x_1(t)$ could be written in the form (12.4.29) by writing $\omega_1 = \sqrt{k/m}$ and $\omega_2 = \sqrt{3k/m}$ as the frequencies, $\phi_1 = \phi_2 = \pi/2$ as the phases, and

$$\mathbf{A}_1 = \begin{pmatrix} a/2 \\ 0 \end{pmatrix} \mathbf{A}_2 = \begin{pmatrix} 0 \\ -a/2 \end{pmatrix} \tag{12.4.30}$$

as the amplitudes.

12.5 CHAPTER SUMMARY

Systems of coupled harmonic oscillators can be analyzed by using either the Lagrangian approach, or alternatively using Newton's second law. The simplest form of these systems consists of two masses connected by springs to each other, with the Lagrangian:

$$\mathcal{L} = T - V = \frac{1}{2}m_1\dot{x_1}^2 + \frac{1}{2}m_2\dot{x_2}^2 - \frac{1}{2}k_1x_1^2 - \frac{1}{2}k_3x_2^2 - \frac{1}{2}k_2(x_1 - x_2)^2$$

This leads to the system of coupled differential equations:

$$m_1\ddot{x}_1 = -k_1x_1 - k_2(x_1 - x_2)$$

$$m_2\ddot{x}_2 = -k_2(x_2 - x_1) - k_3x_2$$

The solution $x_1(t)$ and $x_2(t)$ of such systems are in general complex, and it is not possible to give a simple physical description of the motion of the masses. However, in the special case of equal masses and equal spring constants, we obtain the analytical solution by using the standard matrix techniques of Linear Algebra.

The first possible solution for the motion of the two masses corresponds to the natural frequency $\omega_1 = \sqrt{k/m}$, and both masses move in the same direction and in phase at all times. This type of motion is known as the first normal mode or the *symmetric* mode of oscillation, and can be written as:

$$\begin{pmatrix} x_1(t) \\ x_2(t) \end{pmatrix} = D_1 \begin{pmatrix} 1 \\ 1 \end{pmatrix} \cos(\omega_1 t - \phi_1)$$

The second possible solution for the motion of the two masses corresponds to the natural frequency $\omega_2 = \sqrt{3k/m}$, and in this type of motion the two masses moves in opposite directions, while the center of mass remains stationary. This type of motion is known as the second normal mode or the *antisymmetric* mode of oscillation, and can be written as:

$$\begin{pmatrix} x_1(t) \\ x_2(t) \end{pmatrix} = D_2 \begin{pmatrix} 1 \\ -1 \end{pmatrix} \cos(\omega_2 t - \phi_2)$$

In general, the motion of the system will be a linear combination of the two possible normal modes, corresponding to the frequencies $\omega_1 = \sqrt{k/m}$ and $\omega_2 = \sqrt{3k/m}$.

This system can also be reduced into an eigenvalue problem, by substituting a trial solution of the form $x_1(t) = A_1 e^{i\omega t}$ and $x_2(t) = A_2 e^{i\omega t}$ in the system of coupled equations.

In a second well-known example of coupled harmonic oscillators, we examined the double pendulum, which also has two normal modes. The frequencies of the symmetric and antisymmetric modes of oscillation are $\omega_1 = \sqrt{\left(2 - \sqrt{2}\right) \frac{g}{L}}$ and $\omega_2 = \sqrt{\left(2 + \sqrt{2}\right) \frac{g}{L}}$.

In this chapter we also looked at the general mathematical formalism for the normal modes of coupled oscillators. Here we use generalized coordinates q_j ($j = 1, 2, \ldots, s$) and the corresponding generalized velocities $\dot{q}_j$, where s is the number of degrees of freedom of the system.

The Lagrangian for small oscillations around an equilibrium position can be reduced into a quadratic function of the generalized coordinates and generalized velocities of the system in the form:

$$\mathcal{L} = T - V = \frac{1}{2} \sum_{j,k} a_{jk} \dot{q}_j \dot{q}_k - \frac{1}{2} \sum_{j,k=1}^{n} K_{jk} q_j q_k$$

where a_{jk} and K_{jk} are constants.

For such a system, the Euler-Lagrange equations become:

$$\sum_{k=1}^{n} a_{jk} \ddot{q}_k = - \sum_{k=1}^{n} K_{jk} q_k$$

This system of coupled differential equations can be solved using standard techniques of Linear Algebra.

Any solution $\mathbf{q} = q_j (j = 1, 2, \ldots, s)$ of these equations of motion can be written as the linear combination of s independent oscillating terms, each of which has its own characteristic natural frequency ω_j:

$$\mathbf{q} = \sum_{j=1}^{s} b_j \sin\left(\omega_j t - \phi_j\right) \boldsymbol{A}_j$$

These oscillating terms are known as the *normal coordinates* of the system.

12.6 END−OF−CHAPTER PROBLEMS

The symbol ⌨ indicates a problem which requires some computer assistance, in the form of graphics, numerical computation, or symbolic evaluation.

Sections 12.1-12.2: Coupled Oscillations of Spring-Mass Systems -Normal Mode Analysis

1. Two masses m_1 and m_2 are connected to each other with a spring of spring constant k, and they are placed on a horizontal frictionless table. Show that the natural frequency of the system is $\omega = \sqrt{k/\mu}$, where $\mu = m_1 m_2/(m_1 + m_2)$ is the reduced mass of the system.

2. Consider the two-mass, three-spring system studied in this chapter, with different masses m_1, m_2 and identical spring constants $k_1 = k_2 = k_3$.

a. Transform the Lagrangian by introducing two new variables $q_1 = x_1 + x_2$ and $q_2 = x_1 - x_2$. Obtain new equations of motion by using the Euler-Lagrange equations for this transformed Lagrangian.

b. Solve the new equations and obtain $q_1(t)$ and $q_2(t)$, and the natural frequencies of the system. This is an example where by using a new set of coordinates, we can uncouple the differential equations of motion. These new coordinates $q_1(t)$, $q_2(t)$ represent the *normal coordinate*s of the system.

3. In the two-mass, three-spring system studied in this chapter, obtain the analytical solution $x_1(t)$ and $x_2(t)$, by using the following initial conditions: the first mass is at rest in the equilibrium position, and the second mass is moved a distance a from equilibrium, and released from rest.

4. Obtain an approximate analytical solution for the case of two weakly coupled oscillators described in Example 12.4. This system is weakly coupled if the spring constant k_2 of the middle spring is much smaller than the spring constants of the two end springs ($k_2 << k_1 = k_3$). Use $m_1 = m_2$ and the following initial conditions: the first mass is at rest in the equilibrium position, and the second mass is moved a distance a from equilibrium, and released from rest. Find the frequency ω at which energy is transferred back and forth between the two coupled oscillators.

5. ⌨ Consider two equal masses m attached to one wall and to each other with springs of spring constant k, as shown in Figure 12.9. Show that the natural frequencies of the system are $\omega = \frac{\sqrt{5} \pm 1}{2} \sqrt{k/m}$.

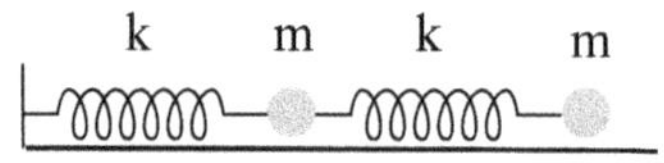

Figure 12.9: Two equal masses attached to each other and to a wall by springs, Problem 12.5.

6. ⌨ Consider the three different situations shown in Figure 12.10, where three identical masses m are attached to springs with the same spring constant k . Write the equations of motion for these three cases.

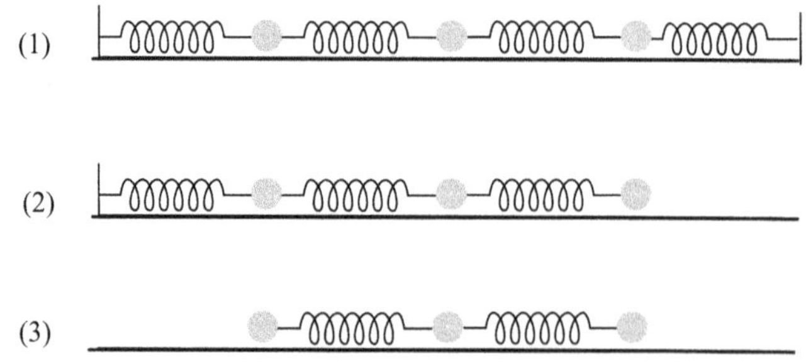

Figure 12.10: Three identical masses connected by identical springs, Problem 12.6.

7. 🖥 Three equal masses m are connected to each other with identical springs of constant k, as shown in Figure 12.11. Find and describe the normal modes of oscillation (assume no friction), by using a CAS.

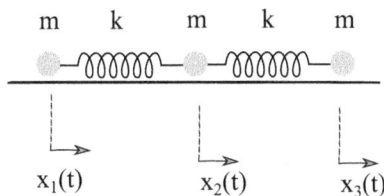

Figure 12.11: Three identical masses connected by two identical springs, Problem 12.7.

8. 🖥 Three equal masses m are connected with springs of the same spring constant k. The masses are connected to each other and to two end walls, as shown in Figure 12.12. Calculate the frequencies of the normal-modes of oscillations of this system and describe the corresponding motion of the three masses in each normal mode.

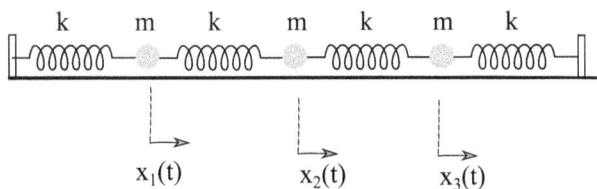

Figure 12.12: Three identical masses connected to four identical springs, Problem 12.8.

9. Find the effective spring constants and natural frequencies for each of the oscillating systems shown in Figure 12.13 below. In Figure 12.13a, two springs with spring constants k_1 and k_2 are attached in parallel to a mass m and to one wall. In Figure 12.13b, two springs with spring constants k_1 and k_2 attached in series to a mass m and to one wall.

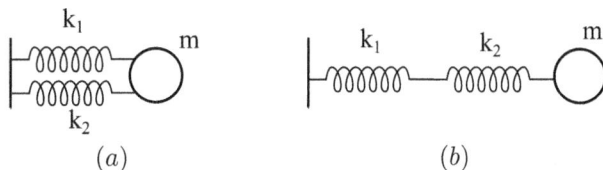

Figure 12.13: A mass connected to a support by two springs, Problem 12.9.

10. 🖥 Two identical masses are connected with identical springs with spring constant, k, and are restricted to move on a circle as shown in Figure 12.14. Find expressions for the natural frequencies and describe the motion of the system.

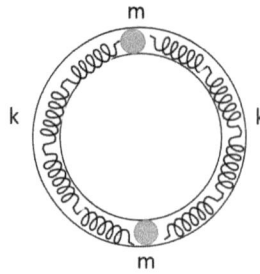

Figure 12.14: Two masses connected along a circle, Problem 12.10.

11. ⌨ Repeat Problem 10, with three identical masses connected with identical springs with spring constant k. The equilibrium position is when the masses are equally spaced.

12. ⌨ Four identical masses are connected by four identical springs, and are constrained to move on a frictionless circle of radius R. The equilibrium position is when the masses are equally spaced. How many normal modes of small oscillations are there and what are the frequencies of small oscillations?

13. ⌨ Three bodies of equal mass m are connected with springs of constant, k, and are placed in an equilateral shape as shown in Figure 12.15. The masses are constrained to move on the xy-plane. Construct the Lagrangian of this system and find the natural frequencies.

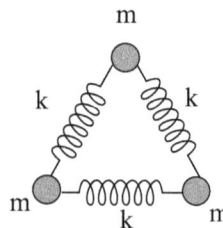

Figure 12.15: Three masses connected in a triangle, Problem 12.13.

14. ⌨ A simple classical model of the CO_2 molecule would be a linear structure of three masses, with the electrical forces between the ions represented by two identical springs of equilibrium length L and force constant k. Assume that only motion along the original equilibrium line is possible, i.e., ignore rotations. Let m be the mass of each oxygen and M be the mass of the carbon atom located between the oxygen atoms.

   a. How many vibrational degrees of freedom does this system have?

   b. Find the normal modes and calculate the natural frequencies.

15. 🖳 Two equal masses m are suspended from a ceiling by springs with constants k, as shown in Figure 12.16.

    a. Find the equations of motion of the two masses by using a coordinate system (y_1, y_2) as measured from the ceiling.

    b. Transform these equations of motion by making a substitution $y_1 = x_1 + u_1$ and $y_2 = x_2 + u_2$, where u_1 and u_2 are the equilibrium positions of the two masses. Solve the new system of equations and describe the normal modes of oscillations of this system in the vertical direction.

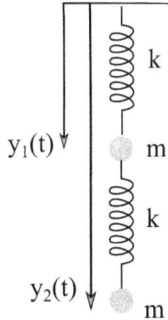

Figure 12.16: Two masses hanging by springs from a vertical support, Problem 12.15.

Section 12.3: The Double Pendulum

16. A double pendulum consists of two equal masses m, and the upper and lower strings have lengths $L_1 = 3L$ and $L_2 = 4L$, respectively. Find the normal modes for small oscillations of this system.

17. 🖳 Two unequal masses m and M (with $M > m$) hang from a support by strings of equal lengths l, with the respective angles with the vertical θ_1 , θ_2. The masses are coupled by a spring of spring constant K and of unstretched length equal to the distance L between the support points, as shown in Figure 12.17. Assume that the distance between the supports $L \ll l$. Find the normal mode frequencies for the small oscillations along the line between the two masses, and describe the motion of the two masses in each mode. Write down the most general solution.

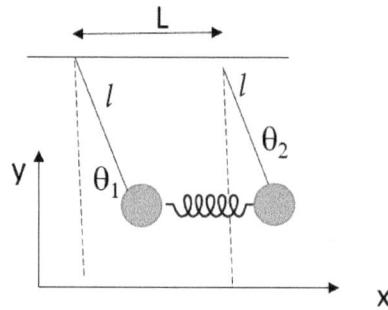

Figure 12.17: Two unequal mass pendula masses m, M connected by a spring, Problem 12.17.

Section 12.4: General Theory of Small Oscillations and Normal Coordinates

18. A mass M moves along the x-axis on a frictionless surface as shown in Figure 12.18. The mass m is connected to M by a massless string.

   a. Write a Lagrangian for this system for a small angle approximation.

   b. Find $x(t)$ and $\theta(t)$.

   c. Find the frequency of small oscillations.

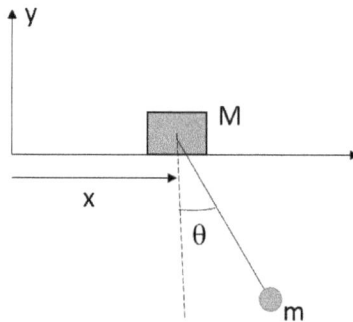

Figure 12.18: Pendulum attached to a sliding mass, Problem 12.18.

19. ⌨ A mass M moves along the x-axis on a frictionless surface, and is attached to a wall with a spring with spring constant k as shown in Figure 12.19. The mass m is connected to M by a massless string.

   a. Write a Lagrangian for this system for a small angle approximation.

   b. Find the frequencies of small oscillations.

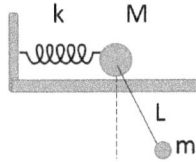

Figure 12.19: Pendulum attached to an oscillating support, Problem 12.19.

20. 🖥 A thin uniform bar of mass m and length $3L/2$ is suspended by a string of length L and negligible mass as shown in Figure 12.20. Find the normal frequencies for small oscillations in the xy-plane.

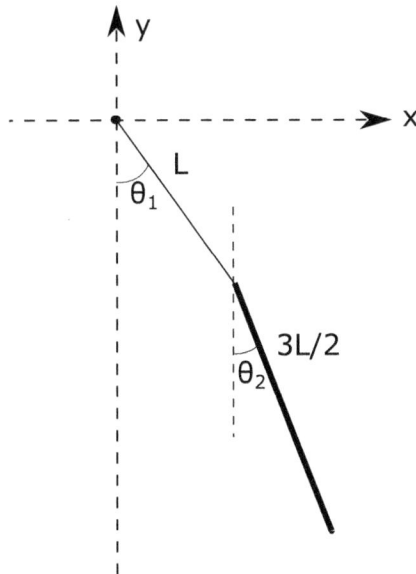

Figure 12.20: A thin uniform bar suspended by a string, Problem 12.20.

21. A mass m moves in a gravitational field g pointing in the z-direction, and on the inside wall of a frictionless axially symmetric vessel given by $z = b(x^2 + y^2)/2$, where a is a constant and z is in the vertical direction, as shown in Figure 12.21. The particle is moving in a circular orbit at height $z = z_0$.

  a. Obtain the angular frequency of the circular motion in terms of m, z_0, b, and g.

  b. Find the total energy and angular momentum in terms of m, z_0, b, and g.

  c. The particle is pushed slightly downwards in the horizontal circular orbit. Obtain the frequency of oscillation about the unperturbed orbit, in the case of small oscillations around the stable circular motion.

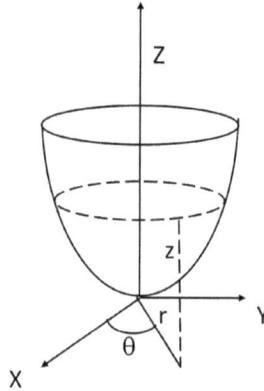

Figure 12.21: An axially symmetric vessel given by $z = b(x^2 + y^2)/2$, Problem 12.21.

22. A particle of mass m is constrained to move on the parabola $z = x^2/a$ in the xz-plane, where a is a constant length, and there is a constant gravitational force acting in the negative z direction.

    a. Define a suitable generalized coordinate system, and write the Lagrangian in terms of these generalized coordinates.

    b. Where is the equilibrium position for the particle?

    c. Write and solve the equation for small oscillations about this equilibrium point.

23. A hemisphere with a non-uniform density is placed on a table as shown in Figure 12.22. The radius of the hemisphere is R, and the center of mass is located a distance a above the table.

    a. Find the potential energy when the hemisphere is slightly misplaced by an angle θ from equilibrium. Under what conditions will the object perform oscillations around a stable equilibrium point?

    b. Find the frequency of small oscillations around this stable equilibrium point, when the moment of inertia of the object around the center of mass is I_{CM}. Assume that the object is only allowed to roll around the pivot contact point.

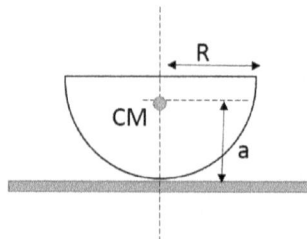

Figure 12.22: A hemisphere placed on a table, Problem 12.23.

24. A uniform sphere of radius a rests in equilibrium inside a uniform hemispherical shell of radius R, as shown in Figure 12.23.

   a. Find the relationship between a and R in order to have a stable equilibrium situation for small oscillations.

   b. Find the natural frequency for small oscillations of the system.

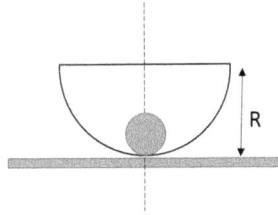

Figure 12.23: A uniform sphere resting inside a hemispherical shell, Problem 12.24.

13 Nonlinear Systems

Nonlinear systems are everywhere! In physics, one of the most common examples of a nonlinear system is an oscillator undergoing large amplitude oscillations. However, nonlinear systems occur in other sciences as well. Examples of nonlinear systems in other fields include predator-prey models, population dynamics, disease propagation, arms-race models, economic models, and the list goes on. In fact, nonlinear systems are the norm, and linear systems are often approximations. While the diversity and difficulty of nonlinear systems may be concerning, there are some basic methods of analysis which are applicable to most of the nonlinear systems you would encounter. In this chapter, we will learn some of those methods. Along the way, we will discover some interesting new behaviors that only nonlinear systems can exhibit, such as bifurcations and chaos. At the end of the chapter, you will have an opportunity to apply what you have learned, by analyzing systems from a variety of natural and social sciences.

13.1 LINEAR VS. NONLINEAR SYSTEMS

So far, most of the systems studied in this book have been linear. In other words, the equations that describe the motion of an object have a linear dependence on the coordinate $x(t)$ and its derivatives. For example, we generally have not seen terms such as $\dot{x}x$, x^2, or $\sin(x)$ in the equations of motion we have studied. Of course, there are exceptions, the quadratic drag force and the simple plane pendulum contained a nonlinear term in their equations of motion. In the case of quadratic drag force, we could solve the equations of motion in closed form. In the case of the simple plane pendulum, we needed to perform a small angle approximation in order to solve the equation of motion. However, it is often the case that we cannot find closed-form solutions to equations of motion with nonlinear terms. Small-quantity approximations, such as the small angle approximation, are not always appropriate for an analysis.

Consider a system that is described by a coordinate $x(t)$. Systems whose behavior are determined by differential equations with only a linear dependence on the dependent variable x and its derivatives are called *linear systems*. If nonlinear terms in $x(t)$ and its derivatives are present in the system's equation of motion, then the system is called a *nonlinear system*. Note this classification does not include the independent variable t. For example, the equation $\ddot{x} + \omega_0^2 x = t^2$ is still considered linear, because all of the terms are linear in the dependent variable x and its derivatives. While the difference between linear and nonlinear systems may seem minor, nonlinearity has serious mathematical consequences.

One of the first consequences of nonlinearity is that the principle of linear superposition no longer holds. Consider the equation of motion for a simple harmonic oscillator, $\ddot{x} + \omega_0^2 x = 0$, studied in Chapter 6. Suppose we find two solutions for this equation, $x_1(t)$ and $x_2(t)$. Then, according to the principle of linear superposition, the general solution to the harmonic oscillator equation is: $c_1 x_1(t) + c_2 x_2(t)$, where c_1 and c_2 are constants as demonstrated by (13.1.1).

$$\begin{aligned} \ddot{x}_1 + \omega_0^2 x &= (c_1 \ddot{x}_1 + c_2 \ddot{x}_2) + \omega_0^2 (c_1 x_1 + c_2 x_2) \\ &= c_1 \left(\ddot{x}_1 + \omega_0^2 x_1 \right) + c_2 \left(\ddot{x}_2 + \omega_0^2 x_2 \right) \\ &= c_1 (0) + c_2 (0) \\ &= 0 \end{aligned} \tag{13.1.1}$$

Now, let us change the differential equation to the form: $\ddot{x} + x^2 = 0$. Suppose we find two solutions to this equation, x_1 and x_2. Is it true that $c_1 x_1(t) + c_2 x_2(t)$ is still a solution? We insert the superposition into the nonlinear differential equation:

$$\begin{aligned} \ddot{x} + x^2 &= (c_1 \ddot{x}_1 + c_2 \ddot{x}_2) + (c_1 x_1 + c_2 x_2)^2 \\ &= c_1 \left(\ddot{x}_1 + x_1^2 \right) + c_2 \left(\ddot{x}_2 + x_2^2 \right) + 2 c_1 c_2 x_1 x_2 \qquad (13.1.2) \\ &= 2 c_1 c_2 x_1 x_2 \end{aligned}$$

Notice that now, we do not get zero. The result is that unlike linear systems, nonlinear systems cannot be broken into parts, solved individually, and then have those solutions combined for a solution to the whole system. Recall the topic of the driven damped harmonic oscillator. If there are multiple sinusoidal drive terms on the right-hand side of the equation, we could solve the differential equation for one right-hand term at a time, then add the individual solutions to get a particular solution. That cannot be done for nonlinear systems. This makes nonlinear systems more difficult to solve in closed-form. In fact, closed-form solutions for nonlinear differential equations often cannot be found.

Many systems from the physical, biological, and social sciences are nonlinear, and linear descriptions of behavior are often approximations. For example, Hooke's Law, $F(x) = -kx$, holds for an oscillator undergoing small amplitude oscillations. Once the amplitude of the oscillations is large enough, then the restoring force is no longer linear, and nonlinear terms need to be included in order for the equation of motion to accurately describe the oscillation. One of the results is that the period of oscillation is no longer independent of the amplitude of oscillation.

As we will see, nonlinear systems can display a much more diverse set of behaviors than linear systems. The limited types of behavior and mathematical simplicity makes linear systems easier to solve, and allows us to predict future states of the system with a high degree of accuracy. Nonlinear systems, however, are often not exactly solvable and display complex behaviors such as bifurcations and chaos, which can make long-term predictions of behavior impossible.

Linear approximations can shed some light on the solution of a nonlinear system. These linear approximations, often done using Taylor expansion, will allow us to use some of the tools for linear differential equations in order to provide some understanding of the nonlinear system.

One of the powerful tools we have for analyzing nonlinear systems is the phase space plot mentioned in Chapter 6. Phase space plots will provide graphical methods of understanding the long-term behavior of a nonlinear system. Recall, that for the damped driven harmonic oscillator, we were mostly interested in the steady-state solution and less interested in the transient behavior. The same will generally be true for nonlinear systems, and phase space plots will allow us to identify the nature of the steady-state solution.

Nonlinear systems appear in many fields besides physics. The mathematical techniques we will learn in this chapter are applicable to systems from a wide range of natural and social sciences. Some of the most exciting developments in science are the applications of "traditional physics techniques" to problems in the biological and social sciences.

13.2 DAMPED HARMONIC OSCILLATOR, REVISITED

In this section, we return to the damped harmonic oscillator, a linear system, to expand upon the concept of phase space diagrams. Recall that in a phase space diagram, we plot velocity vs. position. The behavior of the damped harmonic oscillator is modeled by the equation

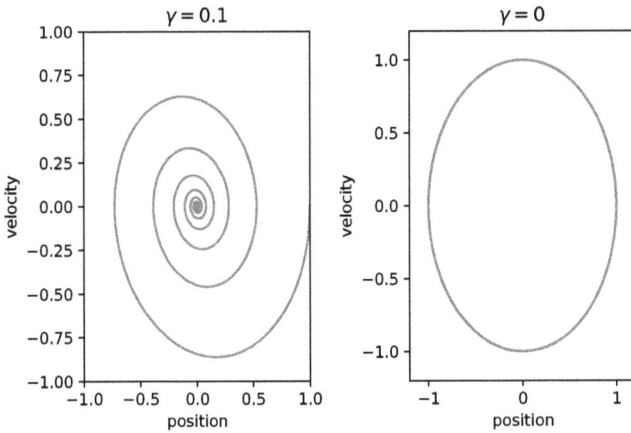

Figure 13.1: The phase space plot for (13.2.1) with $\gamma = 0.1$ (left) and $\gamma = 0$ (right).

$$\ddot{x} + 2\gamma\dot{x} + \omega_0^2 x = 0 \tag{13.2.1}$$

where x is the displacement of the mass m from the spring's equilibrium, γ is the damping parameter, and $\omega_0^2 = k/m$ depends on the spring constant k.

The phase portrait of (13.2.1) is shown in Figure 13.1 where $m = 1$, $\omega_0 = 1$, and $x(0) = 1$, $\dot{x}(0) = 0$. For the undamped case, $\gamma = 0$ and for the damped case $\gamma = 0.1$.

Pretend, for a moment, that we did not know how to solve this problem in closed form, but instead all we had were the phase space plots that were generated numerically; this is often true for nonlinear systems. What can we learn from Figure 13.1 about the behavior of (13.2.1)?

Let us first consider the left graph in Figure 13.1, where $\gamma = 0.1$ and, hence, there is a small amount of damping in the system. To help visualize the behavior of the system, we will continue to think of (13.2.1) as describing a mass on a spring with damping. The path in phase space, called the *phase space trajectory*, or simply *trajectory*, represents the solution to (13.2.1).

To understand the behavior of the system, follow the trajectory in phase space! As we follow the trajectory from its starting point, we see that the position decreases towards zero, while the velocity becomes more negative. Thinking of a mass on a spring, the trajectory represents a mass starting at rest at a position of 1.0 m to the right of equilibrium ($x = 0$), and when released, it moves to the left towards equilibrium, gaining speed. The mass reaches equilibrium where it has its maximum (negative) speed, overshoots equilibrium, and begins to slow down (velocity becomes less negative). The mass continues moving to the left to a distance of about 0.6 meters left of equilibrium and, at that point, the velocity is zero. The mass, however, has reached a turning point and moves to the right (positive velocity) gaining speed as it approaches equilibrium again. The mass reaches its highest speed at equilibrium and begins to slow down. However, notice that the mass does not return to its initial position, instead ending up at a distance of about 0.5 m to the right of equilibrium before turning around. This decaying oscillation continues until the mass is at rest at equilibrium (trajectory ends at the origin).

We can use Figure 13.1 to answer our question about the steady-state of the system (the mass is at rest at the equilibrium position), and we could even discuss the transient behavior, all without having a closed-form solution! The final point, in this case the origin, is a point where both $\dot{x} = 0$ and $\dot{v} = 0$. Points in phase space that satisfy the conditions, $\dot{x} = 0$ and $\dot{v} = 0$, are called *fixed points* of the system. Fixed points correspond to equilibrium points of the system and take on different forms. In the case of $\gamma = 0.1$, the fixed point is called a *stable spiral*, and it represents a decaying oscillation onto the equilibrium position.

We can get the same qualitative information about the system's behavior in the case of $\gamma = 0$, which corresponds to the right graph in Figure 13.1. Again, starting with an initial condition of $(x = 1, v = 0)$, and continuing with the mass-spring analogy, we can follow the trajectory counterclockwise to see that the mass moves towards equilibrium with an increasing speed (to the left), overshoots equilibrium, slows down and stops a distance of 1.0 m to the left of equilibrium. At that point, the mass begins moving to the right with increasing speed until it overshoots equilibrium and returns to its starting position. The closed curve trajectory represents an oscillation, the system returning to its initial state. The origin is still an equilibrium position and is still a fixed point, but the mass does not come to rest there. In this case, the fixed point at the origin is referred to as a *center* (as in center of oscillation).

As we can see, it is possible to get a lot of valuable information from phase space plots. We can ask further questions. For example, do all initial conditions have trajectories that spiral into the origin when $\gamma = 0.1$? The answer, of course, is yes because damping removes energy from the system and causes it to come to rest at equilibrium. However, we could find this out by selecting several initial conditions and numerically solving (13.2.1). We would find that all points have trajectories that spiral into the origin, and therefore the origin is a *global attractor*, a fixed point which "attracts" trajectories starting from all points in the phase space. Likewise, we could repeat the exercise for $\gamma = 0$ and find that the resulting phase space graph consists of nested closed trajectories. In this case, the origin neither attracts nor repels trajectories and is considered to be neutrally stable. Hence, we could answer the question about the behavior of the system for any initial condition, even without solving the system in closed form.

In the above example, we saw two types of fixed points, the stable spiral and a center, representing a stable and neutral equilibrium, respectively. There are other types of fixed points which repel trajectories, corresponding to unstable equilibrium, as well as fixed points that both attract and repel trajectories (also corresponding to unstable equilibrium). In the next section, we will discuss fixed points more generally. We will look into how to find them and how to classify the stability of their equilibrium states. The next section is mathematical in nature and a bit abstract, but we will follow with two examples which will help demonstrate the key ideas of the section.

13.3 FIXED POINTS AND PHASE PORTRAITS

Although Newton's second law motivates us to study second-order ordinary differential equations in classical mechanics, we know that we can rewrite them as a pair of coupled first-order equations. Hence, for this section, we will consider pairs of first-order equations of the form:

Generic First-Order Equations of Motion

$$\left.\begin{aligned} \dot{x} &= f(x,y) \\ \dot{y} &= g(x,y) \end{aligned}\right\}$$

(13.3.1)

where $f(x, y)$ and $g(x, y)$ are functions of the variables x and y. For example, x and y could represent position and velocity for a physical system, or they could represent the numbers of rabbits and foxes in a predator-prey model. We can recast (13.2.1) into a form similar to (13.3.1) by choosing $y = v$ and (13.2.1) becomes,

$$\left. \begin{aligned} \dot{x} &= y \\ \dot{y} &= -2\gamma y - x \end{aligned} \right\} \tag{13.3.2}$$

where $f(x, y) = y$ and $g(x, y) = -2\gamma y - x$. Equations (13.3.1) are sometimes referred to as a two-dimensional system of equations because it depends on two variables, x and y. In addition, (13.3.1) is called the *equations of motion* for our system. Recall that the term, equations of motion, for a physical system describes the actual motion of a particle. However, it is common to refer to (13.3.1) as equations of motion even if we are using them to describe the populations of animals, or any other type of system that is not necessarily moving (but is changing). The equations of motion describe the behavior of a given system regardless of the nature of the system.

While nonlinear systems are diverse and often difficult to analyze, there are some general strategies on how to approach such systems. For example, a general approach that we like to recommend to our students is the following:

The General Approach to Analyzing Nonlinear Systems

1. Find the fixed points (equilibrium points) of the system, points $(x^*,\ y^*)$ such that $f(x^*,\ y^*) = 0$ and $g(x^*,\ y^*) = 0$.

2. Identify the stability of the fixed points.

3. Using initial conditions near the fixed points, plot the *phase portrait*, y vs. x, which contains a representative selection of trajectories.

4. Identify other nonlinear structures in the phase portrait such as limit cycles and strange attractors (see below).

5. Identify bifurcations (see below).

In this section, we will focus on the first two steps, and then, in later sections, we will demonstrate the final two steps by example.

To find the fixed points, $(x^*,\ y^*)$ for the system (13.3.1), we simply solve the equations:

$$\left. \begin{aligned} f(x^*,\ y^*) &= 0 \\ g(x^*,\ y^*) &= 0 \end{aligned} \right\} \tag{13.3.3}$$

for $(x^*,\ y^*)$. In our damped harmonic oscillator example, there is one fixed point $(x^* = 0, y^* = 0)$, found by solving $f(x^*,\ y^*) = y^* = 0$ and $g(x^*,\ y^*) = -2\gamma y^* - x^* = 0$. Systems may have multiple fixed points. Solving for fixed points can be done either by hand, or by using a computer algebra system.

Once we have found the fixed points, we need to find the stability of those fixed points. In other words, are the equilibrium states stable, unstable, or neutrally stable? To find the stability, we must know how the system behaves near each equilibrium state, i.e., we need to find the mathematical form of the equations of motion near the equilibrium point.

Taylor series expansions provide a polynomial approximation of equations near a particular point. Hence, by Taylor expanding f and g, we can get a simple mathematical form for the equations of motion near the equilibrium states. The Taylor-expanded f and g will result in simple linear differential equations that can be solved in closed-form. The closed-form solutions will tell us the stability of the system near the equilibrium point. In summary, in order to get the stability of the fixed point, Taylor expand the equations of motion about the fixed point and solve the resulting differential equations.

To perform the Taylor series expansion, we perturb the system about the fixed point. The perturbation is done via a change of coordinates: $x = x^* + u$ and $y = y^* + v$, where u and v are small compared to x^* and y^*, respectively. Note that v is not being used to denote velocity. Furthermore, the origin of our new coordinates, (u, v), is the fixed point (x^*, y^*). Next, we substitute $x = x^* + u$ and $y = y^* + v$ into the equations of motion (13.3.1) and Taylor expand about the point (x^*, y^*). Note that $\dot{x} = \dot{u}$ because x^* is constant, similarly for $\dot{y}$.

After substitution, the Taylor expansion of f in (13.3.1) becomes:

$$\dot{u} = f\left(x^* + u, y^* + v\right) \tag{13.3.4}$$

$$= f\left(x^*,\ y^*\right) + u \left.\frac{\partial f}{\partial x}\right|_{(x^*, y^*)} + v \left.\frac{\partial f}{\partial y}\right|_{(x^*, y^*)} + \mathcal{O}\left(u^2, v^2\right) \tag{13.3.5}$$

$$= u \left.\frac{\partial f}{\partial x}\right|_{(x^*, y^*)} + v \left.\frac{\partial f}{\partial y}\right|_{(x^*, y^*)} \tag{13.3.6}$$

Notice that some terms appear to be lost from (13.3.5) to (13.3.6). The term $f\left(x^*,\ y^*\right) = 0$ by the definition of fixed points, while the term $\mathcal{O}\left(u^2, v^2\right)$ contains second-order and higher terms. Recall that u and v are small, so u^2 and v^2 are very small and negligible for the dynamics. While there are cases where those nonlinear terms need to be retained, we will focus this text on systems for which those terms can be neglected. The same procedure can be repeated for $g(x, y)$:

$$\dot{v} = u \left.\frac{\partial g}{\partial x}\right|_{(x^*, y^*)} + v \left.\frac{\partial g}{\partial y}\right|_{(x^*, y^*)} \tag{13.3.7}$$

The equations for $\dot{u}$ and $\dot{v}$ tell us how the perturbations from the fixed point (x^*, y^*) evolve in time. By Taylor expanding (13.3.1) about the fixed point, we obtain a system of equations that is linear in u and v. The above process is sometimes called the *linearization* of the equations of motion. We can solve these *linearized* equations in closed form and get an understanding of how the trajectories behave near the fixed point. In other words, we will be able to find the stability of the equilibrium point. It is helpful to cast the linearized equations into matrix form:

Linearized Equations of Motion

$$\begin{pmatrix} \dot{u} \\ \dot{v} \end{pmatrix} = \begin{pmatrix} \frac{\partial f}{\partial x} & \frac{\partial f}{\partial y} \\ \frac{\partial g}{\partial x} & \frac{\partial g}{\partial y} \end{pmatrix} \begin{pmatrix} u \\ v \end{pmatrix} \tag{13.3.8}$$

where we have removed the notation for the evaluation of each partial derivative for simplicity in notation.

The 2×2 matrix in (13.3.8) is called the *Jacobian matrix* **A** and, for simplicity, we can recast (13.3.8) in the form,

$$\dot{\mathbf{u}} = \mathbf{A}\mathbf{u} \tag{13.3.9}$$

where $\mathbf{u} = (u, v)$. Equation (13.3.9) represents a system of first-order differential equations which can be solved with the solution $\mathbf{u} = \mathbf{w}e^{\lambda t}$, where $\mathbf{w}$ is a constant vector. Inserting our trial solution into (13.3.9) yields,

$$\lambda \mathbf{w} = \mathbf{A}\mathbf{w} \tag{13.3.10}$$

which is the eigenvector equation for the matrix **A**. Hence, in our trial solution, $\mathbf{u} = \mathbf{w}e^{\lambda t}$, λ is the eigenvalue of A associated with the eigenvector, $\mathbf{w}$. We have previously studied eigenvalues and eigenvectors in Chapters 11 and 12. Here we will study them from a geometric perspective.

Returning to our analysis, we can rewrite (13.3.10) in the form,

$$(\mathbf{A} - \lambda \mathbf{1})\,\mathbf{w} = \mathbf{0} \tag{13.3.11}$$

where $\mathbf{1}$ is the 2×2 identity matrix (with 1's on the diagonal and 0's otherwise) and $\mathbf{0}$ is the zero vector. Linear algebra tells us that if the matrix, $\mathbf{A} - \lambda \mathbf{1}$, is invertible, then the solution to (13.3.11) is $\mathbf{w} = \mathbf{0}$, which is the trivial solution. For a nontrivial solution, we need the matrix $\mathbf{A} - \lambda \mathbf{1}$ to be singular (i.e., not invertible). Linear algebra also tells us that a matrix is singular if its determinant is zero. Hence, in order to get a nontrivial solution, $\det (\mathbf{A} - \lambda \mathbf{1}) = 0$ must be true. The condition $\det (\mathbf{A} - \lambda \mathbf{1}) = 0$ produces a polynomial in λ, called the characteristic polynomial, which can be solved to obtain the eigenvalues of **A**. Because the matrix **A** is a 2×2 matrix, it will produce two eigenvalues and each eigenvalue can be plugged into (13.3.11) to get its associated eigenvector. Using the principle of linear superposition, the solution to (13.3.9) is therefore,

$$\mathbf{u} = c_1 \mathbf{w}_1 e^{\lambda_1 t} + c_2 \mathbf{w}_2 e^{\lambda_2 t} \tag{13.3.12}$$

where λ_i is the eigenvalue of **A** associated with the eigenvector $\mathbf{w}_i$, and c_i are constants determined by initial conditions.

Before moving on with examples, it is important to pause and try to understand the implications of (13.3.10) and (13.3.12). These next three paragraphs are abstract, but worth the effort needed in understanding them. When a vector is multiplied by a matrix, the result is a vector. Consider the matrix equation, $\mathbf{M}\mathbf{v}_1 = \mathbf{v}_2$ where **M** is a $N \times N$ matrix and $\mathbf{v}_1$ and $\mathbf{v}_2$ are N-dimensional vectors ($N \times 1$ matrices). We can think of the equation $\mathbf{M}\mathbf{v}_1 = \mathbf{v}_2$ as saying that when multiplied by the matrix **M**, the vector $\mathbf{v}_1$ is transformed into the vector $\mathbf{v}_2$. The new vector, $\mathbf{v}_2$, may have a different magnitude and direction compared to $\mathbf{v}_1$. The point here is that matrices change vectors. Now, let us look at (13.3.10). Notice that in (13.3.10), the vector $\mathbf{w}$, when multiplied by the matrix **A**, essentially does not change direction. The length of $\mathbf{w}$ is changed by a factor of λ, but the new vector $\lambda \mathbf{w}$ still lies along the same line as $\mathbf{w}$ (if $\lambda < 0$, the new vector $\lambda \mathbf{w}$ now points in the opposite direction of $\mathbf{w}$). This makes eigenvectors special, they essentially define directions that are unchanged by the transformation **A**.

Why do we care about how matrices change vectors? Recall that the proposed solution to (13.3.9) is $\mathbf{u} = \mathbf{w}e^{\lambda t}$, and $\mathbf{u}$ is a location in the phase space near the fixed point. The magnitude of $\mathbf{u}$ grows or decays exponentially in time depending on the sign of λ. The geometric interpretation of eigenvalues and eigenvectors tells us that a solution that starts near the fixed point along the direction of $\mathbf{w}$ will either move towards the fixed point along the direction of $\mathbf{w}$ if $\lambda < 0$, or away from the fixed point along the direction of $\mathbf{w}$ if $\lambda > 0$. At the end of this section, we will briefly address the case of $\lambda = 0$.

The sign of the eigenvalue λ gives the stability of the fixed point along the direction of $\mathbf{w}$. Because we are working with problems in two dimensions, (13.3.12) tells us that there are two directions of stability for the fixed point. One direction is along $\mathbf{w}_1$ and the other is along $\mathbf{w}_2$, these directions are eigenvectors of A. Each direction has its own stability determined by λ_1 and λ_2, respectively. Hence, if $\lambda_1 < 0$ and $\lambda_2 > 0$, then a point $\mathbf{u}$ starting near the fixed point will initially move towards the fixed point along the direction of $\mathbf{w}_1$ but, as time progresses, it will eventually start to move away from the fixed point along the direction of $\mathbf{w}_2$. The fixed point in this case represents an unstable equilibrium since perturbations from the equilibrium (the fixed point) grow exponentially as $t \to \infty$.

In the next two examples, we will demonstrate the procedure of finding and classifying fixed points using the outline developed above. Example 13.1 examines the damped harmonic motion and demonstrates how to find and classify fixed points analytically.

Example 13.1: The damped harmonic oscillator, revisited, again

Consider the equations of motion for the damped harmonic oscillator with $m = 1$:

$$\dot{x} = y$$

$$\dot{y} = -2\gamma y - \omega_0^2 x$$

Find the Jacobian matrix and compute its eigenvalues and its eigenvectors.

Solution:

We have already noted that $f = y$ and $g = -2\gamma y - \omega_0^2 x$ and that the fixed point is located at the origin. We can use (13.3.8) to get the Jacobian,

$$A = \begin{pmatrix} \frac{\partial f}{\partial x} & \frac{\partial f}{\partial y} \\ \frac{\partial g}{\partial x} & \frac{\partial g}{\partial y} \end{pmatrix} \Bigg|_{(0,0)} = \begin{pmatrix} 0 & 1 \\ -\omega_0^2 & -2\gamma \end{pmatrix}$$

Next, we begin the process of computing the eigenvalues,

$$\det(A - \lambda\mathbf{1}) = \det\left(\begin{pmatrix} 0 & 1 \\ -\omega_0^2 & -2\gamma \end{pmatrix} - \lambda \begin{pmatrix} 1 & 0 \\ 0 & 1 \end{pmatrix} \right) = \begin{vmatrix} -\lambda & 1 \\ -\omega_0^2 & -2\gamma - \lambda \end{vmatrix} = 0$$

Computing the determinant gives:

$$-\lambda(-2\gamma - \lambda) + \omega_0^2 = \lambda^2 + 2\gamma\lambda + \omega_0^2 = 0$$

which is solved for λ:

$$\lambda = -\gamma \pm \sqrt{\gamma^2 - \omega_0^2}$$

Does this look familiar? It should! Next, we will use (13.3.11) to find the eigenvector associated with $\lambda = -\gamma + \sqrt{\gamma^2 - \omega_0^2}$, (we will let the reader do the other one):

$$(A - \lambda\mathbf{1})\mathbf{w} = \left(\begin{pmatrix} 0 & 1 \\ -\omega_0^2 & -2\gamma \end{pmatrix} - \left(-\gamma + \sqrt{\gamma^2 - \omega_0^2}\right) \begin{pmatrix} 1 & 0 \\ 0 & 1 \end{pmatrix} \right) \cdot \begin{pmatrix} w_1 \\ w_2 \end{pmatrix} = \begin{pmatrix} 0 \\ 0 \end{pmatrix}$$

which simplifies to:

$$\begin{pmatrix} \gamma - \sqrt{\gamma^2 - \omega_0^2} & 1 \\ -\omega_0^2 & -\gamma - \sqrt{\gamma^2 - \omega_0^2} \end{pmatrix} \cdot \begin{pmatrix} w_1 \\ w_2 \end{pmatrix} = \begin{pmatrix} 0 \\ 0 \end{pmatrix}$$

or

$$\left(\gamma - \sqrt{\gamma^2 - \omega_0^2}\right) w_1 + w_2 = 0$$

$$-\omega_0^2 w_1 + \left(-\gamma - \sqrt{\gamma^2 - \omega_0^2}\right) w_2 = 0$$

resulting in

$$\begin{pmatrix} w_1 \\ w_2 \end{pmatrix} = \begin{pmatrix} -\frac{1}{\omega_0^2}\left(\gamma + \sqrt{\gamma^2 - \omega_0^2}\right) \\ 1 \end{pmatrix}$$

What does all of this say about the stability of the equilibrium at the origin? We will return to that later in this section. For now, this example is intended to simply demonstrate the procedure outlined so far in this section.

Example 13.2 demonstrates how to use Python and Mathematica to find and classify fixed points. The system in Example 13.2 does not describe a particular physical system. However, it is a simple system of equations that allows for a direct demonstration of the procedure and is more easily extended to a discussion of stability than Example 13.1.

Example 13.2: A system of two first-order ordinary differential equations
Consider the following system of ordinary differential equations:

$$\dot{x} = y(x + 1)$$
$$\dot{y} = x(y + 1)$$

Using Mathematica and Python, find the fixed points and solve the corresponding linearized system of equations.

Solution:
The code appears below. We find there are two fixed points, the origin and the point $(-1, -1)$. After normalizing the eigenvectors, the solution near the origin is

$$\begin{pmatrix} u \\ v \end{pmatrix} = c_1 \begin{pmatrix} 1/\sqrt{2} \\ 1/\sqrt{2} \end{pmatrix} e^t + c_2 \begin{pmatrix} -1/\sqrt{2} \\ 1/\sqrt{2} \end{pmatrix} e^{-t}$$

where c_1 and c_2 are constants that cannot be found without initial conditions for our linearized system. We will leave it to the reader to normalize the solution near the point $(-1, -1)$.

Python Code
After defining **f** and **g**, we solved for the fixed points. Note that Python put the origin as the last fixed point. We then used **Matrix** to create the Jacobian matrix A and found its eigenvalues and eigenvectors. We used list comprehensions in **eigenvects1** and **eigenvects2** so that the **print** command would print the eigenvectors as lists.

```
from sympy import symbols, diff, solve, Matrix

print('-'*28,'CODE OUTPUT','-'*29,'\n')

x, y = symbols('x,y')

f = y*(x + 1)
g = x*(y + 1)

#find the fixed points
fps = solve([f,g],[x,y])

fps1 = fps[1] #identify the first fixed point as origin
fps2 = fps[0]

print('The fixed points are, ', fps)

#compute the Jacobian matrix
A = Matrix([[diff(f,x), diff(f,y)],\
    [diff(g,x),diff(g,y)]])

A_fps1 = (A.subs([(x,fps1[0]),(y,fps1[1])]))
A_fps2 = (A.subs([(x,fps2[0]),(y,fps2[1])]))

eigenvals1 = list(A_fps1.eigenvals().keys())
eigenvals2 = list(A_fps2.eigenvals().keys())

eigenvects1 = [list(ev[2][0]) for ev in A_fps1.eigenvects()]
eigenvects2 = [list(ev) for ev in A_fps2.eigenvects()[0][2]]

print('The eigenvalues for the origin are:', eigenvals1)
print('The eigenvalues for (-1,-1) are:', eigenvals2)

print('The eigenvectors for the origin are ', eigenvects1)
print('The eigenvectors for the (-1,-1) are ', eigenvects2)


---------------------------- CODE OUTPUT ----------------------------
The fixed points are,  [(-1, -1), (0, 0)]
The eigenvalues for the origin are: [-1, 1]
The eigenvalues for (-1,-1) are: [-1]
The eigenvectors for the origin are  [[-1, 1], [1, 1]]
The eigenvectors for the (-1,-1) are  [[1, 0], [0, 1]]
```

Mathematica Code
After defining f and g, we solved for the fixed points. Note that Mathematica put the origin as the last fixed point. We then created the Jacobian matrix A and found its eigenvalues and eigenvectors using the `Eigenvalues` and `Eigenvectors` commands.

```
f = y * (x + 1);

g = x * (y + 1);

(* Find the fixed points *)

fps = Solve[{f == 0, g == 0}, {x, y}]

{{x → -1, y → -1}, {x → 0, y → 0}}

(* Compute the Jacobian matrix *)

A = {{D[f, x], D[f, y]}, {D[g, x], D[g, y]}};

(* We will label the origin as fps1 *)

Afps1 = A /. fps[[2]];

Afps2 = A /. fps[[1]];

Print["The eigenvalues of the origin are ", Eigenvalues[Afps1]]

The eigenvalues of the origin are {-1, 1}

Print["The eigenvalues of (-1,-1) are ", Eigenvalues[Afps2]]

The eigenvalues of (-1,1) are {-1, -1}

Print["The eigenvectors of the origin are ", Eigenvectors[Afps1]]

The eigenvectors of the origin are {{-1, 1}, {1, 1}}

Print["The eigenvectors of (-1,-1) are ", Eigenvectors[Afps2]]

The eigenvectors of (-1,-1) are {{0, 1}, {1, 0}}
```

Now let us examine the solution near the origin from Example 13.2. We see that along the direction of $\mathbf{w}_1 = (1, 1)$, there is an exponential growth away from the origin since $\lambda > 0$. However, along the direction of $\mathbf{w}_2 = (-1, 1)$, there is exponential decay towards the origin, since $\lambda < 0$. This fixed point has two different stabilities which depend on direction. The unstable direction $\mathbf{w}_1$ is tangent to a curve called the *unstable manifold* of the fixed point.

The stable direction $\mathbf{w}_2$ is tangent to a curve called the *stable manifold*. Each vector, $\mathbf{w}_1$ and $\mathbf{w}_2$, is tangent to its respective manifold at the fixed point. The origin in Example 13.2 is called a *saddle point* (or simply, saddle) and is ultimately unstable, as we will see below. To better understand what is going on here, let us examine the phase portrait near the origin in Example 13.3.

Example 13.3: The phase portrait of the system from Example 13.2
Plot the phase portrait near the origin for the system of equations in Example 13.2.

Solution:
The code below creates the phase portraits shown in Figure 13.2. In each case, the solid black lines represent the stable and unstable manifold. The arrowheads on the stream plots show the flow of trajectories near the origin. Trajectories starting near the stable manifolds approach the origin, but they eventually turn toward the unstable manifold and move away from the origin.

Python Code
The stable and unstable manifolds are found by integrating the equations near the origin. The command **streamplot** is used to show the vector field around the origin. Trajectories will flow along the arrows created by the stream plot.

```python
import numpy as np
import matplotlib.pyplot as plt
from scipy.integrate import odeint

times = np.linspace(0,8,100)

def deriv(z,t):
    x, y = z
    dzdt = [y*(x+1), x*(y+1)]
    return dzdt

ic1 = [-0.001,0.0009]
ic2 = [0.001, -0.001]
ic3 = [-0.001, 0.0011]
ic4 = [-0.001, -0.001]

soln1 = odeint(deriv, ic1, -times)
soln2 = odeint(deriv, ic2, -times)
soln3 = odeint(deriv, ic3, times)
soln4 = odeint(deriv, ic4, times)

X, Y = np.meshgrid(np.linspace(-0.1,0.1,100), \
    np.linspace(-0.1,0.1,100))

f_vec = Y*(X + 1)
g_vec = X*(Y + 1)

plt.streamplot(X, Y, f_vec, g_vec)
plt.plot(soln1[:,0],soln1[:,1],'k')
plt.plot(soln2[:,0],soln2[:,1],'k')
plt.plot(soln3[:,0],soln3[:,1],'k')
plt.plot(soln4[:,0],soln4[:,1],'k')
plt.xlabel('x')
plt.ylabel('y')
plt.xlim(-0.1,0.1)
plt.ylim(-0.1,0.1)
plt.show()
```

Mathematica Code
We begin by computing trajectories that will be the stable and unstable manifolds. We create multiple plots using `ParametricPlot`, one for each manifold. We then plot the stream plot using `StreamPlot` where the inputs are $f(x,y)$ and $g(x,y)$. The `Show` command displays all of the graphs as one plot.

```
tmax = 6;

soln1 = NDSolve[{x'[t] == y[t](x[t] + 1), y'[t] == x[t](y[t] + 1), x[0] == -0.001,
y[0] == 0.001}, {x, y}, {t, 0, -tmax}];
soln2 = NDSolve[{x'[t] == y[t](x[t] + 1), y'[t] == x[t](y[t] + 1), x[0] == 0.001,
y[0] == -0.001}, {x, y}, {t, 0, -tmax}];
soln3 = NDSolve[{x'[t] == y[t](x[t] + 1), y'[t] == x[t](y[t] + 1), x[0] == 0.001,
y[0] == 0.0011}, {x, y}, {t, 0, tmax}];
soln4 = NDSolve[{x'[t] == y[t](x[t] + 1), y'[t] == x[t](y[t] + 1), x[0] == -0.001,
y[0] == -0.001}, {x, y}, {t, 0, tmax}];

stable1 = ParametricPlot[{x[t], y[t]}/.soln1, {t, 0, -tmax}, PlotStyle->Black];
stable2 = ParametricPlot[{x[t], y[t]}/.soln2, {t, 0, -tmax}, PlotStyle->Black];
unstable1 = ParametricPlot[{x[t], y[t]}/.soln3, {t, 0, tmax}, PlotStyle->Black];
unstable2 = ParametricPlot[{x[t], y[t]}/.soln4, {t, 0, tmax}, PlotStyle->Black];

stream = StreamPlot[{y(x + 1), x(y + 1)}, {x, -0.1, 0.1}, {y, -0.1, 0.1},
StreamColorFunction->Blue];

Show[stream, stable1, stable2, unstable1, unstable2, Frame->True,
Axes -> False, BaseStyle -> {Bold, FontSize -> 16}, FrameLabel -> {x, y}]
```

As we can see in the above example, the eigenvalues give the stability of the fixed point. There are three simple types of fixed points associated with real eigenvalues in two-dimensional systems: stable node, unstable node, and a saddle.

- The *stable node* has two negative eigenvalues and trajectories that start near the stable node and approach it exponentially. A stable node corresponds to a stable equilibrium position. For each stable node there is a region in the phase space called its *basin of attraction*. The basin of attraction for a stable node is the set of all initial conditions whose trajectories approach the stable node as $t \to \infty$.

- The *unstable node*, sometimes called a repellor, has two positive eigenvalues. It is called a repellor because trajectories starting near the node move away from the node at an exponential rate. An unstable node corresponds to an unstable equilibrium.

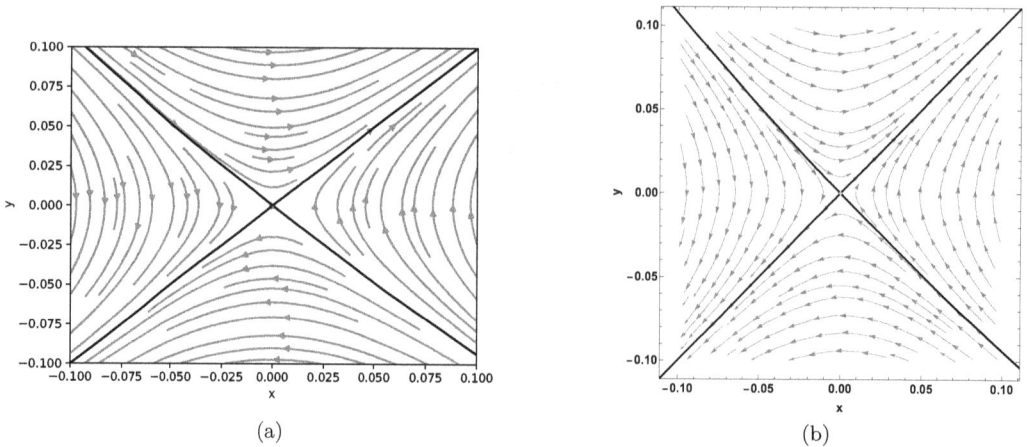

(a)   (b)

Figure 13.2: Phase portrait for Example 13.3 created by Python (a) and Mathematica (b). In both graphs, the X-shaped black lines are the stable and unstable manifold. The blue arrows (in the e-book) show the flow of the trajectories.

- The *saddle*, explored in Examples 13.2 and 13.3, has one positive and one negative eigenvalue. It corresponds to an unstable equilibrium. The stable manifolds of the saddle often serve as borders for the basin of attraction for stable nodes and other stable structures in the phase space (see below).

In addition to the points listed above, there are three types of fixed points with complex eigenvalues. Suppose $\lambda = \alpha + i\beta$, then $e^{\lambda t} = e^{\alpha t}(\cos(\beta t) + i\sin(\beta t))$. In the case of complex eigenvalues we see that there is a combination of exponential growth or decay and oscillation, similar to the damped harmonic oscillator case. The possible fixed points are:

- A *stable spiral* occurs when $Re(\lambda) < 0$, and the trajectory spirals onto the fixed point. A stable spiral appears in the phase portrait of the underdamped harmonic oscillator with $\beta < \omega_0$.

- An *unstable spiral* occurs when $Re(\lambda) > 0$, and the trajectory spirals away from the fixed point. This would correspond to the damped harmonic oscillator with a negative damping parameter.

- A *center* occurs when $Re(\lambda) = 0$ and there is neither exponential growth away from the fixed point, nor is there exponential decay onto the fixed point. The point is neutrally stable and the trajectories circle the center. This corresponds to the undamped simple harmonic oscillator.

What happens when there is a zero eigenvalue? One typically uses numerical methods to analyze the system. Finally, it should be mentioned that there are cases that are called *degenerate*, because they have only one eigenvalue and/or one eigenvector. These lead to other types of fixed points which are beyond the scope of this book. The interested reader is directed to Strogatz [11] for a thorough and readable review of even more types of fixed points. Next, we will use our newfound knowledge of fixed points to analyze the simple plane pendulum.

13.3.1 SIMPLE PLANE PENDULUM, REVISITED

Consider a simple plane pendulum shown below in Figure 13.3. The pendulum consists of a massless rod of length ℓ, bob of mass m, and makes an angle with the vertical θ.

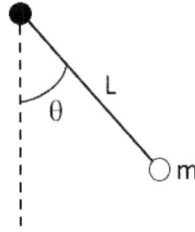

Figure 13.3: Simple plane pendulum.

The equation of motion of the pendulum has been found many times in this book,

$$\ddot{\theta} + \omega_0^2 \sin\theta = 0 \tag{13.3.13}$$

where the natural frequency $\omega_0^2 = g/\ell$. Equation (13.3.13) can be solved in closed form using elliptical integrals; however, those solutions are not very insightful. Let us instead proceed with the analysis developed in Section 13.3, by breaking up (13.3.13) into two first-order equations:

$$\begin{aligned} \dot{\theta} &= y \\ \dot{y} &= -\omega_0^2 \sin\theta \end{aligned} \tag{13.3.14}$$

where the variable y is the angular velocity. Notice that fixed points occur at $(\theta^* = 0, y = 0)$ and $(\theta^* = \pm\pi, y = 0)$; these make sense physically because they correspond to the very bottom and top of the pendulum's path. Based on the way we have defined our coordinate system, $\theta = \pm\pi$ are the same point, the very top of the pendulum's path. Note that $\theta = +\pi$ when the pendulum approaches the top moving counterclockwise, and $\theta = -\pi$ when it approaches the top moving clockwise. The Jacobian of the system is:

$$A = \begin{pmatrix} 0 & 1 \\ -\omega_0^2 \cos\theta & 0 \end{pmatrix} \tag{13.3.15}$$

At the origin, the Jacobian (13.3.15) becomes:

$$\begin{pmatrix} 0 & 1 \\ -\omega_0^2 & 0 \end{pmatrix} \tag{13.3.16}$$

with eigenvalues $\lambda = \pm i\omega_0$. The origin, corresponding to the bottom of the pendulum's path, is a center. The center is representing the oscillation of the pendulum about the stable equilibrium position.

At $(\pm\pi, 0)$, the Jacobian (13.3.15) becomes

$$\begin{pmatrix} 0 & 1 \\ \omega_0^2 & 0 \end{pmatrix} \tag{13.3.17}$$

which has eigenvalues $\lambda = \pm\omega_0$. The fixed point corresponding to the unstable equilibrium is a saddle. Were you expecting an unstable node? The topmost point cannot be an unstable mode because an unstable mode repels all trajectories. A saddle, on the other hand, has a stable direction. We know that the pendulum can approach the top before turning around, and hence, the topmost point must be a saddle. Now that we know the location and stability of the fixed points, we are ready to draw the phase portrait.

Figure 13.4 illustrates the phase portrait for the simple pendulum using $\omega_0 = 1$. Recall that the points $\theta = \pm\pi$ correspond to the same physical location. Hence, the two saddle

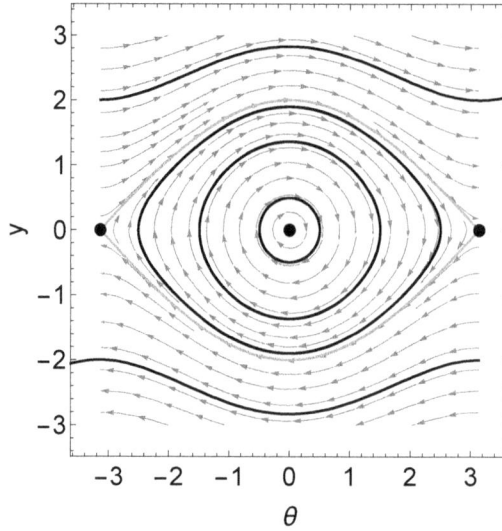

Figure 13.4: Phase portrait for (13.3.13) with $\omega_0 = 1$.

points in the phase portrait are actually the same point. The red line (in the e-book) in Figure 13.4 is both the stable and unstable manifold of the saddle, which is sometimes referred to as a *homoclinic orbit*, a trajectory that connects a saddle point to itself. Physically, the homoclinic orbit represents the pendulum starting with an angular displacement of $\theta = \pi$ (vertical position) and an angular velocity of 0, where the pendulum then swings down in a clockwise direction through the stable equilibrium and returns to the vertical position, coming to a momentary rest and swinging back in a clockwise direction. The homoclinic orbit represents the largest possible amplitude of oscillation of the pendulum.

Initial conditions that start inside the homoclinic orbit correspond to oscillatory behavior. Notice that these inside trajectories (black lines) in Figure 13.4 form closed loops, which we know to mean that the system eventually returns to its initial state after a time τ, the period of oscillation. Notice how the trajectories begin to take the shape of the homoclinic orbit as the amplitude of oscillation gets bigger. The graphs of $\theta(t)$ for each of the trajectories inside the homoclinic orbit are shown in Figure 13.5. As the amplitude increases in size, the period of oscillation increases. Recall that in the simple harmonic oscillator, the period of oscillation was independent of the amplitude of oscillation. In nonlinear oscillators, there is typically a relationship between amplitude and period.

Finally, notice that the trajectories outside of the homoclinic orbit in Figure 13.4 do not form closed loops. Those orbits correspond to the pendulum going over the top of its motion and continually rotating in one direction, either clockwise (trajectories below the homoclinic orbit) or counterclockwise (trajectories above the homoclinic orbit). Hence, we see that the homoclinic orbit separates two distinct behaviors of the pendulum (oscillatory versus continual rotation in one direction). The homoclinic orbit is sometimes called a *separatrix*, because it separates two distinct types of behavior.

13.3.2 DOUBLE-WELL POTENTIAL, REVISITED

Recall the double-well potential,

$$V(x) = -\frac{1}{2}kx^2 + \frac{1}{4}\epsilon x^4 \qquad (13.3.18)$$

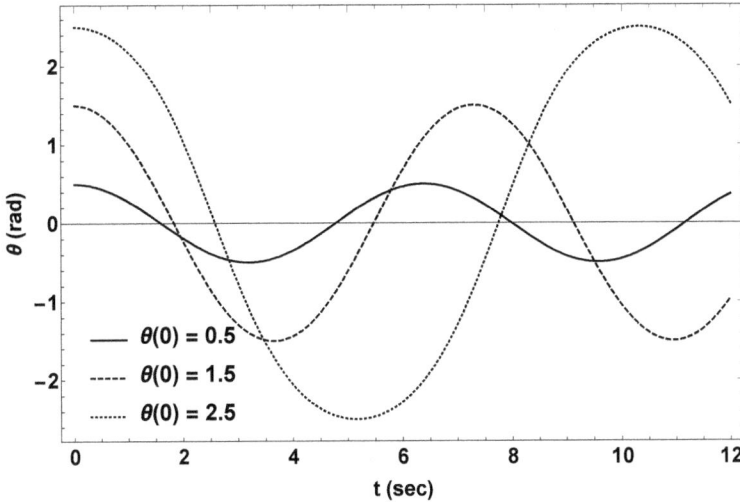

Figure 13.5: Graphs of θ (in radians) versus time (seconds) for each of the oscillatory trajectories shown in Figure 13.4.

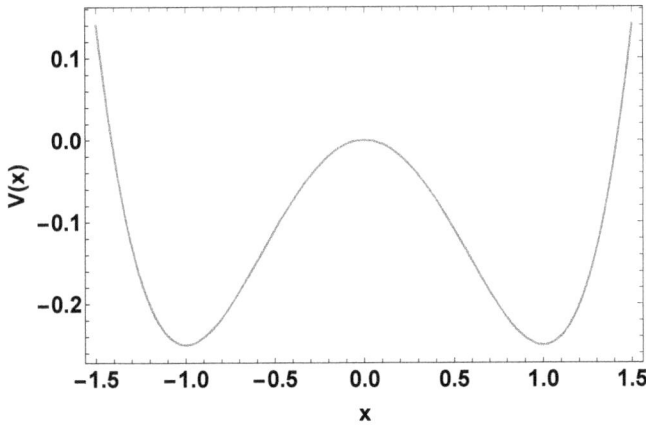

Figure 13.6: Double-well potential (13.3.18) with $k = \epsilon = 1$.

An example of the double-well potential is graphed in Figure 13.6 with $k = \epsilon = 1$. We can see in Figure 13.6 that there are two stable equilibria at $x = \pm 1$ and one unstable equilibrium at $x = 0$. The double-well potential is associated with a force of the following form, $F(x) = kx - \epsilon x^3$. The resulting equation of motion is then,

$$m\ddot{x} = kx - \epsilon x^3 \tag{13.3.19}$$

Next, using $m = k = \epsilon = 1$, we will find the fixed points, identify their stability, and plot the phase portrait. In order to plot the phase portrait, we will need to recast (13.3.19) into a two-dimensional system,

$$\left.\begin{aligned} \dot{x} &= y \\ \dot{y} &= x - x^3 \end{aligned}\right\} \tag{13.3.20}$$

We see from (13.3.20) that fixed points exist at $(0,0)$, $(1,0)$, and $(-1,0)$. The Jacobian matrix is,

$$A = \begin{pmatrix} 0 & 1 \\ 1 - 3x^2 & 0 \end{pmatrix} \tag{13.3.21}$$

which provides eigenvalues of ± 1 for $(0,0)$ and $\pm i\sqrt{2}$ for the points $(1,0)$ and $(-1,0)$. Hence, the origin is a saddle, and the other two points are centers. In Example 13.4, we plot the phase portrait for the double well.

> ### Example 13.4: Double-well potential
> Plot the phase portrait for the double-well potential (13.3.20).
>
> ### Solution:
> Figure 13.7 shows the output of the Mathematica code used to create the phase portrait. Notice the two homoclinic orbits, one in red (in the e-book) and the other in black (in the e-book), which surround the two centers. Inside each homoclinic orbit are closed-loop trajectories corresponding to oscillations about each stable equilibrium point. Outside of the homoclinic orbit are larger closed-loop trajectories which correspond to an oscillation that passes through each equilibrium point.

Python Code
This code follows that of Example 13.3 in structure. Note the variables `right_hc` and `left_hc` are the solutions for the left and right homoclinic orbits, respectively.

```python
import numpy as np
import matplotlib.pyplot as plt
from scipy.integrate import odeint

times = np.linspace(0,20,200)

def deriv(z,t):
    x, y = z
    dzdt = [y, x - x**3]
    return dzdt

ic1 = [0.001,0.001]
ic2 = [-0.001, -0.001]


right_hc = odeint(deriv, ic1, times)
left_hc = odeint(deriv, ic2, times)


X, Y = np.meshgrid(np.linspace(-2,2,100), \
    np.linspace(-2,2,100))

f_vec = Y
g_vec = X - X**3

plt.streamplot(X, Y, f_vec, g_vec)
plt.plot(right_hc[:,0],right_hc[:,1],'r')
plt.plot(left_hc[:,0],left_hc[:,1],'k')
plt.xlabel('x')
plt.ylabel('y')
plt.xlim(-2,2)
plt.ylim(-2,2)
plt.show()
```

Mathematica Code

This code is similar to that of Example 13.3. The variables leftHomoClinicSolution and rightHomoclinicSolution contain the solutions for the left and right homoclinic orbits, respectively.

```
sp = StreamPlot[{y, x − x^3}, {x, −2, 2}, {y, −2, 2}, FrameLabel → {x, y},

BaseStyle → {Bold, FontSize → 16}, StreamColorFunction->Blue];


leftHomoclinicSolution =

NDSolve[{x′[t] == y[t], y′[t] == x[t] − x[t]^3, x[0] == 0.01, y[0] == 0}, {x, y},

{t, 0, 100}];

leftHomoclinicPlot = ParametricPlot[Evaluate[{x[t], y[t]}/.leftHomoclinicSolution],

{t, 0, 100}, BaseStyle → {Bold, FontSize → 22}, PlotStyle → Black];


rightHomoclinicSolution =

NDSolve[{x′[t] == y[t], y′[t] == x[t] − x[t]^3, x[0] == −0.01, y[0] == 0}, {x, y},

{t, 0, 100}];

rightHomoclinicPlot = ParametricPlot[Evaluate[{x[t], y[t]}/.rightHomoclinicSolution],

{t, 0, 100}, BaseStyle → {Bold, FontSize → 22}, PlotStyle → Red];


Show[sp, rightHomoclinicPlot, leftHomoclinicPlot]
```

13.3.3 DAMPED DOUBLE-WELL

As another example of a phase portrait, we will examine a damped double-well potential. The force acting on the particle is,

$$F(x) = kx - \epsilon x^3 - b\dot{x} \tag{13.3.22}$$

where the first two terms of (13.3.22) are from the double-well potential studied in the previous section and the final term is a linear damping term. If we assume the particle has a unit mass and set $k = \epsilon = 1$ and $b = 0.1$, then we can rewrite the equation of motion as,

$$\left.\begin{array}{l} \dot{x} = y \\ \dot{y} = x - x^3 - 0.1y \end{array}\right\} \tag{13.3.23}$$

We can see that (13.3.23) has three fixed points, $(0,0)$ and $(\pm 1, 0)$. The Jacobian matrix of (13.3.23) is,

$$A = \begin{pmatrix} 0 & 1 \\ 1 - 3x^2 & -0.1 \end{pmatrix} \tag{13.3.24}$$

and has eigenvalues of ± 1 at the origin and $-0.05 \pm 1.41i$ for the other two points. Hence, we see that the origin is a saddle and the other two points are a stable spiral. This should

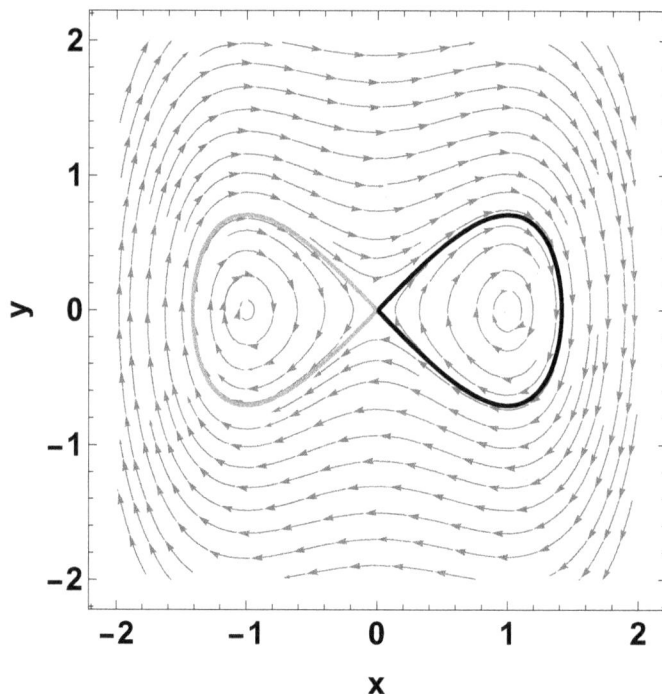

Figure 13.7: Mathematica-generated phase portrait for the double-well potential from Example 13.4.

not be surprising: the damping is causing decaying oscillation onto the equilibria at $(\pm 1, 0)$. To better see this, we will study the phase portrait in Example 13.5.

Example 13.5: Damped double-well potential
Plot the phase portrait for the damped double-well potential (13.3.20).

Solution:
Figure 13.8 shows the Python-generated phase portrait. We did not include the stream plot so that the trajectories would stand out better in the graph. The stable manifolds are black (in e-book), and they connect the saddle to each stable spiral. The stable spirals represent stable equilibria the oscillator will decay onto. The red (in e-book) curves are the unstable manifolds of the saddle. All initial conditions (besides those starting on the stable manifold) will eventually spiral onto one of the stable spirals.

Python Code
This code follows that of Example 13.4 in structure. The natures of the manifolds produced by each solution are described by their variable name.

```python
import numpy as np
import matplotlib.pyplot as plt
from scipy.integrate import odeint

times = np.linspace(0,80,2000)

def deriv(z,t):
    x, y = z
    dzdt = [y, x - x**3 - 0.1*y]
    return dzdt

ic1 = [0.01, 0.0]
ic2 = [-0.01, 0.0]

unstable_1 = odeint(deriv, ic1, times)
unstable_2 = odeint(deriv, ic2, times)
stable_1 = odeint(deriv, ic1, -times)
stable_2 = odeint(deriv, ic2, -times)

plt.plot(unstable_1[:,0],unstable_1[:,1],'k')
plt.plot(unstable_2[:,0],unstable_2[:,1],'k')
plt.plot(stable_1[:,0],stable_1[:,1],'r')
plt.plot(stable_2[:,0],stable_2[:,1],'r')
plt.xlabel('x')
plt.ylabel('y')
plt.xlim(-2,2)
plt.ylim(-2,2)
plt.show()
```

Mathematica Code

This code is similar to that of Example 13.4. We identified the stable and unstable manifolds in the variable names for each plot.

```
soln1 = NDSolve[{x'[t]==y[t], y'[t]==x[t](1 − x[t]^2) − 0.1 ∗ y[t],

x[0] == 0.01, y[0] == 0}, {x, y}, {t, 0, 100}];


unstable1 = ParametricPlot[Evaluate[{x[t], y[t]}/.soln1],

{t, 0, 100}, PlotStyle → Black,

PlotRange->{{−2.1, 2.1}, {−2.1, 2.1}}];


soln2 = NDSolve[{x'[t]==y[t], y'[t]==x[t](1 − x[t]^2) − 0.1 ∗ y[t],

x[0] == 0.01, y[0] == 0}, {x, y}, {t, 0, −40}];


stable1 = ParametricPlot[Evaluate[{x[t], y[t]}/.soln2],

{t, 0, −20}, PlotStyle → Red, PlotRange->{{−2.1, 2.1}, {−2.1, 2.1}}];


soln3 = NDSolve[{x'[t]==y[t], y'[t]==x[t](1 − x[t]^2) − 0.1 ∗ y[t],

x[0] == −0.01, y[0] == 0}, {x, y}, {t, 0, 100}];


unstable2 = ParametricPlot[Evaluate[{x[t], y[t]}/.soln3],

{t, 0, 100}, PlotStyle → Black,

PlotRange->{{−2.1, 2.1}, {−2.1, 2.1}}];


soln4 = NDSolve[{x'[t]==y[t], y'[t]==x[t](1 − x[t]^2) − 0.1 ∗ y[t],

x[0] == −0.01, y[0] == 0}, {x, y}, {t, 0, −40}];


stable2 = ParametricPlot[Evaluate[{x[t], y[t]}/.soln4],

{t, 0, −20}, PlotStyle → Red, PlotRange->{{−2.1, 2.1}, {−2.1, 2.1}}];


Show[unstable1, unstable2, stable1, stable2, FrameLabel → {x, y},

BaseStyle → {FontSize → 16}, Frame->True, Axes->False]
```

Examining the code in Example 13.5 provides a good opportunity to discuss how to find stable manifolds. Notice that the stable manifolds, with variable names `stable_1` and `stable_2` (in Python), are found by integrating (13.3.23) backwards in time. The stable manifold is the set of points that moves towards the saddle as $t \to \infty$. The stable manifold can be difficult to find integrating forward in time. Hence, in order to find the stable

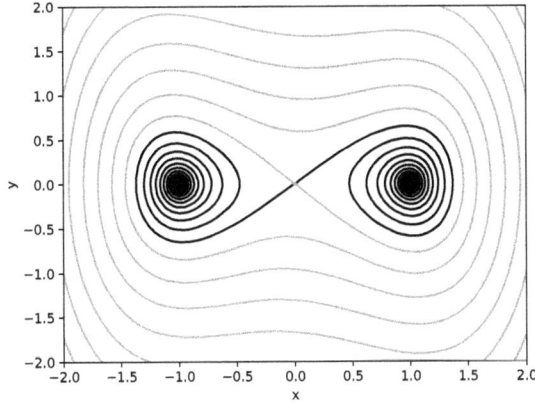

Figure 13.8: Python-generated phase portrait for the double-well potential from Example 13.4.

manifold, one chooses an initial condition near the saddle point and integrates the equations backwards in time. Points on the stable manifold move away from the saddle as $t \to -\infty$. The unstable manifold is the set of points that moves towards the saddle as $t \to -\infty$ (or away as $t \to \infty$), and it is found by solving the equations forward in time, using an initial condition near the saddle.

13.3.4 BIFURCATIONS OF FIXED POINTS

An interesting phenomenon that can occur in nonlinear systems is the so-called *bifurcation* of fixed points. If we think of the parameters of a nonlinear system as being variables, we see that the location and stability of fixed points (and therefore equilibria) can change. Consider again the damped double-well potential,

$$\left.\begin{array}{l} \dot{x} = y \\ \dot{y} = kx - \epsilon x^3 - by \end{array}\right\} \tag{13.3.25}$$

The fixed points in (13.3.25) occur at the origin and at,

$$(x^*, \, y^*) = \left(\pm\sqrt{\frac{k}{\epsilon}}, 0\right) \tag{13.3.26}$$

If we think of k and ϵ as parameters that can be varied, then we see that the location of the fixed points in (13.3.26) can change. For example, as $k \to 0$ the fixed points (13.3.26) collide at the origin. This collision is an example of a bifurcation, and it tells us the condition for which two stable equilibria can exist in the system. In this case, we use the term *stable equilibrium* in the sense of Chapter 5, i.e., as a local minimum in the potential. Furthermore, the eigenvalues of the Jacobian matrix evaluated at (13.3.26) are,

$$\lambda = \frac{1}{2}\left(-b \pm \sqrt{b^2 - 8k}\right) \tag{13.3.27}$$

When $0 < b < \sqrt{8k}$, we see that the fixed points located at (13.3.26) are stable spirals. However, if $b \to 0$, the previously stable fixed points become centers. This is an example of another type of bifurcation where the fixed point changes stability. There are many types of bifurcations between fixed points, and a complete overview of them is beyond the scope of this book. However, the interested reader should consult Refs. [11, 12, 13]. In the next paragraph, we will briefly mention two important bifurcations.

One of the most common bifurcations is the so-called *saddle-node bifurcation*. In a saddle-node bifurcation, a saddle point and a stable node collide and both points disappear from the phase portrait. Physically speaking, the system goes from having two equilibria, one stable and one unstable, to having no equilibrium points (sometimes there is another attracting structure elsewhere in the phase portrait). Another common bifurcation of fixed points is the so-called *transcritical bifurcation*. In a transcritical bifurcation, a saddle and a stable node collide, and pass through each other. However, as the collision occurs, the fixed points exchange stabilities, so the saddle becomes a stable node and the stable node becomes a saddle. Physically speaking, the equilibria exchange stabilities, but both equilibrium states continue to exist after the bifurcation. Of course, for each bifurcation, the equilibrium state is changing because the fixed point is moving, therefore changing the position and speed of the equilibrium state.

We might imagine, that if we were designing a system, we would want to know of possible bifurcations in the system in order to prevent problems such as a desirable stable equilibrium suddenly becoming unstable. For example, we might want to engineer an airplane wing that stays relatively flat during flight. All airplane wings move a little during flight. However, we would want to avoid the flat state from becoming unstable and the system undergoing galloping, where the amplitude of oscillation of the wing gets large. A bifurcation in which the galloping state becomes stable would be a dangerous situation, as can be seen in videos of the famous Tacoma Narrows Bridge collapse.

Next, we are going to study more complicated behaviors unique to nonlinear systems, namely limit cycles and chaos.

13.4   LIMIT CYCLES

We are now going to eliminate the restriction of small amplitude oscillations and allow for larger amplitudes. So far we have learned that nonlinear oscillators can have simple, sinusoidal-like oscillations, just like the harmonic oscillator. However, as the amplitude of oscillation gets larger, the period begins to depend on the amplitude. In some nonlinear systems, the oscillations can get complicated and look nothing like a sinusoid. These nonlinear oscillations appear in the phase portrait as a limit cycle, an isolated closed-loop trajectory. The limit cycle's isolation makes them different from centers, which have multiple closed orbit trajectories around the fixed point. Like fixed points, limit cycles can be stable, unstable, and even take on saddle-like stability. The stability of limit cycles is beyond the scope of this text, but Strogatz [11] provides a good introduction, and we will provide some other references at the end of this book.

13.4.1   DUFFING EQUATION

Recall that in Chapter 5, we made some assumptions about the potential associated with a restoring force. Namely, the potential needed to have a local minimum at the equilibrium position x_0, which in this Chapter we will set to 0. We then found that if we considered small amplitude deviations from the equilibrium, then we could approximate the potential energy using a parabola, $V = 1/2kx^2$, which results in Hooke's Law, $F = -kx$. All of this was done by Taylor expanding the potential energy about the equilibrium. If necessary, go back to chapter 5 and reread that section. Because we are now looking at larger amplitude oscillations, we will need to include more terms in the Taylor expansion of the potential energy $V(x)$ about the equilibrium at the origin,

$$V(x) \approx V(0) + \frac{dV}{dx}\bigg|_{x=0} x + \frac{1}{2!}\frac{d^2V}{dx^2}\bigg|_{x=0} x^2 + \frac{1}{3!}\frac{d^3V}{dx^3}\bigg|_{x=0} x^3 + \frac{1}{4!}\frac{d^4V}{dx^4}\bigg|_{x=0} x^4 + \cdots \qquad (13.4.1)$$

We have already made the argument that $\frac{dV}{dx} = 0$ at the equilibrium, and that argument continues to hold for the large amplitude case. The derivatives produce a constant value when evaluated at the origin, so we will rewrite (13.4.1) in the form:

$$V(x) \approx V(0) + \frac{1}{2!}kx^2 + \frac{1}{3!}\alpha x^3 + \frac{1}{4!}\delta x^4 + \cdots \qquad (13.4.2)$$

Next, we apply a restriction to the force which is common in physics, even in nonlinear systems, that $V(x)$ is symmetric about the equilibrium. Hence, the potential energy of the object at a position x is the same as that at a position $-x$. In other words, we would not expect the oscillator to have an energy that is dependent upon which side of equilibrium it is on. In order to satisfy this restriction, we must require that $\alpha = 0$. We can then find F by calculating $F = -dV/dx$ to get the force:

$$F(x) = -kx - \epsilon x^3 \qquad (13.4.3)$$

where $\epsilon = \delta/6$. The resulting equation of motion is,

$$m\ddot{x} + kx + \epsilon x^3 = 0 \qquad (13.4.4)$$

using $F(x) = m\ddot{x}$. We see that the result of examining large amplitude oscillations is that we must include a cubic nonlinearity in our equation of motion. If we include a damping term and a drive term, the new net force acting on the system is $m\ddot{x} = F_{net}(x) = -kx - \epsilon x^3 - \beta \dot{x} + f\cos(\omega t)$, and our equation of motion becomes:

$$\ddot{x} + c_1\dot{x} - c_2 x + c_3 x^3 = d\cos(\omega t) \qquad (13.4.5)$$

where $c_1 = \beta/m$, $c_2 = k/m$, $c_3 = \epsilon/m$, and $d = f/m$. Equation (13.4.5) is called the *Duffing equation* and is a standard model equation used in nonlinear dynamics to study a wide range of nonlinear phenomena. With the right choice of parameters, c_i and d, the Duffing equation can be made to display a large variety of nonlinear behaviors. We will use the Duffing equation to demonstrate limit cycles, period-doubling bifurcations, and chaos.

13.4.2 LIMIT CYCLES AND PERIOD-DOUBLING BIFURCATIONS

To demonstrate limit cycles and what is called *period-doubling bifurcations,* we will use equation (13.4.5) with $c_i = 1$, for all i, $\omega = 1$, and vary the value of d. In each case we will solve the Duffing equation with initial conditions $x(0) = 0$ and $\dot{x}(0) = 0$ and plot the phase portrait along with a short segment of $x(t)$. All units can be considered to be SI units.

Figures 13.9, 13.10, and 13.11 show sample solutions of (13.4.5) for various values of d, after the transients have decayed and the system settled onto its attracting limit cycle. We will demonstrate how these figures are made in Example 13.6 in the next section.

Notice that in Figure 13.9 when $d = 0.68$, the Duffing equation produces a limit cycle, an isolated closed trajectory corresponding to an oscillation. The solution $x(t)$ (on the right), may look sinusoidal, but it is not; notice how narrow the peaks and troughs are compared to a sinusoid. Furthermore, the limit cycle is not elliptical, and sinusoidal oscillations are the result of elliptical trajectories in the phase portrait.

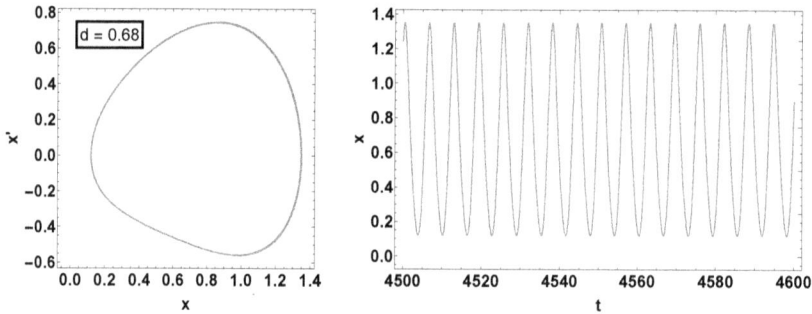

Figure 13.9: Phase portrait (left) and $x(t)$ (right) of (13.4.5) with $c_i = 1$, for all i, $\omega = 1$, and $d = 0.68$.

The graphs of Figure 13.10 shows the phase portrait (left) and solution (right) when d is increased to 0.69. The limit cycle appears to intersect itself; however, it does not. The apparent intersection is due to a three-dimensional phase portrait being projected into two dimensions. Although beyond the scope of this chapter, the Duffing equation is actually a three-dimensional system due to the explicit dependence of time in this system. The solution for $d = 0.69$ is called a *period-2 solution* because, loosely speaking, the oscillator has to go through two oscillations before returning to its initial state, once the oscillator has settled onto the limit cycle behavior.

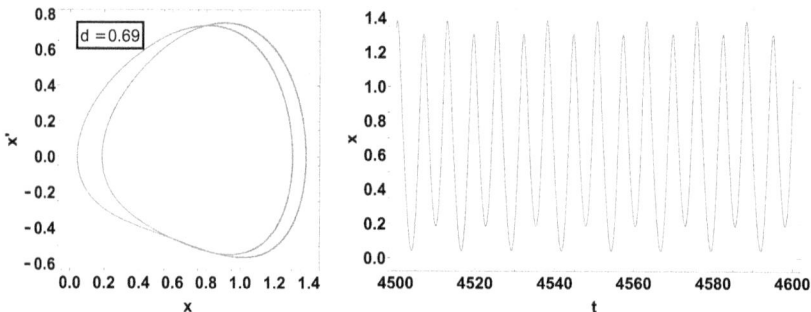

Figure 13.10: Phase portrait (left) and $x(t)$ (right) of (13.4.5) with $c_i = 1$, for all i, $\omega = 1$, and $d = 0.69$.

Figure 13.11 shows the phase portrait (left) and solution (right) when d is increased to 0.75. There are even more apparent intersections in the phase portrait and the solution $x(t)$ appears even more complicated. Figure 13.11 illustrated a *period-4 solution* to (13.4.5). We are seeing an interesting phenomenon that can occur in nonlinear oscillators called a *period-doubling bifurcation*, whereas the parameter d in (13.4.5) varies, the period of the oscillator (and therefore the limit cycle) doubles. One method of illustrating period-doubling bifurcations is by using something called a *Poincaré section*, which we will discuss below.

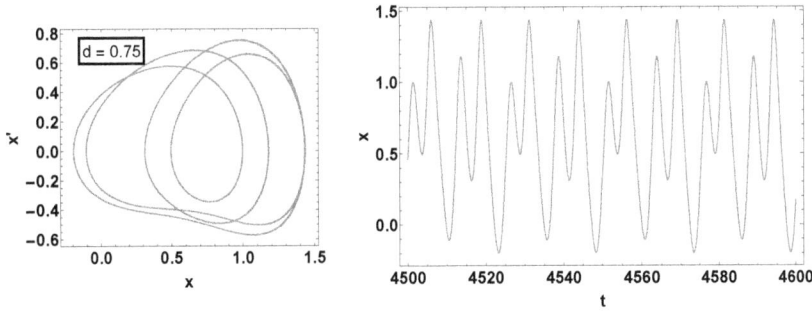

Figure 13.11: Phase portrait (left) and $x(t)$ (right) of (13.4.5) with $c_i = 1$, for all i, $\omega = 1$, and $d = 0.75$.

The *Poincaré section* samples the solution once a period and then plots the phase space. Since the drive frequency is $\omega = 1$, we can sample $x(t)$ and $\dot{x}(t)$ at times $t = 2n\pi$, where n is an integer. The code for producing Poincaré sections is shown in Example 13.6 in the next section.

The Poincaré sections for the solutions in Figures 13.9 through 13.11 are shown in Figure 13.12.

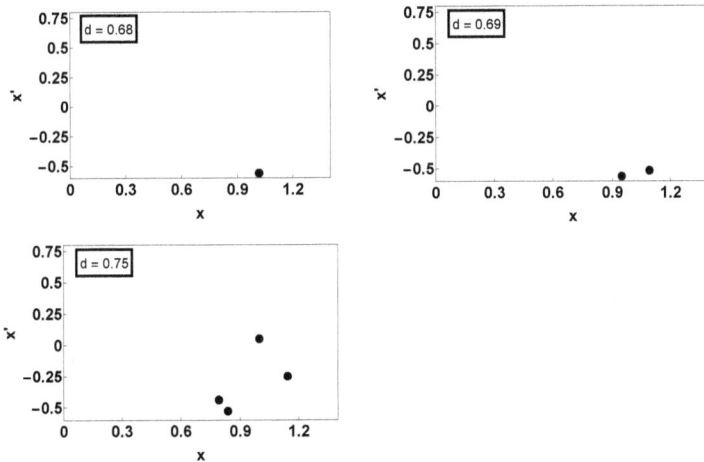

Figure 13.12: The Poincaré sections for the solutions shown in Figures 13.9 through 13.11.

Notice how in Figure 13.12, for $d = 0.68$, there is only one point. That is because, after settling onto the limit cycle, the solution repeats its state once every period. We are not sampling the solution at its maximum, i.e., the x-coordinate of the point in the Poincaré section does not correspond to the maximum value of the right graph in Figure 13.9. The fact that the point in the Poincaré section does not match either the maximum or minimum value of the solution does not matter. What matters is that every $t = 2n\pi$, the system returns to the same state. However, we see that when $d = 0.69$, the oscillator appears to jump between two points in its Poincaré section. It takes $4\pi = 2 \times 2\pi$, or twice the period to repeat its state. Continuing this reasoning, we see that when $d = 0.75$, the system is exhibiting period-4 behavior.

As mentioned previously, Figures 13.9 through 13.12 are demonstrating period-doubling bifurcations. We see a period-doubling bifurcation in the limit cycle between $d = 0.68$ and $d = 0.69$. We see another period-doubling bifurcation between $d = 0.69$ and $d = 0.75$. As we continue to increase d, we would see several more period-doubling bifurcations, the series of which is called a *period-doubling cascade*, which will eventually lead to chaotic behavior once $d = 0.8$. However, to illustrate chaos, we will change the parameters used in (13.4.5) in order to better visualize the attractor corresponding to chaotic behavior.

13.5 CHAOS

In addition to interesting oscillations, nonlinear systems can also display a behavior called *chaos*. To explore the phenomenon of chaos, we will use the Duffing equation, but with a specific set of parameter values:

$$\ddot{x} + 0.15\dot{x} - x + x^3 = 0.3\cos(t) \tag{13.5.1}$$

It is worth mentioning that there are many different values of c_i and d from (13.4.5) that will produce chaotic behavior in the Duffing equation, including those used in Section 13.4.2 with $d = 0.8$. However, we use the parameters as in (13.5.1) because they provide a good view of the attractor associated with the chaotic dynamics. Example 13.6 will demonstrate the chaotic solution of (13.5.1) and also show how Figures (13.9)-(13.12) are created. Let us start by plotting the solution, phase portrait, and Poincaré section of the solution to (13.5.1)

> **Example 13.6: Chaotic solution to the Duffing equation**
> Solve the Duffing equation
>
> $$\ddot{x} + 0.15\dot{x} - x + x^3 = 0.3\cos(t) \tag{13.5.2}$$
>
> using the initial conditions $x(0) = 0$ and $\dot{x}(0) = 0$. Plot $x(t)$, the phase portrait, and the Poincaré section after allowing transients to decay.
>
> *Solution:*
> The solution is presented in Figure 13.13, which shows the output of the Python code. Notice how in Figure 13.13, the phase portrait does not contain a limit cycle, but some unusual trajectory which seems to oscillate about each of the stable equilibrium in the double-well potential. Those oscillations are clear in the graph of $x(t)$, where the solution oscillates about the equilibrium at $x = 1$, before switching and oscillating about $x = -1$. Notice that the duration of time spent near each equilibrium varies and the switching between one equilibrium to the next seems unpredictable. This is strange behavior!
>
> Like in Section 13.4.2, we can use a Poincaré section to understand the structure of the attractor. Notice that the Poincaré section is not a simple collection of points. In fact, it has its own complex structure. You might wonder if we solved (13.5.1) for a longer period of time, we would find that any of the points in the Poincaré section repeat themselves, hence demonstrating a periodicity. The answer is no, there is no repetition, and therefore the system is not periodic. The Poincaré section is even more interesting if we zoom into it and look at its structure. Doing so would reveal that the bands that are visible in the Poincaré section of Figure 13.13 persist when zoomed in at any scale provided enough points are plotted, which can get computationally expensive. The presence of structure at arbitrary scale means that the attractor is a fractal!

> **Python Code**
> To use `odeint`, we need to rewrite (13.5.2) as
>
> $$\left.\begin{array}{l} \dot{x} = v \\ \dot{v} = -0.15y + x - x^3 + 0.3\cos(t) \end{array}\right\} \tag{13.5.3}$$

We set a transient time of 4800 using the variable `start`. Notice that `start` appears in the plot command as the starting point for the slices used to create each plot (except for the Poincaré section). The transient time is much shorter than 4800; however, we choose a large value to make the graphs easier to read.

To create the Poincaré section, we used a slice in the `scatter` command. The stride of the slice is `tau = 2*np.pi`, which matches the period of the drive term in (13.5.2). To properly visualize the Poincaré section, we needed a lot of data. Therefore, we used the full solution for the Poincaré section.

```python
import numpy as np
import matplotlib.pyplot as plt
from scipy.integrate import odeint

dt = 0.001 #step size
times = np.arange(0,5000,dt)

def deriv(z,t):
    x, y = z
    dzdt = [y, -0.15*y + x - x**3 + 0.3*np.cos(t)]
    return dzdt

ics = [0,0]

soln = odeint(deriv,ics,times)

start = int(4800/dt) #starting index after transient

tau = int(2*np.pi/dt)

fig, ax = plt.subplots(nrows = 2, ncols = 2)

#plot x(t)
ax[0,0].plot(times[start:],soln[start:,0])
ax[0,0].set_xlabel('time')
ax[0,0].set_ylabel('x')
ax[0,0].set_title('x(t)')

#plot phase portrait
ax[0,1].plot(soln[start:,0],soln[start:,1])
ax[0,1].set_xlabel('x(t)')
ax[0,1].set_ylabel('v(t)')
ax[0,1].set_title('Phase Portrait')

#plot Poincare section
ax[1,0].scatter(soln[::tau,0],soln[::tau,1],s = 1)
ax[1,0].set_xlabel('x')
ax[1,0].set_ylabel('v')
ax[1,0].set_title('Poincare Section')

#delete unneeded graph
ax[1,1].set_axis_off()

plt.tight_layout()
plt.show()
```

Mathematica Code

We use NDSolve to solve (13.5.2). The plots of $x(t)$ and the phase portrait are similar to that done earlier in this chapter. To create the Poincaré section, we sampled the solution at time interval of $t = 2\pi$ and stored the result as the table psData, which was then plotted with ListPlot.

```
SetOptions[{Plot, ListPlot, ParametricPlot}, Axes->False, Frame->True,

BaseStyle->{FontSize->16}, ImageSize->Large];

soln = NDSolve[{x"[t] + 0.15 * x'[t] - x[t] + x[t]^3==0.3 * Cos[t],

x[0]==0, x'[0]==0}, x, {t, 0, 5000}];

p1 = Plot[x[t]/.soln, {t, 4500, 5000}, FrameLabel->{"time", "x(t)"},

PlotLabel->"x(t)"];

p2 = ParametricPlot[{x[t], x'[t]}/.soln, {t, 4500, 5000},

FrameLabel->{x, v}, PlotLabel->"Phase Portrait"];

psData = Table[{First[x[t]/.soln], First[x'[t]/.soln]}, {t, 1000, 5000, 2 * π}];

p3 = ListPlot[psData, FrameLabel->{x, v}, PlotLabel->"Poincare Section"];

GraphicsGrid[{{p1, p2}, {p3}}]
```

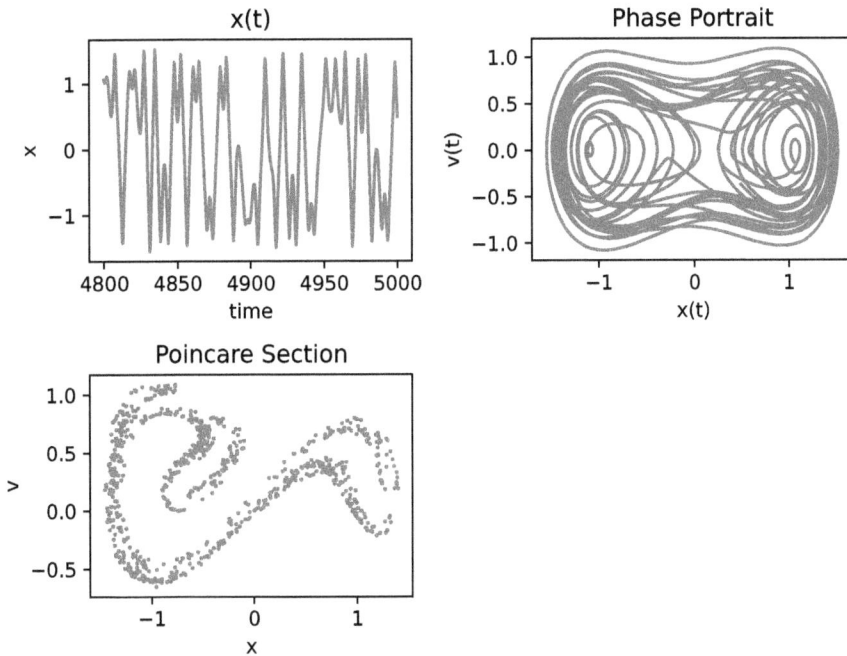

Figure 13.13: Python-generated plots of $x(t)$, the phase portrait, and Poincaré section of (13.5.2) from Example 13.6.

What have we learned in our study so far from Figure 13.13? We learned that chaotic systems are not periodic. The trajectory of a chaotic system's solution in the phase portrait is attracted to a structure with fractal properties. Furthermore, the behavior of the solution is not random; the system is obeying the Duffing equation. This is getting strange! But wait, there is more...

13.5.1　CHAOS AND INITIAL CONDITIONS

Let us look at the behavior of two solutions of (13.5.1) with different, but similar, initial conditions. Do we expect the behaviors to significantly differ? Consider two children, Chris and Gail, each on a swing side-by-side. The swings are identical in length, and because we are physicists, let us assume they are both wearing space suits and are swinging in vacuum, under the influence of a gravitational field similar to that of the Earth's. Chris starts by displacing his swing 10 degrees from equilibrium. Gail starts by displacing hers 11 degrees from equilibrium. In each case, the small angle approximation is satisfied, and both swings behave like simple harmonic oscillators. What are the resulting behaviors? We know the frequencies of oscillation are the same (simple harmonic motion, with same length swings), so we would expect that Chris and Gail would swing in sync, but Gail would go a little higher than Chris because her initial state had a slightly larger initial amplitude. What happens if their swings are chaotic? The ride would be much more interesting! After a period of time, Chris and Gail's swings will fall out of sync with each other.

To illustrate this, let us compare two solutions of (13.5.1), one with an initial condition of $x(0) = 0$, $\dot{x}(0) = 0$, and the other with the same initial velocity but with $x(0) = 0.00001$, 0.01, and 0.1. The results are shown in Figure 13.14 with a plot of the simple harmonic oscillator (SHO) with two initial conditions $x(0) = 1.1$ and $x(0) = 1.0$, each with $\dot{x}(0) = 0$, for comparison.

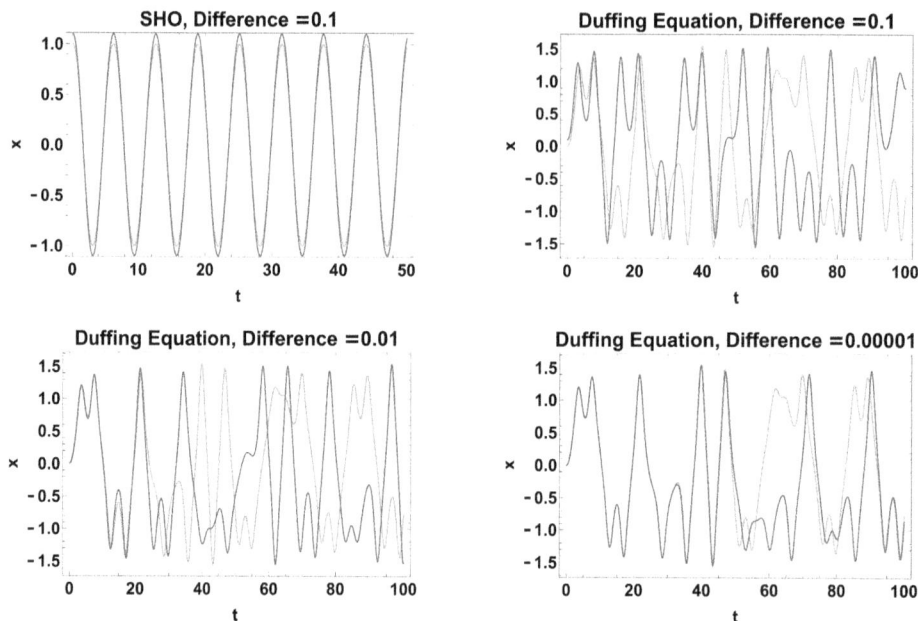

Figure 13.14: Plots comparing similar initial conditions for the simple harmonic oscillator (SHO) and the Duffing equation. The value of *Difference* in the title of each plot describes the difference in the initial positions of the two oscillators. The black solid curve has an initial position of $x(0) = 0$ and the red dashed curve (in the e-book) has an initial position of $x(0) = 0 + \text{Difference}$. In all cases $\dot{x}(0) = 0$.

In Figure 13.14, the black line (in the e-book) for the SHO is the solution for $x(0) = 1$ and $\dot{x}(0) = 0$. For the Duffing equation, the black line is the solution for $x(0) = 0$ and $\dot{x}(0) = 0$. The red dashed line (in the e-book) for each graph has a zero initial velocity, but an initial position that is off by the value of "Difference" (the value of which is in each plot's title) from the red line's initial condition (e.g., for the SHO, the red dashed line has an initial position of $x = 1.1$). Notice that the SHO behaves exactly as described in the above paragraph. The two solutions track each other with only a small difference in their amplitudes (as expected from the initial condition). However, the Duffing equation solutions are very different from the SHO solutions. Notice that over time, the small difference in the initial conditions is magnified. Eventually, the two solutions are not similar at all. For example, there are times where the red solution is near $x = 1$, but the black solution is near the other equilibrium at $x = -1$, and vice versa. We see in Figure 13.14 that the smaller the initial difference, the longer the solutions are similar, but even extreme initial similarity (the lower right graph in Figure 13.14) eventually yields a divergence in the solutions. This phenomenon is referred to as *sensitive dependence on initial conditions*, and is a critical component of chaotic behavior. Attractors on which the trajectories exhibit sensitive dependence on initial conditions are called *strange attractors*.

Is the sensitive dependence on initial conditions just a fancy way of saying that the system is random? No, the system is not random. The Duffing equation is an ordinary differential equation. The solution of the Duffing equation is dictated by its initial conditions, and therefore is a deterministic system. In other words, future values of $x(t)$ and $\dot{x}(t)$ depend on the initial conditions $x(0)$ and $\dot{x}(0)$. In a random system, past states do not dictate future states (otherwise, there would be no lottery!). The fact that the Duffing equation is deterministic means that the interesting behavior arises due to the nonlinearity of the system, not due to some random elements.

Other than being a mathematical curiosity, why is sensitive dependence on initial conditions important? As an example, it is helpful to consider the case where chaos was first presented in its modern form by meteorologist Edward Lorenz (1917-2008), who was creating weather models.

Lorenz's weather model demonstrated chaotic behavior. To illustrate sensitive dependence on initial conditions, suppose that the weather perfectly follows Lorenz's model (in reality, it does not). Does that mean we can perfectly predict the weather at any point in the future? No. In order to make predictions, we need to input an initial condition into the model. The initial condition comes from instruments used to measure various atmospheric conditions. The instruments have a certain degree of accuracy; they may only measure the true value of a quantity to, for example, one part in a thousand. The actual degree of accuracy is irrelevant, the point is that the measured initial conditions will be different from the true state of the atmosphere. Since, in our example, the weather perfectly obeys a chaotic model, then the slight difference in measured versus true values will mean that after a period of time, the model's predictions will diverge from the actual weather. The chaotic system is sensitive to initial conditions! The result of a chaotic system's sensitive dependence on initial conditions is that there is no accurate long-term prediction possible for chaotic systems.

13.5.2 LYAPUNOV EXPONENTS

We can quantify the sensitivity to initial conditions by looking at the rate at which initially close trajectories diverge. Consider two trajectories on the chaotic attractor, $\mathbf{x}(t)$, and $\mathbf{x}(t) + \boldsymbol{\delta}(t)$, where $\boldsymbol{\delta}(t)$ is a vector that measures the distance between the two trajectories and $\delta(0)$, the initial separation, is very small, $\delta(0) << 1$. Numerical studies show that,

$$\|\boldsymbol{\delta}(t)\| \approx \|\boldsymbol{\delta}(0)\| \, e^{\lambda t} \qquad (13.5.4)$$

where λ is called the *maximum Lyapunov exponent*. The maximum Lyapunov exponent measures the rate of exponential divergence between two initially nearby trajectories on the chaotic attractor. Chaotic systems have $\lambda > 0$. The maximum Lyapunov exponent is typically difficult to calculate for a system; however, there are methods of finding λ from measured data called *time series*. A review of such methods can be found in Kantz and Schreiber [14].

So what have we learned? First, chaotic systems are not random; they are deterministic. Second, their trajectories live on strange attractors in the phase portrait, and these attractors have fractal structure. This means that chaotic systems are not periodic; they do not settle onto fixed points. They are also not *quasiperiodic*, an oscillator consisting of a sum of incommensurate frequencies. Third, trajectories on the chaotic attractor exhibit sensitive dependence on initial conditions. This means that there is no long-term predictability for a chaotic system. So what is chaos? Chaos has eluded a formal definition since it was first formally presented by Lorenz in 1963. However, for our purposes, Strogatz's definition [11], will suffice:

> *Chaos is aperiodic long-term behavior in a deterministic system that exhibits sensitive dependence on initial conditions.*

If the authors may be so bold, we suggest that of all the factors, it is the sensitivity to initial conditions that is truly the hallmark of chaos. That sensitivity is what makes chaotic systems difficult to predict and very interesting!

13.6   A FINAL WORD ON NONLINEAR SYSTEMS

Mathematician Stanislaw Ulam (1909-1984) once said, "Using a term like nonlinear science is like referring to the bulk of zoology as the study of non-elephant animals." Nonlinear systems are everywhere, and understanding how to study them is an important part of a scientist's preparation. There is an exciting world out there where physics intersects the biological and social sciences, and nonlinear systems often take center stage in those arenas. By focusing only on linear systems, one would miss an opportunity to explore exciting new fields such as econophysics. The study of nonlinear systems can open up new worlds of problems outside of traditional physics, such as modeling disease propagation, population dynamics, social networks, and much more.

In this chapter, we discussed that nonlinear systems can take on a wide variety of behaviors. We have not even scratched the surface on the variety of interesting phenomena that can occur in nonlinear systems. As promised, we will include a few extra suggestions on how to extend the reader's study of nonlinear systems. For a more general discussion of nonlinear systems, we recommend chapter 12 of our book, *Mathematical Methods using Python* (Pagonis and Kulp [3]). The next book we recommend reading is Strogatz [11], as the book is easily understandable and enjoyable to read. It also covers the topics in this chapter more in depth and has applications outside of physics and engineering.

After reading Strogatz [11], one can take many pathways to learning nonlinear systems. We frequently recommend Enns [13] because it contains some very interesting applications of nonlinear systems to a wide variety of fields in the natural and social sciences. Another book commonly used in nonlinear dynamics courses is Hilborn [12]. If the reader is interested in analyzing data measured from nonlinear systems, both Kantz and Schreiber [14] and Abarbanel [15] are recommended. Finally, if the reader is interested in developing models of systems (especially complex systems), we recommend Wilensky and Rand [16].

If one spends enough time working with (and learning about) nonlinear systems, eventually, one will come to start thinking of problems not as linear versus nonlinear, or even as physics versus not physics. Instead, they will likely see the system they are working with as just that, an interesting system to be analyzed with one of many tools the reader will have in his/her mathematical and computational toolbox. In the end, breaking up problems in the natural world by discipline is a human-imposed construct. Real systems are neither biology versus physics versus sociology, they are just systems to be studied. While the construction of silos for problems is useful for the purposes of education, once you no longer need it, we recommend discarding it and focusing on studying problems that are interesting to you.

13.7   CHAPTER SUMMARY

In this chapter, we studied systems whose equations of motion have nonlinear terms. These equations have the form

$$\dot{x} = f(x, y)$$
$$\dot{y} = g(x, y)$$

Although the equations governing nonlinear systems are not usually solvable in closed form, we can use tools such as phase portraits to understand possible steady state behaviors of the system as well as identify the stability of equilibria.

The fixed points of the system can be found by solving

$$f(x^*, y^*) = 0$$
$$g(x^*, y^*) = 0$$

Although we cannot solve the whole system analytically, we can linearize the system about the fixed points, which results in an easily solvable system of equations:

$$\begin{pmatrix} \dot{u} \\ \dot{v} \end{pmatrix} = \begin{pmatrix} \frac{\partial f}{\partial x} & \frac{\partial f}{\partial y} \\ \frac{\partial g}{\partial x} & \frac{\partial g}{\partial y} \end{pmatrix} \begin{pmatrix} u \\ v \end{pmatrix}$$

where the eigenvalues of the Jacobian matrix A give us information about the stability of the equilibrium. The eigenvalues can be found using the characteristic polynomial,

$$\det (A - \lambda \mathbf{1}) = 0$$

In nonlinear systems, the equilibrium states can change (both location in phase space and stability).

Finally, we also explored a behavior called *chaos*, which is unique to nonlinear systems. A hallmark of chaos is that the solutions of a chaotic system display sensitive dependence on initial conditions. In other words, two initially similar states will evolve to very different behaviors, thus inhibiting one's ability to make long-term predictions on chaotic systems.

13.8   END-OF-CHAPTER PROBLEMS

The symbol ⌨ indicates a problem which requires some computer assistance, in the form of graphics, numerical computation, or symbolic evaluation.

Section 13.1: Linear vs. Nonlinear Systems

1. Consider the differential equation

$$\frac{d^3 x}{dt^3} + x = 0$$

   Is this equation a linear or a nonlinear differential equation? Justify your answer. Solve the equation for $x(t)$.

2. Are the following differential equations linear or nonlinear? Justify your answers.

   a. $\frac{d^3 x}{dt^3} + t^3 x = 0$.
   b. $\ddot{x} + t^3 \dot{x} x = 0$
   c. $\ddot{x} + t^2 x = A \cos (t)$
   d. $\dot{x} + x \sin t = t^3$

3. Consider the differential equation

$$\ddot{x} = t^2 x - tx^2$$

   Suppose you have found two solutions to this equation, $x_1(t)$ and $x_2(t)$. Is it true that $c_1 x_1 + c_2 x_2$ is also a solution? Prove your answer.

Section 13.2: Damped Harmonic Oscillator, Revisited

4. 🖥 Using a computer, plot the phase portrait for the damped harmonic oscillator with $\gamma < 0$ and $\omega_0^2 = 1$. You may use $x(0) = 1$ and $\dot{x}(0) = 0$. Do you get what you expect? Why or why not? What does it mean for $\gamma < 0$?

5. 🖥 Using a computer, plot the phase portrait for the damped pendulum, $\ddot{x} + \omega_0^2 \sin x + \gamma \dot{x} = 0$ for $\omega_0^2 = 1$ and three values of γ, less than, equal to, and greater than ω_0^2. For each case, use $x(0) = 1$ and $\dot{x}(0) = 0$. Discuss your results.

Section 13.3: Fixed Points and Phase Portraits

6. Consider a one-dimensional system, $\dot{x} = f(x)$. We can assign a potential function to the system, $V(x)$, such that

$$f(x) = -\frac{dV}{dx}$$

which makes the potential function similar to the potential energy discussed in Chapter 5. In this chapter, we discussed that the condition for fixed points (i.e., equilibrium points), x^*, $f(x^*) = 0$. How can we use the potential to find fixed points using $V(x)$? In other words, how does one find fixed points using $V(x)$? How can one use $V(x)$ to identify the stability of a fixed point?

7. Consider the system $\dot{x} = 1 - x^2$. Find the potential corresponding to this system (see Problem 6). Using the potential, find the fixed points and identify their stability.

8. In this chapter, we discussed how to identify the stability of fixed points for two-dimensional systems. The logistic equation, $\dot{x} = rx(1 - x/k)$, is used to model population sizes, and is a one-dimensional system. Note that r is a positive constant called the *linear growth rate* and k is a positive constant called the *carrying capacity*. Find the fixed points for the logistic equation and give them a physical interpretation. Find the stability of each fixed point by perturbing the fixed point, $x = x^* + u$ (for $u << 1$) and expanding the logistic equation in powers of u. The result will be a differential equation for u which you can solve.

9. 🖥 Find the eigenvalues and eigenvectors for the following matrices. Solve this problem analytically and by using a computer algebra system.

$$\text{(a)} \quad \begin{pmatrix} 0 & 1 \\ 3 & 0 \end{pmatrix} \qquad \text{(b)} \quad \begin{pmatrix} 2 & 1 \\ 7 & 3 \end{pmatrix} \qquad \text{(c)} \quad \begin{pmatrix} 1 & 1 \\ 1 & 1 \end{pmatrix}$$

10. 🖥 Write a computer program that recreates Figure 13.4.

11. 🖥 Solve the pendulum equation (13.3.13) numerically with an initial condition of $\theta(0) = 2.7$, 2.9, and 3.0 (all with $\dot{\theta}(0) = 0$ and $\omega_0 = 1$). Keeping in mind Figure 13.5, why does the solution, $\theta(t)$, change as the initial angular displacement approaches the vertical $(x = \pi)$?

12. 🖵 Strogatz [11] discusses the mathematics of love affairs, an interesting and fun way to explore two-dimensional linear systems. Let $R(t)$ be Romeo's love (positive) or hate (negative) for Juliet at time t, and $J(t)$ be Juliet's love or hate for Romeo at time t. The love affair can be modeled using the equations:

$$\dot{R} = aR + bJ$$
$$\dot{J} = cJ + dR$$

Interesting relationships can be created by choosing different values of the parameters a, b, c, and d. Following the methods of Section 13.3, classify the fixed point at the origin for an eager Romeo ($a = 1$, $b = 2$) and a very cautious Juliet ($c = -1$, $d = -3$) by analytically finding the eigenvalues of the Jacobian matrix. Plot the phase portrait, and comment on the possible outcomes for various initial conditions of the relationship.

13. 🖵 In this problem, we will build on Problem 12 with a nonlinear romance! We will have both Romeo and Juliet be very sensitive to their own emotions:

$$\dot{R} = -2R - 2J(1 - |J|)$$
$$\dot{J} = J + R(1 - |R|)$$

The nonlinear term $J(1 - |J|)$ is sometimes called a *repair nonlinearity* (Sprott [17]), and the number 1 in the second term on the right-hand side is a measure of when Juliet's love becomes counterproductive. Plot the phase portrait for this system and discuss the nature of the relationship between Romeo and Juliet.

14. Find the Jacobian matrix for (13.3.25) and compute its eigenvalues for each of the fixed points.

15. 🖵 The *Lotka-Volterra model of competition* is a mathematical model for two species X and Y who compete for the same resources. Let x be the population size of species X and y be the population size of species Y, then one choice of a Lotka-Volterra model would be:

$$\dot{x} = 4x - 3x^2 - xy$$
$$\dot{y} = 2y - 4y^2 - 3xy$$

For this model, identify the fixed points and their stability. Can the two species ever coexist? Note that this kind of equation-based model tends to work well for large populations. For small populations, agent-based models tend to be more useful. For a discussion of agent-based versus equation-based modeling, the interested reader should consult Wilensky and Rand [16].

16. 🖵 Plot the phase portrait for the Lotka-Volterra competition model in Problem 15. Discuss the final state of the system for various initial conditions.

17. Lotka-Volterra equations can also be used to describe predator-prey models where species Y (predator) eats species X (prey). Consider the predator-prey model

$$\dot{x} = 3x - 7xy$$
$$\dot{y} = -2y + 5xy$$

where like the competition model, we will assume $x(t)$ and $y(t)$ measure the number of each species in the hundreds. Find analytically the fixed points of this predator-prey model and identify their stability. Comment on the nature of the equilibrium state corresponding to each fixed point.

18. 🖥 Using a computer, draw the phase portrait for the predator-prey model in Problem 17.

19. The SIS (susceptible-infected-susceptible) disease transmission model can be used for diseases that do not remove individuals from the population, either through immunity or death. SIS models are sometimes appropriate for some bacterial diseases such as streptococcal pharyngitis (also known as sore throat). Consider two populations of individuals $s(t)$ and $i(t)$, representing the number of susceptible and infected individuals, respectively. The SIS model is:

$$\dot{s} = \mu n - \mu s + \gamma i - \beta si$$
$$\dot{i} = -\mu i - \gamma i + \beta si$$

where the coefficients represent rates, and $n = s + i$ is the total population. Comment on the physical meaning of each coefficient: μ, γ, and β. Find the fixed points for the system and identify their stability. Note that negative values of s and i are meaningless.

Section 13.4: Limit Cycles

20. 🖥 Consider the Duffing equation:

$$\ddot{x} - x + x^3 + 0.1\dot{x} = d\cos\omega t \tag{13.8.1}$$

where $\omega = 1.4$. Show that the Duffing equation, (13.8.1), undergoes a series of period-doubling bifurcations by changing the value of d. Start with $d = 0.1$ and gradually increase its value until a period-doubling bifurcation is observed. Use $x(0) = 0$ and $\dot{x}(0) = 0$ as initial conditions.

21. 🖥 The *logistic map*, $x_{n+1} = rx_n(1 - x_n)$, is a discrete form of the logistic equation (see Problem 8). After choosing an initial value x_0 and parameter r, it is easy to compute successive values. For example, $x_1 = rx_0(1 - x_0)$ and $x_2 = rx_1(1 - x_1)$, and so on. Using $x_0 = 0.1$, plot x_n vs. n for $r = 2.9, 3.2, 3.5$, and 3.55. Identify the periodicity of each solution. You can produce a Poincaré section by plotting x_{n+1} vs. x_n. Finally, try $r = 3.91$. What kind of behavior is the logistic map exhibiting when $r = 3.91$?

22. 🖥 Weakly driven nonlinear oscillators can be excited to large amplitudes through a phenomenon called *autoresonance*. The oscillator is driven by a periodic forcing term whose amplitude is small, but its frequency changes (typically decreasing) with time. As the drive frequency decreases through the oscillator's resonant frequency, the oscillator becomes phase-locked with the driving force, resulting in the oscillator maintaining resonance even though the drive frequency is no longer the oscillator's natural frequency. Consider the following driven pendulum:

$$\ddot{x} + \omega_0^2 \sin(x) = \epsilon \cos\left(\omega_0 t - 1/2\alpha t^2\right)$$

Notice that the drive frequency (the time derivative of the drive phase) is $f_d = \omega_0 - \alpha t$, so that at $t = 0$, the drive frequency is equal to the natural frequency of the pendulum. Starting with initial conditions, $x(-1000) = 0$ and $\dot{x}(-1000) = 0$, compute $x(t)$ for $\epsilon = 0.0459$ and $\epsilon = 0.0461$ for the pendulum with $\omega_0 = 2\pi$, $\alpha = 0.001$. For which value of ϵ is the pendulum displaying autoresonance? For more information on autoresonance and the pendulum, check out Fajans and Friédland [18].

23. 🖥 Electrical circuits provide excellent examples of nonlinear systems. One of the most famous circuit-inspired equations is the so-called *Van der Pol equation*

$$\ddot{x} - \epsilon\left(1 - x^2\right)\dot{x} + x = 0$$

which is a simple harmonic oscillator with a nonlinear damping term. Notice that the sign of the damping actually changes with the value of x. For example if $\epsilon > 0$, then the damping is negative for $x < 1$ and is positive for $x > 1$. Thus small oscillations will grow due to negative damping, but once the oscillations are large enough, they will dampen out as the damping term becomes positive. The Van der Pol oscillator can also display so-called *relaxation oscillations*, a common form of oscillations in nonlinear systems where rapid changes of x are followed by slower variations. To see relaxation oscillations in action, solve the Van der Pol oscillator for $\epsilon = 8.0$. Plot both $x(t)$ and the phase portrait. Describe the solution.

24. It can sometimes be convenient to describe limit cycles using polar coordinates. Consider the following system in polar coordinates:

$$\dot{r} = r(r - 1)(r - 2)$$
$$\dot{\theta} = 1$$

where r and θ are the usual polar coordinates with $r \geq 0$. Notice that one can find values of r that make $\dot{r} = 0$, but there are no values of θ or r for which $\dot{\theta} = 0$. Give a description of the phase portrait in this case. There are limit cycles in this system. Where are they and what is their stability? The stability can be found by using the method outlined in Problem 8 for the $\dot{r}$ equation.

Section 13.5: Chaos

25. 🖥 The Lorenz equations:

$$\dot{x} = \sigma(y - x)$$
$$\dot{y} = rx - y - xz$$
$$\dot{z} = xy - bz$$

display chaotic behavior with $\sigma = 10$, $r = 28$, and $b = 8/3$. Plot the phase portrait (x, y, z) for the Lorenz equations.

26. ⌨ The Lorenz equations in Problem 25 are a chaotic system. Show that they demonstrate sensitive dependence on initial conditions.

27. ⌨ Chua's circuit is a electrical circuit with a piecewise linear negative resistance N_R. It is one of the simplest circuits to exhibit a variety of nonlinear behaviors including a period-doubling route to chaos. The circuit schematic is shown in Figure 13.15. Let x, y, and z be proportional to the voltages across the capacitors C_1 and C_2, and the current in the inductor L, respectively. The equations governing the circuit can be obtained from Kirchoff's rules and simplified to be:

$$\dot{x} = \alpha(y - x - g(x))$$
$$\dot{y} = x - y + z$$
$$\dot{z} = -\beta y$$

where $g(x) = m_1 x + \frac{1}{2}(m_0 - m_1)(|x + 1| - |x - 1|)$ with $\alpha = C_2/C_1$, $\beta = R^2 C_2/L$, and m_0 and m_1 being the slopes of the inner and outer segments of the voltage-current curve, respectively. For more information the interested student should consult Enns [13]. Create a phase portrait (x-y-z space) for the equations of motion using $\alpha = 15.6$, $\beta = 25.58$, $m_0 = -8/7$, $m_1 = -5/7$. The result will be the famous double-scroll attractor.

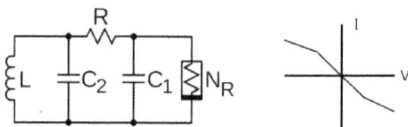

Figure 13.15: Chua's circuit, Problem 13.28.

28. ⌨ A period-doubling route to chaos can be observed in the equations for Chua's circuit (see Problem 27) by using the same parameters as used in Problem 27, but with $\beta = 50$, 35, 33.8, etc. For each value of β, create a graph of $x(t)$ and identify the periodicity of each solution. For what value of β is there a period-8 solution?

One Final Problem for Nonlinear Systems

29. ⌨ As a final problem for this chapter, we will explore the idea of bifurcations of fixed points. Consider the following predator-prey model:

$$\dot{x} = x(1 - x) - xy$$
$$\dot{y} = -\epsilon y + xy$$

   a. Which species is the predator? Which is the prey?

   b. Find the fixed points. Assume that $\epsilon > 0$.

c. Notice that for a certain value of ϵ, fixed points collide in the phase portrait (i.e., two or more fixed points occupy the same location in the phase portrait). Those collisions are bifurcations of fixed points. For what value of ϵ do the bifurcations occur?

d. Find the Jacobian of the system and evaluate the Jacobian at each fixed point.

e. Find the stability of each fixed point by finding the eigenvalues of the Jacobian associated with each fixed point. You may want help from a computer algebra system for this.

f. Describe how the stability of each fixed point changes with ϵ. Besides collisions between fixed points, bifurcations can also involve changes in the stability of a fixed point. Plotting the eigenvalues as a function of ϵ is often a useful trick to identify whether or not the eigenvalues are positive or negative.

g. Draw phase portraits for the system before and after the bifurcation(s) by choosing a value of ϵ less than and greater than the bifurcation conditions you found in parts c and e. Notice that the phase portraits are different.

h. Provide a qualitative description of what happens to the system as ϵ increases. What does ϵ represent in our system? Does the increasing the value of ϵ have the effect on the phase portrait that you expected? Why or why not?

A Introduction to Python

A.1 DATA TYPES AND VARIABLES IN PYTHON

The main numeric types in Python are integers, floating point numbers and complex numbers. A floating point number (or float) is a real number written in decimal form. Python stores floats and integers internally in different ways.

In Example A.1 we assign a value to a variable using the assignment operator $=$, for example a = 0.0003 assigns the value of 0.0003 to the variable a. We can also define variables by using scientific notation to represent floats, for example b = 1.64e-4. Python uses the letter j to represent $\sqrt{-1}$, and the built-in function complex(a,b) creates the complex number a+b*j. We can use the built-in function type() to identify the type of a Python object, and the built-in function print(a) is used to print the value of a variable. Similar to several other computer languages, the line e = 'time' defines a string variable e.

Example A.1 shows how the function type(a) can be used to find out what class of object the variable a belongs to. In most codes of this book you will see the line of code print('-'*28,'CODE OUTPUT','-'*29), which is used to separate the Python code from the output generated by the code.

Example A.1: Assigning Python variables
Assign Python variables to each of the quantities 0.00007, 1.64e-4, 10, 3+4i, time. Use the Python function type(a) to determine the type of each variable.

Python Code

```
#define variables
a, b, c, d = 0.00007, 1.64e-4, 10, complex(3,4)
e = 'time'

print('-'*28,'CODE OUTPUT','-'*29)

print('Variable a = 0.0003 belongs to type:',type(a))
print('Variable b = 1.64e-4 belongs to type:',type(b))
print('Variable c = 10 belongs to type:',type(c),'\n')

print('A complex number created with complex(3,4)=',d)
print('Variable d = complex(3,4) belongs to type:',type(d),'\n')

print("Variable e = 'time' belongs to type:",type(e))

----------------------- CODE OUTPUT ---------------------------
Variable a = 0.0003 belongs to type: <class 'float'>
Variable b = 1.64e-4 belongs to type: <class 'float'>
Variable c = 10 belongs to type: <class 'int'>
A complex number created with complex(3,4)= (3+4j)
Variable d = complex(3,4) belongs to type: <class 'complex'>
Variable e = 'time' belongs to type: <class 'str'>
```

Multiple variable assignments can be compressed to one line. For example, the code `a,b,c = 1, 2, 3` is a compressed version of the three individual lines of code `a = 1`, `b = 2`, and `c = 3`.

The common arithmetic operators in Python are: addition +, subtraction −, multiplication *, division / and exponentiation **. In addition, the % operator is used to find the remainder (or modulo), and the // is used for integer division.

In Python we can use *f-strings* to format the printing of variables. Within the f-string, any variable can be enclosed inside curly brackets, for example `print(f'{a}')` will print the numerical value of the variable a, and `print(f'{a**2}')` prints the numerical value of a^2. By using the f-string formats `print(f'{a:.3}')`, `print(f'{a:.2e}')`, `print(f'{a:.2f}')` and `print(f'{a:g}')`, we print the variable a using two decimals, using scientific notation, and as a float in generic notation, respectively.

In Example A.2 we use *f-strings* to format the printing of variables.

Example A.2: Assign Python variables a, b to the numbers 5.0/3 and 1.645.
(a) Use f-strings to print the variable a using scientific notation with two significant figures.
(b) Print the variable a^2 without any formatting and by using generic notation.
(c) Print the variable b using two decimal points in the output of the code.
(d) What does `round(a,2)` and `round(b,1)` produce in the code?

Python Code

```
a = 5.0/3                 #define variables
b = 1.645

print('-'*28,'CODE OUTPUT','-'*29,'\n')

print(f'Without formatting, the variable a={a}','\n')
print(f'Using scientific notation, the variable a={a:.2e}')

print(f'Without formatting, the square of variable a is {a**2}')
print(f'Using generic notation, the square of variable a is \
{a**2:g}','\n')

print(f'Using two decimals, the variable b=1.645 is      {b:.3}')

print('\nround(a,2) produces ',round(a,2))
print('round(b,1) produces ',round(b,1))

------------------------- CODE OUTPUT ---------------------------
Without formatting, the variable a=1.6666666666666667
Using scientific notation, the variable a=1.67e+00
Without formatting, the square of variable a is 2.777777777777778
Using generic notation, the square of variable a is 2.77778
Using two decimals, the variable b=1.645 is      1.65

round(a,2) produces  1.67
round(b,1) produces  1.6
```

A variable name cannot begin with a number, and there are reserved words in Python which should not be used as variables. Table A.1 shows a partial list of reserved words.

Table A.1
Partial list of reserved words in Python.

False	None	def	elif	True
class	for	from	or	if
finally	lambda	nonlocal	else	sum
return	try	and	import	min
is	not	del	break	max
continue	sum	global	list	array

A.2 SEQUENCES IN PYTHON

The main sequence types in Python are *lists*, *tuples* and *range* objects. The main differences between these sequence objects are:

Lists are mutable, i.e., their elements can be modified, and are usually homogeneous (i.e. objects of the same type create a list of similar objects).

Range objects are efficient sequences of integers (commonly used in for loops). They use a small amount of memory and yield items only when needed.

Tuples are immutable, i.e. their elements cannot be modified, and their elements are usually heterogeneous (i.e., objects of different types create a tuple, describing a single structure).

In the next three subsections, we look at examples of these types of sequences.

A.2.1 LISTS

We create a Python list using square brackets, with items separated by commas. Lists can contain data of any type and even other lists. For example, a=[2.0,[3,0],5j,[1,1,2],'s',1] defines a list containing floats, integers, complex numbers, other lists, and strings.

We can access the elements of a list by their index, and it is important to remember that lists are indexed starting at 0, not at 1. For example print(a[0]) prints the first element in the list a, and print(a[2]) prints the third element. We can also use *negative indices* to access elements starting from the end of the list, so that print(a[-1]) prints the last element in the list, print(a[-2]) prints the second to last element, etc.

Since lists are mutable, they can be altered, so we can redefine any of the elements in the list, for example a[-1] = -2 sets the last element of the list to -2. We can use *multiple indices* to access several entries in a list of lists, for example print(a[1][0]) will print the first element in the element a[1]. In the above list a, the element print(a[1][0]) is 0.

We can define sub-lists called *slices*, by using the syntax `a[start:end:step]`. For example, the slice `a[2:5:2]` starts at index 2 and increases by a step of 2 taking every second element in the list, but does *not* include the ending value 5 of the index. We can omit one of the indices or the step parameter in a slice, for example `a[:5]` creates a list of the first five elements of `a`, and `a[2:]` creates a list of all the elements of `a` after and including `a[2]`. Likewise, `a[::2]` creates a list which contains every other element in `a`.

Lists can be *concatenated* using the addition operator +, for example, `a[1]+a[3]` creates a new list with the elements from both `a[1]`, `a[3]`.

Example A.3 presents how to use these general properties of lists and how to access and modify parts of a list.

Example A.3: Consider the Python list
`a = [2,[3,0],5,[1,1,2],'s',[1,4],2,5]`
(a) Write a code to add the first and last element in this list.
(b) Modify the third element in the list so that it is equal to -2.
(c) What is the `a[3][2]` element of the list?
(d) Create a new list which starts from the third element in the list, `a`, and contains every third element of `a`.
(e) Create a new list that contains the first five elements of `a`.
(f) What does the code `a[1]+a[3]` produce?
(g) What does the code `a[1]*3` produce?

Python Code
(a) The first and last elements in this list are added by `a[0]+a[-1]`.
(b)-(c) `a[2]=-2` sets the third element in the list equal to -2. Similarly, `a[3][2]` obtains the third part of the fourth element of the list.
(d)-(e) The line `a[2:len(a):2]` creates a new list which starts from the third element `a[2]` in the list, up to the length of the list which is specified by `len(a)`.
(f)-(g) The code `a[1]+a[3]` produces a new list by concatenating the list $[1, 1, 2]$ to the end of the list $[3, 0]$. Finally, `a[1]*3` produces a new list by repeating the elements of the list `a[1]` three times.

```
print('-'*28,'CODE OUTPUT','-'*29,'\n')

a = [2,[3,0],5,[1,1,2],'s',[1,4],2,5] # define list a

print('list a=',a)

print('the sum of the first and last element is: ',a[0]+a[-1])

a[2] = -2
print('\na[2] = -2 modifies list a into \na =',a,'\n')

print('The a[3][2] element is =', a[3][2],'\n')
print('The a[2:len(a):2] sequence is =',a[2:len(a):2])
print('The a[:5] sequence is =',a[:5])

print('\nThe a[1]+a[3] sequence is =',a[1]+a[3])
print('The a[1]*3 sequence is =',a[1]*3)

---------------------------- CODE OUTPUT ----------------------------
list a= [2, [3, 0], 5, [1, 1, 2], 's', [1, 4], 2, 5]
the sum of the first and last element is:  7

a[2] = -2 modifies list a into
a = [2, [3, 0], -2, [1, 1, 2], 's', [1, 4], 2, 5]
The a[3][2] element is = 2
The a[2:len(a):2] sequence is = [-2, 's', 2]
The a[:5] sequence is = [2, [3, 0], -2, [1, 1, 2], 's']

The a[1]+a[3] sequence is = [3, 0, 1, 1, 2]
The a[1]*3 sequence is = [3, 0, 3, 0, 3, 0]
```

Individual elements can be assigned separate variables in a list. For example, a, b, c = [1,2,3] stores the value of 1 in the variable a, the value of 2 in b, and the value of 3 in c. This is sometimes referred to as unpacking a list.

Lists *cannot* be copied like numeric data types by using for example a statement like b = a. Specifically a statement like b = a does *not* create a new list b from list a, but simply makes a *reference* to a. This confusion can be avoided by using the b = a.copy() command, which creates two isolated objects, whose contents share the same reference. This is referred to as *shallow copying* of an object. One can also use the b = copy.deepcopy(a) command, which creates two isolated objects, whose contents and structure are completely isolated from each other. This is referred to as deep copying of an object. The deep copy is an independent copy of the original object *and all its nested objects*. In other words, if we make changes to any *nested* objects in the original object, we will see no changes to the deep copy.

A.2.2   RANGE SEQUENCES AND LIST COMPREHENSIONS

The second important type of sequence in Python is a *range* object, created with the built-in function `range(a:b:step)` where a, b and step are integers. This function creates an object representing the sequence of integers from `a` to `b` (excluding `b`), incremented by the variable `step`.

An important feature of Python that we use frequently in this book is a list comprehension, with the general syntax `[expression for item in sequence]`. Here `sequence` is a sequence object (e.g., a range, a list, a tuple, etc.), `item` is a variable name which takes each value in the sequence, and `expression` is a Python expression which is calculated for each value of `item`. For example `[u**2 for u in [1,2,3]]` produces a new sequence with the squares of the integer sequence $[1,2,3]$.

As another example, we can use the remainder operator % to create a periodic sequence of $(0,1,2,3)$ and length 12 with the line of code `[x%4 for x in range(0,12)]`. Example A.4 presents various properties of the `range()` function. Notice that when we print a range object, this does not display the elements when printed. This is because a range object yields values only when they are needed. However, the function `list()` can convert a range object into a list object, whose elements can then be printed. This example also demonstrates list comprehensions.

Example A.4:
(a) What does the line `print(range(1,7,2))` produce in the output of the Python code?
(b) Change the object `range(1,7,2)` into a list, by using the Python function `list()`.
(c) Find the maximum element, the sum of the elements and the length of the object `range(1,7,2)`.
(d) What will be the result of the code `[u**3 for u in [1,2,3]]`?
(e) What will be the result of the code `[x%3 for x in range(0,12)]`?

Python Code

```
print('-'*28,'CODE OUTPUT','-'*29,'\n')

a = range(1,7,2)
print('Define the range sequence a = ',a,'\n')
# print(a) does not print the elements of the range() object

# Use list(a) to convert the range into a list, then print elements
print('list(a) gives: ',list(a),'\n')

print('The length of the range sequence a is = ',len(a))
print('The sum of elements in the range sequence a is = ',sum(a))
print('The maximum of the range sequence a is = ',max(a))

c, d = [3,'s']                    # unpacking a sequence
print("\nUnpack the sequence [3,'s'], to get: c = ",c,' d = ',d,'\n')

e = [u**3 for u in [1,2,3]]    # a list comprehension
print('The list comprehension e = ',e)

f = [x%3 for x in range(0,12)]  # another list comprehension
print('The list comprehension f = ',f,'\n')

-------------------------- CODE OUTPUT ----------------------------
Define the range sequence a =  range(1, 7, 2)
list(a) gives:  [1, 3, 5]
The length of the range sequence a is =  3
The sum of elements in the range sequence a is =  9
The maximum of the range sequence a is =  5

Unpack the sequence [3,'s'], to get: c =  3    d =  s
The list comprehension e =  [1, 8, 27]
The list comprehension f =  [0, 1, 2, 0, 1, 2, 0, 1, 2, 0, 1, 2]
```

A.2.3 TUPLE SEQUENCES

The third type of Python sequence is a *tuple*. We create a tuple by using *parentheses* instead of square brackets. For example a = (1,3,'s'). The major difference between tuples and lists is that tuples are immutable, i.e., their elements cannot be altered. However, the *indexing*, *slicing* and *concatenating* operations for tuples are essentially the same as for lists. In this book we will be using list and range objects in most examples, and will occasionally encounter tuples.

A.2.4 FUNCTIONS ON SEQUENCES

Python has several built-in functions for computing with sequences. For example, len(a), sum(a), max(a) and min(a), evaluate the *length*, *sum*, *maximum* and *minimum* element in the list. We can also *sort* the list using sorted(a). We also note a major difference between string and list types in Python. Lists are mutable but strings are not, i.e., we can modify the value of an element in a list, but not for a string. However, parts of a string can be extracted using slices, for example 'hi'[1] will extract the second letter in the string 'hi'.

A.3 FUNCTIONS, FOR LOOPS AND CONDITIONAL STATEMENTS

This section is a brief introduction to *functions, for loops,* and *conditional statements* in Python. We will demonstrate these over the next four examples. All programming languages have their own way of setting up functions, for loops, and conditionals. Python uses indentations and colons for these statements. In the next four examples, pay close attention to how Python uses indentations and colons to delimit blocks of code. There should be no extra white spaces in the beginning of any line, and the line before any indented block must end with a colon character.

We begin with Example A.5 where we wish to calculate the position of a particle starting at the origin with initial speed v_0, moving under a constant acceleration a at a time t using:

$$y = v_0 \, t + \frac{1}{2} a \, t^2 \tag{A.3.1}$$

Because we wish to evaluate (A.3.1) several times in our code, we will define (A.3.1) as a function in Python and we will use a `for` loop for our calculation.

The first three lines of code in Example A.5 define the function f, by using the `def` keyword. A *function* is generally a named unit of code which can be called from other parts of a program. In Python a function may have one or more variables as arguments, which receive their values from the calling program. The function f has three arguments vo, a, t, which are listed in parentheses in the line defining f. All three arguments must be provided when the function is called.

Inside the indented code of the function definition, the evaluation of (A.3.1) is stored in the variable y. The variable y is a *local variable* used by the function, and its value is not known outside the definition of the function f. By contrast, the variable ypos is a *global variable*, whose numerical value is known to the entire code.

The `return y` statement sends the variable y as the result of the calculation to the calling code. One can specify more than one variable in the return statement by separating them with commas.

Suppose we wanted to calculate the value of the function f for vo = 2, a = 1, and t = 3. This can be done using the syntax f(2,1,3). For example, p = f(2,1,3) would store the value 10.5 in the variable p.

However, Example A.5 asks us to calculate the particle's position at multiple times. One can construct *for loops* in Python which execute blocks of code repeatedly. In Example A.5 we create a *for loop* for such a purpose. However, before we begin the *for loop*, the code line ypos = [] initializes an empty list and which will be used to store the position of the particle at various times. Notice that this line of code is unindented because, as a global variable, ypos is defined outside of the function f.

To define a *for loop*, we need two things, a variable for iteration and a range for that iteration. In Example A.5, the variable of iteration is t which takes on the values in the sequence `range(4)`. The indented block of code following the *for loop* will be repeated several times. In the first iteration of the loop, the variable t has the value of 0, the first element of the sequence `range(4)`. The function f is evaluated using f(1,2,t) which, in this iteration of the loop, is the same as f(1,2,0). The value of f(1,2,0) is then appended to the end of the list ypos by using the ypos.append(f(1,2,t)) command.

In the second iteration of the loop, the variable t has the value of 1 and the process is repeated. Afterwards, the loop has two more iterations, t = 2 and t = 3. The integer 3 is the last element of range(4) and, therefore the loop stops after that iteration.

Example A.5:
Create a function f(vo,a,t) which evaluates the position y of a particle starting at the origin, moving with initial speed v_0, under a constant acceleration a, and at time t:

$$y = v_0 t + \frac{1}{2} a t^2 \qquad (A.3.2)$$

Use a *for loop* so that the function f is called repeatedly to evaluate the position of the particle at times $t = 0, 1, 2$, and 3 seconds.

Python Code
As outlined above, we construct the *for loop* using range(4). The function f is called repeatedly inside the loop, to evaluate the position variable ypos. The value of f is added (appended) to the list ypos by using the ypos.append(f(1,2,t)) command.

```
print('-'*28,'CODE OUTPUT','-'*29,'\n')

def f(vo,a,t):            # define function f
    y = vo*t+a*t**2/2     # functions require indentation
    return y              # return the variable y to the calling code

ypos = []                 # create the empty list ypos
for t in range(4):        # for loops require indentation
    ypos.append(f(1,2,t)) # add the value of f to the list ypos

print('position y(t) = ',ypos)

--------------------------- CODE OUTPUT ----------------------------
position y(t) =   [0.0, 2.0, 6.0, 12.0]
```

In some cases, the functions we need in our code are simple and they depend only on one variable, so that we can use a simple function structure called a lambda function. Example A.6 defines a lambda named function f, which evaluates the position ypos of the particle based on (A.3.1).

Example A.6:
Repeat the previous example by using a lambda function.

Python Code
Note that when using this simpler type of function, the variables v_0 and a are treated as *global* variables with fixed values $v_0 = 1$ and $a = 2$.

```
print('-'*28,'CODE OUTPUT','-'*29,'\n')

vo, a = 1, 2                    # global variables vo, a
f = lambda t: vo*t+a*t**2/2     # define lambda function f(t)

ypos = []                       # empty list ypos

for t in range(4):              # for loops require indentation
    ypos.append(f(t))           # add value of f to the list ypos

print('position y(t) = ',ypos)

--------------------------- CODE OUTPUT ----------------------------
position y(t) =  [0.0, 2.0, 6.0, 12.0]
```

In many occasions we need to execute parts of the code only if certain conditions are true. In Python such conditional statements are implemented using the if, elif and else keywords, as in Example A.7. The conditional statement also uses indentation, just like the functions and for loops.

Example A.7
Repeat the previous example; however, this time use an if statement inside the for loop, so that the code stores only values of the position which are smaller than 7.

Python Code
In this example the ypos variable is evaluated and stored only if the conditional statement f(t)<7 is true, otherwise, the code line ypos.append(f(t)) is ignored.

```
print('-'*28,'CODE OUTPUT','-'*29,'\n')

vo, a = 1, 2                    # global variables xo, vo, a
f = lambda t: vo*t+a*t**2/2     # define lambda function f(t)

ypos = []                       # empty list ypos

for t in range(4):              # for loops require indentation
    if f(t)<7:                  # if statements require indentation
        ypos.append(f(t))       # if f(t)<7 is true, then add value
                                # of f(t) to the list ypos
                                # if f(t)>=7, then ignore this statement

print('position y(t) = ',ypos)

--------------------------- CODE OUTPUT ----------------------------
position y(t) =  [0.0, 2.0, 6.0]
```

A.4 IMPORTING PYTHON LIBRARIES AND PACKAGES

One of the major advantages of Python is the availability of modules, libraries and packages for various scientific applications. In this text, we'll use the term *library* as a generic term for all three.

There are several different ways to import Python libraries, and the most common methods are explained in Example A.8, by using the numpy library and the cosine function cos() as an example.

The simplest way to import functions from the NumPy library is using numpy.cos(0.5), where the function is invoked using the form library_name.function_name() .

In the second method, we use an *alias* for the module name, so that we do not have to type repeatedly long module names. In this example import numpy as np, and call the cosine function in the form np.cos(0.5). In the third method of importing a function, we use from numpy import cos and it is understood that the cos() function in the code will refer to the numpy library.

In the fourth method we use the character * as a wild card for importing all available functions with the code line from numpy import *. In this last method, we do not need to type the name of the alias or library. However, one must be careful when using the * method, since this can cause trouble when more than one modules is imported, which could be using the same name for different functions.

Other common libraries we will work with are SymPy (Symbolic Python), SciPy (Scientific Python) and the graphics library Matplotlib. These libraries are discussed later in this chapter.

Example A.8:
Write a Python code to demonstrate the four methods described above for importing the numpy package, and then evaluate the cosine function cos(0.5) in each method.

Python Code
The comments in this code explain the differences between the four methods of loading and using the package.

```
print('-'*28,'CODE OUTPUT','-'*29,'\n')

import numpy          # method 1: use name.function() syntax
print('Using NumPy function, result is: ', numpy.cos(0.5))

import numpy as np    # method 2: use alias np, instead of numpy
print('Using np shorthand notation, result is: ', np.cos(0.5))

from numpy import cos  # method 3: import only function cos() from numpy
print('Importing just the NumPy function, result is: ', cos(0.5))

from numpy import *    # method 4: import all necessary from numpy
print('Importing all NumPy functions, result is: ', cos(0.5))

-------------------------- CODE OUTPUT ----------------------------
Using NumPy function, result is:  0.8775825618903728
Using np shorthand notation, result is:  0.8775825618903728
Importing just the NumPy function, result is:  0.8775825618903728
Importing all NumPy functions, result is:  0.8775825618903728
```

Libraries are sometimes organized into multiple modules. For example, the module name A.B indicates a *module* B contained within a *library* named A. For example, `numpy.random.normal()` refers to the function `normal()` contained within the module `random` of the library `numpy`. The general format for this type of function is `package.module.function()`.

A.5 THE NUMPY LIBRARY

NumPy is the core Python library for numerical computing. NumPy supports operations on compound data types like arrays and matrices. The first thing to learn is how to create arrays and matrices using the NumPy library. Python lists can be converted into multidimensional arrays.

In this book we usually adopt the standard convention and import NumPy using the alias name np, with the code line `import numpy as np`. As mentioned above, one can also import NumPy functions using the syntax `from numpy import *`. If NumPy is the only package being used in the code, then there is no possibility of any function name conflicts.

A.5.1 CREATING NUMPY ARRAYS

The fundamental object provided by the NumPy library is the `ndarray`. We can think of a 1D (one-dimensional) ndarray as a list, a 2D (two-dimensional) ndarray as a matrix, a 3D (three-dimensional) ndarray as a 3-tensor, and so on.

Using the function `np.array()` we can create a NumPy array from a Python sequence, such as a list, a tuple or a list of lists. For example, `a = np.array([1,5,2])` creates a 1D NumPy array from a Python list.

There are several other functions that can be used for creating different types of arrays and matrices. Some examples of such array functions are shown in Table A.2.

Table A.2
NumPy functions for creating arrays and matrices.

Function	Description
np.array(a)	Creates N-dimensional NumPy array from sequence a.
np.arange(start, stop, step)	Creates an evenly spaced 1D array from *start* to *stop*, excluding the *stop* value.
np.linspace(start, stop, N)	Creates a 1D array with length equal to N, including the *start* and *stop* value.
np.zeros(shape)	Creates array of given shape and type, filled with zeros.
np.ones(shape)	Creates an array of given shape and type, filled with ones.
np.random.random(shape)	Creates an array of given shape and type, filled with random float numbers from 0 to 1.
np.reshape(array, newshape)	Changes the dimensions of a 1D array.

Example A.9 applies some of these array functions to create various types of arrays. In the first two lines of code, notice that when we print a NumPy array it looks a lot like a Python list, except the elements are separated by *spaces*, while in a list the elements are separated by *commas*.

The function np.arange(start,stop,step) is very similar to the list function range(start,stop,step) that we saw before when discussing lists. In Example A.9, the code line np.arange(0,3.5,.5) creates a 1D NumPy array with values from 0 to 3.5 in steps of .5, *excluding* the end value of 3.5.

The function np.linspace(start,stop,N) is slightly different, since we specify the number of points N between the *start* and *end* values, instead of the *step* in the array. In the same example, np.linspace(0,8,6) creates an array with exactly six elements between 0 and 8, *including* the end value 8.

Similarly np.zeros(5) creates a 1D NumPy array of zeros of length 5. Similarly, np.zeros(2,1) creates a 2D NumPy array of zeros with 2 rows and 1 column. The code line np.random.random([1,2]) will generate a matrix with one row and two columns, with random values between 0 and 1.

Example A.9 demonstrates various types of arrays, and how to use the np.arange(), np.linspace(), np.zeros() and np.random.random() functions.

Example A.9
Write a simple code to show how to create arrays using the np.arange(), np.linspace(), np.zeros() and np.random.random() functions.

Python Code
The comments in the code explain the various commands.

```
import numpy as np
print('-'*28,'CODE OUTPUT','-'*29,'\n')
print('printing the list [1,2,3]:                ',[1,2,3])
print('printing the array np.array([1,2,3]): ',np.array([1,2,3]),'\n')

print('np.arange(5) gives: ',np.arange(5))
print('np.arange(0,3.5,.5) gives: ',np.arange(0,3.2,.5),'\n')
# array from 0 to 3.5 steps 0.5 (excluding 7)

print('np.linspace(0,8,6) gives: ',np.linspace(0,8,6),'\n')
# create array with 6 values between 0 and 8

print('array with zeros: ',np.zeros(5))  # generate array with 5 zeros
print('array with ones: ',np.ones([2,1]),'\n')
# generate two rows and one column, with ones

print('array with random values: ',np.random.random([1,2]))
# generate random numbers in one row and two columns, in interval (0,1)

--------------------------- CODE OUTPUT -----------------------------
printing the list [1,2,3]:               [1, 2, 3]
printing the array np.array([1,2,3]):  [1 2 3]
np.arange(5) gives:  [0 1 2 3 4]
np.arange(0,3.5,.5) gives:  [0.  0.5 1.  1.5 2.  2.5 3. ]
np.linspace(0,8,6) gives:  [0.  1.6 3.2 4.8 6.4 8. ]
array with zeros:  [0. 0. 0. 0. 0.]
array with ones:  [[1.]
 [1.]]
array with random values:  [[0.32685396 0.76130921]]
```

A.5.2 ARRAY FUNCTIONS, ATTRIBUTES, AND METHODS

In this section we present a brief overview and examples of working with Python arrays. We will look at how we can access the properties (or *attributes*) of an array, and will examine some of the functions that can be used with NumPy arrays.

Example A.10 applies various functions on arrays and also shows how we can extract some of the important properties of arrays. Some examples of useful commands for extracting the properties of arrays are shown in Table A.3.

Note that the syntax in Table A.3 is `A.dtype`, `A.size` etc. These are examples of using methods objects in Python, as opposed to the functions objects we have looked at so far in this book. Recall also that we use the function `type(A)` for a list `A`, but we use the similar command `A.dtype` for a NumPy array `A`. This is an example of using a *function* versus using a *method*.

Functions are called by placing argument expressions in parentheses after the function name as in `type(a)`, where `a` is an object (list, string, float, etc.). The functions that we define in this book will always be called using the function name first, followed in parenthesis by all of the arguments of the function.

Methods are somewhat similar to functions, but they are called using the dot notation, such as `A.dtype`. Loosely speaking, methods are always attached to a specific object, while functions are isolated. Another example of a function versus a method is the implementation of the dot and cross products of two vectors A and B in SymPy and NumPy. In SymPy the dot product is implemented as a method `A.dot(B)`, while in NumPy the dot product is represented as a function `dot(A,B)`. We will later see several more different types of functions and methods within the NumPy, SymPy and SciPy packages used in this book.

Table A.3

Useful commands (*methods*) for extracting the properties or attributes of an array.

Command	Description
A.dtype	Prints the data type for NumPy array A.
A.ndim	Prints the number of dimensions for NumPy array A.
A.size	Prints the size (total number of elements) for NumPy array A.
A.shape	Finds the number of rows and columns for NumPy array A.

Example A.10 shows how we can use some of these functions with arrays. All entries in a NumPy array are of the *same* data type. We will mostly work with numeric arrays which contain integers, floats, complex numbers or Booleans. We will also mostly be working with the default integer type numpy.int64 and the default float type numpy.float64. We can access the datatype of a NumPy array A by its `A.dtype` attribute. We create a 2D NumPy array from a list of lists of integers using the lines of code `A = np.array([[1,2,3],[4,5,6]])` and `A.dtype` prints out `dtype('int32')` i.e, the data is of integer type and is represented in the computer's memory by 32-bits. Similarly, `A.ndim` tells us that A has two dimensions, with the first dimension corresponding to the vertical direction counting the *rows*, and the second dimension corresponds to the horizontal direction counting the *columns*. Finally, we can find out the total elements in array A with `A.size`.

In the next section we see how to carry out arithmetic operations with various types of arrays.

A.5.3 ARITHMETIC OPERATIONS WITH NUMPY ARRAYS

Mathematical functions in NumPy are *vectorized*, and a partial list of functions we can use to compute with NumPy arrays is given in Table A.4.

Vectorized functions operate elementwise on arrays and produce new arrays as output. For example `np.sin(2*np.pi*x)` computes the values for each elements of the array x = np.arange(0,1.25,0.25). These vectorized functions compute values across arrays very quickly. NumPy also provides mathematical constants such as π (`np.pi`), and e (`np.e`).

We can modify the contents of a NumPy array using *indexing* and *slicing*, just as in the case of lists.

Table A.4

Partial list of useful array functions in Python.

Array functions in NumPy

np.sum	np.argmax	np.min	np.std
np.max	np.argmin	np.mean	np.prod

Mathematical functions in NumPy

np.sin	np.exp	np.arcsin
np.cos	np.log	np.arccos
np.tan	np.log10	p.arctan

Mathematical constants in NumPy

np.pi	np.e

Arithmetic operations are applied to NumPy arrays element-by-element: these include addition +, subtraction -, multiplication *, division / and exponentiation **. See Example A.10 for a demonstration of the addition of two NumPy arrays.

Normally we can only add vectors or matrices of the same size. However, NumPy has a set of rules called *broadcasting*, which allows the combination of arrays of different sizes when the process makes sense. For example, we can add a constant to a 1D NumPy array, by using the expression `1+np.array([5,6,7])**2`. In this example of broadcasting, the array `np.array([5,6,7])` is squared, and then we add the array of ones [1. 1. 1.] which has the same shape. Here we are adding the scalar number 1 to a 1D NumPy array of length 3, and the broadcasting rule allows us to use a very simple syntax.

Example A.10 demonstrates various array operations. For example, to compute the average or mean of the values in an array, we use `np.mean([5,6,7])`, and to find the index of the maximum element in the array, we use `np.argmax([5,6,7])` which gives the value of the index 2, i.e., it identifies the third element in the array as the max value.

Array functions apply to multi-dimensional arrays as well, but we have to choose along which axis of the array to apply the function. For example, for the array M = `np.array([[2,4,2],[2,1,1],[3,2,0]])`, the function `np.sum(M,axis=0)` will sum the *columns* of M. However, the function `np.sum(M,axis=1)` will sum the *rows* of M. In a similar manner, a three-dimensional array of size $3\times3\times3$ can be summed over each of its three axes.

Example A.10:
Define two NumPy arrays, a and b from the lists [1, 2, 3] and [4, 5, 6], respectively. In addition, define a 3×3 matrix M of your choosing as a NumPy array.
Write a simple code to demonstrate the following:

1. The use of the functions `mean(a)`, `argmax(a)`, `std(a)` for the NumPy array a.

2. The result of a + b and a * b.

3. Broadcasting, using 1 + a**2.

4. The use of the functions `np.sum(M,axis=0)`, `np.sum(M,axis=1)`.

Python Code

```
import numpy as np
print('-'*28,'CODE OUTPUT','-'*29,'\n')


a = np.array([1,2,3])
b = np.array([4,5,6])
M = np.array([[2,4,2],[2,1,1],[3,2,0]])


#Part 1
print('mean of elements in array a = ',np.mean(a))
print('std dev of elements in array a = ',np.std(a))
print('max of elements in array a = ',np.max(a))
print('sum of elements in array a = ',np.sum(a),'\n')


#Part 2
print('a + b = ' , a+b)
print('a * b = ', a*b, '\n')


#Part 3
print('1 + a**2 = ', 1 + a**2,'\n')


#Part 4
print('matrix M =',M,'\n')
print('sum of elements of M along axis=0 (rows) = ',np.sum(M,axis=0))
print('sum of elements of M along axis=1 (columns) = ',np.sum(M,axis=1))

------------------------- CODE OUTPUT -----------------------------
mean of elements in array a =  2.0
std dev of elements in array a =  0.816496580927726
max of elements in array a =  3
sum of elements in array a =  6
a + b =  [5 7 9]
a * b =  [ 4 10 18]
1 + a**2 =  [ 2  5 10]
matrix M = [[2 4 2]
 [2 1 1]
 [3 2 0]]
sum of elements of M along axis=0 (rows) =  [7 7 3]
sum of elements of M along axis=1 (columns) =  [8 4 5]
```

A.5.4 INDEXING AND SLICING OF NUMPY ARRAYS

NumPy arrays can be indexed, sliced and copied, just like Python Lists.

Example A.11 shows how to use slicing and indexing in multidimensional NumPy array. For example, the fourth element in array v is accessed with v[3] and we can use two indexes v[1,2] to find the element at row index 1 and column index 2. Similarly v[-1,-1] accesses the last element. The syntax v[2,:] addresses the *row* with index 2, and v[:,3] addresses the *column* with index 3.

Example A.11:
(a) Define a 3×3 matrix v of your choosing and demonstrate the use of indexing with v[3], v[1,2] and v[-1,-1].
(b) What elements of the matrix correspond to v[2,:] and v[:,3] ?

Python Code

```
import numpy as np
print('-'*28,'CODE OUTPUT','-'*29,'\n')


# define 3x3 matrix using NumPy array
v = np.array([[1,2,3],[4,5,6],[7,8,9]])


print('The third element v[2] = ',v[2],'\n')

print('The element at row index 1 and column index 2=',v[1][2])
print('The element v[1,0] = ',v[1,0])
print('The last element v[-1,-1] = ',v[-1,-1],'\n')

print('The column with index 0 = ',v[:,0])
print('The row with index 2 = ',v[2,:],'\n')

-------------------------- CODE OUTPUT -----------------------------
The third element v[2] =  [7 8 9]
The element at row index 1 and column index 2= 6
The element v[1,0] =  4
The last element v[-1,-1] =  9
The column with index 0 =  [1 4 7]
The row with index 2 =  [7 8 9]
```

A.6   THE MATPLOTLIB MODULE

Matplotlib is a Python library that can generate many different types of plots and graphical representations of data, using just a few lines of code. The user has full control of line styles, font properties, axes properties, etc. Often data points for the plotting functions are supplied as Python lists or as NumPy arrays.

In the next two subsections we will see how to create two- and three-dimensional plots using Matplotlib. We will also briefly discuss two different methods of using Matplotlib, namely object-oriented programming versus procedural programming.

A.6.1   2D PLOTS USING MATPLOTLIB

The general procedure to create a 2D plot is to first import the Python library *Matplotlib* and its submodule *pyplot*. Typically we will import the matplotlib.pyplot submodule using the short-hand alias plt. Next we create two sequences of x-values and y-values, and then use the general plot command plt.plot(x,y,[fmt],**kwargs). Here [fmt] is an optional string format and **kwargs are optional keyword arguments specifying the properties of the plot. To improve the appearance of the plot, we use `pyplot()` functions to add a figure title, legend, grid lines, etc. Finally, the line `plt.show()` displays the figure.

Let's begin with a basic example, of solving a simple physics problem in which we throw a ball straight up with an initial speed $v_0 = 5$ m/s. We want to plot the position $y(t)$ as a function of time, and also find numerically the maximum height reached by the ball. The kinematic equations for this problem are:

$$y = v_0\, t - \frac{1}{2} g\, t^2 \tag{A.6.1}$$

$$v(t) = v_0 - g\, t \tag{A.6.2}$$

where $g = 9.8$ m/s^2 is the acceleration of gravity. Example A.12 shows how we would solve this problem using the numerical capabilities of Python.

Example A.12
Write a code to evaluate and plot the kinematic equation (A.6.1), from $t = 0$ to $t = 1$ s with $vo = 5$ m/s.

Python Code
The first line in the code `import numpy as np` causes Python to import the NumPy library, with the alias `np`. We import the Python library *Matplotlib* and its submodule *pyplot*, in order to plot the position $y(t)$. The code imports the `matplotlib.pyplot` submodule using the short-hand alias `plt`. After the libraries are imported, we define the initial condition variable vo and the acceleration of gravity g, in a single line `vo, g = 5, 9.8`.
We need to tell Python for which values of time t we will be computing the position $y(t)$. In this case, we use `t=np.linspace(0,1,100)` to define the array with the values of times t, which will be used to compute $y(t)$. We next use a *list comprehension* in the form `y=[vo*u-g*u**2/2 for u in t]` to obtain the array y containing the values of $y(t)$.
The pyplot commands `plot`, `title`, `ylabel`, `xlabel` are used to improve the appearance of the plot in Figure A.1, and finally `plt.show()` prints the plot. The last four lines in the code show how to evaluate numerically the maximum height $ymax$ and the corresponding time $tmax$, by applying the functions `max()` and `argmax()`, respectively.

```
print('-'*28,'CODE OUTPUT','-'*29,'\n')
import numpy as np
import matplotlib.pyplot as plt
vo, g = 5, 9.8                    # define values of vo, g

t = np.linspace(0,1.03,100)    # create sequence of times
y = [vo*u-g*u**2/2 for u in t] # create sequence of positions, plot y(t)

plt.plot(t,y)

plt.title('Ball thrown straight up')  # add title to plot
plt.ylabel('Vertical distance y(t)')  # add labels for x and y axes
plt.xlabel('Time t [s]')

ymax = max(y)                        # find position ymax
tmax = t[np.argmax(y)]               # find time tmax

print('Max height reached = ',f'{ymax:.4f}', ' m')
print('Time to reach ymax = ',f'{tmax:.4f}',' s')

plt.show()

---------------------------- CODE OUTPUT ----------------------------
Max height reached =  1.2755  m
Time to reach ymax =  0.5098  s
```

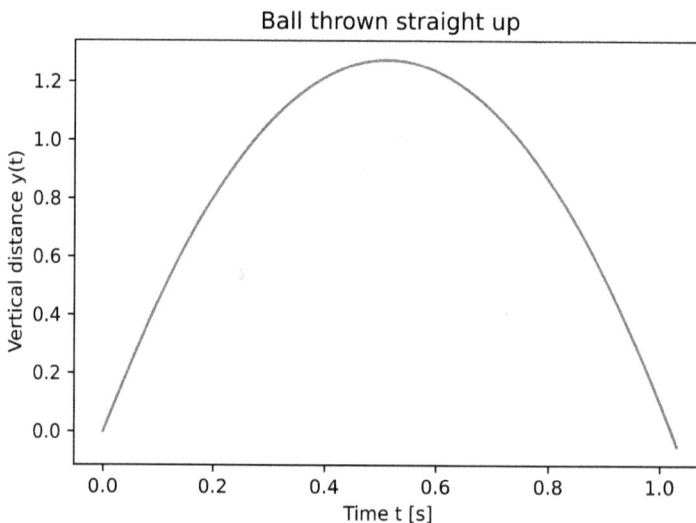

Figure A.1: Plot of the kinematic equation (A.6.1), from $t = 0$ to $t = 1$ s with $vo = 5$ m/s.

We can improve further the plots by specifying various options inside the `plt.plot()` function. For example, we specify the *color* , e.g., by name as in `color='red'`, or by a RGB tuple such as (1,0,1). We can also specify the type of line to be used with `linestyle='dashed'`, the type of marker for data points with `marker='o'`, the linewidth with `linewidth=2`, etc.

It is very convenient to use combined format strings as in the command `plt.plot(x,y,'rd--')` where the string `'rd--')` signifies a red line ('r'), a diamond marker ('d') and a dashed line (--).

Examples of *colors* are: b blue, g green, r red, c cyan, m magenta, y yellow, k black, w white. Examples of character *Markers* are: . point, o circle, v triangle down, ^ triangle up, s square, p pentagon, * star, + plus, x x, D diamond. Finally, examples of *linestyles* are: - solid line style, -- dashed line, -. dash-dot line, : dotted line.

The graphical commands in this example are an example of procedural programming. Next, we will use the Matplotlib module to produce 3D plots using object oriented programming.

A.6.2   3D PLOTS USING MATPLOTLIB

Example A.13 demonstrates how to plot a surface $z = f(x, y)$ in 3D.

Example A.13
Write a Python code to plot the surface function $z = f(x, y) = x + y + 3$ in 3D, when $-3 \leq x \leq 3$ and $-3 \leq y \leq 3$.

Python Code
The 3D plot in the example is created using the `plot_surface()` function within the `matplotlib` library. The lines `fig = plt.figure()` and and `fig.add_subplot(projection ='3d')` create the 3D plot, and the wireframe style is used for the plot.
The NumPy command `meshgrid(x, x)` creates double arrays `X,Y` from the single NumPy arrays `x,y,z`. It is necessary to created double arrays which are required as inputs for the function `plot_surface(X,Y,Z)`.
The plotting function `plot_wireframe(X,Y,Z)` plots the 3D surface of the plane $z = f(x, y) = 3 + x + y$ with the result shown in Figure A.2.

```
import numpy as np
import matplotlib.pyplot as plt

# surface plot for z=3+x+y
x = np.arange(-3, 3, 0.6)      # grid of points on x-axis
X, Y = np.meshgrid(x, x)       # grid of points on xy-plane
Z= 3+X+Y                       # values of z for points on the xy-grid

# plot 3D surface using object oriented programming commands
# define the objects fig and axes for 3D plotting
fig = plt.figure()
axes = fig.add_subplot(projection ='3d')

# plot 3D surface using wireframe style
axes.plot_wireframe(X, Y, Z, color='skyblue')

axes.set_xlabel('X')           # labels for x,y,z axes
axes.set_ylabel('Y')
axes.set_zlabel('Z')

axes.text(-2.6,2,6,'z=f(x,y)=3+x+y')  # add text to the plot

plt.show()
```

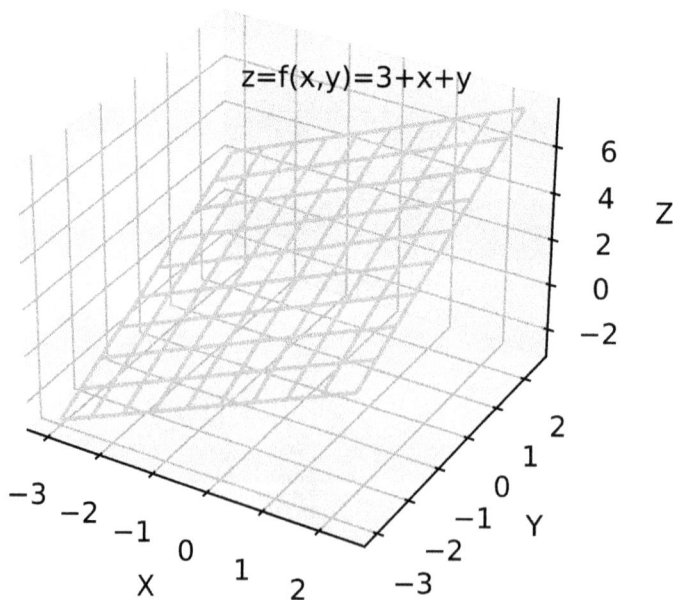

Figure A.2: Plot of a 3D surface $z = f(x, y)$ from Example A.13 using MatPlotLib.

A.7 SYMBOLIC COMPUTATION WITH SYMPY

Symbolic Computation involves using a computer to help find closed-form solutions to differential equations, integrals, eigenvalues, and many more types of symbolic evaluations. For example, we can enter the equation or integral we want solved, and the computer program returns a closed-form solution, if it exists.

Python cannot perform symbolic manipulations by itself, so we need to import the SymPy library in order to expand Python's capabilities. This is sometimes done in the first line of the code with `import sympy as sym`, which causes Python to import the SymPy library with the alias `sym`. Most often, we will import from SymPy only the functions we need.

As an example of a symbolic type of computation, we will use SymPy to solve again *symbolically* the simple physics problem of throwing a ball straight up with an initial speed $v_0 = 5$ m/s. We want to evaluate symbolically the maximum height y_{max} reached by the ball (neglecting air resistance), and how long it takes to reach this maximum height. The kinematic equations for this problem are:

$$y = v_0\, t - \frac{1}{2} g\, t^2 \tag{A.7.1}$$

$$v(t) = v_0 - g\, t \tag{A.7.2}$$

By setting the speed $v(t) = v_0 - g\, t = 0$ at the maximum height reached by the ball, we find the corresponding time $t = v_0/g$. Substituting this value of t into (A.7.1) we find the maximum height:

$$y_{\max} = v_0 \left(\frac{v_0}{g}\right) - \frac{1}{2} g \left(\frac{v_0}{g}\right)^2 = \frac{v_0^2}{2g} \tag{A.7.3}$$

Example A.14 shows how we would solve this problem using the symbolic capabilities of Python.

Example A.14:
Use SymPy to solve again the problem of throwing a ball straight up with an initial speed $v_0 = 5$ m/s. Evaluate symbolically the maximum height $y_{\max}$ reached by the ball (neglecting air resistance), and how long it takes to reach this maximum height.

Python Code
The following Python code evaluates the time $t_{\max}$ and the maximum height $y_{\max}$, starting from (A.7.1) and (A.7.2). In this example, the SymPy library includes the function `solve` and the method `.subs` which can help us solve algebraic equations.
Notice that we tell Python that the variables `vo, t, g` need to be defined, using the `symbols` command, and that we also include the option `real=True` to signify that they are to be treated as real variables.
The code line `solve(vo-g*t,t)` solves symbolically the equation $v_0 - g\, t = 0$ for the variable t, while the line `y.subs(t,tmax)` substitutes symbolically the value of t with `tmax`.

```
from sympy import  symbols, solve
print('-'*28,'CODE OUTPUT','-'*29,'\n')


vo, t,  g = symbols('vo, t,  g ',real=True)    # define variables


# solve equation for tmax symbolically using solve() function
tmax = solve(vo-g*t,t)[0]
print('Time to reach max height=',tmax)


y = vo*t-g*t**2/2
# substitute tmax in y(t) to find ymax, using method .subs
print('Max height reached = ',y.subs(t,tmax))


--------------------------- CODE OUTPUT ----------------------------
Time to reach max height= vo/g
Max height reached =  vo**2/(2*g)
```

In the next section we provide an example which shows how to use the `lambdify` and `diff` functions in Python.

A.8 LAMBDIFY() AND DIFF() FUNCTIONS IN PYTHON

Example A.15 shows how one can combine the results from NumPy and SymPy libraries, by using the `lambdify()` function. The `lambdify()` function can calculate numerical values from SymPy expressions, and the `diff` function is used to obtain the symbolic derivative of a function.

In Example A.15, we are asked to evaluate a symbolic derivative and then plot its result. We use SymPy to evaluate the symbolic derivative of a function. However, `lambdify` must then be used, so that we can numerically evaluate the result before we create the plot.

Example A.15:
Use SymPy to evaluate the symbolic derivative of the function $f = a\sin(bt)$ with respect to time t, and evaluate this derivative in the interval $t = 0$ to $t = 6$. Use the numerical values $a = 1$ and $b = 2$.

Python Code
The Python code imports the `lambdify` function with the code line `from sympy.utilities.lambdify import lambdify`. Notice that we tell Python that the variables a, b, t need to be defined using the `symbols` command, and that we also include the option `real=True` to signify that they are to be treated as real variables. The `diff(f,t)` command is used to evaluate the symbolic derivative of the function $f = a\sin(b\,t)$ and the result is stored in the variable `deriv`.

Next, `f.subs(a:1,b:2)` substitutes the numerical values $a = 1$ and $b = 2$ into the symbolic derivative, and code line `y = lambdify(t,f.subs(a:1,b:2))` creates a function $y = f(t)$.

```
import numpy as np
from sympy import symbols, sin, diff
from sympy.utilities.lambdify import lambdify


print('-'*28,'CODE OUTPUT','-'*29,'\n')


a, b, t = symbols('a,b,t',real=True) # define symbols


f = a*sin(b*t)
deriv = diff( f,t)                       # evaluate symbolic derivarive of f
print('Symbolic Derivative v=dy/dt: ',deriv)


tims = np.linspace(0,6,5)               # sequence of times tims


# substitute a=1 and b=2 in derivative, using .subs method
y = lambdify(t,deriv.subs({a:1,b:2}))


# print y(t)
print('Derivative values = \n',y(tims))

-------------------------- CODE OUTPUT ----------------------------
Symbolic Derivative v=dy/dt:   a*b*cos(b*t)
Derivative values =
  [ 2.          -1.97998499  1.92034057 -1.82226052  1.68770792]
```

A.9 OVERVIEW OF INTEGRATION METHODS IN PYTHON

This section provides a brief overview of different symbolic and numerical integration methods available in Python.

In general, there are two broad types of integration functions available in Python. The first type of Python function is used to integrate analytical functions $f(x)$, and several examples are given in this chapter from various areas of science. This type of integration is carried out using two different methods:

- *Method #1:* Using the `sympy.integrate` library, we carry out symbolic integration of both proper and improper integrals. In addition, by using the `scipy.special` library we can evaluate integrals of special functions frequently encountered in physics (Error function, Bessel function, Hermite polynomials, etc).

- *Method #2:* Using the `scipy.integrate` library, we can carry out numeric integration of functions from other libraries, or functions defined by the user (see Table A.5), or samples of a function which are often stored as NumPy arrays (see Table A.6).

Table A.5
Integration functions available in the `scipy.integrate` library for given functions.

Python function	Description
quad	General purpose integration.
dblquad	General purpose double integration.
tplquad	General purpose triple integration.
fixed_quad	Integrate func(x) using Gaussian quadrature of order n.
quadrature	Integrate with given tolerance using Gaussian quadrature.
romberg	Integrate func using Romberg integration.

Of the methods in Table A.5, we will use the `quad` command in this book to evaluate 1D integrals, while the functions `dblquad` and `tplquad` will be used to evaluate two-dimensional and three-dimensional integrals.

Table A.6
Integration functions available in the `scipy.integrate` library for sampled functions.

Python function	Description
trapezoid	Use the trapezoidal rule to compute an integral
cumulative_trapezoid	Use the trapezoidal rule to cumulatively compute an integral
simpson	Use Simpson's rule to compute an integral from samples of the function
romb	Use Romberg Integration to compute integral from $(2^k + 1)$ evenly-spaced samples

For a more complete listing of integration routines in SciPy, and details on how each algorithm works, see the online manual and tutorial for `scipy.integrate`.

Example A.16 evaluates an indefinite integral which appears often in physics, using the `sympy.integrate` command.

Example A.16:
Calculate these integrals using SymPy.

$$\text{(a)} \quad \int x\cos(n\,x)\,dx \qquad \text{(b)} \quad \int x\sin(n\,x)\,dx$$

Python Code
Using integration by parts, we obtain:

$$\int x \cos(n\,x)\,dx = \frac{x\sin(n\,x)}{n} + \frac{\cos(n\,x)}{n^2} + C$$

and

$$\int x \sin(n\,x)\,dx = -\frac{x\cos(n\,x)}{n} + \frac{\sin(n\,x)}{n^2} + C \qquad\qquad \text{(A.9.1)}$$

where C is an arbitrary constant of integration. The following Python code to evaluate the definite integral gives the same result. Notice that we tell Python the variable the variable x needs to be defined using the **symbols** command, and that the variable n is not zero, by including the option **positive=True**. We also use the symbolic form of **cos()** for the cosine function, instead of the numerical form of **numpy.cos()**.

```
from sympy import symbols, integrate, cos, sin
x = symbols('x')
n = symbols('n', positive=True)  # define symbols

print('-'*28,'CODE OUTPUT','-'*29,'\n')
print("The indefinite cosine integral = ",integrate(x*cos(n*x),x))
print("The indefinite sine integral = ",integrate(x*sin(n*x),x))

--------------------------- CODE OUTPUT ----------------------------
The indefinite cosine integral =   x*sin(n*x)/n + cos(n*x)/n**2
The indefinite sine integral =   -x*cos(n*x)/n + sin(n*x)/n**2
```

A.10 DSOLVE() COMMAND IN SYMPY

In Python we obtain symbolic general solutions of a differential equation using the SymPy command **dsolve**. This command has typically three arguments: the differential equation that we are trying to solve, the function for which the differential equation is being solved, and the initial conditions of the problem. Let's look at the details of the Python code.

Example A.17:
   Consider a particle that is moving along a line with an acceleration proportional to time $a\,t$, where a is a constant. If at time $t = 0$ the particle's velocity is v_0, find the formula for the particle's velocity as a function of time, by solving the differential equation $dv/dt = a\,t$.

Python Code
   The first line tells Python to import several commands from the Symbolic Python (SymPy) library. Libraries are written by experienced programmers, and they include functions which can be used by any code that imports the library. In this case, the SymPy library includes functions like **dsolve** which solve differential equations in closed form.

All of the symbols need to be specified in Python. We use `Symbol('t', real=True, positive=True)` to indicate that `t` is a real and positive variable. Similarly, `symbols` is used to indicate that `a,v0` are real variables. We use the `Function` command to indicate to Python that `v[t]` is a function.

The general solution of the differential equation is given by the SymPy command `dsolve`, whose arguments are the differential equation, the function for which the differential equation is being solved, and the initial conditions of the problem. In this example, the structure of this command is

    dsolve(diff(v(t),t)-a*t,v(t),ics=initconds).

Notice that in Python, the differential equation is written such that the right-hand side is equal to zero; in this example `diff(v(t),t)-a` represents the equation $dv/dt = a\,t$, and only the left-hand side of the differential equation is entered. Here `v(t)` is the desired solution, and `ics=initconds` specifies the initial conditions of the problem. These conditions are specified in the line `initconds = v(0):v0` which represents the given initial condition $v(0) = v_0$.

```
from sympy import Symbol, symbols, Function, dsolve, diff

print('-'*28,'CODE OUTPUT','-'*29,'\n')

# define symbols and the function v
v = Function('v')
t = Symbol('t', real=True, positive=True)
a, v0 = symbols('a, v0', real=True)

# initial conditions for the ODE
initconds = {v(0):v0}

# use dsolve to obtain solution of ODE   dv/dt=a
solution = dsolve(diff(v(t),t)-a*t,v(t),ics=initconds).rhs

print('The solution is   v(t) = ',solution)

--------------------------------- CODE OUTPUT ----------------------------------
The solution is   v(t) =   a*t**2/2 + v0
```

A.11   NUMERICAL INTEGRATION: THE ODEINT() COMMAND IN SCIPY

In Python we obtain numerical solutions of a differential equation using the SciPy command `odeint()`.

Example A.18

Using an initial velocity of $v(0) = v_0 = 1$ m/s, find and graph the numerical solution to the differential equation:

$$\frac{dv}{dt} = a\,t \qquad\qquad\qquad (A.11.1)$$

where the constant $a = 2$ m/s^2.

Python Code

We import three libraries, NumPy, SciPy, and the graphics library Matplotlib. From the SciPy library we import only the function `odeint` which is used to numerically solve the differential equation (A.11.1). After the libraries are imported, we assign the values of the initial condition variable v0 and the acceleration `a`.

To solve the differential equation numerically, we create a new function in Python that we called `velderiv(v,t)`, which contains the differential equation we are solving. In Python, function definition is done using the command `def`. The arguments of the function are `t` and `t`, and they are included in parentheses following the function name.

The next few lines contain the actual calculation of the function. Notice we included a local variable `dvdt`, which stays within the function and is equal to the first derivative of `v`. The last line of the function is `return dvdt`, which returns the value of the variable `dvdt`.

We also need to tell Python for which values of `t` we will be computing $v(t)$. In this case, `times` is an array which contains the list $[t_0, t_1, \ldots, t_{30}] = [0, 0.1, \ldots, 3.0]$, and hence we will be solving for $v(0), v(0.1), \ldots, v(3.0)$. Next, we solve the differential equation using the command `odeint`, which is imported from the `scipy.integrate` library. The `odeint` command in this example is:

   `velocity = odeint(velderiv,v0,times).`

This command has three arguments: the differential equation (as the user-defined function `velderiv`), the initial value (`v0`), and the list of times `times`. The result is an array which we called `velocity`, and contains values $[v_0, v_1, \ldots]$, where $v_i = v(t_i)$. The last group of lines in the program set up the graph of `v` versus `t`.

```python
import numpy as np
from scipy.integrate import odeint
import matplotlib.pyplot as plt

# initial condition v(0)=1, and acceleration a =9.8 m/s^2
v0 = 1
a = 2

# define function velderiv to be called by odeint
def velderiv(v,t):
    dvdt = a*t
    return dvdt

# times at which to evaluate the numerical solution
times = np.linspace(0,3,30)

# solve the ODE with initial v(0)=v0
velocity = odeint(velderiv,v0,times)

# plot v(t) and label the axes
plt.plot(times,velocity)
plt.ylabel('v(t), m/s')
plt.xlabel('Time, s')
plt.show()
```

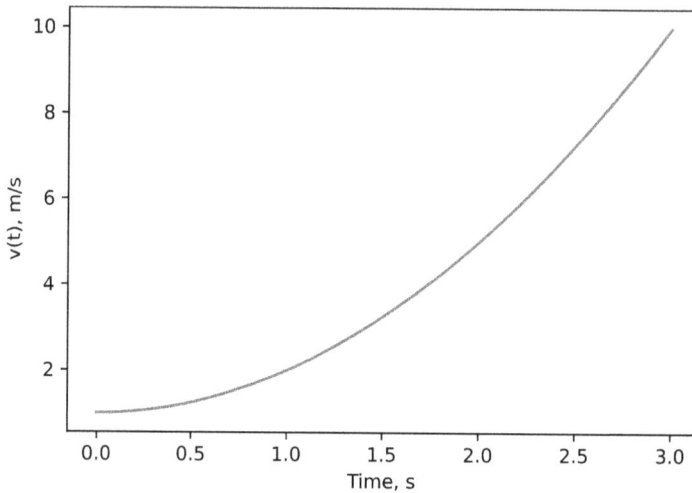

Figure A.3: Plot of the numerical solution of the differential equation A.11.1 from Example A.18.

A.12   VECTOR REPRESENTATION IN SYMPY

The Python library `sympy.vector` is used in this book to represent vectors and vector operations. .Some examples of the relevant vector functions and methods are shown in Table A.7.

Table A.7

Partial list of vector functions and methods in the package `sympy.vector`.

Function/Method	Description
`R = CoordSys3D('R')`	Define Cartesian coordinate system named R.
`R.base_vectors()`	The unit base vectors of the coordinate system R.
`v.components`	components of vector **v**.
`v.magnitude()`	magnitude of vector **v**.
`v.normalize()`	the normalized version of vector **v**.
`v.projection(u)`	The vector projection of vector **v** on vector **u**,
`v.projection(u, scalar=True)`	The scalar magnitude of the projection of vector **v** on vector **u**.
`v.cross(u)`	cross product of vectors **v**, **u**.
`v.dot(u)`	dot product of vectors **v**, **u**.
`v.to_matrix(C)`	Modify vector **v** into a matrix form C.

For several more examples of Python codes covering many of the topics in this book, the readers are referred to our recently published book *Mathematical Methods using Python* (Pagonis and Kulp, [3]).

B Introduction to Mathematica

B.1 VARIABLES IN MATHEMATICA

Although Mathematica is best known for its symbolic capabilities, it is a programming language in and of itself. Therefore, it can store floats, integers, complex numbers, and strings as variables. In general, variable names in Mathematica should be lowercase (functions begin with uppercase letters); however, there is nothing preventing you from using upper case letters in Mathematica. The `Print` command can be used to display a variable's contents. Note that Mathematica can also print contents without using the `Print` command. By default, Mathematica displays the output when commands are executed, or variables are assigned. Lines ending with a semicolon (;) suppress the output.

> **Example B.1: Assigning and displaying variables in Mathematica**
> Assign Mathematica variables to each of the quantities 0.0004, 3×10^4, $1 - 2i$, and the string "Hello world!" Note that $i = \sqrt{-1}$.
>
> **Solution:**
> Notice that we used ^ for exponentiation and that I is used for $\sqrt{-1}$. Furthermore, double quotes are used for strings. For the variables a, b, and c, we suppressed the output with a ; at the end of the line in which the variables are defined. By leaving out the ; when s is defined, we allowed Mathematica to display the contents of s immediately below its definition. We also demonstrated the use of the `Print` command. The last line shows that variables need not be assigned a value. Mathematica can work with the x without it being assigned beforehand.

Mathematica Code

$a = 0.0004;$

$b = 3 * 10^\wedge 4;$

$c = 1 - 2I;$

$s = $ "Hello World!"

Hello World!

$\text{Print}[\text{"a} = \text{"}, a]$

a = 0.0004

$\text{Print}[\text{"b} = \text{"}, b]$

b = 30000

$\text{Print}[\text{"c} = \text{"}, c]$

c = 1 - 2i

$\text{Print}[\text{"s} = \text{"}, s]$

s = Hello World!

$\text{Print}[\text{"We stored nothing in x. x} = \text{"}, x]$

We stored nothing in x. x = x

Variable names in Mathematica cannot start with a number. It is often good to use variable names that are clear. However, Mathematica does not allow spaces or underscores in variable names. Words can be separated in a variable's name using capitalization. For example, suppose we wish to calculate a projectile's time of flight. We can store that time in a variable named `timeOfFlight` where we capitalized the O and the F to separate words.

Variable names can not be the same as function names. For example, you cannot use `Sin` as a variable name since that is Mathematica's reserved command for the trig function. However, `sin` can be used as a variable name. The Mathematica naming convention (of using capitalization for functions and not variables) can help prevent you from trying to assign variables to protected names.

The Mathematica notebook will also give color clues as to variables which have already been defined. For example, if you were to type the letter **a** later in the code used in Example B.1, then it would appear black because Mathematica knows the value of **a**. Likewise, if you type the name of a command such as `Sin`, it would also appear black, because Mathematica knows that function. However, if you were to type `sin` in the code, it would appear blue. The variable `sin` is not defined and Mathematica colors variables and terms blue when it doesn't understand them. Coloration can be a useful way of debugging code in Mathematica. Finally, some characters may appear in the color teal. These are variables which Mathematica intends to use in a function. For example in the command `Solve[x^2 == 1,x]`, we are asking Mathematica to solve the equation $x^2 = 1$ for x. The variable x is not previously defined, but Mathematica is using it in the `Solve` command.

B.2 LISTS IN MATHEMATICA

In Mathematica, a list contains a series of objects (of any kind, string, integer, float, etc.). Lists are defined using curly brackets {}. For example, a = {1,2,3} defines the list {1,2,3} in the variable a. Lists can contain more than one type of object. For example, we could define a new list b = {1, {3,4,5}, "Hello World!",2I} .

Let us return to the list a as defined in the paragraph above. We can access an element of a list by its index (the numerical position in the list). In Mathematica, the first element has an index of 1. Hence, if we want to ask Mathematica for the first element of a list, we type a[[1]]. Note the double square brackets. Likewise, if we want to change the value of an element in the list, we can use the element's index. For example, a[[-1]] = 5 changes the last value of the list to the number 5.

Sublists can be defined using ;;. This is shown in Example B.2.

Example B.2: Lists in Mathematica
Define the list $a = \{2, \{1, 2, 3\}, 3 * I, \{1, 4\}, "Hello"\}$.
a. What does the line a[[2;;4]] produce?
b. Change the last value of the list to "world."
c. What does a[[4]] + a[[5]] produce?
d. What does 3*a[[2]] produce?

Mathematica code
Notice that a[[2;;4]] creates a list that starts with the second element of a, and ends with the fourth element of a. The ranges are inclusive and the resulting list has three elements.

In part (b), note that a[[-1]] indicates the last element of the array a.

When computing a[[4]] + a[[5]], we find that we get a new list where the string "world" is added to each element of a[[4]]. Note that Mathematica is okay with combining variable types in this way. Finally 3*a[[2]] produces a list where each element of a[[2]] is multiplied by the first element of a (the number 3). Further, notice that we did not use the Print command and let Mathematica output the result of each calculation immediately below each line that doesn't end with a semicolon.

```
(*Part (a)*)

a = {2, {1, 2, 3}, 3 * I, {1, 4}, "Hello"};

a[[2;;4]]


{{1, 2, 3}, 3i, {1, 4}}


(*Part (b)*)

a[[-1]] = "world";

a


{2, {1, 2, 3}, 3i, {1, 4}, world}


(*Part (c)*)

a[[5]] + a[[4]]


{1 + world, 4 + world}


(*Part (d)*)

3 * a[[2]]


{3, 6, 9}
```

B.3 FUNCTIONS, LOOPS, AND CONDITIONAL STATEMENTS

Mathematica has thousands of predefined functions. However, there is still a need for user-defined functions. For example, suppose you want to calculate the position of a particle $x(t)$ moving with a constant acceleration a. We know from basic physics that the position of the particle x is described by

$$x = x_0 + v_0 t + \frac{1}{2} a t^2 \tag{B.3.1}$$

where x_0 and v_0 are the position and velocity of the particle at time t, respectively. By creating a user-defined function for (B.3.1), we can evaluate it several times. Example B.3 shows how to create and evaluate a user-defined function.

Example B.3: Functions and lists in Mathematica
Create a function X[v0,a,t] which evaluates (B.3.1) to find the position of a particle at times $t = 0, 1, 2$, and 3 seconds with $v_0 = 3$ m/s and $a = 2$ m/s^2.

Mathematica Code

Following Mathematica's naming convention, we began with defining the function X using a capital letter. In the square brackets, we included each argument for X, in this case the initial velocity v0, acceleration a, and time t. To define the function, we use `SetDelayed` (`:=`) which assigns the right-hand side to the function X, but does not evaluate the definition until it is called in the code. In the second line of the code, we show how to evaluate the function. Finally, we create a list of positions using the `Table` command. `Table` creates a list where the first argument is the object being computed (in this case, the function X with specified v0 and a). The second argument of `Table` is a list that contains the variable being iterated, the starting value for the iteration, and the final value (both values are inclusive). A step size can be included as a fourth value in the list. For example, the command `Table[X[3, 2, t], {t, 0, 3,2}]` would create a list starting at $t = 0$ and ending at $t = 3$, but calculate only every other element. The result would be `{0,10}`.

```
X[v0_, a_, t_] := v0 * t + 1/2 * a * t^2;

X[3, 2, 1]

4

Table[X[3, 2, t], {t, 0, 3}]

{0, 4, 10, 18}
```

The `Table` command is one method of performing a loop in Mathematica that is particularly useful if you have a simple calculation, or if you want the output of the loop to be a list. More complicated loops can be done in Mathematica using the command `Do`. In the next example, we will repeat Example B.3 using a Do-loop.

Example B.4: Functions and lists in Mathematica

Create a function x[v0,a,t] which evaluates (B.3.1) to find the position of a particle. Use a Do loop to calculate x[3,2,t] at times $t = 0, 1, 2$, and 3 seconds.

Mathematica Code

We defined the function X as in Example B.3. We then created an empty list Xlist which will store the position of the particle as a function of time. Notice that the Do loop has several lines. Before the comma in Do are the lines that are iterated during the loop. In this case, we evaluate the function X[3,2,t] and store the result in the variable xpos. The AppendTo command appends the value of xpos to the end of the list. Finally, we display the value of the variable Xlist.

$X[\text{v0}\_,\text{a}\_,\text{t}\_]:=\text{v0}*t+1/2*a*t{\wedge}2;$

$\text{xlist} = \{\};$

$\text{Do}[\text{xpos} = X[3,2,t];$

$\text{AppendTo}[\text{xlist},\text{xpos}],$

$\{t,0,3\}];$

xlist

$\{0,4,10,18\}$

Conditional statements are used when we need to execute parts of the code, but only if certain conditions are true. In Mathematica, we use the If command as demonstrated in Example B.5.

Example B.5: Functions and lists in Mathematica
Create a function x[v0,a,t] which evaluates (B.3.1) to find the position of a particle. Use a Do loop to calculate X[3,2,t] at times $t = 0, 1, 2$, and 3 seconds, but store the value in a list only if X[3,2,t] < 10 is true.

Mathematica Code
This time we included an If command in the Do loop. We used white space to separate the if-statement for clarity. The first argument of If is xpos < 10 and it states the condition that must be true in order for the second argument AppendTo[xlist, xpos] to be executed. In this case if the particle's position is less than 10 meters, then the position's value is appended to xlist. A third argument can be included in If, which would be the calculation executed if the condition were false.

$X[\text{v0\_}, \text{a\_}, \text{t\_}] := \text{v0} * t + 1/2 * a * t\hat{} 2;$

$\text{xlist} = \{\};$

$\text{Do}[\text{xpos} = X[3, 2, t];$

$\text{If}[\text{xpos} < 10,$

$\text{AppendTo}[\text{xlist}, \text{xpos}]]$

$, \{t, 0, 3\}];$

xlist

$\{0, 4\}$

B.4 VECTORS AND MATRICES IN MATHEMATICA

Vectors and matrices are important mathematical structures in physics, and Mathematica treats both as lists. In Example B.6, we show how Mathematica stores vectors and matrices as well as to perform basic calculations. Instruction on more detailed calculations will be provided throughout the text.

Example B.6: Vectors and matrices in Mathematica
Consider the vectors
$$\mathbf{r}_1 = 3\hat{\mathbf{i}} - 2\hat{\mathbf{j}} + \hat{\mathbf{k}}$$
$$\mathbf{r}_2 = -\hat{\mathbf{i}} + \hat{\mathbf{j}} + 4\hat{\mathbf{k}}$$
and the matrices
$$A = \begin{pmatrix} 1 & 2 \\ 3 & 4 \end{pmatrix} \quad B = \begin{pmatrix} a & b \\ c & d \end{pmatrix}$$
Calculate: $\mathbf{r}_1 + \mathbf{r}_2$, $\mathbf{r}_1 \cdot \mathbf{r}_2$, $A + B$, and AB.

Mathematica Code
Notice that in Mathematica, vectors are stored as lists. The first element is the x-component, the second is the y-component and so on. In general, an n-dimensional vector is stored as a list of length n. The command . or `Dot[r1,r2]` is used for the dot product.

Matrices are stored as a two-dimensional array in Mathematica. Notice that the elements of the array are row vectors. Matrix multiplication is done using . just like the dot product. Furthermore, we demonstrate the `MatrixForm` command which produces easier to read output for matrices. The double slashes tells Mathematica to perform the `MatrixForm` command on the output.

```
r1 = {3, -2, 1};
r2 = {-1, 1, 4};

A = {{1, 2}, {3, 4}};
B = {{a, b}, {c, d}};

r1 + r2

{2, -1, 5}

r1.r2

-1

A + B

{{1 + a, 2 + b}, {3 + c, 4 + d}}

A.B//MatrixForm
```

$$\begin{pmatrix} a + 2c & b + 2d \\ 3a + 4c & 3b + 4d \end{pmatrix}$$

B.5 CREATING PLOTS IN MATHEMATICA

The graphical representation of functions is important in physics. Mathematica has many different plotting functions. However, the four most important in this text are:

- Plot is used to plot functions, both Mathematica and user-defined functions, including the solutions to differential equations.

- ListPlot is used to plot lists such as data points.

- ParametricPlot is used to plot two functions which depend on the same independent variable. For example, if you needed to plot the path of a projectile, you would plot the projectile's horizontal coordinate $x(t)$ and its vertical coordinates $y(t)$ in a parametric plot.

- Plot3D is used to plot functions of two independent variables. The result is the surface of a function in 3D.

Each type of plot is demonstrated in Example B.7.

Example B.7: Plotting functions in Mathematica

Consider the functions:

$$x(t) = 1 + 2t \tag{B.5.1}$$

$$y(t) = 2 + 3t - 4.5t^2 \tag{B.5.2}$$

$$f(x,y) = 3\exp\left(-\frac{x^2+y^2}{3}\right) \tag{B.5.3}$$

(a) Use `Plot` to plot $y(t)$ for $t = 0\ldots1$.
(b) Use `ListPlot` to plot the ordered pairs $\{x(t), y(t)\}$ for $t = 0, 1, 2, 3$, and 4.
(c) Use `ParametricPlot` to plot $\{x(t), y(t)\}$ for $t = 0\ldots3$.
(d) Use `Plot3D` to plot $f(x,y)$ for $x \in [-2, 2]$ and $y \in [-2, 2]$.

Mathematica Code

(a) We begin with the `SetOptions` command which forces certain options on the commands that appear in its first argument. In this case, we are forcing the options `Axes -> False` which removes the axes from the plot, `Frame -> True` which encloses the plot in a frame, and `BaseStyle -> {FontSize -> 16}`, which forces the font size to be 16 for the commands `Plot`, `ListPlot`, and `ParametricPlot`. The command `Plot3D` does not have a `Frame` option, hence the additional line of code.

For the `Plot` command, the first argument is the function to be plotted, and the second argument is the range of independent coordinate values. The range is inputted as a list. The first argument of the list is the independent variable, the second argument is the starting value, and the third argument is the final value. The `FrameLabel` option is used to label the plot's axes. The output of `ListPlot` is shown in Figure B.1(a).

(b) Before we could use `ListPlot`, we needed to define a list of ordered pairs. That is done using the `Table` command. In this case, the first argument is the ordered pair. The second argument contains the options. The output of `ListPlot` is shown in Figure B.1(b).

(c) `ParametricPlot` is used in the same way as `Plot`; however, an ordered pair is plotted instead. The option `AspectRatio -> Full` is used to prevent Mathematica from using the same scale to display the vertical and horizontal axes. The reader is encouraged to download the code from this chapter and change the `AspectRatio` to see how the plot is changed. The output of `ListPlot` is shown in Figure B.1(c).

(d) Finally `Plot3D` is used like `Plot`, except ranges for both independent variables must be given. The output is a wireframe surface. If you are using the Mathematica notebook, you can mouse over the plot and click and drag to change its orientation. The output of `ListPlot` is shown in Figure B.1(d).

```
SetOptions[{Plot, ListPlot, ParametricPlot}, Axes->False,

Frame->True, BaseStyle->{FontSize->16}];

SetOptions[Plot3D, BaseStyle->{FontSize->16}];

x[t_]:=1 + 2 * t;

y[t_]:=2 + 3 * t - 4.5 * t^2;

f[x_, y_]:=3 * Exp[-(x^2 + y^2)/3]

Plot[y[t], {t, 0, 1}, FrameLabel->{"time", "y(t)"}]

list = Table[{x[t], y[t]}, {t, 0, 4}];

ListPlot[list, FrameLabel->{x, y}]

ParametricPlot[{x[t], y[t]}, {t, 0, 3}, FrameLabel->{"x(t)
", "y(t)"}, AspectRatio->Full]

Plot3D[f[x, y], {x, -2, 2}, {y, -2, 2}, AxesLabel->{x, y, f}]
```

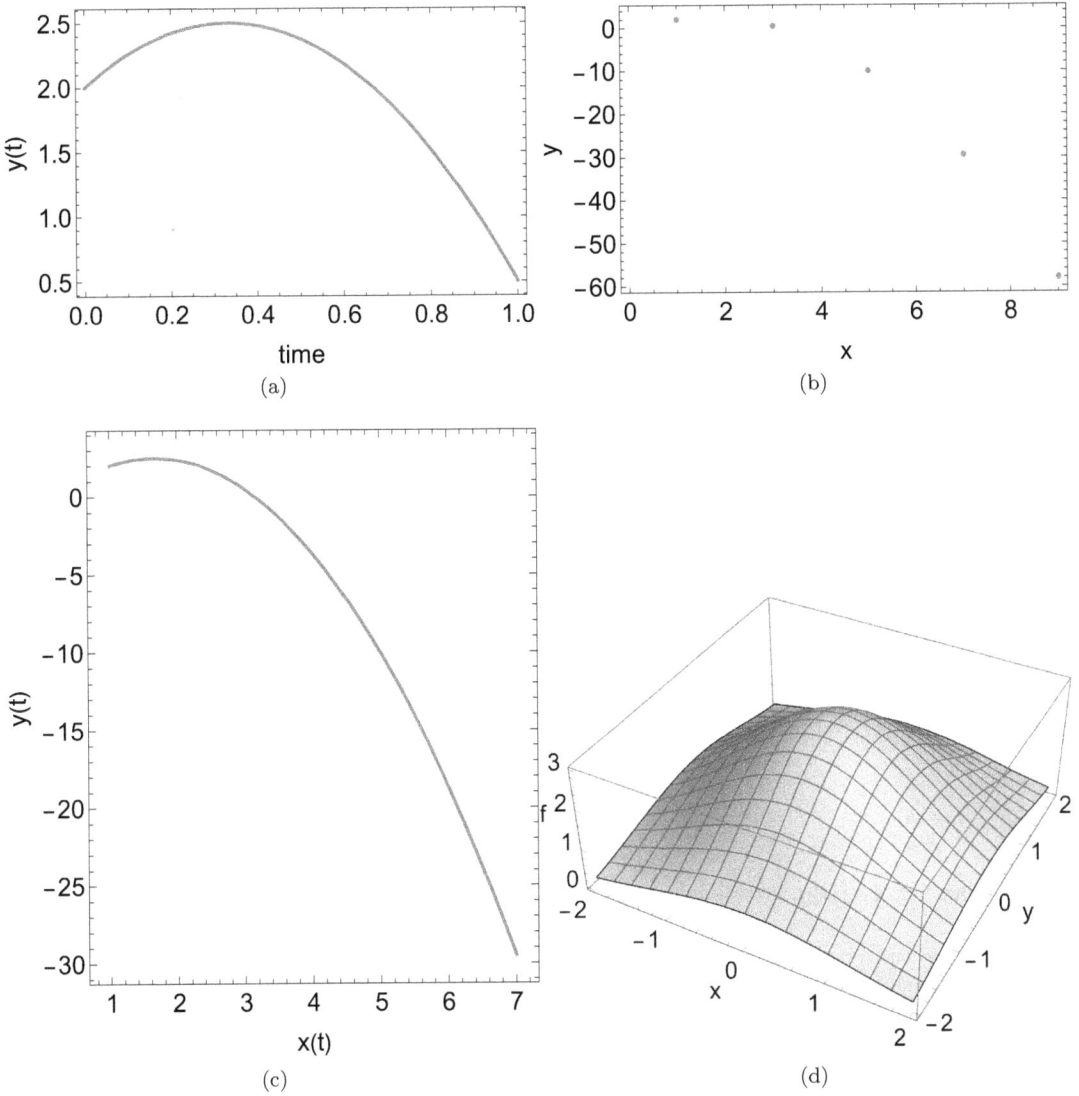

Figure B.1: Output of `Plot` (a), `ListPlot` (b), `ParametricPlot` (c), and `Plot3D` (d) from Example B.7.

B.6 BASIC SYMBOLIC CALCULATIONS IN MATHEMATICA

Mathematica is a powerful computer algebra system. In this book, we will demonstrate many of its capabilities. For now, we are going to present four basic calculations:

- `Solve` allows Mathematica to symbolically solve an equation for a single variable, or a system of equations for multiple variables.

- `D` calculates symbolic derivatives.

- `Integrate` calculates symbolic integrals.

- `DSolve` solves ordinary differential equations symbolically.

We will demonstrate each of these in the next three examples.

Example B.8: Symbolic manipulation and differentiation
The height of a projectile is

$$y(t) = y_0 + v_0 t - \frac{1}{2}g t^2 \tag{B.6.1}$$

where y_0 is the projectile's initial vertical position, v_0 is the initial velocity in the y-direction, and g is the acceleration due to gravity. Using Mathematica, find the maximum height of the projectile.

Mathematica Code
This problem involves finding the maximum of the function $y(t)$, which is done by finding $dy/dt = 0$, solving for t, and inserting that value of t into $y(t)$.

We begin by defining (B.6.1) as a variable (not a function) y. The command D[y, t] computes dy/dt (the result is displayed below the command). We then use the Solve command to find the value of t when $dy/dt = 0$. In the Solve command, note the use of ==, which is used as equivalence. A single equal sign is used for variable assignment. If one needs to test equality or set two sides of an equation equal, one must use ==.

Finally, we use the replacement command /. to substitute the value of tmax into y.

$y = y0 + v0*t - 1/2*g*t^2;$

$D[y, t]$

$-gt + v0$

$tmax = Solve[D[y, t] == 0, t]$

$\left\{\left\{t \to \frac{v0}{g}\right\}\right\}$

$y/.tmax[[1]]$

$\frac{v0^2}{2g} + y0$

Example B.9: Symbolic integration
Find the work done by the force $F = -kx$ exerted from the point $x = x_1$ to the point $x = x_2$.

Mathematica Code
We know that this problem involves integrating the force. After defining the variable, force, we use the command Integrate to perform the integration. The first argument of Integrate is the function being integrated. The second argument can be one of two

things. Had we simply put x as the second argument, Mathematica would have given an indefinite integral (without the constant of integration). However, we included a list of the variable of integration, its lower limit, and its upper limit. The result is a definite integral.

---

force $= -k * x$;

Integrate[force, $\{x, \mathrm{x1}, \mathrm{x2}\}$]//Simplify

$\frac{1}{2} k \left(\mathrm{x1}^2 - \mathrm{x2}^2 \right)$

Example B.10: Symbolic solution of ordinary differential equations
Solve the differential equation

$$\frac{d^2 x}{dt^2} = -kx \qquad\qquad (B.6.2)$$

with the initial conditions $x(0) = x_0$ and $\dot{x}(0) = 0$.

Mathematica code

The DSolve command is used to solve (B.6.2). The first element is a list which includes the ODE and the initial conditions. Note the use of apostrophes to denote differentiation, and square brackets to denote that x is a function of time t. The second argument of DSolve is the function being solved for, and the third element is the independent variable. If we wanted a general solution, we could have omitted the initial conditions from the first argument of DSolve and Mathematica would have given a solution with arbitrary constants. Finally, we display the solution using x[t]/. soln[[1]], which tells Mathematica to display the function x[t] stored in the variable soln.

---

soln $=$ DSolve[$\{x"[t] == - k * x[t], x[0] == \mathrm{x0}, x'[0] == 0\}, x, t$];

$x[t]/.\mathrm{soln}[[1]]$

$\mathrm{x0}\,\mathrm{Cos}\left[\sqrt{k}\,t\right]$

B.7 NUMERICAL SOLUTIONS TO INTEGRALS AND DIFFERENTIAL EQUATIONS IN MATHEMATICA

Sometimes, closed-form solutions to ODEs or integrals are either impossible to find, or not helpful. In those cases, we can use numerical methods which provide a list of values as a function of an independent variable (or variables). Mathematica often uses interpolating functions as numerical solutions, so the solutions can be evaluated just like any other function.

Numerical integration is done using the command `NIntegrate` which is used exactly like `Integrate` except that the limits of integration need numerical values. Furthermore, all parameters in the function (such as k in Example B.10) must be given numerical values.

We demonstrate numerical integration of ODEs in the next example.

Example B.11: Numerical solution of ordinary differential equations
Solve the differential equation

$$\frac{d^2x}{dt^2} = -kx \tag{B.7.1}$$

where $k = 2$ N/m and the initial conditions $x(0) = x_0 = 1$ and $\dot{x}(0) = 0$. Plot the result for $t = 0 \ldots 10$.

Mathematica Code

The `NDSolve` command is used to solve (B.7.1). Like `DSolve`, the first element is a list which includes the ODE and the initial conditions. The second argument of `NDSolve` is the function being solved for, and the third element is the numerical range of independent variable. The result of `NDSolve` is an interpolating function. We plot the resulting function using the `Plot` command. Note again the use of `x[t] /. soln` to specify the function `x[t]`, which is stored in the variable `soln`.

```
SetOptions[Plot, Frame->True, Axes->False,

BaseStyle->{FontSize->16}];

soln = NDSolve[{x''[t]== - 2 * x[t], x[0]==1, x'[0]==0}, x,

{t, 0, 10}];

Plot[x[t]/.soln, {t, 0, 10}, FrameLabel->{"time", "x(t)"}]
```

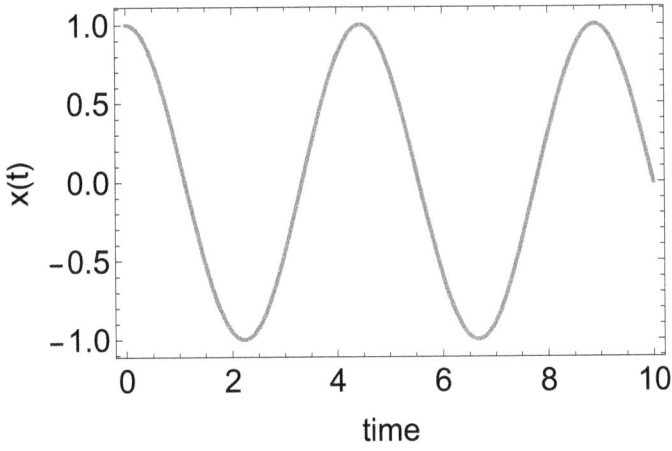

Figure B.2: A graph of the solution to (B.7.1) from Example B.11.

Bibliography

1. J. Taylor, Classical Mechanics, University Science Books, Sausalito, CA, 2005.

2. A. Hadhazy, Fact or fiction: The days (and nights) are getting longer, Scientific American (6 2001).

3. V. Pagonis, C. W. Kulp, Mathematical Methods using Python, CRC Press, 2024.

4. W. Hamilton, On a general method in dynamics; by which the study of the motions of all free systems of attracting or repelling points is reduced to the search and differentiation of one central relation, or characteristic function, Philosophical Transactions of the Royal Society, Part 2 (1834) 247–308.

5. W. Hamilton, Second essay on a general method in dynamics, Philosophical Transactions of the Royal Society, Part 1 (1835) 95–144.

6. S. Thornton, J. Marion, Classical Dynamics of Particles and Systems, 5th Edition, Thomson - Brooks/Cole, Belmont, CA, 2004.

7. E. Noether, Invariante variationsprobleme, Nachrichten von der Gesellschaft der Wissenschaften zu Göttingen, Mathematisch-Physikalische Klasse 1918 (1918) 235 – 257.

8. H. Goldstein, C. Poole, J. Safko, Classical Mechanics, 3rd Edition, Pearson, London, 2001.

9. V. Arnold, Mathematical Methods of Classical Mechanics, 2nd Edition, Springer, New York, 1997.

10. J. Worthington, A study of the planar circular restricted three body problem and the vanishing twist, Ph.D. thesis, University of Sydney (10 2012).

11. S. Strogatz, Nonlinear Dynamics and Chaos: With Applications to Physics, Biology, Chemistry, and Engineering, Chapman and Hall/CRC, Cambridge, UK, 2024.

12. R. Hilborn, Chaos and Nonlinear Dynamics: An Introduction for Scientists and Engineers, 2nd Edition, Oxford University Press, New York, 2001.

13. R. Enns, It's a Nonlinear World, Springer, New York, 2011.

14. H. Kantz, T. Schreiber, Nonlinear Time Series Analysis, 2nd Edition, Cambridge University Press, Cambridge, UK, 2004.

15. H. Abarbanel, Analysis of Observed Chaotic Data, Springer, New York, 1996.

16. U. Wilensky, W. Rand, An Introduction to Agent-Based Modeling, MIT Press, Cambridge, MA, 2015.

17. J. Sprott, Dynamical models of love, Nonlinear Dynamics, Psychology, and Life Sciences 8 (8) (2004) 303 – 313.

18. J. Fajans, L. Friédland, Autoresonant (nonstationary) excitation of pendulums, plutinos, plasmas, and other nonlinear oscillators, American Journal of Physics 69 (2001) 1096 –1102.

Index

For Product Safety Concerns and Information please contact our EU
representative GPSR@taylorandfrancis.com
Taylor & Francis Verlag GmbH, Kaufingerstraße 24, 80331 München, Germany

www.ingramcontent.com/pod-product-compliance
Lightning Source LLC
Chambersburg PA
CBHW060941210326
41598CB00031B/4693